AF389505

Fiducial Reference Measurements for Satellite Ocean Colour

Fiducial Reference Measurements for Satellite Ocean Colour

Editors

Andrew Clive Banks
Christophe Lerebourg
Kevin Ruddick
Gavin Tilstone
Riho Vendt

MDPI • Basel • Beijing • Wuhan • Barcelona • Belgrade • Manchester • Tokyo • Cluj • Tianjin

Editors
Andrew Clive Banks
Institute of Oceanography,
Hellenic Centre for Marine
Research (HCMR)
Greece

Christophe Lerebourg
ACRI-ST
France

Kevin Ruddick
Royal Belgian Institute for
Natural Sciences (RBINS)
Belgium

Gavin Tilstone
Plymouth Marine Laboratory
(PML)
UK

Riho Vendt
Tartu Observatory,
University of Tartu
Estonia

Editorial Office
MDPI
St. Alban-Anlage 66
4052 Basel, Switzerland

This is a reprint of articles from the Special Issue published online in the open access journal *Remote Sensing* (ISSN 2072-4292) (available at: https://www.mdpi.com/journal/remotesensing/special_issues/2nd_ocean_color_RS).

For citation purposes, cite each article independently as indicated on the article page online and as indicated below:

LastName, A.A.; LastName, B.B.; LastName, C.C. Article Title. *Journal Name* **Year**, *Article Number*, Page Range.

ISBN 978-3-03943-064-2 (Hbk)
ISBN 978-3-03943-065-9 (PDF)

Cover image courtesy of Copernicus Sentinel data (2015)/European Space Agency, CC BY-SA 3.0 IGO.

Contents

About the Editors

Andrew Clive Banks is a senior researcher in satellite oceanography. He has more than 20 years of research experience in Earth observation/remote sensing since his first publication in the field in 1997 and is now author of more than 70 papers in peer-reviewed journals and international conference proceedings. His academic qualifications are from premier United Kingdom institutions in the field of Earth observation: B.S. (Honours) in Physical Geography from the University of Reading in 1993 (concentrating on remote sensing in collaboration with NUTIS, the NERC Unit for Thematic Information Systems); M.S. in Remote Sensing from University College London (UCL) and Imperial College in 1994; and Ph.D. in remote sensing and environmental physics from the NERC Environmental Systems Science Centre in 2000 (formerly NUTIS, now part of the University of Reading Department of Meteorology). Dr. Banks was a satellite oceanography researcher at the Hellenic Centre for Marine Research (HCMR) for 12 years (2000–2012) before gaining wider international experience, first at the Joint Research Centre of the European Commission as a scientific officer for satellite oceanography for three years (2012–2015) and then as a senior research scientist in Earth observation and climate at the National Physical Laboratory (NPL) of the U.K. for a further three years (2015–2018). In 2018, Dr. Banks returned to HCMR as the principal research scientist for satellite oceanography and marine optics at the Institute of Oceanography. During his career, he has worked on more than 30 international projects including as a PI in FRM4SOC. His research interests include: Earth observation, satellite oceanography (biological and physical), ocean color, marine optics, radiative transfer modelling, climate change and essential climate variables, validation and vicarious calibration of satellite data, accuracy of satellite and in situ data (uncertainty and SI-traceability), fiducial reference measurements, and open ocean and coastal remote sensing of the Hellenic Seas and East Mediterranean. He has chaired a number of international remote sensing conference and workshop sessions, has been an expert projects evaluator for the European Commission under different framework programmes, is a reviewer for journals such as *Remote Sensing of Environment*, the *International Journal of Remote Sensing*, and *Frontiers in Marine Science*, and is currently an Editorial Board Member for MDPI Remote Sensing.

Christophe Lerebourg was awarded a Ph.D. in marine optics from the University of Plymouth (U.K.). He then worked for two years at the Federal University of Rio Grande (Brazil) on in situ radiometric measurements and EO data processing in the context of ocean color and SST studies. He joined ACRI-ST in 2006 to work on various projects requiring a strong expertise in ocean sciences, optical remote sensing, and bio-optics. His activities have included: (1) Copernicus GBOV service manager (https://land.copernicus.eu/global/gbov); (2) OLCI A & B vicarious calibration and operational Cal/Val; (3) Vicarious calibration for MERIS 3rd and 4th data reprocessing (algorithm development and operational implementation); (4) MERIS and Sentinel-3 Cal/Val activities; (5) MERMAID facility specification and development; (6) ODESA processing platform development and organization of training for the OC users community; (7) Organization and implementation of ocean color workshops as president of the GIS COOC executive committee; (8) In situ data collection and quality control to populate MERMAID matchup database; and (9) Algorithm development and data validation in the GlobColour project. As part of the MERIS maintenance activities, he has been deeply involved in the development of the MERIS vicarious adjustment procedure. With the MERIS mission, he is currently involved in the Sentinel-3 Mission Performance Centre for the specification, development, and exploitation of multi-sensor Cal/Val tools for Sentinel-3. In particular, he has extended the MERMAID facility for OLCI validation activities. His expertise in MERIS vicarious calibration is of prime interest for the development of the OLCI vicarious calibration procedures. In addition, he has been involved in various OC projects for CNES, such as the development of the Kalicôtier platform, the development and the assessment of new generation OC algorithms, and in-flight calibration of OC sensors. Finally, Dr. Lerebourg is involved in the definition of the Geo-OCAPI mission and is president of the GIS-COOC Executive Committee (French scientific group on ocean color).

Kevin Ruddick is the Remote Sensing and Ecosystem Monitoring (REMSEM) team leader at the Operational Directorate for the Natural Environment (ODNature) of the Royal Belgian Institute for Natural Sciences (RBINS). His academic training involved obtaining a Ph.D. in Oceanography (Docteur en Sciences) from the Université de Liège, Belgium, in 1995, after a M.S. in Marine Environmental Modelling (1992) from the Université de Liège, Belgium, a M.S. in Fluid Dynamics (1988) from the von Karman Institute, Belgium, and a B.A. in Mathematics (1987) from the University of Cambridge, United Kingdom. Dr. Ruddick has a general interest in understanding how the physical environment (light, temperature, currents, etc.) affects life in the sea with more specific interests and expertise in marine optics and remote sensing, including: algorithms for total suspended matter, chlorophyll-a, diffuse attenuation, turbidity, (harmful) algal blooms, etc.; atmospheric correction, especially for turbid waters; in situ validation measurements, especially recorded the most accurate possible reflectance measurements and understanding their uncertainty; geostationary ocean color, the next ocean color revolution; marine reflectance spectra from the blue to the short wave infrared (400–3000 nm); coastal water applications of remote sensing data; and the underwater light climate. Dr. Ruddick has participated in many cruises and field campaigns using above water radiometry with a large number of other teams from a large diversity of platforms, both fixed and ship-based.

Gavin Tilstone is a U.K. Natural Environment Research Council (NERC) Merit Scientist with over 25 years' experience in the fields of optics, phytoplankton, photosynthesis, primary production, and remote sensing ocean color. He leads research on satellite ocean color validation and algorithm development and on water quality, including detecting harmful algal blooms and the use of satellite ocean color in quantifying the marine carbon cycle. He impacts policy implementation using primary production derived from space observations for OSPAR. He leads a team of optical and remote sensing scientists who develop and apply his research within international, European, and UKRI research projects. He leads the Sentinel-3 EUROHAB web alert system project, and three European Space Agency (ESA) projects on the validation, uncertainty quantification, and development of CO2 flux products using Sentinel data. He has been PI and Co-I on four ESA contracts and two European Union (EU) INTERREG contracts. He has published over 85 peer-reviewed papers, with an H-index of 30. Education 1993–1996 University of Exeter, U.K., Ph.D., Botany. Multiple Metal Tolerance in Mimulus guttatus. 1990–1992 University of Swansea, U.K., B.S. (Hons.) Environmental Biology 2:1. Professional Experience 1997–1999 PDRA, Instituto Investigaciones Marinas CSIC, Vigo, Spain. 1999–2000 PDRA, Universite Libre de Bruxelles / MUMM, Brussels, Belgium. 2000–2004 NERC Scientist (Band 6) at PML, U.K., on NERC MARPROD 1 (NER/T/S/1999/00070). 2004–2018 NERC Senior Scientist (Band 5), Plymouth Marine Laboratory, U.K. 2018–present Merit Scientist

(NERC Band 4), Plymouth Marine Laboratory, U.K. Successful Funding Applications PI on: ESA AMT4OceanSatFlux (ESRIN/RFQ/3-14457/16/I-BG); EU INTERREG Va S-3EUROHAB; ESA AMT4SentinelFRM (ESRIN/RFQ/3-14457/16/I-BG); NERC-FAPESP SemSAS (NE/P00878X/1); ESA Extreme Case 2 waters (C2X) (4000113691/15/I-LG); EU Marie Curie EIF "UVphytoMAA" for Kadija Oubelkheir; NERC small grant "Backscatter for ERSEM" (NE/C514215/1); EU FP5, REVAMP (EVG1–CT–2001–00049); ESA COASTDRIVE, (ESRIN AO/1-4240/02/I-LG). Co-I on: ESA FRM4SOC (ESA/AO/1-8500/15/I-SBo); EU Horizons JMP EUNOSAT (11.0661/2017/750678/SUBIENV.C2); INTERREG IVA 2SEAS Information System on the Coastal Eutrophication (ISECA); NERC consortium grant "ECOMAR" (NE/C513018/1); NERC small grant "CASI-PP in estuaries" (NE/D007232/1); NERC MarProd Phase II (NE/C508385/1). WP Lead on: NERC UK SOLAS grant ICON (NE/C517176/2) Oct 2005–Sept 2010; NERC Oceans 2025, Theme 2, Mar Biogeochem in a high CO2 world. Supervisor of 8 PhDs: NE/L50189X/1, Shelf Seas Biogeochemistry Ph.D. Curran; NERC CASE Studentship; Johanna Gloel, PML Studentship Barnes, Robinson (0711560248004), Jackson, Arctic Marine productivity, Martinez-Vicente, Uni. Bangor; 2 Current Ph.D.s: Ford, GW4+ Ph.D., NERC Industrial Ph.D., Courticuisse.

Riho Vendt is a senior researcher in metrology and space technology. Dr. Vendt earned his Ph.D. in physics from the University of Tartu, Estonia, in 2013. For almost 20 years, he worked for the National Metrology Institute of Estonia (Metrosert, 1994–2013) in various positions: research scientist, head of section, and technical manager, and was responsible for the national standards for thermometry (2004–2012). Currently, he is a Senior Research Fellow at Tartu Observatory, University of Tartu with the Department of Space Technology. Dr. Vendt has a high level of expertise in quality assurance of tests and measurements. He has designed and piloted several intercomparison measurement campaigns. He was a member of the Technical Committee for Thermometry (TC-Therm) with the European Association of National Metrology Institutes (EURAMET) as the national representative for Estonia (2002–2013) and a member of the consultative committee for metrology at the Estonian National Accreditation Board (2000–2013). He has been an active assessor of quality management systems since 2004, the Head of the Testing, Calibration, and Space Technology Laboratories at Tartu Observatory since 2012, a member of the Scientific Council of Tartu Observatory since 2015, and a member of EURAMET's European Metrology Networks (EMN) expert group for Climate and Ocean Observation since 2019. He was also responsible for the quality assurance and testing campaign of the first satellite of Estonia, the ESTCube-1 (2012–2013). His research interests include metrology, calibration, and testing; SI traceability and uncertainty evaluation; characterization of measurement instruments; quality assurance; data validation; accuracy of in situ data; fiducial reference measurements; design of intercomparison measurements and analysis of comparison data; optical radiometry; thermometry, thermal effects and modelling; Earth observation; and space technology. He has been teaching courses and supervising students at the undergraduate and graduate levels since 2006 and has worked in several European Space Agency (ESA), EUMETSAT, European Metrology Research Program (EMRP), and other EU-funded projects. Dr. Vendt was the coordinator of the FRM4SOC project.

Preface to "Fiducial Reference Measurements for Satellite Ocean Colour"

Ocean colour's fundamental measure is water-leaving radiance, with chlorophyll estimates derived therefrom providing a proxy for phytoplankton occurrence. This is why ocean colour measured by satellite-mounted optical sensors is an essential climate variable that is routinely used as a central element in assessing the health and productivity of marine ecosystems and the role of oceans in the global carbon cycle. For satellite ocean colour to be trustworthy and used in these and other important environmental applications, its data must be reliable and of the highest quality.

Pre-flight and on-board calibration is conducted for satellite ocean colour sensors; however, once in orbit, their data quality can only be fully assessed via independent calibration and validation activities using surface measurements. These measurements therefore need to be at least as high quality as the satellite data, which necessitates SI-traceability and an uncertainty budget. This is the basis of fiducial reference measurements (FRM) that are: a suite of independent ground measurements that provide the maximum return on investment for a satellite mission by delivering to users the required confidence in data products. This is in the form of independent validation results and satellite measurement uncertainty estimation, over the entire end-to-end duration of a satellite mission. The FRM must: have documented traceability to SI units (via an unbroken chain of calibrations and comparisons); be independent from the satellite retrieval process; be accompanied by a complete estimate of uncertainty, including contributions from all FRM instruments and all data acquisition and processing steps; follow well-defined protocols/community-wide management practices and; be openly available for independent scrutiny.

Within this context, the European Space Agency (ESA) funded a series of projects targeting the validation of satellite data products (atmosphere, land, and ocean) and set up the framework, standards, and protocols for future satellite validation efforts. The Fiducial Reference Measurements for Satellite Ocean Colour (FRM4SOC) project was structured to provide support for evaluating and improving the state of the art in ocean colour radiometry (OCR) through a series of comparisons under the auspices of the Committee on Earth Observation Satellites (CEOS) working group on calibration and validation and in support of the CEOS ocean colour virtual constellation.

The objectives of FRM4SOC were to establish and maintain SI-traceable ground-based FRM for satellite OCR with the relevant protocols and uncertainty budgets for an ongoing international reference measurement system supporting the validation of satellite ocean colour. This was undertaken to ensure the high quality and accuracy of Copernicus satellite mission data and, in particular, the Sentinel-2 MSI and Sentinel-3 OLCI ocean colour products, providing and continuing to provide a fundamental contribution to the European system for monitoring the Earth (Copernicus).

This Special Issue of MDPI Remote Sensing was designed to showcase this essential Earth observation work through the publication of the project's main achievements and results, accompanied by other select relevant articles. It covers the following topics:

- FRM4SOC project overview and scientific roadmap for the future of ocean colour validation (Article 1);

- Measurement requirements and protocols when operating FRM OCR for satellite validation (Articles 2 and 3);

- Improvements in ocean colour radiometers (Article 4);

- OCR calibration source inter-comparisons (Article 5);

- Laboratory-based OCR inter-comparisons (Article 6);
- Field-based OCR inter-comparisons (Articles 7–9);
- SI traceability and end-to-end uncertainty budgets (Article 10);
- FRM in the context of ocean colour system vicarious calibration (Articles 11 and 12);
- Improvements to satellite ocean colour time series (Articles 13 and 14); and
- Improvements in satellite ocean colour application to lakes and coastal areas (Articles 15 and 16).

We sincerely thank everyone that contributed to FRM4SOC, both to the project and this Special Issue.

Andrew Clive Banks, Christophe Lerebourg, Kevin Ruddick, Gavin Tilstone, Riho Vendt
Editors

 *remote sensing*

Project Report

Fiducial Reference Measurements for Satellite Ocean Colour (FRM4SOC)

Andrew Clive Banks [1,*], Riho Vendt [2], Krista Alikas [2], Agnieszka Bialek [3], Joel Kuusk [2], Christophe Lerebourg [4], Kevin Ruddick [5], Gavin Tilstone [6], Viktor Vabson [2], Craig Donlon [7] and Tania Casal [7]

[1]   Institute of Oceanography, Hellenic Centre for Marine Research (HCMR), Former US Base Gournes, 71500 Hersonissos, Crete, Greece
[2]   Tartu Observatory (TO), University of Tartu, Observatooriumi 1, EE-61602 Tõravere, Estonia; riho.vendt@ut.ee (R.V.); krista.alikas@ut.ee (K.A.); joel.kuusk@ut.ee (J.K.); viktor.vabson@ut.ee (V.V.)
[3]   National Physical Laboratory (NPL), Teddington, Middlesex TW11 0LW, UK; agnieszka.bialek@npl.co.uk
[4]   ACRI-ST, 260 Route du Pin Montard, 06904 Sophia Antipolis, France; christophe.lerebourg@acri-st.fr
[5]   Royal Belgian Institute for Natural Sciences (RBINS), 29 Rue Vautierstraat, 1000 Brussels, Belgium; kruddick@naturalsciences.be
[6]   Plymouth Marine Laboratory (PML), Prospect Place, Plymouth PL1 3DH, UK; ghti@pml.ac.uk
[7]   European Space Agency (ESA-ESTEC), Keplerlaan 1, 2201 AZ Noordwijk, The Netherlands; craig.donlon@esa.int (C.D.); tania.casal@esa.int (T.C.)
*   Correspondence: andyb@hcmr.gr; Tel.: +30-2810-337815

Received: 29 February 2020; Accepted: 16 April 2020; Published: 22 April 2020

Abstract: Earth observation data can help us understand and address some of the grand challenges and threats facing us today as a species and as a planet, for example climate change and its impacts and sustainable use of the Earth's resources. However, in order to have confidence in earth observation data, measurements made at the surface of the Earth, with the intention of providing verification or validation of satellite-mounted sensor measurements, should be trustworthy and at least of the same high quality as those taken with the satellite sensors themselves. Metrology tells us that in order to be trustworthy, measurements should include an unbroken chain of SI-traceable calibrations and comparisons and full uncertainty budgets for each of the in situ sensors. Until now, this has not been the case for most satellite validation measurements. Therefore, within this context, the European Space Agency (ESA) funded a series of Fiducial Reference Measurements (FRM) projects targeting the validation of satellite data products of the atmosphere, land, and ocean, and setting the framework, standards, and protocols for future satellite validation efforts. The FRM4SOC project was structured to provide this support for evaluating and improving the state of the art in ocean colour radiometry (OCR) and satellite ocean colour validation through a series of comparisons under the auspices of the Committee on Earth Observation Satellites (CEOS). This followed the recommendations from the International Ocean Colour Coordinating Group's white paper and supports the CEOS ocean colour virtual constellation. The main objective was to establish and maintain SI traceable ground-based FRM for satellite ocean colour and thus make a fundamental contribution to the European system for monitoring the Earth (Copernicus). This paper outlines the FRM4SOC project structure, objectives and methodology and highlights the main results and achievements of the project: (1) An international SI-traceable comparison of irradiance and radiance sources used for OCR calibration that set measurement, calibration and uncertainty estimation protocols and indicated good agreement between the participating calibration laboratories from around the world; (2) An international SI-traceable laboratory and outdoor comparison of radiometers used for satellite ocean colour validation that set OCR calibration and comparison protocols; (3) A major review and update to the protocols for taking irradiance and radiance field measurements for satellite ocean colour validation, with particular focus on aspects of data acquisition and processing that must be considered in the estimation of measurement uncertainty and guidelines for good practice; (4) A technical comparison of the main radiometers used globally for satellite ocean colour validation bringing radiometer manufacturers together around the same table for the first time to discuss instrument characterisation and its documentation, as needed for measurement uncertainty estimation; (5) Two major international side-by-side field intercomparisons of multiple ocean colour radiometers, one on the Atlantic Meridional Transect (AMT) oceanographic cruise, and the other on the Acqua Alta oceanographic tower in the Gulf of Venice; (6) Impact and promotion of FRM within the ocean colour community, including a scientific road map for the FRM-based future of satellite ocean colour validation and vicarious calibration (based on the findings of the FRM4SOC project, the consensus from two major international FRM4SOC workshops and previous literature, including the IOCCG white paper on in situ ocean colour radiometry).

Keywords: satellite ocean colour; fiducial reference measurements (FRM); calibration and validation; SI traceability and uncertainty; Copernicus; European Space Agency (ESA); Committee for Earth Observation Satellites (CEOS)

---

1. Introduction

Copernicus [1] is the European system for monitoring the Earth. It includes earth observation satellites, notably the Sentinel series developed by ESA [2], ground-based measurements and data processing to provide users with reliable and up-to-date information delivered through a set of Copernicus Services related to environmental and security issues. The Copernicus Marine Environmental Monitoring Service (CMEMS, [3]) provides critical marine information in near-real time to the various levels of the user community. Copernicus satellite missions are designed to serve CMEMS by providing systematic measurements of the Earth's oceans to monitor and understand large-scale global dynamics as well as providing data for coastal and inland water applications including eutrophication monitoring, sediment transport and environmental impact assessments [4].

The Committee for Earth Observation Satellites (CEOS) defines calibration as "the process of quantitatively defining a system's responses to known, controlled signal inputs". Validation, on the other hand, is "the process of assessing, by independent means, the quality [uncertainty] of the data products derived from those system outputs" [5,6]. Validation is a core component of a satellite mission (and should be planned for accordingly), starting at the moment satellite instrument data begin to flow until the end of the mission. Without adequate validation, the geophysical retrieval methods, algorithms, and geophysical parameters derived from satellite measurements cannot be used with confidence because meaningful uncertainty estimates cannot be provided to users.

The societal benefits of Ocean Colour Radiometry (OCR) are well-articulated [7–10] and include management of the marine ecosystem and the role of the ocean ecosystem in climate change, aquaculture, fisheries, coastal zone water quality, and the mapping and monitoring of harmful algal blooms. Consequently, Copernicus has developed two relevant satellite families (Sentinel-2 and Sentinel-3) that carry two complementary payload instruments that can measure ocean colour to support the CMEMS service. These are the Multi Spectral Instrument or MSI [11]; and the Ocean and Land Colour Instrument or OLCI [12]. Once in orbit, the uncertainty characteristics of (a) the satellite instruments established during pre-launch laboratory calibration and characterisation activities and (b) the end-to-end geophysical measurement retrieval process, can only be assessed via independent calibration and validation activities. Ground reference measurements are therefore essential to the Sentinel-2 MSI and Sentinel-3 OLCI OCR but were not adequately covered in the operational Copernicus system plan.

Within this context, the European Space Agency (ESA) has funded a series of Fiducial Reference Measurements (FRM) projects [13] targeting the validation of satellite data products of the atmosphere, land, and ocean, and setting the framework, standards, and protocols for future satellite validation efforts. Fiducial reference measurements, as originally defined by [14,15], are a suite of independent ground measurements that provide the maximum return on investment for a satellite mission by delivering to users the required confidence in data products. This is in the form of independent validation results and satellite measurement uncertainty estimation, over the entire end-to-end duration of a satellite mission. The FRM must: have documented traceability to SI units (via an unbroken chain of calibrations and comparisons); be independent from the satellite retrieval process; be accompanied by a complete estimate of uncertainty, including contributions from all FRM instruments and all data acquisition and processing steps; follow well-defined protocols/community-wide management practices and; be openly available for independent scrutiny.

Following the recommendations from the International Ocean Colour Coordinating Group's white paper on in situ ocean colour radiometry [16], and in support of the CEOS ocean colour virtual constellation [9], the main aim of FRM4SOC [17] was therefore to establish and maintain SI traceable ground-based fiducial reference measurements for ocean colour with the relevant protocols and uncertainty budgets for an ongoing international reference measurement system supporting the validation of satellite ocean colour. This paper details how the FRM4SOC project achieved this and showcases the most important results, including the community consensus-driven scientific road map for the future of satellite ocean colour validation based on fiducial reference measurements.

The paper is structured according to the sections listed below, which follow the project's organisation. SI-traceability and uncertainty budgets are essential for FRM and a focus on these was maintained throughout (see Figure 1). Section 1 is this introduction section, with the following sections being numbered in order starting with Section 2:

- FRM and the future of system vicarious calibration of satellite OCR.
- Measurement requirements and protocols when operating FRM OCR for satellite validation
- Review of the most common ocean colour radiometers used for satellite validation.
- Comparisons of irradiance and radiance reference sources used in the calibration of ocean colour radiometry.
- Laboratory and controlled outdoor comparisons to verify the performance of ocean colour radiometers used for satellite validation.
- Field intercomparison experiments to verify the performance of ocean colour radiometers used for satellite validation.
- End-to-end uncertainty.

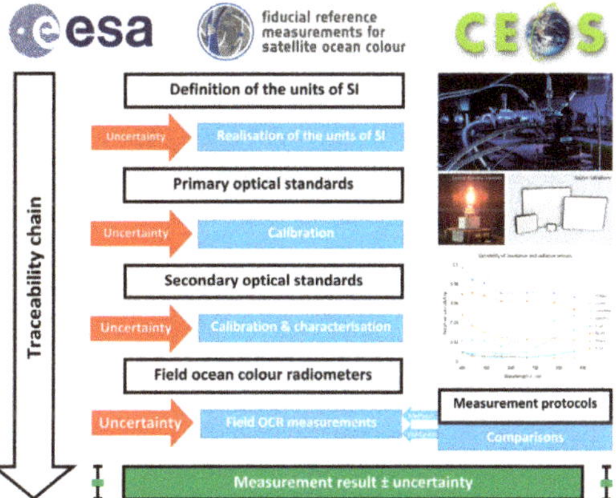

Figure 1. FRM4SOC project overview: a focus on SI traceability and the addition of uncertainty at each step of the FRM process.

2. FRM and the Future of System Vicarious Calibration of Satellite OCR

Post launch system vicarious calibration (SVC) using highly precise and accurate ground radiometric measurements is an essential step in the process of achieving sufficient satellite ocean colour product quality to meet the needs of Copernicus [1] and the Global Climate Observing System (GCOS, [18]). At present there is only one fully operational dedicated ocean colour SVC facility run by NASA and NOAA off the coast of Hawaii, USA (MOBY, [19,20]); and only one other site in the world (BOUSSOLE, [21]), which, although it has reached the requirements and high standard of data quality expected for SVC purposes, is at pre-operational status due to a lack of long-term investment.

From an operational perspective, it is crucial that SVC is implemented as early as possible in an ocean colour satellite mission's lifetime as it is the key to public product release (ideally SVC infrastructure should be operational before launch to ensure continuity of long-term data records in a multi-mission perspective). Past experience has demonstrated that approximately two high quality matchups per month are produced by a permanent mooring for the purpose of SVC [22]. At this rate, several years can pass before consolidated vicarious gains can be derived from a single infrastructure. In an operational context, it is, therefore, crucial to increase the number of operational SVC systems to reduce this delay. Furthermore, the EC, ESA and EUMETSAT have put a significant amount of investment into the Sentinel series of satellites and the OLCI and MSI sensors to provide ocean colour products. Value for money from this investment, in terms of good quality ocean colour data and products, is potentially at serious risk if the European SVC infrastructure is not upgraded and supported in the long term.

With the above in mind, between the 21st and 23rd of February 2017, the FRM4SOC project organised an international workshop at ESA entitled "Options for future European satellite OCR vicarious adjustment infrastructure for the Sentinel-3 OLCI and Sentinel-2 MSI series" [22]. The primary objective of this workshop was to evaluate the options and approaches for the long-term vicarious calibration of the Sentinel-3 OLCI and Sentinel-2 MSI series of satellite sensors. This evaluation was performed with the support and active participation of the world's experts in ocean colour SVC and ocean colour radiometry fields. Presentations were given covering all major aspects of ocean colour SVC globally; and open debates were held to discuss lessons learned, to analyse strengths and weaknesses of the different approaches, and to review the cost and requirements to implement, operate, and maintain SVC infrastructure, in order to clearly establish Copernicus' needs in the short and long term. Drawing from the current status of ocean colour SVC the workshop concluded with a consensus for the development of Copernicus' capacity. The key recommendations of this consensus can be summarised as follows:

- Copernicus does not directly support either MOBY or BOUSSOLE. The risk of losing one or both and their associated expertise, and therefore losing the capacity to deliver robust EO products, must be taken into consideration. Supposing that the US MOBY infrastructure is secured in the long term, Copernicus should consider maintaining two operational SVC sites, resulting in a minimum of three sites globally. This will ensure system redundancy and robustness of ocean colour SVC as recommended by the Committee on Earth Observation Satellites (CEOS). Maintaining two sites in Europe will also: secure the existing expertise, knowledge and knowhow in Europe; develop new expertise; and stimulate technical, scientific and industrial innovation. From a risk mitigation perspective, it is also essential that Copernicus maintain control over its vicarious calibration capacity to ensure Sentinel-2 and Sentinel-3 product quality for the next two decades.
- For the development of these two proposed Copernicus operational SVC sites, it is clear that building upon existing systems and expertise (namely BOUSSOLE and MOBY) would be most cost effective. Consequently, the final community recommendation for SVC development within the framework of Copernicus was: to maintain BOUSSOLE in the long term and upgrade it to full operational status for SVC purposes and also support the development and long-term operation

of a second new European infrastructure in a suitable location to ensure the required Copernicus operational system for SVC including operational redundancy.

- As was implemented for MOBY, and now for BOUSSOLE, for any SVC infrastructure a good metrological foundation with 'hands-on' involvement of National Metrological Institutes (NMIs) at all stages of development and operation is a key component. This fiducial reference measurement (FRM) ethos ensures SI traceability, full uncertainty characterisation and the best possible accuracy and precision for the SVC measurements and process.
- In situ radiometry for SVC should be of high spectral resolution, exceptionally high quality, and of an SI-traceable FRM nature, with a full uncertainty budget and regular SI-traceable calibration [23].
- For the second European SVC infrastructure, the results of studies to date [24,25] clearly point to a site located in the Eastern Mediterranean Sea, near the island of Crete, as the best candidate in European waters, although other options (for example in non-European waters) were not excluded at this stage.
- A MOBY-Net system [26], which includes the transportable modular optical system developed by NASA and the MOBY team, was recommended for the new site. It offers a technologically proven system within a realistic timeframe for Copernicus' needs and its use reinforces collaboration of world-class experts and centres of excellence. However, it was also recommended that, in parallel, steps should be taken within the framework of Copernicus to develop a European solution for the mid and long-term.

3. Measurement Requirements and Protocols When Operating FRM OCR for Satellite Validation

One of the key achievements of the FRM4SOC project has been to review the state of the art of protocols for the measurement of downwelling irradiance [27] and water-leaving radiance [28]. This builds on heritage from the NASA Ocean Optics protocols series [29], recently updated in [30], but: (a) broadens the scope from oceanic waters [31] to all waters where satellite data products are used, including coastal and inland waters [32]; (b) takes account of the many protocol refinements since 2004, including input from the MERIS optical measurement protocols [33]; (c) focuses particularly on the estimation of uncertainties from the data acquisition and processing steps, as required in the FRM context.

3.1. Measurement Requirements

As regards the measurement requirements for satellite OCR validation in the FRM context, it is necessary to:

- Measure in situ the water-leaving radiance and downwelling irradiance in order to derive water-leaving radiance reflectance (or a similar product such as normalised water-leaving radiance), the essential parameter used for comparison with corresponding satellite data products.
- For above-water measurement systems, to distribute and archive the intermediate measurements of downwelling irradiance and upwelling radiance from water and sky radiance and the effective Fresnel reflectance used in processing.
- For underwater measurement systems, to distribute and archive the upwelling radiance just beneath the water surface and the diffuse attenuation coefficient used for extrapolation of upwelling radiance to the surface or complete information on the extrapolation method if the vertical profile is not assumed to be exponential.
- Collect mandatory ancillary data for geographical position (preferably according to the WGS84 datum) and altitude of the air-water interface and UTC date and time (expressed as start, centre and finish times of the measurement).
- Collect, where possible, highly desirable ancillary data, including total water depth, significant wave height, wind speed and direction, surface atmospheric pressure, water salinity and temperature, air temperature, cloud cover and type (e.g., according to World Meteorological codes

27000 and 0508), and photographs of water state (showing water colour, waves and any floating material), sky conditions (full sky, using fish-eye lens) and the radiometers themselves (showing any fouling or possible obstructions).

- Estimate the uncertainty of each measurement based on documented methodology and taking account of all possible sources of uncertainty.
- Provide traceability of the measurement to the SI system of units, using published data acquisition and processing protocols.
- Provide quality control and associated measurement and processing flags along with the measurements themselves.
- Facilitate full traceability of data processing, e.g., by open publication of data processing software.
- Ensure that data is archived for long-term curation in an open access data repository.

Many of these requirements are described in the CEOS/IOCCG White Paper [16], to which the reader is referred for more detail. A detailed description of auxiliary optical and biogeochemical parameters can be found in [29], and further considerations on relevant metadata can be found in [34].

3.2. FRM4SOC Review of Data Acquisition and Processing Protocols

There have been several major developments over the period 2004-2017 since the last revision of the NASA Ocean Optics protocols [29] that helped shape the FRM4SOC protocols, including:

1. A maturing of methods for above water radiometry (although significant diversity still exists particularly for the skyglint correction).
2. A growing consensus that downwelling irradiance should be measured above water, even for protocols that derive water-leaving radiance from the vertical extrapolation of underwater measurements. This allows significant simplification and restructuring with respect to the NASA Ocean Optics protocols by splitting the FRM4SOC protocols reviews into two papers, one dealing with downwelling irradiance and one dealing with water-leaving radiance.
3. A move away from supervised measurements, typified by individual seaborne cruises, to unsupervised measurements (e.g., BOUSSOLE [21], MOBY [19,20], AERONET-OC [35] and potential future drifting systems) because of obvious advantages in terms of measurements/year and the economies of scale for automated acquisition and processing.
4. A growing availability of high spatial resolution satellite data for inland and coastal water applications and the need for validation of such data. Conceptually there are no fundamental differences between the application of protocols for oceanic or inland waters, although different circumstances may occur more frequently in the latter that will impact the choice and/or performance of protocols, e.g., bottom reflectance, very high vertical attenuation, very shallow water, optical impacts of surrounding trees/buildings/terrain, fetch limited surface gravity wave field, etc.
5. Reinforcement of the need for measurements to be accompanied by a full uncertainty budget with traceability to SI standards, introduction of the terminology of Fiducial Reference Measurements [13–15] and the detailed set of recommendations of the IOCCG/CEOS INSITU-OCR White Paper [16]. The FRM4SOC protocols in fact focus on describing elements that should be considered in an uncertainty budget rather than prescribing exactly how measurements should be made.

The essential methods described in the FRM4SOC protocols for measuring downwelling irradiance (three generic methods—see Figure 2) and water-leaving radiance (four generic methods—see Figure 3) can be considered to have reached a reasonable degree of maturity in that they have existed for at least 10–15 years in some form. However, it is clear that there are many incremental improvements still occurring and still possible because of improved understanding/modelling of optical processes and new instruments and measuring platforms.

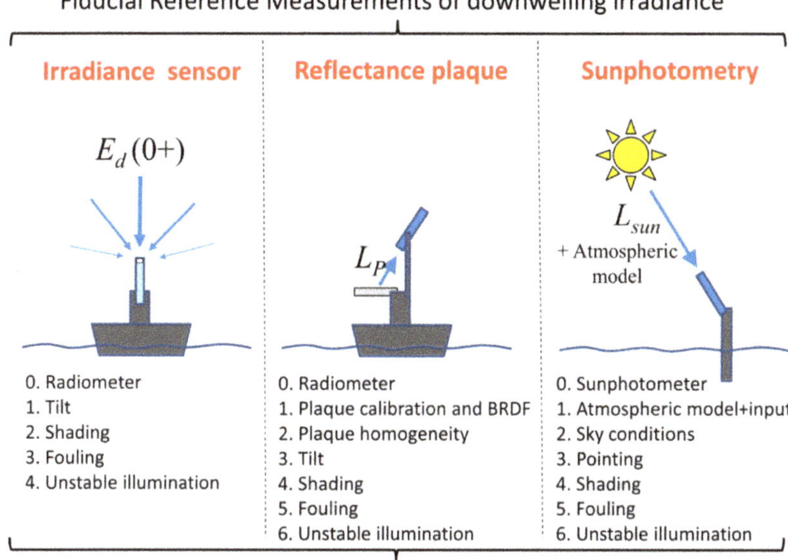

Figure 2. Summary of sources of uncertainty for the three generic families of method for measurement of downwelling irradiance. Reproduced with permission from [27].

Fiducial Reference Measurements of water-leaving radiance

Figure 3. Summary of sources of uncertainty for the four generic families of method for measurement of water-leaving radiance. Reproduced with permission from [28].

The FRM4SOC protocols review papers for downwelling irradiance [27] and for water-leaving radiance [28] discuss in detail the different measurement approaches and the sources of uncertainty that need to be considered and provide guidelines on best practice for making these measurements.

3.3. Recommendations from FRM4SOC Protocols Review

In addition to the guidelines provided by the protocols themselves, there are some key recommendations from them for teams participating in satellite ocean colour validation activities that need to be considered when attempting to achieve FRM status for their measurements:

- Analyse carefully their present measurement protocol and construct an uncertainty budget including minimally the elements listed in the corresponding sections of the FRM4SOC protocols [27,28].
- Participate in intercomparison exercises to validate their uncertainty estimates against those of other methods/scientists.
- Consider the IOCCG/CEOS INSITU-OCR White Paper [16] and the FRM4SOC protocols [27,28] and provide comments for their improvement.

Furthermore, it is recommended to ESA and other space agencies to:

- Facilitate discussion and adoption of best practice and uncertainty estimation by sponsoring intercomparison exercises with appropriate funding for post-measurement analysis of results.
- In the medium term encourage and stimulate the adoption of FRM requirements and in the long term, when sufficient progress and consensus is achieved, use only FRM for the routine validation of satellite ocean colour data.

Finally, it is recommended to the IOCCG:

- To adopt a terminology that reflects the generic nature of aquatic optical processes: "air-water interface" instead of "sea surface", "water colour/reflectance" instead of "ocean colour", "aquatic/water optics protocols" instead of "ocean optics protocols", etc.

4. Review of the Most Common FRM OC Radiometers Used for Satellite OCR Validation

As mentioned in the previous section on protocols, the type of instrument used and its calibration are also major components of a validation measurement uncertainty budget. Therefore, the FRM4SOC project has undertaken a review of the most common ocean colour radiometers used for the purpose of taking validation measurements. The main objectives in carrying out this review were to:

i.   Document the different designs and performance of Ocean Colour Radiometers (OCR) commonly used for satellite OCR validation including a review of their known characterisation and identify significant issues to address.

ii.   Highlight the technical strengths/weaknesses of each system.

iii.   Build on available material and include a dedicated section on instrument characterisation and identify issues that must be addressed for each OCR system.

iv.   Conclude with a justified set of actions to ensure that each OCR used for satellite validation attains FRM status.

The review therefore focused on the radiometers used for in situ measurement and, in particular, on establishing traceable documentation on their characterisation, including factors such as immersion factor, cosine response, linearity, stray light/out of band response, spectral response, temperature sensitivity, dark currents, radiometric noise and polarisation sensitivity. It also contains some information on radiometric calibration and wavelength calibration of the instruments, although calibration aspects were dealt with in more detail in other parts of the FRM4SOC project and this paper (see Sections 5 and 6).

The list of the radiometers reviewed can be seen in Table 1. The full report, which gives further details of the characteristics of each of the instruments listed, is publicly available from the project website [36].

Table 1. Summary of key characteristics of the instruments and systems described in the review (L represents radiance, and E_d represents downwelling irradiance).

Manufacturer/Instrument	Type	Deployment	Wavelength Range	Spectral Resolution (FWHM)	Radiance FOV in Air (FWHM)
Biospherical/C-OPS	Multispectral underwater system, L_u, E_d	Underwater, ship-tethered, slow free-fall	(305 … 1100) nm, (1100…1650) nm, available with InGaAs detectors	10 nm	7°
CIMEL/SeaPRISM	Multispectral system, sun/L_{sky}/L_{water}	Above water (fixed platform)	(412 … 1020) nm	8 … 10 nm	1.2°
IMO/DALEC	Hyperspectral system, E_d/L_{sky}/L_{water}	Above water (ship)	(305 … 1050) nm, calibrated in (400 … 900) nm	10 nm	5°
Satlantic/HyperOCR	Hyperspectral instrument, L or E_d	Above water, Underwater	(305 … 1100) nm, calibrated in (350 … 800) nm	10 nm	6° and 23°
Satlantic/OCR500	Multispectral instrument, L or E_d	Above water, Underwater	(380 … 865) nm, optional from 305 nm	10 nm (or 20 nm)	28°
TriOS/RAMSES	Hyperspectral instrument, L or E_d	Above water, Underwater	(320 … 950) nm	10 nm	7°
WaterInsight/WISP	Hyperspectral system, E_d/L_{sky}/L_{water}	Above water (handheld)	(380 … 800) nm	4.9 nm	3°

To our knowledge, this report is the first attempt that has been made to compile information on all commonly used OCR to the level of detail that is required to construct a full uncertainty budget for instrument-specific aspects. This level of detail far surpasses the information that is generally made publicly available, e.g., on manufacturer websites, and should in any case be available for individual instrument units and not just for an instrument family. In many cases, sufficient information is just not available. In some cases, radiometer manufacturers have performed characterisation tests, but the information is not publicly available and/or is considered confidential, which is contrary to FRM requirements. It is not the intention, and in fact would be neither feasible nor ethically acceptable, to recommend a "best" OCR nor, a fortiori, a "best value for money" OCR. It is for the OCR users, as customers, to make such decisions. However, it is hoped that the FRM4SOC survey and report will help understand what information is or is not currently available for preparation of an FRM uncertainty budget, so that these users will be able to make informed purchase decisions and request the relevant information on radiometer characterisation from their suppliers. Similarly, this process should reward the efforts of the most conscientious instrument manufacturers, who perform careful characterisation tests and provide this information to their customers and to the scientific public and space agencies that use data from these instruments for satellite validation purposes.

To ensure the reliability of measurement results, i.e., traceability to the units of SI with the associated uncertainty evaluation, the review recommended to instrument manufacturers:

- To characterise new types of instruments in well-equipped optics laboratories under stable reference conditions as well as under varied conditions similar to in-field measurements.
- To provide further public information on instrument performance and characterisation where necessary to fill gaps in present knowledge.

The review recommended to instrument users:

- To order regularly the radiometric calibration of instruments in well-equipped calibration labs, collect and carefully analyse the results.

- To request, as customers, detailed performance information from the instrument manufacturers.
- To verify specifications of instrument performance by performing independent tests. For scientists with access to a well-equipped optics laboratory these tests could be quite detailed, e.g., measurement of cosine response of irradiance sensors, measurement of thermal sensitivity, measurement of stray light/out of band response, etc., although it is fully recognised that such tests may be very time-consuming and will generally require specific funding. For scientists without access to a well-equipped optics laboratory it is still possible to verify certain aspects of instrument performance, e.g., by intercomparison of measurements made by different instruments pointing at a uniform target such as a cloudless sky or by participation in multi-partner intercomparison activities (such as the activities of the laboratory and field comparison experiments of the FRM4SOC project detailed in Sections 6 and 7 of this paper).

The review recommended to ESA and other space agencies or entities, including Copernicus Services, requiring Fiducial Reference Measurements for satellite validation, to fund and encourage:

- Preparation of a guide document setting minimum requirements for the most important properties of OCR instruments (like temporal stability, linearity, thermal stability, angular response, stray light/out of band response, etc.).
- Activities to test radiometers from all manufacturers according to a standardised methodology.
- Further development of OCR instruments, including a requirement that such developments provide FRM-compatible information on radiometer characterisation.

5. Comparisons of Irradiance and Radiance Reference Sources Used in the Calibration of Ocean Colour Radiometry

From the International Vocabulary of Metrology (VIM, [37]), metrological traceability is the property of a measurement result whereby the result can be related to a reference through a documented unbroken chain of calibrations, each contributing to the measurement uncertainty. For FRM4SOC optical radiometry this traceability is to SI where the primary standard/reference is provided by the NPL cryogenic radiometer [38]. The traceability chain for OCR for satellite validation can be seen in Figure 4.

What is not evident from this diagram is that the calibrated irradiance and radiance sources are usually the first part of the chain that is distributed outside an NMI such as NPL. Therefore, these sources provide the foundation of testing the performance of any international network of calibration laboratories and satellite validation.

The main objectives of this initial phase of FRM4SOC were therefore to design and document protocols and procedures and implement a laboratory-based (round-robin) comparison experiment to verify the performance of reference irradiance and radiance sources (i.e., lamps, plaques, etc.) used to maintain the calibration of FRM OCR radiometers traceable to SI. The protocols used to implement both the irradiance and radiance source comparisons are publicly available from the project website [39,40].

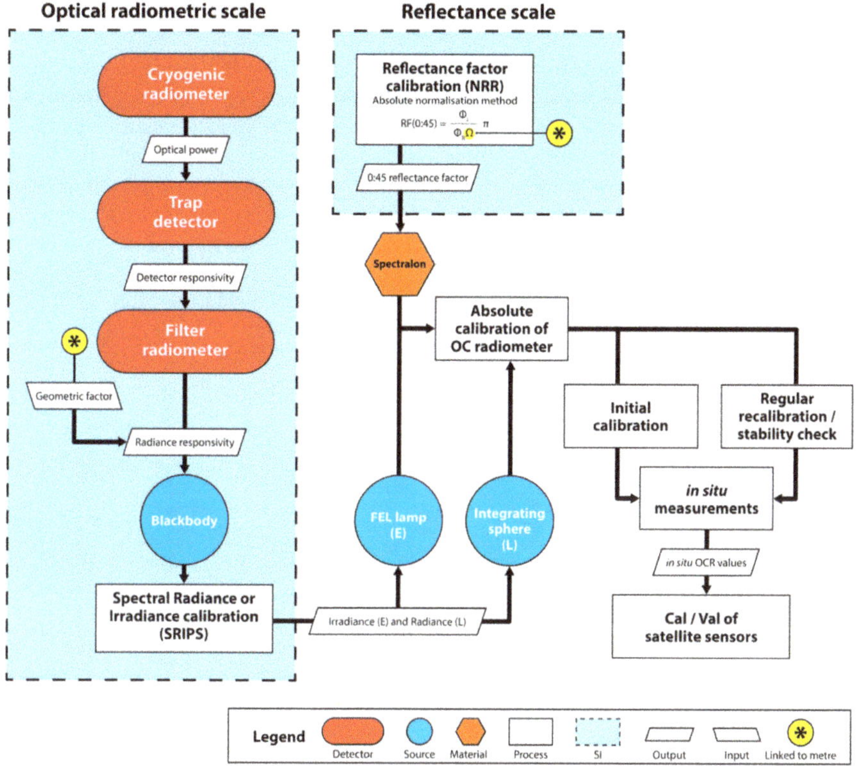

Figure 4. Simplified SI traceability chain at NPL for satellite ocean colour validation.

5.1. Irradiance Reference Source Comparisons

These international comparisons took place at the National Physical Laboratory (NPL) of the UK between the 3rd and 7th of April 2017. The main aim was to verify the performance of reference irradiance sources that are used in the calibration of ocean colour radiometers. Participants were from the following organisations and countries: NPL, UK (pilot); Tartu Observatory, Estonia; Laboratoire d'Océanographie de Villefranche-sur-Mer (LOV), France; Satlantic, Canada; Commonwealth Scientific and Industrial Research Organisation (CSIRO), Australia; Natural Environment Research Council (NERC) Field Spectroscopy Facility, UK; and the National Oceanic and Atmospheric Administration (NOAA), USA. All participants were required to bring or send three (minimum of two) FEL lamps that are used as reference irradiance sources in their calibration laboratories. It was mandatory that each of the participant's lamps had an SI traceable certificate from its last calibration and information about burn time since that calibration (less than 50 h).

At NPL the Spectral Radiance and Irradiance Primary Scales (SRIPS, [39,41,42]) facility is used to transfer the scale from the NPL primary reference standard for spectral emission, a high-temperature blackbody, to lamp and integrating sphere sources. These sources are then used as secondary spectral radiance and irradiance standards further down the chain. For the FRM4SOC irradiance comparison each participant lamp was measured against such an NPL secondary standard lamp obtaining irradiance values under the carefully controlled conditions of the SRIPS and Reference Spectroradiometer System (RefSpec) facilities at NPL.

The results of these SI-traceable comparisons, as represented in Figure 5, show a comparison of irradiance sources used for OCR calibration. To achieve these results a somewhat more complex analysis than a simple difference to the NPL scale was required. The results of the comparison were expressed in terms of the difference between the spectral irradiance values measured by each participant and the mean spectral irradiance values measured by all participants. Since the participants all measured different lamps (i.e., their own lamps), the required differences between them were determined via measurements at NPL of all lamps. The mean ratio between the participants' measurements and those made at NPL was calculated and results for each lamp were then expressed relative to this mean ratio, so showing the degree to which the individual measurements agree with one another. This was necessary for a couple of important reasons: (1) the participants had various different SI-traceability routes for their lamps, i.e., several different NMIs providing their calibration; and (2) a few of the lamps were recently calibrated at NPL, which with a simple difference to the NPL scale would have shown them performing almost perfectly and thus giving a misleading and biased comparison.

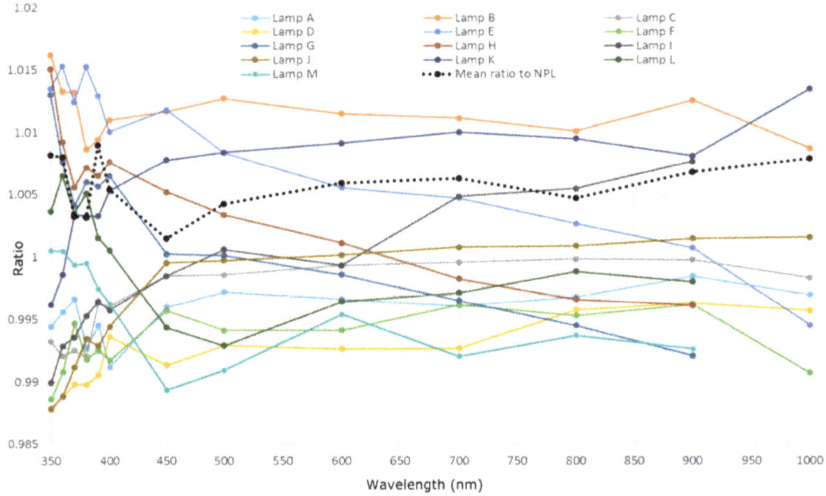

Figure 5. Comparison between irradiance sources from several OCR calibration laboratories, including the NPL spectral irradiance scale as a reference. Reproduced with permission from [43].

All participants' lamps are traceable to SI, and so the results show how the lamps compare with this realisation of the SI irradiance scale, i.e., the mean of all. The comparison shows that they agree among each other to within ±1 to 1.5%. Also included is the difference between the mean of all and the NPL spectral irradiance scale, which shows an agreement for the entire set of lamps to within 1% across all wavelengths. Uncertainties were calculated for each lamp's ratio (not shown) and these generally ranged from 1.6%–1.7% in the UV wavelengths down to 0.9% in the NIR.

These comparisons should not be misinterpreted to mean that there is anything wrong with any of the lamps at particular wavelength ranges. The trend of any single lamp will be due to a combination of factors that may include, for example, the trends of all the other lamps and whether any of them are suffering from the effects of ageing since their last calibration. The likely reason for lamps following the same trend is probably related to whether calibration for more than one has been transferred from the same primary lamp. Again, there is nothing wrong with doing this and this does not indicate anything wrong with the calibration reference lamp. These results are in fact comparable to differences seen between different NMI realisations of the irradiance scale [41,42].

5.2. International Transfer Radiometer Round Robin for Radiance Reference Source Comparison

The radiance round robin comparison took place between June 2017 and May 2018. Its main aim was to verify the performance of radiance sources used to calibrate ocean colour radiometers. The comparison was conducted by NPL as pilot through the round-robin circulation of two ocean colour transfer radiometers. The transfer radiometers used were 7-band multispectral Satlantic ocean colour radiometers (OCR-200) on loan to the pilot from the Joint Research Centre (JRC) of the European Commission. Satlantic had customised these two particular instruments for JRC in terms of their angular characteristics to provide a narrower (~3°) field of view than standard. Initial characterisation measurements to confirm this FOV were carried out by NPL in air, and found to be 2.5° ± 0.3° at FWHM, with a close to Gaussian profile.

The most commonly used radiance source for ocean colour radiometer calibration was used for these comparisons, i.e., an FEL lamp and reflectance panel combination. The FEL lamps were by design the same ones included in the irradiance comparison (see previous section).

The two ocean colour transfer radiometers were sent to each participant according to the schedule shown in Figure 6 in order for them to take at least two sets of radiance measurements of their in-house radiance source (lamp-panel combination) according to NPL protocols that accompanied the transfer radiometers [40].

Month / Dates	Location	Round Robin Leg
June, 2017	NPL, UK	Transfer radiometer preparation
July, 2017	TO, Estonia	
August, 2017	DLR, Germany	European leg (part 1)
September, 2017	NERC, UK	
October, 2017	NIVA, Norway	
December 04-15, 2017	NPL, UK	Transfer radiometer measurement check
December 16-22, 2017	LOV, France	European leg (part 2)
January 08-29, 2018	NOAA, USA	North American leg
February 05-26, 2018	Satlantic, USA & Canada	
March 05-09, 2018	NPL, UK	Transfer radiometer measurement check
March 19-April 02, 2018	CSIRO (Perth), Australia	Australian leg
May 01-08, 2018	NPL, UK	Transfer radiometer measurement check
May 14-18, 2018	JRC, Italy	Final EC leg

Figure 6. Schedule of the FRM4SOC international radiance round robin.

The transfer radiometers were checked by NPL before and after each round of measurements by the participants. The round robin measurements were directly traceable to the NPL primary reference standards, using well-characterised facilities, and were supported by full uncertainty budgets. This direct link to SI not only provided a stringent test of the reliability of the various traceability routes used by the participants, but also allowed the uncertainties associated with the comparison to be evaluated. As in the irradiance comparison, use of the calibration certificates of each participant's lamp and panel was also essential because they are a critical part of the SI traceability and uncertainty evaluation of each participant's radiance measurements.

Each participant was requested to evaluate uncertainties associated with their radiance source operating in their own laboratory for these measurements. This included all the additional uncertainty components related to the alignment of the lamp, panel and radiometer, distance measurements, and other relevant laboratory specific factors such as power supply stability and accuracy. The pilot had discussed all these aspects with participants and trained them in order to facilitate the correct compiling and reporting back of this uncertainty budget evaluation using pre-agreed templates.

The comparison measurand was the calibration factor determined for the transfer radiometer using each participant's own spectral radiance reference (i.e., a lamp-panel combination in a 0°: 45° arrangement with the lamp set at a known distance from the panel). For each participant a separate calibration factor was determined for each of the seven specified wavebands of the transfer radiometers and each waveband was treated independently for the purposes of this analysis. It was recognised that participants may be using different types of reflectance panels to the 46 cm (18 inch) Spectralon panel used by NPL. Thus, it was essential that participants supplied the pilot with the technical details and history of the artefacts along with SI-traceable calibration certificates, the uncertainty evaluation according to the pre-defined and agreed format, see [40], and as much additional information on their laboratory conditions as possible, in order to aid the pilot in carrying out this comparison.

An example of the results from one of the two transfer radiometers of the comparison are presented in Figure 7 in terms of differences between each participant's measurements and the mean value of all of them. The other radiometer showed similar results and in general, they all agree within ±4%, a slightly higher difference than expected. The majority of the results forms a group located at around the 0% line and below on the y-axis values. A second group of 4 entries is located at the level of +3% difference from the mean comparison value. The majority of participants exhibited a range of differences across the channels of within 1% to 2.5%, with the notable exception of one of the participants blue (412 nm) channel measurements.

Additional investigation showed that the reason for these differences and groupings may be caused by a combination of the size of source effect and instrument effective FOV that affected the results of the smaller group. If these effects could be corrected for, or the measurements repeated with different settings, an agreement within 2.5% might be expected. Furthermore, each participant's uncertainty budget for the radiance measurements gave values of between 1.8% to 2.0% for low uncertainty participants, 2.1% to 2.4% for medium uncertainty participants and 2.5% to 3.1% for high uncertainty participants. Full details can be found in [43].

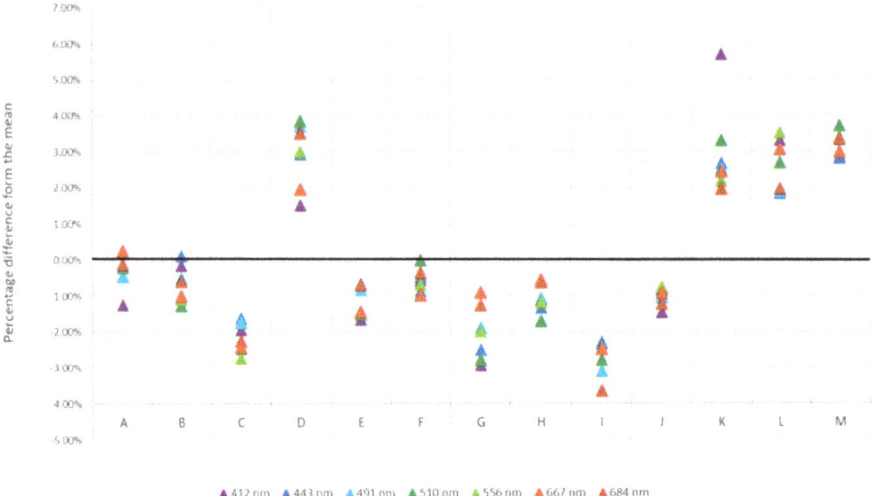

Figure 7. Comparison between radiance sources of participant OCR calibration laboratories (A–M) showing mean coefficient percentage difference across all distances. Reproduced with permission from [43].

6. Comparisons to Verify the Performance of Ocean Colour Radiometers Used for Satellite Validation

The main aim of these comparisons was to link the ocean colour field measurements to the radiometers' SI-traceable calibrations and verify whether different instruments measuring the same light source in the lab, or the same patch of water or sky outdoors, will provide consistent results within the expected uncertainty limits. As an outcome, methodologies used by participants for the measurements and data handling were also critically reviewed.

The laboratory and outdoor comparisons took place at Tartu Observatory (TO) in Estonia and at a lake nearby (Lake Kääriku) between the 8th and 13th of May 2017 with the calibration of all participants' radiometers taking place just prior to this between the 2nd and 7th of May 2017. This was an international event with participants and their radiometers taking part from several different organisations and countries: TO (Tõravere, Estonia) as pilot; Alfred-Wegener-Institut (AWI, Bremerhaven, Germany); Royal Belgian Institute of Natural Sciences (RBINS, Brussels, Belgium); National Research Council of Italy (CNR, Rome, Italy); University of Algarve (CIMA, Faro, Portugal); University of Victoria (UVIC, Victoria, BC, Canada); Sea Bird Scientific (Halifax, NS, Canada); Plymouth Marine Laboratory (PML, Plymouth, UK); Helmholtz-Zentrum Geesthacht (HZG, Geesthacht, Germany); University of Tartu (UT, Tartu, Estonia); Cimel Electronique S.A.S. (Paris, France).

The comparison exercise therefore consisted of three sub-tasks: an SI-traceable radiometric calibration of participating radiometers just before the intercomparison; a laboratory intercomparison of the measurement of stable lamp sources in a controlled environment; and an outdoor intercomparison of the measurement of natural radiation sources at a lake. Altogether, 44 radiometric sensors from 11 institutions were involved: 16 TriOS RAMSES, 2 Satlantic OCR-3000, 4 Satlantic HyperOCR, 4 WaterInsight WISP-3, 1 Cimel SeaPRISM and 1 Spectral Evolution SR-3500 radiance sensors, and 10 TriOS RAMSES, 1 Satlantic OCR-3000, 2 Satlantic HyperOCR, 2 WaterInsight WISP-3, and 1 Spectral Evolution SR-3500 irradiance sensors.

6.1. Laboratory SI-Traceable Radiometric Calibrations

Before the comparisons could take place the first task was the SI-traceable absolute radiometric calibration of the 44 participating radiometers. The calibrations were performed in the optical radiometry laboratory of Tartu Observatory (TO), Estonia. Calibration measurements were performed at the room temperature of 21.5 °C $\pm$ 1.5 °C in an EN ISO 14644 Class 8 equivalent cleanroom environment.

NPL provided two Gigahertz-Optik BN9101-2 FEL-type irradiance calibration standard lamps for the calibrations and comparison exercise. The lamps were calibrated by NPL and had not been used since the last calibration. Differences in responsivity, in the range of 340 nm to 980 nm, were less than $\pm$0.5%. The drift of the irradiance values (at 500 nm) measured during the calibration campaign was ~0.1%, which is close to the detection limit of the filter radiometer. In certificates issued for the radiometers from these calibrations, the arithmetic mean of the responsivity measured by the two lamps was used. Radiance calibration was performed using the same lamps and a Sphere Optics calibrated white reflectance panel. Normal incidence for the illumination and 45° from normal for viewing were used. The panel had been previously calibrated in the same illumination and viewing conditions by NPL.

Additionally, a large number of the sensors involved in the comparisons were recalibrated at TO a year later for the FRM4SOC field intercomparison on the Acqua Alta Oceanographic Tower (AAOT) in the Gulf of Venice (see below) allowing the evaluation of the stability of the sensors. Most of these sensors (>80%) changed less than $\pm$1% during this year.

6.2. Laboratory Intercomparison of Measurements

The main set of laboratory comparisons took the form of carefully controlled measurements of irradiance and radiance using stable lamp sources. These were a seasoned but uncalibrated FEL lamp for irradiance and a Bentham ULS-300 integrating sphere with internal illumination as a stable radiance source. Minimum sets of 30 measurements were taken by each radiometer with overall results seen in the graphs below. Consensus values were calculated as the median [37] of all presented comparison values. Reference values were applicable only for the indoor irradiance measurements, when the measurand used for this exercise was, during the comparison, also measured with a precision filter radiometer serving as a reference.

Despite different sensor types, as the radiation sources used for indoor comparison were spectrally very similar to calibration sources, agreement between sensors was reasonably close for radiance and for irradiance sensors (see Figures 8 and 9). No outliers were present, after correction of data by participants and unified data handling (especially harmonisation of spectral interpolation) by the pilot.

Relative uncertainty budget tables for irradiance and radiance, based on the spread of individual sensors measuring the same source during the indoor comparisons were also produced. Effects of different characteristics of the radiometers, such as temperature dependency, stray light, non-linearity, cosine responsivity and field of view, on the calibration and measurement uncertainty are discussed in detail in [44,45]. In summary though, from the indoor experiment, when conditions were similar to calibration conditions, a high effectiveness of the SI-traceable radiometric calibration has been demonstrated, and a large group of different types of radiometers operated by different scientists achieved a reasonably close consistency giving low standard deviations between radiance (27 in total) or irradiance (15 in total) results ($s < 1\%$). This was, however, only achieved after some unification of measurement and data processing, e.g., alignment of sensors, structuring of collected data, and application of unified wavelength bands, a spectral interpolation method and non-linearity corrections. Nevertheless, variability between sensors may be insufficient for complete quantification of uncertainties in the measurements. For example, standard deviation of nonlinearity estimates versus the mean effect demonstrated that differences are not able to reveal the full size of systematic errors common for all the instruments. Therefore, all radiometers should be individually tested for all significant systematic effects that may affect the results, as this is the only way to get a full estimate of the effects degrading traceability to the SI scale.

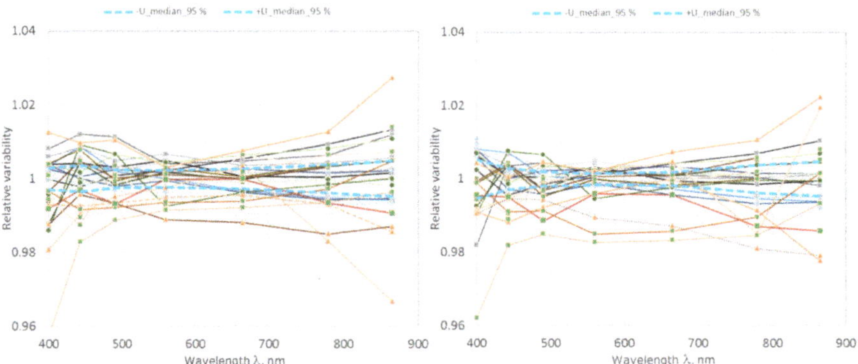

Figure 8. Comparison of low intensity (**left**) and high intensity radiance (**right**) sources as measured by each participating instrument compared to the median over all instruments; after reviewing data by pilot, corrections submitted by participants and/or unified data handling by pilot. Blue dotted lines—expanded uncertainty of the median consensus value. Reproduced with permission from [45].

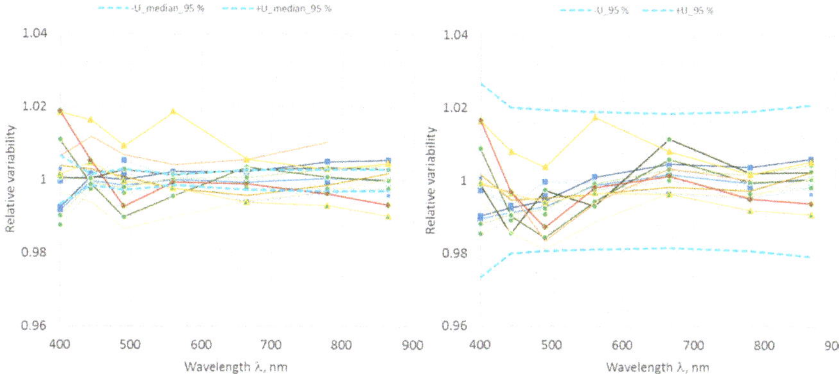

Figure 9. (**left**) Comparison of each participating irradiance sensor measuring an irradiance source compared to the median over all instruments; after reviewing data by pilot, corrections submitted by participants and/or unified data handling by pilot with blue dotted lines the expanded uncertainty of the median consensus value; (**right**) compared to the reference values of the filter radiometer with blue dashed lines the expanded uncertainty covering 95% of all data points and uncertainty of radiometric calibration included. Reproduced with permission from [45].

6.3. Outdoor Intercomparison of Measurements

The outdoor comparisons were conducted as a direct intercomparison of the downwelling irradiance E_d, the downwelling sky radiance L_d, and the total upwelling water radiance L_u from a diving platform on the end of the 50 m pier at the southern shore of Lake Kääriku in Estonia, as shown in Figure 10. The physical and optical characteristics of the part of the lake measured were characteristically eutrophic, well known to the pilot, and are detailed in [44,46].

The outdoor measurements were performed in 5-min casts. Between the pilot announced beginning and end times of casts, all participants recorded the radiance and irradiance data at their usual fieldwork data acquisition rate. Thirty casts were recorded in total, but only seven of them were included in the intercomparisons. The selection of casts was based on the time series of the 550 nm spectral band. The pilot received the 550 nm time series data for 16 radiance and 10 irradiance sensors. Only the casts with the most stable signal and least missing data were selected for further analysis. All the selected casts were measured on May 12 2017—the second day of the outdoor experiment—due to adverse environmental conditions on the first programmed day, forcing the comparisons to be limited to E_d, L_d, and L_u, rather than also including the remote sensing reflectance R_{rs} and the water-leaving radiance L_w derived from simultaneously measured E_d, L_d, and L_u. Consensus values of irradiance and radiance were assigned as the median of the valid casts (C) for each of the conditions measured (Figure 11).

The measurement results for the field casts are presented in Figures 12 and 13 as the deviation from the above consensus values as this was considered the most appropriate way to show differences between the radiometers.

Relative uncertainty budget tables for the downwelling irradiance and water leaving radiance were also produced, based on the spread of individual sensors measuring the same target during the outdoor comparison [44,46]. Investigations to try to explain the marked differences between radiometers and types of radiometer shown in Figures 12 and 13 and these uncertainty tables were also undertaken within the framework of these comparisons and the FRM4SOC project.

Figure 10. Pier and diving platform at the southern coast of Lake Kääriku with all the radiance (**right**) and irradiance radiometers (**left**) mounted in common frames for the outdoor experiment.

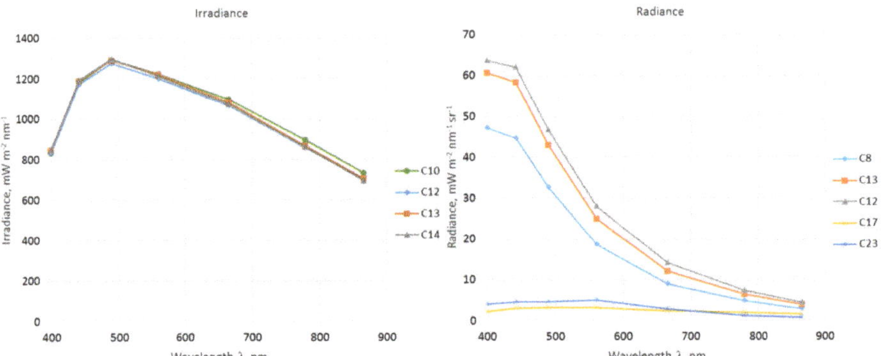

Figure 11. Irradiance and radiance consensus values in the outdoor experiment. C8, C10, C12, C13, C14—blue sky (radiance) or direct sunshine (irradiance); C17—water in cloud shadow; C23—sunlit water. Reproduced with permission from [46].

For irradiance, the difference in cosine response was the main source of differences between different sensor groups revealed during the field experiment. Variability between irradiance sensors was about five times larger than that observed during the indoor laboratory exercise. This large variability between sensors during the outdoor exercise cannot be explained simply by the poor stability of sensors, as a stability check in lab conditions a year later has shown smaller changes than during the outdoor measurements some days after calibration. Variability cannot be fully explained by factors such as temperature, nonlinearity, and stray light either, as one could expect a smaller difference between radiance and irradiance sensors in this case. Most likely, the different behaviours of RAMSES and HyperOCR sensors are largely due to a different construction of input optics of these sensors and imperfect cosine response [47]. This hypothesis was supported by the angular response characterisation of 5 RAMSES irradiance sensors and comparing the integral cosine error values to the deviations from the consensus value in the outdoor experiment [46].

For radiance, the angular response (different fields of view) and spatial non-uniformity of the targets provides the main difference between different sensor groups. In the case of a spatially heterogeneous target (sky with scattered clouds, water at an oblique viewing angle) the large differences of FOV of different sensors will likely cause significant discrepancies between sensors. The variability between radiance sensors was about two times larger than during the indoor exercise. This can be partly explained by the larger effects of factors like temperature, stray light and nonlinearity that were not corrected for during the field experiment. For example, dependence of the calibration coefficients on temperature can cause significant deviation from the SI-traceable result. For a maximum temperature difference of about 20 °C between calibration and later measurements (typically between 0 °C and 40 °C) a responsivity change of more than 10% may be possible [48,49]. This feeds back to the calibration procedure, which may be improved if its conditions are designed to cover situations possible during the use of an instrument in the field. For example, if it is known that the radiometer has a linear response with temperature [49], the responsivity of the radiometer can be evaluated when calibration is performed at three different temperatures covering the possible range of temperature variations during its later use in the field.

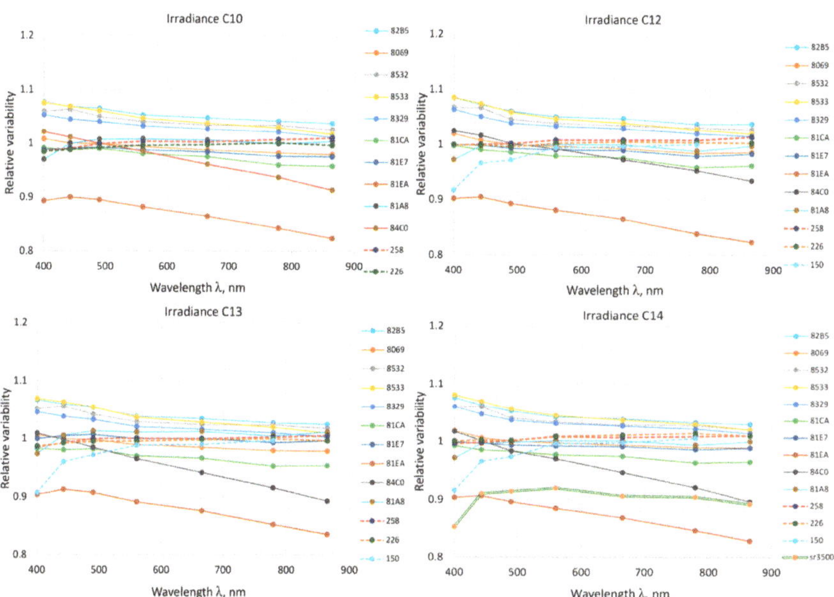

Figure 12. Irradiance sensors compared to the consensus value. Solid lines—RAMSES sensors; dashed lines—HyperOCR sensors; double line—SR-3500. Reproduced with permission from [46].

The different behaviours of RAMSES and HyperOCR sensor groups were also clearly revealed during the comparisons. For the RAMSES group, the variability of radiance sensors during indoor and outdoor exercises was very similar and the HyperOCR and WISP-3 sensors mainly caused the larger variability for the outdoor measurements. For irradiance measurements, the deviation of HyperOCR sensors from the consensus value of the group was very small, and the group of RAMSES sensors was the main cause of an increase in variability.

The spread of irradiance and radiance results from the comparison, with differences between the sensors due to their calibration state before the experiment, is summarised in Figure 14. All standard deviations of laboratory measurements were smaller than 1%. Standard deviations of the field results are substantially higher (1%–5%), but still much smaller than the variability due to the calibration state of the sensors before the experiment (5%–10%), i.e., the calibration that each participant would have used if the radiometers were not freshly calibrated just before the start of the intercomparison exercise. It must be noted, however, that some instruments had not been used for fieldwork in recent years and their calibration coefficients were several years old.

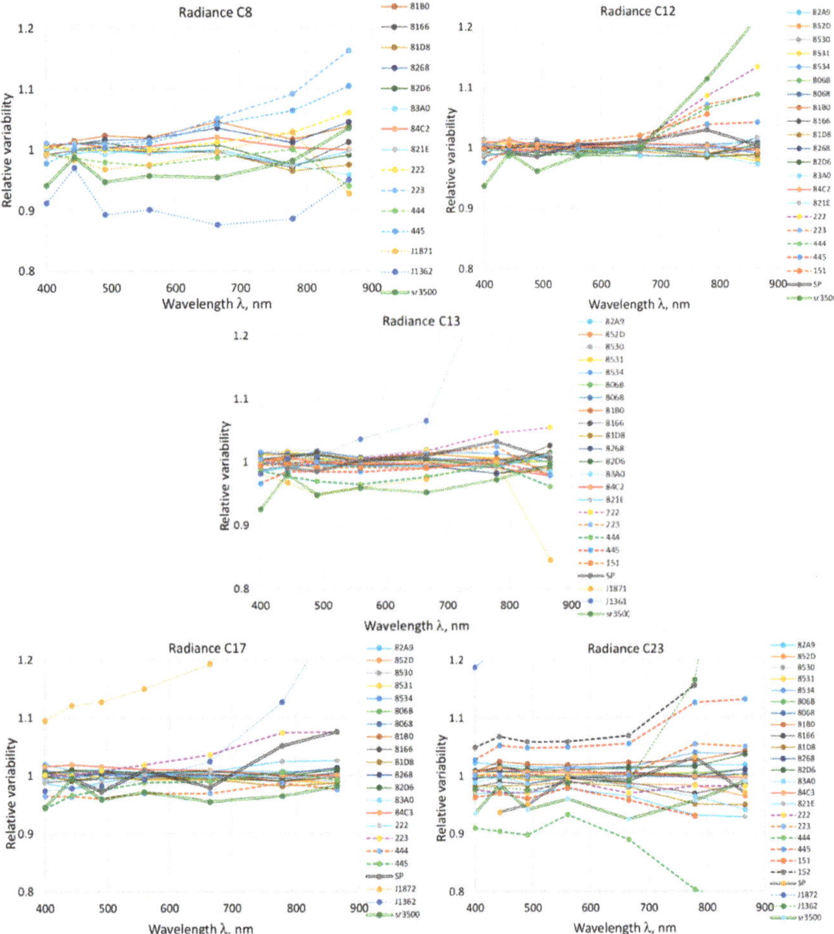

Figure 13. Radiance sensors compared to the consensus value in the outdoor experiment. C8, C12, C13—blue sky; C17—water in cloud shadow at 139° VZA; C23—sunlit water at 130° VZA. Solid lines—RAMSES sensors; dashed lines—HyperOCR sensors; double lines—SeaPRISM (SP) and SR-3500; dotted lines—WISP-3. Reproduced with permission from [46].

In the frame of the outdoor experiment when conditions for calibration and in the field are very different from each other, the variability between freshly calibrated individual sensors did increase substantially. This demonstrated a limitation of typical OC field measurements, even for sensors having recent SI-traceable radiometric calibration. Including laboratory intercomparison in the comparison of OCR sensors has clearly shown that a further reduction of the uncertainty of radiometric calibration of sensors will not improve the agreement between field results significantly. More relevant for achieving better SI-traceability and lower uncertainties in field measurements are improved specifications of radiometers, additional characterisation of individual sensors accounting for specific field conditions, and unified data handling.

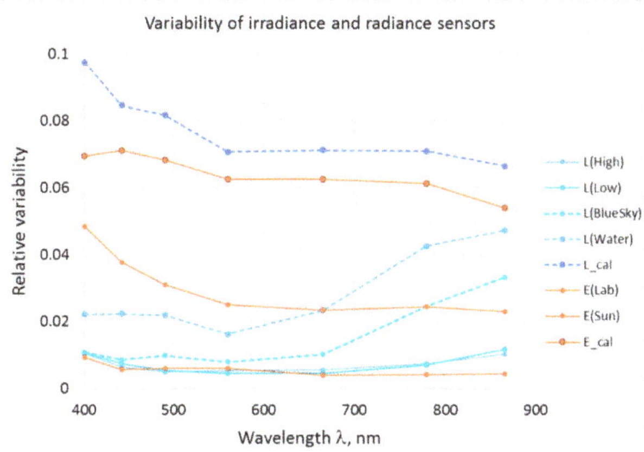

Figure 14. Variability between irradiance and radiance sensors. *E*_cal and *L*_cal—due to calibration state; *E*(Lab), *L*(Low) and *L*(High)—variability in laboratory intercomparison; *E*(Sun), *L*(BlueSky) and *L*(Water) variability in the field. Reproduced with permission from [46].

The indoor experiment demonstrated the effectiveness of performing the radiometric calibration at the same laboratory just before intercomparison measurements [45,46] in obtaining consistent results. However, besides regular calibration, a sufficient individual characterisation of radiometers by testing them for all significant systematic effects is suggested from these comparisons as the best way to enable reduction of biases in outdoor intercomparisons. This should lead to a smaller variability between measurements from different instruments in the field, and a more realistic and complete quantification of uncertainties in measurement. To help in the interpretation of the results and in future outdoor intercomparison campaigns, the following further suggestions were proposed:

- The instruments' internal (photodetector) temperatures should be logged whenever possible;
- During the responsivity calibration, different ambient temperatures should be used;
- Acquisition of the data for all instruments should start synchronously within ±1 s and sampling intervals should be the same, to make it possible to compare the individual spectra instead of temporal averages;
- Characterisation of the angular response of the radiance radiometers is important, especially in the case of variable sky conditions;
- Irradiance measurements under clear sky conditions covering a large span of solar zenith angles are necessary to assess the uncertainties caused by irradiance entrance optics (cosine response);
- Intercomparisons should be done in varying water optical property conditions;
- The calibration history for each participating radiometer is vital in order to detect possible instrument misbehaviour and remove outliers;
- It is highly recommended to use a well-characterised reference instrument;
- An aligned photo or video camera should be used to continuously record the measurement scene during outdoor experiments;
- The data processing algorithms should be well defined and agreed between the participants.

7. Field Intercomparison Experiments to Verify the Performance of Ocean Colour Radiometers Used for Satellite Validation

The overall objective of these field intercomparison experiments was to design and document protocols and procedures and implement field comparisons of FRM OCR radiometers, as well as build a database of OCR field radiometer performance knowledge over several years.

7.1. The Atlantic Meridional Transect (AMT) Cruise Field Intercomparison Experiment

Plymouth Marine Laboratory (PML), in collaboration with the National Oceanography Centre (NOC) Southampton, has operated the AMT since 1995 [50]. The cruise is conducted between the UK and the sparsely sampled South Atlantic during the annual passage from October to November of a NERC ship (RRS James Clark Ross, RRS James Cook or RRS Discovery). The transect covers several ocean provinces where key physical and biogeochemical variables such as chlorophyll, primary production, nutrients, temperature, salinity and oxygen are measured. The stations sampled are principally in the North and South Atlantic Gyres, but also the productive waters of the Celtic Sea, Patagonian Shelf and Equatorial upwelling zone are visited, which therefore offered a wide range of variability in which to conduct field intercomparisons for FRM4SOC.

The results from the AMT cruises have enabled the intercomparison of simultaneous measurements of water leaving radiance and reflectance. The differences observed between these measurements form a key component of estimating errors and uncertainties resulting from environmental variability, as well as instrument deployment methodology, instrument specifications and calibration.

The main AMT comparison for FRM4SOC was conducted from 23rd September to 4th November 2017 from Southampton, UK to South Georgia and the Falkland Islands on AMT-27, to compare along track measurements of L_w and $R_{rs}(\lambda)$ between PML and Tartu Observatory (TO) radiometers. Measurements were carried out in various solar zenith angle, water and weather conditions. The ambient temperature varied from 1 °C to 28 °C. Altogether, data was collected from ~30 stations.

The AMT-27 cruise data consists of synchronised measurements of water leaving reflectance with two sets of hyperspectral radiometers both consisting of three radiometers in order to measure the upwelling radiance $L_u(\lambda)$, downwelling radiance from the sky $L_d(\lambda)$, and downwelling solar irradiance $E_d(\lambda)$. The PML set consisted of three Satlantic HyperSAS sensors and the TO set of three TriOS RAMSES sensors. All radiance and irradiance sensors were SI-traceably calibrated at the Tartu Observatory before and after the campaign. All of these sensors were involved a year before in the laboratory calibration intercomparison campaign (Section 6.1) and demonstrated differences less than ±1% both for radiance and irradiance results during indoor measurements (Section 6.2). However, during the outdoor exercise, the PML irradiance sensors showed up to 6% higher values in the blue part of the spectrum, and the PML radiance sensors showed up to approximately 10% higher values in the red and IR parts of the spectrum when compared to the respective TO sensors.

The radiance sensors $L_d(\lambda)$ and $L_u(\lambda)$ were mounted side by side on a common steel frame positioned at the front of the ship using 40° zenith and nadir viewing angles, respectively. The downwelling irradiance sensors were mounted on another steel frame positioned on the mast at the front of the ship, to avoid any ship shadows. Positioning of sensors ensured nearly identical measurement conditions for both 3-sensor radiometric systems (see Figure 15).

Figure 15. The route of AMT-27 through the Atlantic and the position of the FRM4SOC radiometers on RRS Discovery in operation during AMT-27.

The intercomparison allowed the analysis of the variability of responsivity between different types of freshly calibrated sensors with respect to the environmental and illumination conditions. As an example, the difference in the results of downwelling irradiance between PML and TO, as a function of ambient temperature and solar zenith angle, are shown in Figure 16.

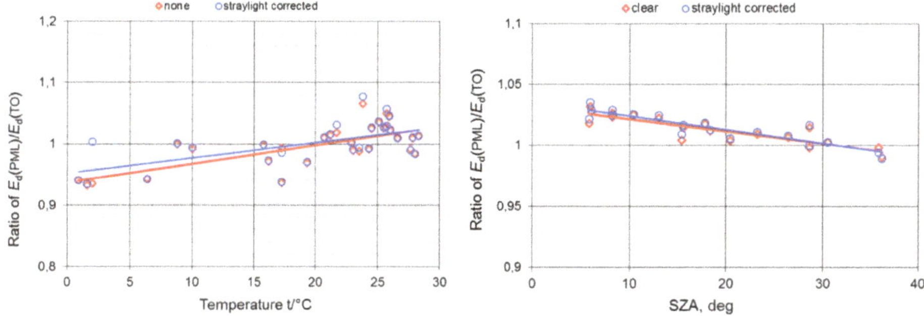

Figure 16. Difference in downwelling irradiance between PML and TO sensors as a function of ambient temperature (**left**) and solar zenith angle (SZA, **right**).

With regard to ambient temperature, radiometric calibration of the sensors was performed in lab conditions at 21 °C and no temperature correction factors were applied for the field results. Responsivity change for both sensors was larger (and unknown) compared to the change of the signal ratio shown. The differences varied from approximately −5 to +5% in the temperature range of 1 to 30 °C. However, the sensors recorded similar irradiance values around 21 °C which corresponds to the calibration temperature. This result clearly shows the need for characterisation of field radiometers for thermal effects.

For solar zenith angle, the variation is in agreement with known or expected errors of the cosine collectors of compared sensors, evaluated to be within ±2% [45,46]. The stray light correction effect is negligible and shown in Figure 16 for reference only.

The comparison of HyperSAS and RAMSES measured water-leaving reflectance after applying stray light correction showed a very high agreement over all wavelengths. The systematic biases were negligible (see Figure 17).

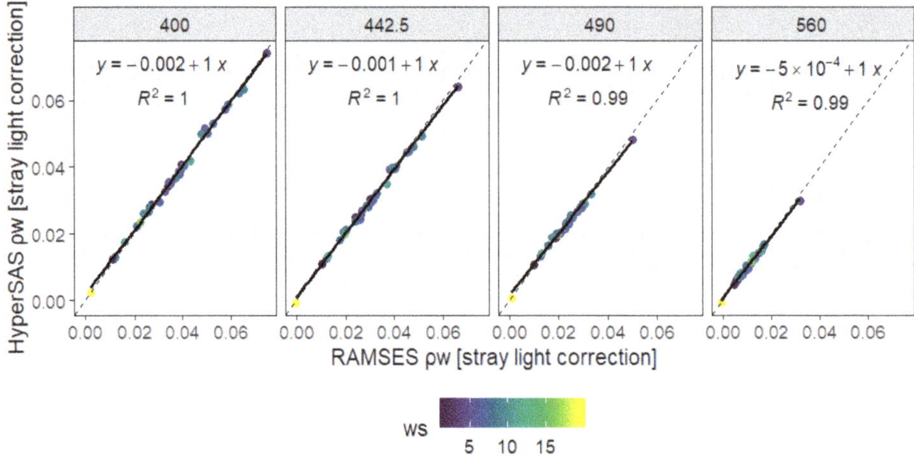

Figure 17. Correlation between HyperSAS and RAMSES measured water-leaving reflectance after stray light correction on selected wavelengths. Colour is wind speed (m s^{-1}) during the measurement.

The comparison between the OLCI-derived and in situ water-leaving reflectance, either by RAMSES (Figure 18A) or HYPERSAS (Figure 18B), showed a very good correlation in the blue to green wavelengths. For these wavelengths, the correlation with OLCI-derived water-leaving reflectance was even better after applying the NIR similarity correction [51,52] (Figure 19).

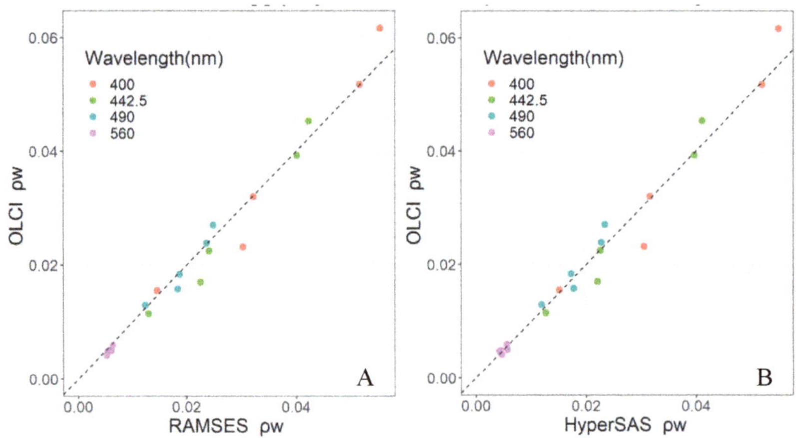

Figure 18. Correlation between OLCI-to-RAMSES (**A**) and OLCI-to-HyperSAS (**B**) water-leaving reflectances at selected wavelengths.

The above summary analysis shows that by comparing results to ancillary instrument data during the cruise (with regards to environmental conditions), the sources of any differences can begin to be established. From these results, recommendations can be made to adjust processing methodology (e.g., applying appropriate filtering thresholds), future instrument deployment methodology, and calibration processes. Furthermore, these comparisons contribute to the Type B estimates in an uncertainty budget [53]. A complete comparison analysis, including uncertainties, is being published using data collected during AMT-27 but nevertheless these initial results are promising, especially given the large differences in environmental conditions experienced during the AMT cruise.

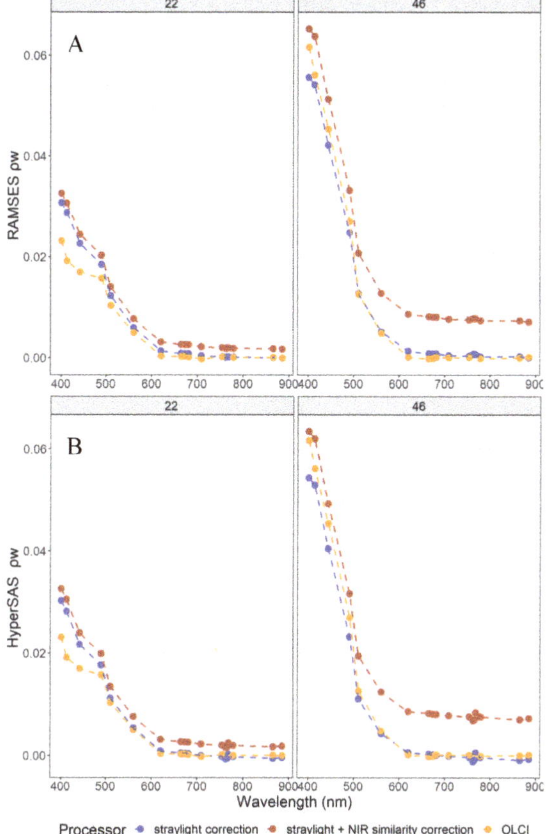

Figure 19. Comparison of RAMSES (**A**) and HyperSAS (**B**) radiometer-derived water-leaving reflectance after stray light correction (blue) and after stray light+NIR similarity correction (brown) compared to OLCI's derived water-leaving reflectance at two stations.

7.2. *The Acqua Alta Oceanographic Tower (AAOT) Field Intercomparison Experiment*

The main aim of the AAOT intercomparison was to assess differences in radiometric quantities determined using a range of above-water and in-water radiometric systems (including both different instruments and processing protocols). Specifically, we evaluated the differences among:

1. Hyperspectral sensors (five above-water TriOS-RAMSES, two Seabird-HyperSAS, one Pan-and-Tilt System with TriOS-RAMSES sensors - PANTHYR, one in-water TriOS-RAMSES system) and multispectral sensors (one in-water Biospherical-C-OPS).
2. In-water and above-water measurement systems.

The field intercomparison was conducted at the Acqua Alta Oceanographic Tower (AAOT) which is located in the Gulf of Venice, Italy, in the northern Adriatic Sea at 45.31°N, 12.50°E during July 2018. The AAOT is a purpose-built steel tower with a platform containing an instrument house to facilitate the measurement of ocean properties under exceptionably stable conditions (Figure 20). In total nine institutes participated in the international intercomparison: University of Algarve (UAlg, Faro, Portugal); Tartu Observatory, University of Tartu (UTar, Tartu, Estonia); Helmholtz-Zentrum Geesthacht (HZG, Geesthacht, Germany); Alfred Wegener Institute (AWI, Bremerhaven, Germany); Royal Belgian Institute of Natural Sciences (RBINS, Brussels, Belgium); Plymouth Marine Laboratory

(PML, Plymouth, United Kingdom); University of Victoria (UVic, Victoria, BC, Canada); Flanders Marine Institute (VLIZ, Ostend, Belgium); Laboratoire d'Oceanographique de Villefranche-sur-Mer (LOV, Villefranche-sur-Mer, France). This enabled the comparison of ten measurement systems comprising 29 radiometers.

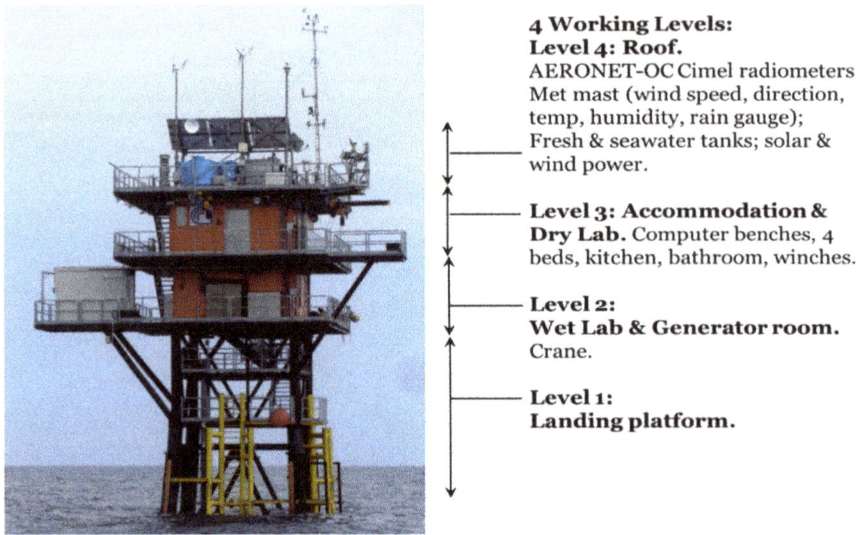

Figure 20. Layout of the Acqua Alta Oceanographic Tower (AAOT). Reproduced with permission from [54].

To rule out any differences arising from absolute radiometric calibration, all of the sensors used during the campaign were calibrated at the University of Tartu (UT), under the same conditions, within ~1 month of the campaign. Measurements were then performed at the AAOT under near ideal conditions, on the same deployment platform and frame, under clear sky conditions, relatively low sun zenith angles and moderately low sea state.

All above-water radiometers except the PANTHYR system were located on the same purpose-built frames. The radiance sensors were located on the deployment platform on level 3 on a 6 m pole that situated them above the solar panels on level 4 (Figure 20). The frame was fabricated from aluminium to position the sensors side by side at 12.3 m from the sea surface (Figure 21A). All L_{sky} and L_t sensors were installed on this frame with identical viewing zenith angles and the deployment frame was adjusted for each measurement sequence to reduce sun glint. The radiance mast was positioned at the same level as the SeaPRISM AERONET-OC system (Figure 21B,C). For irradiance measurements, a telescopic (Fireco) mast was used on level 4 to minimise interference from the tower super-structure and other overhead equipment (Figure 21E,F). The mast and sensors were installed in the eastern corner of the platform at a height of 18.9 m above the sea surface (Figure 21E). The in-water deployment of a TRIOS profiler was carried out using an extendable boom from level 4 of the tower, whereas the C-OPS in water system was deployed from the CNR Research Vessel *Litus*.

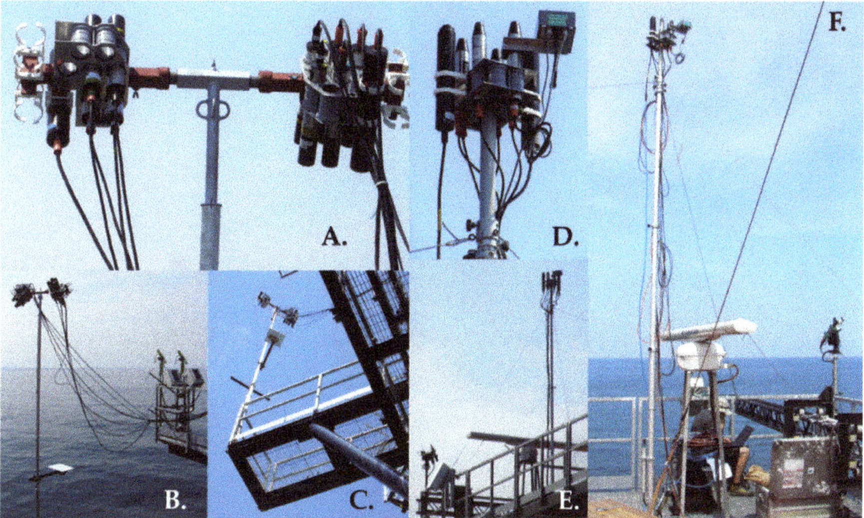

Figure 21. Configuration of sensors on the AAOT platform showing (**A**) the mounting for L_{sky} and L_t radiometers, (**B**) location of radiance sensors next to the AERONET-OC SeaPRISM, (**C**) location of the radiance sensors on level 3 of the AAOT, (**D**) location of the irradiance sensors on the mounting block, (**E**) telescopic mast with irradiance sensors at the eastern corner of the AAOT, (**F**) proximity of the telescopic mast with irradiance sensors and the PANTHYR system just above the railings below. Reproduced with permission from [54].

Measurements were made from the 13th to 17th July 2018. All above water measurements were conducted every 20 min from 08:00 to 13:00 GMT over a discrete measurement period of 5 min (known as casts). In water C-OPS were also coordinated to these times and in water TRIOS measurements were made directly after the above water casts. Only casts with wind speeds < 5 m s^{-1} and clear skies (no cloud) were accepted. Using these criteria, 35 casts were valid from the campaign. Each institute used their standard processing to compute downwelling irradiance (E_d), sky radiance (L_{sky}), radiance from the water surface (L_t) and remote sensing reflectance (R_{rs}). Mean, median and standard deviation values of these parameters over each 5-min cast were submitted. These were compared to the weighted mean of above-water systems that were submitted by the 'blind' submission date, and subsequently used as a reference.

For downwelling irradiance (E_d), there was generally good agreement between sensors with differences of <6% for most of the sensors over the spectral range 400 nm–665 nm. One sensor exhibited a systematic bias, of up to 11%, due to poor cosine response. For L_{sky}, the spectrally averaged difference between optical systems was <2.5% and for L_t the difference was <3.5%. For R_{rs}, the differences between above-water TriOS RAMSES were <3.5% and <2.5% at 443 and 560 nm, respectively, and were <7.5% for some systems at 665 nm. Seabird HyperSAS sensors were on average within 3.5% at 443 nm, 1% at 560 nm, and 3% at 665 nm. The differences between the weighted mean of the above-water and in-water systems was <16.5% across visible bands (Figure 22).

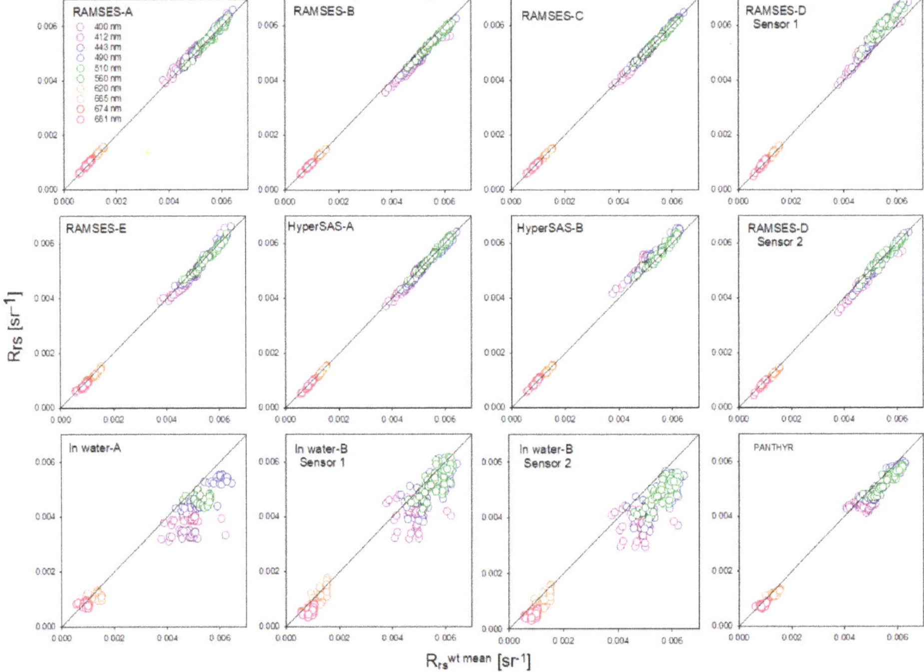

Figure 22. Scatter plots of R_{rs} from the different above- and in-water systems versus weighted mean R_{rs} from above-water systems (RAMSES-A, -B, -C, HyperSAS-A, -B). Reproduced with permission from [54].

These results give an indication of the importance and need for similar regular comparisons in the future highlighting errors in or differences between sensor systems and methods and helping characterise possible uncertainties. A more detailed analysis can be found in [54].

7.3. The FRM4SOC Field Intercomparison Database of OCR

During the course of the project PML designed and built a database for FRM4SOC. Essentially this is a PostgreSQL database with a GIS web portal interface. It provides a web interface to remotely sensed, modelled and in situ data. Its functionality includes the ability to carry out simple analysis and plotting, as well as at all stages of analysis the ability to download data for local processing if preferred. Figure 23 shows the overall design.

The portal uses the Open Geospatial Consortium (OGC) Web Map Service for displaying imagery data and the OGC Web Feature Service (WFS) and Sensor Observation Service (SOS) interface standards for interacting with in situ data. The analysis and plotting capabilities include: time series; latitude or longitude Hovmöller plots; scatter/regression; compositing; animations; and match-ups from CSV files. Data from the AMT cruises and the AAOT experiment have been included along with the calibration and traceability information for the OCR radiometers that were used throughout the FRM4SOC intercomparisons.

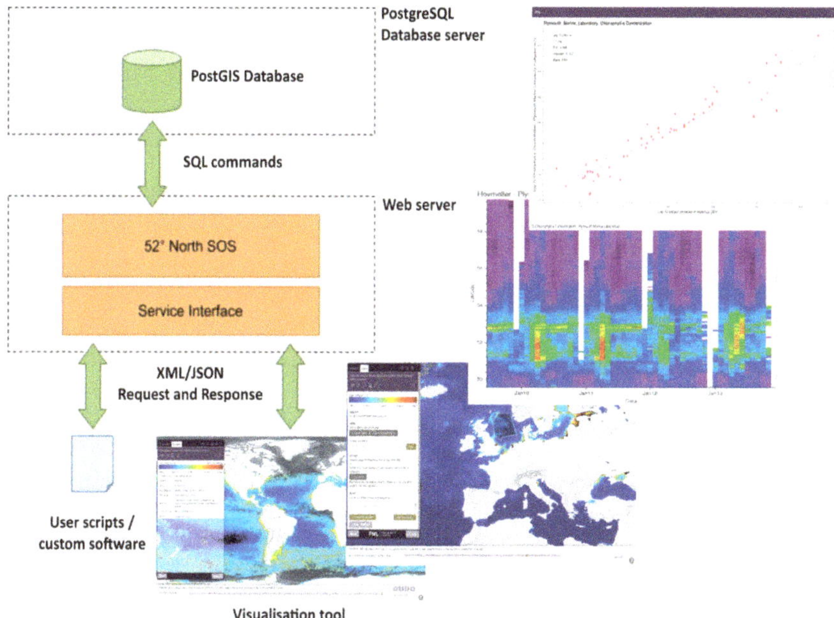

Figure 23. The architecture and functionality of the FRM4SOC field intercomparison database. Graphs and plots are included only as examples of the functionality of the visualisation tool.

8. End-to-End Uncertainty

Having an uncertainty estimate for a measurement is crucial for objectively and numerically gauging how much trust we can place in that measurement. Furthermore, an uncertainty estimate or budget for a field OCR measurement should be constructed and calculated from uncertainty estimates from an unbroken chain of calibrations back to a primary reference standard (preferably SI), in order for this measurement to be considered as an FRM. This concept of end-to-end uncertainty for FRM4SOC meant using NMI agreed protocols to conduct a derivation and specification of uncertainty budgets for FRM OCR field measurements used for satellite OCR validation that had been collected as part of FRM4SOC.

NPL therefore developed a methodology that was based on the guide to the expression of uncertainty in measurement (GUM) [53]. This was based on the Monte Carlo method of uncertainty evaluation GUM supplement [55] and calculated this uncertainty budget for three TriOS RAMSES instruments, one ACC-VIS measuring irradiance and two ARC-VIS measuring radiance, supplied by the University of Tartu [56].

These radiometers were used throughout FRM4SOC, i.e., they were calibrated, characterised and used as part of the laboratory intercomparison measurements, the controlled outdoor intercomparison measurements and the FRM4SOC field intercomparison experiment at the Acqua Alta Oceanographic Tower (AAOT) in the Gulf of Venice (see previous sections). It is these AAOT measurements that were used as the example where uncertainty is propagated from the preceding FRM4SOC calibrations and characterisations. Two sets of observations of irradiance and radiance were used from the AAOT, one from 13th July 2018 between 11:00 and 11:04 ('cast 1') and another from 14th July 2018 between 11:40 and 11:44 local time ('cast 2'). At these times, downwelling irradiance, downwelling radiance and upwelling radiance were all measured simultaneously. Measurements were performed at the AAOT under near ideal conditions, on the same deployment platform and frame (see previous section), under clear sky conditions, sun zenith angles of approximately 24° and moderately low sea state with wind speed of 3.1 m s^{-1} and 0.5 m s^{-1} for each cast, respectively. The average chlorophyll content was Chl = 0.77 mg m^{-3} and absorption of the coloured dissolved organic matter was CDOM (442 nm) = 0.12 m^{-1}.

A Monte Carlo approach was chosen for this uncertainty propagation because the analytical method can become difficult to apply on complex functions with many correlated input parameters where the calculation of sensitivity coefficients is not straightforward. Monte Carlo Methods (MCM) for uncertainty estimation are recognised, accepted and summarised in the GUM supplement [55]. MCM is a numerical method that requires a distinct probability distribution function (PDF) for each of the input components; if input components are correlated then the joint PDF and the measurement equation are required. The MCM will then run a large number of numerical calculations of the measurement equation and with each iteration will use a random choice of each of the inputs from the available range defined by the relevant PDF. The large number of output values calculated using different input values at each iteration, provides the uncertainty of the output value with its PDF.

The true value of a measurement can never be exactly known; only an estimate can be made which is as good as the instrument and method used. Therefore, an error (bias) will always exist between the measured and best estimate value. Figures 24 and 25 illustrate the error (bias) contributions for the measurement equations for downwelling irradiance and water-leaving radiance respectively. These diagrams were first designed in the Horizon 2020 FIDUCEO project [57] to show the sources of uncertainty from their origin through to the measurement equation. The outer labels describe the effects that cause the corresponding uncertainty.

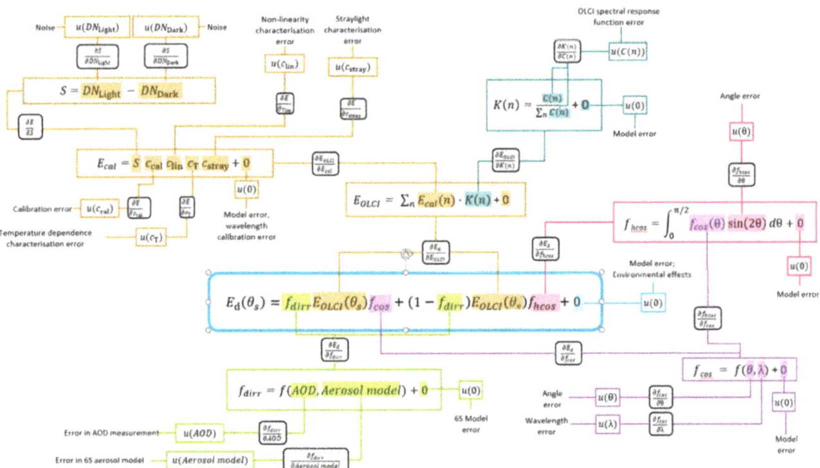

Figure 24. Uncertainty tree diagram for downwelling irradiance (E_d). Reproduced with permission from [56].

To propagate uncertainty for the measurands of interest for FRM4SOC (E_d and L_w) the following Monte Carlo approach was applied:

1. Measurement functions were defined based on the uncertainty tree diagrams that include all inputs defined as quantities that can have an influence on the measurand.
2. All inputs had their standard uncertainty identified in terms of magnitude (value) and PDF shape.
3. The measurement equations were run a large number of times (10^4 in this case).
4. The correlation between some input quantities (for example, the absolute radiometric calibration coefficients of the different instruments) was handled by treating them as systematic contributions, thus the draws from that distribution are not randomised.

5. The final uncertainty value is derived from the resultant PDF.
6. All uncertainties are reported with a $k = 1$ coverage factor.

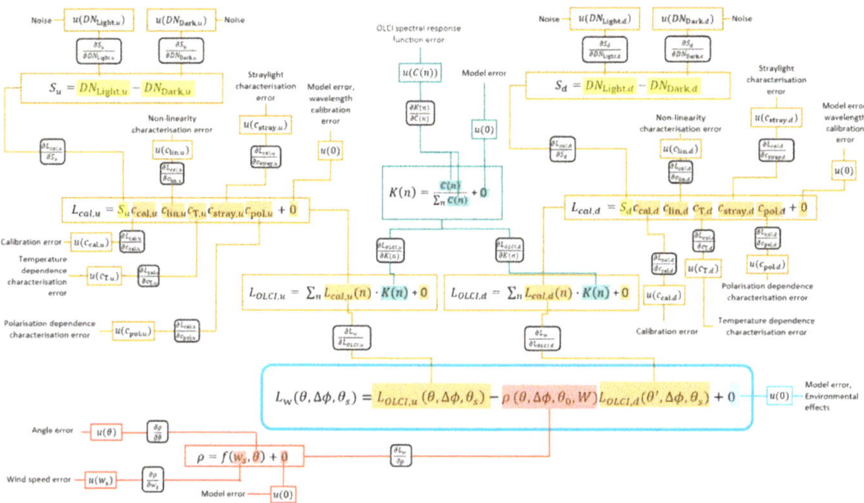

Figure 25. Uncertainty tree diagram for water leaving radiance (L_w). Reproduced with permission from [56].

Specifically, two scenarios were investigated in which the known biases are corrected (the ideal case), and the known biases are not corrected but treated as an uncertainty contributor (the non-ideal case). In addition, we present how the non-ideal case shows an under-estimation of measurement uncertainty because the biases are not corrected and the errors, due to a lack of that correction, are not accounted for. The required data for this activity included downwelling irradiance, downwelling radiance and upwelling radiance as well as all correction factors, the Fresnel reflectance of the water surface, and the fraction of diffuse to direct radiation at the time of measurement.

The resultant outputs of the uncertainty analysis are therefore for the ideal and non-ideal cases, as well as a corrected case where an extra correction is applied to show the true resultant uncertainty when not corrected. The MCM for downwelling irradiance and water-leaving radiance was run over two casts and results in Tables 2 and 3 are presented for the seven OLCI bands of interest (400, 442.5, 490, 560, 665, 778.8, 865 nm). It should be noted that environmental uncertainty is not included, and this may be the major limiting factor since it is likely to be larger than the absolute calibration uncertainty. An evaluation of how to correctly estimate environmental uncertainty for a range of conditions is yet to be completed.

Table 2. The mean and standard uncertainty as a percentage of the mean of the downwelling irradiance, E_d [mWm^{-2} nm^{-1}], presented for the ideal and non-ideal cases. Reproduced with permission from [56].

	Mean E_d			Standard Uncertainty $u(E_\mathrm{d})$ [%]		
Band (nm)	Ideal	Non-Ideal	Bias of E_d	Ideal	Non-Ideal	Corrected Non-Ideal
400.0	1150.0	1080.0	73.9	1.00	4.44	11.30
442.5	1540.0	1480.0	60.8	0.85	2.62	6.72
490.0	1650.0	1590.0	53.8	0.87	2.51	5.89
560.0	1550.0	1510.0	43.5	1.23	2.73	5.61
665.0	1370.0	1320.0	51.9	1.43	3.04	6.99
778.8	1090.0	1050.0	44.6	1.76	3.41	7.67
865.0	899.9	876.0	23.6	2.90	3.90	6.60

Table 3. The mean and standard uncertainty as a percentage of the mean of water-leaving radiance L_w [mWm^{-2} nm^{-1} sr^{-1}] presented for the ideal and non-ideal cases. Reproduced with permission from [56].

	Mean L_w			Standard Uncertainty $u(L_w)$ [%]		
Band (nm)	Ideal	Non-Ideal	Bias of L_w	Ideal	Non-Ideal	Corrected Non-Ideal
400.0	5.90	5.77	0.1340	1.28	3.33	5.65
442.5	7.26	7.22	0.0353	0.94	1.58	2.07
490.0	9.10	9.09	0.0097	0.89	0.89	1.00
560.0	7.35	7.31	0.0442	1.40	1.48	2.08
665.0	0.84	0.86	0.0218	3.89	4.37	6.91

This part of FRM4SOC therefore demonstrated how to conduct an end-to-end uncertainty analysis for in situ radiometers of ocean colour measurements. The results of the three scenarios (ideal, non-ideal and corrected non-ideal) in Tables 2 and 3 highlight the importance and benefits of carrying out instrument characterisations before campaigns and performing instrument corrections in addition to absolute radiometric calibration. It is recommended that the sources of uncertainty that are likely to dominate over the absolute calibration uncertainty (or other more dominant uncertainty contributors which cannot be corrected for) should be characterised before campaigns so that these can be corrected for. This will produce results with reduced uncertainties as demonstrated in the ideal scenario (Tables 2 and 3). The most likely parameters that will need prior characterisations are stray light, cosine response, temperature and non-linearity corrections. Full details can be found in [56] and following these guidelines will support compliance with the FRM requirements of in situ ocean colour measurements for use in satellite product validation.

9. Conclusions and the Road Map for the FRM-Based Future of Satellite Ocean Colour Validation and Vicarious Calibration

The work and results of FRM4SOC highlighted in this paper is already having a significant impact on the earth observation and ocean colour community. In particular, FRM4SOC played a prominent role in the two previous Sentinel-3 validation team meetings at EUMETSAT [58,59], and the FRM4SOC international workshop report [22] on ocean colour system vicarious calibration (OC-SVC) is being used as one of the main requirements reference documents for the future of Copernicus OC-SVC infrastructure. The project has also inspired a sibling in the form of the amt4sentinelfrm project run by the Plymouth Marine Laboratory of the UK specifically for following FRM principles in the measurements taken on the yearly Atlantic Meridional Transect cruises [50,60].

Even though the FRM4SOC developed measurement protocols and uncertainty budgets have been thoroughly tested in several laboratory and in-field comparison exercises, and the space agencies are beginning to demand FRM for satellite product validation, there remains considerable effort required before FRM in ocean colour has gained widespread adoption within the ocean colour validation community. Considering that this continued effort is in support of ensuring high quality and accuracy Copernicus satellite mission data, in particular Sentinel-2 MSI and Sentinel-3 OLCI ocean colour products, and contributes directly to the work of ESA and EUMETSAT to ensure that these instruments are validated in orbit, FRM4SOC produced a scientific road map for the FRM-based future of satellite ocean colour validation and vicarious calibration [61]. Therefore, along with the main project conclusions in the form of recommendations, this paper concludes with the main associated FRM4SOC scientific road map recommended actions (Figure 26).

IMPLEMENTING FRM

C1 Measurement results collected for EO data validation shall have metrological traceability to the units of SI with related uncertainty evaluation.

C2 Space agencies should: *i.* in the medium term, encourage and stimulate the adoption of FRM requirements, and *ii.* in the long term, when sufficient progress and consensus is achieved, use only FRM for the routine validation of satellite ocean colour data. In the near term, use of non-FRM quality data for satellite calibration or validation should only be done with great care.

C3 Space agencies and National Metrology Institutes should consider forming a symbiotic relationship in order to harmonise approaches, methodologies and implement the principles of FRM worldwide.

C4 Financial support from ESA and other space agencies or entities shall be ensured for implementing the principles of FRM.

A1 International collaboration is needed to agree on establishing and implementing FRM requirements.

METHODS, PROTOCOLS AND PROCEDURES AND UNCERTAINTY BUDGETS

C5 International worldwide cooperation on all levels (e.g., agencies, research institutes, experts, etc.) is imperative in order to ensure high quality data for global climate and coastal and inland water environmental monitoring. Different protocols existing for OCR data validation all over the world shall be harmonised, understood and applied in a consistent manner to ensure global uniformity of measurements.

C6 Data (including appropriate metadata) and expertise collected over years by the international community shall be acknowledged, preserved and passed on to the next generations.

C7 Principles of good practice in performing measurements shall be documented and their application encouraged.

C8 Practical consolidated examples on compiling uncertainty budgets shall be provided.

C9 Established methods, principles of good practice, and uncertainty budgets shall be validated in comparison measurements.

C10 Definition, adoption and validation of the principles of good practice and uncertainty budgets shall be supported with appropriate funding from ESA and other space agencies or entities.

A2 International co-operation is needed on all levels to:

    a.   harmonise measurement protocols;

    b.   agree and establish principles of good practice in performing measurements, particularly to estimate and document measurement uncertainties;

    c.   identify, harmonise and establish requirements for measurement and correction of gains and assess their uncertainty levels;

    d.   provide consolidated examples on compiling uncertainty budgets

    e.   provide training on good practice and building uncertainty budgets.

A3 Ensure appropriate funding to define, adopt and validate the principles of good practice and uncertainty budgets.

PROPERTIES OF OCR

C11 The performance of OC radiometers must reflect the needed accuracy for satellite OCR data validation and correspond to requirements as identified and established by the international community in the field. Community consensus on practically feasible requirements is needed. However, the principles of metrology—SI traceability and uncertainty - must be followed.

C12 A document, setting minimum requirements for the most important properties of radiometric instruments used for satellite OCR validation, is needed. Preparation of such a document should be

Figure 26. *Cont.*

encouraged and funded by ESA and other space agencies or entities.

C13 Vital components and specifications for new generation instruments shall be identified and characterisation capabilities of required metrology infrastructure shall be developed accordingly.

C14 ESA and other space agencies or entities should encourage further development of OCR instruments, including a requirement that such developments provide FRM-compatible information on radiometer characterisation.

C15 Characterisation and regular calibration of OCR is needed in order to ensure traceability to the units of SI and evaluate the instrument related uncertainty contributions.

C16 ESA and other space agencies or entities should fund and encourage activities to test radiometers from all manufacturers according to a standardised methodology.

A4 Identify and document requirements and expected specifications (e.g., measurement range, maximum permissible errors, uncertainties, etc.) for Ocean Colour Radiometry (OCR) instruments to meet the requirements for validation of mission data (A2. c.)

A5 Identify, document, map existing and develop missing metrology infrastructure and its capabilities required for calibration and characterisation of OCR (incl. new generation e.g., hyperspectral) instruments.

A6 Identify, document and implement a recommended (standardised) plan for initial and periodic calibration and characterisation of OCR instruments.

A7 Establishment and intercomparison of regional reference laboratories for calibration and characterisation of OCR.

A8 Ensure appropriate funding to identify and document requirements for specifications of OCR instruments and their calibration and characterisation.

COMPARISON EXPERIMENTS AND DATABASE OF OCR FIELD RADIOMETER PERFORMANCE

C17 Periodic comparison experiments are needed for validation of established methods and uncertainty budgets at all levels of the traceability chain.

C18 Comparison experiments also serve the purpose of training, sharing experience, and facilitating common understanding and interpretation of the measurement protocols.

C19 Application of unified data handling or a community processor will reduce overall uncertainty and improve agreement between individual datasets, although care not to limit innovation must be ensured.

C20 Worldwide international participation of agencies and research organisations in comparison exercises shall be aimed for.

C21 ESA and other space agencies or entities shall encourage and support implementing of comparison experiments with appropriate funding.

A9 Organise periodic comparison experiments on all levels of the traceability chain:
    a. reference standards (NMI and OCR calibration laboratory level);
    b. calibration and characterisation methods of OCR (calibration laboratory level);
    c. in situ field measurements:
        • understanding, interpretation, and following established protocols;
        • competence and experience of personnel (all levels).

A10 Development and application of unified data handling/ community processor.

A11 Ensure appropriate funding to organise comparison experiments for validation of established methods and uncertainty budgets on all levels of the traceability chain.

OPTIONS FOR LONG-TERM FUTURE EUROPEAN SATELLITE OCR VICARIOUS ADJUSTMENT

C22 Operational FRM infrastructures to underpin SVC with SI traceability, full uncertainty characterisation and the best possible accuracy and precision are mandatory. Such FRM infrastructure of the quality needed for SVC shall be redundant in order to ensure steady and sufficient data provision.

C23 BOUSSOLE as the existing unique SVC site in Europe must be maintained in the long term and upgraded to full operational status.

C24 Development and long-term operation of a second new European infrastructure for OC-SVC in a suitable

Figure 26. *Cont.*

> location to gain ideal SVC conditions and ensure operational redundancy is needed.
> **A12** Upgrade BOUSSOLE to fully operational status.
> **A13** Develop a new infrastructure based on MOBY-Net and/or new European technology in a suitable location, e.g., the Eastern Mediterranean near Crete.
> **A14** Involvement of National Metrological Institutes (NMIs) at all stages of development of an SVC infrastructure.
> **A15** Train a new group to operate a second SVC.
> **A16** Support long-term interaction of the different SVC operations groups globally.
> **A17** Support scientific and research activities on SVC sites.
> **A18** Ensure long-term investments for both SVC sites.

Figure 26. Conclusions (**C**) from the FRM4SOC project and their recommended actions (**A**).

Author Contributions: Conceptualisation, A.C.B, R.V., C.D., C.L., K.R., G.T., T.C.; methodology, A.C.B., R.V., C.D., K.A., A.B., J.K., C.L., K.R., G.T., V.V.; writing—original draft preparation, ACB; writing—review and editing, A.C.B., R.V., C.D., K.A., A.B., J.K., K.R., G.T., V.V.; visualisation, A.C.B., R.V., K.A., A.B., K.R., G.T., J.K., V.V.; project administration, R.V., C.D., T.C. All authors have read and agreed to the published version of the manuscript.

Funding: This research was funded by the European Space Agency project Fiducial Reference Measurements for Satellite Ocean Colour (FRM4SOC), contract No. 4000117454/16/I-Sbo."

Acknowledgments: Input and comments from numerous scientists and experts are gratefully acknowledged. The FRM4SOC team, G. Zibordi, D. Antoine, all participants of the project events, and manufacturers of the ocean colour radiometers are kindly thanked for the valuable support and contribution to the project outcomes and this paper.

Conflicts of Interest: The authors declare no conflict of interest.

References

1. Copernicus. Europe's Eyes on Earth. Available online: http://www.copernicus.eu (accessed on 13 January 2020).
2. ESA Sentinel Online. Available online: http://sentinel.esa.int (accessed on 13 January 2020).
3. Copernicus Marine Environment Monitoring Service (CMEMS). Available online: http://marine.copernicus.eu (accessed on 13 January 2020).
4. Vanhellemont, Q.; Ruddick, K. Landsat-8 as a Precursor to Sentinel-2: Observations of Human Impacts in Coastal Waters. In Proceedings of the Sentinel-2 for Science Workshop, Frascati, Italy, 20–23 May 2014. ESA Special Publication SP-726.
5. CEOS Working Group on Calibration and Validation (WGCV). Available online: http://www.ceos.org/wgcv (accessed on 13 January 2020).
6. CEOS Cal/Val Portal. Available online: http://calvalportal.ceos.org (accessed on 13 January 2020).
7. Platt, T.; Hoepffner, N.; Stuart, V.; Brown, C. (Eds.) Why Ocean Colour? The Societal Benefits of Ocean-Colour Technology, Reports of the International Ocean-Colour Coordinating Group, No. 7. IOCCG: Dartmouth, Canada, 2008. Available online: https://ioccg.org/wp-content/uploads/2015/10/ioccg-report-07.pdf (accessed on 13 January 2020).
8. IOCCG Brochure, "Why Ocean Colour? The Societal Benefits of Ocean-Colour Radiometry". Available online: http://www.ioccg.org/reports/WOC_brochure.pdf (accessed on 13 January 2020).
9. CEOS Virtual Constellations: Ocean Colour Radiometry. Available online: http://ceos.org/ourwork/virtual-constellations/ocr/ (accessed on 13 January 2020).
10. The International Ocean Colour Coordinating Group (IOCCG). Available online: http://www.ioccg.org/ (accessed on 13 January 2020).
11. Drusch, M.; Bello, U.D.; Carlier, S.; Colin, O.; Fernandez, V.; Gascon, F.; Hoersch, B.; Isola, C.; Laberinti, P.; Martimort, P.; et al. Sentinel-2: ESA's Optical High-Resolution Mission for GMES Operational Services. *Remote Sens. Environ.* **2012**, *120*, 25–36. [CrossRef]
12. Donlon, C.; Berruti, B.; Buongiorno, A.; Ferreira, M.H.; Féménias, P.; Frerick, J.; Goryl, P.; Klein, U.; Laur, H.; Mavrocordatos, C.; et al. The Global Monitoring for Environment and Security (GMES) Sentinel-3 mission. *Remote Sens. Environ.* **2012**, *120*, 37–57. [CrossRef]

13. ESA Fiducial Reference Measurements: FRM. Available online: https://earth.esa.int/web/sppa/activities/frm (accessed on 13 January 2020).

14. Donlon, C.J.; Wimmer, W.; Robinson, I.; Fisher, G.; Ferlet, M.; Nightingale, T.; Bras, B. A Second-Generation Blackbody System for the Calibration and Verification of Seagoing Infrared Radiometers. *J. Atmos. Ocean. Technol.* **2014**, *31*, 1104–1127. [CrossRef]

15. Donlon, C.; Goryl, P. Fiducial Reference Measurements (FRM) for Sentinel-3. In Proceedings of the Sentinel-3 Validation Team (S3VT) Meeting, ESA/ESRIN, Frascati, Italy, 26–29 November 2013.

16. IOCCG, International Network for Sensor Inter-Comparison and Uncertainty assessment for Ocean Color Radiometry (INSITU-OCR). Available online: http://www.ioccg.org/groups/INSITU-OCR_White-Paper.pdf (accessed on 13 January 2020).

17. FRM4SOC–Fiducial Reference Measurements for Satellite Ocean Colour. Available online: https://frm4soc.org (accessed on 13 January 2020).

18. Global Climate Observing System (GCOS). Available online: https://gcos.wmo.int/en/home (accessed on 13 January 2020).

19. Clark, D.K.; Gordon, H.R.; Voss, K.J.; Ge, Y.; Broenkow, W.; Trees, C.C. Validation of atmospheric corrections over oceans. *J. Geophs. Res.* **1997**, *102*, 17209–17217. [CrossRef]

20. Clark, D.K.; Yarbrough, M.A.; Feinholz, M.; Flora, S.; Broenkow, W.; Kim, Y.S.; Johnson, B.C.; Brown, S.W.; Yuen, M.; Mueller, J. *MOBY, A Radiometric Buoy for Performance Monitoring and Vicarious Calibration of Satellite Ocean Color Sensors: Measurement and Data Analysis Protocols. In Ocean Optics Protocols for Satellite Ocean Color Sensor Validation, Revision 4*; Mueller, J.L., Fargion, G.S., McClain, C.R., Eds.; NASA Goddard Space Flight Center: Greenbelt MD, USA, 2003; Volume 6, pp. 3–34.

21. Antoine, D.A.; Guevel, P.; Deste, J.-F.; Becu, G.; Louis, F.; Scott, A.J.; Bardey, P. The "BOUSSOLE" Buoy-A New Transparent to Swell Taut Mooring Dedicated to Marine Optics: Design, Tests, and Performance at Sea. *J. Atmos. Ocean. Technol.* **2008**, *25*, 968–989. [CrossRef]

22. Lerebourg, C.; Deliverable, D. FRM4SOC D-240, Proceedings of WKP-1 (PROC-1). Report of the International Workshop, 2017. Available online: https://frm4soc.org/wp-content/uploads/filebase/FRM4SOC-WKP1-D240-Workshop_Report_PROC-1_v1.1_signedESA.pdf (accessed on 13 January 2020).

23. Zibordi, G.; Mélin, F.; Voss, K.; Johnson, B.C.; Franz, B.A.; Kwiatkowska, E.; Huot, J.P.; Wang, M.; Antoine, D. System vicarious calibration for ocean color climate change applications: Requirements for in situ data. *Remote Sens. Environ.* **2014**, *159*, 361–369. [CrossRef]

24. Zibordi, G.; Mélin, F. An evaluation of marine regions relevant for ocean color system vicarious calibration. *Remote Sens. Environ.* **2017**, *190*, 122–136. [CrossRef]

25. Zibordi, G.; Mélin, F.; Talone, M. *JRC Technical Report EU 28433 EN: System Vicarious Calibration for Copernicus Ocean Colour Missions: Requirements and Recommendations for a European Site*; European Union Publications Office: Rue Mercier, Luxembourg, 2017.

26. Voss, K.; Johnson, B.C.; Yarbrough, M.; Flora, S.; Feinholz, M.; Houlihan, T.; Peters, D.; Gleason, A. Overview and status of MOBY-Net concept. International Ocean Colour Symposium, Lisbon, Portugal, May 2017. Available online: https://iocs.ioccg.org/wp-content/uploads/2017/05/mon-1445-bo3-voss.pdf (accessed on 29 January 2020).

27. Ruddick, K.G.; Voss, K.; Banks, A.C.; Boss, E.; Castagna, A.; Frouin, R.; Hieronymi, M.; Jamet, C.; Johnson, B.C.; Kuusk, J.; et al. A Review of Protocols for Fiducial Reference Measurements of Downwelling Irradiance for the Validation of Satellite Remote Sensing Data over Water. *Remote Sens.* **2019**, *11*, 1742. [CrossRef]

28. Ruddick, K.G.; Voss, K.; Boss, E.; Castagna, A.; Frouin, R.; Gilerson, A.; Hieronymi, M.; Johnson, B.C.; Kuusk, J.; Lee, Z.; et al. A Review of Protocols for Fiducial Reference Measurements of Water-Leaving Radiance for Validation of Satellite Remote-Sensing Data over Water. *Remote Sens.* **2019**, *11*, 2198. [CrossRef]

29. Mueller, J.L.; Fargion, G.S.; McClain, C.R. (Eds.) *Ocean Optics Protocols for Satellite Ocean Color Sensor Validation, Revision 5*; NASA Goddard Space Flight Center: Greenbelt, MD, USA, 2004.

30. Zibordi, G.; Voss, K.J.; Johnson, B.C.; Mueller, J.L. *Protocols for Satellite Ocean Colour Data Validation: In Situ Optical Radiometry (v3.0) in IOCCG Ocean Optics and Biogeochemistry Protocols for Satellite Ocean Colour Sensor Validation, Volume 3.0*; IOCCG: Dartmouth, NS, Canada, 2019. [CrossRef]

31. McClain, C.R.; Cleave, M.L.; Feldman, G.C.; Gregg, W.W.; Hooker, S.B.; Kuring, N. Science Quality SeaWiFS Data for Global Biosphere Research. *Sea Technol.* **1998**, *9*, 10–16.

32. Mouw, C.B.; Greb, S.; Aurin, D.; DiGiacomo, P.M.; Lee, Z.; Twardowski, M.; Binding, C.; Hu, C.; Ma, R.; Moore, T.; et al. Aquatic Color Radiometry Remote Sensing of Coastal and Inland Waters: Challenges and Recommendations for Future Satellite Missions. *Remote Sens. Environ.* **2015**, *160*, 15–30. [CrossRef]

33. MERIS Optical Measurement Protocols Part A: In-situ Water Reflectance Measurements, CO-SCI-ARG-TN-008 Issue 2.0, August 2011 ESA/ARGANS. Available online: http://mermaid.acri.fr/dataproto/CO-SCI-ARG-TN-0008_MERIS_Optical_Measurement_Protocols_Issue2_Aug2011.pdf (accessed on 14 January 2020).

34. Barker, K.; Mazeran, C.; Lerebourg, C.; Bouvet, M.; Antoine, D.; Ondrusek, M.; Zibordi, G.; Lavender, S. MERMAID: The MEris MAtchup In-Situ Database. In Proceedings of the 2nd MERIS/(A)ATSR User Workshop, ESA ESRIN, Frascati, Italy, 22–26 September 2008.

35. Zibordi, G.; Holben, B.; Slutsker, I.; Giles, D.; D'Alimonte, D.; Mélin, F.; Berthon, J.-F.; Vandemark, D.; Feng, H.; Schuster, G.; et al. AERONET-OC: A Network for the Validation of Ocean Color Primary Products. *J. Atmos. Ocean. Technol.* **2009**, *26*, 1634–1651. [CrossRef]

36. Ruddick, K.G.; Banks, A.C.; Donlon, C.; Kuusk, J.; Tilstone, G.H.; Vabson, V.; Vendt, R. 'FRM4SOC D-70: Technical Report TR-2, A Review of Commonly Used Fiducial Reference Measurement (FRM) Ocean Colour Radiometers (OCR) used for Satellite Validation', 2018. Available online: https://frm4soc.org/wp-content/uploads/filebase/FRM4SOC-TR2_TO_signedESA.pdf (accessed on 14 January 2020).

37. Joint Committee for Guides in Metrology (JCGM). JCGM 200:2012 (JCGM 200:2008 with Minor Corrections): International Vocabulary of Metrology–Basic and General Concepts and Associated Terms, VIM 3rd ed. Bureau International des Poids et Mesures (BIPM): Paris, France, 2012. Available online: https://www.bipm.org/utils/common/documents/jcgm/JCGM_200_2012.pdf (accessed on 14 January 2020).

38. Martin, J.E.; Fox, N.P.; Key, P.J. A Cryogenic Radiometer for Absolute Radiometric Measurements. *Metrologia* **2005**, *21*. [CrossRef]

39. Banks, A.C.; Bialek, A.; Servantes, W.; Goodman, T.; Woolliams, E.R.; Fox, N.P. FRM4SOC D-80a, Protocols and Procedures to Verify the Performance of Reference Irradiance Sources used by Fiducial Reference Measurement Ocean Colour Radiometers for Satellite Validation (TR-3a), 2017. Available online: https://frm4soc.org/wp-content/uploads/filebase/D80a_LCE1_TR3a_NPL_signedTO.pdf (accessed on 14 January 2020).

40. Banks, A.C.; Goodman, T.; Greenwell, C.; Bialek, A.; Scott, B.H.G.; Woolliams, E.R.; Fox, N.P. FRM4SOC D-80b: Protocols and Procedures to Verify the Performance of Reference Radiance Sources used by Fiducial Reference Measurement Ocean Colour Radiometers for Satellite Validation (TR-3b), 2017. Available online: https://frm4soc.org/wp-content/uploads/filebase/D80b_LCE1_TR3b_NPL_signedTO.pdf (accessed on 14 January 2020).

41. Woolliams, E.R.; Fox, N.P.; Cox, M.G.; Harris, P.M.; Harrison, N.J. Final report on CCPR K1-a: Spectral irradiance from 250 nm to 2500 nm. *Metrologia* **2006**, *43*. Technical Supplement. Full report. Available online: https://www.bipm.org/utils/common/pdf/final_reports/PR/K1/CCPR-K1.a.pdf (accessed on 14 January 2020). [CrossRef]

42. Goodman, T.; Servantes, W.; Woolliams, E.; Sperfeld, P.; Simionescu, M.; Blattner, P.; Källberg, S.; Khlevnoy, B.; Dekker, P. Final report on the EURAMET.PR-K1.a-2009 comparison of spectral irradiance 250 nm—2500 nm. Metrologia 2015, 52, Technical Supplement; Full report. Available online: https://www.bipm.org/utils/common/pdf/final_reports/PR/K1/EURAMET-PR-K1.a-Final-Report.pdf (accessed on 14 January 2020).

43. Bialek, A.; Goodman, T.; Woolliams, E.; Brachmann, J.F.S.; Schwarzmaier, T.; Kuusk, J.; Ansko, I.; Vabson, V.; Lau, I.C.; MacLellan, C.; et al. Results from Verification of Reference Irradiance and Radiance Sources Laboratory Calibration Experiment Campaign. *Remote Sens.* **2020**, in press.

44. Kuusk, J.; Vabson, V.; Ansko, I.; Vendt, R.; Alikas, K. FRM4SOC D-170: Results from the First FRM4SOC Field Ocean Colour Radiometer Verification Round Robin Campaign (TR-6), 2018. Available online: https://frm4soc.org/wp-content/uploads/filebase/FRM4SOC-D-170-TR-6-TO_signed.pdf (accessed on 15 January 2020).

45. Vabson, V.; Kuusk, J.; Ansko, I.; Vendt, R.; Alikas, K.; Ruddick, K.; Ansper, A.; Bresciani, M.; Burmester, H.; Costa, M.; et al. Laboratory Intercomparison of Radiometers Used for Satellite Validation in the 400–900 nm Range. *Remote Sens.* **2019**, *11*, 1101. [CrossRef]

46. Vabson, V.; Kuusk, J.; Ansko, I.; Vendt, R.; Alikas, K.; Ruddick, K.; Ansper, A.; Bresciani, M.; Burmester, H.; Costa, M.; et al. Field Intercomparison of Radiometers Used for Satellite Validation in the 400–900 nm Range. *Remote Sens.* **2019**, *11*, 1129. [CrossRef]

47. Mekaoui, S.; Zibordi, G. Cosine error for a class of hyperspectral irradiance sensors. *Metrologia* **2013**, *50*, 187. [CrossRef]

48. Antoine, D.; Schroeder, T.; Slivko, M.; Klonowski, W.; Doblin, M.; Lovell, J.; Boadle, D.; Baker, B.; Botha, E.; Robinson, C.; et al. IMOS Radiometry Task Team, Final Report, 2017. Available online: http://imos.org.au/fileadmin/user_upload/shared/IMOS%20General/documents/Task_Teams/IMOS-RTT-final-report-submission-30June2017.pdf (accessed on 15 January 2020).

49. Zibordi, G.; Talone, M.; Jankowski, L. Response to Temperature of a Class of In Situ Hyperspectral Radiometers. *J. Atmos. Ocean. Technol.* **2017**, *34*, 1795–1805. [CrossRef]

50. Atlantic Meridional Transect. Available online: https://www.amt-uk.org (accessed on 15 January 2020).

51. Ruddick, K.; De Cauwer, V.; Park, Y.; Moore, G. Seaborne measurements of near infrared water-leaving reflectance: The similarity spectrum for turbid waters. *Limnol. Oceanogr.* **2006**, *51*, 1167–1179. [CrossRef]

52. Ruddick, V.; De Cauwer, V.; Van Mol, B. Use of the near infrared similarity reflectance spectrum for the quality control of remote sensing data. *Remote Sens. Coastal Ocean. Environ.* **2005**, *5885*, 588501.

53. Joint Committee for Guides in Metrology (JCGM). *JCGM 100:2008, GUM 1995 with Minor Corrections, Evaluation of Measurement Data-Guide to the Expression of Uncertainty in Measurement*; Bureau International des Poids et Mesures (BIPM): Paris, France, 2008.

54. Tilstone, G.; Dall'Olmo, G.; Hieronymi, M.; Ruddick, K.; Beck, M.; Ligi, M.; Costa, M.; D'Alimonte, D.; Vellucci, V.; Vansteenwegen, D.; et al. Field Intercomparison of Radiometer Measurements for Ocean Colour Validation. *Remote Sensing* **2020**. in prass.

55. Joint Committee for Guides in Metrology (JCGM). *JCGM 101:2008. Evaluation of Measurement Data-Supplement 1 to the "Guide to the Expression of Uncertainty in Measurement"-Propagation of Distributions Using a Monte Carlo Method*; Bureau International des Poids et Mesures (BIPM): Paris, France, 2008.

56. Białek, A.; Douglas, S.; Kuusk, J.; Ansko, I.; Vabson, V.; Vendt, R.; Casal, T. Example of Monte Carlo Method Uncertainty Evaluation for Above-Water Ocean Colour Radiometry. *Remote Sens.* **2020**, *12*, 780. [CrossRef]

57. FIDUCEO FIDelity and Uncertainty in Climate data Records from Earth Observations. Available online: http://www.fiduceo.eu (accessed on 15 January 2020).

58. Kwiatkowska, E.; Nieke, J. Ocean Colour Subgroup Conclusions and Recommendations. In Proceedings of the 4th Sentinel 3 Validation Team (S3VT) Meeting, EUMETSAT, Darmstadt, Germany, 13–15 March 2018.

59. Kwiatkowska, E.; Nieke, J. Ocean Colour Subgroup Conclusions and Recommendations. In Proceedings of the 5th Sentinel 3 Validation Team (S3VT) Meeting, EUMETSAT, Darmstadt, Germany, 7–9 May 2019.

60. amt4sentinelfrm: Validating Copernicus Sentinel data products using Fiducial Reference Measurements. Available online: https://amt4sentinelfrm.org/Home (accessed on 15 January 2020).

61. Vendt, R.; Banks, A.C.; Bialek, A.; Casal, T.; Donlon, C.; Fox, N.; Kuusk, J.; Lerebourg, C.; Ruddick, K.; Tilstone, G.; et al. FRM4SOC D-280: FRM4SOC Scientific and Operational Roadmap. Available online: https://frm4soc.org/wp-content/uploads/filebase/D-280-FRM4SOC-SOR_-signed_UT_ESA.pdf (accessed on 16 January 2020).

Review

A Review of Protocols for Fiducial Reference Measurements of Downwelling Irradiance for the Validation of Satellite Remote Sensing Data over Water

Kevin G. Ruddick [1,*], Kenneth Voss [2], Andrew C. Banks [3], Emmanuel Boss [4], Alexandre Castagna [5], Robert Frouin [6], Martin Hieronymi [7], Cedric Jamet [8], B. Carol Johnson [9], Joel Kuusk [10], Zhongping Lee [11], Michael Ondrusek [12], Viktor Vabson [10] and Riho Vendt [10]

[1] Royal Belgian Institute of Natural Sciences (RBINS), Operational Directorate Natural Environment, 29 Rue Vautierstraat, 1000 Brussels, Belgium
[2] Physics Department, University of Miami, Coral Gables, FL 33124, USA
[3] Institute of Oceanography, Hellenic Centre for Marine Research (HCMR), Former US Base Gournes, 71500 Hersonissos, Crete, Greece
[4] University of Maine, Orono, ME 04469, USA
[5] Protistology and Aquatic Ecology Research Group, Gent University, Krijgslaan 281, 9000 Gent, Belgium
[6] Scripps Institution of Oceanography, University of California San Diego, 8810 Shellback Way, La Jolla, CA 92037, USA
[7] Institute of Coastal Research, Helmholtz-Zentrum Geesthacht (HZG), Max-Planck-Str. 1, 21502 Geesthacht, Germany
[8] Université du Littoral-Côte d'Opale, CNRS, Université de Lille, UMR8187, F62930 Wimereux, France
[9] National Institute of Standards and Technology (NIST), 100 Bureau Drive, Gaithersburg, MD 20899, USA
[10] Tartu Observatory, University of Tartu, 61602 Tõravere, Estonia
[11] School for the Environment, University of Massachusetts Boston, 100 Morrissey Blvd., Boston, MA 02125-3393, USA
[12] National Oceanic and Atmospheric Administration (NOAA), Center for Weather and Climate Prediction, 5830 University Research Court, College Park, MD 20740, USA
* Correspondence: kruddick@naturalsciences.be

Received: 17 June 2019; Accepted: 18 July 2019; Published: 24 July 2019

Abstract: This paper reviews the state of the art of protocols for the measurement of downwelling irradiance in the context of Fiducial Reference Measurements (FRM) of water reflectance for satellite validation. The measurement of water reflectance requires the measurement of water-leaving radiance and downwelling irradiance just above water. For the latter, there are four generic families of method, using: (1) an above-water upward-pointing irradiance sensor; (2) an above-water downward-pointing radiance sensor and a reflective plaque; (3) a Sun-pointing radiance sensor (sunphotometer); or (4) an underwater upward-pointing irradiance sensor deployed at different depths. Each method—except for the fourth, which is considered obsolete for the measurement of above-water downwelling irradiance—is described generically in the FRM context with reference to the measurement equation, documented implementations, and the intra-method diversity of deployment platform and practice. Ideal measurement conditions are stated, practical recommendations are provided on best practice, and guidelines for estimating the measurement uncertainty are provided for each protocol-related component of the measurement uncertainty budget. The state of the art for the measurement of downwelling irradiance is summarized, future perspectives are outlined, and key debates such as the use of reflectance plaques with calibrated or uncalibrated radiometers are presented. This review is based on the practice and studies of the aquatic optics community and the validation of water reflectance, but is also relevant to land radiation monitoring and the validation of satellite-derived land surface reflectance.

Keywords: downwelling irradiance; satellite validation; Fiducial Reference Measurements; water reflectance

---

1. Introduction

The objective of this paper is to review the state-of-the-art of protocols for the measurement of downwelling irradiance, as used for the validation of satellite remote sensing data over water.

1.1. The Need for Fiducial Reference Measurements for Satellite Validation

Satellite remote sensing data is now used routinely for many applications, including the monitoring of oceanic phytoplankton in the context of global climate change, the detection of harmful algal blooms in coastal and inland waters, the management of sediment transport in coastal water, estuaries, and ports, the optimization and monitoring of dredging operations, etc. [1]. To be able to trust and use the remote sensing data, this must be validated, usually by a "matchup" comparison of simultaneous measurements by satellite and in situ. The terminology of "Fiducial Reference Measurements (FRM)" was introduced to establish the requirements on the in situ measurements that can be trusted for use in such validation. Using the definition proposed by [2] in the context of sea surface temperature measurements, the defining mandatory characteristics of a "Fiducial Reference Measurement (FRM)" are:

- An uncertainty budget for all FRM instruments and derived measurements is available and maintained, and is traceable where appropriate to the *International System of Units/Système International d'unités* (SI), ideally through a National Metrology Institute;
- **FRM measurement protocols and community-wide management practices (measurement, processing, archive, documents, etc.) are defined and adhered to;**
- FRM measurements have documented evidence of SI traceability that is validated by an intercomparison of instruments under operational-like conditions;
- FRM measurements are independent from the satellite retrieval process.

The second term above, given in bold, situates the current review, which should provide such a definition of measurement protocols for the downwelling irradiance measurement.

1.2. Scope and Definitions

This review is focused on the validation of satellite data products for water reflectance at the bottom of the atmosphere. In the present review, the terminology of "remote sensing reflectance", R_{rs}, is used as shown in Equation (1):

$$R_{rs}(\lambda, \theta, \phi) = \frac{L_w(\lambda, \theta, \phi)}{E_d^{0+}(\lambda)} \tag{1}$$

where $E_d^{0+}(\lambda)$ is the above-water downwelling irradiance, which is also called the "spectral downward plane irradiance", and $L_w(\lambda, \theta, \varphi)$ is the water-leaving radiance [3], after the removal of the air–water interface reflection, just above the water in the upward direction measured by the radiance sensor and defined by nadir viewing angle θ and azimuth angle φ. The conventions used for these angles are defined in Figure 1.

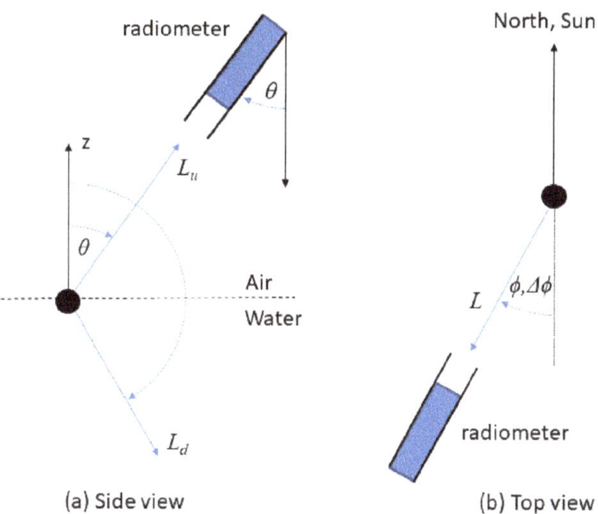

Figure 1. Nadir and azimuth viewing angle conventions illustrated for a reference system centered on the water surface (black dot). (**a**) Viewing nadir angle, θ, is measured from the downward vertical axis: upward radiances are viewed at $\theta < \pi/2$, downward radiances (from sky and Sun) are viewed at $\theta > \pi/2$. (**b**) Azimuth viewing angle, ϕ, and relative azimuth viewing angle, $\Delta\phi$, are measured for viewing directions clockwise from the north and Sun respectively: radiance viewed by a radiometer pointing toward north has an azimuth viewing angle of 0, and radiance viewed by a radiometer pointing toward and away from the Sun have relative azimuth viewing angles of 0 and π, respectively.

$E_d^{0+}(\lambda)$ is itself defined [3] as the integral of radiance, $L(\lambda, \theta, \phi)$, over the downward hemisphere of solid angles (giving the geometric factor $Sin\theta$) and weighted by $|Cos(\theta)|$ (since this is plane irradiance) and is measured in Wm^{-2} nm^{-1}:

$$E_d^{0+}(\lambda) = \int_{\phi=0}^{2\pi} \int_{\theta=\pi/2}^{\pi} L(\lambda, \theta, \phi) \, |Cos(\theta)| Sin\theta d\theta d\phi \tag{2}$$

In the following text, λ, θ, and ϕ are omitted in the notations for brevity.

The θ integral limits from $\pi/2$ to π in Equation (2) correspond to the nadir viewing angle convention defined in Figure 1, but are different from the integration limits from 0 to $\pi/2$ found in some references, e.g., Equation (2.9) of [4], which defines θ as the incidence angle of photons from air. While there is diversity in the nadir/zenith angle terminology in different references, and Figures 2.1 and 2.4 of [4] are themselves quite ambiguous in the use of θ, in practice it is not difficult to follow a consistent angle convention. Similarly, for azimuth angles, these may be defined in some references for the light propagation direction or for the direction toward which the radiometer is pointing (or, in satellite metadata, for the azimuth of the satellite/Sun as seen from the ground location). These azimuth angle conventions can easily be understood and converted provided that they are well defined.

Thus, the validation of R_{rs} is based on simultaneous measurement of two parameters: E_d^{0+} and L_w. A companion paper [5] focuses on the measurement of $L_w(\lambda)$. The present review focuses on the measurement of E_d^{0+}, reviewing the state-of-the-art of measurement protocols in the FRM context, particularly as regards components of the measurement uncertainty budget relating to the measurement protocol.

In addition to the use of E_d^{0+} to enable the validation of satellite-derived reflectance, E_d^{0+} measurements can also be used to validate separately the E_d^{0+} (or equivalently the atmospheric transmittance) calculated as an intermediate product in satellite data-processing chains.

In some references, E_d^{0+} may be called "surface irradiance"—typically with notation E_s—or more ambiguously "reference irradiance". The parameter is most completely described as "above-water spectral downward horizontal plane irradiance".

E_d^{0+} is composed of photons that reach the surface directly from the Sun ("direct irradiance") and of photons that reach the surface from the sky after scattering in the atmosphere ("diffuse irradiance"). The latter may also include some photons that have interacted with the surrounding surface and subsequently been backscattered in the atmosphere—see page 12 of [6].

Thus, E_d^{0+} spectra are related to: (a) the extraterrestrial solar irradiance, (b) the Sun zenith angle, (c) atmospheric scattering and absorption from molecules, aerosols, and clouds, and (d) to a lesser extent, surface reflectance. Some typical E_d^{0+} are plotted in Figure 2 for different Sun zenith angles and atmospheric conditions.

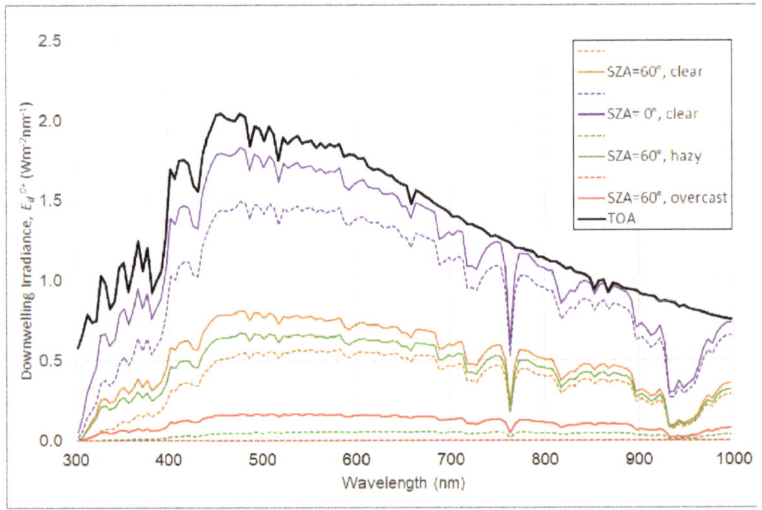

Figure 2. E_d^{0+} for four combinations of Sun zenith angle (SZA) and atmospheric conditions, averaged over 5-nm bands. Solid colored lines are total E_d^{0+}; dashed lines are the corresponding direct component. The solid black line is the band-averaged extraterrestrial solar downwelling irradiance for comparison. Redrawn from [7].

In sunny, low to moderate Sun zenith angle conditions where direct irradiance is greater than diffuse irradiance, E_d^{0+} varies over the day approximately according to the cosine of the Sun zenith angle. This temporal variability is greatest just after sunrise and just before sunset. The time averaging of replicate E_d^{0+} measurements can be simple mean averaging with reference to a central time if the total duration for replicates is short or can be normalized by the cosine of the Sun zenith angle before averaging.

The present paper is focused on aquatic applications, including the full range, size, and diversity of water bodies from deep oceans through coastal and estuarine waters to ports and inland lakes. The measurement of E_d^{0+} is required also for the radiometric validation of surface reflectance over land—such applications are not the focus of the present paper, although there are in principle no major differences between the measurement of E_d^{0+} over land and over water. Measurements of E_d^{0+} without simultaneous L_w are also relevant, outside the R_{rs} validation context, for a variety of applications, including monitoring the Earth's radiation budget for climate applications [8,9], ground-level ultraviolet radiation [10,11] for health-related and ecosystem-related applications, photosynthetically available radiation for biological applications [12,13], solar energy and building applications [14], etc. These applications are not specifically covered here, although many considerations of the measurement protocols described here are valid for all such applications.

Using the terminology of [15], the spectral ranges of primary interest here are the visible (380 nm to 760 nm) and near infrared (760 nm to 1400 nm) ranges. The considerations for measurement of E_d^{0+} given here should be valid also for the near ultraviolet (300 nm to 400 nm) and middle infrared (1400 nm to 3000 nm), although the importance of the various uncertainty sources may be different because of the different intensity and angular distribution of downwelling irradiance, and the equipment (irradiance/radiance sensor, reflectance plaque) may have different properties in these ranges.

The protocols described here are relevant for the validation of a vast range of optical satellites, including the dedicated medium resolution "ocean color" missions, such as AQUA/MODIS, Sentinel-3/OLCI, NPP/VIIRS, etc., but also the operational high spatial resolution missions such as Landsat-8/OLI and Sentinel-2/MSI, as well any other optical mission from which water reflectance can be derived, including the geostationary COMS/GOCI-1 and MSG/SEVIRI, the extremely high resolution Pléiades and PlanetDove constellations, etc.

The current document does not try to identify a "best" protocol; it cannot provide typical uncertainty estimates if good practice is followed (that depends on many factors) and does not aim to prescribe mandatory requirements on specific aspects of a measurement protocol such as "acceptable tilt" or "minimum distance for ship shadow avoidance". While such prescriptions have great value in encouraging convergence of methods and challenging scientists to make good measurements, the diversity of aquatic and atmospheric conditions where validation is required, the diversity of radiometers and platforms, and the corresponding diversity of measurement protocols suggests that more flexibility is needed. This flexibility is acceptable, provided that each measurement is accompanied by an SI-traceable uncertainty budget that is: (a) based on a full analysis of the protocol, and (b) that is itself validated, e.g., by measurement intercomparison exercises [16–18]. Then, the data user can accept or reject such measurements by applying their own threshold for "acceptable" measurement uncertainty.

The present review does aim to provide an overview of all the relevant protocols, including guidelines for radiometer deployment and the quality control of data and an overview of elements that should be considered in the complete uncertainty analysis of a measurement protocol. The approach is structured as follows: for each aspect of the measurement protocol contributing to measurement uncertainty, the perfect situation is summarized in a single sentence in boldface, e.g., "**the irradiance sensor should be vertical**". This is followed by a discussion of techniques to achieve or monitor this (e.g., gimballing, measurement of tilt, removal of tilted data), practical considerations and problems (e.g., changes to ballasting of ships), and approaches to estimate uncertainty when this perfect situation is not achieved (e.g., model studies, experiments). While this highly structured approach may seem over-rigorous or even trivial (isn't it obvious that an irradiance sensor should be vertical?), we do feel that it is necessary to be complete and rigorous in the FRM context (is it obvious to all measurement scientists that a reflectance plaque should be perfectly horizontal?).

For a general treatment of uncertainties in measurements, including a recommended terminology (e.g., "expanded uncertainty") and generic methods for estimating each component uncertainty and combining uncertainties to achieve a total uncertainty, the reader is referred to the Guide to the Expression of Uncertainty in Measurement (GUM) [19].

The present review covers only aspects of the measurement relating to the protocol, including radiometer deployment, data acquisition, and processing aspects, but excluding any uncertainties arising from radiometer imperfections, such as calibration, thermal sensitivity, spectral response (straylight/out-of-band effects), non-linearity, and angular (cosine) response. These radiometer-related aspects deserve a review paper of their own; the reader is referred to Volume II of the NASA Ocean Optics Protocols [20], Section 3 of [21], Chapter 2 of [22] and to the papers in this volume, e.g., [23,24].

In the satellite validation context covered by this review, the focus is on clear sky conditions. There is no clear consensus regarding an objective definition of "clear sky" conditions, although Web Appendix 1 of [25] proposes for moderate Sun zenith angles the test $L_d/E_d^{0+}(750\ nm) < 0.05$, where L_d

is sky radiance at a 135° relative viewing azimuth to the Sun and a 140° viewing nadir angle. This test will detect clouds in front of the Sun because of the consequent increase in $1/E_d^{0+}$, and will detect clouds in the specified sky-viewing direction because clouds have greater L_d values than blue sky at 750 nm. A more complete test for "clear sky" conditions could involve the use of hemispherical camera photos, but would need automated image analysis for an objective test.

1.3. Previous Protocol Reviews

Most of the pre-2004 in situ measurements of water reflectance were made for the purpose of oceanic applications, and most aquatic optics investigators base their measurement protocol in some way on the NASA Ocean Optics Protocols [20] and the references contained within that multi-volume publication. While there are no fully new methods for the measurement of E_d^{0+} since the NASA 2004 protocols collection, the current review aims to better reflect the current practices. The main evolutions since 2004 include:

- more frequent use of unsupervised measurements for validation, e.g., AERONET-OC [26] and Bio-ARGO [27], instead of shipborne supervised measurements
- greater need for validation measurements in coastal and inland waters rather than the prior focus on oceanic waters
- preference for above-water measurement of E_d^{0+} rather than extrapolation from underwater profiles
- reduction in the cost of radiometers facilitating use of an irradiance sensor (instead of a radiance sensor and a reflectance plaque), and better availability of hyperspectral radiometers.

1.4. Overview of Methods

Protocols for measurement of E_d^{0+} are grouped into three broad families of method:

- Direct above-water measurement of E_d^{0+} with an upward-pointing irradiance sensor ("Irradiance sensor method")
- Estimation of E_d^{0+} using a downward-pointing radiance sensor and a reflective plaque ("Reflectance plaque method")
- Estimation of E_d^{0+} from direct sunphotometry and a clear sky atmospheric model ("Sunphotometry method")

A fourth family of method, estimating E_d^{0+} from underwater measurements of downwelling irradiance at differences depths, $E_d(z)$, is now considered obsolete for measurement of E_d^{0+}—see Section 5.

For each family of method, the measurement equation is defined, and the measurement parameters are briefly described in Sections 2–4, respectively. The elements that should be included for the estimation of total protocol-related measurement uncertainty are discussed with some key considerations, guidelines, and recommendations. The "protocol-related" measurement uncertainty includes both known imperfections in the protocol (e.g., atmospheric models used in sunphotometry) and deployment-related imperfections (e.g., the tilting of sensors/plaques).

2. Direct above-Water Measurement of E_d^{0+} with an Upward-Pointing Irradiance Sensor

2.1. Measurement Equation

Since E_d^{0+} can be measured directly using radiometers that are designed to measure plane irradiance, the measurement equation here simply relates the electrical output of a radiometer to calibrated irradiance. Imperfections in such radiometers (angular response, spectral response, non-linearity, thermal sensitivity, etc.) contribute, of course, to the total uncertainty budget of the measurement, and the imperfect cosine response is an important consideration for the measurement of E_d^{0+}, e.g., [24,28].

The direct measurement of E_d^{0+}, which is sketched in Figure 3, can be made from various platforms including ships, small inflatable boats, buoys, fixed offshore structures, and underwater profiling platforms that contain a floating element or the ability to surface. These measurements can be either supervised or unsupervised. In all cases, it is recommended to mount the E_d^{0+} radiometer as high as possible, above any superstructure elements and passing humans, in order to avoid the optical contamination of the measurement from the shading of both Sun and sky light. This can be achieved by the use of a fixed or telescopic mast, e.g., [29].

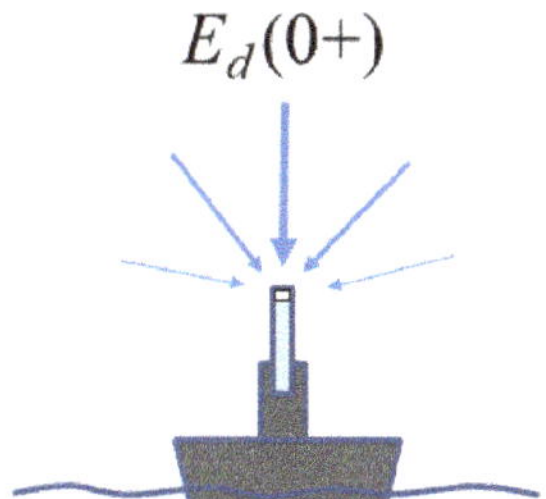

Figure 3. Schematic (not drawn to scale) of (shipborne) direct above-water measurement of E_d^{0+} with an irradiance sensor (pale blue with flat white cosine collector).

2.2. Protocol-Dependent Sources of Uncertainty

In addition to the radiometer-related sources of uncertainty that arise from imperfections in the radiometers themselves, including the angular (cosine) response of the radiometer, the direct measurement of above-water downwelling irradiance has a number of sources of uncertainty relating to the deployment conditions. These protocol-related sources of uncertainty are described in Sections 2.2.1–2.2.4.

2.2.1. Tilt Effects

The irradiance sensor should be vertical.

The non-verticality of the E_d^{0+} radiometer, e.g., caused by imprecise installation, wave-tilting of floating structures (buoys, ships), wind-tilting of offshore structures, including masts, and even ballast changes for ships (shifts in fuel, water, large equipment), will result in a bias in the measurement of E_d^{0+}. Therefore, it is necessary to measure the tilt of radiometers at sufficiently high frequency and perform the appropriate filtering of non-vertical data and/or averaging of data to reduce tilt effects.

For E_d^{0+}, the effect of tilt may be particularly strong in sunny (satellite validation) conditions because of the highly anisotropic light field. The main effect of tilt is similar to a change in the effective Sun zenith angle, and is strongest for tilt in the solar plane. The passive gimballing of an E_d^{0+} sensor, if sufficiently well designed, may help to reduce tilt, as implemented in the DALEC system [30,31]. Active gimballing of an E_d^{0+} sensor, using electric motors to correct for tilt, may now be feasible, although at the time of writing, the authors are not aware of documentation on the use of such hardware for E_d^{0+} measurement.

The impact of tilt on measurement uncertainty can be estimated if the two angles of tilt with respect to the Sun are measured and the approximate angular variation of sky radiance is known, e.g., from imaging cameras, or estimated from atmospheric properties. At high tilt, an E_d^{0+} sensor may also measure some light from the underlying water/land/platform surface instead of the sky, although grazing angle incident light has a low contribution to the cosine-weighted integral for E_d^{0+}.

Obviously, minimization of tilt can be a consideration in the design [32] or in the location (e.g., low waves) of validation measurement structures. Floating buoys and small ships may be particularly subject to high tilt.

2.2.2. Shading from Superstructure

The irradiance sensor should be deployed above the height of all the other structures or objects (including humans).

The light field that is being measured may itself be perturbed by the presence of solid objects such as the superstructure of the platform used to mount them. This may be especially problematic on ships, where practical considerations may prevent mounting the E_d^{0+} sensor above all other structures, particularly if regular inspection by humans of the fore-optics is required.

The process of sky shading can be easily understood from fish-eye photographs taken vertically upwards at the location of an E_d^{0+} sensor, as illustrated in Figures 4 and 5. Any part of the upward hemisphere that is not sky represents optical contamination of the measurement, and this contamination will be related to the solid angle of sky that is replaced by the object with near-zenith objects contributing more than near-horizontal objects to the cosine integral of radiances. Of course, it is best to make such photos with a calibrated fully hemispherical sky radiance camera [33]. However, even photos from simple cameras with a wide-angle lens and without any radiometric calibration can rapidly identify a major contamination of measurements from superstructures and/or other objects.

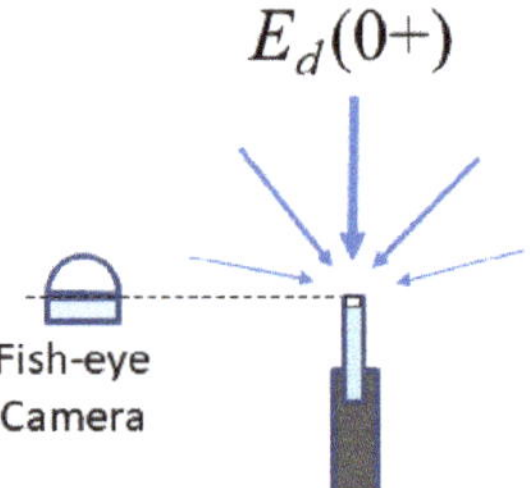

Figure 4. Schematic showing how a fish-eye camera, preferably fully hemispherical, can be used to qualitatively check for the superstructure contamination of E_d^{0+} measurements.

While direct Sun shadowing of the E_d^{0+} sensor is generally avoided by design of the deployment method and can easily be identified and removed from data, the impact of more subtle optical contaminations of sky radiance can be more difficult to identify and estimate.

It is obvious that humans should remain fully below the level of an E_d^{0+} sensor at all times during measurements. It is not unknown for resting birds to contaminate unsupervised E_d^{0+} measurements [34], and measures may be taken to avoid this, e.g., the use of spikes below the field of view, but sufficiently close to threaten discomfort. Unusual contaminations may be identified by time series analysis or video camera monitoring of unsupervised installations.

On some platforms, optical contamination may also arise from atmospheric steam or smoke emissions from ship engine funnels and other exhaust gases (air conditioning, etc.).

Fixed offshore structures with limited access (e.g., oil and gas platforms, wind farm structures, navigational structures) as well as large ships with tall masts may be particularly subject to superstructure shading. Improvements in the stability of telescopic masts [35], which allow high mounting but easy inspection of fore-optics, and reductions in the price of such equipment should facilitate the adoption of deployment techniques with greatly reduced or zero superstructure shading.

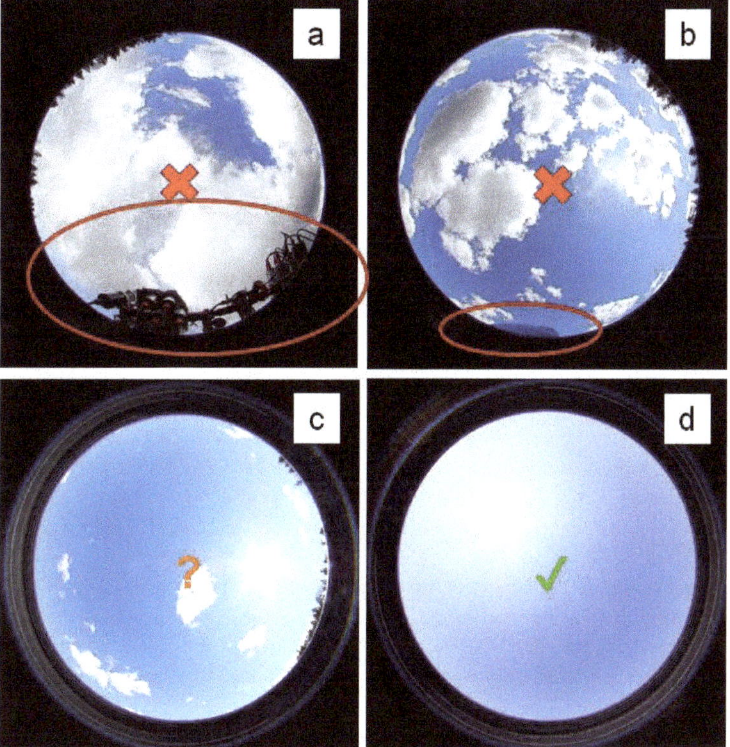

Figure 5. Example fish-eye photos taken to check for contamination of E_d^{0+} measurements. (**a**) Contamination of field of view by other radiometers; (**b**) Contamination of field of view by a scientist in the bottom of the photo; (**c**) No contamination of field of view, partly cloudy sky; (**d**) No contamination of field of view, clear sky. The trees visible in the bottom-left photo, typical of inland water or very nearshore measurements, do affect the measurement, but are not considered as "contamination" in the context of this review. The impact of such far-field objects contributes to the natural downwelling irradiance at the measurement location, and should be measured as such.

For supervised shipborne E_d^{0+} measurements, the use of a floating platform to carry the E_d^{0+} radiometer away from the ship will clearly minimize—to possibly a negligible amount—the superstructure-related perturbations. This may be conveniently combined in a floating/profiling platform used for underwater profiling of upwelling radiance.

Measures to reduce and/or estimate the uncertainties associated with superstructure shading may include redundant measurements by multiple sensors located in different positions, and hence subject to different shading effects, or experiments with sensors at different heights/locations, etc. Three-dimensional (3D) radiative transfer modeling may also be used to estimate uncertainties in E_d^{0+} measurements associated with superstructure effects.

2.2.3. Fouling

The fore-optics of the irradiance sensor should be kept clean.

Upward-facing sensors needed for measuring E_d^{0+} are prone to fouling of the fore-optics, especially during long-term unsupervised deployments.

Fouling may occur because of sea spray, the atmospheric deposition of particles (which may even embed within the structure of some diffuser materials used as fore-optics [36]), rain droplets, bird feces, etc. This can be mitigated by cleaning the fore-optics, and can be monitored by frequent calibration checks, e.g., with portable relative calibration devices [37].

Fouling is generally kept negligible for supervised deployments by regular inspection and, when necessary, the cleaning of fore-optics and protection by lens caps when not measuring (e.g., at night and between "stations" for discrete measurements).

Exposure to ultraviolet light can lead to the photodegradation of materials used as diffusers.

For unsupervised deployments, fouling and photodegradation can be minimized by the protection of fore-optics when not measuring by the use of external mechanical shutters [38] or the rotation of sensors to point downwards (typified by the "parking" function of the CIMEL CE-318 sunphotometer when not measuring).

Major fouling events can be identified by time series analysis of data and/or video camera imagery.

The uncertainty related to fouling can be estimated by comparing post-deployment calibrations before and after cleaning, although it is also noted that fouling may vary non-monotonically in time because of the cleaning effect of rain water. To separate the effects of fouling from intrinsic sensitivity changes (e.g., long-term drift or short-term changes typically caused by mechanical shock), these measurements must be done immediately before and after cleaning, e.g., in the field (using a stable light source such as a clear sky) or in a calibration laboratory (which must be provided with the uncleaned radiometer).

2.2.4. Fast Natural Fluctuations

Measurements should be used only during periods of stable illumination.

In clear sky conditions, the natural variability of E_d^{0+} over a typical measurement time scale (~1 to 10 min) is low, and may be easily estimated from a clear sky irradiance model, e.g., [39], using as input the temporal variation of the Sun zenith angle and an estimation/measurement of aerosol optical thickness.

If measurements are made during partially cloudy conditions, in addition to the tilt-induced fluctuations described in Section 2.2.1, the natural variability of E_d^{0+} may be non-negligible, particularly if there are clouds or haze near the Sun. In such cases, careful quality control of data is necessary to remove individual measurements or complete sets of measurements that cannot be used for satellite validation. Quality control will typically include tests on temporal variability including second derivative "spike/jump" analysis and min/max/standard deviation analysis, and may also include the comparison of data with a clear sky model.

A full sky imager can be used to provide detailed information on sky conditions for quality control [40].

It is suggested here that FRM for satellite validation should not be made during fully cloudy conditions or when the Sun is obscured by clouds or haze. In situ measurements can be made at a slightly different times from the satellite overpass, e.g., 1 to 6 h depending on natural variability, and so a cloud-free satellite image could theoretically correspond with an in situ reflectance measurement made during cloudy conditions within an acceptable time window. However, many factors, including the very different bidirectional reflectance of water under a sunny or a cloudy sky, suggest that this should be avoided in the FRM satellite validation context. In other contexts, such as the simultaneous measurement of reflectance and chlorophyll *a* for algorithm calibration/validation, it may be acceptable to use measurements made in cloudy conditions, particularly fully overcast conditions, provided that the corresponding measurement uncertainties are sufficiently quantified and limited.

The question of whether FRM can be made in partially cloudy conditions is relevant. It can be argued that only the best measurements should be used, and this requires perfectly clear sky conditions. On the other hand, if a measurement scientist is able to estimate the uncertainties associated with partially cloudy conditions, then the data user could later decide whether to use or reject such measurements for their specific application on the basis of a threshold on measurement uncertainty. There is no clear consensus on this question at present, but perhaps the debate requires first a more objective definition of "cloudiness" and/or "clear sky" conditions—see Section 1.2. Isolated clouds with small solid angles, away from the Sun and low on the horizon, so with low zenith cosine weighting, have little impact on E_d^{0+}.

Uncertainties associated with fast natural fluctuations can be estimated from the standard deviation of replicate measurements made over a certain interval of time. High uncertainty may lead to simple rejection of the measurement.

2.3. Variants on the Method of Direct above-Water Measurement of E_d^{0+} with an Upward-Pointing Irradiance Sensor

Underwater drifting floats used for satellite radiometry validation [27] may lack a permanently above-water E_d^{0+} sensor, and make only occasional E_d^{0+} measurements when surfacing. There is no fundamental difference between the "surfacing" E_d^{0+} sensor and the permanently above-water E_d^{0+} sensors considered in the rest of this review. However, it is noted that there may be different designs of E_d^{0+} sensors for in-water and in-air measurements; the time and horizontal space differences between E_d^{0+} and L_w measurements must be considered; and the presence of water, as already mentioned in Section 2.2.3, and aquatic algae on the fore-optics may be more problematic.

With an additional moving "shadowband" accessory, it is possible to combine full Sun and sky E_d^{0+} with a direct Sun-obscured measurement, thus giving the diffuse sky component of E_d^{0+}, which is termed E_d^{dif}. This is not commonly used for the validation of satellite data over water, since the primary radiometric product from satellites, e.g., the reflectance product given in Equation (1), does not require a decomposition of E_d^{0+} into direct and diffuse components. However, this additional information does provide the additional opportunity to validate the satellite data processing for direct and diffuse atmospheric transmittance, and does potentially allow improving the bidirectional reflectance distribution functions (BRDF) corrections. The measurement of direct and diffuse components of E_d^{0+} can also be used to improve self-shading corrections when making underwater measurements of upwelling radiance. The measurement of E_d^{dif} in addition to the total E_d^{0+} is of major importance for other applications such as earth radiation budget monitoring, agriculture, solar energy, etc. A discussion of E_d^{dif} data acquisition and processing with the shadowband technique can be found in [41].

3. Estimation of E_d^{0+} Using a Downward-Pointing Radiance Sensor and a Reflective Plaque

3.1. Measurement Equation

E_d^{0+} can also be calculated indirectly by measuring the exitant radiance, L_P, from a horizontally deployed reflectance plaque of known reflectance, ρ_P—see Figure 6. If the plaque is perfectly Lambertian, then:

$$E_d^{0+} = \frac{\pi * L_P}{\rho_P} \tag{3}$$

where all the terms may vary with wavelength, but the wavelength variation is dropped for brevity throughout this section. If the plaque is not perfectly Lambertian, then the downwelling light field can be approximated as a collimated beam of light from the Sun direction [42], giving the measurement equation:

$$E_d^{0+} = \frac{L_P(\theta_v, \phi_v)}{f_r(\theta_i, \phi_i, \theta_v, \phi_v)} \tag{4}$$

where $f_r(\theta_i, \phi_i, \theta_v, \phi_v)$ is the plaque bidirectional reflectance distribution function (BRDF), θ_v, ϕ_v are the viewing nadir and azimuth angles and θ_i, ϕ_i are the zenith and azimuth angles of the incident collimated beam, which are generally assumed to correspond to the Sun beam direction.

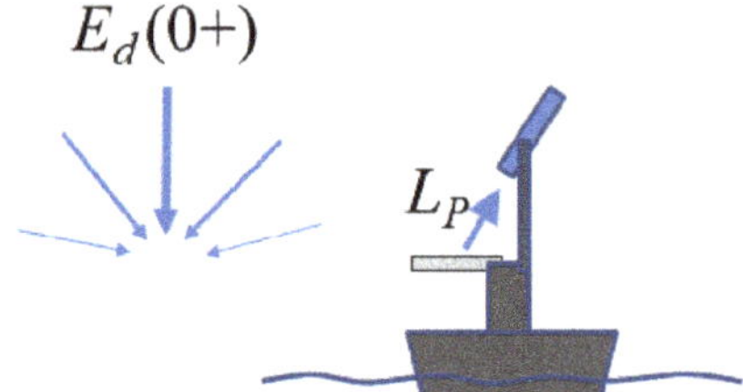

Figure 6. Schematic showing indirect measurement of E_d^{0+} using a downward-pointing radiance sensor and a reflective plaque (sensor, plaque, and holder not to scale).

A common material for such plaques is sintered polytetrafluorethylene (PTFE), which is typically sold under the product name Spectralon™ (see disclaimer at the end before the references), which can be manufactured to give near 100% reflectance ($\rho_P \approx 1.0$) for "white" plaques with low spectral variation of reflectance, low departure from the perfect Lambertian angular response [43], low spatial heterogeneity, and reasonable temporal stability. Lower reflectance "grey" plaques, e.g., $\rho_P \approx 0.18$, can also be used, although they have less Lambertian angular response. Other diffusive materials have been used in this method, including grey "cards" that are used traditionally in photography. All the materials used in the FRM context need to be adequately characterized as regards bidirectional, spectral, spatial and temporal variability.

Historically, the measurement of E_d^{0+} using a downward-pointing radiance sensor and a Lambertian reflective plaque was adopted for cost considerations, allowing all the measurements to be made with a single radiance sensor. This method also allows the reduction of some calibration-related uncertainties, since only one sensor is used. Moreover, if only R_{rs} is required, this method may be implemented with an uncalibrated sensor (but see the discussion in Section 3.1.1).

The reflectance plaque method is popular in the land remote sensing community, possibly because the measurements for some middle infrared wavelengths (1.4 μm to 2.5 μm) are important, which very significantly raises the cost of a radiometer and increases the uncertainty relating to cosine response for an irradiance sensor with a transmissive diffuser.

Measurements with a reflective plaque are often supervised, although it is possible to automate such measurements, e.g., [44].

Outside the FRM satellite validation context, the educational value of measurements made using this protocol, e.g., with very simple and inexpensive optical radiometers [45], is clearly recognized.

3.1.1. Is It Necessary to Use a Calibrated Radiance Sensor?

The preparation of this review generated much discussion within the community regarding the question of whether an uncalibrated radiance sensor can be used to acquire measurements for satellite validation. This method was suggested in the NASA Ocean Optics protocols 2003 version "Method 2" [46] as being appropriate for the measurement of reflectance using an uncalibrated sensor. Indeed R_{rs} can be calculated via Equation (1) from measurements of L_w and E_d^{0+} made by the same radiance sensor, even if this sensor is not calibrated, i.e., providing data for L_w and E_d^{0+} in (dark-corrected) digital counts rather than in SI-traceable units. While it is essential to characterize the sensor, e.g., for straylight, non-linearity, thermal effects, etc., it is not necessary to calibrate the sensor to perform radiometer-related corrections and uncertainty estimates. In fact, some radiometer-related uncertainties are best treated before calibration, e.g., non-linear effects may depend directly on the digital count data [47,48] (as compared to the maximum possible, saturated, digital counts), but not on the calibrated radiance.

There is formally nothing in the FRM definition that would require a calibrated radiance sensor to be used for the measurement of R_{rs}. However, the use of a calibrated radiance sensor does have two advantages:

- A calibrated radiance sensor will provide a calibrated E_d^{0+}, which can then be compared with clear sky models [39] for quality control purposes, and can be compared to satellite data to validate the computations of atmospheric transmittance (in addition to the more important R_{rs} products).
- The interpretation of in situ measurement intercomparison exercises [17], as required by the FRM process, necessitates a separation of uncertainties arising from L_w and E_d^{0+} measurements, e.g., comparing E_d^{0+} measurements from a vertically-mounted irradiance sensor (impacted by cosine angle uncertainties, etc.) with E_d^{0+} measurements deduced from a radiance sensor viewing a reflectance plaque (impacted by BRDF uncertainties, etc.).

Moreover, it is noted [42] that the simple cancellation of unknown calibration factors used to calculate $R_{rs} = \pi L_w / E_d^{0+}$ in native spectral resolution no longer works precisely when spectrally convolving L_w and E_d^{0+} with a spectral response function, as needed for the validation of R_{rs} for individual spectral bands of satellite sensors.

3.1.2. What Nadir Angle Should Be Used for Viewing a Reflectance Plaque?

The NASA 2003 protocols (Volume III, Section 3.3) recommended that measurements of E_d^{0+} with a reflective plaque should be made with a vertical downward (nadir) pointing radiance sensor and a plaque with BRDF calibration for varying downwelling light distributions (typically characterized by Sun zenith angle) and vertical upwelling reflected radiance. However, off-nadir viewing with the same nadir angle as water-viewing L_w measurements, typically 40°, has often been adopted for practical reasons, e.g., for easy switching between plaque and water-viewing modes for certain deployments. It is noted that [49] provides the scientific basis for a water-viewing nadir angle of 40° (and relative azimuth to Sun of 135°) as a good geometry for sunglint avoidance, but does not give a scientific basis for a plaque-viewing nadir angle of 40°—the latter is merely suggested as practically convenient. On the other hand, an off-nadir plaque-viewing geometry may indeed be desirable for scientific reasons, since the radiometer shading of the plaque will be greater with nadir-viewing when the Sun zenith angle is low [42]. For off-nadir plaque viewing, there seems to be no standardization of the viewing azimuth angle, although the same azimuth angle as used for L_w measurements (90° or 135° with respect to the Sun) would be a typical choice for both practical and shadow-avoidance reasons.

Optimal plaque-viewing geometry was investigated in [42], who recommend, for moderate Sun zenith angles between 20–60°, a plaque-viewing nadir angle of 40° for a ~100% reflective white plaque, to minimize operator/radiometer shading/reflection, but a nadir view for less reflective, grey plaques, where reflectivity may vary strongly with the viewing nadir angle. For both types of plaque, a viewing azimuth angle of 90° with respect to the Sun was recommended.

The FRM context does not prescribe a single viewing geometry (or any other specific aspect of a measurement protocol), but "simply" requires that, for whatever plaque-viewing geometry is adopted, the related uncertainties (radiometer and superstructure shading of plaque, plaque BRDF) be quantified.

3.2. Protocol-Dependent Sources of Uncertainty

In addition to the radiometer-related sources of uncertainty that arise from imperfections in the radiometers themselves, the measurement of E_d^{0+} using a reflectance plaque has a number of sources of uncertainty relating to the deployment conditions. These protocol-related sources of uncertainty are described in Sections 3.2.1–3.2.7.

3.2.1. Plaque Calibration

The reflectance plaque must be calibrated.

Clearly, the reflectance of the plaque used for this measurement must be calibrated with traceability to an SI standard and an uncertainty associated with this calibration. Optical contamination/degradation of the plaque and bidirectional effects are further considered in Sections 3.2.5 and 3.2.7.

3.2.2. Plaque Homogeneity and Sensor Field of View

The reflectance plaque should be homogeneous and should fill the radiance sensor field of view.

It is known that plaques do have spatial and azimuthal inhomogeneities, and so it is assumed that the measurement area on the plaque corresponds sufficiently well to the area on the plaque used during plaque calibration, taking account of the surface average of any inhomogeneities.

Clearly, the plaque must fully fill, and preferably exceed, the sensor field of view (FOV) so that the measurement of E_d^{0+} will not be contaminated by the background around the reflectance plaque. This can be facilitated by small FOV radiometers. In any case, the angular response of the radiance sensor should be checked for any residual response outside the manufacturer-specified FOV, e.g., by occulting the plaque partially with a black material moved from each edge of the plaque towards the center until an impact is detected

Uncertainties associated with the sensor field of view and plaque inhomogeneity can be assessed by experiments deploying the radiometers at different heights and at different horizontal locations above the reflectance plaque, and by changing the background around the reflectance plaque (since the radiometer shading effects will also vary with radiometer height—see Section 3.2.4).

3.2.3. Tilt Effects

The reflectance plaque should be horizontal.

The non-horizontality of the reflectance plaque that is used for measurements of E_d^{0+} will give uncertainty in the measurement of E_d^{0+} in the same way as the non-verticality of an irradiance sensor used to directly measure E_d^{0+}, as discussed previously in Section 2.2.1. Tilting of the plaque can be caused by a number of factors, including imprecise leveling and, if measuring from a ship, ship roll during measurements. Therefore, it is necessary to measure the tilt of the plaque (not just the ship) at sufficiently high frequency and perform the appropriate filtering of non-horizontal data and/or averaging of data to reduce tilt effects.

Although digital inclinometers are now readily available for integration with radiometric data streams, they seem to not yet be used for shipborne measurement of E_d^{0+} using a reflectance plaque.

For E_d^{0+}, the effect of tilt may be particularly strong in sunny (satellite validation) conditions because of the highly anisotropic light field, and the effect of a non-horizontal plaque is similar to a change in Sun zenith angle, and is strongest for tilt in the solar plane. At high tilt, the measurement may also measure some light from the water/land/platform instead of the sky, although the grazing angle incident light has a low contribution to the cosine-weighted integral for E_d^{0+}.

The impact of tilt on measurement uncertainty can be estimated if the two angles of tilt with respect to the Sun and approximate angular variation of the sky radiance (from imaging cameras or estimated from atmospheric properties) are known—see Section 2.2.1.

The minimization of tilt should be a consideration in the choice of measurement platform, taking account of expected wave conditions. Small ships may be particularly subject to high tilt because of larger ship roll.

3.2.4. Shading from Superstructure and Radiometers and Mounting Equipment

The reflectance plaque should be deployed above the height of all other structures or objects (including humans).

The light field that is being measured is itself perturbed by the presence of solid objects anywhere above the level of the reflectance plaque. This includes, necessarily, the radiometer itself, which is used for measurements, but also any superstructure elements of the ship/platform as well as any equipment related to fixing the radiometer above the reflectance plaque.

The shading problems associated with this method are conceptually similar to those already described for direct measurement of E_d^{0+} (Section 2.2.2), but are significantly worse:

- Firstly, there will always be some shading of sky radiance onto the plaque from the radiometer itself. The radiometer must be held above the plaque at a height that is sufficiently small so that the plaque fills the whole field of view of the radiometer. The exact height depends on the radiometer and the size of the plaque. Shading from the radiometer (and any associated fixations) will be related to the zenith cosine-weighted solid angle of sky filled by the radiometer, as seen from any point on the reflectance plaque, and will be worse for radiometers held close to the plaque or that have a large diameter.
- Secondly, while it is typical to mount irradiance sensors high on poles/masts (Section 2.2.2) and certainly above head height, measurements with a reflectance plaque are nearly always made much lower on a ship/platform for practical reasons: it is generally necessary to manipulate the radiometer (e.g., to then point to water and sky) and the plaque (e.g., to protect it when not measuring). Optical contamination from ship/platform sides, upper decks, masts, and even humans (often including those making the measurement) can be significant and difficult to quantify.

The process of sky shading can be easily understood from fish-eye photographs taken vertically upwards at the location of a reflectance plaque – see Figure 7. Any part of the upward hemisphere that is not sky represents the optical contamination of the measurement, and this contamination will be related to the zenith cosine-weighted solid angle of sky that is replaced by the object with near-zenith objects contributing more than near-horizontal objects to the cosine integral of radiances.

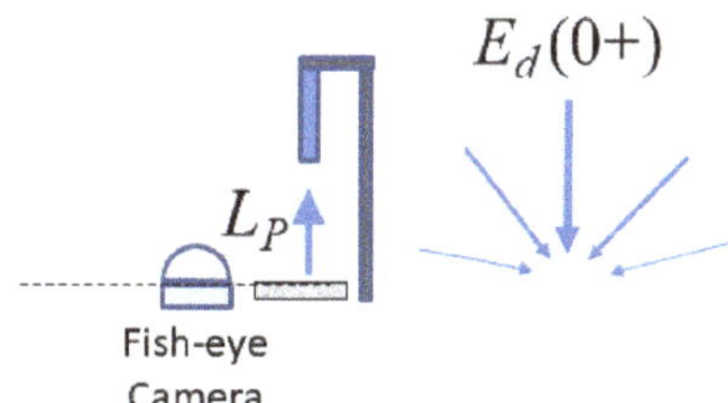

Figure 7. Location of fish-eye camera used for qualitative checking of shading of reflectance plaque, for comparison with Figure 4 for the direct measurement of E_d^{0+} using an irradiance sensor, as described in Section 2.

Measures to estimate the uncertainties associated with shading/reflection could include experiments made with irradiance sensors, with well-characterized cosine response, located (a) alongside the plaque, and (b) on a mast above the possible optical contamination and/or experiments combining optimal and non-optimal locations [50]. Such an experiment is reported by [51] for land remote sensing applications, but the issues are clearly the same as for water remote sensing. In that study, the height of the sensor above the plaque and the position of a human observer were varied. The shading (but not reflection) effects from radiometer and observer are analyzed in detail in the model simulations of [42], for different Sun zenith angles and aerosol conditions, with the conclusion that a plaque-viewing nadir angle of 40° and relative azimuth to the Sun of 90° is recommended when viewing a ~100% reflectance plaque.

3.2.5. Fouling

The radiometer fore-optics and the reflectance plaque should be kept clean.

When measurements made with a reflectance plaque are supervised, there should be negligible contamination of the radiance sensor fore-optics, provided that it is cleaned whenever necessary following the manufacturers' recommendations.

Optical contamination of the plaque itself may be a significant problem because of the atmospheric deposition of particles (which may embed within the structure of some diffuser materials) of both natural and ship-related origin, marks from contact with any objects including materials used to protect the plaque during storage, etc. For example, it is recommended to keep plaques away from plastics and hydrocarbons (diesel fumes) and to build a storage box that holds the plaque fixed in a way such that the reflective surface is not in contact with anything. Obviously, humans, especially those with greasy fingers, should not touch the diffusive surface itself. The cleaning of dirty plaques is, of course, recommended, but should be accompanied by recalibration or pre/post-cleaning calibration checks.

In addition to optical contamination, plaques may change naturally from photodegradation processes related to ultraviolet exposure. For example, the reflectivity of Spectralon™, a proprietary form of sintered polytetrafluoroethylene (PTFE) produced by Labsphere Inc., USA, and used for both spaceborne calibration diffusers and many ground-based measurements, may change at short wavelengths because of absorption from organic impurities [52,53], which can only be removed by vacuum baking. The careful handling and storage of plaques is required to limit such degradation.

The uncertainty estimate related to fouling can be validated by comparing post-deployment calibrations before and after cleaning a plaque.

3.2.6. Fast Natural Fluctuations

Measurements should be used only during periods of stable illumination.
Considerations and uncertainties associated with fast natural fluctuations of E_d^{0+} over a typical measurement time scale (~1 min to 10 min) are identical to those already discussed in Section 2.2.4, except that the asynchronicity of E_d^{0+} and L_w measurements is inevitable for this method. In the latter context, replicate measurements, e.g., E_d^{0+} before and after L_w, can be used.

3.2.7. Bidirectional Reflectance of Plaques

The bidirectional reflectance of the plaque should be known.
In general, a plaque calibration is made for unidirectional illumination (typically 8°) and with hemispherical collection, using an integrating sphere, which is termed "8/h" calibration. Whereas the cosine response of irradiance sensors must be considered for the direct measurement of E_d^{0+} (Section 2), the bidirectional reflectance of a plaque (from all illuminating directions to the single viewing direction) must be considered in the uncertainty estimate for the reflectance plaque method. This data is reported in some cases for typical white Spectralon™ plaques [53] and for grey Spectralon™ plaques [42,54], but they may be unknown for other materials, including grey cards. A full characterization of the optical properties of a plaque will include polarization sensitivity in the calibration process [55]. The full four-dimensional and reciprocal Mueller matrix bidirectional reflectance distribution function of sintered polytetrafluoroethylene is reported at four wavelengths in [56]. The uncertainty associated with the imperfect Lambertian response of a plaque can be validated by comparison, for a range of Sun zenith angles, with a zenith-pointing irradiance sensor, if the latter has a sufficiently characterized cosine response and is associated with a full uncertainty analysis.

3.3. Variants on the Method for Measurement of E_d^{0+} Using a Downward-Pointing Radiance Sensor and a Reflectance Plaque

Multiple measurements can be made with different plaques [18], e.g., of different reflectivity, to reduce/validate the uncertainties associated with individual plaques (calibration, optical contamination/degradation, bidirectionality, etc.).

Although not used for the measurement of E_d^{0+} as such, it is interesting to note the use of a "blue tile" reported by B.C. Johnson in Section 7.10 of [18]. This specially-manufactured reflectance plaque has spectral properties similar to those of blue water, and so provides an intercomparison target, which allows the testing of some aspects of above-water L_w protocols with some aspects of radiometer characterization, such as straylight.

4. Estimation of E_d^{0+} from Direct Sunphotometry and a Clear Sky Atmospheric Model

As an alternative to the direct measurement of E_d^{0+} using a vertically-pointing irradiance sensor as described in Section 2, it is possible to estimate aerosol optical thickness by measuring the direct Sun radiance with a sunphotometer and estimate the total atmospheric transmittance with this and other inputs—see Figure 8. This method was originally developed for satellite validation measurements using the hand-held SIMBAD(A) radiometer [57], and has the interesting feature for satellite validation studies of providing more information on atmospheric parameters than just the E_d^{0+} measurement described in Sections 2 and 3. In the hand-held SIMBAD(A) protocol, only aerosol optical thickness is measured, but for automated Sun/sky radiometers, such as those of the AERONET-OC network [26] with many other pointing scenarios, many extra atmospheric parameters can be estimated, including aerosol size distribution and phase function [58].

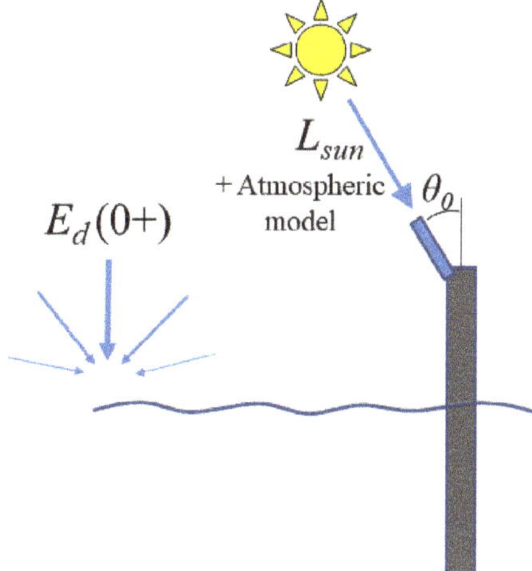

Figure 8. Schematic of direct Sun measurement for the estimation of E_d^{0+}.

This method was described in the NASA Ocean Optics Protocols [46] as above-water radiometry "Method 3", in combination with measurements of water-leaving radiance using a vertical polarizer, as implemented for the SIMBAD(A) radiometer. However, this method for estimating E_d^{0+} may be combined with different methods for estimating L_w, e.g., above-water methods without a vertical polarizer, and so is described here as a generic method for estimating E_d^{0+}.

The pointing accuracy required for direct Sun measurements generally requires a very stable platform, such as a fixed offshore structure as in the AERONET-OC protocol [26], for unsupervised measurements, or can be achieved by a hand-held sunphotometer, e.g., SIMBAD(A) radiometer [57]. However, the feasibility of making direct Sun measurements from a moving platform has been demonstrated for an airborne radiometer [59], so it is conceivable that such measurements may be made in the future from structures with some movement, e.g., buoys.

4.1. Measurement Equation

The full measurement equations for this method are described in [57] using a notation typical for atmospheric radiative transfer studies, which does not explicitly mention E_d^{0+}. For compatibility with the rest of this review, these equations are rewritten here in a form that facilitates the identification of E_d^{0+} itself.

Thus, the total (direct and diffuse) downward (Sun to water) atmospheric transmittance, T_0, is defined by:

$$T_0 = \frac{E_d^{0+}}{E_d^{TOA}} \tag{5}$$

and the downwelling irradiance at Top of Atmosphere, E_d^{TOA}, is estimated from:

$$E_d^{TOA} = F_0 Cos\vartheta_0 \left(\frac{d_0}{d}\right)^2 \tag{6}$$

where F_0 is the extraterrestrial solar irradiance for mean Sun–Earth distance d_0, e.g., tabulated by [60], ϑ_0 is the Sun zenith angle, and d is the Sun–Earth distance at the time of the measurement, which can be easily calculated from position and date/time using earth orbital models.

Combining Equations (5) and (6) gives:

$$E_d^{0+} = T_0 F_0 Cos\vartheta_0 \left(\frac{d_0}{d}\right)^2 \tag{7}$$

T_0 is estimated using a clear sky radiative transfer model, e.g., [61], which takes as input vertically integrated ozone amounts (obtained from extraneous data such as Total Ozone Mapping Scanner satellite data and/or meteorological models or climatologies), ϑ_0, surface atmospheric pressure (which influences Rayleigh optical thickness and may be obtained from simultaneous surface measurements or from appropriate meteorological models), and aerosol optical thickness, $\tau_a(\lambda)$—see Equation (7) of [57]. The impact of other absorbing gases and absorbing aerosols and other parameters such as surface reflectance may be included in the atmospheric radiative transfer model, if necessary.

In the estimation of T_0, the effects of multiple scattering from surface to atmosphere back to surface are generally neglected. These effects can be important over reflective waters and nearby land, especially at short wavelengths, where the spherical albedo of the atmosphere becomes large; for a more complete treatment, see [6].

The aerosol optical thickness $\tau_a(\lambda)$ is deduced from direct Sun measurements taking account of sunphotometer calibration, Earth–Sun distance variation d/d_0, Sun zenith angle ϑ_0, and including corrections for molecular scattering and gaseous absorption, which is considered to be mainly due to ozone—see Section 4.1 of [57], including Equations (5) and (6). In theory, sky radiance information (in the principal plane and almucantar, especially aureole), in addition to direct sunlight measurements, could be used to better determine the aerosol type, and therefore better estimate the atmospheric transmittance. In practice, only aerosol optical thickness is used to estimate atmospheric transmittance from AERONET-OC and SIMBAD(A) measurements, because the anisotropy factor of the aerosol phase function is quite constant for most aerosol models [62]. However, when aerosols are absorbing, the impact of absorption can be significant [63].

The Ångström exponent for the spectral variation of $\tau_a(\lambda)$ can also be computed, and in the SIMBAD protocol it is used in the skyglint correction for L_w, but is not needed for the computation of E_d^{0+}.

The calculation of T_0 required for this E_d^{0+} measurement protocol is comparable to the computation of E_d^{0+} made in satellite data processing software, e.g., SeaDAS.

4.2. Protocol-Dependent Sources of Uncertainty

In addition to the radiometer-related sources of uncertainty that arise from imperfections in the radiometers themselves, including the Bouguer–Langley calibration, the measurement of above-water downwelling irradiance from direct Sun radiometry and atmospheric modeling has a number of sources of uncertainty relating to the measurement equation and deployment conditions. These protocol-related sources of uncertainty are described in Sections 4.2.1–4.2.6.

4.2.1. Atmospheric Radiative Transfer Model

The atmospheric radiative transfer model and its inputs (extraterrestrial solar irradiance, absorbing gases, atmospheric pressure, Sun zenith angle, etc.) should be accurate.

The atmospheric radiative transfer model used to estimate T_0 has both intrinsic uncertainties, which are associated with models and simplifications of many complex atmospheric optical processes, as well as uncertainties in the various input parameters (aerosol parameters, absorbing gas amounts, atmospheric pressure, Sun zenith angle, etc.) and which propagate through the model. The extraterrestrial solar irradiance also includes some uncertainty; ideally, the same solar irradiance data will be used for in situ and satellite data processing.

The estimation of uncertainty from all these sources is complex and is described in detail in Section 5 of [57], except for the adjacency effect of multiple surface–atmosphere scattering, which was mentioned in Section 4.1.

An intercomparison of atmospheric radiative transfer codes and discussion of issues can be found in [64].

4.2.2. Sky Conditions

The atmosphere should be cloud-free and horizontally homogeneous.

The atmospheric radiative transfer model used to estimate T_0 assumes that the atmosphere is horizontally homogeneous and, in particular, contains no clouds. This assumption is valid for the design conditions of clear sky satellite validation, but significant and difficult-to-estimate uncertainties will arise if this assumption is violated, e.g., for a partially cloudy sky. In the SIMBAD(A) and AERONET-OC protocols, automated quality control steps identify when the direct Sun measurement is affected by clouds or haze near the Sun, and remove such data from processing. In the SIMBAD(A) protocol, the human observer can also identify suboptimal conditions, such as clouds somewhere else in the sky, and quality flag such data accordingly.

4.2.3. Pointing Effects

The sensor FOV should contain entirely the Sun and be centered on the Sun.

While high pointing accuracy is crucial for direct Sun measurements, this can be well achieved by both robotic and handheld systems allowing for fine pointing adjustments. The field of view of sunphotometers is by design small, e.g., 1° to 3°, and typically not much larger than the Sun's linear angle of about 0.53°, to minimize the contribution of atmospheric scattering yet completely cover the Sun disk.

Inadequate pointing accuracy can be identified from replicate measurements and/or very high apparent optical thickness and corresponding measurements removed during quality control steps.

Uncertainties associated with direct Sun pointing may be grouped with other uncertainties in the measurement of aerosol optical thickness.

4.2.4. Shading

The direct path from Sun to sensor should be free of obstructions.

Shading of the direct Sun measurement by the presence of solid objects is generally not a problem because—in contrast to the direct measurement of E_d^{0+} with an irradiance sensor where the whole upward hemisphere should be free of obstructions—for direct Sun measurements, only the direct Sun path must be free of obstructions. For unsupervised measurements, most structure shading will be very obvious in direct Sun measurements, and can be automatically removed either a priori, by defining a range of acceptable viewing azimuth angles, or a posteriori, by eliminating very low radiance values. Minor obstructions such as wires and cables potentially in the field of view should be eliminated during deployment, and other occasional obstructions (birds, humans) can be monitored by video camera. For supervised measurements, any structural shading can easily be identified and avoided.

On some platforms, there may be a risk of optical contamination from atmospheric steam or smoke emissions and other exhaust gases (air conditioning, etc.).

4.2.5. Fouling

The sensor fore-optics should be clean.

Sunphotometers are always associated with a pointing mechanism that is either robotic or human, and so can generally be protected from most fouling mechanisms when not measuring.

Nevertheless, some fouling of the fore-optics may occur for long-term unsupervised deployments because of sea spray, rain droplets, and/or spiders and insects, etc.

Major fouling events can be identified by time series analysis of data and/or video camera imagery.

The uncertainty estimate related to fouling can be validated by comparing post-deployment calibrations before and after cleaning [26].

4.2.6. Fast Natural Fluctuations

Measurements should be used only during periods of stable illumination.

This method for E_d^{0+} can only be used in ideal clear sky conditions, where fast natural fluctuations of E_d^{0+} do not occur. The latter can easily be detected by replicate measurements, and the corresponding measurement sequence can be eliminated.

4.3. Variants on the Method of Measurement of E_d^{0+} from Direct Sunphotometry and a Clear Sky Atmospheric Model

As mentioned previously, this protocol can be used with human or robotic pointing systems. Since this protocol has very different assumptions and very different sources of uncertainty from the protocol using a vertically-pointing irradiance sensor (Section 2), there is significant added value to combine the sunphotometric estimation of E_d^{0+} with the direct measurement of E_d^{0+} using an irradiance sensor, as proposed in the OSPREY system [65].

5. Estimation of E_d^{0+} from Underwater Measurements

It is common for underwater radiometric measurements of the profile with depth, z, of nadir upwelling radiance, $L_{un}(z)$, to be accompanied by underwater measurements of downwelling irradiance, $E_d(z)$. Historically, E_d^{0+} was often estimated from these underwater measurements by extrapolation to just beneath the surface and transmission across the air–water interface. However, the temporal variability of $E_d(z)$ associated with wave focusing/defocusing is particularly difficult to remove, and this method for estimating E_d^{0+} has been replaced by the direct above-water E_d^{0+} measurement, and will not be discussed further in this review. A detailed description of protocols for measuring $E_d(z)$, the spectral diffuse attenuation coefficient of downwelling irradiance, $K_d(\lambda, z)$, and, if considered useful, E_d^{0+}, can be found in the NASA Ocean Optics protocols [66].

Outside the satellite validation context, underwater measurements of $E_d(z)$ are still relevant for the estimation of optically and biologically important parameters such as $K_d(\lambda, z)$, and related parameters such as euphotic depth.

6. Conclusions

6.1. Summary of the State of the Art

This paper reviews the current state of the art of protocols for the measurement of downwelling irradiance for the validation of satellite remote sensing data over water. In the FRM context, particular attention is paid to the protocol-related elements of the measurement uncertainty budget. These aspects of the protocol are discussed with reference to documented studies, and guidelines are provided on how to estimate such uncertainties, e.g., design of experiments and/or model studies.

Three basic measurement protocols have been identified:

- Direct above-water measurement of E_d^{0+} with an upward pointing irradiance sensor
- Estimation of E_d^{0+} using a downward pointing radiance sensor and a reflective plaque
- Estimation of E_d^{0+} from direct sunphotometry and a clear sky atmospheric model

A fourth measurement method that was previously used, estimating E_d^{0+} from the underwater vertical profiles of $E_d(z)$, is now considered inappropriate, and is no longer recommended. This method remains relevant for the measurement of $E_d(z)$ and related parameters such as diffuse attenuation coefficient, but not E_d^{0+}.

The main body of this paper is summarized in Table 1, which lists the equipment needed, method variants, and any special issues, and in Table 2. The latter summarizes the components of the uncertainty estimation giving ideal conditions, recommendations for best practice, and approaches to estimating uncertainty, but excludes any uncertainties arising from radiometer imperfections, such as calibration, thermal sensitivity, spectral response (straylight/out of band effects), non-linearity, and angular (cosine) response.

Table 1. Summary of the three measurement methods as regards equipment, method variants, and special issues.

	Upward-Pointing Irradiance Sensor	Radiance Sensor and Reflective Plaque	Direct Sunphotometry
Equipment	Irradiance sensor (cosine response) Inclinometer	Radiance sensor Reflective plaque Inclinometer	Sunphotometer (radiance) sensor Pointing mechanism Atmosphere radiative transfer model
Variants	Surfacing of underwater drifting floats. Shadowband for diffuse/direct.	White/grey plaques	Hand-held or robotic pointing
Other notes		Uncalibrated radiometers? (see Section 3.1.1) Plaque viewing nadir angle? (see Section 3.1.2)	

For the "irradiance sensor" and the "reflectance plaque" methods, the main challenge is to deploy the radiometer/plaque sufficiently high enough to avoid any shading. In this context, "shading" does not only refer to the obvious shadowing of direct Sun, but also refers to the difference between the unobstructed hemisphere of Sun and sky radiance and the reality of measuring in situations where the radiometer/plaque are not higher than all the other structures. For the "irradiance sensor" method, it is also a major challenge to have a sensor that is sufficiently well-designed and well-characterized as regards angular (cosine) response [28].

Table 2. Summary of the three measurement methods, including components that must be considered for the uncertainty estimation. BRDF: bidirectional reflectance distribution functions; I = Ideal conditions; R = Recommendations; U = Uncertainty estimation; Cal = calibration; FOV = field of view; AOT = aerosol optical thickness; r/t = radiative transfer; S.D. = standard deviation; N/A = Not Applicable. See text for more details on each topic.

Method	Upward-Pointing Irradiance Sensor	Radiance Sensor and Reflective Plaque	Direct Sunphotometry
Plaque cal and characterization	N/A	I: BRDF-calibrated, homogeneous plaque fills FOV R: Tests to check FOV U: Plaque certificate including BRDF, experiments for homogeneity and height above plaque/FOV	N/A
Tilt/pointing	I: Deploy vertical R: Monitor with inclinometer U: Modeling/experiments	I: Deploy horizontal R: Monitor with inclinometer U: Modeling/experiments	I: Sensor FOV contains and centered on Sun R: Small FOV, accurate pointing, check AOT U: Via estimation of AOT
Superstructure shading	I: Deploy above all structures R: Use mast and fish-eye photos U: Experiments (different heights/locations) and modeling	I: Deploy above all structures (except radiometer) R: Use mast and fish-eye photos U: Experiments (different heights/locations) and modeling	I: Clear radiometer–direct Sun path R: Check with video surveillance and data QC U: N/A (if not rejected)
Fouling	I: Keep fore-optics clean R: Inspect/clean/protect, monitor with portable cal devices U: Pre-/post-cleaning cal of radiometer	I: Keep radiometer fore-optics and plaque clean R: Inspect/clean/protect, monitor radiometer with portable cal devices U: Pre-/post-cleaning cals for radiometer and plaque	I: Keep fore-optics clean R: Inspect/clean/protect U: Pre-/post-cleaning cals
Fast natural fluctuations	I: Reject if unstable illumination R: Compare replicates/time series U: S.D. of accepted measurements	I: Reject if unstable illumination R: Compare replicates/time series U: S.D. of accepted measurements	I: Reject if unstable illumination R: Compare replicates/time series U: S.D. of accepted measurements
Sky conditions and atmospheric r/t model	N/A	N/A	I: Perfectly cloud-free sky, horizontally homogeneous atmosphere and surface. Perfect r/t model and inputs R: Reject if clouds detected. Intercompare r/t models, check inputs U: Modeling. See Section 4.2.1

6.2. Irradiance Sensor or Reflectance Plaque?

The preparation of this review stimulated considerable discussion within the community on the pros/cons of the reflectance plaque method as compared to the irradiance sensor method in addition to the question of whether the reflectance plaque method radiance sensor needs to be calibrated (see Section 3.1.1). When correctly applied, the reflectance plaque method can clearly meet the criteria expected of an FRM. However, in practice, this method has often been associated with less rigorous implementation. Specifically, recognizing that the reflectance plaque is performing the same function as the fore-optics of an irradiance sensor, which collects light from the upward hemisphere according to a zenith cosine weighting and directs that light to a photodetector, it is necessary that:

1. There be no humans above the level of the reflectance plaque/irradiance sensor (and thereby affecting the sky radiance contributing to E_d^{0+} in a way that is highly variable and essentially not quantifiable in an uncertainty estimate),

2. The reflectance plaque/irradiance sensor be mounted as high as possible on the ship/platform, typically higher than any superstructure elements with significant solid angle as viewed from the plaque/sensor,

3. The reflectance plaque/irradiance sensor be mounted on a fixed structure, not hand-held, and associated with an inclinometer allowing the estimation of uncertainties associated with non-horizontal/vertical measurements,

4. The measurements made using the reflective plaque/irradiance sensor be supported by experiments and/or simulations to estimate the measurement uncertainties associated with any superstructure shading of the plaque/irradiance sensor.

6.3. Future Perspectives

In contrast to the more difficult L_w measurement, where there has been considerable evolution and diversity since the publication of the NASA Ocean Optics Protocols [20], measurement protocols for E_d^{0+} seem now to be quite mature and stable.

Future improvements to E_d^{0+} measurements are expected to come from the following developments:

- Improvements in the design and usage of calibration monitoring devices, which can be used in the field, are likely to improve the identification of fore-optics fouling and radiometer sensitivity changes.
- Model simulations of the 3D light field and experiments for deployments with structures above the irradiance sensor/reflectance plaque are likely to improve estimations of related uncertainties.
- Improvements in the stability and reduction in the cost of telescopic masts may reduce superstructure shading effects.
- Reduction in the cost of pointing systems, thanks to the video camera surveillance industry, should improve the protection ("parking") of irradiance sensors when not in use, and thus reduce fouling for long-term deployments.
- Improvements in automatic gimballing systems might reduce the tilt effects for the irradiance sensor method.
- Greater use of full sky imaging cameras, whether calibrated (expensive) or not (inexpensive), will allow the better identification of suboptimal measurement conditions.

As regards the future for the validation of water reflectance more generally:

- The tendency to move to highly automated systems with long-term, e.g., one year, essentially maintenance-free deployments is likely to significantly improve the quantity of data available for validation.
- The advent of operational satellite missions such as NPP/VIIRS, Sentinel-3/OLCI, Sentinel-2/MSI, and Landsat-8/OLI with the need for a guaranteed long-term validated data stream will increase the need for FRM.
- The huge increase in optical satellite missions used for aquatic remote sensing will also increase the need for highly automated measurement systems.
- As regards the needs of the validation community, it is recommended to:
- Update this review, e.g., on a 10-year time frame, to take account of developments in the protocols, particularly in the estimation of uncertainties. Such an update is best preceded by community discussion at an international workshop.
- Organize regular, e.g., on a two-year time frame, intercomparison exercises to ensure that measurement protocols and scientists remain state of the art (as required by the FRM context).

Although not targeted by this review, it is possible that the considerations developed here may be useful for other applications where E_d^{0+} measurements are needed, including the validation of satellite-derived photosynthetically available radiation products [67], the validation of surface reflectance over land, and the monitoring of solar irradiance for the solar energy industry, for agriculture, for the building industry, for the estimation of the Earth's radiation budget, and absorbing atmospheric gases, etc.

Author Contributions: Conceptualization, K.G.R.; methodology, K.G.R.; writing—original draft preparation, K.G.R.; writing—review and editing, K.V., A.C.B., E.B., A.C., R.F., M.H., C.J., B.C.J., J.K., Z.L., M.O., V.V., R.V.

Funding: The collection of information for and the writing of this study were funded by the European Space Agency, grant number ESA/AO/1-8500/15/I-SBo, "Fiducial reference measurements for satellite ocean color (FRM4SOC)" project. EB's contributions are supported by the NASA Ocean Biology and Biochemistry program.

Acknowledgments: Colleagues from the FRM4SOC project, the Sentinel-3 Validation Team, and the NOAA/VIIRS cal/val team are acknowledged for helpful discussions on measurement protocols during project meetings and teleconferences. Craig Donlon and Tânia Casal are gratefully acknowledged for stimulating discussions and constructive support throughout the FRM4SOC project, from conception to implementation. Giuseppe Zibordi is thanked for constructive comments.

Conflicts of Interest: The first author is involved in the design of systems based on the "irradiance sensor" method and in the acquisition of data using the "irradiance sensor" and "sunphotometer" methods. Other co-authors are involved in the acquisition of data using the "irradiance sensor", the "reflectance plaque" and the "sunphotometer" methods. The funders had no role in the design of the study other than the drafting of the FRM4SOC project Statement of Work, which states that such a protocol review should be performed. The funders had no role in the collection, analyses, or interpretation of data or in the writing of the manuscript other than general project monitoring and encouragement. The funders did encourage publication of this review in order to ensure that high quality measurements of known uncertainty are available for the validation of the satellite missions they design and operate. None of these interests is considered to be conflictual or to inappropriately influence the recommendations expressed here.

Disclaimer: Certain commercial equipment, instruments, or materials are identified in this paper in order to specify the experimental procedure adequately. Such identification is not intended to imply recommendation or endorsement by the National Institute of Standards and Technology or any other organization involved in the writing of this paper, nor is it intended to imply that the materials or equipment identified are necessarily the best available for the purpose.

References

1. International Ocean Colour Coordinating Group (IOCCG). *Why Ocean Colour? The Societal Benefits of Ocean-Colour Technology*; IOCCG Report 7; IOCCG: Dartmouth, NS, Canada, 2008; p. 141.

2. Donlon, C.J.; Wimmer, W.; Robinson, I.; Fisher, G.; Ferlet, M.; Nightingale, T.; Bras, B. A second-generation blackbody system for the calibration and verification of seagoing infrared radiometers. *J. Atmospheric Ocean. Technol.* **2014**, *31*, 1104–1127. [CrossRef]

3. Mobley, C.D. *Light and Water: Radiative Transfer in Natural Waters*; Academic Press: London, UK, 1994.

4. Mueller, J.L.; Morel, A. Fundamental definitions, relationships and conventions. In *Ocean Optics Protocols for Satellite Ocean Color Sensor Validation, Revision 4, Volume I: Introduction, Background and Conventions*; NASA/TM; Goddard Space Flight Space Center: Greenbelt, MD, USA, 2003; Chapter 2; pp. 11–30.

5. Ruddick, K.; Voss, K.; Boss, E.; Castagna, A.; Frouin, R.J.; Gilerson, A.; Hieronymi, M.; Johnson, B.C.; Kuusk, J.; Lee, Z.; et al. A review of protocols for Fiducial Reference Measurements of water-leaving radiance for validation of satellite remote sensing data over water. *Remote Sens. (submitted to this volume).*

6. Vermote, E.; Tanré, D.; Deuzé, J.L.; Herman, M.; Morcette, J.J.; Kotchenova, S.Y. Second Simulation of the Satellite Signal in the Solar Spectrum-Vector (6SV). 6S User Guide Version 3. 2006, p. 134. Available online: http://6s.ltdri.org/pages/manual.html (accessed on 23 July 2019).

7. Mobley, C. Light from the Sun. Available online: http://www.oceanopticsbook.info/view/light_and_radiometry/level_2/light_from_the_sun (accessed on 17 May 2019).

8. Ohmura, A.; Dutton, E.G.; Forgan, B.; Fröhlich, C.; Gilgen, H.; Hegner, H.; Heimo, A.; König-Langlo, G.; McArthur, B.; Müller, G.; et al. Baseline Surface Radiation Network (BSRN/WCRP): New Precision Radiometry for Climate Research. *Bull. Am. Meteorol. Soc.* **1998**, *79*, 2115–2136. [CrossRef]

9. Wild, M. Global dimming and brightening: A review. *J. Geophys. Res. Atmos.* **2009**, *114*, 1984–2012. [CrossRef]

10. Smith, R.C.; Prezelin, B.B.; Baker, K.S.; Bidigare, R.R.; Boucher, N.P.; Coley, T.; Karentz, D.; MacIntyre, S.; Matlick, H.A.; Menzies, D.; et al. Ozone depletion: ultraviolet radiation and phytoplankton biology in antarctic waters. *Science* **1992**, *255*, 952–959. [CrossRef]

11. Eerme, K.; Veismann, U.; Lätt, S. Proxy-based reconstruction of erythemal UV doses over Estonia for 1955–2004. *Ann. Geophys.* **2006**, *24*, 1767–1782. [CrossRef]

12. Balzarolo, M.; Anderson, K.; Nichol, C.; Rossini, M.; Vescovo, L.; Arriga, N.; Wohlfahrt, G.; Calvet, J.-C.; Carrara, A.; Cerasoli, S.; et al. Ground-based optical measurements at European flux sites: a review of methods, instruments and current controversies. *Sensors* **2011**, *11*, 7954–7981. [CrossRef]

13. Frouin, R.; Ramon, D.; Boss, E.; Jolivet, D.; Compiègne, M.; Tan, J.; Bouman, H.; Jackson, T.; Franz, B.; Platt, T.; et al. Satellite radiation products for ocean biology and biogeochemistry: Needs, state-of-the-art, gaps, development priorities, and opportunities. *Front. Mar. Sci.* **2018**, *5*, 3. [CrossRef]

14. Loutzenhiser, P.G.; Manz, H.; Felsmann, C.; Strachan, P.A.; Frank, T.; Maxwell, G.M. Empirical validation of models to compute solar irradiance on inclined surfaces for building energy simulation. *Sol. Energy* **2007**, *81*, 254–267. [CrossRef]

15. International Standards Organisation (ISO). *Space Environment (Natural And Artificial)—Process for Determining Solar Irradiances*; ISO Report 21348:2007; International Standards Organisation (ISO): Geneva, Switzerland, 2007.

16. Hooker, S.B.; Zibordi, G.; Maritorena, S. The Second SeaWiFs Ocean Optics DARR (DARR-00). In *Results of the Second SeaWiFS Data Analysis Round Robin, March 2000 (DARR-00)*; SeaWiFS Postlaunch Technical Report Series; NASA Goddard Space Flight Space Center: Greenbelt, MD, USA, 2001; Volume 15, pp. 4–45.

17. Zibordi, G.; Ruddick, K.; Ansko, I.; Moore, G.; Kratzer, S.; Icely, J.; Reinart, A. In situ determination of the remote sensing reflectance: an inter-comparison. *Ocean Sci.* **2012**, *8*, 567–586. [CrossRef]

18. Ondrusek, M.; Lance, V.P.; Arnone, R.; Ladner, S.; Goode, W.; Vandermeulen, R.; Freeman, S.; Chaves, J.E.; Mannino, A.; Gilerson, A.; et al. *Report for Dedicated JPSS VIIRS Ocean Color December 2015 Calibration/Validation Cruise*; NOAA Technical Report 148; U.S. Department of Commerce, National Oceanic and Atmospheric Administration, National Environmental Satellite, Data, and Information Service: Washington, DC, USA, 2016; p. 66.

19. *International Standards Organisation (ISO) Evaluation of Measurement Data—Guide to the Expression of Uncertainty in Measurement*; JCGM 101: 2008; International Standards Organisation (ISO): Geneva, Switerland, 2008.

20. Mueller, J.L.; Fargion, G.S.; McClain, C.R. *Ocean Optics Protocols for Satellite Ocean Color Sensor Validation*; Technical Memorandum TM-2003-21621/Revision 5; NASA Goddard Space Flight Space Center: Greenbelt, MD, USA, 2004.

21. Zibordi, G.; Voss, K.J. *In situ optical radiometry in the visible and near infrared. In Optical Radiometry for Ocean Climate Measurements*; Zibordi, G., Donlon, C., Parr, A.C., Eds.; 2014; pp. 248–305.

22. Zibordi, G.; Voss, K. *Protocols for Satellite Ocean Color Data Validation: In situ Optical Radiometry*; IOCCG Protocols Document; IOCCG: Dartmouth, Canada, 2019.

23. Vabson, V.; Kuusk, J.; Ansko, I.; Vendt, R.; Alikas, K.; Ruddick, K.; Ansper, A.; Bresciani, M.; Burmester, H.; Costa, M.; et al. Laboratory intercomparison of radiometers used for satellite validation in the 400–900 nm range. *Remote Sens.* **2019**, *11*, 1101. [CrossRef]

24. Vabson, V.; Kuusk, J.; Ansko, I.; Vendt, R.; Alikas, K.; Ruddick, K.; Ansper, A.; Bresciani, M.; Burmester, H.; Costa, M.; et al. Field intercomparison of radiometers used for satellite validation in the 400–900 nm range. *Remote Sens.* **2019**, *11*, 1129. [CrossRef]

25. Ruddick, K.; De Cauwer, V.; Park, Y.; Moore, G. Seaborne measurements of near infrared water-leaving reflectance: The similarity spectrum for turbid waters. *Limnol. Oceanogr.* **2006**, *51*, 1167–1179. [CrossRef]

26. Zibordi, G.; Holben, B.; Slutsker, I.; Giles, D.; D'Alimonte, D.; Mélin, F.; Berthon, J.-F.; Vandemark, D.; Feng, H.; Schuster, G.; et al. AERONET-OC: A network for the validation of ocean color primary product. *J. Atmospheric Ocean. Technol.* **2009**, *26*, 1634–1651. [CrossRef]

27. Claustre, H.; Bernard, S.; Berthon, J.-F.; Bishop, J.; Boss, E.; Coaranoan, C.; D'Ortenzio, F.; Johnson, K.; Lotliker, A.; Ulloa, O. *Bio-Optical Sensors on Argo Floats*; IOCCG Report; IOCCG: Dartmouth, NS, Canada, 2011.

28. Mekaoui, S.; Zibordi, G. Cosine error for a class of hyperspectral irradiance sensors. *Metrologia* **2013**, *50*, 187–199. [CrossRef]

29. Hlaing, S.; Harmel, T.; Ibrahim, A.; Ioannou, I.; Tonizzo, A.; Gilerson, A.; Ahmed, S. Validation of ocean color satellite sensors using coastal observational platform in Long Island Sound. In *Remote Sensing of the Ocean, Sea Ice, and Large Water Regions 2010*; International Society for Optics and Photonics: Bellingham, WA, USA, 2010; pp. 782504-1–782504-8.

30. Brando, V.E.; Lovell, J.L.; King, E.A.; Boadle, D.; Scott, R.; Schroeder, T. The potential of autonomous ship-borne hyperspectral radiometers for the validation of ocean color radiometry data. *Remote Sens.* **2016**, *8*, 150. [CrossRef]

31. Slivkoff, M.M. Ocean Colour Remote Sensing of the Great Barrier Reef waters. PhD Thesis, Curtin University, Bentley, Australia, 2014.

32. Antoine, D.; Guevel, P.; Desté, J.-F.; Bécu, G.; Louis, F.; Scott, A.J.; Bardey, P. The "BOUSSOLE" buoy-a new transparent-to-swell taut mooring dedicated to marine optics: Design, tests, and performance at sea. *J. Atmospheric Ocean. Technol.* **2008**, *25*, 968–989. [CrossRef]

33. Voss, K.J.; Chapin, A.L. Upwelling radiance distribution camera system, NURADS. *Opt. Express* **2005**, *13*, 4250. [CrossRef]

34. Vansteenwegen, D.; Ruddick, K.; Cattrijsse, A.; Vanhellemont, Q.; Beck, M. The pan-and-tilt hyperspectral radiometer system (PANTHYR) for autonomous satellite validation measurements – prototype design and testing. *Remote Sens.* **2019**, *11*, 1360. [CrossRef]

35. Hooker, S.B. The telescopic mount for advanced solar technologies (T-MAST). In *Advances in Measuring the Apparent Optical Properties (AOPs) of Optically Complex Waters*; NASA/TM; NASA Goddard Space Flight Space Center: Greenbelt, MD, USA, 2010; Volume 215856.

36. Vaskuri, A.; Greenwell, C.; Hessey, I.; Tompkins, J.; Woolliams, E. Contamination and UV ageing of diffuser targets used in satellite inflight and ground reference test site calibrations. *J. Phys. Conf. Ser.* **2018**, *972*, 012001. [CrossRef]

37. Hooker, S.B.; Maritorena, S. An evaluation of oceanographic radiometers and deployment methodologies. *J. Atmospheric Ocean. Technol.* **2000**, *17*, 811–830. [CrossRef]

38. Kuusk, J.; Kuusk, A. Hyperspectral radiometer for automated measurement of global and diffuse sky irradiance. *J. Quant. Spectrosc. Radiat. Transf.* **2018**, *204*, 272–280. [CrossRef]

39. Gregg, W.W.; Carder, K.L. A simple spectral solar irradiance model for cloudless maritime atmospheres. *Limnol. Oceanogr.* **1990**, *35*, 1657–1675. [CrossRef]

40. Kalisch, J.; Macke, A. Estimation of the total cloud cover with high temporal resolution and parametrization of short-term fluctuations of sea surface insolation. *Meteorol. Z.* **2008**, *17*, 603–611. [CrossRef]

41. Batlles, F.J.; Olmo, F.J.; Alados-Arboledas, L. On shadowband correction methods for diffuse irradiance measurements. *Sol. Energy* **1995**, *54*, 105–114. [CrossRef]

42. Castagna, A.; Johnson, B.C.; Voss, K.; Dierssen, H.M.; Patrick, H.; Germer, T.; Sabbe, K.; Vyverman, W. Uncertainty in global downwelling plane irradiance estimates from sintered polytetrafluoroethylene plaque radiance measurements. *Appl. Opt.* **2019**, *58*, 4497–4511. [CrossRef] [PubMed]

43. Early, E.A.; Barnes, P.Y.; Johnson, B.C.; Butler, J.J.; Bruegge, C.J.; Biggar, S.F.; Spyak, P.R.; Pavlov, M.M. Bidirectional Reflectance Round-Robin in Support of the Earth Observing System Program. *J. Atmos. Ocean. Technol.* **2000**, *17*, 1077–1091. [CrossRef]

44. Niemi, K.; Metsämäki, S.; Pulliainen, J.; Suokanerva, H.; Böttcher, K.; Leppäranta, M.; Pellikka, P. The behaviour of mast-borne spectra in a snow-covered boreal forest. *Remote Sens. Environ.* **2012**, *124*, 551–563. [CrossRef]

45. Leeuw, T.; Boss, E. The hydrocolor app: above water measurements of remote sensing reflectance and turbidity using a smartphone camera. *Sensors* **2018**, *18*, 256. [CrossRef]

46. Mueller, J.L.; Davis, C.; Arnone, R.; Frouin, R.; Carder, K.; Lee, Z.P.; Steward, R.G.; Hooker, S.; Mobley, C.D.; McLean, S. Above-water radiance and remote sensing reflectance measurements and analysis protocols. In *Ocean Optics Protocols for Satellite Ocean Color Sensor Validation Revision 4*; National Aeronautical and Space Administration: Greenbelt, MD, USA, 2003; Volume III, Chapter 3; pp. 21–31.

47. Salim, S.G.R.; Fox, N.P.; Theocharous, E.; Sun, T.; Grattan, K.T.V. Temperature and nonlinearity corrections for a photodiode array spectrometer used in the field. *Appl. Opt.* **2011**, *50*, 866–875. [CrossRef]

48. Yoon, H.W.; Butler, J.J.; Larason, T.C.; Eppeldauer, G.P. Linearity of InGaAs photodiodes. *Metrologia* **2003**, *40*, S154–S158. [CrossRef]

49. Mobley, C.D. Estimation of the remote-sensing reflectance from above-surface measurements. *Appl. Opt.* **1999**, *38*, 7442–7455. [CrossRef] [PubMed]

50. Doxaran, D.; Cherukuru, N.C.; Lavender, S.J.; Moore, G.F. Use of a Spectralon panel to measure the downwelling irradiance signal: case studies and recommendations. *Appl. Opt.* **2004**, *43*, 5981. [CrossRef]

51. Behnert, I.D.; Deadman, A.J.; Fox, N.P.; Harris, P.M.; Gürol, S.; Özen, H.; Bachmann, M.; Boucher, Y.; Lachérade, S. *Measurement Report CEOS WGCV Pilot Comparison of Techniques and Instruments Used for the Vicarious Calibration of Land Surface Imaging through a Ground Reference Standard Test Site 2009*; NPL Report OP5; National Physical Laboratory: Teddington, UK, 2010.

52. Stiegman, A.E.; Bruegge, C.J.; Springsteen, A.W. Ultraviolet stability and contamination analysis of Spectralon diffuse reflectance material. *Opt. Eng.* **1993**, *32*, 799–804. [CrossRef]

53. Georgiev, G.T.; Butler, J.J. Long-term calibration monitoring of Spectralon diffusers BRDF in the air-ultraviolet. *Appl. Opt.* **2007**, *46*, 7892–7899. [CrossRef] [PubMed]

54. Georgi, G.T.; James, J. *Butler BRDF Study of Gray-Scale Spectralon*; SPIE Proceedings 7081; SPIE: Bellingham, WA, USA, 2008; p. 9.

55. Georgiev, G.T.; Butler, J.J. *The Effect of Incident Light Polarization on Spectralon BRDF Measurements*; SPIE Proceedings 5570; SPIE: Bellingham, WA, USA, 2004; p. 492.

56. Germer, T.A. Full four-dimensional and reciprocal Mueller matrix bidirectional reflectance distribution function of sintered polytetrafluoroethylene. *Appl. Opt.* **2017**, *56*, 9333–9340. [CrossRef]

57. Deschamps, P.-Y.; Fougnie, B.; Frouin, R.; Lecomte, P.; Verwaerde, C. SIMBAD: A field radiometer for satellite ocean color validation. *Appl. Opt.* **2004**, *43*, 4055–4069. [CrossRef]

58. Holben, B.; Eck, T.F.; Slutsker, I.; Tanre, D.; Buis, J.P.; Setzer, A.; Vermote, E.; Reagan, J.A.; Kaufman, Y.J.; Nakajima, T.; et al. AERONET—A federated instrument network and data archive for aerosol characterization. *Remote Sens. Environ.* **1998**, *66*, 1–16. [CrossRef]

59. Segal-Rosenheimer, M.; Russell, P.B.; Schmid, B.; Redemann, J.; Livingston, J.M.; Flynn, C.J.; Johnson, R.R.; Dunagan, S.E.; Shinozuka, Y.; Herman, J.; et al. Tracking elevated pollution layers with a newly developed hyperspectral Sun/Sky spectrometer (4STAR): Results from the TCAP 2012 and 2013 campaigns. *J. Geophys. Res. Atmos.* **2014**, *119*, 2013JD020884. [CrossRef]

60. Thuillier, G.; Herse, M.; Labs, D.; Foujols, T.; Peetermans, W.; Gillotay, D.; Simon, P.C.; Mandel, H. The solar spectral irradiance from 200 to 2400 nm as measured by the SOLSPEC spectrometer from the ATLAS and EURECA missions. *Sol. Phys.* **2003**, *214*, 1–22. [CrossRef]

61. Deuzé, J.L.; Herman, M.; Santer, R. Fourier series expansion of the transfer equation in the atmosphere-ocean system. *J. Quant. Spectrosc. Radiat. Transf.* **1989**, *41*, 483–494. [CrossRef]

62. Tanré, D.; Herman, M.; Deschamps, P.-Y.; de Leffe, A. Atmospheric modeling for space measurements of ground reflectances, including bidirectional properties. *Appl. Opt.* **1979**, *18*, 3587–3594. [CrossRef] [PubMed]

63. Gordon, H.R.; Du, T.; Zhang, T. Remote sensing of ocean color and aerosol properties: resolving the issue of aerosol absorption. *Appl. Opt.* **1997**, *36*, 8670–8684. [CrossRef] [PubMed]

64. Emde, C.; Barlakas, V.; Cornet, C.; Evans, F.; Korkin, S.; Ota, Y.; Labonnote, L.C.; Lyapustin, A.; Macke, A.; Mayer, B.; et al. IPRT polarized radiative transfer model intercomparison project—Phase A. *J. Quant. Spectrosc. Radiat. Transf.* **2015**, *164*, 8–36. [CrossRef]

65. Hooker, S.B.; Bernhard, G.; Morrow, J.H.; Booth, C.R.; Comer, T.; Lind, R.N.; Quang, V. *Optical Sensors for Planetary Radiant Energy (OSPREy): Calibration and Validation of Current and Next-Generation NASA Missions*; NASA Goddard Space Flight Space Center: Greenbelt, MD, USA, 2012.

66. Mueller, J.L. In-water radiometric profile measurements and data analysis protocols. In *Ocean Optics Protocols for Satellite Ocean Color Sensor Validation, Revision 4, Volume III: Radiometric Measurements and Data Analysis Protocols*; NASA/TM; NASA Goddard Space Flight Space Center: Greenbelt, MD, USA, 2003; Chapter 2; pp. 7–20.

67. Frouin, R.; Pinker, R.T. Estimating Photosynthetically Active Radiation (PAR) at the earth's surface from satellite observations. *Remote Sens. Land Surf. Stud. Glob. Chang.* **1995**, *51*, 98–107. [CrossRef]

 *remote sensing*

Review

A Review of Protocols for Fiducial Reference Measurements of Water-Leaving Radiance for Validation of Satellite Remote-Sensing Data over Water

Kevin G. Ruddick [1,*], Kenneth Voss [2], Emmanuel Boss [3], Alexandre Castagna [4], Robert Frouin [5], Alex Gilerson [6], Martin Hieronymi [7], B. Carol Johnson [8], Joel Kuusk [9], Zhongping Lee [10], Michael Ondrusek [11], Viktor Vabson [9] and Riho Vendt [9]

[1] Royal Belgian Institute of Natural Sciences (RBINS), Operational Directorate Natural Environment, 29 Rue Vautierstraat, 1000 Brussels, Belgium
[2] Physics Department, University of Miami, Coral Gables, FL 33124, USA
[3] School of Marine Sciences, University of Maine, Orono, ME 04469, USA
[4] Protistology and Aquatic Ecology Research Group, Gent University, Krijgslaan 281, 9000 Gent, Belgium
[5] Scripps Institution of Oceanography, University of California San Diego, 9500 Gilman Drive #0224, La Jolla, CA 92093-0224, USA
[6] Department of Electrical Engineering, The City College of New York, 160 Convent Avenue, New York, NY 10031, USA
[7] Institute of Coastal Research, Helmholtz-Zentrum Geesthacht (HZG), Max-Planck-Str. 1, 21502 Geesthacht, Germany
[8] National Institute of Standards and Technology (NIST), 100 Bureau Drive, Gaithersburg, MD 20899, USA
[9] Tartu Observatory, University of Tartu, 61602 Tõravere, Estonia
[10] School for the Environment, University of Massachusetts Boston, 100 Morrissey Blvd., Boston, MA 02125-3393, USA
[11] National Oceanic and Atmospheric Administration (NOAA), Center for Weather and Climate Prediction, 5830 University Research Court, College Park, MD 20740, USA
* Correspondence: kruddick@naturalsciences.be

Received: 24 July 2019; Accepted: 6 September 2019; Published: 20 September 2019

Abstract: This paper reviews the state of the art of protocols for measurement of water-leaving radiance in the context of fiducial reference measurements (FRM) of water reflectance for satellite validation. Measurement of water reflectance requires the measurement of water-leaving radiance and downwelling irradiance just above water. For the former there are four generic families of method, based on: (1) underwater radiometry at fixed depths; or (2) underwater radiometry with vertical profiling; or (3) above-water radiometry with skyglint correction; or (4) on-water radiometry with skylight blocked. Each method is described generically in the FRM context with reference to the measurement equation, documented implementations and the intra-method diversity of deployment platform and practice. Ideal measurement conditions are stated, practical recommendations are provided on best practice and guidelines for estimating the measurement uncertainty are provided for each protocol-related component of the measurement uncertainty budget. The state of the art for measurement of water-leaving radiance is summarized, future perspectives are outlined, and the question of which method is best adapted to various circumstances (water type, wavelength) is discussed. This review is based on practice and papers of the aquatic optics community for the validation of water reflectance estimated from satellite data but can be relevant also for other applications such as the development or validation of algorithms for remote-sensing estimation of water constituents including chlorophyll *a* concentration, inherent optical properties and related products.

Keywords: water reflectance; satellite validation; Fiducial Reference Measurements; water-leaving radiance

---

1. Introduction

The objective of this paper is to review the state of the art of protocols for the measurement of water-leaving radiance, as used for the validation of satellite remote-sensing data over water.

1.1. The Need for Fiducial Reference Measurements for Satellite Validation

Satellite remote-sensing data is now used routinely for many applications, including monitoring of oceanic phytoplankton in the context of global climate change, detection of harmful algae blooms in coastal and inland waters, management of sediment transport in coastal water, estuaries and ports, the optimization and monitoring of dredging operations, etc. [1]. To be able to trust and use the remote-sensing data, these must be validated, usually by "matchup" comparison of simultaneous measurements by satellite and in situ. The terminology of "fiducial reference measurements (FRM)" was introduced to establish the requirements on the in situ measurements that can be trusted for use in such validation. Using the definition proposed in the context of sea surface temperature measurements [2], the defining mandatory characteristics of a FRM are:

- An uncertainty budget for all FRM instruments and derived measurements is available and maintained, traceable where appropriate to the *International System of Units/Système International d'unités* (SI), ideally through a national metrology institute.
- **FRM measurement protocols and community-wide management practices (measurement, processing, archive, documents, etc.) are defined and adhered to**
- FRM measurements have documented evidence of SI traceability and are validated by intercomparison of instruments under operational-like conditions.
- FRM measurements are independent from the satellite retrieval process.

The second term above, given in bold, situates the current review, which should provide such a definition of measurement protocols for the water-leaving radiance measurement.

1.2. Scope and Definitions

This review is focused on measurements of the water-leaving radiance as necessary for the validation of satellite data products for water reflectance at the bottom of the atmosphere. In the present review, the terminology of "remote-sensing reflectance", R_{rs}, is used where

$$R_{rs}(\lambda, \theta, \phi) = \frac{L_w(\lambda, \theta, \phi)}{E_d^{0+}(\lambda)} \tag{1}$$

where $E_d^{0+}(\lambda)$ is the spectral downward plane irradiance, also called "above-water downwelling irradiance", and $L_w(\lambda, \theta, \varphi)$ is the water-leaving radiance, defined, e.g., see [3], as the component of above-water directional upwelling radiance that has been transmitted across the water–air interface in the upward direction measured by the sensor and defined by viewing nadir angle θ and azimuth angle φ. The conventions used for these angles are defined in Figure 1. In other words, and as illustrated in Figure 2, L_w is the above-water directional upwelling radiance, L_u^{0+}, just above the air–water interface, after removal of radiance from air–water interface reflection, L_r:

$$L_w = L_u^{0+} - L_r \tag{2}$$

The latter term is called hereafter "skyglint" but may include also sunglint reflected from wave facets.

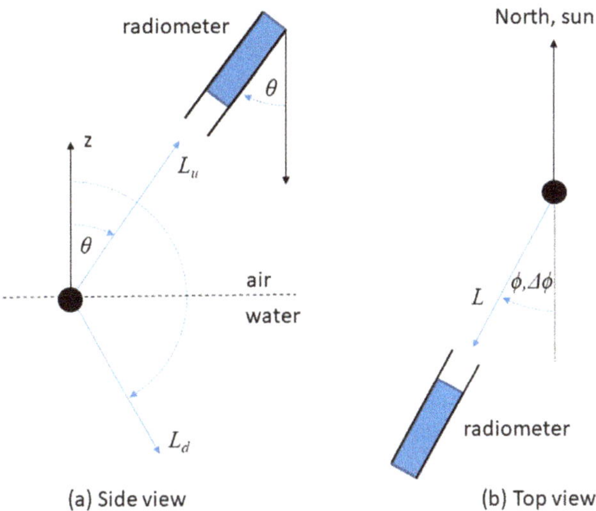

Figure 1. Nadir and azimuth viewing angle conventions illustrated for a reference system centred on the water surface (black dot). (**a**) Viewing nadir angle, θ, is measured from downward vertical axis: upward radiances are viewed at $\theta < \pi/2$, downward radiances (from sky and sun) are viewed at $\theta > \pi/2$. (**b**) Azimuth viewing angle, ϕ, and relative azimuth viewing angle, $\Delta\phi$, are measured for viewing direction clockwise from North and sun respectively: radiance viewed by a radiometer pointing towards North has azimuth 0 and radiance viewed by a radiometer pointing towards and away from sun have relative azimuth 0 and π respectively.

L_w is generally measured for nadir viewing geometry by under water or on water approaches (see Sections 2, 3 and 5) and generally measured for an off-nadir geometry by above-water approaches (see Section 4). When measured for (or extrapolated by a suitable model to) the nadir viewing geometry, the term nadir water-leaving radiance will be used where $L_{wn}(\lambda) = L_w(\lambda, \theta = 0°)$.

All radiometric quantities in this review are assumed to vary spectrally but for brevity the dependence on wavelength, λ, is generally omitted in the terminology.

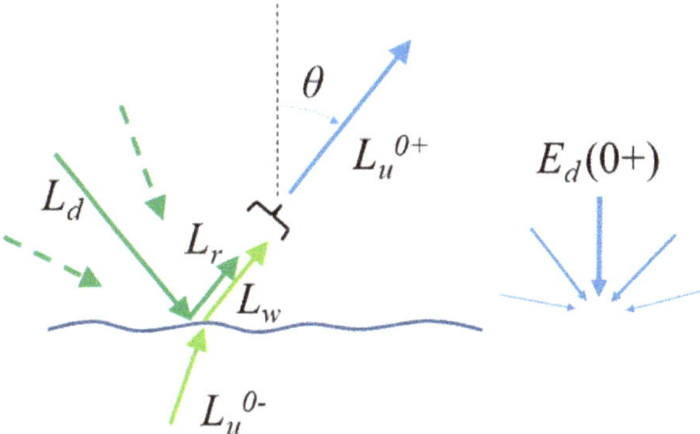

Figure 2. Illustration of definitions of water-leaving radiance, L_w, above and below water upwelling radiances, L_u^{0+} and L_u^{0-}, above-water downwelling (sky) radiance in the specular reflection direction, L_d, above-water upwelling radiance from reflection at the air–water interface ("skyglint"), L_r, and downwelling irradiance, E_d^{0+}. See also [4]. The widths of the arrows for E_d^{0+} represent the zenith cosine weighting for the different incident angles.

The validation of R_{rs} thus requires simultaneous measurement of two parameters: $E_d^{0+}(\lambda)$ and $L_w(\lambda, \theta, \varphi)$, although an alternative approach is to validate only $L_w(\lambda, \theta, \varphi)$. A companion paper [5] focuses on measurement of $E_d^{0+}(\lambda)$. The present review focuses on measurement of $L_w(\lambda, \theta, \varphi)$, reviewing the state of the art of measurement protocols in the FRM context, particularly as regards components of the measurement uncertainty budget relating to the measurement protocol.

The focus here is on aquatic applications, including the full range and diversity of water bodies from deep oceans through coastal and estuarine waters to ports and inland lakes.

Measurements of R_{rs} and hence $L_w(\lambda, \theta, \varphi)$ are also relevant outside the satellite validation context, for example when simultaneous in situ measurements are made of R_{rs} and in-water properties such as chlorophyll *a* concentration or inherent optical properties (IOPs) (without simultaneous satellite data) for algorithm calibration/validation purposes [6] or when in situ R_{rs} is used on its own for monitoring [7]. These applications are not specifically covered here, although many considerations of the measurement protocols described here are valid for all such applications.

Using the terminology of the International Standards Organisation (ISO, 2007) the spectral range of primary interest here is the visible (380 nm to 760 nm) and the lower wavelength part of the near infrared (760 nm to 1400 nm) ranges [8]. The considerations for measurement of L_w given here should be valid also for the near ultraviolet (300 nm to 400 nm) and middle infrared (1400 nm to 3000 nm), although the importance of the various uncertainty sources may be different because of the different intensity and angular distribution of downwelling irradiance and upwelling radiance and the instrumentation (radiance sensor detector and fore-optics) may have different properties in these ranges. Although L_w is measurably non-zero in the range 1000 nm to 1100 nm in extremely turbid waters [9], L_w will be effectively negligible for the longer near infrared from 1100 nm to 1400 nm and the middle infrared (1400 nm to 3000 nm) wavelengths because of the very high pure water absorption at these wavelengths. The need for L_w measurements in the range 1100 nm to 3000 nm is very limited, because satellite R_{rs} data will typically be set to zero during atmospheric correction. However, there may be some interest in this range for quality control of above-water L_w measurements, with non-zero measurement indicating a data quality problem, e.g., skyglint or sunglint contamination or floating material, for the whole spectrum. Also, there may be some interest in the range 1100 nm to 3000 nm for applications such as measurement of floating aquatic vegetation, although this is not strictly speaking L_w and should be measured only using above-water radiometry and without a skyglint/sunglint correction for the percentage of surface covered by vegetation [10].

The protocols described here are relevant for validation of a vast range of optical satellites including the dedicated medium resolution "ocean colour" missions, such as AQUA/MODIS, Sentinel-3/OLCI, JPSS/VIIRS, etc., but also the operational high resolution missions such as Landsat-8/OLI and Sentinel-2/MSI, as well any other optical mission from which water reflectance can be derived, including the geostationary COMS/GOCI-1 and MSG/SEVIRI, the extremely high resolution Pléiades and PlanetDove satellite constellations, airborne data, etc.

The current document does not try to identify a "best" protocol, nor does it aim to prescribe mandatory requirements on specific aspects of a measurement protocol such as "best nadir and azimuth angles for above-water radiometry" or "minimum distance for ship shadow avoidance". While such prescriptions have great value in encouraging convergence of methods and in challenging scientists to make good measurements, the diversity of aquatic and atmospheric conditions where validation is required, the diversity of radiometers and platforms and the corresponding diversity of measurement protocols suggests that more flexibility is needed. This flexibility is acceptable in the FRM context provided that each measurement is accompanied by a SI-traceable uncertainty budget that is (a) based on a full analysis of the protocol and (b) that is itself validated, e.g., by measurement intercomparison exercises [11–13] or by optical closure with inherent optical property measurements and radiative transfer modelling [14,15].

The present review aims to provide an overview of all relevant protocols, including guidelines for radiometer deployment and quality control of data and an overview of elements that should be considered in the complete uncertainty analysis of a measurement protocol. The approach is structured as follows: for each aspect of the measurement protocol contributing to measurement uncertainty the ideal situation is summarized in a single sentence in bold face, e.g., "**The radiance sensor should be vertical**" when making underwater radiance measurements. This is followed by a discussion of techniques to achieve or monitor this (e.g., slow descent free-fall platforms, measurement of tilt, removal of tilted data), practical considerations and problems (e.g., need for multiple deployments to reduce uncertainties for fast free-fall deployments) and approaches to estimate uncertainty when this ideal situation is not achieved (e.g., model studies, experiments).

For a general treatment of uncertainties in measurements, including a recommended terminology (e.g., "expanded uncertainty") and generic methods for estimating each component of uncertainty and combining uncertainties to achieve a total uncertainty the reader is referred to the Guide to the Expression of Uncertainty in Measurement (GUM) of the ISO [16].

The present review covers only aspects of the measurement relating to the protocol, including radiometer deployment, data acquisition and processing aspects but excluding any uncertainties arising from radiometer imperfections, such as calibration (including immersion coefficients for underwater radiometry), thermal sensitivity, spectral response (straylight/out of band effects) and spectral interpolation, non-linearity and angular response and polarization sensitivity. The decomposition of measurements into "protocols" (deployment, data acquisition and processing methods) and "radiometers" is adopted here in order to conveniently represent the wide diversity of possible combinations of methods and radiometers in a synthetic and generic way. However, it is fully recognised that "protocol" and "radiometer" must be coupled for the assessment of the uncertainty of any specific measurement. For example, the uncertainty associated with the skyglint correction in above-water radiometry or the uncertainty associated with wave-focusing effects in underwater radiometry depend on the speed (integration time) of the radiometer used (as well as the number of replicate measurements and the temporal processing and quality control processes). These radiometer-related aspects deserve a review paper of their own—the reader is referred to Volume II of the National Aeronautics and Space Administration (NASA) Ocean Optics Protocols [17] and Section 3 of [18] and Chapters 2 and 3 of [19].

The present review is limited in scope to the measurement of $L_w(\lambda, \theta, \varphi)$ in a single viewing geometry and does not discuss bidirectional reflectance distribution function (BRDF) corrections that can be applied to data to facilitate in situ vs. satellite comparisons. For example, a BRDF correction may be applied to the satellite data (and to off-nadir above-water in situ measurements) to estimate the nadir-viewing water-leaving radiance from the off-nadir viewing geometry. Alternatively, a BRDF correction may be applied to the in situ measurement to estimate water-leaving radiance in the satellite viewing geometry. This and other topics relating to the use of $L_w(\lambda, \theta, \varphi)$ measurements for satellite validation, including the impact of the different space and time scales [20,21], should be reviewed in a separate paper. The measurement of $E_d^{0+}(\lambda)$, as needed to calculate R_{rs}, and as needed for temporal correction and/or quality control of $L_w(\lambda, \theta, \varphi)$ in some protocols is reviewed in [5].

In the satellite validation context covered by this review, the focus is on clear sky conditions. There is no clear consensus regarding an objective definition of "clear sky" conditions, although Web Appendix 1 of [22] proposes for moderate sun zenith angles the test $L_d/E_d^{0+}(750\ nm) < 0.05$ where L_d was sky radiance at 135° relative viewing azimuth to sun and 140° viewing nadir angle. This test will detect clouds in front of the sun because of the consequent increase in $1/E_d^{0+}$ and will detect clouds in the specified sky-viewing direction because clouds have greater L_d than blue sky. A more complete test for "clear sky" conditions could involve use of hemispherical camera photos but would need automated image analysis for an objective test.

1.3. Previous Protocol Reviews

Most of the pre-2004 in situ measurements of water reflectance were made for the purpose of oceanic applications and most aquatic optics investigators base their measurement protocol in some way on the NASA Ocean Optics Protocols [17] and the references contained within that multi-volume publication. While the methods for measurement of L_w from underwater radiometry using fixed-depth measurements or vertical profiles were already well established at the time of that protocol collection, there has been considerable evolution of methods for above-water radiometry and development of the "skylight-blocked approach (SBA)". Current practices have also been affected by technological evolutions since 2004 including:

- More frequent use of unsupervised measurements for validation, e.g., AERONET-OC [23] and Bio-Argo [24], instead of shipborne supervised measurements;
- greater need for validation measurements in coastal and inland waters rather than the prior focus on oceanic waters;
- reduction in cost and size of radiometers, e.g., facilitating multi-sensor above-water radiometry and reducing self-shading problems for underwater radiometry; and
- increased availability of hyperspectral radiometers.

A draft of new Protocols for Satellite Ocean Color Data Validation [19] has been released within the framework of the International Ocean Colour Coordinating Group (IOCCG), providing many updates on the previous NASA-2004 collection.

1.4. Overview of Methods and Overview of This Paper

Protocols for measurement of L_w are grouped into four broad families of methods:

- Underwater radiometry using fixed-depth measurements ("underwater fixed depths")
- Underwater radiometry using vertical profiles ("underwater profiling")
- Above-water radiometry with sky radiance measurement and skyglint removal ("above-water")
- On-water radiometry with skylight blocked ("skylight-blocked")

For each family of method, the measurement equation is defined and the measurement parameters are briefly described in Sections 2–5 respectively. The elements that should be included for estimation of total protocol-related measurement uncertainty are discussed with some key considerations, guidelines and recommendations. The "protocol-related" measurement uncertainty includes both known imperfections in the protocol (e.g., models for reflectivity of the air–water interface) and deployment-related imperfections (e.g., tilting of sensors). Finally, the question of which protocol is best adapted to which water types and wavelengths is considered and some directions for probable future evolution of protocols are outlined in Section 6.

2. Underwater Radiometry—Fixed-Depth Measurements

2.1. Measurement Equation

In fixed-depth underwater radiometry, as typified by BOUSSOLE [25,26] and MOBY [27–29], radiometers are deployed underwater and attached to permanent floating structures, to measure nadir upwelling radiance, $L_{un}(z)$, at two or more depths, $z = z_1, z_2, \ldots$ —see Figure 3. A further measurement is made above water of downwelling irradiance, E_d^{0+}, to allow for calculation of R_{rs} via Equation (1) and to monitor for possible variation of illumination conditions during the measurement. In the case of MOBY these $L_{un}(z)$ measurements are made with $z_1 = 1$ m, $z_2 = 5$ m and $z_3 = 9$ m, while the BOUSSOLE system makes measurements at $z_1 = 4$ m, and $z_2 = 9$ m. Strictly speaking, these are fixed *nominal* depths because actual depth varies with tilt of structures and waves—see Section 2.2.5.

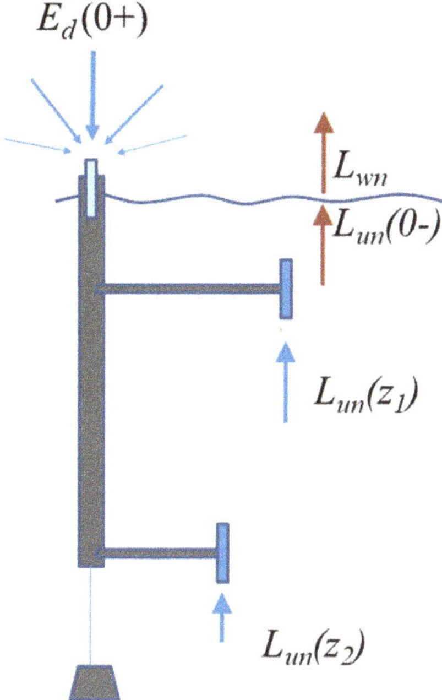

Figure 3. Schematic of fixed-depth underwater measurements.

The nadir water-leaving radiance, L_{wn}, is calculated by first estimating the nadir upwelling radiance just beneath the water surface, $L_{un}(0^-)$, by extrapolating from, preferably, the two shallowest depth measurements z_1 and z_2 assuming that the depth variation of $L_{un}(z)$ between the surface, $z = 0$, and $z = z_2$, is exponential with constant diffuse attenuation coefficient for upwelling radiance, K_{Lu}. Thus, using the convention that depths beneath the water surface are considered as positive (but retaining the notation 0^- for radiance just beneath the water surface),

$$L_{un}(0^-) = L_{un}(z_1, t_1)exp[K_{Lu}z_1] \tag{3}$$

with,

$$K_{Lu} = \frac{1}{z_2 - z_1}ln\left[\frac{L_{un}(z_1, t_1)}{L_{un}(z_2, t_2)}\frac{E_d^{0+}(t_2)}{E_d^{0+}(t_1)}\right] \tag{4}$$

where $E_d^{0+}(t_1)$ and $E_d^{0+}(t_2)$ represent the downwelling irradiance measured at times t_1 and t_2, corresponding to the times of measurement of $L_{un}(z_1)$ and $L_{un}(z_2)$. If these radiances are measured at precisely the same time, as is the case for most such implementations, then Equation (4) simplifies to:

$$K_{Lu} = \frac{1}{z_2 - z_1}ln\left[\frac{L_{un}(z_1)}{L_{un}(z_2)}\right] \tag{5}$$

Finally, the water-leaving radiance is obtained from $L_{un}(0^-)$ by propagating the latter across the water–air interface using,

$$L_{wn} = \frac{T_F}{n_w^2}L_{un}(0^-) \tag{6}$$

where T_F is the Fresnel transmittance of radiance from water to air and n_w is the refractive index of water. The refractive index of air, n_{air}, is here assumed equal to unity. T_F, which depends also on n_w, can be easily calculated from Fresnel's equations in the case of a flat water–air interface, e.g., [3] chapter 4.2, and has a typical value of 0.975 at normal incidence for oceanic water. T_F/n_w^2 takes a typical value of 0.543 for oceanic water [30]. In the case of a wave-roughened interface, by combining the reciprocity condition between radiance reflectance and transmittance coefficients [31] and the simulations of Figure 18 of [32], it was established that there is negligible (much less than 1%) difference for T_F between a flat interface and a wave-roughened interface for wind speeds up to 20 m/s (neglecting the whitecaps and breaking waves that occur already at wind speeds much less than 20 m/s) [33]. However, for a more precise calculation of T_F/n_w^2 it is necessary to take account of wavelength, salinity and temperature variations of the refractive index, n_w [34], both for oceanic waters [33] and for inland waters.

The choice of depth, z_1, for the shallowest measurement is determined by the competing interests of a shallow depth to reduce errors due to propagation to the surface and reducing the chances of the shallow depth measurement broaching the surface. This choice is then dependent on the sea-state expected at the measurement location. The choice of depth, z_2, for the second measurement is likewise a compromise between increasing $z_2 - z_1$, which reduces the uncertainty in the derived K_{Lu}, the possibility of an inhomogeneous water column over the measurement depth thus not being representative of K_{Lu} from z_1 to the surface, the natural variation in K_{Lu} due to inelastic processes [35], possible increased signal to noise because K_{Lu} is different at each wavelength, and an increase in overall length of the structure.

In addition to the time variation of illumination conditions due to time-varying sun zenith angle and diffuse atmospheric transmission (aerosols, clouds) which is accounted for in $E_d^{0+}(t_1)$ and $E_d^{0+}(t_2)$, it is necessary to account for the temporal variation of underwater radiances $L_{un}(z_1)$ and $L_{un}(z_2)$ associated with waves at the air–water interface. Wave focusing and defocusing effects [36–39] and wave shadowing [40] may have very fast time scales, less than 1 s, and very short length scales, less than 1 cm, giving a time-varying 3D light field. These effects are reduced by averaging for $L_{un}(z_1)$ and $L_{un}(z_2)$ over a large number of measurements and making the extrapolation to depth 0^- with the time-averaged values $\overline{L_{un}}(z_1)$ and $\overline{L_{un}}(z_2)$ or $\overline{L_{un}(z_1)/E_d^{0+}(t_1)}$ and $\overline{L_{un}(z_2)/E_d^{0+}(t_2)}$ (performing time-averaging on each parameter before taking the ratio). The probability density functions for $E_d^{0+}(t_1)$ and $L_{un}(z,t)$ are skewed near the surface and approach normal distributions with depth [39,41]. For BOUSSOLE data, median averaging is used [26]. For MOBY mean averaging is used as defined in p21 of [28].

At high wind speed and wave height various problems may occur affecting measurement quality or usability. For example, whitecaps and/or breaking waves may affect the water-air Fresnel transmittance. Tilt may become high. Depth measurement may become uncertain or sensors may even emerge from water. Such conditions are usually excluded from satellite data products and validation analyses anyway because the air–water interface correction of satellite data is also not suited for high whitecap coverage and/or breaking wave conditions. There is no clear consensus on acceptable wind speed for the L_w measurements, and this will clearly be dependent on the specific deployment equipment. A limit of 10 m/s would give an estimated whitecap coverage of 1% for fully-developed wind waves [42].

2.2. Protocol-Dependent Sources of Uncertainty

The protocol-related sources of uncertainty are described in the following subsections.

2.2.1. Non-Exponential Variation of Upwelling Radiance with Depth

The vertical variation of upwelling radiance between the lowest measurement depth and the air–water interface should be known

The essential assumption of exponential variation of $L_{un}(z)$ used to extrapolate measurements from two fixed depths to just beneath the water surface is only an approximation of reality. Firstly, the water inherent optical properties themselves may vary with depth [43], for example because of vertical variability related to thermal stratification including a "Deep Chlorophyll Maximum", or related to resuspended or river plume particles in coastal waters. Secondly, inelastic processes such as Raman scattering and fluorescence [35] cause non-exponential variation of radiance, particularly in the red and near infrared for Raman scattering. Thirdly, while for a homogeneous aquatic medium the attenuation with distance of a collimated beam of light can indeed be expected to be exponential the same does not hold for a diffuse light field. The angular distribution of upwelling light varies with depth, e.g., [44], and K_{Lu} depends on the angular distribution of light and so may be expected to vary with depth even for a homogeneous water column and without inelastic scattering—see Figures 9.5 and 9.6 of [3].

If a more appropriate non-exponential functional form can be found to represent the vertical variation of radiance with depth, e.g., by characterising vertical variability from profile measurements or from radiative transfer modelling [45], it is possible to modify Equation (3) to improve accuracy of the extrapolation, as suggested by using Case 1 models in Appendix A of [26] and [46].

The difficulties of non-exponential variation of upwelling radiance with depth become greater in waters or at wavelengths where the diffuse attenuation coefficient is high compared to the reciprocal of the measurement depths, e.g., in turbid waters and/or at red and near infrared wavelengths.

The uncertainty estimate associated with K_{Lu} can be validated by measuring K_{Lu} at high vertical resolution and close to the surface, e.g., from occasional shipborne campaigns.

2.2.2. Tilt Effects

The radiance sensors should be deployed vertically

Non-verticality of radiometers, e.g., caused by wave- or current-tilting of floating structures, will give uncertainty in the measurements of both E_d^{0+} and $L_{un}(z)$ because of the anisotropic nature of the down- and up-welling light fields respectively. Therefore, it is necessary to measure the tilt of radiometers using fast response inclinometers and perform appropriate filtering of non-vertical data and/or averaging of data to reduce tilt effects.

The impact of tilt on E_d^{0+} measurements is discussed in [5].

Tilt can also affect the effective underwater radiance measurement depths, z_i, which should therefore be measured continuously, e.g., using pressure sensors close to the optical sensors.

Obviously, minimisation of tilt can be a consideration in the design or in the location of validation measurement structures. As an example, the BOUSSOLE structure was designed to have low sensitivity to swell. The mean tilt of the buoy was measured as 4° (with 4° of pitching) for a 4.6 m swell of period 5.2 s [25] and data is rejected for tilt greater than 10° [26].

2.2.3. Self-Shading and/or Reflection from Radiometer and/or Superstructure

The light field should not be perturbed by the measurement radiometer and platform

In practice, the light field that is being measured is itself perturbed by the presence of solid objects such as the radiometers and the superstructure used to mount them. These perturbations are most pronounced when the water volume being measured (roughly defined horizontally by radiometer field of view and vertically by the diffuse attenuation coefficient, K_{Lu}) is in some way shadowed from direct sun, although shadowing of downwelling skylight and side/back-reflection of down/upwelling light also contribute to optical perturbations.

Shading can lead to either under- or over-estimation of K_{Lu} depending on relative impacts at the depths z_1 and z_2.

As regards the radiometers, self-shading can be minimised by using a sensor with fore-optics of small diameter compared to the mean free path of photons. This requirement becomes more challenging at longer wavelengths, such as in the near-infrared where the water absorption coefficient is high. A partial correction for self-shading effects for a radiometer with idealised geometry was proposed [47] for a concentric sensor, tested experimentally [48] and further generalized, including shallow water effects [49]. This correction requires measurement or estimation of IOPs.

As regards the superstructure, self-shading can be minimised by limiting the cross-section of the structure above the radiometers, e.g., by a sub-surface buoy [25] rather than surface buoy, and by increasing the distance between structure and radiometer, e.g., by the use of horizontal arms. The use of multiple redundant radiometers at the same depth but differently affected by superstructure and/or the measurement of superstructure azimuth and the identification/correction [50] of possible superstructure effects can also reduce superstructure shading uncertainty and/or be used to validate uncertainty estimates.

2.2.4. Bio-Fouling

The fore-optics of the radiance sensors should be kept clean

In addition to sensitivity changes inherent to the radiometer, modification of the transmissivity of the fore-optics can occur because of growth of algal films, particularly for long-term underwater deployments. Such bio-fouling can be mitigated: (a) by the use of shutters and/or wipers (provided the latter do not themselves scratch optical surfaces), (b) by use of copper surfaces and/or release of anti-fouling compounds close to the optical surface, e.g., p15 of [28], or by ultraviolet (UV-C) irradiation [51] (c) by limiting the duration of deployments between maintenance [26], (d) by monitoring optical surfaces in some way, e.g., occasional diver-operated underwater calibration lamps, e.g., p15 of [28], and (e) by regular diver cleaning of optics during the deployment.

In general, downward facing-sensors used to measure L_u are much less prone to bio-fouling than upward-facing sensors used to measure E_d [52].

An accumulation of bubbles on the horizontal surface of the L_u fore-optics would also affect data and radiometers should be designed to avoid trapping of bubbles, e.g., by removal of any concave shields or collimators used for some above-water radiance sensors.

Fouling of the above-water upward-facing E_d^{0+} sensor is described in [5].

Residual uncertainty related to bio-fouling (taking account of any biofouling corrections, e.g., linear drift) can be estimated by comparing post-deployment calibrations before and after cleaning and by comparing pre-/post-cleaning operations by divers using a portable calibration source or by using L_u time series in stable conditions [53].

2.2.5. Depth Measurement

The depth of radiance measurements should be accurately known

The measurement equation implies that the depth of measurement is accurately known. For large and permanent structures such as MOBY and BOUSSOLE, measurement of depth can be achieved quite precisely using pressure sensors (including a simultaneous above-water measurement of atmospheric pressure [54]) accounting for any time variation because of tilt and wave and current effects. If fixed-depth measurements are used at shorter vertical length scales, e.g., in shallow lakes or for measurement in high attenuation waters or wavelengths, depth measurements should be made sufficiently accurate so as to not contribute significantly to overall measurement uncertainty.

2.2.6. Fresnel Transmittance

The Fresnel transmittance for upwelling radiance should be accurately calculated

The Fresnel transmittance, T_F, used to propagate upwelling nadir radiance across the water surface in Equation (6), is often assumed to have a constant value of 0.543 in sea water, but does vary with wavelength, salinity and temperature via the index of refraction of water—see also Section 2.1 and [33] where improvements on use of a constant value and uncertainties associated with T_F are discussed.

2.2.7. Temporal Fluctuations

Temporal fluctuations associated with surface waves should be removed

Measurements are averaged over a certain interval of time (see Section 2.1) to remove as far as possible the fast variations associated with wave focusing/defocusing effects. Simulations can be performed [39,41] to assess the effectiveness of different averaging approaches/time intervals and any associated residual uncertainty.

If measurements from all sensors are not simultaneous the corresponding time corrections should be made and residual uncertainty estimated.

2.3. Variants on the Fixed-Depth Underwater Radiometric Method

Section 2 has been written primarily for MOBY/BOUSSOLE-style systems where radiometers are deployed at fixed underwater depths attached to a structure tethered to the sea bottom in an approximately constant geographical location (notwithstanding possible small horizontal movements associated with currents). Variants on this method, which are based on the same essential measurement equation, are briefly discussed here.

While the MOBY/BOUSSOLE superstructures are designed with small optical cross-section to minimise optical perturbations, buoys/platforms designed for other purposes, e.g., hydrographic measurements or navigation-related structures, may also be used for underwater radiometric measurements. The essential measurement equation and checklist of elements to be included in the uncertainty budget remain the same, although measurement uncertainties associated with superstructure shading will need to be very carefully assessed and will generally be much more significant.

Fixed-depth measurements may also be made from ships, e.g., when using radiometers with too slow a response time for fast vertical profiling. Again, the essential measurement equation and checklist of elements to be included in the uncertainty budget remain the same, although measurement uncertainties associated with ship shading/reflection will need to be very carefully assessed and will generally be much more significant unless the radiometers are somehow deployed at a sufficient distance from the ship.

At the time of writing, there are no known cases of multiple fixed-depth radiometric validation measurements being made from a horizontally moving platform. In general, horizontally moving platforms [24] (BioArgo, PROVAL, HARPOONS/Waveglider – see disclaimer at end before references) can also move vertically and so use a measurement technique based on high vertical resolution profiling, as described in Section 3.

The tethered attenuation chain colour sensors (TACCS) [55] is a variant on the fixed-depth measurement, where a single underwater L_{un} measurement, made at 0.5 m depth, is supplemented by a vertical chain of four downwelling irradiance sensors measuring $E_d(z)$ at multiple depths, in addition to the usual above-water E_d^{0+} measurement. The diffuse attenuation coefficient, K_{Ed}, that is derived from these $E_d(z)$ measurements is then used as an approximation of the K_{Lu}, that is needed to extrapolate $L_{un}(-0.5\,\text{m})$ to $L_{un}(0^-)$. In one implementation [12] the $E_d(z)$ measurements are made at a lower spectral resolution that the L_{un} measurements, and K_{Ed} must, therefore, be interpolated/extrapolated spectrally. In other respects this variant on the fixed-depth underwater radiometry method has the same sources of uncertainty as listed in Section 2.2, except that further uncertainties must be assessed relating to the modelling of K_{Lu} from K_{Ed}, and the spectral interpolation/extrapolation of K_{Ed}.

In some implementations a single measurement of upwelling radiance is made close to the air–water interface [56]. The K_{Lu} required to extrapolate to the surface is then not measured but is either assumed zero or estimated using a model which takes the L_{un} spectrum as input (potentially repeated iteratively), giving a measurement uncertainty in both cases. In the optical floating system [57], measurements were made within 2 cm of the surface in very calm conditions. Vertical extrapolation of single depth near-surface measurements are discussed in Section 3E of [35].

3. Underwater Radiometry—Vertical Profiles

Water-leaving radiance can also be measured using underwater radiometry based on vertical profiling—see Figure 4. This method has frequently been used in supervised deployments from ships [58] and can also be made from fixed platforms [43]. Theoretically, vertical profiling from a fixed platform could also be automated and unsupervised, although in practice long-term deployments of radiometers with moving underwater parts are vulnerable to mechanical failures. As an alternative, unsupervised vertical profiles can be carried out from horizontally drifting platforms or "floats" [59,60], as further described in Section 3.3.

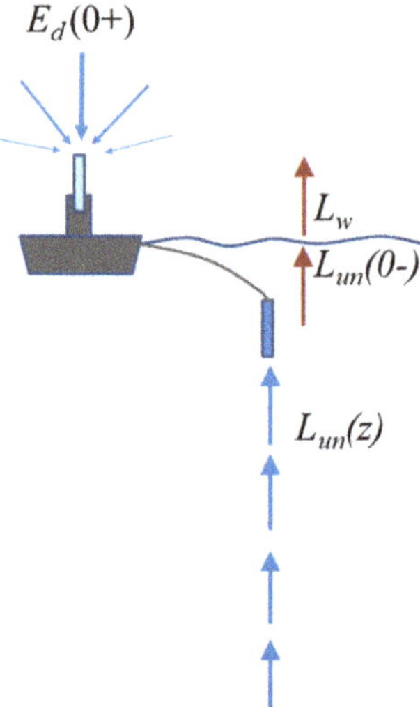

Figure 4. Schematic of underwater vertical profile measurements. This sketch shows deployment typical of a free-fall radiometer tethered to a ship, although the method is generic and does not need to be ship-tethered, e.g., could be tethered to a fixed offshore platform or moored buoy, or could be untethered and horizontally drifting, while profiling.

The first vertical profile radiometric measurements were generally made from winches attached to ships [61]. However, it is clearly important to avoid as far as possible optical (shadow/reflection) [62] and hydrographic perturbations (ship wake, ship hull and propeller-induced mixing, bow wave, etc.) from the ship as well as vertical motion of optical sensors due to ship motion. It has been recommended to make measurements from the stern of a ship with the sun's relative bearing aft of the beam at a minimum distance of $1.5/K_{Lu}$ from the ship or at greater minimum distance when deploying off the beam of a large vessel—see Section 2.2, p8 of [63].

A popular method for getting radiometers away from ship perturbations is to float radiometers away a few tens of metres and then profile vertically using a specially-designed rocket-shaped free-fall platform [64]. More recently a new "kite" free-fall design allows slower profiling, closer to the water surface [54]. Remotely operated vehicles can also be used [65].

In view of such improvements in deployment hardware that have become commercially available over the last 15 years it is likely that fiducial reference measurements will generally not be made from shipborne winch deployments, although this is not formally precluded provided that the measurement is supported by a careful uncertainty analysis covering all perturbations specific to the ship/deployment method/water type combination, including, for example, measurements made at different distances from the ship and/or 3D optical model studies.

Vertical profiles can also be made from offshore structures, including fixed platforms, e.g., the WISPER system on the Aqua Alta Oceanographic Tower (AAOT) [43], or moored buoys with a vertical wire-mounted package. These structures have the advantage over shipborne winches of reduced tilt of radiometers and reduced hydrodynamic perturbations, although optical perturbations still need to be evaluated, e.g., by measurements made at different distances from the platform [66] and/or 3D optical model studies [67].

3.1. Measurement Equation

The fundamental measurement equation is similar to that used for fixed-depth measurements, except that measurements are now available for a range of depths $z_1 \leq z \leq z_2$ for estimation of the vertical variation of $L_{un}(z)$.

By definition of K_{Lu}, the diffuse attenuation coefficient for L_{un}:

$$L_{un}(z, t_0) = L_{un}(0^-, t_0)e^{-\int_0^z K_{Lu}(z')dz'} \tag{7}$$

where z is positive underwater and increases with depth beneath the surface (but retaining the notation 0^- for radiance just beneath the water surface) and t_0 is the time to which measurements are referred. This gives, after natural logarithm transformation and reorganisation:

$$\ln[L_{un}(z, t_0)] = ln[L_{un}(0^-, t_0)] - \int_0^z K_{Lu}(z')dz' \tag{8}$$

If it is assumed that K_{Lu} is constant with depth over the depth range of measurements and up to the water surface, then this simplifies to:

$$\ln[L_{un}(z, t_0)] = ln[L_{un}(0^-, t_0)] - K_{Lu}z \tag{9}$$

$L_{un}(0^-, t_0)$ is then obtained from vertical profile measurements as the exponential of the intercept of a linear regression of $\ln[L_{un}(z, t_0)]$ against z over a specified depth range.

Since measurements at different depths are made at slightly different times, t, the radiance measurements are first corrected for any variations in above-water downwelling irradiance by:

$$L_{un}(z, t_0) = L_{un}(z, t)\frac{E_d^{0+}(t_0)}{E_d^{0+}(t)} \tag{10}$$

Finally, the water-leaving radiance is obtained from $L_{un}(0^-, t_0)$ by propagating the latter across the water-air interface as in Equation (6).

A number of deployment and data-processing factors influence the quality of $L_{un}(0^-, t_0)$ derived from measurements of $L_{un}(z, t)$:

- Measurements should be made as close as possible to the air–water interface to minimise the uncertainties associated with extrapolation from depth, particularly if there are vertical gradients of inherent optical properties or for wavelengths/waters with high vertical attenuation. Very near-surface measurements are complicated by waves, which affect radiometer tilt and vertical positioning as well as the radiance field itself (focusing/defocusing). To deal with this, new profiling platforms have been designed for very slow and stable sampling close to the surface [54].
- Sufficient measurements are needed for each depth (interval) to ensure that wave focusing and defocusing effects can be removed, implying that profiling speed should be sufficiently slow, adding to the time required to make a cast, a practical consideration, and the possibility of temporal variation of illumination conditions, a data quality consideration.

- The vertical profiling speed should be matched to the acquisition rate of the radiometers to ensure that the depth z of each measurement can be determined with sufficient accuracy.
- The depth range $z_1 \leq z \leq z_2$ chosen for data processing is "the key element in extracting accurate subsurface data from in-water profiles" [68]. z_1 should be chosen sufficiently large to avoid problems of near-surface tilt, wave focusing/defocusing and bubbles, but sufficiently small to limit uncertainties associated with extrapolation to the surface, particularly for high attenuation waters/wavelengths. Any depth interval with significant ship/superstructure shadowing must also be avoided. In practice, the choice of depth range is generally made subjectively [11] because of the difficulty to automate such thinking.
- The depth range $z_1 \leq z \leq z_2$ used in data processing can be wavelength-dependent (unlike for the fixed-depth method of Section 2), e.g., using optical depth to set z_2 differently at each wavelength.
- Different mathematical methods used to perform the regression analysis for Equation (9) and different methods for filtering outliers [69] may give quite different results. Such considerations were analysed in detail in the Round Robin experiments documented by [11].
- For measurements with significant temporal variability of $E_d^{0+}(t)$, some time filtering of $E_d^{0+}(t)$ may be needed before application of Equation (10). For example, $E_d^{0+}(t_0)$ may be chosen as the median of $E_d^{0+}(t)$ over the measurement interval or, for ship-induced periodic variability, $E_d^{0+}(t)$ may be first linearly fitted as function of t.

For profiling systems where the upcast is made by applying tension to a wire, only downcast ("free-fall") data is used to avoid irregular motion and high tilt.

3.2. Protocol-Dependent Sources of Uncertainty

The protocol-related sources of uncertainty are described here for the case of a profiling system that is supposed to be fixed, or almost fixed, in horizontal space, e.g., tethered to a ship or an offshore platform. Additional considerations to account for significant horizontal movements, e.g., from glider platforms, are summarised in Section 3.3.

3.2.1. Non-Exponential Variation of Upwelling Radiance with Depth

The vertical variation of upwelling radiance between the highest measurement depth and the air–water interface should be known

The essential assumption of exponential variation of $L_{un}(z)$ from the measurement depth range $z_1 \leq z \leq z_2$ to just beneath the air–water interface is clearly an approximation of reality. This assumption will cause uncertainties in conditions of near-surface optical stratification, inelastic scattering (Raman, fluorescence) and variability of the angular distribution of upwelling radiance, as already described in Section 2.2.1 for fixed-depth radiometry.

The uncertainty associated with non-exponential variation of $L_{un}(z)$ can be assessed for the measurement range $z_1 \leq z \leq z_2$ by considering the goodness-of-fit of Equation (8), after suitable filtering of temporal variability and taking account of realistic uncertainties. For $0 \leq z \leq z_1$, between the measurement range and the surface, potential non-exponential variation of $L_{un}(z)$ can be assessed by model studies [45]. If this non-exponential variation is already considered in the fitting methodology, then the uncertainty is reduced to the residual uncertainty associated with the difference between the true non-exponential variation of $L_{un}(z)$ and the estimated non-exponential variation.

Clearly z_1 should be kept as shallow as possible, within constraints of deployment, tilt contamination and temporal variability, particularly if there may be near-surface stratification of the water column.

3.2.2. Tilt Effects

The radiance sensor should be deployed vertically

Non-verticality of radiometers, e.g., caused by wave-tilting of free-fall platforms or ship winch-deployed frames, gives uncertainty in the measurements of $L_{un}(z, t)$ because of the anisotropic nature of upwelling light fields. It is, therefore, necessary to measure the tilt of radiometers using fast response inclinometers and perform appropriate filtering of non-vertical data and/or averaging of data to reduce tilt effects [69].

The uncertainty associated with tilt effects can be estimated by reprocessing of oversampled vertical profile measurements with different thresholds for removal of non-vertical data and by 3D optical model simulations.

The impact of tilt on E_d^{0+} measurements is discussed in [5].

Obviously, minimisation of tilt should be a consideration in the design of deployment hardware. Vertical profiles carried out from fixed platforms suffer less from such tilt effects. The "rocket-shaped" free fall platforms may suffer from high tilt, particularly in near-surface waters and high wave conditions. The new designs of "kite-shaped" profilers [70] and autonomous profiling floats [60] have significantly reduced tilt.

3.2.3. Self-Shading from Radiometers and/or Superstructure

The light field should not be perturbed by the measurement radiometers and platform

In practice, the light field that is being measured is itself perturbed by the presence of solid objects such as the radiometers and the superstructure used to mount them, as discussed previously in Section 2.2.3 for fixed-depth underwater radiometry. For free-fall radiometer platforms, the considerations and corrections discussed in Section 2.2.3 as regards self-shading from the radiometer collector and from the mounting frame are relevant also for vertical profiling. For ship-tethered free-falling radiometers with an off-centre L_u sensor, azimuthal rotation should be controlled to have the L_u sensor on the sunny side.

Redundant deployment of two sensors at the same depth but on different sides of a profiling platform can help identify and remove the data worst affected by platform shading. Knowledge of platform azimuth with respect to sun can help assess such effects [60].

For ship- or fixed platform-deployed vertical profiling radiometers, superstructure shading/reflection effects may be considerable and should be carefully limited, by maximising horizontal distance from the structure. Uncertainties should be estimated, e.g., by radiative transfer modelling [67,71] and/or by in situ measurements at different distances from the structure.

3.2.4. Bio-Fouling

The fore-optics of the radiance sensor should be kept clean

Supervised underwater radiometric measurements generally do not suffer from bio-fouling provided that fore-optics are kept clean between deployments.

Fouling of the above-water upward-facing E_d^{0+} sensor is described in [5].

Unsupervised fixed location vertical profiling measurements are rare but would suffer from similar problems to those described in Section 2.2.4 for fixed-depth measurements.

Horizontally drifting vertical profiling systems (Section 3.3) may arrange to spend most time at great depth to minimise bio-fouling [24]. Residual bio-fouling uncertainties (after any biofouling correction, e.g. linear drift) can be estimated by comparing pre- and post-deployment calibrations, although recovery of horizontally drifting systems is not always possible.

3.2.5. Depth Measurement

The depth of radiance measurements should be accurately known

The measurement equation implies that the depth of measurement is precisely known by a fast response and appropriately calibrated pressure sensor located close to the optical sensor. Any permanent vertical shift between depth sensor and optical sensor must be corrected and any tilt-induced vertical difference between depth and optical measurements must be included in the uncertainty estimate. Accurate measurement of depth and associated uncertainties is needed, including referencing to surface atmospheric pressure at the moment of profiling (pressure "taring") and temperature-sensitivity of pressure transducers—see Section 5.2 of [54].

3.2.6. Fresnel Transmittance

The Fresnel transmittance for upwelling radiance should be accurately calculated

As in Section 2.2.6.

3.2.7. Temporal Fluctuations

Temporal fluctuations associated with surface waves should be removed

The removal of temporal fluctuations in $L_{un}(z,t)$, e.g., from wave focusing/defocusing is complicated for vertical profile measurements because both the light field and the measurement depth, z, vary with t, and because measurements may be affected by both natural variability (wave effects, water variability) and by deployment-related variability (e.g., tilt and vertical wave motions).

If all other factors (above-water illumination, water optical properties) are assumed invariant in time during the measurements, or suitably corrected, and $L_{un}(z,t)$ is assumed to be tilt-free after filtering, then natural variability caused by wave effects [72] can be minimised by performing sufficient measurements to allow adequate averaging. This can be achieved by slow profiling [54,73] or, if this is not possible, by multicasting [68].

The uncertainty associated with all sources of temporal fluctuations must be estimated, e.g., by testing alternative data processing options on oversampled measurements and by 4D optical simulations [45]. Uncertainty estimates should be validated, e.g., by measurement intercomparison exercises [12].

3.3. Variants on the Vertical Profiling Underwater Radiometric Method

Following on from the success of the Argo float network designed for physical oceanography, a number of horizontally-drifting vertical-profiling radiometer platforms have been designed for long-term unsupervised measurement of optical properties [24,59,60]. Such floats, when suitably networked, allow for much better spatial coverage of the oceans (but not shallow seas or inland waters). Typically, the radiometer will park at great depth during most of the day and night (to reduce bio-fouling) and perform one or more vertical profiles per day (rising at about 4 cm/s to 10 cm/s or slower), potentially timed to match the acquisition times of specific ocean colour sensors. Such systems can also combine vertical profiling with near-surface fixed-depth "drifting buoy" measurements, thus falling within both Sections 2 and 3 of this document and allowing the vertical profile K_{Lu} measurements to be used for the near-surface single fixed-depth measurements.

The essential measurement equation and sources of uncertainty for such measurements are the same as for other vertically profiling radiometers. As for all unsupervised measurements, biofouling, particularly for the upward-facing E_d^{0+} measurement [5], may be a significant source of uncertainty, especially if the radiometer cannot be recovered for post-deployment calibration. On the other hand, the possibility of diving deep limits exposure to biofouling.

In contrast to vertical profile measurements made from ships or fixed offshore structures, drifting floats generally do not have a permanent above-water radiometer for $E_d^{0+}(t)$ and so there will be an additional uncertainty associated with possible time variation of illumination conditions during the vertical profile, although the latter may also be reduced by analysis of the $E_d(z, t)$ profile data [74].

Floats can also accommodate radiometers on horizontal arms and redundant radiometers to provide additional constraints on sensor drift and shading by platform [60].

4. Above-Water Radiometry with Sky Radiance Measurement and Skyglint Removal

4.1. Measurement Equation

In above-water radiometry one or two radiometers are deployed above water from a ship or fixed structure to measure (a) upwelling radiance, $L_u(0^+, \theta_v, \Delta\varphi)$, at a suitable viewing nadir angle, $\theta_v < 90°$, and viewing azimuth angle relative to sun, $\Delta\varphi$, and (b) downward (sky) radiance, $L_d(0^+, 180° - \theta_v, \Delta\varphi)$, in the "mirror" direction which reflects at the air–water interface into the water-viewing direction—see Figure 5.

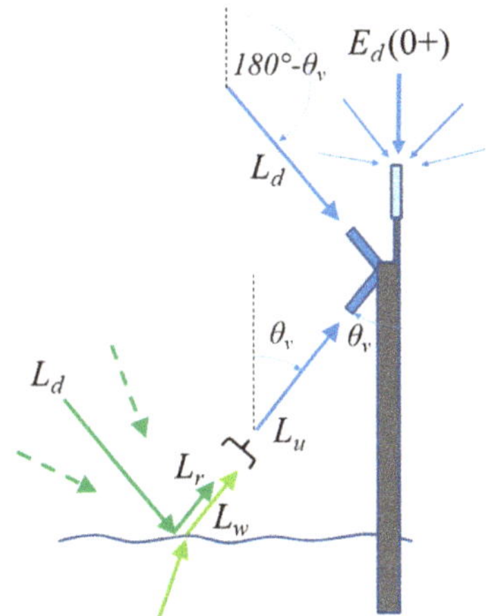

Figure 5. Schematic of above-water radiometry with measurement of sky radiance, L_d, and removal of skyglint radiance, L_r. Dashed arrows indicate that contributions to the skylight reflected at the air–water interface come from directions that are not directly measured by the L_d radiance sensor, including possible contributions from the direct sunglint direction.

Then the water-leaving radiance in the water-viewing direction is estimated from the measurement equation:

$$L_w(\theta_v, \Delta\varphi) = L_u\big(0^+, \theta_v, \Delta\varphi\big) - L_r(\theta_v, \Delta\varphi) \tag{11}$$

where the skyglint radiance, L_r, which cannot be measured directly, is typically estimated as a multiple of the downwelling sky radiance, L_d, by

$$L_r(\theta_v, \Delta\varphi) = \rho_F L_d\big(0^+, 180° - \theta_v, \Delta\varphi\big) \tag{12}$$

where ρ_F is a coefficient that represents the fraction of incident skylight that is reflected back towards the water-viewing sensor at the air–water interface and is the Fresnel reflectance coefficient for a flat water surface, or is called here the "effective Fresnel reflectance coefficient" for a roughened water surface.

The second part of this measurement Equation (12), which forms the basis of this protocol, is adopted as a pragmatic way of estimating and removing the upwelling radiance that originates from reflection at the air–water interface. However, it is well understood that such radiance may originate from portions of the sky dome other than the portion that is actually measured, as defined by $(180° - \theta_v, \Delta\varphi)$ and the field of view of the L_d radiometer. L_r may include reflection of direct sun glint—see Figures 1 and 2 of [75] and Equation (1) of [76]. This is discussed further in Section 4.2.1. In reality, the right hand side of (12) is an approximation of the convolution of sky radiances for the full hemisphere with the wave slope statistics, defining the probability of encountering a part of the air–water interface that reflects specularly into the direction $(\theta_v, \Delta\varphi)$, and the Fresnel reflectance coefficient for the corresponding incidence angle—see Chapter 4 and Equation (4.3) of [3] or Equation (3) of [77] for a complete description.

In the case of a flat water surface with only specular reflection processes (i.e., no whitecaps or other diffuse reflection processes) and with unpolarised downwelling light, and for an infinitesimally small sensor field of view, ρ_F is simply given by the Fresnel reflectance equation and is plotted in Figure 6:

$$\rho_F(\theta_v) = \frac{1}{2}\left\{ \left[\frac{sin(\theta_v - \theta_t)}{sin(\theta_v + \theta_t)}\right]^2 + \left[\frac{tan(\theta_v - \theta_t)}{tan(\theta_v + \theta_t)}\right]^2 \right\} \tag{13}$$

where θ_v is the viewing nadir angle ("above-water incidence angle") and θ_t is the angle of light transmitted to below water after refraction:

$$\theta_t = 180° - sin^{-1}(sin\theta_v / n_w) \tag{14}$$

where n_w is the index of refraction of water with respect to air and is often approximated by the value 1.34 but does also vary with salinity, temperature and wavelength [3].

For nadir-viewing, $\theta_v = 0$, and Equation (13) is replaced by:

$$\rho_F(0) = \left(\frac{n_w - 1}{n_w + 1}\right)^2 \tag{15}$$

The nadir viewing angle variation of ρ_F is illustrated for this flat-water surface and for modelled wavy water surfaces in Figure 6.

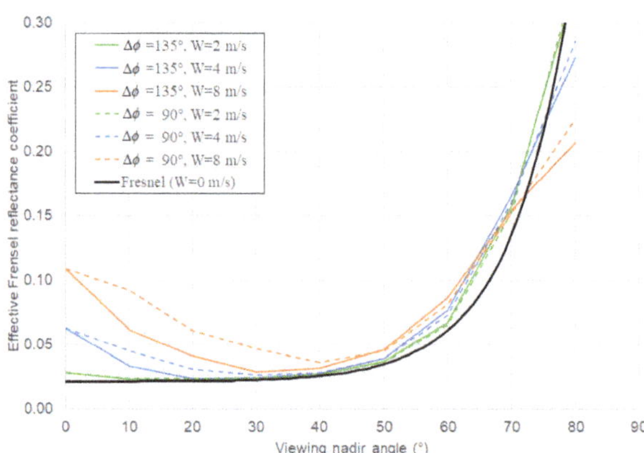

Figure 6. Effective Fresnel reflectance coefficient, ρ_F, as function of viewing nadir angle, θ_v, for the flat water case (Fresnel reflectance given by Equation (13)) and for a wind-roughened surface, modelled [75] at 10° intervals for $\lambda = 550$ nm, $\theta_0 = 30°$, and various wind speeds, W, for L_r with relative viewing azimuth angles, $\Delta\varphi_v$.

In reality:

- The water surface is not flat but is a wavy surface [32] implying that (a) the portion of sky reflected into the water-viewing direction may come from directions other than $L_d(0^+, 180° - \theta_v, \Delta\varphi)$ [75], and that (b) the incidence angle required for calculation of the Fresnel coefficient is different from

θ_v, with spatial variation of the incidence angle within the sensor field of view that increases with wave inclination.

- The downwelling light is not unpolarised, but, particularly for the molecularly scattered "Rayleigh" component at 90° scattering angle from the sun, may be strongly polarised [78].
- Some radiometers have a field of view that can be quite significant, e.g., >10°, meaning that the measurements $L_u^{0+}(0^+,\ \theta_v,\Delta\varphi)$ and $L_d(0^+,180°-\theta_v,\Delta\varphi)$ are weighted averages over a range of viewing angles $(\theta_v,\Delta\varphi)$ and the model for ρ_F may need to account for different incidence angles even for a flat water surface.

These considerations are dealt with in detail in the following Sections and their references.

As regards the classification of methods for measuring L_w, it is suggested here to drop the Method1/2/3 above-water radiometry classification used in the NASA Ocean Optics 2003 protocols [79] mainly for the E_d^{0+} measurement and in future classify the above-water L_w measurements according to viewing geometry, measuring radiance with:

- Viewing nadir angle, e.g., $\theta_v = 0°$ (pointing towards nadir) or $\theta_v = 40°$ or "other".
- Viewing relative azimuth angle to sun for off-nadir measurements, e.g., $\Delta\varphi = 90°$ or $\Delta\varphi = 135°$ or "other".

and

- The method used to estimate skylight reflected at the air–water interface.

In general nadir-viewing is avoided because of the high uncertainties associated with skyglint removal in geometries close to sunglint [75] and because of difficulties in avoiding optical perturbation from the ship/platform. However, there may be situations where nadir-viewing can be acceptable (e.g., mirror-flat lakes, sensors deployed well above water surface from an optically small structure, high sun zenith angle) provided that uncertainties are careful assessed and validated.

The measurement of polarized upwelling radiance [80,81] is considered as a variant of the above-water L_w method – see Section 4.3.

In view of the quite different measurement uncertainties, the skylight-blocked approach (SBA) [76,82] is treated in the separate Section 5.

Temporal Processing of Radiance Measurements

Measurement of both sky radiance and water radiance involves time integration for each individual measurement and replicate measurements which are subsequently processed to yield a single value for $\overline{L_u}(0^+,\ \theta_v,\Delta\varphi)$ and $\overline{L_d}(0^+,180°\ -\theta_v,\Delta\varphi)$ where the overbar represents the multitemporal measurement, typically called "time-average", although the temporal processing may be different from a mean average and will generally involve prior outlier removal or time series based quality control.

The integration time depends on the radiometer concept and the brightness of the target. Filter-wheel radiometers generally measure fast, typically at many hertz, whereas spectrometer-based systems may be fast, e.g., 8 ms to 32 ms, for bright targets such as the sky, but much slower, e.g., integration time of 1 s to 4 s, for darker targets such as water.

For the sky radiance measurement, $\overline{L_d}(0^+,\ 180°-\theta_v,\Delta\varphi)$, a small number of replicate measurements should be sufficient. If the sky conditions are good (clear blue sky) then 3 to 5 replicates should be sufficient to establish this and provide a mean average and standard deviation for this parameter. If the sky conditions are not good (e.g., scattered clouds and/or partially obscured sun) then this will also be immediately apparent from even a low, e.g., 3 to 5, number of replicates either in the standard deviation or in the magnitude of $\overline{L_d}/E_d^{0+}$ at 750 nm, which will be much higher than that of an ideal sky model, see Web Appendix 1 of [22].

For the water radiance measurement, $\overline{L_u}(0^+,\ \theta_v,\Delta\varphi)$, a much larger number of replicate measurements is needed because of the rapid and large temporal variations associated with surface gravity waves. These variations include the darkening/brightening effect of large surface gravity waves oriented towards/away from the sensor (because of air–water interface reflectance differences and/or reflection of brighter/darker portions of the sky) as well as the very bright, small and fast sunglint "flashes" from specular reflectance of direct sun at suitably oriented capillary wave facets, particularly when viewing at low $\theta_v - \theta_0$, low $\Delta\varphi$ and for high wave amplitudes. The temporal processing of $L_u(0^+, \theta_v, \Delta\varphi)$ measurements should also depend on the integration time of each measurement and may be linked to the method for estimation of ρ_F. For example, a temporal processing method has been used for a rapidly sampling, small field of view radiometer that retains the minimal values of $L_u(0^+, \theta_v, \Delta\varphi)$ over a number of replicates and uses a flat sea model for ρ_F using the principle that sunglint flashes and brighter waves can be resolved and eliminated by the minimum filter [83]. A different approach was suggested [75] for the case effectively of a slowly sampling radiometer where the contributions of different wave facets cannot be isolated but are effectively averaged in time (and possibly space, depending on the field of view and distance from the water surface) for each individual $L_u(0^+, \theta_v, \Delta\varphi)$ measurement. In the latter case a quite different value of ρ_F may be required from that of the flat water surface model of Equation (13)—see Figure 2 of [84].

4.2. Protocol-Dependent Sources of Uncertainty

The protocol-related sources of uncertainty are described in the following subsections.

4.2.1. Estimation of Reflected Skylight

Upwelling radiance from reflection at the air–water interface (skyglint/sunglint) should be removed

The most critical aspect of above-water measurements of L_w lies in the removal of skylight reflected at the air–water interface, represented by the coefficient ρ_F in Equation (11). For waters or wavelengths where R_{rs} is low, the right-hand side of (11) can be the difference of two values which are much larger than the left hand side. For example, in clear waters in the near infrared, L_w may be negligibly small whereas $L_u(0^+, \theta_v, \Delta\varphi)$ and $\rho_F L_d(0^+, 180° - \theta_v, \Delta\varphi)$ are not. Any uncertainty in ρ_F is then greatly amplified when taking the difference. It is important to note that the uncertainty on $\rho_F L_d(0^+, 180° - \theta_v, \Delta\varphi)$ is an absolute uncertainty for L_w [22] that is unrelated to the value of L_w itself and so becomes more important in relative terms as L_w decreases. This is in contrast to most radiometer-related uncertainties (calibration, E_d^{0+} cosine response, radiometer thermal sensitivity, etc.) which are relative uncertainties that can be expressed as a percentage of the desired parameter, L_w or R_{rs}.

In view of the importance of estimating L_r or the product $\rho_F L_d(0^+, 180° - \theta_v, \Delta\varphi)$ there is quite large diversity of approaches. In the crudest approach, ρ_F is simply taken from the flat sea Equation (13) and therefore generates large uncertainties that may be strongly positively biased for L_w. For waters with low red or near infrared reflectance, a further "residual" correction may be applied [85], assuming that $L_w = 0$ for a suitable wavelength, λ_0, and that $L_r(\theta_v, \Delta\varphi)$ has spectral variation given by $L_d(0^+, 180° - \theta_v, \Delta\varphi)$.

Such an approach may also be used in highly absorbing waters at both ultraviolet and near infrared wavelengths to provide two fixed points at each extreme of the spectrum for a full spectrum construction of $L_r(\theta_v, \Delta\varphi)$ [86].

For brighter waters, a wavelength λ_0 with negligible L_w may not exist and, in an approach analogous to turbid water aerosol correction algorithms, a "turbid water" residual correction was proposed [87] based on measurements at 715 nm and 735 nm. This approach was generalised for any pair of near infrared wavelength [88], but was suggested for use in quality control/uncertainty estimation rather than data correction.

Scalar radiative transfer simulations were carried out [75] to establish ρ_F as function of sun and viewing geometry $(\theta_0, \theta_v, \Delta\varphi_v)$ and wind speed at a height of 10 m above the water, W, assuming a Cox-Munk relationship [89] between surface wave field and wind speed. In general, the directionality of the wave field (in particular the azimuth angle between wind direction and sun) is not accounted for when applying such corrections, although variability with wind direction has been observed [89] and this directionality may affect data [40]. In the case of fetch-limited inland waters W will typically be set to zero or a small value, since the Cox–Munk relationship will not apply. Similarly in overcast conditions (not very relevant for satellite validation) the dependence on surface wave field and/or W is also less strong and a constant value of $\rho_F = 0.028$ has been proposed [75]. The table of values calculated for ρ_F as function of $(\theta_0, \theta_v, \Delta\varphi_v)$ and W is provided for download at [90], together with an updated table including polarisation effects [91], as described below.

It has been noted [76] that, since contributions to $L_r(\theta_v, \Delta\varphi)$ arise from different portions of the sky (including direct sun) when the surface is not perfectly flat, these will have different spectral shapes from the $L_d(0^+, 180° - \theta_v, \Delta\varphi)$ that is measured. This effect is not accounted for in the simulations of [75] where the model assumes the same colour of the sky in all directions.

Sky radiance measured over small inland waters may include a component of light which has been scattered by land and then further backscattered in the atmosphere, giving, near vegetated land, a stronger near infrared contribution than typical oceanic skies [92].

For measurements made in inland waters very close to trees or in the vicinity of steep mountains, the sky radiance measurement may even include directly light from objects that are not sky—such problems could be mitigated by choosing the most favourable of the two possible relative azimuth angles (left or right of sun) although it will clearly be very challenging to make good measurements in such circumstances of highly anisotropic downwelling "skydome" hemisphere.

It has been shown that ρ_F is, in reality, significantly lower than that in the simulations of [75] because the downward radiance is not unpolarized [93]. This effect is particularly strong when viewing near the Brewster angle of about $53°$. Further simulations do take account of such polarisation effects [91,94] and the impact of aerosols, showing the further dependency of ρ_F on aerosol optical thickness [95]. Other simulations take account of polarisation effects and also demonstrate that quite different mean surface slopes and hence quite different surface reflectance factors can arise from a single wind speed [40].

In one study, also taking account of polarization, the sunglint and skyglint components of light reflected at the air–water interface are treated separately [77]. In that formulation, the reflected light is still modelled as a multiple of the measured incident skylight in the sky-viewing direction, $L_d(0^+, 180° - \theta_v, \Delta\varphi)$, but the air–water interface reflection coefficient, ρ_F, is split into two reflection coefficients, $\rho_{sun}(\lambda)$, and $\rho_{sky}(\lambda)$ representing respectively the sunglint and skyglint contributions. Although these coefficients are considered as "spectrally varying" in that paper it is noted that this "spectral variation" is a model to correct for the fact that the $L_d(0^+, 180° - \theta_v, \Delta\varphi)$ measurement is not representative of the spectral variation of sky radiances from all portions of sky (including direct sun) that are reflected towards the water-viewing sensor. The true spectral variation of the flat sea Fresnel coefficient, because of salinity and temperature related variation of the refractive index of water, is less significant (but also accounted for in that study). Using this decomposition of $L_r(\theta_v, \Delta\varphi)$ into skyglint and direct sunglint components [77], the spectral variation of the latter follows the spectral radiance of the direct sun radiance, which is clearly different from the measured sky radiance $L_d(0^+, 180° - \theta_v, \Delta\varphi)$ and may be closer in spectral variation to that of the measured downwelling irradiance, E_d^{0+}.

The effective air–water interface reflection coefficient, ρ_F, has been modelled for a continuum of viewing nadir and azimuth angles, sun zenith angles and wind speeds [84]. The impact of aerosol optical thickness on ρ_F was demonstrated and it was recommended that above-water radiometric measurements be accompanied by measurements of aerosol optical thickness.

In a way that is analogous with the development of full spectrum coupled ocean-atmosphere modelling in satellite data atmospheric correction algorithms, more complex schemes have been proposed for taking account of the expected spectral shapes of L_w and $\rho_F L_d(0^+, 180° - \theta_v, \Delta\varphi)$. e.g., [96].

For hyperspectral measurements it has been proposed [97] to use the fact that R_{rs} can be expected to be spectrally quite smooth whereas both $L_u(0^+, \theta_v, \Delta\varphi)$ and $\rho_F L_d(0^+, 180° - \theta_v, \Delta\varphi)$ are affected by atmospheric absorption features. Thus ρ_F can be constrained or estimated as the value that will yield a spectrally smooth R_{rs}.

While there have been many recent and diverse developments for the removal of skyglint in data post-processing, the acquisition geometry of $\theta_v = 40°$ viewing angle for the water and $180° - \theta_v = 140°$ viewing angle for the sky observations, as proposed in [75] and endorsed by [79], remains a very robust and practical approach: viewing angles below 40° are more often associated with the impact of sunglint effects [84], while at viewing angles larger than 40° the reflectance coefficient becomes more sensitive to the small changes of the viewing angle as clearly follows from Figure 6. In addition, for moderate wind speeds the impact of aerosol optical thickness and polarization on the reflectance coefficient is typically smaller than for other viewing angles [84]. The azimuth angle for the water and sky observations should be closely monitored and should be the same for both measurements because of the significant azimuthal gradient of the sky radiance [84].

Using a hyperspectral imaging camera, relative uncertainties for L_w have been estimated arising from L_r correction for the spectral range 450 nm to 900 nm and for viewing angles 20° to 60° as a function of wind speed [84]. These uncertainties are most critical at blue wavelengths for waters with low blue reflectance, typical of coastal waters, where L_r/L_w is greatest. That study [84] also showed that both water and sky radiance measurements are not sensitive to the field of view (FOV) of the optics for FOV between 4° and 31.2° for measurements made at between 6 m and 8 m above water level with integration time 20 ms to 50 ms for a wind speed of 5.6 m/s.

If L_u and L_d are measured with different radiometers, e.g., as in the implementation of [22], then the differences between the radiometer sensitivities as a function of wavelength will add some measurement uncertainty for the spectrally-binned L_w—this is often visible in hyperspectral measurements where narrow and strong atmospheric absorption features, such as oxygen absorption near 762 nm, lead to "blips" in L_w or R_{rs} spectra.

In view of the wide diversity of approaches for estimation of ρ_F [98] and continued research into methodological improvements, the present document does not intend to prescribe a single protocol for estimating $L_r(\theta_v, \Delta\varphi)$ or ρ_F in FRM measurements. In fact, for most data acquisition protocols, different methods for estimating ρ_F or $L_r(\theta_v, \Delta\varphi)$ can be applied in post-processing and could be applied to historical data. Rather the approach of the current document is merely to insist that the uncertainties of any approach be thoroughly estimated and validated.

One method for estimation of uncertainties associated with $L_r(\theta_v, \Delta\varphi)$ removal is to consider the spectral consistency of $R_{rs}(\theta_v, \Delta\varphi)$ in the near infrared. For clear waters and at sufficiently long wavelength R_{rs} can be assumed zero and any offset in measurements can be used as an estimator of total measurement uncertainty, provided this information has not already been used to perform a "residual correction" of data—this approach was suggested by [99], although in their study the uncertainty was expected to come more from ship perturbations (Section 4.2.3) than from $L_r(\theta_v, \Delta\varphi)$ removal. The approach was extended [88] for moderately turbid waters, where R_{rs} is non-zero in the near infrared, but adopts a spectral shape determined primarily by the pure water absorption coefficient [22].

4.2.2. Tilt and Heading Effects

Radiance measurements should be made at exactly the prescribed viewing nadir and relative azimuth angles

The uncertainty in the pointing angle of radiometers used for measuring both $L_u(0^+, \theta_v, \Delta\varphi)$ and $L_d(0^+, 180° - \theta_v, \Delta\varphi)$ must be propagated through to give an uncertainty for $L_w(\theta_v, \Delta\varphi)$.

When operating from boats inaccuracies in pointing angle may arise from (a) the initial setup and levelling of radiometers for the "at rest" balancing of the boat, and any resetting that is required during a campaign, e.g., because of changes in boat balance (ballasting, fuel and water tanks, deployment of equipment overboard by crane, etc.) and; (b) pitch and roll, which may easily reach 10° or more in heavy sea states or for small boats. Above-water radiometry from most fixed platforms is not significantly affected by wave- or wind-induced tilt and angular accuracy of <1° is easily achieved with a rigid structure, but can be exceeded for a flexible mast.

The impact of tilt can be estimated and reduced by: (a) measuring the inclination of the radiometers or the mounting platform/ship with a fast response well-calibrated inclinometer and removing all data where tilt exceeds a user-defined threshold; and (b) calculating the mean average and standard deviation of a time series of replicate measurements.

For the $L_u(0^+, \theta_v, \Delta\varphi)$ measurement, tilt, particularly any setup angle error, will affect the effective angle of data for $L_w(\theta_v, \Delta\varphi)$ and hence any bidirectional corrections that may subsequently be applied to reproject data to nadir-viewing or to the satellite-viewing geometry. However, the related uncertainties will generally be low provided that data are sufficiently tilt-thresholded before processing. Tilt will also affect the effective incidence angle for calculation of the effective Fresnel reflectance, particularly for high wave conditions and when viewing at high viewing nadir angle such as >40°.

While pointing away from the sun azimuth minimizes the azimuthal variation of effective Fresnel reflectance, the deviation between nominal $\Delta\varphi$ and actual $\Delta\varphi$ provides an additional source of uncertainty. The actual $\Delta\varphi$ should therefore be measured, typically using a magnetic compass and modelled sun azimuth angle for shipborne measurements. For unsupervised deployments a reference azimuth is generally set during installation by sun-pointing and is regularly checked.

For the $L_d(0^+, 180° - \theta_v, \Delta\varphi)$ measurement, tilt will result in a different portion of the sky being measured from the sky that is effectively reflected by the air–water interface into the water-viewing sensor.

4.2.3. Self-Shading from Radiometers and/or Superstructure

The light field should not be perturbed by the measurement platform

Measurements from boat- and platform-mounted water-viewing radiometers may be contaminated by optical perturbations from the boat/platform. These perturbations are most pronounced when the water volume being measured is in some way shadowed from direct sun, although shadowing of downwelling skylight and reflection of downwelling light from structures also contribute to optical perturbations.

For the above-water optical perturbations to E_d, one can imagine operating a fish-eye camera pointing vertically upwards from the water surface at the centre of the radiometer field of view—see Figures 2 and 3 of [5] except that, in the context of impact on the L_w measurement, the location for such photos is the water surface target. Anything in the hemispherical picture that is not the sun/sky represents an optical perturbation, that will be wavelength-dependent and may be either positive or negative, e.g., blue sky replaced by part of the ship. This effect is most important for objects close to zenith because of their greater contribution to the cosine-weighted integral of E_d (see Equation 2 of [5]), for objects close to the sun where sky radiance is greatest and for objects which occupy a large solid angle of the sky.

The ship/platform may also throw a shadow (or reflections) that affect the underwater light field and hence $L_w(\theta_v, \Delta\varphi)$, particularly in clear waters and/or for wavelengths with low diffuse attenuation coefficient.

Optical perturbations from the ship/platform are generally reduced in the system design by:

1. Mounting the water-viewing radiometer as high as possible, e.g., on a telescopic mast [100,101];
2. Choosing the radiometer mounting position to limit optical perturbations, e.g., at the prow of a ship, facing forward [22,102] or at a corner of a fixed offshore platform [103];
3. Viewing at a moderate nadir angle, because low nadir angle viewing generally implies that the ship/platform will be closer to the water target and will occupy a larger solid angle of the sky as

seen from the water surface (but too large nadir angle will increase uncertainties associated with effective Fresnel reflectance calculation); and

4. Considering the viewing azimuth angle as a compromise between avoiding sunglint (need high $\Delta\varphi$—see Section 4.2.1) and avoiding direct shadow (need not too high $\Delta\varphi_v$).

Finally, the ship/platform may also affect the surface roughness and effective ρ_F described in Section 4.2.1 by wind-shadowing so that the measured wind speed no longer represents the wave field producing sunglint/skyglint.

Optical perturbations caused by the radiometers themselves are generally not a problem unless the radiometers are operated very close to the water surface, e.g., within 1 m.

Uncertainties associated with optical perturbations can be assessed by 3D optical simulations [67], by making measurements at different distances from the ship/platform and/or by very high resolution satellite/aircraft/drone measurements.

4.2.4. Bio-Fouling and Other Fore-Optics Contamination

The fore-optics of the radiance sensor(s) should be kept clean

In addition to sensitivity changes inherent to the radiometer, modification of the transmissivity of the fore-optics can occur because of deposition of atmospheric particles and/or water (rain, salty sea spray) and/or bio-fouling from animals (spiders, insects, birds, etc.) on the fore-optics or associated collimator tubes.

Such contamination can be easily avoided by regular checking and cleaning of the fore-optics in supervised deployments, but may be problematic for long-term unsupervised deployments, particularly for the upward facing $L_d(0^+, \theta_v, \Delta\varphi)$ sensor. Sea spray can leave a salty deposit on fore-optics and can be reduced by mounting sensors sufficiently high above the sea surface.

For long-term unsupervised deployments fore-optics contamination can be significantly reduced by parking the radiometer facing downwards (e.g., CIMEL/Seaprism approach) when not measuring and during periods of rain, as detected by a humidity sensor. Collimator tubes or other concave shielding of the fore-optics may also help reduce fore-optics contamination, e.g., from sea spray, but may provide attractive shelter to spiders and insects.

The uncertainty related to bio-fouling and other foreoptics contamination can be estimated by comparing post-deployment calibrations before and after cleaning.

4.2.5. Temporal Fluctuations

Temporal fluctuations associated with surface waves should be removed

Measurements are averaged over a certain interval of time to remove as far as possible the temporal variations associated with surface gravity waves—see Section 4.2.1. Variations in illumination conditions, e.g., clouds/haze passing near the sun, or in cloudiness of the portion of sky that reflects into the water-viewing sensor, can be detected in time series of replicates and the associated data can be rejected if a user-defined threshold of variation is reached.

If $L_d(0^+, 180° - \theta_v, \Delta\varphi)$ and $L_u(0^+, \theta_v, \Delta\varphi)$ are measured with the same radiometer then illumination changes between these two measurement times should be monitored, e.g., via continuous $E_d(0^+)$ measurements.

Uncertainties associated with any temporal fluctuations of illumination conditions (both the direct sun and the sky in the sky-viewing direction) that pass the time series quality control can be quantified by simple model simulations.

4.2.6. Bidirectional Effects

The viewing geometry (nadir and relative azimuth angle to sun) should be accurately known

The difference between satellite and in situ viewing directions and associated BRDF corrections, as mentioned in Section 1.2 is outside the scope of the present study and warrants a study of its own, although it is noted here that off-nadir angles, e.g., $\theta_v = 40°$, are generally used in above-water radiometry. BRDF corrections from off-nadir to nadir-viewing geometries are more significant in optically shallow waters.

4.2.7. Atmospheric Scattering between Water and Sensor

The atmospheric path length for scattering between water and sensor should be negligible

Atmospheric scattering (or absorption) can occur between the water surface and the radiance sensor introducing an error in the L_u measurement. In practice this is often ignored because the deployment height is typically only a few metres. However, for completeness in the FRM context and particularly when deployments are made from high masts (to avoid superstructure and shading effects), the uncertainty associated with atmospheric scattering between water and sensor should be estimated.

4.3. Variants on the Above-Water Radiometric Method

In addition to the various viewing geometries that have been used for above-water radiometry, one important protocol variant was introduced [80] and further developed [81], for the SIMBAD/SIMBADA radiometers with a vertically polarising filter placed as fore-optics and a measurement protocol with $\theta_v = 45°$ and $\Delta\varphi = 135°$. This design allows dramatic reduction of the magnitude of $L_r(\theta_v, \Delta\varphi)$ and hence associated uncertainties, provided that the polarising filter can be adequately calibrated and the residual polarised component of reflected skyglint can be adequately modelled.

Above-water measurements could also be made for multiple nadir and azimuth angles, e.g., from a robotic pointing system or from an imaging camera system [104].

It is entirely feasible to combine both polarised and unpolarised measurements of $L_u(0^+, \theta_v, \Delta\varphi)$, e.g., in a filter-wheel radiometer or by mounting in parallel radiometers with and without polarising filters [105]. The main component of skyglint can be effectively removed for a range of viewing angles by use of a vertical polarizer [106]. However, small background noise still exists because of different orientations of the wave facets and the sunglint is not well removed by a vertical polariser because polarization is in a different plane. The partial polarization of L_w itself needs to be considered in such techniques.

Theoretically above-water radiometric measurements could be made for satellite validation from low altitude airborne platforms such as tethered balloons or drones, which would have advantages in terms of reducing optical perturbation by increasing distance from the water surface. However, in practice, the control of viewing geometry (platform stability) and logistical considerations (power supply, cleaning, maintenance) seems to preclude significant use of such platforms for unsupervised measurements at present.

5. Skylight-Blocked Approach

5.1. Measurement Equation

In view of the potentially large uncertainties which may arise from the skyglint correction of above-water radiometry (Section 4.2.1), the SBA was suggested [76,107,108] and further developed [82]. In this approach the upwelling radiance measurement is made with a radiance sensor to which an extension cone or cylinder is added so that the tip of the cone/cylinder lies fully beneath the air–water interface but the sensor fore-optics remains in air—see Figure 7. A photograph of an actual deployment can be found in Figure 2 of [82].

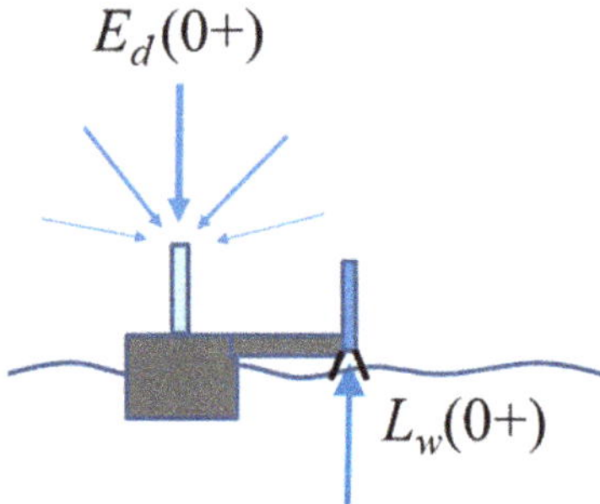

Figure 7. Schematic of above-water radiometry with skylight-blocked approach. Note that the radiometer fore-optics are in air, but the radiometer body is extended with a cone or shield (black lines) that extends below the water surface, ensuring blocking of skylight reflection.

With this approach there should be no skyglint reflected at the air–water interface and the measurement equation is simply given by:

$$L_w(\theta_v, \Delta\varphi) = L_u\left(0^+,\ \theta_v, \Delta\varphi\right) \tag{16}$$

This measurement can be made for the nadir-viewing direction, $\theta_v = 0$, typically from a buoy which is floated away from a ship or tethered to a mooring, but other configurations are possible (see Section 5.3).

Measurement of water radiance involves time integration for each individual measurement and replicate measurements which are subsequently processed to yield a single value for $\overline{L_u}(0^+,\ \theta_v, \Delta\varphi)$ where the overbar denotes the multitemporal measurement, typically called "time-average", although the temporal processing may be different from a mean average.

The integration time depends on the radiometer design and the brightness of the target. Filter-wheel radiometers generally measure fast, typically at many hertz, whereas spectrometer-based systems may be much slower, e.g., integration time of 1 s to 4 s, for dark targets such as water.

5.2. Protocol-Dependent Sources of Uncertainty

The protocol-related sources of uncertainty are described in the following subsections.

5.2.1. Self-Shading from Radiometers and/or Superstructure

The underwater light field should not be perturbed by the measurement radiometer, sky-blocking cone and platform

The skylight blocking cone/shield is designed to fully block all downward radiance at the air–water interface so that the reflection of skylight from the air–water interface is zero with zero uncertainty provided that there are no internal reflections within the cone and from the sensor fore-optics. However, in practice the cone/shield and radiometer will also block sun and skylight illuminating the water volume that is being measured. This (spectrally-dependent) uncertainty, also called self-shading, needs to be evaluated and will depend on:

- Diameter of the cone/shield (preferably small);
- Angular variation of downwelling radiance (preferably high sun zenith angle);
- Inherent optical properties of the water (preferably low absorption);

- Distance of the cone beneath the air–water interface (preferably very small compared to a vertical attenuation length scale).

The first three parameters are similar as for the process of radiometer self-shading for underwater radiometry [47]. Minimisation of the distance of the cone beneath the air–water interface depends on surface wave height and stability of the deployment platform and should be measured or estimated. Outliers caused by waves can be removed during data processing.

The uncertainties associated with self-shading using this protocol have been estimated by [109], who propose also a scheme for correcting for these effects.

Further contamination of measurements may arise from optical perturbations from the deployment platform, typically a buoy floated away from a ship to a distance sufficient to ensure no optical contamination from the ship itself. Clearly the water volume being measured should not be in the direct sun shadow of any deployment platform (buoy). This can be achieved by duplicate radiometers on opposite sides of a buoy, one of which will always be outside the direct sun shadow. Measurement of the azimuthal rotation of the deployment structure with respect to sun will facilitate estimation of the uncertainty relating to optical contaminations. Figure 4 of [109] shows, from 3D Monte Carlo simulations of the structure, that azimuthal dependence of self-shading is low provided that direct sun shadow is avoided.

Even if outside the direct sun shadow the deployment structure will to some extent modify the downwelling radiance field illuminating the water volume. Consequent uncertainties can be estimated, as for the other methods (Section 2.2.2), by 3D optical modelling, by high-resolution imagery (e.g., from drone-mounted cameras) or by experiments with radiometers held at different distance from the deployment structure.

5.2.2. Tilt Effects

The radiance sensor should be deployed vertically

Any variation in the pointing angle of the radiometer ("tilt") must be considered to give an uncertainty for L_{wn} as for fixed-depth underwater measurements—Section 2.2.2.—but using here the above-water angular variability of L_w. Typically a tilt threshold will be set for acceptable measurements and the associated uncertainty can be assessed from model simulations.

5.2.3. Bio-Fouling and Other Fore-Optics Contamination

The fore-optics of the radiance sensor should be kept clean

Since this protocol involves a downward-facing sensor with shadowed fore-optics, bio-fouling from algae is not expected to be a major problem, even for unsupervised deployment—see also Section 2.2.4 for fixed-depth underwater radiometry.

More problematic may be the possibility of water droplets reaching the fore-optics, which is supposed to be in air. For seaborne deployments, salt water reaching the fore-optics may leave a salty deposit. This can be particularly problematic in high sea state, but can be limited by choice of a stable deployment platform [82] and a sufficiently long and air-tight cone/shield (subject to radiometer field of view constraints). In addition, for supervised deployments, a small brush can be used to clean the fore-optics regularly.

The uncertainty related to any foreoptics contamination can be estimated by comparing post-deployment calibrations before and after cleaning.

5.2.4. Temporal Fluctuations

Temporal fluctuations associated with surface waves should be removed

Measurements are averaged (after quality control) over a certain interval of time to remove as far as possible the fast variations associated with natural variability (wave focusing/defocusing—see also

Section 2.2.7), and with surface gravity waves, which may affect the depth of water in the shield/cone (Section 5.2.1).

Variations in illumination conditions, e.g., clouds/haze passing near the sun, can be detected in time series of L_w / E_d^{0+} or E_d^{0+} and the associated data can be rejected if a user-defined threshold of variation is reached. Uncertainties associated with any rapid fluctuations of illumination conditions that pass the time series quality control can be quantified by simple model simulations.

5.3. Variants on the Skylight-Blocked Approach

The SBA protocol could be used with various radiometers, shields/cones and deployment methods (buoys, etc.). The preceding subsections are thought to be sufficiently generic to cover these variants.

6. Conclusions

6.1. Summary of the State of the Art

This paper reviews the current state of the art of protocols for the measurement of water-leaving radiance for validation of satellite remote-sensing data over water in the FRM context. This review focusses particularly on protocol-related elements of the measurement uncertainty budget. These aspects of the protocol are discussed with reference to documented studies and guidelines are provided on how to estimate such uncertainties, e.g., design of experiments and/or model studies.

Four basic measurement protocols have been identified:

- Underwater radiometry using fixed-depth measurements ("underwater fixed depths");
- Underwater radiometry using vertical profiles ("underwater profiling");
- Above-water radiometry with sky radiance measurement and skyglint removal ("above-water"); and
- On-water radiometry with optical blocking of skylight ("skylight-blocked").

These protocols are summarized in Table 1 as regards equipment, protocol maturity, automation aspects, and challenging waters/wavelengths.

In this review we have tried to cover a very wide range of potential environmental conditions and a rather generic consideration of the four basic protocol families. For example, the MOBY and BOUSSOLE systems are obvious models for the underwater fixed-depth method and are both operating from floating platforms in deep, oligotrophic "case 1" waters with high performance and high cost infrastructure and instrumentation. However, the fixed-depth protocol can be applied in very different circumstances such as in very shallow inland waters (with much closer vertical spacing of radiometers) or from fixed platforms (with negligible tilt). Similarly, the AERONET-OC system is an obvious model for above-water radiometry and is characterised by fixed, offshore platforms with negligible tilt and no azimuthal rotation (of the platform itself). However, the above-water protocol can be applied in very different circumstances, e.g., from ships, or even small boats, with tilt and azimuthal rotation. The overview of protocol-related uncertainties given in Table 2, therefore, refers to the generic protocol rather than to any of these specific implementations.

Table 1. Summary of the four measurement methods as regards: equipment; standard (S) and variant (V) methods; viewing geometry; protocol maturity/diversity; automation maturity; automation challenges; and challenging waters/wavelengths/conditions (see Section 6.2 for more details). The automation challenges refers to the protocol-specific challenges and excludes common challenges such as the logistics of maintenance visits, power supplies, hardware failures, radiometer calibration requirements, protection from damage, etc. CDOM and NAP are abbreviations for coloured dissolved organic matter and non-algae particles, respectively.

	Underwater Fixed Depths	Underwater Profiling	Above-Water	Skylight-Blocked
Equipment (in addition to ship/platform/buoy)	2 radiance sensors Inclinometer Pressure/depth sensor	Radiance sensor and profiling platform Inclinometer Pressure/depth sensor	Radiance sensor and robotic/human pointing or 2 radiance sensors Inclinometer, Compass/protractor	Radiance sensor Sky-blocking cone/shield Inclinometer
Standard (S) and Variants (V)	S: tethered buoy, at least two fixed depths V: Single very near-surface radiometer; single radiometer successively at different depths	S: free-fall away from ship V: platform/mooring-tethered vertical wire; Horizontally drifting platforms	S: unpolarised radiometer V: vertical polarizer option	S: tethered buoy V: boats and other platforms
Viewing geometry	Nadir	Nadir	Off-nadir, usually $\theta_v = 40°$ and $\Delta\varphi = 90°$ or $135°$	Nadir (or off-nadir)
Protocol maturity/diversity	Mature	Mature	Mature basis but also diverse and evolving skyglint corrections	Mature
Automation maturity	Operational	Prototype	Operational	Feasible
Automation challenges	Fore-optics contamination	Fore-optics contamination Mechanical reliability of profiling (fixed location systems)	Fore-optics contamination	Fore-optics contamination
Challenging water types/wavelengths/conditions	High K_{Lu} (high CDOM/NAP blue, red, near infrared) High waves Very shallow or stratified waters	High K_{Lu} (high CDOM/NAP blue, red, near infrared) High waves Very shallow or stratified waters	Low reflectance (high CDOM blue, low backscatter red/near infrared) High waves Scattered clouds in sky-viewing direction	High waves

Table 2. Summary of the four measurement methods as regards protocol-related uncertainty estimation. I = Ideal conditions; R = Recommendations; U = Uncertainty estimation. Cal = calibration. N/A = not applicable. See text for more details on each topic. Depth measurement and Fresnel transmittance should also be included in the uncertainty budget for the underwater fixed-depth and profiling methods, but are not included in the table. Radiometer-related uncertainties must also be estimated for all methods but are beyond the scope of this review.

	Underwater Fixed Depths	Underwater Profiling	Above-Water	Skylight-Blocked
Non-exponential vertical variation	I: Known (e.g., exponential) variation R: Extra depths, profiles and modelling U: as R.	I: Known (e.g., exponential) variation R: Measure close to surface U: Goodness-of-fit tests, modelling	N/A	N/A
Tilt	I: Deploy vertical R: Monitor inclination and pressure U: Modelling, time series analysis	I: Deploy vertical R: Stable free-fall or wire-guided, Monitor inclination and pressure U: Modelling, time series analysis	I: Accurate pointing, stable platform R: Monitor inclination U: Modelling	I: Stable platform R: Monitor inclination U: Modelling, time series analysis
Self-shading from radiometer	I: Negligible size radiometer R: Small diameter radiometer U: Modelling	I: Negligible size radiometer R: Small diameter radiometer U: Modelling	N/A (in general)	I: Negligible size cone/shield R: Small diameter cone/shield U: Modelling
Self-shading from structure/platform	I: Negligible size superstructure R: Limit cross-section, horizontal arms, redundant radiometers U: Modelling, comparison of redundant radiometers	I: Negligible size superstructure R: Limit cross-section, deploy away from ship, redundant radiometers U: Modelling, comparison of redundant radiometers	I: Negligible size superstructure R: Target away from platform (masts) or ship (forward from prow), azimuth filtering to avoid shadow U: Modelling, experiments (different heights/positions/azimuths)	I: Negligible size platform R: Limit cross-section, horizontal arms, redundant radiometers U: Modelling, comparison of redundant radiometers
Fore-optics contamination	I: Keep fore-optics clean (in water) R: Inspect/clean/protect, monitor with portable cal devices U: Pre-/post-cleaning cal of radiometer	I: Keep fore-optics clean (in water) R: Inspect/clean/protect, monitor with portable cal devices U: Pre-/post-cleaning cal of radiometer	I: Keep fore-optics clean (in air) R: Inspect/clean/protect, monitor with portable cal devices U: Pre-/post-cleaning cal of radiometer	I: Keep fore-optics clean (in air, close to water) R: Inspect/clean/protect, monitor with portable cal devices U: Pre-/post-cleaning cal of radiometer
Temporal fluctuations	I: Clear sky, flat water R: Time series analysis U: Modelling, time series analysis	I: Clear sky, flat water R: Time series analysis, multi-casting U: Modelling, time series and multi-cast analysis	(here for sky, see below for waves) I: Clear, stable sky R: Replicates U: Standard deviation of replicates	I: Clear sky, flat water R: Time series analysis U: Modelling, time series analysis
Skylight reflection correction	N/A	N/A	I: Flat sea R: Very diverse, see text U: Very diverse, see text	N/A

6.2. Underwater or Above-Water Measurement?

So which is the best approach to use? A newcomer to the field of water radiance measurements will typically be confronted with important decisions for:

- purchasing radiometers and associated equipment;
- purchasing, renting or arranging access to a deployment platform such as a fixed structure (offshore platform, jetty, pier, buoy, etc.), a ship (research vessel, small boat, passenger ferry "ship of opportunity", etc.), a drifting underwater platform, or even a low-altitude airborne vehicle (tethered balloon, drone, etc.); and
- training and financially supporting staff to make the measurements (if supervised) or to setup and maintain and monitor the measurement system (if unsupervised), including radiometer checks, calibration and characterisation and data processing, quality control, archiving and distribution.

The choice of protocol will affect both the quality and quantity of data and the setting and running costs of acquiring data. The choice of protocol will obviously be driven by the objectives of the measurement program and the environmental conditions (type of water: brightness, colour, depth, vertical homogeneity) as well as by any cost constraints and/or cost-sharing opportunities (such as the existence of platforms or other measurement programs).

The main fundamental differences in data quality that can be expected between the two underwater methods and the above-water (skyglint corrected) method, in their most generic implementations, can be related to the need for vertical extrapolation in the underwater methods and the need for skyglint correction in the above-water method:

- Uncertainties associated with vertical extrapolation in underwater methods will be highest for situations (water types, wavelengths) where the diffuse attenuation coefficient length scale, $1/K_{Lu}$, is small compared to the depth of the highest usable upwelling radiance measurement, z_1. Thus, the requirement for underwater measurements close to the surface becomes more and more demanding for waters/wavelengths with high K_{Lu}, including blue wavelengths in waters with high coloured dissolved organic matter (CDOM) or high non-algae particle (NAP) absorption and red and, a fortiori, near infra-red wavelengths in all waters. Self-shading also increases for high attenuation waters.
- Uncertainties associated with skyglint correction in above-water methods will be highest for low reflectance waters/wavelengths and for high sun zenith angle (as well as for cloudy and partially cloudy skies although these are supposed to be removed by quality control in the FRM context) and for blue wavelengths. Thus, the requirement for a highly accurate skyglint correction method becomes more and more demanding for blue wavelengths in waters with high CDOM absorption (and to a lesser extent high non-algae particle absorption) and for red and near infrared wavelength in low particulate backscatter waters.

It is interesting to note that these two challenging conditions, high K_{Lu} and low reflectance, generally correlate in highly absorbing waters/wavelengths but anticorrelate in highly scattering waters.

Both the underwater methods and the above-water methods have uncertainties that increase with surface wave conditions because of wave focusing/defocusing effects and skyglint removal respectively.

The skylight-blocked approach has quite different sensitivity to the water type and wavelength of measurement from the underwater and above-water approaches, because it requires neither vertical extrapolation nor skyglint removal. The most challenging conditions for this method will probably be practical deployment in high wave conditions and self-shading correction for low sun zenith and high K_{Lu} conditions.

6.3. Future Perspectives

In contrast to the simpler E_d^{0+} measurement [5], there has been considerable evolution and diversity of the L_w measurement since the publication of the NASA Ocean Optics Protocols [17].

Future improvements to L_w measurements are expected to come in the future from the following developments:

- Improvements in the design and usage of calibration monitoring devices, which can be used in the field, are likely to improve identification of fore-optics fouling and radiometer sensitivity changes.
- Model simulations (with polarisation) of the 3D light field and dedicated experiments for all four protocols are likely to improve estimation of related uncertainties.
- Improvements in the stability and reduction in the cost of telescopic masts may reduce superstructure shading effects for above-water radiometry.
- Reduction in the cost of pointing systems, thanks to the video camera surveillance industry, should facilitate multi-directional above-water radiometry [110] and improve the protection ("parking") of radiometers when not in use and thus reduce fouling for long-term deployments.
- Greater use of full sky imaging cameras [111], whether calibrated (expensive) or not (typically inexpensive), potentially coupled with automated image analysis techniques, will allow better identification of suboptimal measurement conditions.
- Above-water imaging cameras may allow better characterisation of the air–water interface (wave field) and hence better removal of L_r in above-water radiometric measurements [104,106].

As regards the future for validation of water reflectance more generally:

- The tendency to move to highly automated systems with long-term, e.g., one year, essentially maintenance-free deployments is likely to improve significantly the quantity of data available for validation. Networks of such systems further increase the power and efficiency for validation purposes. Networks of automated systems are now already operational or in advanced prototype testing phases for systems based on the above-water, underwater profiling and underwater fixed-depth methods and are conceptually feasible for the skylight-blocked approach.
- The advent of operational satellite missions such as VIIRS and Sentinel-3/OLCI, Sentinel-2/MSI and Landsat-8/OLI with the need for a guaranteed long-term validated data stream will increase the need for FRM.
- The huge increase in optical satellite missions used for aquatic remote-sensing will also increase the need for highly automated measurement systems and the economy of scale for such deployments—one in situ radiometer system can validate many, many satellite instruments.

As regards the needs of the validation community, it is recommended to:

- Update this review, e.g., on a 10-year time frame, to take account of developments in the protocols, particularly in the estimation of uncertainties and for the above-water family of methods, where evolution and innovations in basic methodology are continuing. Such an update is best preceded by community discussion at an international workshop.
- Organise regular intercomparison exercises, e.g., on a two-year time frame, covering the full diversity of methods, to ensure that measurement protocols and scientists, remain state of the art (as required by the FRM context).

Although not targeted by this review it is possible that the considerations developed here may be useful for other applications where L_w measurements are needed, including calibration/validation data for IOP retrieval algorithms.

Author Contributions: Conceptualization, K.G.R.; methodology, K.G.R.; writing—original draft preparation, K.G.R.; writing—review and editing, K.V., E.B., A.C., R.F., A.G., M.H., B.C.J., J.K., Z.L., M.O., V.V. and R.V.

Funding: The collection of information for and the writing of this study were funded by the European Space Agency, grant number ESA/AO/1-8500/15/I-SBo, "Fiducial reference measurements for satellite ocean colour (FRM4SOC)" project. EB's and RF's contributions are supported by the NASA Ocean Biology and Biochemistry program.

Acknowledgments: Colleagues from the FRM4SOC project, the Sentinel-3 Validation Team and the NOAA/VIIRS cal/val team are acknowledged for helpful discussions on measurement protocols during project meetings and teleconferences. Craig Donlon and Tânia Casal are gratefully acknowledged for stimulating discussions and constructive support throughout the FRM4SOC project, from conception to implementation. Giuseppe Zibordi is thanked for constructive comments. An anonymous reviewer is thanked for a very careful reading of the text and for many useful suggestions.

Conflicts of Interest: The first author is involved in the design of systems and in the acquisition of data based on the "above-water radiometry" method of Section 4. Other co-authors are involved in the design of systems and/or acquisition of data of other methods, covering all protocols presented here in Sections 2–5. The funders had no role in the design of the study other than the drafting of the FRM4SOC project Statement of Work, which states that such a protocol review should be performed. The funders had no role in the collection, analyses, or interpretation of data or in the writing of the manuscript other than general project monitoring and encouragement. The funders did encourage publication of this review in order to ensure that high quality measurements of known uncertainty are available for the validation of the satellite missions they design and operate. None of these interests is considered to be conflictual or to inappropriately influence the recommendations expressed here.

Disclaimer: Certain commercial equipment, instruments, or materials are identified in this paper in order to specify the experimental procedure adequately. Such identification is not intended to imply recommendation or endorsement by the National Institute of Standards and Technology or any other organization involved in the writing of this paper, nor is it intended to imply that the materials or equipment identified are necessarily the best available for the purpose.

References

1. International Ocean Colour Coordinating Group (IOCCG). *Why Ocean Colour? The Societal Benefits of Ocean-Colour Technology*; Technical Report No. 7; IOCCG: Dartmouth, NS, Canada, 2008; p. 141.
2. Donlon, C.J.; Wimmer, W.; Robinson, I.; Fisher, G.; Ferlet, M.; Nightingale, T.; Bras, B. A second-generation blackbody system for the calibration and verification of seagoing infrared radiometers. *J. Atmos. Ocean. Technol.* **2014**, *31*, 1104–1127. [CrossRef]
3. Mobley, C.D. *Light and Water: Radiative Transfer in Natural Waters*; Academic Press: London, UK, 1994.
4. Mobley, C. Overview of Optical Oceanography—Reflectances. Available online: http://www.oceanopticsbook. info/view/overview_of_optical_oceanography/reflectances (accessed on 24 July 2019).
5. Ruddick, K.G.; Voss, K.; Banks, A.C.; Boss, E.; Castagna, A.; Frouin, R.; Hieronymi, M.; Jamet, C.; Johnson, B.C.; Kuusk, J.; et al. A review of protocols for fiducial reference measurements of downwelling irradiance for validation of satellite remote sensing data over water. *Remote Sens.* **2019**, *11*, 1742. [CrossRef]
6. Gitelson, A. The peak near 700 nm on radiance spectra of algae and water: Relationships of its magnitude and position with chlorophyll concentration. *Int. J. Remote Sens.* **1992**, *13*, 3367–3373. [CrossRef]
7. Randolph, K.; Wilson, J.; Tedesco, L.; Li, L.; Pascual, D.L.; Soyeux, E. Hyperspectral remote sensing of cyanobacteria in turbid productive water using optically active pigments, chlorophyll a and phycocyanin. *Remote Sens. Environ.* **2008**, *112*, 4009–4019. [CrossRef]
8. International Standards Organisation (ISO). *Space Environment (Natural and Artificial)—Process for Determining solar Irradiances*; Technical Report No. 21348:2007; International Standards Organisation (ISO): Geneva, Switzerland, 2007.
9. Knaeps, E.; Dogliotti, A.I.; Raymaekers, D.; Ruddick, K.; Sterckx, S. In situ evidence of non-zero reflectance in the OLCI 1020 nm band for a turbid estuary. *Remote Sens. Environ.* **2012**, *120*, 133–144. [CrossRef]
10. Dogliotti, A.; Gossn, J.; Vanhellemont, Q.; Ruddick, K. Detecting and quantifying a massive invasion of floating aquatic plants in the Río de la Plata turbid waters using high spatial resolution ocean color imagery. *Remote Sens.* **2018**, *10*, 1140. [CrossRef]
11. Hooker, S.B.; Zibordi, G.; Maritorena, S. *The Second SeaWiFs Ocean Optics DARR (DARR-00)*; SeaWiFS Postlaunch; Technical Report No. 5; NASA Technical Memorandum 2001-206892: Greenbelt, MD, USA, 2001; pp. 4–45.
12. Zibordi, G.; Ruddick, K.; Ansko, I.; Moore, G.; Kratzer, S.; Icely, J.; Reinart, A. In situ determination of the remote sensing reflectance: An inter-comparison. *Ocean Sci.* **2012**, *8*, 567–586. [CrossRef]
13. Ondrusek, M.; Lance, V.P.; Arnone, R.; Ladner, S.; Goode, W.; Vandermeulen, R.; Freeman, S.; Chaves, J.E.; Mannino, A.; Gilerson, A.; et al. *Report for Dedicated JPSS VIIRS Ocean Color December 2015 Calibration/Validation Cruise*; U.S. Department of Commerce, National Oceanic and Atmospheric Administration, National Environmental Satellite, Data, and Information Service: Washington, DC, USA, 2016; p. 66.

14. Mobley, C.D.; Sundman, L.K.; Boss, E. Phase function effects on oceanic light fields. *Appl. Opt.* **2002**, *41*, 1035–1050. [CrossRef]

15. Tzortziou, M.; Herman, J.R.; Gallegos, C.L.; Neale, P.J.; Subramaniam, A.; Harding, L.W.; Ahmad, Z. Bio-optics of the Chesapeake Bay from measurements and radiative transfer closure. *Estuar. Coast. Shelf Sci.* **2006**, *68*, 348–362. [CrossRef]

16. International Standards Organisation (ISO). *Evaluation of Measurement Data—Guide to the Expression of Uncertainty in Measurement*; Technical Report No. JCGM 100:2008; International Standards Organisation (ISO): Geneva, Switzerland, 2008.

17. Mueller, J.L.; Fargion, G.S.; McClain, C.R. *Ocean Optics Protocols for Satellite Ocean Color Sensor Validation*; Technical Report No. TM 2003 21621/Revision 5; NASA: Greenbelt, MD, USA, 2004.

18. Zibordi, G.; Voss, K.J. In situ optical radiometry in the visible and near infrared. In *Optical Radiometry for Ocean Climate Measurements*; Zibordi, G., Donlon, C., Parr, A.C., Eds.; Academic Press: Oxford, UK, 2014; pp. 248–305.

19. Zibordi, G.; Voss, K. *Protocols for Satellite Ocean Color Data Validation: In situ Optical Radiometry*; IOCCG Protocols Series; IOCCG: Dartmouth, NS, Canada, 2019.

20. Minnett, P.J. Consequences of sea surface temperature variability on the validation and applications of satellite measurements. *J. Geophys. Res. Oceans* **1991**, *96*, 18475–18489. [CrossRef]

21. Loew, A.; Bell, W.; Brocca, L.; Bulgin, C.E.; Burdanowitz, J.; Calbet, X.; Donner, R.V.; Ghent, D.; Gruber, A.; Kaminski, T.; et al. Validation practices for satellite-based Earth observation data across communities. *Rev. Geophys.* **2017**, *55*, 779–817. [CrossRef]

22. Ruddick, K.; De Cauwer, V.; Park, Y.; Moore, G. Seaborne measurements of near infrared water-leaving reflectance: The similarity spectrum for turbid waters. *Limnol. Oceanogr.* **2006**, *51*, 1167–1179. [CrossRef]

23. Zibordi, G.; Holben, B.; Slutsker, I.; Giles, D.; D'Alimonte, D.; Mélin, F.; Berthon, J.-F.; Vandemark, D.; Feng, H.; Schuster, G.; et al. AERONET-OC: A network for the validation of ocean color primary product. *J. Atmos. Ocean. Technol.* **2009**, *26*, 1634–1651. [CrossRef]

24. Claustre, H.; Bernard, S.; Berthon, J.-F.; Bishop, J.; Boss, E.; Coaranoan, C.; D'Ortenzio, F.; Johnson, K.; Lotliker, A.; Ulloa, O. *Bio-Optical Sensors on Argo Floats*; Technical Report No. 11; IOCCG: Dartmouth, NS, Canada, 2011.

25. Antoine, D.; Guevel, P.; Desté, J.-F.; Bécu, G.; Louis, F.; Scott, A.J.; Bardey, P. The BOUSSOLE buoy—A new transparent-to-swell taut mooring dedicated to marine optics: Design, tests, and performance at sea. *J. Atmos. Ocean. Technol.* **2008**, *25*, 968–989. [CrossRef]

26. Antoine, D.; d'Ortenzio, F.; Hooker, S.B.; Bécu, G.; Gentili, B.; Tailliez, D.; Scott, A.J. Assessment of uncertainty in the ocean reflectance determined by three satellite ocean color sensors (MERIS, SeaWiFS and MODIS-A) at an offshore site in the Mediterranean Sea (BOUSSOLE project). *J. Geophys. Res.* **2008**, *113*. [CrossRef]

27. Clark, D.K.; Gordon, H.R.; Voss, K.J.; Ge, Y.; Broenkow, W.; Trees, C. Validation of atmospheric correction over the oceans. *J. Geophys. Res.* **1997**, *102*, 17209–17217. [CrossRef]

28. Clark, D.K.; Yarbrough, M.A.; Feinholz, M.; Flora, S.; Broenkow, W.; Kim, Y.S.; Johnson, B.C.; Brown, S.W.; Yuen, M.; Mueller, J.L. MOBY, a radiometric buoy for performance monitoring and vicarious calibration of satellite ocean color sensors: Measurement and data analysis protocols. Chapter 2. In *Ocean Optics Protocols for Satellite Ocean Color Sensor Validation: Special Topics in Ocean Optics Protocols and Appendices*; Technical Report No. TM-2003-211621/Rev4; NASA: Greenbelt, MD, USA, 2003; Volume 6, pp. 3–34.

29. Brown, S.W.; Flora, S.J.; Feinholz, M.E.; Yarbrough, M.A.; Houlihan, T.; Peters, D.; Kim, Y.S.; Mueller, J.L.; Johnson, B.C.; Clark, D.K. *The Marine Optical Buoy (MOBY) Radiometric Calibration and Uncertainty Budget for Ocean Color Satellite Sensor Vicarious Calibration.*; SPIE Proceedings: Bellingham, WA, USA, 2007; Volume 67441.

30. Austin, R.W.; Halikas, G. *The Index of Refraction of Seawater*; Technical Report 76-1; Scripps Institution of Oceanography: San Diego, CA, USA, 1976.

31. Gordon, H.R. Normalized water-leaving radiance: Revisiting the influence of surface roughness. *Appl. Opt.* **2005**, *44*, 241. [CrossRef] [PubMed]

32. Preisendorfer, R.W.; Mobley, C.D. Albedos and glitter patterns of a wind-roughened sea surface. *J. Phys. Oceanogr.* **1986**, *16*, 1293–1316. [CrossRef]

33. Voss, K.J.; Flora, S.J. Spectral dependence of the seawater-air radiance transmission coefficient. *J. Atmos. Ocean. Technol.* **2017**, *34*, 1203–1205. [CrossRef]

34. Röttgers, R.; Doerffer, R.; McKee, D.; Schönfeld, W. *The Water Optical Properties Processor (WOPP): Pure Water Spectral Absorption, Scattering and Real Part of Refractive Index Model*; Technical Report No WOPP-ATBD/WRD6; HZG: Geesthaacht, Germany, 2011; p. 20. Available online: http://calvalportal.ceos.org/data_access-tools (accessed on 24 July 2019).

35. Li, L.; Stramski, D.; Reynolds, R.A. Effects of inelastic radiative processes on the determination of water-leaving spectral radiance from extrapolation of underwater near-surface measurements. *Appl. Opt.* **2016**, *55*, 7050. [CrossRef]

36. Zaneveld, J.R.; Boss, E.; Hwang, P. The influence of coherent waves on the remotely sensed reflectance. *Opt. Express* **2001**, *9*, 260. [CrossRef] [PubMed]

37. D'Alimonte, D.; Zibordi, G.; Kajiyama, T.; Cunha, J.C. Monte Carlo code for high spatial resolution ocean color simulations. *Appl. Opt.* **2010**, *49*, 4936–4950. [CrossRef] [PubMed]

38. Darecki, M.; Stramski, D.; Sokólski, M. Measurements of high-frequency light fluctuations induced by sea surface waves with an underwater porcupine radiometer system. *J. Geophys. Res. Oceans* **2011**, *116*. [CrossRef]

39. Hieronymi, M.; Macke, A. On the influence of wind and waves on underwater irradiance fluctuations. *Ocean Sci.* **2012**, *8*, 455–471. [CrossRef]

40. Hieronymi, M. Polarized reflectance and transmittance distribution functions of the ocean surface. *Opt. Express* **2016**, *24*, A1045. [CrossRef] [PubMed]

41. Gernez, P.; Stramski, D.; Darecki, M. Vertical changes in the probability distribution of downward irradiance within the near-surface ocean under sunny conditions. *J. Geophys. Res.* **2011**, *116*. [CrossRef]

42. Monahan, E.C.; Muircheartaigh, I. Optimal power-law description of oceanic whitecap coverage dependence on wind speed. *J. Phys. Oceanogr.* **1980**, *10*, 2094–2099. [CrossRef]

43. Zibordi, G.; Berthon, J.-F.; D'Alimonte, D. An evaluation of radiometric products from fixed-depth and continuous in-water profile data from moderately complex waters. *J. Atmos. Ocean. Technol.* **2009**, *26*, 91–106. [CrossRef]

44. Voss, K.J. *Radiance distribution measurements in coastal water*; SPIE Proceedings: Bellingham, WA, USA, 1988; Volume 0925, pp. 56–66.

45. D'Alimonte, D.; Shybanov, E.B.; Zibordi, G.; Kajiyama, T. Regression of in-water radiometric profile data. *Opt. Express* **2013**, *21*, 27707–27733. [CrossRef]

46. Voss, K.J.; Gordon, H.R.; Flora, S.; Johnson, B.C.; Yarbrough, M.A.; Feinholz, M.; Houlihan, T. A method to extrapolate the diffuse upwelling radiance attenuation coefficient to the surface as applied to the Marine Optical Buoy (MOBY). *J. Atmospheric Ocean. Technol.* **2017**, *34*, 1423–1432. [CrossRef]

47. Gordon, H.R.; Ding, K. Self-shading of in-water optical instruments. *Limnol. Oceanogr.* **1992**, *37*, 491–500. [CrossRef]

48. Zibordi, G.; Ferrari, G.M. Instrument self-shading in underwater optical measurements: Experimental data. *Appl. Opt.* **1995**, *34*, 2750–2754. [CrossRef] [PubMed]

49. Leathers, R.A.; Downes, T.V.; Mobley, C.D. Self-shading correction for oceanographic upwelling radiometers. *Opt. Express* **2004**, *12*, 4709–4718. [CrossRef] [PubMed]

50. Mueller, J.L. Shadow corrections to in-water upwelled radiance measurments: A status review. In *Ocean Optics Protocols for Satellite Ocean Color Sensor Validation, Revision 5: Special Topics in Ocean Optics Protocols, Part 2*; NASA Technical Memorandum NASA: Greenbelt, MD, USA, 2004; Volume 6, pp. 1–7.

51. Patil, J.S.; Kimoto, H.; Kimoto, T.; Saino, T. Ultraviolet radiation (UV-C): A potential tool for the control of biofouling on marine optical instruments. *Biofouling* **2007**, *23*, 215–230. [CrossRef] [PubMed]

52. Antoine, D.; Curtin University, Perth, Australia. Personal Communication, 2017.

53. Vellucci, V.; Laboratoire Océanographique de Villefranche, Université de Sorbonne, Villefranche-sur-mer, France. Personal communication, 2018.

54. Hooker, S.B.; Morrow, J.H.; Matsuoka, A. Apparent optical properties of the Canadian Beaufort Sea; Part 2: The 1% and 1 cm perspective in deriving and validating AOP data products. *Biogeosciences* **2013**, *10*, 4511. [CrossRef]

55. Beltrán-Abaunza, J.M.; Kratzer, S.; Brockmann, C. Evaluation of MERIS products from Baltic Sea coastal waters rich in CDOM. *Ocean Sci.* **2014**, *10*, 377–396. [CrossRef]

56. Froidefond, J.M.; Ouillon, S. Introducing a mini-catamaran to perform reflectance measurements above and below the water surface. *Opt. Express* **2005**, *13*, 926–936. [CrossRef]

57. Talone, M.; Zibordi, G.; Lee, Z.P. Correction for the non-nadir viewing geometry of AERONET-OC above water radiometry data: An estimate of uncertainties. *Opt. Express* **2018**, *26*, 541–561. [CrossRef]

58. Hooker, S.B.; Maritorena, S. An evaluation of oceanographic radiometers and deployment methodologies. *J. Atmos. Ocean. Technol.* **2000**, *17*, 811–830. [CrossRef]

59. Gerbi, G.P.; Boss, E.; Werdell, P.J.; Proctor, C.W.; Haëntjens, N.; Lewis, M.R.; Brown, K.; Sorrentino, D.; Zaneveld, J.R.V.; Barnard, A.H.; et al. Validation of ocean color remote sensing reflectance using autonomous floats. *J. Atmos. Ocean. Technol.* **2016**, *33*, 2331–2352. [CrossRef]

60. Leymarie, E.; Penkerc'h, C.; Vellucci, V.; Lerebourg, C.; Antoine, D.; Boss, E.; Lewis, M.R.; D'Ortenzio, F.; Claustre, H. ProVal: A new autonomous profiling float for high quality radiometric measurements. *Front. Mar. Sci.* **2018**, *5*, 437. [CrossRef]

61. Smith, R.C.; Booth, C.R.; Star, J.L. Oceanographic biooptical profiling system. *Appl. Opt.* **1984**, *23*, 2791–2797. [CrossRef] [PubMed]

62. Voss, K.J.; Nolten, J.W.; Edwards, G.D. Ship shadow effects on apparent optical properties. In *Ocean Optics VIII*; SPIE Proceedings 0637; SPIE: Bellingham, WA, USA, 1986; pp. 186–190.

63. Mueller, J.L. In-water radiometric profile measurements and data analysis protocols. In *Ocean Optics Protocols for Satellite Ocean Color Sensor Validation, Revision 4, Volume III: Radiometric Measurements and Data Analysis Protocols*; NASA Technical Memorandum 2003-21621: Greenbelt, MD, USA, 2003; Chapter 2; pp. 7–20.

64. Waters, K.J.; Smith, R.C.; Lewis, M.R. Avoiding ship-induced light-field perturbation in the determination of oceanic optical properties. *Oceanography* **1990**, *3*, 18–21. [CrossRef]

65. Yarbrough, M.; Feinholz, M.; Flora, S.; Houlihan, T.; Johnson, B.C.; Kim, Y.S.; Murphy, M.Y.; Ondrusek, M.; Clark, D. *Results in Coastal Waters with High Resolution in Situ Spectral Radiometry: The Marine Optical System ROV*; SPIE: Bellingham, WA, USA, 2007; Volume 6680.

66. Zibordi, G.; Doyle, J.-P.; Hooker, S.B. Offshore tower shading effects on in-water optical measurements. *J. Atmos. Ocean. Technol.* **1999**, *16*, 1767–1779. [CrossRef]

67. Doyle, J.P.; Zibordi, G. Optical propagation within a three-dimensional shadowed atmosphere–ocean field: Application to large deployment structures. *Appl. Opt.* **2002**, *41*, 4283–4306. [CrossRef]

68. D'Alimonte, D.; Zibordi, G.; Berthon, J.-F. The JRC data processing system. In *Results of the Second SeaWiFS Data Analysis Round Robin, March 2000 (DARR-00)*; SeaWiFS Postlaunch Technical Report No. DARR-00; NASA: Greenbelt, MD, USA, 2001; Volume 15, pp. 52–56.

69. Maritorena, S.; Hooker, S.B. The GSFC data processing system. In *Results of the Second SeaWiFS Data Analysis Round Robin, March 2000 (DARR-00)*; SeaWiFS Postlaunch Technical Report No. DARR-00; NASA: Greenbelt, MD, USA, 2001; Volume 15, pp. 46–51.

70. Morrow, J.H.; Hooker, S.B.; Booth, C.R.; Bernhard, G.; Lind, R.N.; Brown, J.W. Advances in measuring the apparent optical properties (AOPs) of optically complex waters. *NASA Tech. Memo 215856* **2010**, 42–50.

71. Gordon, H.R. Ship perturbation of irradiance measurements at sea 1: Monte Carlo simulations. *Appl. Opt.* **1985**, *24*, 4172. [CrossRef]

72. Zaneveld, J.R.V.; Boss, E.; Barnard, A. Influence of surface waves on measured and modeled irradiance profiles. *Appl. Opt.* **2001**, *40*, 1442. [CrossRef]

73. Zibordi, G.; D'Alimonte, D.; Berthon, J.-F. An evaluation of depth resolution requirements for optical profiling in coastal waters. *J. Atmos. Ocean. Technol.* **2004**, *21*, 1059–1073. [CrossRef]

74. Organelli, E.; Claustre, H.; Bricaud, A.; Schmechtig, C.; Poteau, A.; Xing, X.; Prieur, L.; D'Ortenzio, F.; Dall'Olmo, G.; Vellucci, V. A novel near-real-time quality-control procedure for radiometric profiles measured by bio-argo floats: Protocols and performances. *J. Atmos. Ocean. Technol.* **2016**, *33*, 937–951. [CrossRef]

75. Mobley, C.D. Estimation of the remote-sensing reflectance from above-surface measurements. *Appl. Opt.* **1999**, *38*, 7442–7455. [CrossRef] [PubMed]

76. Lee, Z.; Ahn, Y.-H.; Mobley, C.; Arnone, R. Removal of surface-reflected light for the measurement of remote-sensing reflectance from an above-surface platform. *Opt. Express* **2010**, *18*, 26313. [CrossRef] [PubMed]

77. Zhang, X.; He, S.; Shabani, A.; Zhai, P.-W.; Du, K. Spectral sea surface reflectance of skylight. *Opt. Express* **2017**, *25*, A1. [CrossRef] [PubMed]

78. Santer, R.; Zagolski, F.; Barker, K.; Huot, J.-P. Correction of the above water radiometric measurements for the sky dome reflection, accounting for polarization. In Proceedings of the MERIS/(A) ATSR & OLCI/SLSTR Preparatory Workshop, ESA Special Publication 711, European Space Agency, Noordwijk, The Netherlands, 15–19 October 2012.

79. Mueller, J.L.; Davis, C.; Arnone, R.; Frouin, R.; Carder, K.; Lee, Z.P.; Steward, R.G.; Hooker, S.; Mobley, C.D.; McLean, S. Above-water radiance and remote sensing reflectance measurements and analysis protocols. In *Ocean Optics Protocols for Satellite Ocean Color Sensor Validation Revision 4, Volume III*; NASA: Greenbelt, MD, USA, 2003; Chapter 3; pp. 21–31.

80. Fougnie, B.; Frouin, R.; Lecomte, P.; Deschamps, P.-Y. Reduction of skylight reflection effects in the above-water measurement of diffuse marine reflectance. *Appl. Opt.* **1999**, *38*, 3844–3856. [CrossRef] [PubMed]

81. Deschamps, P.-Y.; Fougnie, B.; Frouin, R.; Lecomte, P.; Verwaerde, C. SIMBAD: A field radiometer for satellite ocean color validation. *Appl. Opt.* **2004**, *43*, 4055–4069. [CrossRef] [PubMed]

82. Lee, Z.; Pahlevan, N.; Ahn, Y.-H.; Greb, S.; O'Donnell, D. Robust approach to directly measuring water-leaving radiance in the field. *Appl. Opt.* **2013**, *52*, 1693. [CrossRef]

83. Hooker, S.B.; Lazin, G.; Zibordi, G.; McLean, S. An evaluation of above-and in-water methods for determining water-leaving radiances. *J. Atmos. Ocean. Technol.* **2002**, *19*, 486–515. [CrossRef]

84. Gilerson, A.; Carrizo, C.; Foster, R.; Harmel, T. Variability of the reflectance coefficient of skylight from the ocean surface and its implications to ocean color. *Opt. Express* **2018**, *26*, 9615–9633. [CrossRef]

85. Morel, A. In-water and remote measurements of ocean colour. *Bound. Layer Meteorol.* **1980**, *18*, 177–201. [CrossRef]

86. Kutser, T.; Vahtmäe, E.; Paavel, B.; Kauer, T. Removing glint effects from field radiometry data measured in optically complex coastal and inland waters. *Remote Sens. Environ.* **2013**, *133*, 85–89. [CrossRef]

87. Gould, R.W.; Arnone, R.A.; Sydor, M. Absorption, scattering and remote-sensing reflectance relationships in coastal waters: Testing a new inversion algorithm. *J. Coast. Res.* **2001**, *17*, 328–341.

88. Ruddick, K.; Cauwer, V.D.; Van Mol, B. *Use of the Near Infrared Similarity Spectrum for the Quality Control of Remote Sensing Data*; Frouin, R.J., Babin, M., Sathyendranath, S., Eds.; SPIE Proceedings: Bellingham, WA, USA, 2005; Volume 5885.

89. Cox, C.; Munk, W. Measurements of the roughness of the sea surface from photographs of the Sun's glitter. *J. Opt. Soc. Am.* **1954**, *44*, 834–850. [CrossRef]

90. Ocean Optics Web Book—Surface Reflectance Factors. Available online: http://www.oceanopticsbook.info/view/remote_sensing/level_3/surface_reflectance_factors (accessed on 24 July 2019).

91. Mobley, C.D. Polarized reflectance and transmittance properties of windblown sea surfaces. *Appl. Opt.* **2015**, *54*, 4828. [CrossRef] [PubMed]

92. Groetsch, P.M.M.; Gege, P.; Simis, S.G.H.; Eleveld, M.A.; Peters, S.W.M. Variability of adjacency effects in sky reflectance measurements. *Opt. Lett.* **2017**, *42*, 3359–3362. [CrossRef] [PubMed]

93. Harmel, T.; Gilerson, A.; Tonizzo, A.; Chowdhary, J.; Weidemann, A.; Arnone, R.; Ahmed, S. Polarization impacts on the water-leaving radiance retrieval from above-water radiometric measurements. *Appl. Opt.* **2012**, *51*, 8324. [CrossRef]

94. D'Alimonte, D.; Kajiyama, T. Effects of light polarization and waves slope statistics on the reflectance factor of the sea surface. *Opt. Express* **2016**, *24*, 7922. [CrossRef] [PubMed]

95. Foster, R.; Gilerson, A. Polarized transfer functions of the ocean surface for above-surface determination of the vector submarine light field. *Appl. Opt.* **2016**, *55*, 9476. [CrossRef]

96. Groetsch, P.M.M.; Gege, P.; Simis, S.G.H.; Eleveld, M.A.; Peters, S.W.M. Validation of a spectral correction procedure for sun and sky reflections in above-water reflectance measurements. *Opt. Express* **2017**, *25*, A742. [CrossRef]

97. Simis, S.G.H.; Olsson, J. Unattended processing of shipborne hyperspectral reflectance measurements. *Remote Sens. Environ.* **2013**, *135*, 202–212. [CrossRef]

98. Zibordi, G. Experimental evaluation of theoretical sea surface reflectance factors relevant to above-water radiometry. *Opt. Express* **2016**, *24*, A446–A459. [CrossRef]

99. Hooker, S.B.; Morel, A. Platform and environmental effects on above-water determinations of water-leaving radiances. *J. Atmos. Ocean. Technol.* **2003**, *20*, 187–205. [CrossRef]

100. Hlaing, S.; Harmel, T.; Ibrahim, A.; Ioannou, I.; Tonizzo, A.; Gilerson, A.; Ahmed, S. *Validation of Ocean Color Satellite Sensors Using Coastal Observational Platform in Long Island Sound*; SPIE Proceedings: Bellingham, WA, USA, 2010; Volume 7825, p. 782504-1-8.

101. Hooker, S.B. The telescopic mount for advanced solar technologies (T-MAST). In *Advances in Measuring the Apparent Optical Properties (AOPs) of Optically Complex Waters*; Technical Report No. 215856; NASA: Greenbelt, MD, USA, 2010.

102. Hooker, S.B.; Lazin, G. *The SeaBOARR-99 Field Campaign*; Technical Report No. 2000-206892; NASA: Greenbelt, MD, USA, 2000; Volume 8, p. 46.

103. Zibordi, G.; Hooker, S.B.; Berthon, J.F.; D'Alimonte, D. Autonomous above-water radiance measurements from an offshore platform: A field assessment experiment. *J. Atmos. Ocean. Technol* **2002**, *19*, 808–819. [CrossRef]

104. Carrizo, C.; Gilerson, A.; Foster, R.; Golovin, A.; El-Habashi, A. Characterization of radiance from the ocean surface by hyperspectral imaging. *Opt. Express* **2019**, *27*, 1750–1768. [CrossRef] [PubMed]

105. Hooker, S.B.; Bernhard, G.; Morrow, J.H.; Booth, C.R.; Comer, T.; Lind, R.N.; Quang, V. *Optical Sensors for Planetary Radiant Energy (OSPREy): Calibration and Validation of Current and Next-Generation NASA Missions*; Technical Report No. 2012-215872; NASA: Greenbelt, MD, USA, 2012.

106. Gilerson, A.; Carrizo, C.; Foster, R.; Harmel, T.; Golovin, A.; El-Habashi, A.; Herrera, E.; Wright, T. *Total and Polarized Radiance from the Ocean Surface from Hyperspectral Polarimetric Imaging.*; SPIE: Bellingham, WA, USA, 2019; Volume 11014.

107. Olszewski, J.; Sokolski, M. Elimination of the surface background in contactless sea investigations. *Oceanologia* **1990**, *29*, 213–221.

108. Tanaka, A.; Sasaki, H.; Ishizaka, J. Alternative measuring method for water-leaving radiance using a radiance sensor with a domed cover. *Opt. Express* **2006**, *14*, 3099. [CrossRef] [PubMed]

109. Shang, Z.; Lee, Z.; Dong, Q.; Wei, J. Self-shading associated with a skylight-blocked approach system for the measurement of water-leaving radiance and its correction. *Appl. Opt.* **2017**, *56*, 7033. [CrossRef] [PubMed]

110. Vansteenwegen, D.; Ruddick, K.; Cattrijsse, A.; Vanhellemont, Q.; Beck, M. The pan-and-tilt hyperspectral radiometer system (PANTHYR) for autonomous satellite validation measurements—Prototype design and testing. *Remote Sens.* **2019**, *11*, 1360. [CrossRef]

111. Garaba, S.P.; Schulz, J.; Wernand, M.R.; Zielinski, O. Sunglint detection for unmanned and automated platforms. *Sensors* **2012**, *12*, 12545–12561. [CrossRef]

Article

The Pan-and-Tilt Hyperspectral Radiometer System (PANTHYR) for Autonomous Satellite Validation Measurements—Prototype Design and Testing

Dieter Vansteenwegen [1,*], Kevin Ruddick [2], André Cattrijsse [1], Quinten Vanhellemont [2] and Matthew Beck [2]

[1]   Flanders Marine Institute (VLIZ), Wandelaarkaai 7, 8400 Ostend, Belgium; dre@vliz.be
[2]   Royal Belgian Institute of Natural Sciences (RBINS), Operational Directorate Natural Environment, 29 Rue Vautierstraat, 1000 Brussels, Belgium; kruddick@naturalsciences.be (K.R.); qvanhellemont@naturalsciences.be (Q.V.); mbeck@naturalsciences.be (M.B.)
*   Correspondence: dieter.vansteenwegen@vliz.be

Received: 17 May 2019; Accepted: 3 June 2019; Published: 6 June 2019

Abstract: This paper describes a system, named "pan-and-tilt hyperspectral radiometer system" (PANTHYR) that is designed for autonomous measurement of hyperspectral water reflectance. The system is suitable for deployment in diverse locations (including offshore platforms) for the validation of water reflectance derived from any satellite mission with visible and/or near-infrared spectral bands (400–900 nm). Key user requirements include reliable autonomous operation at remote sites without grid power or cabled internet and only limited maintenance (1–2 times per year), flexible zenith and azimuth pointing, modularity to adapt to future evolution of components and different sites (power, data transmission, and mounting possibilities), and moderate hardware acquisition cost. PANTHYR consists of two commercial off-the-shelf (COTS) hyperspectral radiometers, mounted on a COTS pan-and-tilt pointing system, controlled by a single-board-computer and associated custom-designed electronics which provide power, pointing instructions, and data archiving and transmission. The variable zenith pointing improves protection of sensors which are parked downward when not measuring, and it allows for use of a single radiance sensor for both sky and water viewing. The latter gives cost reduction for radiometer purchase, as well as reduction of uncertainties associated with radiometer spectral and radiometric differences for comparable two-radiance-sensor systems. The system is designed so that hardware and software upgrades or changes are easy to implement. In this paper, the system design requirements and choices are described, including details of the electronics, hardware, and software. A prototype test on the Acqua Alta Oceanographic Tower (near Venice, Italy) is described, including comparison of the PANTHYR system data with two other established systems: the multispectral autonomous AERONET-OC data and a manually deployed three-sensor hyperspectral system. The test established that high-quality hyperspectral data for water reflectance can be acquired autonomously with this system. Lessons learned from the prototype testing are described, and the future perspectives for the hardware and software development are outlined.

Keywords: Hyperspectral reflectance; validation; autonomous measurements; ground-truth data; system design

1. Introduction

The objective of this paper is to describe the motivation, design, and prototype testing for an autonomous system of hyperspectral radiometers suitable for validation of satellite-derived water reflectance.

1.1. Motivation and Objective

Satellite imagery of marine, coastal, and inland water reflectance is now routinely used for measuring parameters such as the concentrations of chlorophyll a, a proxy for phytoplankton biomass, and suspended particulate matter, important for sediment transport applications. The satellite data are used for regulatory monitoring of the aquatic environment, e.g., via the European Union Water Framework and Marine Strategy Framework Directives [1,2], and for providing a scientific basis for coastal zone decision-making, e.g., via the assessment of impacts of human activities and constructions [3]. However, end-users of the data require reliable information on data quality, and validation of the satellite data at the level of water reflectance is particularly crucial. This is because of the large errors that may occur during the data calibration and processing, particularly during the atmospheric correction steps [4].

This validation is best achieved by a "match-up" of in situ measurements of water (surface) reflectance made at the same time as the satellite measurement [5], and experience over the last 10 years showed that only autonomous in situ systems can provide sufficient data for this purpose. In particular, AERONET-OC [6], a federated network of multispectral robotically pointed radiometers on offshore platforms all over the world, proved to be the main source of validation data [7] for spaceborne optical missions such as ENVISAT/MERIS, MODIS/AQUA, VIIRS, Sentinel-3/OLCI, Sentinel-2/MSI, Landsat-8/OLI, etc. However, the radiometer adopted within AERONET-OC is only multispectral and cannot adequately cover the spectral bands of all recent and future optical spaceborne missions without spectral interpolation/extrapolation/modelling [8] and associated uncertainties. The WATERHYPERNET network is, therefore, being developed, based closely on the concept of the successful AERONET-OC network [6] but with the essential advantage of a hyperspectral radiometer, thus enabling the validation of all visible and near-infrared bands of all present and future satellite missions providing water reflectance data.

The objective of the present paper is to describe the measurement system, called PANTHYR (pan-and-tilt hyperspectral radiometer) that was developed for use within the WATERHYPERNET network. This measurement system consists of two commercial off-the-shelf (COTS) hyperspectral radiometers, mounted on a COTS pan-and-tilt (PT) pointing system, controlled by a single-board-computer and associated custom-designed electronics which provide power, pointing instructions, and data archiving and transmission.

1.2. Measurement System Requirements

The PANTHYR system was developed to fit the following user requirements:

- Measurement of downwelling irradiance, as well as downward (sky) and upward (water) radiance just above the water surface, at flexible zenith and azimuth (relative to sun) angles for a spectral range covering at least 400–900 nm with full-width half-maximum (FWHM) spectral resolution of 10 nm or better and spectral sampling every 5 nm or better.
- Storage of all measurements and diagnostic logs and regular transmission to a land-based server.
- User interface with flexibility for scientists to easily program pointing and data acquisition scenarios.
- Reliable autonomous operation at remote sites, e.g., offshore platforms, with a typical maintenance frequency of once or twice per year without grid power.
- Resistance to harsh offshore environments, including large temperature ranges (measurement limited to between 2 °V and 40 °C, and survival between −20 °C and 60 °C ambient temperature), rain, salty sea spray, atmospheric deposition, and possible animals (birds, spiders, etc.).
- Modularity to adapt to sites with different possibilities for power (grid/solar/wind), data transmission channels (cabled internet, 3G/4G cellular networks), and mechanical mounting possibilities (rails, masts, etc.), and to cope with future evolution of system components.
- Moderate hardware purchase costs, e.g., typically <10,000 € commercial price excluding taxes for a full system excluding radiometers.

- Pointing accuracy of at least 5° azimuth and 1° zenith.

1.3. Precursor Autonomous Systems

A few autonomous systems with pointable radiometers already exist and are briefly described here.

The most used autonomous system for measurement of water reflectance is the Seaprism version of the CIMEL CE318 multiband photometer system. This system was originally commercialized in the 1990s for the measurement of aerosol properties and includes direct sun, near sun, and principal plane and almucantar sky radiance measurements, and it became the unique instrument of the AERONET network [9]. The Seaprism version was developed in the early 2000s [10] by reprogramming of the pointing system, especially to perform downward pointing measurements of upwelling radiance from water. The system evolved over the years to include improvements of the optical and electronic components, as well as development of new versions, e.g., including the possibility of nighttime lunar measurements [11]. The system consists essentially of an optical head with two fore-optics protected by collimators, containing a filter wheel for multiband optical measurements with a very wide dynamic range (from direct sun to dark water), a robotic pointing system, a control box providing pointing and measurement instructions and managing power and data, and associated auxiliary equipment for power generation (solar panels) and data transmission (METEOSAT satellite network uplink or cellular link). The system is extremely robust, giving reliable maintenance-free performance for long-term (~1 year) deployments in a very wide range of environments including land and water sites from the tropics to the polar regions.

The OSPREY system [12] was designed to be commercialized by Biospherical Instruments as a very-high-performance modular system of radiometers on pointing systems. The radiometers include a hyperspectral spectrometer and numerous single-band microradiometers within a single thermally controlled casing and a filter wheel in front of the spectrometer allowing polarimetric, direct sun, straylight-corrected, and dark measurements. The system, described in detail in Reference [12], has a very high dynamic range and high accuracy pointing, and is capable of water, sun, sky, and moon radiance measurements.

The RFLEX system [13] consists of three hyperspectral TRIOS/RAMSES radiometers measuring downwelling irradiance, downwelling sky radiance, and upwelling water radiance mounted at fixed zenith angles (0°, 40°, and 140°, respectively) on an azimuthally rotating platform with associated control software. This system was designed principally for deployment on moving ships, where the azimuthally rotating platform allows achieving an optimal 90° or 135° relative azimuth to sun for any ship heading. Instructions for system hardware construction using low-cost components (except for the COTS radiometers) were made publicly available via SourceForge (https://sourceforge.net/projects/rflex/) and source code for communication with TRIOS/RAMSES was made available via GitHub (https://github.com/StefanSimis/PyTRIOS). The system was used operationally on two "ferryboxes" mounted on ships of opportunity in the Baltic Sea.

The DALEC system [14] is commercially available from Insitu Marine Optics and consists of three hyperspectral spectrometers measuring downwelling irradiance, and downwelling (sky) and upwelling (water) radiance embedded at fixed zenith angles (180°, 140°, and 40°, respectively) within a compact azimuthally rotating body on a gimballed mount. Control and data are managed by a standard personal computer (PC) using software supplied by the manufacturer. This system was designed principally for deployment on moving ships but can also be used from fixed platforms. The system is deployed on research vessels and an example of data usage is described by Reference [15].

The SAS Solar Tracker system from Seabird Electronics consists of three hyperspectral Seabird/HyperSAS radiometers measuring downwelling irradiance, downwelling sky radiance, and upwelling water radiance mounted at fixed zenith angles (180°, 140°, and 40°, respectively) on an azimuthally rotating platform, with compass-based sun-tracking to ensure optimal relative azimuth angle with respect to sun. Power and data are managed via a deck interface unit and a standard PC. The system is used routinely on a few research vessels and ships of opportunity [16].

The WISP-3 is a handheld system, commercially available from Water Insight, with three embedded radiometers measuring water reflectance from downwelling irradiance, sky radiance, and water radiance. Data for monitoring parameters such as chlorophyll a concentration are logged internally and can be downloaded to a PC as processed data. A variant called WISPstation is under development for autonomous operation. An example of data usage and inter-comparison with other systems is described in Reference [17].

The current PANTHYR design will provide hyperspectral reflectance data not available from the Seaprism system and has advantages over the RFLEX, DALEC, Suntracker, and WISP designs because the zenith pointing flexibility allows sky and water radiance measurements to be made by the same radiometer and allows downward pointing to protect instruments when not measuring. Use of a single radiance sensor provides cost saving and reduces the uncertainties associated with spectral and radiometric calibration differences of the two-radiance-sensor systems. The disadvantage of the PANTHYR sequential measurement of water and sky radiance as compared to the simultaneous measurements possible with the two-radiance-sensor systems is considered to be minimal for the good clear sun and sky illumination conditions needed for satellite validation. If illumination conditions are suboptimal, this will be detected by the PANTHYR replicate measurements and will lead to data rejection. The PANTHYR system does not pretend to achieve the very high performance expected of the OSPREY system, but represents a much lower-cost alternative which should be ideal for deployment at multiple sites worldwide by organizations with moderate budgets.

All of these systems, including PANTHYR, are developing within a context where hyperspectral spectrometers are rapidly evolving with reduction of size, power, and cost (thanks to mass production for medical and industrial applications), and where pointing systems are becoming more affordable and more easily available as COTS items (thanks to mass production for applications such as video surveillance). Other technologies facilitating PANTHYR development include the massive improvements and cost reductions in data transmission (thanks to mobile phone networks) and in microcomputers.

1.4. Overview of Paper

In this paper, the design choices are described in Section 2, together with details of the control electronics, where most of the original developments were made, and the data acquisition protocol and data processing steps, which are strongly based on precursor systems. A first seaborne test of the system on Acqua Alta Oceanographic Tower (AAOT) is described in Section 3, where data acquisition and processing are also described for two precursor systems used for comparison, a manually supervised three-sensor hyperspectral system and the autonomous multispectral AERONET-OC system. Results from the prototype tests are described in Section 4. General system performance is evaluated with description of lessons learned for system improvements. A comparison is made between data acquired with PANTHYR and data from the two precursor systems. Finally, conclusions from these prototype tests and future perspectives for system refinement are outlined in Section 5.

2. System Design

2.1. Top-Level Design Choices

To stay within the design and future purchasing budget, we decided to use mostly COTS components. Self-design of all hardware would allow for greater optimization, but design and manufacturing of all components would add cost and complexity beyond the scope of our goals. Only when commercial hardware was not available or sufficient did we create our own. As availability of the chosen components will change in the next few years, special care was taken to implement a modular design in hardware and software. Individual component changes will require updated software/hardware for only that part, while most of the system remains the same. This also facilitates adding extra hardware and capabilities in the future.

The TRIOS/RAMSES COTS hyperspectral radiometer was chosen because of its relatively high performance, highly robust and mature low-power design, and moderate price. This radiometer provides measurements for a spectral range of 320–950 nm and was used for satellite validation since 2002 [18] by many teams worldwide. The performance of this class of radiometer was extensively studied and characterized with regard to straylight [19], thermal sensitivity [20], polarization sensitivity [21], angular response of the irradiance units [22], and non-linearity. A low-cost, portable, LED-based FieldCAL device is available for rapid checking of radiometric sensitivity in the field.

The FLIR PTU-D48E COTS pointing system was chosen for the PANTHYR system because of its high performance (including good pan and tilt speeds and pointing accuracy), robust design, and moderate price.

While COTS rugged PCs are widely available, a key early decision was to use a small low-power embedded computer board with self-design of additional electronics for managing power and connectivity to components such as the radiometers, GNSS (Global Navigation Satellite System) receiver, and auxiliary sensors. This required more development time, but provided more control over power management and greater flexibility in connecting and supporting additional external devices.

The Linux operating system was chosen to facilitate software portability to next-generation hardware platforms and to improve reliability and security. Linux is open source, easy to customize, and ideal for an offshore system where logistics make on-site intervention difficult and/or expensive.

Python was chosen as the main programming language as it is available under an open-source license, is well known in the scientific community, and produces clear, readable code.

2.2. System Overview and Key Components

The overall PANTHYR system is outlined in Figure 1, showing the main hardware elements and the associated power and data connections. A small embedded computer board controls all components and forms the heart of the system. After start-up, a GNSS receiver provides UTC (Coordinated Universal Time), as well as location, allowing the system to calculate the position of the sun. An optional IP network camera can take still pictures from the measurement target areas, allowing users to check any suspect data for unusual conditions such as floating vegetation or debris, boats, birds, and other obstructions in the field of view. During a measurement cycle, the controller calculates the head position for each measurement step, points the instrument in the required zenith and azimuth angles, and makes a measurement. At the end of a measurement cycle, the head moves to a "park" position where the instruments are pointed downward to prevent fouling. The system then goes into a sleep state, conserving power while waiting for the next cycle.

A myriad of single-board computers with ever increasing performance became available on the market in the last few years. These small (around credit-card size) and cheap (<100 €) boards manage to run an operating system and have on-board storage, as well as a network connection, while consuming as little as 0.5 W of power in standby. For our application, the BeagleBone Black has the following advantages:

- It has five serial ports, which allows it to interface directly with the instruments, without the need for additional external interfaces.
- It was proven to be a reliable option in industrial applications [23] and scientific research [24].
- It is readily available from electronic parts distributors.

To connect and interface the controller to the rest of the system, we designed two electronic boards which are plugged on top of the BeagleBone. The first board translates the voltage levels of the serial ports to RS-232-compatible levels so that the controller can communicate with the instruments and GNSS receiver. The second board provides protection and filtering on the power lines, as well as allowing power saving by cutting the power to external devices when they are not in use. Five protected power outputs with a maximum power draw of 2 A each can be controlled from software. An external solid-state relay is used to control power to the head, which is the only component that needs a switched 24 V direct current supply.

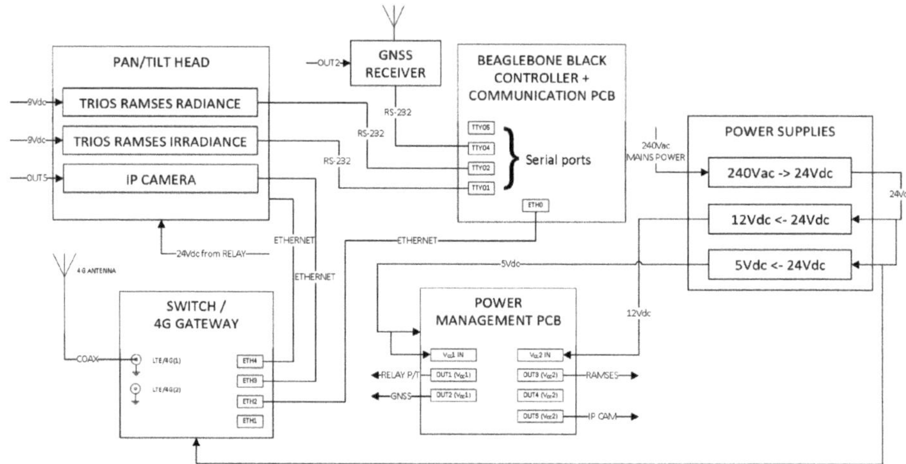

Figure 1. Pan-and-tilt hyperspectral radiometer system (PANTHYR) system diagram.

A number of commercially available PT heads were evaluated. We looked at price, resolution, power consumption, and ruggedness. The FLIR PTU-D48 E matched our requirements most closely. This unit was developed as a rugged solution for demanding applications and has the necessary specifications and certifications such as temperature range, ingress protection 67, and salt spray protection (MIL-810G). Its maximum payload of 7.5 kg leaves margin for extra hardware.

To allow the full pan (+/−174°) and tilt (+90°/−30°) movement, a free loop of cable is required between the rotating instruments and the base of the head. We found no good solution to prevent this loop of cables from getting in front of the instruments at certain positions. The next version of PANTHYR will use the same head but with built-in slip rings, providing internal connections in the head assembly to replace outside cables. This option also adds 360° azimuth movement capability to the head. The limited tilt range of the PTU-D48 E means that the irradiance sensor which needs to be vertical during measurements cannot be parked lower than 30° below the horizon; it is thought that this will provide adequate protection from the elements, but longer-term testing is required to confirm this. The limited tilt range (120°) of the PTU-D48 E also precludes an arrangement with parallel radiance and irradiance sensors since the radiance sensor would need to be tilted through 140° to go from vertical (zenith 180°) to the necessary water-viewing angle (zenith 40°). The radiance sensor was, therefore, fixed at an angle of 40° to the irradiance sensor, giving a zenith angle range for the irradiance sensor of 180° (downwelling irradiance measurement) to 60° (parked) and a zenith angle range for the radiance sensor of 140° (sky radiance measurement) through 40° (water radiance measurement) to 20° (parked).

On average, the PT unit consumes 12.3 W during start-up, 6.2 W during hold, and up to 20 W during combined axial movements. This demonstrates the necessity of keeping the system in a low-power sleep mode as much as possible.

The InSYS MRO-L200 gateway/switch connects all Ethernet devices, as well as providing the gateway to a 4G cellular network. This industrial device can set up and manage a VPN connection to the onshore server, and has a serial port and digital output, which could be used to remotely power cycle PANTHYR in case of problems. Its high power consumption (2.5 W typical) is the main drawback. A low-power network switch and a separate 4G gateway that is switched on only during short intervals would mitigate this, but also remove our option to connect to the installation in case of problems.

Two mechanical structures were designed to mount to existing structures at the measurement sites and install the instruments. A versatile mounting system accommodates different measurement sites, while remaining simple. We started with a 50-mm steel pipe as a range of clamps are commercially available for this standard diameter. Welded on top of this pipe is a triangular plate. The head is mounted on a second similar plate. Three threaded rods connect both plates atop each other. Leveling of the head is achieved by changing the spacing along the three rods. The result is a simple system with a variety of mounting options (Figure 2a).

On top of the head, both sides of a U-shaped bracket accept adapter plates that hold the radiometers. Bolting holes for fixed inclinations of 20° and 60°, as well as a slot that allows a variable 10° to 70° inclination, are available (Figure 2b).

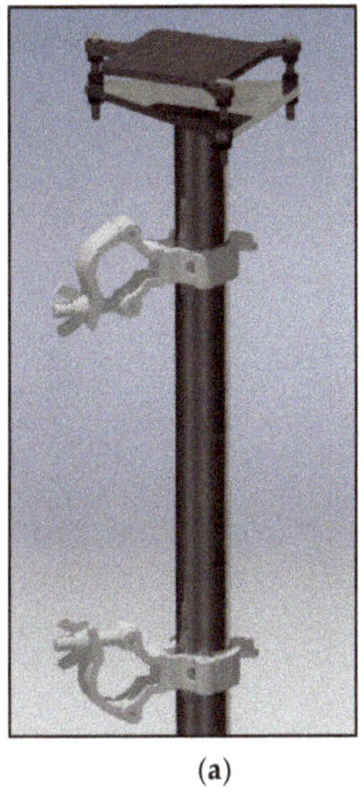

(a)

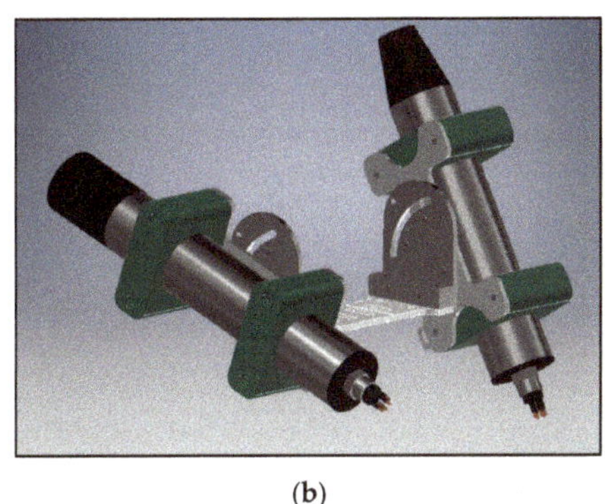

(b)

Figure 2. Engineering drawings of mounting structures. (**a**) Support providing a flexible connection between a platform structure, e.g., horizontal railings or a vertical post fixed through the swivel couplings, and pan/tilt unit to be mounted on uppermost horizontal surface; (**b**) sensor mounting bracket for attaching radiometers to top of pan/tilt unit.

2.3. Software and Usage

The main Python script checks system and user settings and controls all actions. Data and settings are stored in an SQLite database. Software will be released under the GNU GPLv3 (GNU's Not Unix General Public License version 3) at the GitHub page: www.github.com/hypermaq/panthyr.

PANTHYR has one SQLite database which contains the measurement protocol, system settings, logs, and measurement results, as well as a task queue. The user can use a (remotely) connected laptop to access a webpage where the contents of the database can be viewed, changed, and exported (Figure 3). This serves as the main user interface for system configuration.

A "worker script" regularly checks the "queue" database table for tasks that need to be performed. These tasks range from executing a measurement cycle to setting up station parameters.

To perform a measurement cycle, the system gets the necessary settings from the "settings" table and read the first line in the "protocol" table. Each line in this table contains the parameters (instrument, zenith angle, azimuth offset, number of repetitions) describing one sub-cycle in the measurement cycle. Measurement results are written to the "measurements" table. When all scans in the protocol are finished, the system resumes a standby state where it regularly checks for new tasks. In the event of a failure, a log entry is created in the "logs" table. After three failed attempts, a task is ignored to prevent it from blocking the system.

Apart from the database frontend, PANTHYR also hosts an FTP (File Transfer Protocol) server to download log files and pictures, as well as an SSH (Secure SHell) server to allow low-level system maintenance. To prevent provider restrictions enforced on cellular networks, PANTHYR initiates an outgoing OpenVPN connection to an onshore server. Users can connect to the same server and communicate with PANTHYR as if it were next to them.

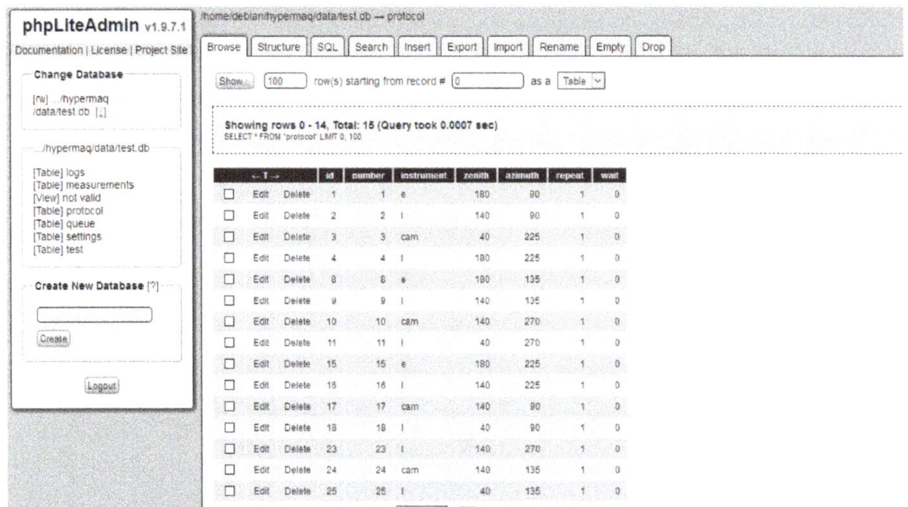

Figure 3. Screenshot of SQLite database interface.

2.4. PANTHYR Data Acquisition Protocol

The measurement protocol presented here is based on various precursors including References [6,25,26] and, as usual for above-water radiometry, gives a water-leaving radiance or reflectance for an off-nadir geometry, which can be matched to satellite viewing geometries by appropriate models of a bidirectional reflectance distribution function. Research to determine an optimal measurement protocol for PANTHYR operations within the WATERHYPERNET network is in progress and some aspects may change in the future, e.g., potential additional zenith and/or azimuth angles and/or a different number of replicate measurements, especially for water viewing. However, the current protocol is considered as already sufficient for demonstrating performance of the hardware.

The PANTHYR system performs automated measurements every 20 min from sunrise until sunset. Each cycle consists of measurements with a 90°, 135°, 225°, and/or 270° relative azimuth to sun. In general, and depending on the installation location, platform geometry, time of day (sun location), and associated platform shading of the water target, only one or two (or sometimes zero) of these azimuth angles are appropriate for measurement of water reflectance; other azimuth angles will be contaminated by platform shading or even direct obstruction of the water target as defined from the instrument field of view. A selection of acceptable azimuth angles is made a priori based on expert judgement (used here for prototype testing; see Section 3) or, better, a study of platform shading effects by modeling or experimentation [27]. The measurement of unacceptable azimuth angles, defined by a "no-go zone", can then be avoided to save power.

For each measurement cycle, the system performs a sub-cycle for each of the configured relative azimuth angles as described in Table 1. Based on the AERONET-OC protocol [6], but with repetition of the E_d and L_d replicates, each azimuthal measurement sub-cycle consists of 2 × 3 replicate scans each of downwelling irradiance, E_d, and sky radiance, L_d, and 11 replicate scans of upwelling radiance, L_u, where "(spectral) scan" refers to acquisition of a single instantaneous spectrum. Firstly, the irradiance sensor is pointed upward, with the radiance sensor offset by 40°, and three replicates of E_d followed by three replicates of L_d are measured. The radiance sensor is then moved to a 40° downward viewing angle to make 11 replicate L_u scans. The irradiance and radiance sensors are then repositioned to make three more replicate scans each of L_d and E_d. Scans are stored as digital counts (DCs), including the embedded dark pixel counts, with instrument serial number and integration time in the metadata.

Table 1. Pan-and-tilt hyperspectral radiometer system (PANTHYR) basic data acquisition cycle. Azimuth is measured clockwise with respect to sun with a relative azimuth of 0°, meaning that the radiance sensor is pointing toward the sun, and a relative azimuth of 90°, meaning that the radiance sensor is pointing to the right of sun. A zenith angle of 180° means that the instrument is pointing vertically upward and, hence, measuring downwelling (ir)radiance. Measurements within a predefined azimuth no-go zone will be skipped.

Sub-Cycle Number	Instrument	Measurement	Zenith Angle (°)	Azimuth Relative to Sun (°)	Replicate Scans
1	Irradiance	E_d	180	90	3
2	Radiance	L_d	140	90	3
3	Camera	Sky photo	140	90	-
4	Radiance	L_u	40	90	11
5	Camera	Water photo	40	90	-
6	Radiance	L_d	140	90	3
7	Irradiance	E_d	180	90	3
8–14	As 1–7	As 1–7	As 1–7	135	As 1–7
15–21	As 1–7	As 1–7	As 1–7	225	As 1–7
22–28	As 1–7	As 1–7	As 1–7	270	As 1–7

2.5. PANTHYR Data Processing

Each of the scans within a sub-cycle are converted from DCs to calibrated (ir)radiances, as described in Equations (1–3) of Reference [20], using the calibration files appropriate for the instrument serial number and date as follows:

- The 16-bit DCs are normalized by dividing by 65,535.
- A long-term dark current correction is performed taking into account the instrument factory characterization and the scan integration time.
- A residual dark signal is subtracted using the mean average from the sensor dark pixels.
- The signal is normalized by the integration time and divided by the calibration sensitivity to retrieve final calibrated (ir)radiances.

Incomplete spectra are removed, as well as L_d and L_u scans with >25% difference between neighboring scans at 550 nm. E_d scans are removed using the same criterion after normalizing E_d by $Cos(\theta_0)$, where θ_0 is the sun zenith angle. The sub-cycle is further processed if sufficient scans pass the quality control: L_u 9/11, and E_d and L_d 5/6 each. The remaining E_d and L_d measurements are then grouped and mean averaged to $\overline{E_d}$ and $\overline{L_d}$. For each L_u scan, the water-leaving radiance, L_w, is computed by removing skyglint radiance as follows:

$$L_w = L_u - \rho_F \overline{L_d}, \tag{1}$$

where ρ_F is the "wind-roughened Fresnel" coefficient that represents the fraction of incident skylight that is reflected back toward the water-viewing sensor at the air–water interface and is given by the look-up table (LUT) of Reference [25]. This LUT describes ρ_F as function of viewing and sun geometry and wind speed. Wind speed is retrieved from ancillary data files or set to a user-defined default value if wind speed data are unavailable. The data in the LUT are linearly interpolated to the current observation geometry and wind speed.

The L_w scans are then converted into ("uncorrected") water-leaving radiance reflectance scans, ρ_{wu}, as follows:

$$\rho_{wu} = \pi \frac{L_w}{\overline{E_d}}. \tag{2}$$

A "near infrared (NIR) similarity spectrum" correction is then applied to remove any white error from inadequate skyglint correction as shown in Equation (3). A spectrally flat measurement error, ε, is estimated using two wavelengths in the NIR [28], where $\lambda_1 = 780$ nm and $\lambda_2 = 870$ nm.

$$\varepsilon = \frac{\alpha \rho_{wu}(\lambda_2) - \rho_{wu}(\lambda_1)}{\lambda_2 - \lambda_1},$$ (3)

where α is the similarity spectrum [26] ratio for the bands used; here, $\alpha(780, 870) = 1/0.523 = 1.912$. Per scan, ε is subtracted from the ρ_w at all wavelengths to give an NIR-corrected water reflectance,

$$\rho_w(\lambda) = \rho_{wu}(\lambda) - \varepsilon,$$ (4)

and all ρ_w scans are mean-averaged to give the final NIR-corrected water reflectance, $\overline{\rho_w}$. This NIR correction is optional and may be turned off, e.g., for extremely turbid waters where it is not valid and/or for situations where the satellite measurement may include a similar constraint in the atmospheric correction which needs to be validated independently.

The final quality control to retain or reject the ρ_w is performed according to Reference [26]; measurements are rejected when $\overline{L_d}/\overline{E_d} > 0.05$ sr^{-1} at 750 nm (indicating clouds either in front of the sun or in the sky-viewing direction), and when the coefficient of variation (CV, standard deviation divided by the mean) of the ρ_w scans is >10% at 780 nm.

3. Prototype Testing

3.1. July 2018 Deployment at the Acqua Alta Oceanographic Tower

After extensive testing of individual components and the full system in the laboratory and outdoors on land, a first prototype system test was performed on the Acqua Alta Oceanographic Tower (AAOT) in July 2018 during the international Field Inter-Comparison Exercise (FICE) organized by the FRM4SOC (Fiducial Reference Measurements for Satellite Ocean Colour) project. AAOT is an oceanographic platform in the Northern Adriatic Sea 15 km offshore from Venice in water of 16 m depth. It was used extensively over the last 20 years for oceanographic data collection including optical oceanography [29] and was used for a number of multi-partner optical radiometry inter-comparison exercises [30,31]. It is the location of the first AERONET-OC site and provides such data since 2002. The above-water platform structure was entirely reconstructed in 2018, but the new structure closely follows the design and layout of the original platform.

The PANTHYR system was deployed on the east corner of the top deck of the platform (Figure 4), with the irradiance sensor collector 2 m above top deck floor and, hence, about 14 m above sea level. The system ran for six days continuously from 12 to 17 July 2018, under the supervision of the developer for the first three days and with non-specialized supervision for the remaining days. Some test conditions were not the most challenging that can be expected in the future; grid power was available, and data were not transmitted over a 4G link but stored onboard. However, the basic functions of autonomous pointing and data acquisition were fully and successfully tested.

This deployment for prototype testing was also installed only at moderate height and not above the height of the top deck masts for practical reasons; for future operational deployments, the PANTHYR system should ideally be positioned above the height of all such masts and structures or with a careful analysis of any shading of downwelling irradiance. For subsequent data comparison, only PANTHYR measurements at relative azimuth angles of 225° or 270° were used.

Using the angular field-of-view data (7° full-width half-maximum) for the radiance instrument supplied by the manufacturer suggests that the water-viewing radiometer sees an elliptic patch of water with about 3 m diameter. Typical integration time for the water-viewing measurement is ≤512 ms during this experiment. Spatio-temporal variability caused by the surface wave field is, thus, mainly averaged and not resolved.

The FICE campaign further involved the deployment of multiple supervised radiometer systems, including seven above-water radiometer systems, one shipborne free-fall underwater radiometer system, and one platform-deployed vertical profiling system. Inter-comparison of all these systems will be the focus of a separate paper. In the current paper, results of the PANTHYR system are compared only to one supervised above-water radiometer system, the manually supervised RBINS (Royal Belgian Institute of Natural Sciences) hyperspectral above-water three-radiometer TRIOS/RAMSES system, hereafter called the "M3TRIOS" system, and the autonomous multispectral AERONET-OC system operated by the Joint Research Centre (JRC) of the European Commission at Ispra. The five TRIOS/RAMSES sensors used here (two for PANTHYR, three for M3TRIOS) were radiometrically calibrated at the Tartu Observatory laboratory just before the FICE campaign, thus minimizing calibration-related differences.

Only measurements from the five days with best sky conditions (13–17 July) are used here. For those days and during the period of day used for comparison in the present study, sun zenith angles ranged from 23.4° to 46.7° and wind speed ranged from 0.1 to 4.3 m/s.

(a)

(b)

(c)

(d)

Figure 4. Photographs of PANTHYR prototype deployment at Acqua Alta Oceanographic Tower (AAOT). (**a**) Irradiance and sky radiance measurement position; (**b**) southwest face of the Acqua Alta Oceanographic Tower; (**c**) upwelling (water + skyglint) radiance measurement position; (**d**) park position.

3.2. Manually Supervised M3TRIOS System Used for Data Comparison

The M3TRIOS system deployed during the FRM4SOC/FICE consists of three TRIOS RAMSES radiometers, one measuring downwelling irradiance, E_d, and two measuring radiance with fixed sky- and water-viewing zenith angles of 140° and 40° for L_d and L_u, respectively. The system was deployed since 2001 at many locations from both large and small ships, and platforms including AAOT [31]. Details of the instrumentation and standard measurement protocol, based on Reference [25], and data processing and quality control can be found in Reference [26]. This is quite similar to the PANTHYR processing described in Section 2.5, except for the following elements:

- The conversion from DCs to calibrated (ir)radiance is performed by the TriOS MSDA_XE software rather than the equivalent Python routines written for PANTHYR.
- Measurements are made simultaneously for E_d, L_d, and L_u with a much larger number of replicate scans, at least 30, with a scan every 10 s for 10 min. The first five scans passing the quality control tests described in Web Appendix 1 of Reference [26] are retained for E_d, L_d, and L_u
- The skyglint correction given as a quadratic function of wind speed by Reference [26] is used as an approximation of the more accurate LUT of Reference [25] described in Section 2.5 for PANTHYR.
- The skyglint correction, Equation (1), and conversion to ρ_{wu}, Equation (2), and subsequent NIR correction, Equations (3) and (4), are applied to each E_d, L_d, and L_u triplet scan individually to give five scans for ρ_w before mean-averaging to yield $\overline{\rho_w}$.

Although the instrument has wider spectral range, data are here limited to the range 400–900 nm where quality was checked by previous inter-comparison exercises. For this specific implementation, the downwelling irradiance sensor was mounted on the top deck of AAOT on a telescopic mast at height 5 m above deck. The sky- and water-viewing radiance sensors were mounted at the southwest face of the platform on a frame tailor-made by Tartu Observatory and Plymouth Marine Laboratory to accommodate many radiometers and to ensure 40° and 140° zenith-angle viewing. On the other hand, 90°, 135°, or 270° relative azimuth angle to sun was achieved by manual rotation of the structure before each measurement. Measurements were made typically every 30 min during daytime between 8:00 a.m. and 1:00 p.m. UTC.

3.3. AERONET-OC Data Used for Comparison

The CIMEL/Seaprism system providing data for AAOT through the AERONET-OC network was installed on the west corner of AAOT on a purpose-built jetty to minimize any shading effects of the platform on the water being measured. Full details on the instrumentation and measurement protocol and data processing and quality control can be found in Reference [6].

Level 1.5 normalized water-leaving radiance data, L_{wn}, were used without f/Q correction [32] as it is not applied to the PANTHYR and M3TRIOS data. This data were downloaded from the AERONET-OC (AOC) website for the "Venise" site located on AAOT [6]. Level 1.5 data are automatically cloud cleared but do not have final post-deployment calibration applied and are, hence, not fully quality assured. The matchup data used in this paper do, however, pass quality control and are likely candidates for level 2.0 data if post-deployment calibration is acceptable [Zibordi, pers. comm. 18 March 2019]. L_{wn} values were converted to water reflectance, ρ_w, by

$$\rho_w = \pi \frac{L_{wn}}{F_0}, \tag{5}$$

where F_0 is the extraterrestrial irradiance [33] resampled to gaussian band-passes with 10-nm full-width half-maximum centered on the reported exact center wavelengths of the CIMEL instrument. Each of the level 1.5 AOC measurements was matched to the closest PANTHYR measurement within 20 min, and the PANTHYR data were resampled to the same band.

4. Results

4.1. System Performance

The system tested at AAOT performed very well and demonstrated its capability of making automated measurements following a predefined protocol and interval. After installation and configuration, there were no blocking issues while the system worked autonomously, and no intervention was required during the six days of operation while the system performed as expected.

Since the system was still in a prototype phase when tested on the AAOT, some functionality is yet to be tested or demonstrated in a real-world scenario. The system was powered from the mains grid and, as the platform was visited daily, there was no need for 3G/4G data transmission. Most measurement sites will require one or both of these functions as mains power or cabled internet may not be available.

With the defined measurement protocol and a measurement interval of 20 min, it became obvious that the system speed needed improvements. The measurement protocol described in Section 2.4 took about 12 min to finish, depending on sun azimuth (since this affects required movements and how many measurements are skipped because they fall within the "no-go" zone). Post-FICE optimization of the control libraries for the pan/tilt head since resulted in faster movements. Additional speed can be gained in the control of the instruments, as well as by implementing multi-threading. Before each measurement cycle, some of the devices need time to start up and/or calibrate. The pan/tilt head, for example, needs to run a calibration routine to find the center point of both axes, and the IP camera needs time to boot and warm up. Implementation of multithreading would allow us to do both at the same time. Shorter measurement cycles result in more low-power standby time in between measurement cycles and lower power consumption, thus achieving better overall efficiency.

Some existing systems such as the CIMEL SeaPRISM hardware rely on the sun to calibrate the system azimuth during installation. While a high accuracy can be achieved with this method, it is only usable on sunny days. An electronic compass that achieves the required accuracy on metal structures, and that has a low power consumption, a marine-grade enclosure, and low cost is yet to be found. The current mode of operation for aligning the system zenith is achieved by placing a spirit level on the top bracket. This allows alignment as accurate as the used spirit level. After this is done, a digital inclinometer is placed on top of the instruments to verify their absolute inclination for different angles. Azimuth calibration is similar to the CIMEL system; the user can order the head to rotate toward the sun, after which the shadows cast by the top bracket serve as a reference while manually performing the alignment.

The images taken as part of each measurement cycle showed an unexpected interest from the local wildlife. Even though the system was (almost) continuously moving, birds seemed not to be deterred but showed a rather close interest in the new technology, both contaminating data by blocking the instrument field of view completely or subtly (Figure 5) and by potentially contaminating the instrument itself by fecal deposits. Major blocking of a radiometer field of view will be easily identified as bad data because the spectral signature will be different from the expected water, sky, and full sky irradiance targets. However, partial blocking of the instrument field of view (Figure 5b) may contaminate data in a way that cannot be automatically detected and rejected. Discussion with experts on bird life suggests that typical visible or audible "scarecrow" devices are not effective deterrents and the best approach is the use of spikes to prevent comfortable resting spots. We, therefore, strapped cable ties around the moving parts (Figure 4c), leaving the ends uncut as deterrent spikes, being careful to avoid the irradiance sensor field of view.

The integration of an IP camera alongside the radiometers proved particularly useful in identifying these unforeseen problems and may be useful for other causes of data contamination such as boats, floating debris/vegetation, or other unusual conditions in the radiometer field of view (Figure 5) since the radiometer gives only spatially integrated information. However, image analysis is presently subjective and is not easy to automate.

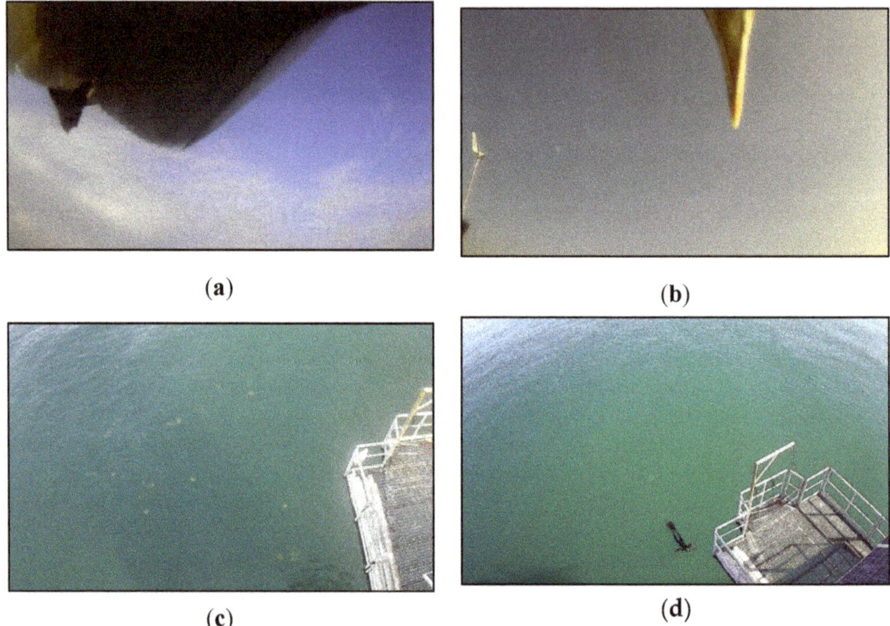

(a) **(b)**

(c) **(d)**

Figure 5. Sample photos taken from the camera aligned with the radiance sensor (for azimuth angles not used for measurements in this paper). (**a**) Bird blocking field of view; (**b**) more subtle measurement contamination; (**c**) floating vegetation and platform shadow on water; (**d**) unexpected water target contamination.

4.2. Water Reflectance Spectra—Mean over Time

For the three clear sky days analyzed here, 30 matchups were recorded between the PANTHYR and M3TRIOS systems and 10 between the PANTHYR and AERONET-OC systems. All 30 water reflectance spectra for PANTHYR and M3TRIOS systems are plotted in Figure 6, with the mean average of the 10 matchup AERONET-OC spectra. Differences between spectra observed on different days/times by a single system in this figure combine both possible temporal variability of the target itself (including possible BRDF effects, which are uncorrected for the PANTHYR and M3TRIOS systems), the random uncertainties associated with the measurement protocol and associated data processing, and certain instrument artefacts (e.g., imperfect irradiance sensor angular response). Since the intra-spectra differences are rather small, we can conclude that the water target was itself rather stable during the three days of analyzed data (spread over a six-day period). One outlier is clearly visible for the M3TRIOS system; detailed analysis of data for that measurement, including the manual log-book and various photos, suggests that this may result from an oily film visible at the water surface within the M3TRIOS system field of view. After exclusion of this outlier, the mean average water reflectance spectrum over the 29 matchups is very similar for PANTHYR and M3TRIOS systems. The mean average water reflectance spectrum over the 10 matchups is also very similar for the PANTHYR and the AERONET-OC systems. For a more detailed understanding of system data quality, including the impact of random uncertainties and removing the differences associated with water target temporal variability, the individual matchups were analyzed as described in the sections below.

4.3. Data Comparison with M3TRIOS System—Matchup Analysis

The 29 matchups between the PANTHYR and M3TRIOS systems, interpolated to typical multispectral wavelengths, are shown via scatterplots of $E_d(\lambda)$, $L_d(\lambda)$, $L_u(\lambda)$, and $\rho_w(\lambda)$, and associated linear regression statistics in Figure 7, including root-mean-square difference (RMSD) and mean average relative difference (MARD).

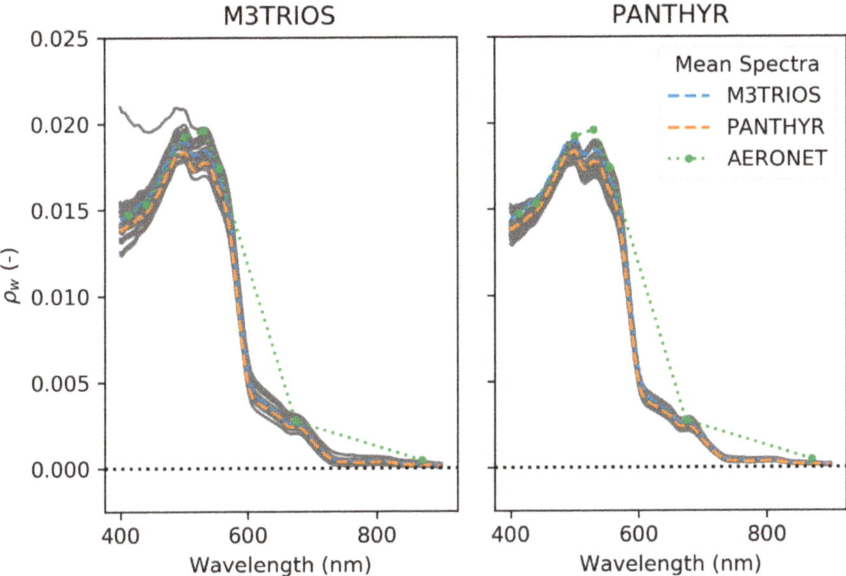

Figure 6. Water reflectance spectra for each of the 30 stations (gray lines) from (left) the three-radiometer TRIOS/RAMSES system (M3TRIOS) system and (right) the PANTHYR system. Mean spectra over all 30 stations are superimposed for the PANTHYR (orange dashed line) and the M3TRIOS system (blue dashed lines) and for all 10 measurements from the AERONET-OC system (green dotted line joining dots where the multispectral data exist).

The MARD of $E_d(\lambda)$ calculated over all wavelengths was 3.1% and the scatterplot (Figure 7a) suggests some spectral variability of differences with lower PANTHYR $E_d(\lambda)$ for 410 nm and 440 nm and higher PANTHYR $E_d(\lambda)$ at 620 nm and 675 nm. A full uncertainty analysis is not yet available for these instruments/systems as deployed at AAOT, and no corrections were applied for instrument artefacts such as imperfect cosine response, straylight, thermal sensitivity, non-linearity or polarization effects, or deployment imperfections, e.g., for optical perturbations from higher mast structures. Operational deployment of the PANTHYR system for fiducial reference measurements will require such an uncertainty analysis but is beyond the scope of this technology-proving paper. However, the observed differences for $E_d(\lambda)$ are not a cause for concern at this stage of testing. The reader is referred also to Reference [34] for a more detailed discussion of differences between measurements of $E_d(\lambda)$ from different instruments in typical field conditions.

The scatterplot (Figure 7b) of $L_d(\lambda)$ sky radiance measurements shows strong correlation but for two distinct groups of data. Analysis of the metadata recorded for both systems revealed that these distinct groups of data correspond to different azimuth angles of the PANTHYR and M3TRIOS systems; they are pointing at very different portions of the sky and, in some cases, the PANTHYR system is pointing at a much brighter sky with cirrus clouds. Since both of the obvious two groups of data corresponded to cases where PANTHYR and M3TRIOS were measuring at different azimuth angles because the systems were deployed on different sides of the AAOT platform and, hence, pointed at the water in different azimuth angles, there is no cause for concern that neither group of points has a slope of one. There were only two matchups where PANTHYR and M3TRIOS were measuring at the same azimuth angle, but this was considered insufficient for detailed analysis.

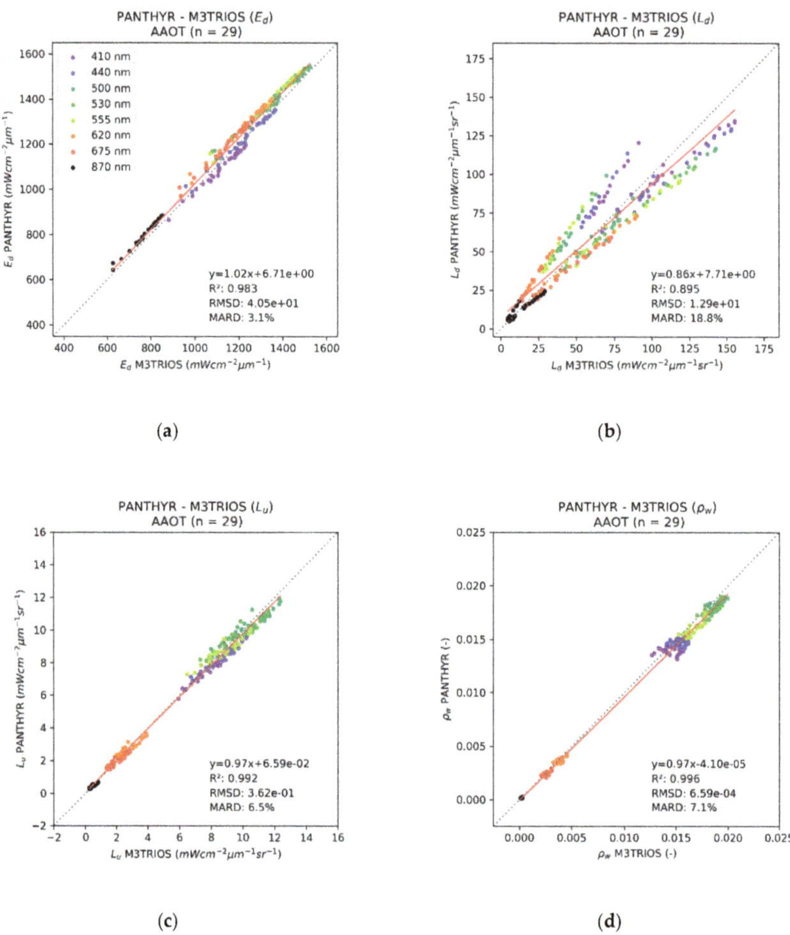

Figure 7. Scatterplot for selected wavelengths comparing PANTHYR data with M3TRIOS system data for (**a**) $\overline{E_d}$, (**b**) $\overline{L_d}$, (**c**) $\overline{L_u}$, and (**d**) $\overline{\rho_w}$ for 29 matchup stations. Linear regression statistics are given, including the root-mean-square difference (RMSD) and the mean average relative difference (MARD).

The scatterplot (Figure 7c) of $L_u(\lambda)$ water (plus reflected skyglint) radiance measurements shows strong correlation ($r^2 = 0.992$) with low systematic differences (slope = 0.97). The MARD of 6.5% was affected by the different azimuth angles of the two systems since this parameter includes reflected skyglint, typically 2.5–3.0% of $L_d(\lambda)$.

The scatterplot (Figure 7d) of $\rho_w(\lambda)$ water reflectance (after skyglint correction) shows high correlation between the PANTHYR and M3TRIOS systems ($r^2 = 0.996$) with low systematic differences (slope = 0.97). The MARD of 7.1% was dominated by the contribution for data at 870 nm, where the water reflectance was very low and absolute differences, e.g., arising from skyglint correction, translated into very large relative differences.

The spectral variations of RMSD and MARD between PANTHYR and M3TRIOS for $E_d(\lambda)$ and $\rho_w(\lambda)$ over these 29 matchups are shown in Figure 8. The short wavelength scale variability for the RMSD of $E_d(\lambda)$ in Figure 8a was related to the similar spectral variability of $E_d(\lambda)$ itself, e.g., the atmospheric oxygen absorption feature at 762 nm or the Frauenhofer lines related to solar photosphere absorption at 516–518 nm, and the way the slightly different central wavelengths of each E_d instrument under-resolved these features. The MARD of $E_d(\lambda)$ between PANTHYR and M3TRIOS (Figure 8b) was <5% for the full spectral range 400–900 nm. The common radiometric calibration of the two E_d instruments just prior to this experiment helped limit MARD; however, a more detailed characterization of these two instruments, particularly including an analysis of their "cosine" angular response and perhaps straylight and non-linear responses, may improve results in the future.

The spectral variation of RMSD between PANTHYR and M3TRIOS for $\rho_w(\lambda)$ shown in Figure 8c suggests an overall spectral shape typical of $L_u(\lambda)$ spectra, i.e., upwelling radiance from green water and reflected "blue" skyglint. Possible causes of this difference are multiple and certainly include imperfect correction of skyglint, but may also include different BRDF at the different azimuth angles, propagation of $E_d(\lambda)$ differences, L radiometer calibration and characterization (straylight, non-linearity, polarization, thermal sensitivity, etc.), etc. A detailed analysis is beyond the scope of this technology-proving paper, which concludes merely that there is no major cause for concern at present.

The MARD between PANTHYR and M3TRIOS shown in Figure 8d for $\rho_w(\lambda)$ was <5% for the spectral range 410–580 nm, with higher MARD around 600 nm. The much higher MARD for wavelengths higher than 700 nm are clearly related to the very low $\rho_w(\lambda)$ for these waters, showing some similarity with the pure water absorption coefficient spectrum for 700–900 nm [26] and with the phytoplankton absorption coefficient at 660–680 nm. In such conditions, the MARD has no practical relevance because a satellite validation analysis would generally consider absolute differences between PANTHYR and satellite or simply not use such data.

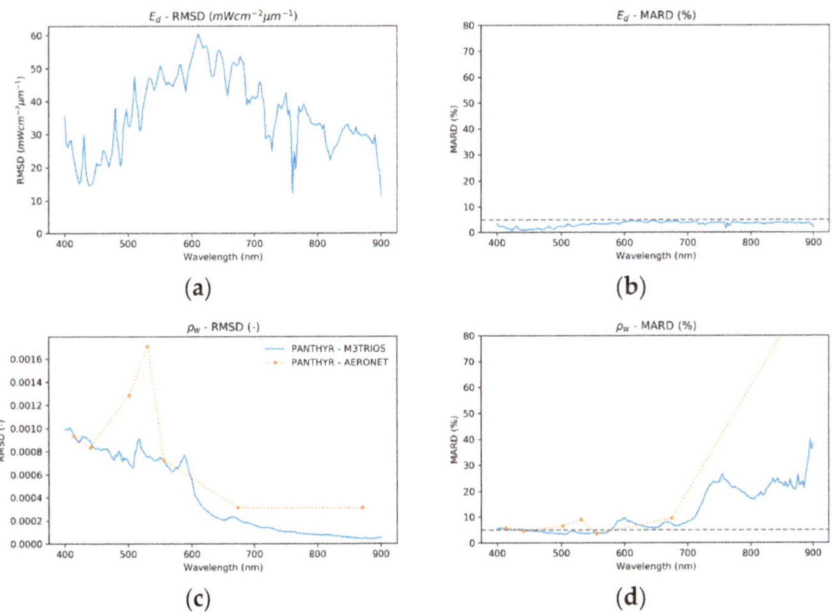

Figure 8. Root-mean-square difference (RMSD) and the mean average relative difference (MARD) between PANTHYR and M3TRIOS systems over 29 stations for (**a**,**b**) $\overline{E_d}$, and (**c**,**d**) $\overline{\rho_w}$. The 5% MARD is shown as a horizontal dashed line.

4.4. Data Comparison with AERONET-OC System—Matchup Analysis

The 10 matchups between the PANTHYR and AERONET-OC systems, with resampling of PANTHYR data to gaussian spectral response functions on the AERONET-OC central wavelengths, are shown via scatterplots of $\rho_w(\lambda)$, and associated linear regression statistics in Figure 9. There is a strong correlation between these datasets ($r^2 = 0.996$) with low systematic differences (slope = 0.95).

The MARD over all wavelengths appeared high (18.3%); however, as seen in the spectral variation of MARD shown in Figure 8d, it was dominated by the data at 870 nm, where RMSD was low but $\rho_w(\lambda)$ itself was very low. For the wavelengths from 413 nm to 555 nm, MARD between PANTHYR and AERONET-OC was between 3.5% and 6.5%, except at 530 nm, where it reached 9.0%. Possible causes of the differences are multiple and may include imperfect correction of skyglint, propagation of $E_d(\lambda)$ differences (not available online for AERONET-OC), radiometer calibration and characterization (straylight and/or spectral response functions, non-linearity, polarization, thermal sensitivity, etc.), etc. A detailed analysis is beyond the scope of this technology-proving paper, which concludes merely that there is no major cause for concern at present.

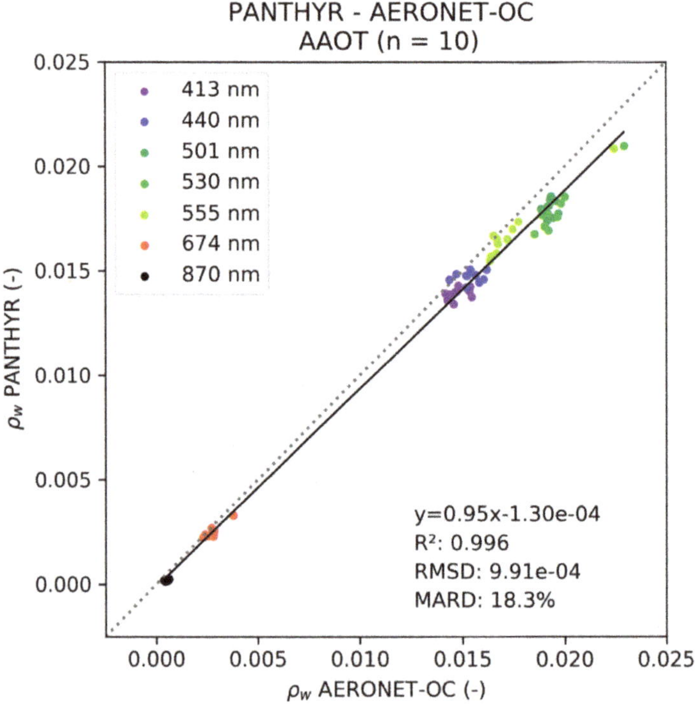

Figure 9. Scatterplot for selected wavelengths comparing PANTHYR data with AERONET-OC Venise Level 1.5 data for $\overline{\rho_w}$ for 10 matchups. Linear regression statistics are given, including the root-mean-square difference (RMSD) and the mean average relative difference (MARD). All PANTHYR data presented in this figure were measured at the 270° relative azimuth. The black line is the reduced major axis regression. Statistics are plotted here for all wavelengths together. For per-band RMSD and MARD, see Figure 8c,d.

5. Conclusions and Future Perspectives

To gather enough high-quality ground-truth data for validation of water reflectance derived from satellite missions, an automated system is necessary. The AERONET-OC federated network of autonomous instrument systems [6] is now the main source of such validation data for all satellite missions measuring water reflectance but provides only multispectral data. The PANTHYR system described here aims to provide hyperspectral water reflectance data for satellite validation.

The design of the PANTHYR system was described in detail here. Two COTS hyperspectral radiometers (one radiance, one irradiance) and an IP camera are mounted on a COTS PT pointing system with original development of control electronics and software providing a low-cost, low-power, but robust, modular, and extendable design. Date/time/location information is received from a GNSS receiver and data are transmitted by a 4G gateway. A flexible mounting bracket was designed to easily fit the system to any suitable platform structure. The prototype was tested during an inter-comparison exercise organized at AAOT, an offshore platform in the Adriatic Sea. The system functioned autonomously over the six-day period without intervention. Data acquired from the system was compared with an established manually deployed hyperspectral system and with the automated AERONET-OC system data deployed simultaneously at AAOT. Data quality from the PANTHYR system was good; water reflectances compared to the two other systems had per-band MARDs within 5.5% for the spectral range 410–550 nm in these good measurement conditions.

Imagery from the IP camera was surprisingly useful and revealed unexpected conditions including birds within the camera field of view, which could contaminate radiometer data in significant (easily detected) or subtle (potentially undetected) ways.

The prototype tests described here were made with mains power supply and without autonomous data transmission over the internet. Use of an autonomous power supply (typically solar panels) with power supply monitoring and automated data transmission over 4G need to be tested in future work.

An upgrade of the PT unit to a version with slip rings is also planned to avoid the risk of cable snags possible with this first prototype.

Future mounting hardware will be machined from 5083 aluminum alloy instead of stainless steel to save weight. Lower overall weight results in lower shipping cost, and limiting the weight on top of the PT head reduces power consumption.

The system can also be extended to function on moving platforms, such as ships or buoys, via the addition of heading and inclination sensors and/or movement compensation mechanisms.

The data acquisition and processing described here are based strongly on precursor work and provide a robust starting point. However, future research may take advantage of the flexible pointing capability to investigate improved data acquisition protocols, e.g., with multiple zenith angle radiance measurements. The skyglint correction for above-water reflectance measurements is also considered to be a major source of measurement uncertainty, and potential improvements were investigated in many recent studies. When considered mature, such improvements can easily be incorporated in the processing software described here.

In conclusion, the PANTHYR system prototype was successfully tested in a basic configuration (without autonomous power supply and without data transmission over the internet) giving robust performance and high-quality hyperspectral data. The system prototype meets the requirements for future worldwide deployment in a network for hyperspectral validation of water reflectance data from satellites.

Author Contributions: Conceptualization, D.V., K.R., and A.C.; methodology, D.V., K.R., and A.C.; software, D.V. and Q.V.; validation, Q.V. and M.B.; formal analysis, K.R., Q.V., and M.B.; writing—original draft preparation, K.R. and D.V.; writing—review and editing, A.C., Q.V., M.B., and D.V.

Funding: This research was funded by BELSPO (Belgian Science Policy Office) in the framework of the STEREO III program project HYPERMAQ (SR/00/335), and by the European Space Agency PRODEX/HYPERNET-OC project. The AAOT fieldwork was partially funded by the European Space Agency FRM4SOC project.

Acknowledgments: The authors would like to thank the Plymouth Marine Laboratory (PML) and Consiglio Nazionale delle Ricerche (CNR), Dr. Gavin Tilstone, Tartu Observatory, and the FRM4SOC project for organizing the FICE at AAOT and for providing mounting frames for the M3TRIOS system radiometers. Tartu Observatory provided the radiometric calibration of all five RAMSES instruments used here. Dr. Giuseppe Zibordi is acknowledged for providing the AERONET-OC data and for kind advice. TRIOS Gmbh is acknowledged for providing details of the serial protocol communication for the RAMSES instruments, allowing us to write the necessary data acquisition software. For the latter, the RFLEX code provided by Stefan Simis provided some useful ideas on how to proceed. Michiel T'Jampens provided appreciated technical feedback. We thank the three anonymous reviewers for many constructive comments on the first version of this paper.

Conflicts of Interest: The authors declare no conflicts of interest. The funders had no role in the design of the study; in the collection, analyses, or interpretation of data; in the writing of the manuscript, or in the decision to publish the results.

References

1. Gohin, F.; Saulquin, B.; Oger-Jeanneret, H.; Lozac'h, L.; Lampert, L.; Lefebvre, A.; Riou, P.; Bruchon, F. Towards a better assessment of the ecological status of coastal waters using satellite-derived chlorophyll-a concentrations. *Remote Sens. Environ.* **2008**, *112*, 3329–3340. [CrossRef]
2. Bresciani, M.; Stroppiana, D.; Odermatt, D.; Morabito, G.; Giardino, C. Assessing remotely sensed chlorophyll-a for the implementation of the Water Framework Directive in European perialpine lakes. *Sci. Total Environ.* **2011**, *409*, 3083–3091. [CrossRef] [PubMed]
3. Vanhellemont, Q.; Ruddick, K.G. Turbid wakes associated with offshore wind turbines observed with Landsat 8. *Remote Sens. Environ.* **2014**, *145*, 105–115. [CrossRef]
4. Giardino, C.; Brando, V.E.; Gege, P.; Pinnel, N.; Hochberg, E.; Knaeps, E.; Reusen, I.; Doerffer, R.; Bresciani, M.; Braga, F.; et al. Imaging Spectrometry of Inland and Coastal Waters: State of the Art, Achievements and Perspectives. *Surv. Geophys.* **2018**, *40*, 401–429. [CrossRef]
5. Bailey, S.W.; Werdell, P.J. A multi-sensor approach for the on-orbit validation of ocean color satellite data products. *Remote Sens. Environ.* **2006**, *102*, 12–23. [CrossRef]
6. Zibordi, G.; Holben, B.; Slutsker, I.; Giles, D.; D'Alimonte, D.; Mélin, F.; Berthon, J.-F.; Vandemark, D.; Feng, H.; Schuster, G.; et al. AERONET-OC: A network for the validation of ocean color primary product. *J. Atmos. Ocean. Technol.* **2009**, *26*, 1634–1651. [CrossRef]
7. Hlaing, S.; Harmel, T.; Gilerson, A.; Foster, R.; Weidemann, A.; Arnone, R.; Wang, M.; Ahmed, S. Evaluation of the VIIRS ocean color monitoring performance in coastal regions. *Remote Sens. Environ.* **2013**, *139*, 398–414. [CrossRef]
8. Mélin, F.; Sclep, G. Band shifting for ocean color multi-spectral reflectance data. *Opt. Express* **2015**, *23*, 2262–2279. [CrossRef]
9. Holben, B. AERONET—A federated instrument network and data archive for aerosol characterization. *Remote Sens. Environ.* **1998**, *66*, 1–16. [CrossRef]
10. Zibordi, G.; Mélin, F.; Hooker, S.B.; D'Alimonte, D.; Holben, B. An autonomous above-water system for the validation of ocean colour radiance data. *IEEE TGARS* **2004**, *42*, 401–415.
11. Barreto, Á.; Cuevas, E.; Granados-Muñoz, M.-J.; Alados-Arboledas, L.; Romero, P.M.; Gröbner, J.; Kouremeti, N.; Almansa, A.F.; Stone, T.; Toledano, C.; et al. The new sun-sky-lunar Cimel CE318-T multiband photometer – a comprehensive performance evaluation. *Atmos. Meas. Tech.* **2016**, *9*, 631–654. [CrossRef]
12. Hooker, S.B.; Bernhard, G.; Morrow, J.H.; Booth, C.R.; Comer, T.; Lind, R.N.; Quang, V. *Optical Sensors for Planetary Radiant Energy (Osprey): Calibration and Validation of Current and Next-Generation Nasa Missions*; NASA Tech. Memo. 2011–215872; NASA Goddard Space Flight Center: Greenbelt, MD, USA, 2012.
13. Simis, S.G.H.; Olsson, J. Unattended processing of shipborne hyperspectral reflectance measurements. *Remote Sens. Environ.* **2013**, *135*, 202–212. [CrossRef]
14. Slivkoff, M.M. Ocean Colour Remote Sensing of the Great Barrier Reef Water. Ph.D. Thesis, Department of Imaging & Applied Physics, School of Science, Curtin University, Curtin, Australia, 2014.
15. Brando, V.; Lovell, J.; King, E.; Boadle, D.; Scott, R.; Schroeder, T. The Potential of Autonomous Ship-Borne Hyperspectral Radiometers for the Validation of Ocean Color Radiometry Data. *Remote Sens.* **2016**, *8*, 150. [CrossRef]
16. Carswell, T.; Costa, M.; Young, E.; Komick, N.; Gower, J.; Sweeting, R. Evaluation of MODIS-Aqua Atmospheric Correction and Chlorophyll Products of Western North American Coastal Waters Based on 13 Years of Data. *Remote Sens.* **2017**, *9*, 1063. [CrossRef]
17. Hommersom, A.; Kratzer, S.; Laanen, M.; Ansko, I.; Ligi, M.; Bresciani, M.; Giardino, C.; Betlrán-Abaunza, J.M.; Moore, G.; Wernand, M.R.; et al. Intercomparison in the field between the new WISP-3 and other radiometers (TriOS Ramses, ASD FieldSpec, and TACCS). *J. Appl. Remote Sens.* **2012**, *6*, 063615. [CrossRef]
18. Ruddick, K.; De Cauwer, V.; Park, Y.; Becu, G.; De Blauwe, J.-P.; Vreker, E.D.; Deschamps, P.-Y.; Knockaert, M.; Nechad, B.; Pollentier, A.; et al. *Preliminary Validation of MERIS Water Products for Belgian Coastal Waters*; European Space Agency: Paris, France, 2003; Volume SP-531.
19. Talone, M.; Zibordi, G.; Ansko, I.; Banks, A.C.; Kuusk, J. Stray light effects in above-water remote-sensing reflectance from hyperspectral radiometers. *Appl. Opt.* **2016**, *55*, 3966. [CrossRef] [PubMed]
20. Zibordi, G.; Talone, M.; Jankowski, L. Response to Temperature of a Class of In Situ Hyperspectral Radiometers. *J. Atmos. Ocean. Technol.* **2017**, *34*, 1795–1805. [CrossRef]

21. Talone, M.; Zibordi, G. Polarimetric characteristics of a class of hyperspectral radiometers. *Appl. Opt.* **2016**, *55*, 10092. [CrossRef] [PubMed]

22. Mekaoui, S.; Zibordi, G. Cosine error for a class of hyperspectral irradiance sensors. *Metrologia* **2013**, *50*, 187–199. [CrossRef]

23. Olesen, D.; Jakobsen, J.; Knudsen, P. Knudsen Low-cost GNSS sampler based on the beaglebone black SBC. In Proceedings of the 2016 8th ESA Workshop on Satellite Navigation Technologies and European Workshop on GNSS Signals and Signal Processing (NAVITEC), Noordwijk, The Netherlands, 14–16 December 2016; pp. 1–7.

24. Mollon, M.; Kaneko, E.H.; Niro, L.; Montezuma, M. *Remote Laboratory for a Servomotor Control System with Embedded Architecture*; IJAREEIE: Chennai, Tamilnadu, India, 2017; Volume 6.

25. Mobley, C.D. Estimation of the remote-sensing reflectance from above-surface measurements. *Appl. Opt.* **1999**, *38*, 7442–7455. [CrossRef] [PubMed]

26. Ruddick, K.; De Cauwer, V.; Park, Y.; Moore, G. Seaborne measurements of near infrared water-leaving reflectance: The similarity spectrum for turbid waters. *Limnol. Oceanogr.* **2006**, *51*, 1167–1179. [CrossRef]

27. Doyle, J.P.; Zibordi, G. Optical propagation within a three-dimensional shadowed atmosphere–ocean field: Application to large deployment structures. *Appl. Opt.* **2002**, *41*, 4283–4306. [CrossRef] [PubMed]

28. Ruddick, K.; Cauwer, V.D.; Van Mol, B. *Use of the Near Infrared Similarity Spectrum for the Quality Control of Remote Sensing Data*; Frouin, R.J., Babin, M., Sathyendranath, S., Eds.; SPIE: Bellingham, WA, USA, 2005; Volume 5885.

29. Zibordi, G.; Berthon, J.F.; Doyle, J.P.; Grossi, S.; van der linde, D.; Targa, C.; Alberotanza, L. *Coastal Atmosphere and Sea Time Series (CoASTS), Part 1: A Long-Term Measurement Program*; Tech Memo TM-2002-206892; NASA Goddard Space Flight Center: Greenbelt, MD, USA, 2002.

30. Hooker, S.B.; Lazin, G.; Zibordi, G.; McLean, S. An evaulation of above- and in-water methods for determining water-leaving radiances. *J. Atmos. Ocean. Technol.* **2002**, *19*, 486–515. [CrossRef]

31. Zibordi, G.; Ruddick, K.; Ansko, I.; Moore, G.; Kratzer, S.; Icely, J.; Reinart, A. In situ determination of the remote sensing reflectance. *Ocean Sci.* **2012**, *8*, 567–586. [CrossRef]

32. Morel, A.; Antoine, D.; Gentili, B. Bidirectional reflectance of oceanic waters: Accounting for Raman emission and varying particle scattering phase function. *Appl. Opt.* **2002**, *41*, 6289–6306. [CrossRef] [PubMed]

33. Thuillier, G.; Herse, M.; Labs, D.; Foujols, T.; Peetermans, W.; Gillotay, D.; Simon, P.C.; Mandel, H. The solar spectral irradiance from 200 to 2400 nm as measured by the SOLSPEC spectrometer from the ATLAS and EURECA missions. *Sol. Phys.* **2003**, *214*, 1–22. [CrossRef]

34. Vabson, V.; Kuusk, J.; Ansko, I.; Vendt, R.; Alikas, K.; Ruddick, K.; Ansper, A.; Bresciani, M.; Burmester, H.; Costa, M.; et al. Field intercomparison of radiometers used for satellite validation in the 400–900 nm range. *Remote Sens.* **2019**, *11*, 1129. [CrossRef]

Article

Results from Verification of Reference Irradiance and Radiance Sources Laboratory Calibration Experiment Campaign

Agnieszka Białek [1,*], Teresa Goodman [1], Emma Woolliams [1], Johannes F. S. Brachmann [2], Thomas Schwarzmaier [2], Joel Kuusk [3], Ilmar Ansko [3], Viktor Vabson [3], Ian C. Lau [4], Christopher MacLellan [5], Sabine Marty [6], Michael Ondrusek [7], William Servantes [1], Sarah Taylor [1], Ronnie Van Dommelen [8], Andrew Barnard [8], Vincenzo Vellucci [9], Andrew C. Banks [1], Nigel Fox [1], Riho Vendt [3], Craig Donlon [10] and Tânia Casal [10]

[1] National Physical Laboratory, Teddington TW11 0LW, UK; teresa.goodman@npl.co.uk (T.G.); emma.woolliams@npl.co.uk (E.W.); william.servantes@npl.co.uk (W.S.); sarah.taylor@npl.co.uk (S.T.); andyb@hcmr.gr (A.C.B.); nigel.fox@npl.co.uk (N.F.)

[2] Deutsches Zentrum für Luft- und Raumfahrt, 82234 Wessling, Germany; brachmann@iabg.de (J.F.S.B.); thomas.schwarzmaier@dlr.de (T.S.)

[3] Tartu Observatory, University of Tartu, 61602 Tõravere, Estonia; joel.kuusk@ut.ee (J.K.); ilmar.ansko@ut.ee (I.A.); viktor.vabson@ut.ee (V.V.); riho.vendt@ut.ee (R.V.)

[4] Commonwealth Scientific and Industrial Research Organisation, Kensington, WA 6151, Australia; ian.lau@csiro.au

[5] Natural Environment Research Council's Field Spectroscopy Facility, Edinburgh EH9 3FE, UK; chris.maclellan@npl.co.uk

[6] Oceanography Group, Norwegian Institute for Water Research, Gaustadalleen 21, 0349 Oslo, Norway; sabine.marty@niva.no

[7] Center for Satellite Applications and Research, National Oceanic and Atmospheric Administration, College Park, MD 20740, USA; michael.ondrusek@noaa.gov

[8] Sea-Bird Scientific, Bellevue, WA 98005, USA; ronnie@xeostech.com (R.V.D.); abarnard@seabird.com (A.B.)

[9] Sorbonne Université, CNRS, Institut de la Mer de Villefranche IMEV, F-06230 Villefranche-sur-Mer, France; enzo@imev-mer.fr

[10] European Space Agency, 2201 AZ Noordwijk, The Netherlands; craig.donlon@esa.int (C.D.); tania.casal@esa.int (T.C.)

[*] Correspondence: agnieszka.bialek@npl.co.uk; Tel.: +44-208-943-6716

Received: 12 June 2020; Accepted: 1 July 2020; Published: 10 July 2020

Abstract: We present the results from Verification of Reference Irradiance and Radiance Sources Laboratory Calibration Experiment Campaign. Ten international laboratories took part in the measurements. The spectral irradiance comparison included the measurements of the 1000 W tungsten halogen filament lamps in the spectral range of 350 nm–900 nm in the pilot laboratory. The radiance comparison took a form of round robin where each participant in turn received two transfer radiometers and did the radiance calibration in their own laboratory. The transfer radiometers have seven spectral bands covering the wavelength range from 400 nm–700 nm. The irradiance comparison results showed an agreement between all lamps within ±1.5%. The radiance comparison results presented higher than expected discrepancies at the level of ±4%. Additional investigation to determine the causes for these discrepancies identified them as a combination of the size-of-source effect and instrument effective field of view that affected some of the results.

Remote Sens. **2020**, *12*, 2220; doi:10.3390/rs12142220   www.mdpi.com/journal/remotesensing

Remote Sens. **2020**, *12*, 2220

Keywords: ocean colour; spectral irradiance comparison; spectral radiance sources comparison; fiducial reference measurements

---

1. Introduction

The National Metrology Institutes (NMIs) are responsible for the International System of Units (SI) which provides the foundation for measurement around seven base units and a system of coherent derived units. There are three key concepts underpinning how the desired multi-century stability and world-wide consistency of these units is achieved: uncertainty analysis, traceability and comparisons [1].

Uncertainty analysis is the systematic review of all sources of uncertainty associated with a particular measurement and the formal propagation of uncertainties through methods defined by the Guide to the Expression of Uncertainty in Measurement (GUM) [2]. Metrological traceability to a measurement unit of the International System of Units [3] is the concept that links all metrological measurements to the SI through a series of calibrations or comparisons. Each step in this traceability chain has rigorous uncertainty analysis, usually peer reviewed or audited and always documented. Comparisons [4] are the process of validating an uncertainty analysis by comparing the measurement of artefacts by different laboratories.

NMIs must participate in regular (usually every 10 years) formal comparisons. Each technical discipline defines a limited number of "key comparisons" and these provide evidence to support uncertainty analysis for a certain number of related quantities in a "Calibration and Measurement Capability Database". For example, the Consultative Committee for Photometry and Radiometry (CCPR) has defined a key comparison for six key measurands (spectral irradiance, spectral responsivity, luminous intensity, luminous flux, spectral diffuse transmittance and spectral regular reflectance). There is no key comparison for spectral radiance, as it is assumed that reliable results (results that are consistent with declared uncertainties) in the spectral irradiance comparison together with results for the comparison of reflectance provide sufficient evidence for spectral radiance measurements as well.

In 2008, the Committee on Earth Observation Satellites (CEOS) established and endorsed the Quality Assurance Framework for Earth Observation (QA4EO) [5] which set general principles for Earth Observation (EO) data quality assurance and which follows the same metrological principles of the NMIs. Although QA4EO does not explicitly require traceability to SI and allows "or [to] a community-agreed reference", it does state "preferably [to] SI". To apply these principles in practice a concept of Fiducial Reference Measurements (FRM) was established and defined as [6]:

> *Fiducial Reference Measurements (FRM) are a suite of independent, fully characterized, and traceable ground measurements that follow the guidelines outlined by the Group on Earth Observations (GEO)/CEOS Quality Assurance framework for Earth Observation (QA4EO). These FRM provide the maximum Return On Investment (ROI) for a satellite mission by delivering, to users, the required confidence in data products, in the form of independent validation results and satellite measurement uncertainty estimation, over the entire end-to-end duration of a satellite mission* [7].

The Fiducial Reference Measurements for Satellite Ocean Colour (FRM4SOC) project was established and funded by the European Space Agency (ESA) to provide support for evaluating and improving the state of the art in satellite ocean colour validation through a series of comparisons under the auspices of the CEOS. The project makes a fundamental contribution to the Copernicus Earth Observation system, led by the European Commission, in partnership with ESA, by ensuring high quality ground-based measurements for ocean colour radiometry (OCR) for use in validation of ocean colour products from missions like Sentinel-3 Ocean Colour and Land Instrument (OLCI) [8] and Sentinel-2 Multi Spectral Imager (MSI) [9].

In the past, a dedicated program to support the quality of Sea-viewing Wide Field-of-view Sensor (SeaWIFS) [10] products was conducted. The Seventh SeaWiFS Intercalibration Round-Robin Experiment (SIRREX) [11] and the Second Intercomparison and Merger for Interdisciplinary Ocean Studies (SIMBIOS) [12–14] included comparisons of irradiance and radiance sources as well as radiometers by the teams participating in validation activities of the SeaWIFS. For the Medium Resolution Imaging Spectrometer (MERIS) [15] a comparison of in situ measurements was performed at the Acqua Alta Oceanographic Tower (AAOT) [16] and showed the discrepancies between the measurements mostly explained by the combined uncertainties of the compared measurements with a few exceptions.

From the metrological point of view, it is important to repeat such comparison exercises at regular time frames; firstly to achieve the measurements agreement, then to ensure that the consistency between the organisations is held stable and, finally, to enable new participants to verify their measurements capability.

This paper presents results from the first step in the OCR measurement chain and includes a comparison intended to verify the performance of the irradiance and radiance sources used to calibrate ocean colour radiometers.

2. Methods

2.1. Layout and Organisation of the Comparisons

Public announcements were made to invite all laboratories involved in the satellite Ocean Colour (OC) validation activities. To participate, laboratories had to hold working standards with spectral irradiance and radiance values traceable to SI. Irradiance and radiance measurements comparisons were addressed separately.

First National Physical Laboratory (NPL), the UK NMI, conducted a laboratory comparison of the irradiance sources involving measurements of all participating lamps at NPL in April 2017. Participants were encouraged to attend this comparison in person to hand carry the lamps to and from the comparison and to attend a training course in absolute radiometric calibration and uncertainty evaluation that was given at the same time. Remote participation in irradiance comparison was allowed, however the training course was given only to seven persons present at NPL at the time.

Then, a round-robin of each participant's radiance sources using ocean colour transfer radiometers was performed between May 2017 and October 2018. This involved two calibrated transfer radiometers sent back and forth in turn to each participant to perform radiance measurements. NPL served as pilot and was responsible for inviting participants, circulating the transfer radiometers and for the analysis of data, following appropriate processing by individual participants. The experiment was conducted anonymously. NPL was the only organisation to have access to and was able to view all data from participants.

The list of ten international laboratories that took part in this comparison exercise is shown in Table 1. Note that three of the institutes, Remote Sensing Technology Institute, Deutsches Zentrum für Luft und Raumfahrt (DLR-IMF), Joint Research Centre (JRC) and Norsk Institutt for Vannforskning (NIVA) participated in the radiance round robin only.

Table 1. Laboratories that participated in the measurements (Alphabetic Order).

	Organisation
CSIRO	Commonwealth Scientific and Industrial Research Organisation, Australia
DLR-IMF	Remote Sensing Technology Institute, Deutsches Zentrum für Luft und Raumfahrt, Germany
JRC	European Commission—DG Joint Research Centre
LOV-IMEV	Laboratoire d'Océanographie de Villefranche, France
NERC-FSF	Natural Environment Research Council's Field Spectroscopy Facility, UK
NIVA	Norsk Institutt for Vannforskning, Norway
NOAA	National Oceanic and Atmospheric Administration, USA
NPL	National Physical Laboratory, UK
TO	Tartu Observatory, Estonia
Satlantic	Satlantic Sea Bird Scientific, Canada

2.1.1. Irradiance Comparison

In this comparison 1000 W quartz tungsten halogen (QTH) lamps, so-called FEL lamps (not acronym) according to American National Standard Institute (ANSI) designation, are considered as irradiance sources and were used at the standard calibration distance of 500 mm measured from their reference plane. Absolute spectral irradiance values were determined by reference to the NPL_{2010} spectral irradiance scale; that is the scale that was realised in 2010 and validated in international comparisons, after a major upgrade to its facility prior to 2010. The measurements were made using the NPL Spectral Radiance and Irradiance Primary Scales (SRIPS) [17] or secondary Reference Spectroradiometer (RefSpec) facility. The two facilities are very similar and allowed the comparison of participants' lamps to reference lamps that had been themselves calibrated on SRIPS by reference to a high temperature, high-emissivity blackbody source operated at a temperature of approximately 3050 K. Spectral irradiance measurements were made from 350 nm to 900 nm at 10 nm steps with an instrument bandwidth of approximately 2.8 nm full width at half maximum (FWHM). Ambient temperature during measurements was 22 °C $\pm$ 2 °C. The results of the comparison were expressed in terms of the difference between the spectral irradiance values measured by each participant and the mean spectral irradiance values measured by all participants. Since the participants all measured different lamps (i.e., their own lamps), the required differences between them were determined via measurements at NPL of all lamps. The mean ratio between the participants' measurements and those made at NPL was calculated and results for each lamp were then expressed relative to this mean ratio, so showing the degree to which the individual measurements agree with one another. This approach was taken because:

1. the participants had various different SI-traceability routes for their lamps, i.e., a number of different NMIs providing their calibration. If all results were shown relative to the NPL values, then this might give the impression that traceability to NPL is 'correct' or 'best' whereas traceability to any NMI should be regarded as equally acceptable.
2. A few of the lamps were recently calibrated at NPL and using a simple ratio to the NPL scale would have shown them to be performing almost 'perfectly' and thus give a misleading and biased comparison.
3. Presenting the results in terms of the agreement between each lamp and the mean of all of the lamps shows how well measurements of the different participants agree with each other, regardless of the traceability route. This was the key aim of the comparison and this form of presentation gives the clearest indication of that.
4. the ratio between the NPL scale and the mean of all the participants' lamps is also included, which gives the confidence that the linkage to SI is sound in all cases.

The FEL lamps used in the comparison were sourced from four different commercially available sources: Gooch and Housego (now Optronic Laboratories Inc., Orlando, USA) OL FEL 1000 W, Gigahertz BN-9101 FEL 1000 W, Gamma Scientific Model 5000 FEL 1000 W and L.O.T.-Oriel 63350 FEL 1000 W. In addition, one participant had a modified general purpose Osram Sylvania 1000 W FEL lamp. In total, 14 lamps from the participants and two of the NPL standards were measured. According to the manufacturer's specification and original calibration certification, these lamps were run using their nominal current values and these varied from lamp to lamp between 8.0 A, 8.1 A and 8.2 A. All lamps were within the 50 h of burn time since the last calibration. There were three different traceability chains for the participants. The most common traceability was to the National Institute of Standards and Technology (NIST) scale via Gooch & Housego (now called Optronics Laboratories). The lamps were calibrated by a direct comparison method to the NIST traceable calibration standards. These calibration standards are directly traceable to the NIST irradiance standard. The second group of lamps was traceable to NPL via a direct comparison method to NPL working irradiance standard. Two lamps were calibrated by the Metrology Research Institute (MRI) of Aalto University, Finland. One lamp was calibrated by Tartu Observatory with traceability to MRI. Depending on the lamp type, the appropriate alignment procedure was used following the lamp manufacturer's instructions. The reference plane for a distance measurement of each lamp type was defined by the manufacturer and was followed.

The Gamma Scientific lamps have a dedicated lamp housing enclosure to reduce the stray light in the room during the measurements. During this comparison, one participant provided a lamp with this housing (and the lamp was measured with the housing at NPL), and another participant provided the lamp without the housing and was measured as it was provided. We did not notice a difference in the irradiance measurements performed on NPL facilities (which appropriately shielded stray light) between them, but, of course, the irradiance of an individual lamp will be sensitive to whether or not that lamp housing was included.

Each lamp was ramped up and run for 30 min before measurements commenced. The voltage was monitored during measurement and is given for checking purposes only.

2.1.2. Radiance Comparison

For this comparison, the radiance source was an FEL lamp (the organisations who took part in the irradiance comparison used the same lamps) and a reference reflectance panel (such as Spectralon) set in the 0°:45° (0° incidence angle and 45° viewing angle) setup presented in Figure 1. The recommended distance between the lamp and the panel was 500 mm as this is the default distance for the irradiance calibration. However, it was recognised that some participants could not perform their measurements with a 500 mm lamp-diffuser distance due to laboratory constraints and therefore other distances were allowed as well (as long as the uncertainties were accounted for). Due to the varying source calibration distances or other instrumental constraints, measurements were acquired at lamp-panel distances 500 mm, 750 mm, 1000 mm and 1300 mm by different participants, some participants obtaining measurements at more than one distance. The recommended minimum distance for the radiometer was defined at approximately 250 mm from the panel. A maximum distance was not set as the panel radiance as measured by a radiometer should, within the constraints of source uniformity, be insensitive to the radiometer distance.

Different types and sizes of reflectance standards were allowed, so as not to exclude any of the participants due to the type of the standard used. However, we requested the technical details and calibration history of the artefacts along with SI-traceable calibration certificates. Table 2 contains the details of the lamp-panel distance for each participant/calibration set-up and information about reflectance standard used.

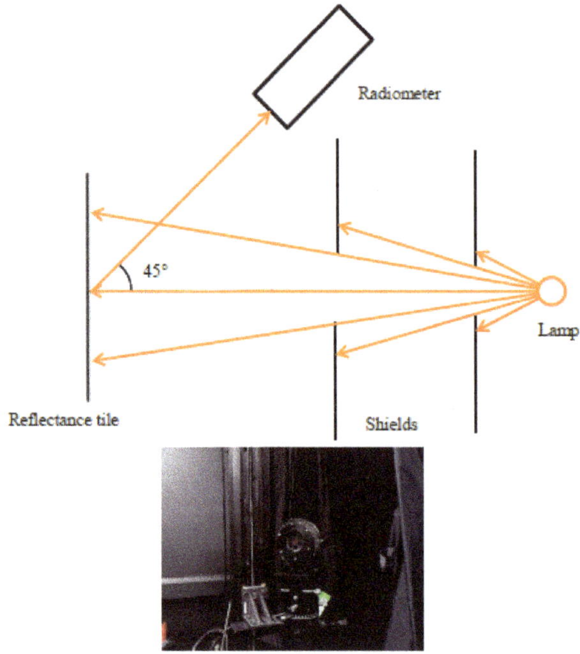

Figure 1. Radiance mode diagram of setup (top), OC filter radiometer (bottom).

Table 2. Detailed information about each participant (A through M) lamp-panel distance setting and type and size of the reflectance standard.

Participant/ Calibration Set-Up	Lamp-Panel Distance in mm	Type	Reflectance Standard Reflectance in %	Size in Inches
A	500	Spectralon	99	10 × 10
B	750	Spectralon	99	10 × 10
C	1000	Spectralon	99	18 × 18
D	500	Spectralon	99	18 × 18
E	1000	Spectralon	99	18 × 18
F	500	Spectralon	99	10 × 10
G	1300	Spectralon	99	18 × 18
H	500	Zenith Lite	95	200 mm × 200 mm
I	500	Spectralon	99	10 × 10
J	1000	Spectralon	99	10 × 10
K	500	Spectralon	99	12 × 12
L	500	Spectralon	99	10 × 10
M	500	Gigahertz-Optik	98	12 × 12

The vast majority of the reflectance standards used by the participants was calibrated for $8°$:hemispherical reflectance by Labsphere laboratory with the NIST traceability using a dual beam spectrophotometer with an integrating sphere accessory. Only two had $0°$:$45°$ reflectance factor calibration traceable to NPL.

To enable a complete radiance source comparison between laboratories, we proposed using a transfer standard radiometer to compare the participant's in-house radiance calibration performance. As NPL does not own an OCR instrument one of the participants, Joint Research Centre of the European Commission (JRC), kindly agreed to provide two of their stable OC multispectral filter radiometers as the transfer radiometers.

On completion of the radiance measurements, each participant sent their results (radiometer readings and spectral radiance data) to the pilot. The pilot calculated radiometer calibration factors using Equation (1) and these data.

$$C_{L_{cal}}(\lambda) = \frac{L(\lambda)}{DN_l(\lambda) - DN_d(\lambda)},\tag{1}$$

where $C_{L_{cal}}$ is absolute radiance calibration coefficient, DN_l and DN_d are radiometer readings during the radiance calibration, with the $_l$ and $_d$ indicating light and dark readings, respectively, L is the radiance value and λ indicates the radiometer spectral channel. Radiance values per each spectral channel are given by Equation (2)

$$L(\lambda) = \frac{E(\lambda)}{\pi} \left(\frac{500}{d}\right)^2 R_{0:45}(\lambda),\tag{2}$$

where E is the lamp irradiance value from the calibration certificate, d is the lamp-panel distance used during the measurement and $R_{0:45}$ is the reflectance panel $0°{:}45°$ reflectance factor from the calibration certificate.

For these participants who did not have their reference reflectance panels calibrated at $0°{:}45°$ reflectance factor geometry, the most common calibration for the $8°$:hemispherical reflectance was allowed. A correction factor of 1.024 was applied to the diffuse reflectance calibration values to correct it for the proper measurement geometry. The value of that correction factor was established based on NPL internal data combined with published data by NIST, the USA National Metrology Institute, [18].

The central wavelength was used to derive the spectral band radiance value. The radiometers used in this comparison have narrow (10 nm) spectral bands, in addition the radiance source does not have any distinct spectral features and is monotonically increasing in the spectral region of interest for this study. All the participants used spectrally similar radiance sources. Therefore, the difference between the spectral band integration values and the central wavelengths are small. The same conclusion was found in SIMBIOS comparisons [13].

Each participant was asked to evaluate uncertainties associated with their radiance source operating in their own laboratory for these measurements. This included all the additional uncertainty components related to the alignment of the lamp, panel and radiometer, distance measurements and other relevant laboratory specific factors such as power supply stability and accuracy. The results of the comparisons are expressed as the percentage difference to the mean calibration coefficients obtained by taking an average of all participants results.

Two Satlantic ocean colour radiometers (OCR-200) were used as transfer radiometers. These are 7 channel multispectral instruments with general technical characteristics of these type of radiometers shown in Figure 2, although the two particular instruments used for FRM4SOC had been customised by Satlantic for JRC in terms of their spatial characteristics to provide a narrower (3°) field of view in air. Initial characterisation measurements to confirm this field of view (FOV) were carried out by NPL in air, and found to be $2.5° \pm 0.3°$ at FWHM, with a close to Gaussian profile (see Figure 3) for both radiometers (serial numbers 051 and 110) used in this comparison.

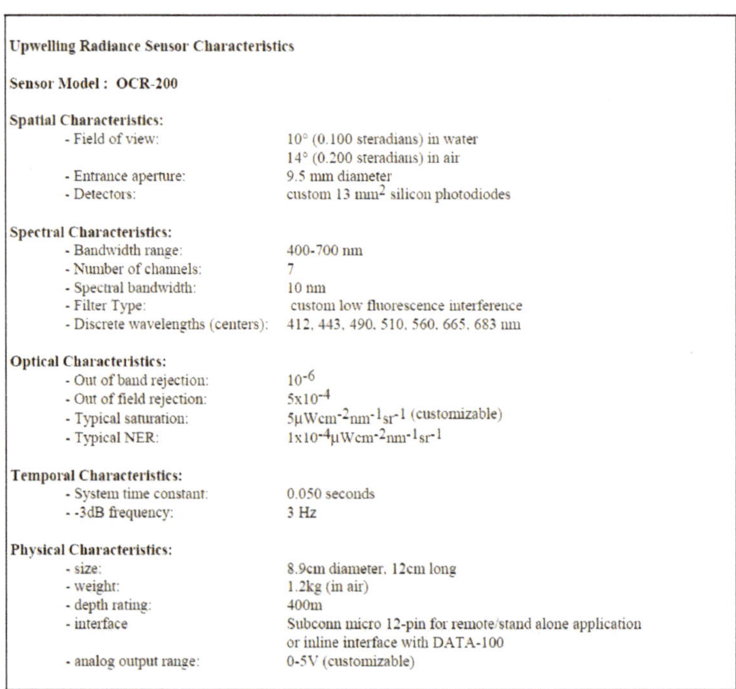

Figure 2. General specifications of the Satlantic ocean colour radiometers (OCR-200).

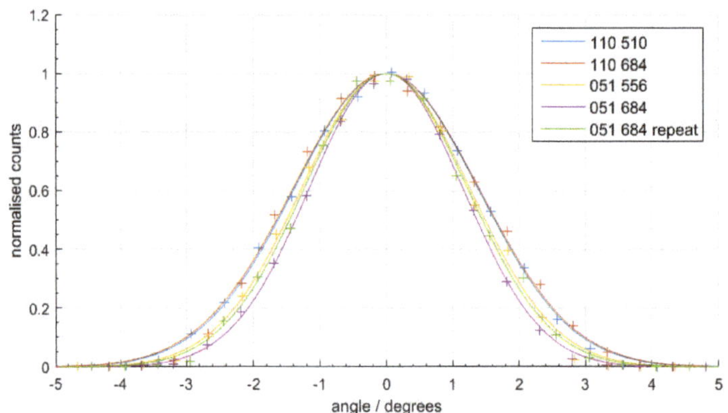

Figure 3. Measurement results to confirm the field of view (FOV) of the transfer radiometers being used in the FRM4SOC radiance round robin. The numbers in plot legend refer to the radiometer serial number and the spectral band (in nm) measured.

3. Results

This section presents the results of the comparisons. The irradiance values are reported at the following wavelengths (350, 360, 370, 380, 390, 400, 450, 500, 600, 700, 800, 900) nm. This selection of wavelengths was dictated by the wavelengths reported in the calibration certificates from the participants. Although currently there are no OC missions that provide data below 400 nm, we wanted to include the ultraviolet spectral region in the comparison. There is a scientific interest to cover shorter wavelengths and this is indeed planned for the Plankton Aerosol Cloud ocean Ecosystem (PACE) [19] EO mission.

The radiance values are reported at the transfer radiometer spectral bands values (412, 443, 491, 510, 560, 667, 684) nm. We have chosen to present results anonymously to keep focus on community consistencies, rather than individual laboratory. Participants were informed which laboratory they were.

3.1. Irradiance

The overall summary result of the irradiance comparison is presented in Figure 4. The data series for each lamp used in the comparison are marked as Lamp A to Lamp N. The black dotted line indicates the mean ratio to NPL.

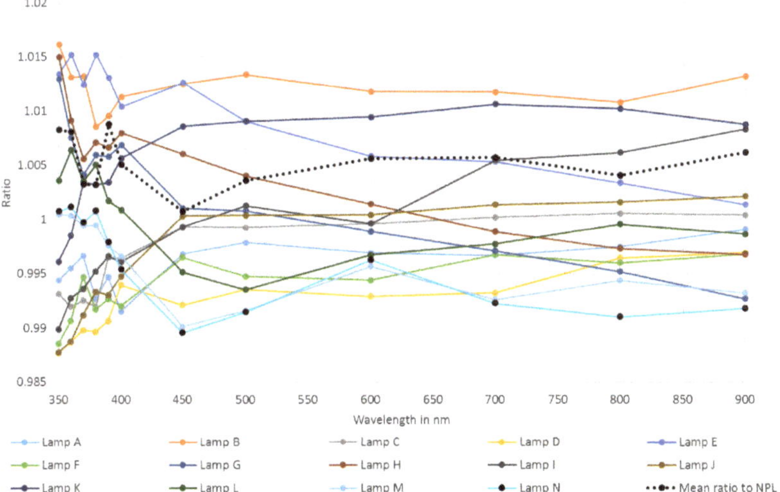

Figure 4. The results of the lamp irradiance comparison. Coloured lines represent the results of each lamp compared to the mean of all lamps. The dotted line compares the mean of all lamps to National Physical Laboratory's (NPL) SI-traceable scale.

The results show an agreement between all measured lamps as all data series above 400 nm lay within the 0.99–1.013 range. The spread in the results is higher for shorter wavelengths as expected due to the higher measurements uncertainty presented in the absolute radiometric calibration for this spectral range.

The uncertainty for the individual lamp ratio is expressed by Equation (3) and was calculated by combining several uncertainty components. All components are relative, thus expressed in percentage form.

$$u(E_r) = \sqrt{\left(u^2\left(\frac{c_{cer}}{2}\right) + u^2(s_{NPL}) + u^2(c_n) + u^2(c_\lambda) + u^2(c_{sl}) + u^2(c_{cur}) + u^2(c_{age}) + u^2(c_{alig})\right)}, \quad (3)$$

where the $u(E_r)$ is the uncertainty in the ratio of the irradiance values from the lamp calibration certificate to measurements performed at NPL, the $u\left(\frac{c_{cer}}{2}\right)$ is uncertainty from the lamp calibration certificate, note that uncertainty value is divided by two to convert it to a standard uncertainty from a coverage factor $k = 2$ used in the certificate. The $u(s_{NPL})$ is the NPL scale uncertainty and the additional components related to the measurements performed at NPL which included noise $u(c_n)$, and the uncertainty contributions to the irradiance value caused by: wavelength setting accuracy $u(c_\lambda)$, room stray light $u(c_{sl})$, the lamp current uncertainty $u(c_{cur})$, ageing of the lamp $u(c_{age})$ and the lamp alignment $u(c_{alig})$. The typical values for the ratio uncertainty expressed in percent are given in Table 3.

Table 3. Example of an FEL comparison uncertainty values.

Wavelength (nm)	Example of a Lamp Ratio Uncertainty ($k = 1$)
350	1.6%
360	1.7%
370	1.7%
380	1.7%
390	1.5%
400	1.5%
450	1.1%
500	1.1%
600	1.0%
700	1.0%
800	1.0%
900	0.9%

3.2. Radiance

The overall summary result of the radiance comparison is presented in Figure 5 for the radiometer with serial number 051 and in Figure 6 for the radiometer with serial number 110. The colour triangles represent the seven spectral bands of the radiometers and the participants/set-ups are marked as letters from A to M. Please note that we present here 13 entries to the summary results that came from 10 participating institutes. The number of entries is higher because some organisations provided results at two different distance settings between the lamp and the reflectance panel, or two different measurements set-up like radiometers alignment to the central channel versus alignment to each channel in turn.

The results for both instruments show the same trends. The difference for any individual participant/set-up from the mean value is up to 4%, which is slightly higher than expected. A clear split into two groups can be seen. The majority of the results forms a group with radiance calibration coefficient values below the mean value and a second group of 4 participants has radiance calibration coefficient values around 3% above the mean value.

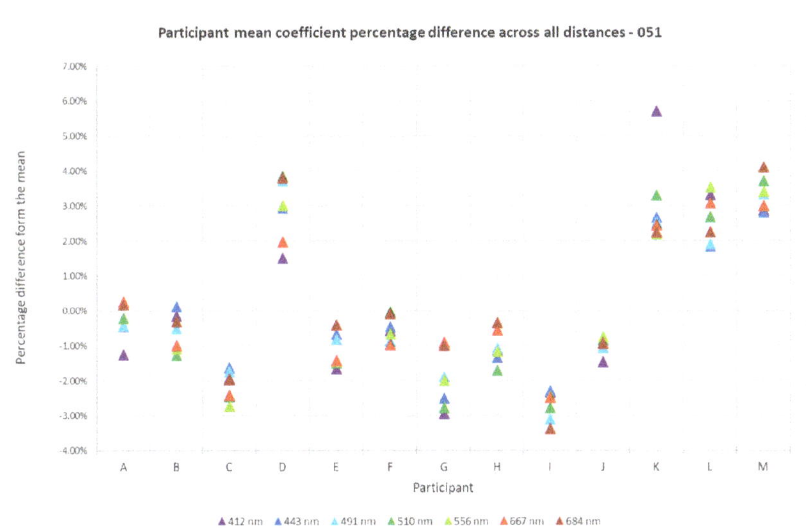

Figure 5. Results of the radiance comparison for radiometer 051. Radiance calibration coefficient compared to mean calibration coefficient for each participant/set-up at each wavelength.

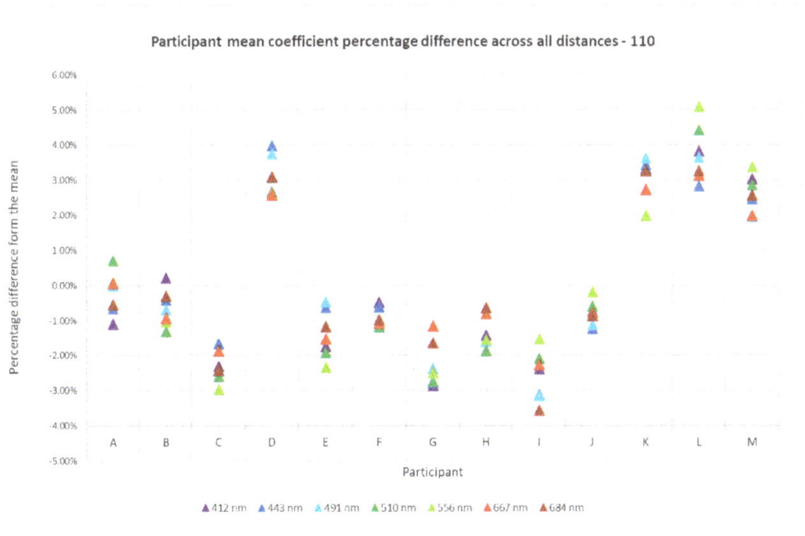

Figure 6. Results of the radiance comparison for radiometer 110. Radiance calibration coefficient compared to mean calibration coefficient for each participant/set-up at each wavelength.

Each participant was asked to provide an uncertainty budget for their calibration of the radiance calibration coefficient values of the radiometers. The uncertainty of radiance measurements is calculated according to Equations (4) and (5). The wavelength dependence is not included in the following equations, but all the values were derived for the central wavelength of each of the radiometer spectral channels and take a similar form for all spectral channels.

$$u_{\text{rel}}(L_{\text{cal}}) = \sqrt{\frac{u^2(DN_{\text{l}}) + u^2(DN_{\text{d}})}{(DN_{\text{l}} - DN_{\text{d}})^2} + u^2(c_{\text{sl}}) + u^2(L_{\text{rel}})}, \tag{4}$$

where $u_{\text{rel}}(L_{\text{cal}})$ is the relative combined uncertainty of radiance calibration coefficient, $u(DN_{\text{l}})$ and $u(DN_{\text{d}})$ are the uncertainty of the radiometer reading during the radiance calibration, with the $_{\text{l}}$ and $_{\text{d}}$ indicating light and dark readings, respectively, $u(c_{\text{sl}})$ is the uncertainty contributor due to the room stray light, expressed as percentage of the radiometer signal after the dark reading subtraction and $u(L_{\text{rel}})$ is the relative uncertainty of the radiance source. The radiance source uncertainty components are listed in Equation (5).

$$u_{\text{rel}}(L) = \sqrt{(u_{\text{rel}}^2(E) + 2^2 u_{\text{rel}}^2(d) + u_{\text{rel}}^2(R_{0:45}) + u^2(c_{\text{cur}}) + u^2(c_{\text{age}}) + u^2(c_{\text{alig}}) + u^2(c_{\text{unif}})}, \tag{5}$$

where $u_{\text{rel}}(E)$ is lamp irradiance absolute calibration uncertainty converted to $k = 1$ from the certificate values, $u_{\text{rel}}(d)$ is the relative uncertainty associated with the distance setting, note that this component has a sensitivity coefficient equal to 2 (from the inverse square law) and, hence, in Equation (5) there is a term 2^2 just before it. In addition, for all measurements at distances different to the 500 mm, the participants were requested to include a filament-offset uncertainty component to account for the difference in the plane of the distance setting and actual lamp filament position. The $u_{\text{rel}}(R_{0:45})$ is the relative uncertainty of the reflectance standard calibration. Please note we use the uncertainty of the reflectance factor calibration at $0°:45°$ geometry at $k = 1$; for the case where a diffuse reflectance calibration value is corrected to $0°:45°$ geometry, an additional uncertainty of that correction has to be included in the equation, NPL recommends a 0.5% value. The remaining terms provide the relative uncertainty associated with radiance due to lamp current uncertainty $u(c_{\text{cur}})$, due to lamp ageing uncertainty, $u(c_{\text{age}})$, due to alignment $u(c_{\text{alig}})$ of the lamp and the reflectance panel at $0°:45°$ configuration and due to target illumination non-uniformity within the FOV $u(c_{\text{unif}})$.

A few examples of participants' radiance measurements uncertainty expressed in percent are presented in Table 4.

Table 4. Examples of participants' radiance calibration relative uncertainty, $k = 2$.

Band (nm)	Participant with Low u	Participant with Middle u	Participant with High u Value
412	2.0%	2.4%	3.1%
443	1.8%	2.4%	2.9%
491	1.8%	2.2%	2.7%
510	1.8%	2.3%	2.7%
556	1.8%	2.2%	2.5%
667	1.8%	2.1%	2.5%
684	1.8%	2.1%	2.5%

4. Discussion

The irradiance comparison results showed a good agreement within the expected uncertainty range. The observed differences are at a similar level to those reported in SIRREX-7 exercise [11], that states 1.3% for the comparisons on the FEL lamps irradiance calibration. However, the comparison presented here included more lamps with different designs and traceability routes. The SIRREX irradiance comparison was expressed as relative percent difference between the NIST and Optronic calibration of the same lamp. Higher differences were observed for wavelengths below 400 nm as this is a more challenging spectral region for radiometric measurements. This indicates that, for future satellite missions the absolute calibration will have higher uncertainty for these new wavelengths.

It is important to note that the same lamps (irradiance standards) used elsewhere in a different laboratory environment, using different power supply or being aligned less carefully, could produce different results. It is also important to note that the spread of results for irradiance lamps would be expected to be higher if the lamps are operated outside the controlled environmental and stray light conditions of a high-quality radiometric laboratory or operated with power supplies that are not as stable as those used at NPL. Therefore, for any measurement using the lamp as the transfer standard for the spectral irradiance it is essential to consider all the uncertainty components given in Equation (3), and not just those quoted on the calibration certificate.

The radiance comparison result, showing a lower level of agreement as compared to the irradiance comparison, led NPL to perform further investigations to explain the cause of the observed difference. The similarly-structured SIMBIOS comparison had results with the absolute value for each participant typically higher than expected by 2% [13,14], but the results were generally within the expected uncertainty range, with few exceptions. The SIMBIOS comparison used a radiometer as a reference and determined the difference in radiance measured by the radiometer and that calculated by each participating laboratory. In our comparison we tried to identify common features for the smaller group of the results that agree well with each other, but are around 3% higher than the top of the "main group". The first common feature for the small group is that those results correspond to a lamp-panel distance of 500 mm; although some of the results in the main group were also made at that lamp-diffuser distance. Thus, the distance setting is not the only cause of the difference. The second common feature was the size of the illuminated patch on the reflectance target from the lamp. This size was influenced by the choice of light shields and other baffles in a particular laboratory setup. For the laboratories that use a lamp in a housing enclosure, the illumination patch size changes with the distance between the lamp and the diffuser.

NPL repeated a set of measurements to accommodate various conditions that participants may have in their own labs. The additional investigation was performed using an 18" Spectralon panel that was illuminated by the lamp at a distance of 500 mm and 1300 mm. The second distance was chosen as this was the longest distance used by a participant during the comparison. The size of the illuminated patch on the panel was varied from the fully illuminated panel, via a patch size with the diameter of around 23 cm to the small patch size of around 15 cm. The top panel in Figure 7a presents the photographs taken for the three different illumination patch sizes. The bottom panel in Figure 7b presents the percentage difference in the calibration coefficients obtained from five scenarios plotted as the data series from A to E. The series A, B and C represent the measurements at 500 mm distance for the fully illuminated panel, patch size 23 cm and the patch size 15 cm, respectively. The series D and E were done at 1300 mm distance for the fully illuminated and 23 cm patch size. The series A is set as the reference in this data set, therefore the percentage difference for this series is 0%.

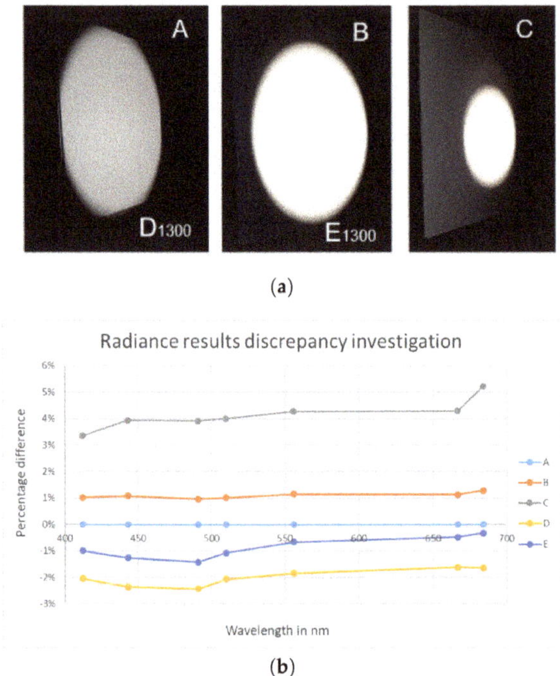

Figure 7. Difference in the radiance measurement due to the distance setting and the size of the source. (**a**) Photographs of illumination patch sizes.The photograph on the left presents cases A and D, the middle one cases B and E, and the right hand side case C. (**b**) Plot with percentage difference between measurements for different patch size and lamp-panel distance setting.

A clearly visible positive bias can be seen for the measurements performed at 500 mm with a smaller size of the source. In addition, a negative bias can be seen for the measurements performed at the larger distance. All participants from the small group had their measurements done at the 500 mm source-panel distance with a relatively small size of their radiance source.

The most likely explanation for this is a combination of the instrument FOV and the size-of-source effect. These will also be influenced, to a small extent, by the distance from the panel to the radiometer, though we believe this is a minor consideration compared to the lamp-panel distance and the baffling that defined the source size Although the radiometers have a FOV defined as 3°, this is a FHWM value and with a Gaussian shape to the FOV, rather than a top-hat. That 3° FOV, when plotted on a logarithmic scale rather than linear, (see Figure 8) shows that there is still light detected at the level of 10^{-3} at 5° and this means that the instrument will see a wider area than expected, and perhaps, therefore, see a less uniform patch. The instrument was not viewing a non-illuminated patch, even for a 5° FOV, but was seeing a less uniform patch, and this would particularly be the case for a 500 mm lamp-panel distance. In addition, scattering on imperfections in the lens will lead to light "lost" from the central field of view and "regained" from the outer parts of the source. With a smaller source, less energy is regained and this can reduce the measured signal. This is known as the "size-of-source effect" [20,21].

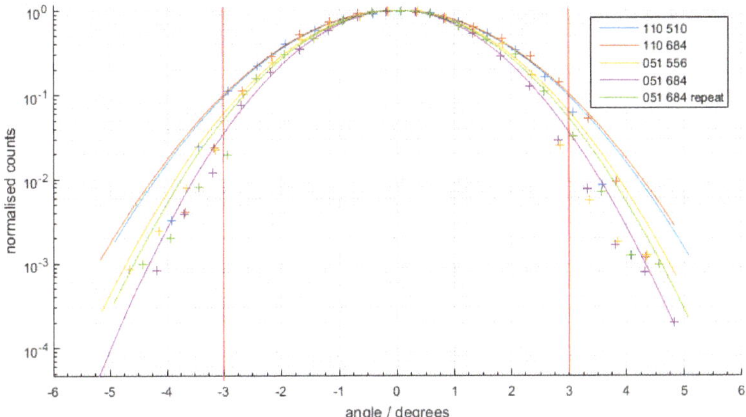

Figure 8. Measurement results to confirm the FOV of the transfer radiometers being used in the FRM4SOC radiance round robin presented on logarithmic scale. The vertical bars indicate ±3°. The numbers in the plot legend refer to the radiometer serial number and the spectral band (in nm) measured.

Thus, the smaller size of the source leads to a reduced measured signal because of the size-of-source effect and a 500 mm lamp-diffuser distance leads to a smaller signal because of the reduced uniformity of the source across the FOV (the source is less bright away from the centre of the filament). When both effects are present, the measured signal is lower and therefore, from Equation (1), the radiance calibration coefficient is higher, as seen in the comparison results. To confirm this hypothesis, we also analysed the data for all participants considering the sensitivity to distance. This effect is not as strong at longer distances, as can be seen in Figure 7, series E, that has a smaller patch size but did not show a positive bias. Thus, here the effect of the size of the source is compensated by a negative bias introduced by the distance setting. We analysed the data of all participants according to their sensitivity to distance. The results of this analysis are presented in Figure 9.

The four data series represent the averaged comparison results for different distances of 500 mm, 750 mm, 1000 mm and 1300 mm, where the 500 mm distance is set as a reference, thus the percentage difference for the 500 mm series is equal to zero. Please note that 750 mm and 1300 mm had one entry to the comparison thus these are not averaged. The 500 mm series contains only the results from the main group and was set as the reference distance for that exercise, thus this data series has 0% difference. A negative bias can be observed with the distance increase for the 1000 mm and 1300 mm distance. For the 750 mm this is not so obvious, however this might be due to the fact that this particular participant has the radiance calibration values provided in radiance units for the whole system, rather than calculated from a lamp irradiance calibrated at 500 nm and a reflectance factor.

The protocol for the comparison encouraged participants to add an uncertainty component to account for set-ups where the source-diffuser distance is different from the 500 mm distance at which the lamps were originally calibrated for spectral irradiance. However, participants did not include an explicit correction for a different lamp-diffuser distance. It is possible that some of the observed differences between the main group and the other group are due to a bias from the filament offset. We would recommend in the future for the filament offset to be evaluated and corrected for [22]. This may also help to reduce the spread within the main group.

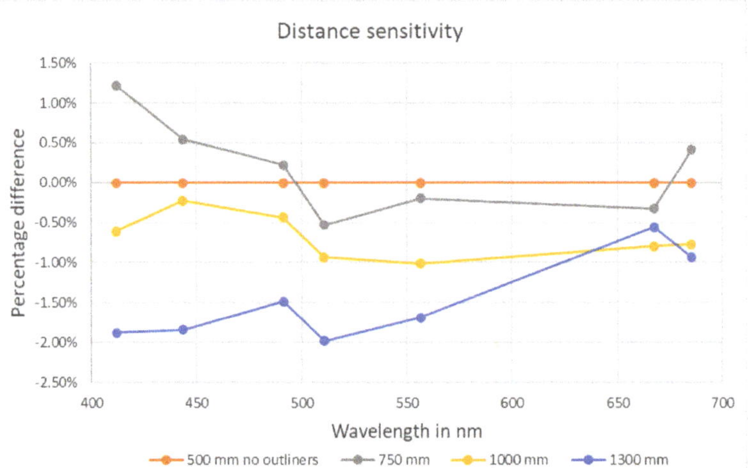

Figure 9. Difference in the radiance calibration coefficient obtained by averaging results from different participants for each lamp-panel distance.

5. Conclusions

We presented here the results of the international irradiance and radiance sources comparisons that were run as a part of the ESA FRM4SOC study activities. Ten international organisations participated in that exercise.

The irradiance comparison was run at NPL, where all participating lamps were measured against NPL standards. The results of that comparison were reported as the difference between the spectral irradiance values measured by each participant and the mean spectral irradiance values measured by all participants to show the degree to which the individual measurements agree with one another, without introducing a bias toward NPL scale. The irradiance comparison values showed good agreements between all lamps.

The radiance comparison took a form of a round robin where two radiometers were sent in turn to each participant to obtain radiance calibration coefficients for the radiometer using their in-house facilities. The results were analysed to compare the calibration coefficients of each participant to that of the mean value of all participants. The results showed the discrepancies between the participants at the level of ±4% and two separate groups with the measurements agreement (see Figures 5 and 6). Additional investigation showed that the reason for this difference was caused by a sensitivity to the size of the illuminated patch (instrument size-of-source effect) and partly because the instrument-effective FOV brought in non-uniform parts of the illumination for a shorter lamp-diffuser distance which affected the results of the smaller group. If these effects could be corrected for or the measurements repeated at different settings we would expect to see all measurements agreeing within ±2.5%, as this is the level of agreement in the results from the majority group. These results were obtained with a modified Satlantic OCR-200 that had a reduced FOV. The sensitivity to size-of-source would be expected to be larger for an unmodified instrument with a larger FOV.

The secondary objective of the comparison exercise was to increase the community awareness of measurement uncertainty evaluation using the GUM methodology. This was achieved via the training course that was provided for the participants being present at NPL during the irradiance comparison exercise week. The participants were given instruction on how to derive the uncertainty components related to their radiance measurements in-house and all the round robin radiance results were reported accompanied by the uncertainty budgets. The results have shown a discrepancy that is larger than the declared uncertainty. In part, this is due to uncorrected sensitivities (e.g., to lamp-panel distance and illuminated-source size) and in part due to the fact that some uncertainty components were not fully investigated. Had the measurements all been made at the same distance, the comparison may not have shown up this sensitivity to the size of the source, and source uniformity. The experience of this comparison provides an opportunity for all participants to improve their thinking of uncertainty analysis and shows the value of a systematic review of all effects during the development of an uncertainty budget.

Space Agencies continue to evolve the OCR constellation. For the first time, a fleet of European Copernicus Earth Observation satellites is now sustained in an operational manner. In the next few years two new Sentinel-2 and two new Sentinel-3 satellites will be launched and commissioned. As Copernicus evolves, the tools, methodology and collaboration developed by FRM4SOC will be applied and further refined to ensure that the best possible OCR measurements are available from space to the user community.

Author Contributions: Conceptualisation, A.B. (Agnieszka Białek), T.G., E.W., N.F.; methodology, A.B. (Agnieszka Białek), T.G., E.W., N.F.; validation, A.B. (Agnieszka Białek); data curation, J.F.S.B., T.S., J.K., I.A., V.V. (Viktor Vabson), I.C.L., C.M., S.M., M.O., W.S., S.T., R.V.D., A.B. (Andrew Barnard), V.V. (Vincenzo Vellucci); writing—original draft preparation, A.B. (Agnieszka Białek); T.G.; A.C.B; writing—review and editing, all; visualisation, A.B. (Agnieszka Białek), S.T.; project administration, R.V., A.C.B., C.D. and T.C. All authors have read and agreed to the published version of the manuscript.

Funding: This work was funded by European Space Agency project Fiducial Reference Measurements for Satellite Ocean Colour (FRM4SOC), contract no. 4000117454/16/I-Sbo.

Acknowledgments: Authors would like to thank Giuseppe Zibordi from JRC for provision of two transfer radiometers used in this comparison and acknowledge his contribution as a radiance comparison participant and data provider. We would like to thank the reviewers for their comments that helped to improve the readability of this paper.

Conflicts of Interest: The authors declare no conflict of interest.

References

1. Mittaz, J.; Merchant, C.J.; Woolliams, E.R. Applying principles of metrology to historical Earth observations from satellites. *Metrologia* **2019**, *56*, 032002, doi:10.1088/1681-7575/ab1705. [CrossRef]

2. JCGM100:2008. Evaluation of Measurement Data—Guide to the Expression of Uncertainty in Measurement; Guidance Document; BIPM. Available online: https://www.bipm.org/utils/common/documents/jcgm/JCGM_100_2008_E.pdf (accessed on 1 March 2019).

3. JCGM200:2012. International Vocabulary of Metrology—Basic and General Concepts and Associated Terms (VIM); Guidance Document; BIPM. Available online: https://www.bipm.org/utils/common/documents/jcgm/JCGM_200_2012.pdf (accessed on 1 March 2019).

4. CIPM. Mutual Recognition of National Measurement Standards and of Calibration and Measurement Certificates Issued by National Metrology Institutes; Guidance Document, BIPM. Available online: https://www.bipm.org/utils/en/pdf/CIPM-MRA-2003.pdf (accessed on 15 October 2019).

5. Quality Assurance Framework for Earth Observation. Available online: www.qa4eo.org (accessed on 10 November 2019).

6. Donlon, C.; Minnett, P.; Fox, N.; Wimmer, W. Strategies for the Laboratory and Field Deployment of Ship-Borne Fiducial Reference Thermal Infrared Radiometers in Support of Satellite-Derived Sea Surface Temperature Climate Data Records. *Exp. Methods Phys. Sci.* **2015**, *47*, 697, ISBN 9780124170117.

7. Fiducial Reference Measurements. Available online: https://earth.esa.int/web/sppa/activities/frm (accessed on 15 November 2019).

8. Donlon, C.; Berruti, B.; Buongiorno, A.; Ferreira, M.H.; Féménias, P.; Frerick, J.; Goryl, P.; Klein, U.; Laur, H.; Mavrocordatos, C.; et al. The Global Monitoring for Environment and Security (GMES) Sentinel-3 mission. *Remote Sens. Environ.* **2012**, *120*, 37–57, doi:10.1016/j.rse.2011.07.024. [CrossRef]

9.   Drusch, M.; Bello, U.D.; Carlier, S.; Colin, O.; Fernandez, V.; Gascon, F.; Hoersch, B.; Isola, C.; Laberinti, P.; Martimort, P.; et al. Sentinel-2: ESA's Optical High-Resolution Mission for GMES Operational Services. *Remote Sens. Environ.* **2012**, *120*, 25–36; doi:10.1016/j.rse.2011.11.026. [CrossRef]

10.   Hooker, S.B.; Esaias, W.E.; Feldman, G.C.; Gregg, W.W.; Mc Clain, C.R. *An overview of SeaWiFS and ocean colour in SeaWiFS Technical Report Series*; NASA Tech. Memo 104566; NASA Goddard Space Flight Centre: Greenbelt, MD, USA, 1992.

11.   Hooker, S.B.; McLean, S.; Sherman, J.; Small, M.; Lazin, G.; Zibordi, G.; Brown, J. *The Seventh SeaWiFS Intercalibration Round-Robin Experiment (SIRREX-7)*; NASA Tech. Memo 2002-206892; NASA Goddard Space Flight Center: Greenbelt, MD, USA, 2002; Volume 17.

12.   McClain, C.R. *SIMBIOS Background*; NASA Tech. Memo TM-1999-208645,1–2; NASA Goddard Space Flight Center: Greenbelt, MD, USA, 1998.

13.   Meister, G.; Abel, P.; Barnes, R.; Cooper, J.; Davis, C.; Godin, M.; Goebel, D.; Fargion, G.; Frouin, R.; Korwan, D.; et al. *The First SIMBIOS Radiometric Intercomparison (SIMRIC-1), April–September 2001*; Technical Report NASA/ TM-2002-210006; NASA Goddard Space Flight Center: Greenbelt, MD, USA, 2002.

14.   Meister, G.; Abel, P.; Carder, K.; Chapin, A.; Clark, D.; Cooper, J.; Davis, C.; English, D.; Fargion, G.; Feinholz, M.; et al. *The Second SIMBIOS Radiometric Intercomparison (SIMRIC-1), March–November 2002*; Technical Report NASA/TM-2002-210006; NASA Goddard Space Flight Center: Greenbelt, MD, USA, 2003; Volume 2.

15.   Rast, M.; Bezy, J.L.; Bruzzi, S. The ESA Medium Resolution Imaging Spectrometer MERIS a review of the instrument and its mission. *Int. J. Remote Sens.* **1999**, *20*, 1681–1702, doi:10.1080/014311699212416. [CrossRef]

16.   Zibordi, G.; Ruddick, K.; Ansko, I.; Moore, G.; Kratzer, S.; Icely, J.; Reinart, A. In situ determination of the remote sensing reflectance: an inter-comparison. *Ocean Sci.* **2012**, *8*, 567–586, doi:10.5194/os-8-567-2012. [CrossRef]

17.   Woolliams, E.R.; Fox, N.P.; Cox, M.G.; Harris, P.M.; Harrison, N.J. The CCPR K1—A key comparison of spectral irradiance from 250 nm to 2500 nm: Measurements, analysis and results. *Metrologia* **2006**, *43*, S98–S104, doi:10.1088/0026-1394/43/2/s20. [CrossRef]

18.   Yoon, H.W.; Allen, D.W.; Eppeldauer, G.P.; Tsai, B.K. The extension of the NIST BRDF scale from 1100 nm to 2500 nm. *Proc. SPIE* **2009**, *7452*, 745204–745204–12, doi:10.1117/12.827293. [CrossRef]

19.   Cetinić, I.; McClain, C.; Werdell, P. *PACE Technical Report Series, Volume 6: Data Product Requirements and Error Budgets*; Nasa Tech. Memo; NASA Goddard Space Flight Space Center: Greenbelt, MD, USA, 2018.

20.   Solorio-Leyva, J.C.; Suarez-Romero, J.G.; Hurtado-Ramos, J.B.; Rodriguez, E.T.; Cortes-Reynoso, J.G.R. The size-of-source effect in practical measurements of radiance. *Proc. SPIE* **2004**, *5622*, 1243–1248, doi:10.1117/12.590882. [CrossRef]

21.   Yoon, H.W.; Proctor, J.E.; Gibson, C.E. FASCAL 2: A new NIST facility for the calibration of the spectral irradiance of sources. *Metrologia* **2003**, *40*, S30–S34, doi:10.1088/0026-1394/40/1/308. [CrossRef]

22.   Manninen, P.; Hovila, J.; Seppälä, L.; Kärhä, P.; Ylianttila, L.; Ikonen, E. Determination of distance offsets of diffusers for accurate radiometric measurements. *Metrologia* **2006**, *43*, S120–S124, doi:10.1088/0026-1394/43/2/s24. [CrossRef]

Article

Laboratory Intercomparison of Radiometers Used for Satellite Validation in the 400–900 nm Range

Viktor Vabson [1,*], Joel Kuusk [1], Ilmar Ansko [1], Riho Vendt [1], Krista Alikas [1], Kevin Ruddick [2], Ave Ansper [1], Mariano Bresciani [3], Henning Burmester [4], Maycira Costa [5], Davide D'Alimonte [6], Giorgio Dall'Olmo [7], Bahaiddin Damiri [8], Tilman Dinter [9], Claudia Giardino [3], Kersti Kangro [1], Martin Ligi [1], Birgot Paavel [10], Gavin Tilstone [7], Ronnie Van Dommelen [11], Sonja Wiegmann [9], Astrid Bracher [9], Craig Donlon [12] and Tânia Casal [12]

[1] Tartu Observatory, University of Tartu, 61602 Tõravere, Estonia; joel.kuusk@ut.ee (J.K.); ilmar.ansko@ut.ee (I.A.); riho.vendt@ut.ee (R.V.); krista.alikas@ut.ee (K.A.); ave.ansper@ut.ee (A.A.); kersti.kangro@ut.ee (K.K.); martin.ligi@ut.ee (M.L.)
[2] Royal Belgian Institute of Natural Sciences, 1000 Brussels, Belgium; kruddick@naturalsciences.be
[3] National Research Council of Italy, 21020 Ispra, Italy; bresciani.m@irea.cnr.it (M.B.); giardino.c@irea.cnr.it (C.G.)
[4] Helmholtz-Zentrum Geesthacht, Institute for Coastal Research, 21502 Geesthacht, Germany; henning.burmester@hzg.de
[5] University of Victoria, Victoria, BC V8P 5C2, Canada; maycira@uvic.ca
[6] Center for Marine and Environmental Research, University of Algarve, 8005-139 Faro, Portugal; davide.dalimonte@gmail.com
[7] Plymouth Marine Laboratory, Plymouth PL1 3DH, UK; gdal@pml.ac.uk (G.D.); GHTI@pml.ac.uk (G.T.)
[8] Cimel Electronique S.A.S, 75011 Paris, France; bahaiddin.damiri@univ-lille1.fr
[9] Alfred Wegener Institute Helmholtz Centre for Polar and Marine Research, D-27570 Bremerhaven, Germany; Tilman.Dinter@awi.de (T.D.); Sonja.Wiegmann@awi.de (S.W.); Astrid.Bracher@awi.de (A.B.)
[10] Estonian marine institute, University of Tartu, 12618 Tallinn, Estonia; birgot.paavel@ut.ee
[11] Satlantic; Sea Bird Scientific, Bellevue, WA 98005, USA; rvandommelen@seabird.com
[12] European Space Agency, 2201 AZ Noordwijk, The Netherlands; craig.donlon@esa.int (C.D.); tania.casal@esa.int (T.C.)
* Correspondence: viktor.vabson@ut.ee; Tel.: +372-737-4552

Received: 25 March 2019; Accepted: 5 May 2019; Published: 8 May 2019

Abstract: An intercomparison of radiance and irradiance ocean color radiometers (The Second Laboratory Comparison Exercise—LCE-2) was organized within the frame of the European Space Agency funded project Fiducial Reference Measurements for Satellite Ocean Color (FRM4SOC) May 8–13, 2017 at Tartu Observatory, Estonia. LCE-2 consisted of three sub-tasks: 1) SI-traceable radiometric calibration of all the participating radiance and irradiance radiometers at the Tartu Observatory just before the comparisons; 2) Indoor intercomparison using stable radiance and irradiance sources in controlled environment; and 3) Outdoor intercomparison of natural radiation sources over terrestrial water surface. The aim of the experiment was to provide one link in the chain of traceability from field measurements of water reflectance to the uniform SI-traceable calibration, and after calibration to verify whether different instruments measuring the same object provide results consistent within the expected uncertainty limits. This paper describes the activities and results of the first two phases of LCE-2: the SI-traceable radiometric calibration and indoor intercomparison, the results of outdoor experiment are presented in a related paper of the same journal issue. The indoor experiment of the LCE-2 has proven that uniform calibration just before the use of radiometers is highly effective. Distinct radiometers from different manufacturers operated by different scientists can yield quite close radiance and irradiance results (standard deviation $s < 1\%$) under defined conditions. This holds when measuring stable lamp-based targets under stationary laboratory conditions with all the radiometers uniformly calibrated against the same standards just prior to the experiment. In addition, some unification of measurement and data processing must be settled. Uncertaint of radiance and irradiance measurement under these conditions largely consists of the sensor's calibration uncertainty and of the spread of results obtained by individual sensors measuring the same object.

Keywords: ocean color radiometers; radiometric calibration; indoor intercomparison measurement; agreement between sensors; measurement uncertainty

---

1. Introduction

Fiducial reference measurements of water reflectance are aimed to validate satellite data with requirement to provide metrological traceability to the SI units with related uncertainty estimates. These measurement uncertainties can arise from instrument specification, calibration and characterization, performance during field measurements due to various conditions of use and different targets, measurement protocol (including any corrections and assumptions), traceability of calibration sources to the primary SI standards. The present study focusses particularly on instrument performance and calibration, assessing whether different instruments freshly calibrated under uniform conditions and methods, but operated by different scientists, can produce consistent estimates within the estimated uncertainty limits when measuring stable radiance and irradiance targets in laboratory conditions. The results of this study are not limited to ocean color (OC) radiometry and are relevant to radiometers operating over 400–900 nm used in air for other applications, including field measurement of land surface reflectance.

An intercomparison of radiance and irradiance ocean color radiometers (The Second Laboratory Comparison Exercise—LCE-2) using stable incandescent lamp sources under controlled indoor conditions was conducted with the aim to provide one link in the chain of traceability from field measurements of water reflectance to the uniform SI-traceable calibration (Figure 1). Intercomparison of data produced by a number of independent radiometric sensors measuring the same object can assess the consistency of different results and their estimated uncertainties depending on the type of the sensor, the spectrum of measured radiation, the environmental conditions, and the particular method used for collecting and handling the measurement data [1,2]. This information can also serve for further elaboration of uncertainty estimation. Additionally, methodologies used by participants for the measurements and data handling can also be critically reviewed. For the LCE-2, a stepwise approach was chosen: first, the radiometric calibration of the sensors was conducted by the same calibration laboratory; second, indoor comparisons using various levels of radiance or irradiance measurements were performed in stable conditions similar to those during radiometric calibration; third, field measurements as described further in [3]. Traceability of the in-situ measurements to SI units is established by regular calibration of field radiometers. Thus, immediately before the comparison, Tartu Observatory (TO) performed consistent calibration of all participating radiometers in order to guarantee that differences in comparison results will not be primarily due to various calibration sources and/or calibration times. Radiometric calibration procedures including respective uncertainties have been, in general, well established over the last decades, tested by several intercomparisons [2,4], and also confirmed by the current experiment. Although during the measurement of stable radiation sources the observations in recorded time series are not always completely independent, their autocorrelation can be accounted for and analysis results (including determination of reference or consensus values) is rather straightforward. Small variability between individual sensors found during the current experiment confirms usefulness of the radiometric calibration performed at the same laboratory just before the comparisons.

According to [5] calibration is an operation that, under specified conditions, in a first step, establishes a relation between the quantity values with measurement uncertainties provided by measurement standards and corresponding indications with associated measurement uncertainties and, in a second step, uses this information to establish a relation for obtaining a measurement result from an indication.

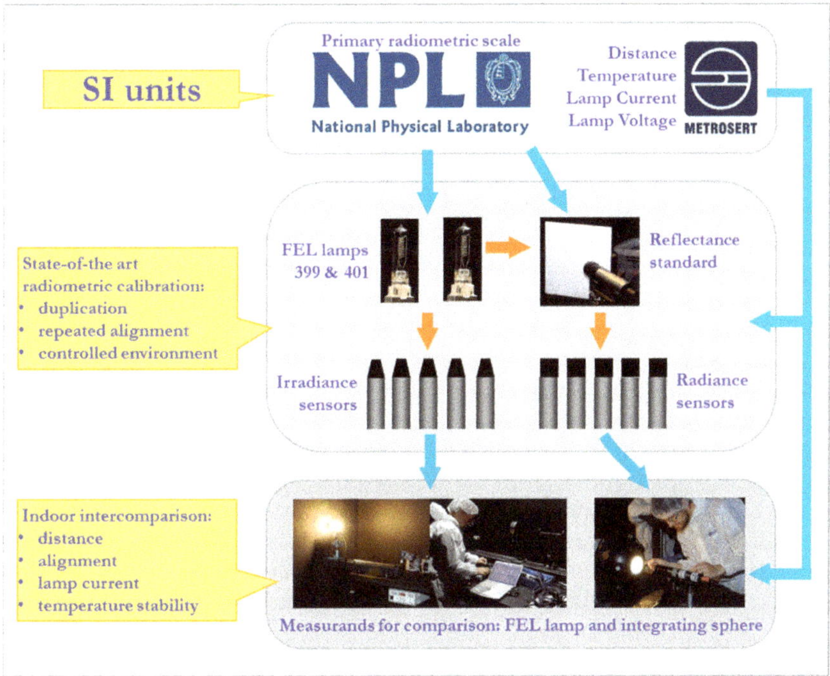

Figure 1. Traceability scheme of the LCE-2 for validation of indoor measurement uncertainties as specific to the present study.

For determination of spectral responsivity of a radiometer, it is usually calibrated against a known source placed at a specified distance from the entrance optics of the radiometer. Such a calibration procedure is well established and validated [6–12]. Unfortunately, specified conditions during the calibration may be quite different from varying conditions which may prevail during later use of the instrument. For radiometric sensors, there can be significant differences between calibration and later use in the field, in regards to the operating temperature, spectral variation of the target (giving different spectral stray light effects), angular variation of the light field (especially for irradiance sensors) and the intensity of the measured radiation. Each of these factors may interact with instrument imperfections to add further uncertainties when an instrument is used in the field and estimation of such uncertainties requires instrument characterization in addition to the well-established absolute radiometric calibration.

Instrument characterization, which can lead to corrections to reduce uncertainties, should include determination of thermal effects, nonlinearity, spectral stray light effects, wavelength calibration, angular response, and polarization effects. Procedures for determination of corrections, including measurement of all relevant influence quantities, are much less studied, and for some instruments often corrections might be not available. For applying corrections, individual testing of radiometers for each effect considered is indispensable. For most of the corrections, tests may be more time consuming than the radiometric calibration. Generally, the corrections should be applied both for calibration spectra and for field spectra calculated using the calibration coefficients, critically increasing the impact of data handling. Fortunately, these individual tests are carried out usually only once in the lifetime of an instrument unit (i.e., a sensor from an instrument family/design with a unique serial number) while radiometric calibration has to be performed on a regular basis at least once a year. Methods for correcting temperature effects [2,4,13–17], spectral stray light effects [4,18–22], nonlinearity [4,15,17,23], and polarization effects [24] are the most studied. Nevertheless, difficulties may arise during the use of a calibrated instrument when some parameters influencing correction should be determined. Some radiometers do not have internal temperature sensors, and therefore, for these instruments the accuracy of temperature corrections is limited even when external temperature sensors are applied during the calibration and later use [2]. Nonlinearity effects present in calibration spectra can be determined rather satisfactorily, but it can be much more difficult to account for nonlinearity when instable natural radiation sources are measured. Effect due to response error of cosine collector [25,26] can be satisfactorily accounted for a well-known radiation source, but in the field conditions the angular distribution of radiation is often not known accurately enough for efficient correction of the cosine error.

This study aims to evaluate techniques and procedures needed for improvement of the traceability of the OC field measurements. In order to improve the consistency of measurements, in this work and in [3] unified and enhanced metrological specification of radiometers, additional individual testing procedures for relevant systematic effects of sensors, and procedures for unified data handling are discussed. Most of the instruments involved in LCE-2 were hyperspectral radiometers having hundreds or thousands of spectral bands and different spectral response functions. Even when the instruments are of the same type, they are not directly comparable to each other due to small manufacturing differences. For instance, each radiometer has individual spectral response function represented by a wavelength scale with different center wavelengths (CWL) of individual bands. Therefore, for comparing the data of all the instruments, a few Sentinel-3 Ocean and Land Color Instrument (OLCI) bands were selected and from the hyperspectral data the OLCI band values were retrieved. The intercomparison analysis was performed using the OLCI band values.

2. Material and Methods

2.1. Participants of the LCE-2

In total, 11 institutions were involved in the LCE-2, see Table 1. Altogether, 44 radiometric sensors from 5 different manufacturers were involved (Figure 2). The list of radiometers reflects the typical selection of instruments used for shipborne validation of satellite-derived water reflectance ('ocean color validation'). However, the number of each type of instrument is not necessarily representative of total validation data usage, since the SeaPRISM instrument is used by a multi-site network of autonomous systems [27], thus providing very significant quantities of validation data. As denoted by the combination "(2L, 1E)" in Table 1, most of the participating teams use an above-water field measurement protocol with three radiometers: two radiance sensors, for upwelling (water) and downwelling (sky) radiances, respectively, and an irradiance sensor, measuring downwelling irradiance. For the RAMSES and HyperOCR this is achieved by three separate devices, while the WISP-3 contains three spectroradiometers integrated into a single device, and the SR-3500 uses a single spectrometer equipped with interchangeable entrance optics for irradiance and radiance (and can, like all radiance sensors, be used sequentially to measure both upwelling and downwelling radiance). The SeaPRISM estimates irradiance (E) from direct sun radiance (L)—see [27]. In the scope of laboratory measurements, the multiple entrance optics of SR-3500 and WISP-3 were treated as separate radiometers. Technical parameters of the participating radiometers are given in Table 2.

Water reflectance can also be measured from underwater radiometers deployed either at fixed depths or during vertical profiles. Indeed, the RAMSES and HyperOCR designs (but not WISP-3, SR-3500, SeaPRISM) may also be used underwater. The present study is fully relevant for the calibration aspects of such radiometers in underwater applications, although extra characteristics, particularly immersion coefficients to transfer in-air calibrations to in-water [28] must also be studied.

Table 1. Institutes and instruments participating in the LCE-2 intercomparison.

Participant	Country	L—Radiance; E—Irradiance Sensor
Tartu Observatory (pilot)	Estonia	RAMSES (2 L, 1 E) WISP-3 (2 L, 1 E)
Alfred Wegener Institute	Germany	RAMSES (2 L, 2 E)
Royal Belgian Institute of Natural Sciences	Belgium	RAMSES (7 L, 4 E)
National Research Council of Italy	Italy	SR-3500 (1 L, 1 E) WISP-3 (2 L, 1 E)
University of Algarve	Portugal	RAMSES (2 L, 1 E)
University of Victoria	Canada	OCR-3000 (OCR-3000 is the predecessor of HyperOCR) (2 L, 1 E)
Satlantic; Sea Bird Scientific	Canada	HyperOCR (2 L, 1 E)
Plymouth Marine Laboratory	UK	HyperOCR (2 L, 1 E)
Helmholtz-Zentrum Geesthacht	Germany	RAMSES (2 L, 1 E)
University of Tartu	Estonia	RAMSES (1 L, 1 E)
Cimel Electronique S.A.S	France	SeaPRISM (1 L)

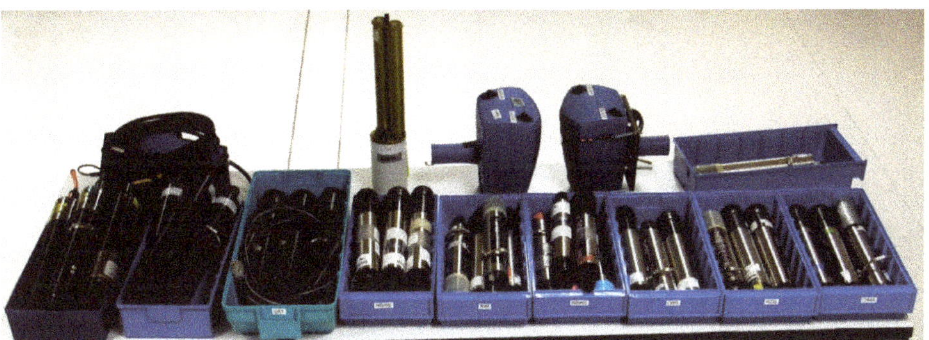

Figure 2. Instruments participating in the LCE-2 intercomparison.

Table 2. Technical parameters of the participating radiometers.

Parameter	RAMSES	HyperOCR	WISP-3	SR-3500	SeaPRISM
Field of View (L/E)	7°/cos	6° (According to the manufacturer, the HyperOCR radiance sensors 444 and 445 have 6° FOV.) or 23°/cos	3°/cos	5°/cos	1.2°/NA
Manual integration time	yes	yes	no	yes	no
Adaptive integration time	yes	yes	yes	yes	yes
Min. integration time, ms	4	4	0.1	7.5	NA
Max. integration time, ms	4096	4096	NA	1000	NA
Min. sampling interval, s	5	5	10	2	NA
Internal shutter	no	yes	no	yes	yes
Number of channels	256	256	2048	1024	12
Wavelength range, nm	320...1050	320 ... 1050	200 ... 880	350 ... 2500	400 ... 1020
Wavelength step, nm	3.3	3.3	0.4	1.2/3.8/2.4	NA
Spectral resolution, nm	10	10	3	3/8/6	10

2.2. Calibration of Irradiance Sensors

A FEL type 1000 W quartz tungsten halogen spectral irradiance standard lamp was used for radiometric calibration of the radiometers [4]. The lamp was operated in constant current mode with a stabilized radiometric power supply Newport/Oriel 69935 ensuring proper polarity as marked on the lamp. A custom designed circuit was used for monitoring the lamp current through a 10 mΩ shunt resistor P310 and providing feedback to the power supply. Lamp current was stabilized to better than ±1 mA. The same feedback unit was used for logging the lamp current and voltage. Voltage was measured with a four-wire sensing method from the connector of the lamp socket. The power supply was turned on and slowly ramped up to the working current of the lamp. Calibration measurements were started after at least a 20-min warm-up time. The voltage across the lamp terminals was compared to the reference value measured during the last calibration of the lamp. A significant change in the operating voltage would have suggested that the lamp was no longer a reliable working standard of spectral irradiance. In addition, the lamp's output was monitored by a two-channel optical sensor to detect possible short-term fluctuations. On completion of the calibration, the lamp current was slowly ramped down to avoid shocking the filament thermally.

The lamp and OC radiometer subject to calibration were mounted on an optical rail that passed through a bulkhead which separated the lamp and radiometer during calibration. A computer-controlled electronic shutter with a Ø60 mm aperture was attached to the bulkhead. The shutter was used for dark signal measurements during the calibration. Two additional baffles with Ø60 mm apertures were placed between the bulkhead and the radiometer at the distance of 50 mm and 100 mm from the bulkhead.

The OC radiometer subject to calibration and a filter radiometer next to it were mounted on a computer-controlled linear translation stage that allowed perpendicular movement with respect to the optical rail. Before calibration the positions of both radiometers were carefully adjusted and the translation stage positions saved in the controlling software. This allowed fast and accurate swapping of the radiometers after the lamp was turned on. In case of two groups of instruments (RAMSES and HyperOCR), several units having a common identical outside diameter in a group allowed a use of a V-block for fast mounting of the radiometers during the calibration. Before the lamp was turned on, the distance between the lamp and sensor was individually measured for each instrument, and a clamp was attached to fix the sensor at the appropriate position. During calibration, the radiometers of the same type were swapped without turning off the lamp. Placing the clamp against the end of the V-block ensured proper distance between the lamp and the radiometer during calibration.

The distance between the lamp and the radiometer was set with a custom designed measurement probe. One end of the probe was placed against the lamp socket reference surface and the other end of the probe had two lasers with beams intersecting at 120° angle. The intersection point defined, in a contactless way, the other end of the probe. Such design allowed distance measurement without touching the diffuser surface of the radiometer. The distance probe was calibrated by using a SI-traceable 500 mm micrometer standard. The uncertainty of distance determined with the probe was better than 0.2 mm.

The filter radiometer was used for monitoring possible long-term drifts in the optical output of the standard lamp. The filter radiometer was based on a three-element trap detector with Hamamatsu S1337-1010Q windowless Si photodiodes and temperature-controlled bandpass filters with peak transmittances at nominal wavelengths of 340 nm, 350 nm, 360 nm, 380 nm, 400 nm, 450 nm, 500 nm, 550 nm, 600 nm, 710 nm, 800 nm, 840 nm, 880 nm, 940 nm, and 980 nm. The photocurrent of the filter radiometer was amplified and digitized with a Bentham 487 current amplifier with integrating analog digital converter (ADC). A Newport 350B temperature controller was used for stabilizing the temperature of the bandpass filters. The filters were changed manually and it took about two minutes for the temperature of the filter to stabilize. Air temperature, relative humidity, and atmospheric pressure in the laboratory were recorded by a device located in the sensor compartment.

At least two different integration times were used for each radiometer (except in the case of the SeaPRISM and WISP-3 instruments for which the manufacturer-provided standard measurement programs were used). After a warm up time, at least 30 spectral measurements were collected measuring the radiation from the lamp. In the case of WISP-3 with internally selected integration time and averaging 10 spectra were collected. Next, the shutter in front of the lamp was closed and the same number of spectral measurements were collected, in order to estimate dark signal and ambient stray light in the laboratory. All measurements were repeated at least twice, including readjustment of the lamp and the sensor.

NPL provided two Gigahertz-Optik BN9101-2 FEL type irradiance calibration standard lamps with S/N 399 and 401 for the LCE-2 exercise in order to relate all measurements performed in the FRM4SOC campaigns to the common traceability source. The lamps were calibrated at the NPL and had not been used since the last calibration. Differences of the spectral flux of the two lamps in the range from 340 nm to 980 nm according to the aforementioned filter radiometer were within ±0.5%. The drift of the irradiance values (at 500 nm) measured during the calibration campaign was ~0.1% which is close to the detection limit of the filter radiometer. In certificates issued for LCE-2 radiometers, arithmetic mean of the responsivities measured by the two lamps was used.

2.3. Calibration of Radiance Sensors

Radiance sensor calibration setup was based on the lamp/plaque method and utilized the components from the irradiance sensor calibration setup [4]. A Sphere Optics SG3151 (200 × 200) mm calibrated white reflectance standard was mounted on the linear translation stage next to the filter radiometer. Normal incidence for the illumination and 45° from normal for viewing were used. The panel was calibrated using the same illumination and viewing geometry at the NPL just before the LCE-2 exercise. A mirror in a special holder and an alignment laser were used for aligning the plaque and radiance sensor. As in the case of irradiance sensors, at least 30 calibration and background spectra were acquired using two different integration times (3 readings for SeaPRISM and 10 spectra for WISP-3). All measurements were repeated at least twice, including readjustment of the lamp, plaque, and radiometers.

2.4. Indoor Experiment of the LCE-2

The indoor experiment took place in the optical laboratory of TO within a few days after the radiometric calibration. The radiance and irradiance experiments were simultaneously set up and running during two days. Measurements were carried out by project participants under the supervision of TO's personnel.

2.4.1. Irradiance Comparison Setup of the LCE-2

The irradiance setup can be seen in Figure 3. A FEL lamp was used as a stable irradiance source for indoor intercomparison. The power supply, current feedback unit, monitor detector, and distance measurement probe were the same as used during the radiometric calibration, but the FEL lamp and measurement distance were different. In order to change and align the radiometers without switching off the lamp, an additional alignment jig was placed between the shutter and the radiometer. When the shutter was closed, it was possible to change and realign the radiometer with respect to the jig. The alignment jig support was fixed to the optical rail during the whole intercomparison experiment and was used as a reference plane for distance measurement. During the intercomparison, the FEL source was switched off only once in the evening of May 9, the first day of the indoor exercise.

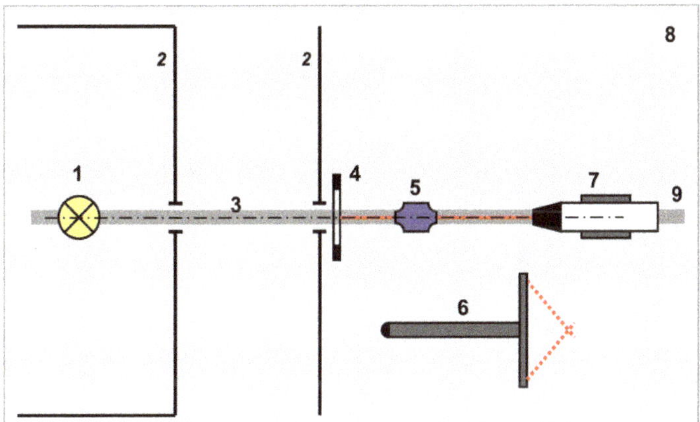

Figure 3. Indoor irradiance comparison. 1—FEL lamp; 2—baffles; 3—main optical axis; 4—alignment jig; 5—alignment laser; 6—distance probe; 7—radiometer on the support; 8—optical table; 9—optical rail.

Each participant measured the irradiance source using two different integration times (with corresponding dark series with the closed shutter) and one series with the instrument rotated by 90° around the optical axis. The measurement series were expected to contain at least 30 readings. As an exception, for the WISP-3 instruments two series (including re-alignment) of 10 readings were recorded.

2.4.2. Radiance Comparison Setup

The radiance setup for indoor intercomparison is depicted in Figure 4. A Bentham ULS-300 integrating sphere with internal illumination was used as the radiance source. ULS-300 is a Ø300 mm integrating sphere with Ø100 mm target port. According to the manufacturer, the uniformity of radiance over the output aperture is ±0.05% independent of the intensity setting. The sphere has a single 150 W quartz tungsten halogen light source (Osram Sylvania HLX 64640) and an eight-branch fiber bundle for transporting the light into the sphere. The sphere has a variable mechanical slit between the light source and the fiber bundle which allows changing the intensity of the light inside the sphere while maintaining the spectral composition of light which corresponds to correlated color temperature of (3100 ± 20) K. The lamp was powered by a Bentham 605 power supply at 6.3 A. A Gigahertz-Optik VL-3701-1 broadband illuminance sensor attached directly to the sphere was used as a monitor detector. The current of the monitor detector was recorded by an Agilent 3458A multimeter and the lamp voltage was measured by a Fluke 45 multimeter. Each participant measured the sphere source at two radiance levels and two distances from the sphere. The current reading of the monitor detector was used for setting the same sphere radiance levels for all the participants. For low radiance measurements, 1 µA monitor current was used corresponding roughly to the typical water radiance at 490 nm during field measurements, whereas 10 µA monitor current was used to simulate typical sky radiance. Obviously, the spectral composition of the incandescent sphere source did not match the field spectra, but was rather similar to the emission of the FEL-type radiometric calibration standard. In addition to sphere radiance, dark measurements were recorded by placing a black screen between the sphere and the radiometer. The sphere radiance was measured at two distances, typically 17 cm and 22 cm from the sphere port. Although the radiance measurement should not depend on measurement distance as long as the sphere port overfills the field-of-view (FOV) of the radiometer, the results measured at two distances were used to estimate the uncertainty component caused by back-reflection from the radiometer into the sphere.

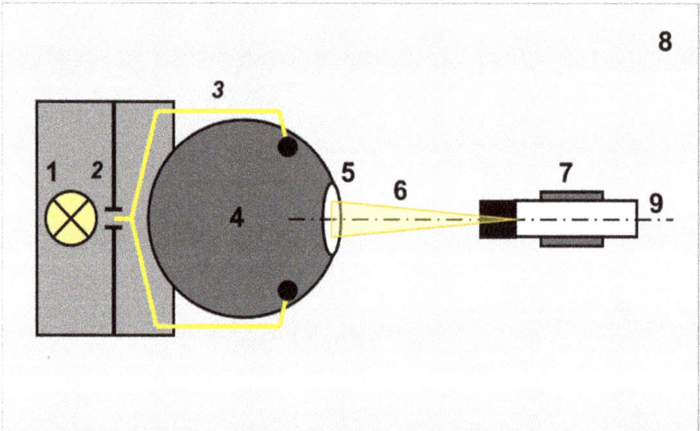

Figure 4. Indoor radiance comparison. 1—quartz tungsten halogen lamp; 2—variable slit; 3—optical fiber; 4—integrating sphere; 5—output port; 6—FOV of the radiometer; 7—radiometer on the support; 8—optical table; 9—main optical axis.

3. Results

3.1. Data Handling

The measurement results, including measurement uncertainty and information about measurement parameters, were reported back to the pilot laboratory in the form of spreadsheet files by most of the participants (for 33 out of the 44 sensors involved). For the rest, the pilot carried out the data analysis based on the raw instrument data. In the case of discrepancies, the pilot repeated the calculations on raw user data applying unified data handling described in the next chapter. Nevertheless, due to differences in hardware and software of the participating radiometers, fully unified data handling was not possible.

The participants were encouraged to perform the data processing for their radiometers and report back the radiance/irradiance values with uncertainty estimates. However, in a few cases, TO accomplished/repeated the calculations for some participants as well. Data processing for the RAMSES, HyperOCR, and WISP-3 instruments was fully automated at TO by purpose-designed computer software. The source code of the software is freely available for the participants.

Data processing of each instrument was performed independently from the others and included the following steps:

- separation of the raw datafiles based on the scene (e.g. low/high radiance, distance), integration time, shutter measurements;
- pairing the raw data with corresponding shutter measurement;
- dark signal subtraction;
- linearity correction whenever applicable;
- division by radiometric responsivity;
- recalculation for the OLCI spectral bands;
- averaging;
- evaluation of the uncertainty.

3.2. Device-Specific Issues

TriOS RAMSES series instruments include both the radiance (ARC) and irradiance (ACC) sensors. The raw spectra are stored in American Standard Code for Information Interchange (ASCII) and/or Microsoft ACCESS database files. Data processing for these radiometers is fully unified based on the measured data (2-byte integer numbers) and calibration files provided by the manufacturer and TO. The detailed procedure to derive the calibrated results is described in [4]. RAMSES instruments are equipped with black-painted pixels on the photodiode array used to compensate for the dark signal and electronical drifts. The background spectrum (with the external shutter closed) was subtracted as well. For subtraction, only the spectra with matching integration times were used. Before division by the responsivity coefficients, linearity correction was applied, see Section 4.1.4.

Satlantic HyperOCR/OCR3000 series instruments include also both the radiance and irradiance sensors with similar data processing chain. The raw spectra stored in binary files were converted to ASCII by participants using the proprietary manufacturer's software. Data processing for the HyperOCR was based on the calibration file provided by TO and is similar to the RAMSES procedure. The HyperOCR radiometers are equipped with an internal mechanical shutter, deployed automatically after every fifth target spectrum. The shutter measurements were detected in the datafiles and the closest shutter measurement was subtracted from each raw spectrum before the next steps.

Water Insight WISP-3 contains a three-channel Ocean Optics JAZ module spectrometer and computer. Two of the input channels are connected to the radiance inputs while the third is attached to the irradiance adaptor. Users can start the acquisition of the spectra by pressing a button, the internal computer is setting the measurement sequence, determining the integration times, and storing the data. All three channels are acquired simultaneously and the data are stored into a single ASCII file. The spectrometers have painted detector array pixels like the RAMSES radiometers. The internal dark signal is subtracted automatically and resulting data are stored in the form of floating point numbers. The only operation needed was the division by the responsivity coefficients determined by TO using the same manual measurement sequence. The linearity correction described in Section 4.1.4 was not used.

Spectral Evolution SR-3500 spectrometer is equipped with an optical fiber input and interchangeable radiance and irradiance fore-optics. Thus, the data processing for the radiance and irradiance measurements are identical. The spectral output is stored in the ASCII files and can contain both the raw and radiometrically calibrated results based on the internal calibration coefficients. The dark signal is subtracted internally using an integrated mechanical shutter. Each target measurement is automatically followed by a dedicated dark measurement. During the radiometric calibration at TO, calibration factors to the existing coefficients were derived. The calibrated data in the files was multiplied by these factors, and finally, the linearity correction as described in Section 4.1.4 was used.

CIMEL SeaPRISM binary output was converted by the owner of the radiometer and was returned to the pilot in the form of ASCII files. Based on these data, TO derived the radiometric calibration coefficients. Neither linearity correction scheme nor re-calculation for the OLCI spectral bands was used for the SeaPRISM at this stage.

3.3. Calculation of Sentinel-3/OLCI Band Values

As the final step of data processing, the radiance and irradiance values were re-calculated for the OLCI spectral bands for each radiometer except for the multispectral SeaPRISM, in which case the initial band values were used. Based on the given CWL of the spectroradiometer λ_n, and the OLCI band definition $O_i(\lambda)$ [29], the weight factors were found for each pixel

$$C_i(n) = O_i(\lambda_n), K_i(n) = \frac{C_i(n)}{\sum_k C_i(k)}, \tag{1}$$

where n is the pixel number with CWL of λ_n, $O_i(\lambda_n)$ is the responsivity of the corresponding ith OLCI band interpolated to λ_n, and $K(n)$ is the normalized weight coefficient for n'th pixel. Finally, the radiance/irradiance value I_i for the corresponding OLCI band was calculated as

$$I_i = \sum_n I(n) \cdot K_i(n), \tag{2}$$

where $I(n)$ denotes the measured radiance/irradiance at the n'th pixel.

3.4. Consensus and Reference Values Used for the Analysis

Consensus values were calculated as median [30] of all presented comparison values. Reference values were applicable only for the indoor irradiance measurements (Figure 8), when the measurand used for this exercise was during comparison measured also with the precision filter radiometer serving as a reference.

3.5. Results of Indoor Experiment

The comparison results are presented as deviation from the consensus value. Despite the different sensor types, as the radiation sources used for indoor comparison were spectrally very similar to calibration sources, agreement between sensors was satisfactory for radiance and for irradiance sensors (Figures 5–8) with no outliers present. In these figures, blue dashed lines show the expanded uncertainty covering 95% of all data points on the right graphs. Solid lines represent RAMSES sensors, dashed lines—HyperOCR sensors, double line—SR-3500, and dotted lines—WISP-3 sensors.

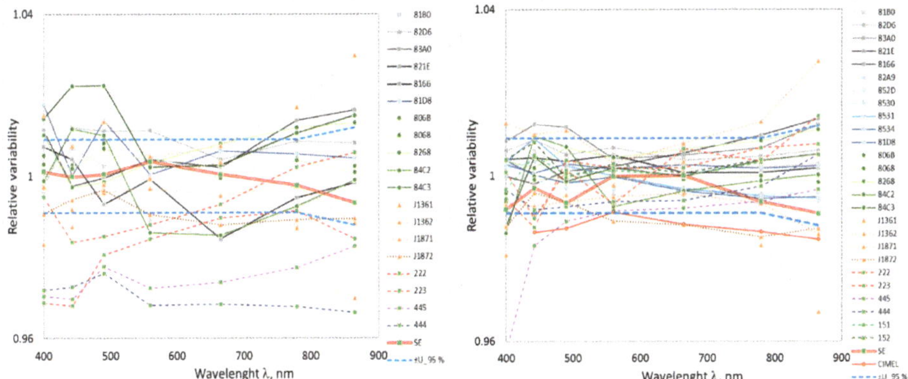

Figure 5. Low intensity radiance; agreement just after receiving data from participants (left), and after reviewing data by pilot, corrections submitted by participants and/or unified data handling by pilot (right). Blue dashed lines—expanded uncertainty covering 95% of all data points on the right graph. Solid lines—RAMSES sensors; dashed lines—HyperOCR sensors; double line—SR-3500; dotted lines —WISP-3 sensors.

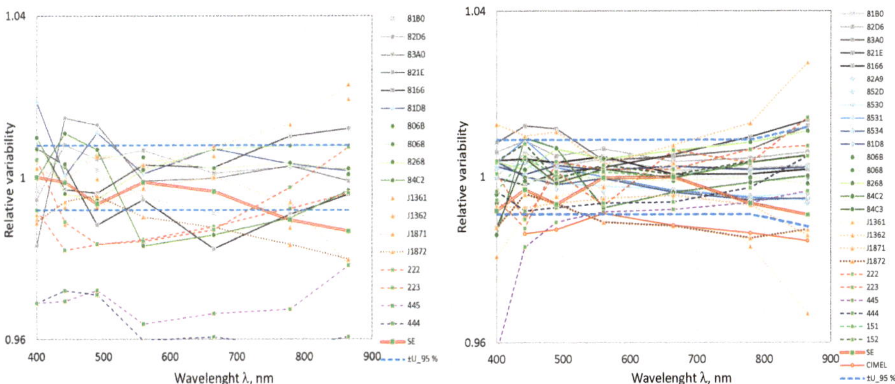

Figure 6. High intensity radiance; agreement just after receiving data from participants (left), and after reviewing data by pilot, corrections submitted by participants and/or unified data handling by pilot (right).

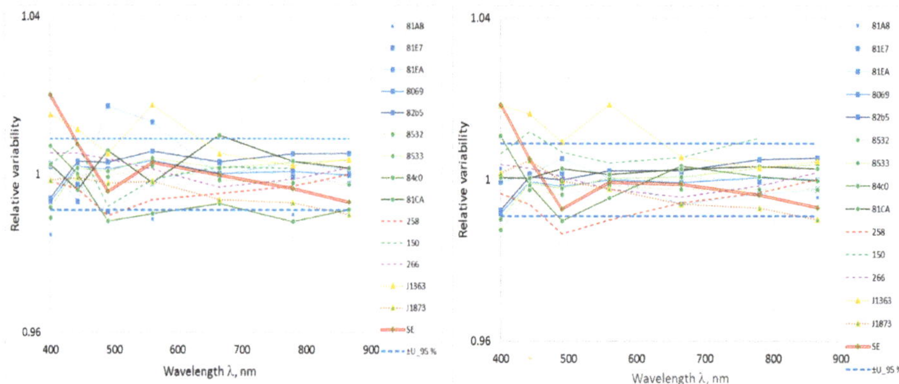

Figure 7. Irradiance sensors; agreement just after receiving data from participants (left), and after reviewing data by pilot, corrections submitted by participants and/or unified data handling by pilot (right).

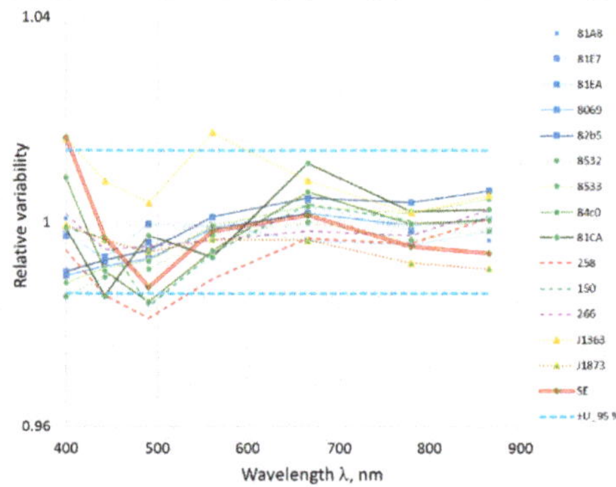

Figure 8. Irradiance sensors; agreement with reference values of the filter radiometer. Blue dashed lines—expanded uncertainty covering 95% of all data points. Uncertainty of radiometric calibration is included.

Larger variability of the results initially reported by participants was caused by applying out-of-date calibration coefficients, by diversely applying or not applying the non-linearity correction (Section 4.1.4) or calculating the OLCI band values differently. For unified data handling carried out by the pilot and described in 3.1 to 3.3, the calibration results obtained during LCE-2 were used, non-linearity correction was applied, OLCI band values were calculated by using individual weights as determined from the wavelength scale of each radiometer. After unified data handling, agreement between comparison results was significantly improved for the radiance sensors (Figures 5 and 6). There was almost no improvement in the case of the irradiance sensors in Figure 7.

4. Measurement Uncertainty

The uncertainty analysis has been carried out according to the ISO Guide to the Expression of Uncertainty in Measurement [31]. The evaluation is based on the measurement model, which describes the output quantity y as a function f of input quantities x_i: $y = f(x_1, x_2, x_3 \dots)$. Standard uncertainty is evaluated separately for each input quantity. There are two types of standard uncertainties: Type A is of statistical origin; Type B is determined by any other means. Both types of uncertainties are indicated as standard deviation and denoted by s and u respectively. The uncertainty component arising from averaging a large number of repeatedly measured spectra of radiation sources by array spectrometers is considered as Type A. Contributions from calibration certificates (lamp, current shunt, multimeter, diffuse reflectance panel, etc.), but also from instability and spatial non-uniformity of the radiation sources are considered of Type B. For all input quantities relative standard uncertainties are estimated. The relative combined standard uncertainty of output quantity is calculated by combining relative standard uncertainty of each input estimate by using Equation (12) in [31]. Uncertainty of the final result is given as relative expanded uncertainty with a coverage factor $k = 2$.

4.1. Effects Causing Variability of the Results

4.1.1. State of Radiometric Calibration

Analysis of the LCE-2 calibration results, comparing them with former calibrations, including the factory calibrations, and also with calibrations carried out on the same set of radiometers by TO one year later (before the FRM4SOC FICE-AAOT intercomparison) demonstrates the importance of radiometric calibration for SI traceable results and reveals interesting information about instability of the sensors. Some uncertainty contributions characteristic to calibration can also be estimated.

The variability of calibration coefficients of radiance and irradiance sensors due to adjustment of the lamps, plaques, and sensors, and due to short-term instability of the lamps and sensors is depicted in Figure 9. All the radiometers were calibrated before LCE-2 using the same pair of lamps (Sections 2.2 and 2.3). Two sets of calibration coefficients were obtained for each sensor and the difference between the lamps was presented as the ratio of these coefficients. The curves in Figure 9 are calculated as standard deviations from the ratios of a whole set of calibration coefficients determined by using the two standard lamps. The systematic difference between lamps (due to small difference in traceability to SI) is neglected and only the other uncertainty components related to individual setting up and measurement of radiometers are accounted for by using standard deviation. Data in Figure 9 include calibration of more than 25 sensors for LCE-2 intercomparison and also for FICE-AAOT intercomparison one year later when a different pair of lamps was used. Remarkable in Figure 9 is the rapid increase of variability between sensors in the UV region.

Figure 10 shows average long-term variability of calibration coefficients of TriOS RAMSES and Satlantic HyperOCR radiance and irradiance sensors. All the radiometers had previous radiometric calibration certificates of various origin and age. The curves in Figure 10 are calculated similarly to Figure 9 as standard deviations of the ratios of previous and the last calibration coefficients. It has to be noted, however, that in this case standard deviation is characterizing dispersion between previous calibrations as these were performed by using various standards and conditions. Many of the RAMSES and HyperOCR radiometers that participated in LCE-2 also took part in the FICE-AAOT field intercomparison experiment one year later. Those sensors were radiometrically calibrated again at TO in June 2018 before the beginning of the field campaign. This gave a good opportunity to estimate the long-term stability of the sensors while minimising other possible factors influencing the calibration result. The sensors were calibrated in the same laboratory by the same operator in similar environmental conditions using the same calibration setup and methodology. Only the calibration standard lamps were exchanged since LCE-2. Nevertheless, the L_1 yr and E_1 yr curves in Figure 10 obtained as standard deviations of the ratios of the calibrations coefficients one year apart exclude the systematic differences between lamps. The two calibrations done in the same lab one year apart showed that over 80% of the sensors have changed less than ±1%. Thus, the inherent long-term stability of the sensors is likely better than 5% to 10% revealed from the previous calibration history, where the differences were likely caused by other factors such as different calibration standards, environmental conditions, calibration setups and methodologies, etc. However, rapid changes in the responsivity of some TriOS RAMSES irradiance sensors may cause even larger deviations which cannot be explained by other factors than the instability of the sensor itself. No quick changes were observed for the RAMSES radiance sensors, however, even after omitting outliers from the stability data of irradiance sensors, the stability of RAMSES radiance sensors is still better.

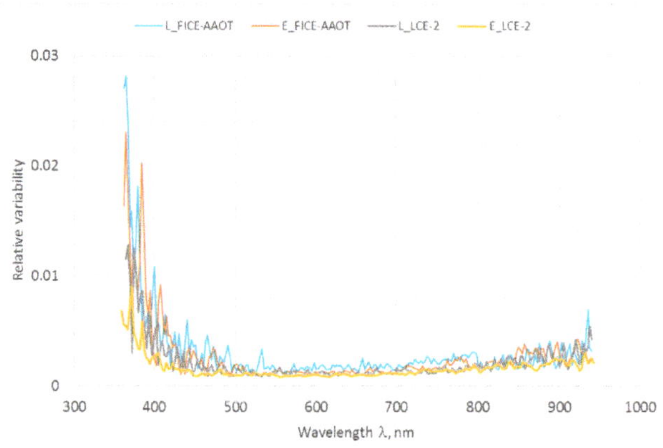

Figure 9. Relative variability of calibration coefficients of radiance (*L*) and irradiance (*E*) sensors with two different lamps used for calibration before LCE-2 and a year later before FICE-AAOT.

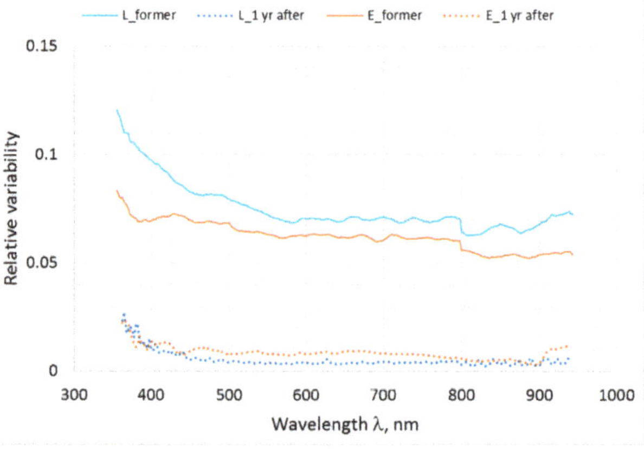

Figure 10. Relative variability of calibration coefficients of radiance (*L*) and irradiance (*E*) sensors: former—difference of previous known calibrations and results of LCE-2 calibration; 1 yr after— changes during one year after LCE-2 calibrations, some extra-large changes excluded.

4.1.2. Abrupt Changes of Responsivity

Factors causing the variability in the responsivity of radiometers were listed in [4]. During the calibration, the uncertainty of the radiation source is the dominant component in the uncertainty budget, assuming that usually the ambient temperature will be within ±1 °C. Based on the experience from LCE-2 and the following FICE activities, differences smaller than ±2% in the wavelength range of (350...900) nm can be observed between different sources used for calibration. Nevertheless, in some cases sharp changes in the responsivity of radiometers were detected, substantially exceeding all possible effects which can cause variability during calibration like the radiation source, alignment of instruments, contamination of fore-optics, temperature effects, etc. Relative change of the spectral irradiance responsivity of the TriOS RAMSES SAM_8329 10 times calibrated during eight years period is depicted in Figure 11. Each calibration in 2016–2018 consists of three repetitions conducted in a short time.

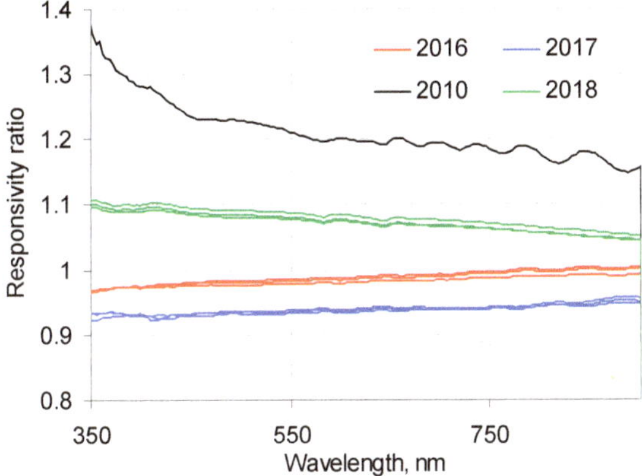

Figure 11. Relative change of responsivity of the SAM 8329. Year of the radiometric calibration is shown with color: 2010 black, 2016 red, 2017, blue, 2018 green.

4.1.3. Temperature Effects

Individual variation of the calibration coefficients as a function of temperature for each radiometer was not determined because of limited time schedule. Temperature effects for the TriOS RAMSES radiometers were evaluated based on [13] instead, see Figure 12.

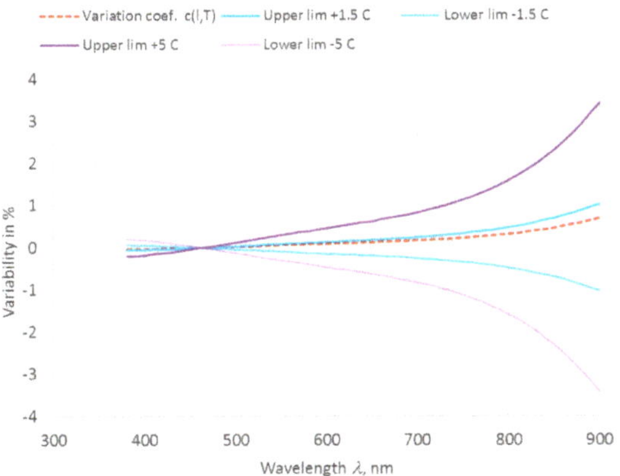

Figure 12. Relative variability of calibration coefficients due to temperature deviations from the reference temperature 21.5 °C.

4.1.4. Nonlinearity Due to the Integration Time

Maximum relative nonlinearity effect due to integration times determined from calibration spectra of TriOS RAMSES radiometers remained in the range of (1.5...3.5)% (Figure 13). Variability between the instruments due to this effect, if not corrected, will mostly be in the range of ±1%. The effect can be corrected down to 0.1% for certain types of radiometers by using the special formula, provided that there are at least two spectra with different integration times available for the same source. Derivation of the correction formula is based on the following assumptions: i) the nonlinearity effect is zero for the dark signal; ii) the effect is proportional with the recorded signal; iii) the effect is wavelength dependent; and iv) the corrected signal does not depend on the initial spectra used for estimation, i.e., it should be of the same size for all possible combinations of initial spectra. Linearity corrected raw spectrum $S_{1,2}(\lambda)$ is calculated as

$$S_{1,2}(\lambda) \;=\; \left[1 - \left(\frac{S_2(\lambda)}{S_1(\lambda)} - 1\right)\left(\frac{1}{t_2/t_1 - 1}\right)\right]S_1(\lambda). \tag{3}$$

Here $S_1(\lambda)$ and $S_2(\lambda)$ are the initial spectra measured with integration times t_1 and t_2. Minimal ratio is usually $t_2/t_1 = 2$, but may be also 4, 8, 16, etc. For large ratios $t_2/t_1 > 8$ the spectrum $S_1(\lambda)$ is close to corrected spectrum $S_{1,2}(\lambda)$ and application of nonlinearity correction is not needed. Uncertainty of corrected spectrum is predominantly determined with the uncertainty of initial spectrum measured with smaller integration time. Therefore, the smallest uncertainty of the corrected spectrum will be obtained if the initial spectra with the largest and with the second-largest non-saturating integration times are used for estimation.

The formula has been found to perform quite effectively for TriOS RAMSES and Satlantic HyperOCR radiometers in the range of (400 … 800) nm. This nonlinearity correction method is not recommended for outdoor measurements, as due to temporal variability of the natural radiation consecutive measurements with different integration times may lead to uncertainty of the corrected spectrum much larger than acceptable 0.2%.

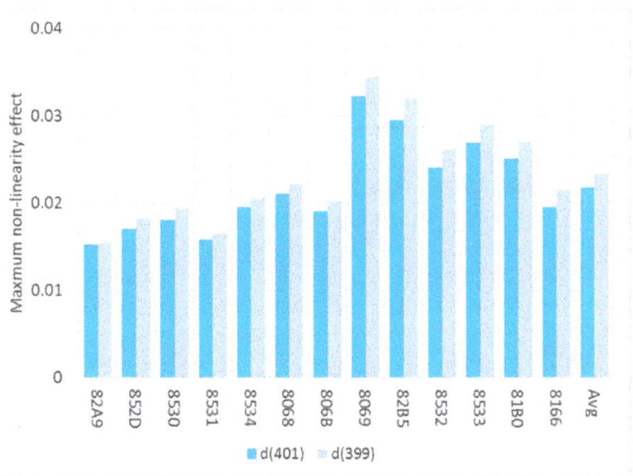

Figure 13. Maximum relative nonlinearity effect determined for 14 RAMSES sensors (both radiance and irradiance) from calibration spectra with FEL lamps 399 and 401.

From the analysis of the calibration data by using the two-spectra formula (3), it became evident that non-linearity errors scaled to full-range value of different radiance sensors behave in similar way. This behavior serves as a basis for derivation of nonlinearity correction applicable to a particular single spectrum that can also be used for the outdoor measurements, see Figure 14.

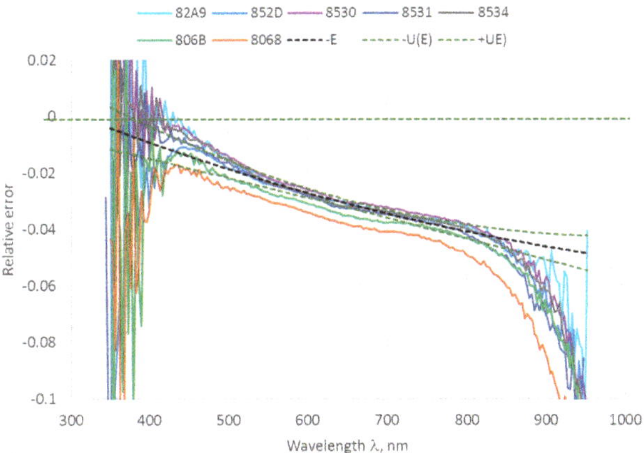

Figure 14. Non-linearity errors of different radiance sensors scaled to full-range value. Dashed lines are fitted model with uncertainty.

Relative nonlinearity correction for the full range signal $\delta x_{\max}(\lambda)$ is

$$\delta x_{\max}(\lambda) = -5.1 \cdot 10^{-8} \lambda^2 + 0.00014 \cdot \lambda - 0.0355. \tag{4}$$

Relative nonlinearity correction $\delta x(\lambda)$ for the signal $x(\lambda)$ is

$$\delta x(\lambda) = \frac{x}{x_{\max}} \delta x_{\max}(\lambda). \tag{5}$$

Corrected signal $x_{\mathrm{cor}}(\lambda)$ can be expressed as

$$x_{\mathrm{cor}}(\lambda) = x(\lambda)\left[1 + \frac{x}{x_{\max}} \delta x_{\max}(\lambda)\right]. \tag{6}$$

The formula has been thoroughly tested on the TriOS RAMSES calibration data, and is effective in the range of (400 . . . 800) nm of correcting nonlinearity mostly better than to 0.2%. The model can be fitted to all the studied RAMSES instruments by adjusting only the constant term.

4.1.5. Spectral Stray Light Effects

For many measurements, spectral stray light can lead to significant distortion of the measured signal and become a significant source of uncertainty [18,28]. Iterative technique [18,32] can be used for the simultaneous correction of bandpass and stray-light effects. When the full spectral stray light matrix (SLM) of a spectrometer is known, the stray light contribution can be removed from the measured signal and the original source spectrum restored. The stray light correction for a remote sensing reflectance measurement made by a common three-radiometer above-water system means that altogether six raw spectra have to be corrected—two for each radiometer, because stray light correction needs to be applied also for the standard source spectrum during the radiometric calibration.

The SLM was known for some radiometers from previous characterization such as for RAMSES sensors of TO, and for HyperOCR sensors of Plymouth Marine Laboratory (PML). Figure 15 presents the impact of stray light correction, evaluated for indoor measurements. The indoor radiance and irradiance sources were spectrally similar to the calibration sources; therefore, the stray light correction has relatively small impact. WISP-3, SR-3500, and SeaPRISM have different optical design, thus, their spectral stray light properties can have different nature compared to the data presented in Figure 15.

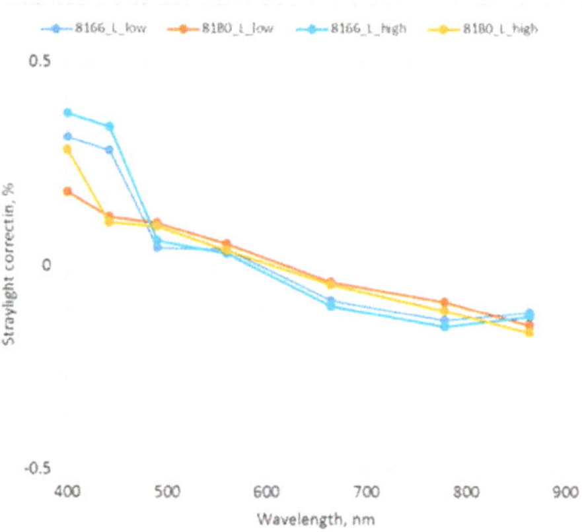

Figure 15. Stray light effects for indoor radiance measurements. Two RAMSES radiance sensors at high and low sphere radiance.

4.2. Uncertainty Budgets for Indoor Comparisons

Uncertainty analysis is made for the laboratory irradiance and radiance measurements. Uncertainty estimates for irradiance sensors measuring an FEL source at approximately 1 m distance are given in Table 3, and for radiance sensors measuring integrating sphere, in Table 4. All the uncertainty estimations of TriOS RAMSES sensors besides experimental data are based on information from [2,24,28,33–36]. For the other radiometer models that took part in the intercomparison, very little publicly available information can be found regarding various instrument characteristics that influence the measurement results [37]. In addition, the RAMSES was the only sensor model that was represented in sufficiently large number for statistical analysis.

The uncertainty is calculated from the contributions originating from the spectral responsivity of the radiometer, including data from the calibration certificate, from interpolation of the spectral responsivity values to the designated wavelengths and/or spectral bands, from instability of the array spectroradiometer, from contribution to the spectral irradiance and/or radiance due to setting and measurement of lamp current, from measurement of the distance between the lamp and input aperture of the radiometer, from the spatial uniformity of the irradiance at 1 m distance, and from reproducibility of the alignment. For the radiometer, uncertainty contributions arising from the non-linearity, temperature effects, spectral stray light, and from dark measurements, as well as from repeatability and reproducibility of averaged signal are included.

The radiometric calibration uncertainty in the following tables is not added to the combined uncertainty because all participating radiometers were calibrated using common standards shortly before the intercomparison and the contribution from calibration when using unified data handling does not affect the relative differences of participants to each other. The following uncertainty budgets describe variability between individual sensors, while uncertainty of radiometric calibration and contributions of systematic effects which influence the instruments in a similar way are not accounted for. Nevertheless, certainly they are relevant for traceability to SI units.

Table 3. Relative uncertainty budget for the irradiance (in percent), based on spread of individual sensors measuring the same lamp during indoor comparison. Data highlighted in green are not used for combined and expanded uncertainties.

	400 nm	442.5 nm	490 nm	560 nm	665 nm	778.8 nm	865 nm
Certificate	0.88	0.68	0.65	0.62	0.59	0.62	0.56
Interpolation	0.5	0.2	0.3	0.2	0.2	0.1	0.1
Instability (sensor)	0.05	0.03	0.04	0.03	0.04	0.03	0.02
Alignment	0.1	0.1	0.1	0.1	0.1	0.1	0.1
Nonlinearity	0.2	0.15	0.15	0.15	0.15	0.15	0.2
Stray light (sensor)	0.2	0.2	0.2	0.2	0.2	0.2	0.2
Temperature	0.02	0.01	0.01	0.03	0.09	0.2	0.38
Instability (source)	0.14	0.14	0.12	0.11	0.1	0.09	0.08
Uniformity	0.1	0.1	0.1	0.1	0.1	0.1	0.1
Stray light (source)	0.1	0.1	0.1	0.1	0.1	0.1	0.1
Signal, type A	0.11	0.04	0.02	0.02	0.01	0.02	0.04
Combined ($k = 1$)	0.63	0.39	0.45	0.38	0.39	0.39	0.52
Expanded ($k = 2$)	1.3	0.8	0.9	0.8	0.8	0.8	1.0

Table 4. Relative uncertainty budget for the radiance (in percent) based on spread of individual sensors measuring the same integrating sphere during indoor comparison. Data highlighted in green are not used for combined and expanded uncertainties.

	400 nm	442.5 nm	490 nm	560 nm	665 nm	778.8 nm	865 nm
Certificate	1.2	0.78	0.76	0.73	0.71	0.73	1.35
Interpolation	0.5	0.2	0.3	0.2	0.2	0.1	0.1
Instability (sensor)	0.04	0.03	0.02	0.01	0.01	0.02	0.01
Back-reflection	0.1	0.1	0.1	0.1	0.1	0.1	0.1
Alignment	0.1	0.1	0.1	0.1	0.1	0.1	0.1
Nonlinearity	0.2	0.15	0.15	0.15	0.15	0.15	0.2
Stray light (sensor)	0.2	0.2	0.2	0.2	0.2	0.2	0.2
Temperature	0.02	0.01	0.01	0.03	0.09	0.2	0.38
Instability (source)	0.14	0.14	0.12	0.11	0.1	0.09	0.08
Uniformity	0.1	0.1	0.1	0.1	0.1	0.1	0.1
Stray light (source)	0.1	0.1	0.1	0.1	0.1	0.1	0.1
Signal, type A	0.12	0.07	0.04	0.02	0.03	0.03	0.06
Combined ($k = 1$)	0.64	0.41	0.46	0.39	0.40	0.40	0.53
Expanded ($k=2$)	1.3	0.8	0.9	0.8	0.8	0.8	1.1

4.3. Uncertainty Components in Tables 3 and 4

4.3.1. Calibration Certificate

Calibration certificate of the radiometer provides calibration points of radiometric responsivity following the individual wavelength scale of the radiometer. This component is presented only for reference and is not included in the combined and expanded uncertainties.

4.3.2. Interpolation

Interpolation of radiometer's data is needed due to differences between individual wavelength scales of the radiometers. Therefore, measured values were transferred to a common scale basis (Sentinel-3/OLCI bands) for comparison, see 3.3. The uncertainty contribution associated with interpolation of spectra is estimated from calculations using different interpolation algorithms. The weights used for binning hyperspectral data to OLCI bands depend on the wavelength scale and exact pixel positions of the hyperspectral sensor. In Table 3, the interpolation components include the contribution of wavelength scale uncertainty estimated from data presented in Figure 14 of [3].

4.3.3. Temporal Instability of Radiometer

Instability of the radiometric responsivity can be estimated from data of repeated radiometric calibrations. For LCE-2, the instruments were calibrated just before the comparisons and only short-term instability relevant for the time needed for the measurements has to be considered. The values are derived from the data collected in calibration sessions of LCE-2 and FICE-AAOT a year later, see Section 4.1.1 and Figure 9. The variability over two weeks was interpolated from the yearly variability data. In addition to instability of the sensors the data in Figure 9 includes other uncertainty components related to the calibration setup (e.g., alignment, short-term lamp instability, etc.).

4.3.4. Back-Reflection

Back-reflection from the radiometer into the integrating sphere was estimated using different distances between the sphere and the radiometer as contribution of radiation reflecting from the radiance sensor back into the integrating sphere.

4.3.5. Polarization

The polarization effect was estimated by indoor irradiance measurements, repeating cast after radiometer was rotated 90° around its main optical axis, and revealing so the combined effect of alignment and polarization. According to [38] the FEL emission is polarized about 3%. As reported in [24], the polarization sensitivity of RAMSES irradiance sensors is varying from (0.05 ... 0.3)% at 400 nm to (0.3 ... 0.6)% at 750 nm. Due to the depolarizing nature of the cosine collector this effect is smaller than polarization sensitivity of RAMSES radiance sensors. Therefore, the observed differences with rotated sensors are mostly caused by other effects like alignment, instability of measured source, etc., and from the indoor irradiance uncertainty budget the polarization component is omitted. Polarization is also not included in the indoor radiance uncertainty budget as the integrating sphere is a strong depolarizer.

4.3.6. Alignment

Evaluation of alignment errors includes determination of distance between the source and the reference plane of the cosine collector, measured along the optical axis. Alignment includes also position errors of the lamp source across optical axes, rotation errors of the lamp [39], and positioning errors of the input optics of the radiometer. Combined alignment and positioning errors are included in variability data of radiometers calibrated with two different lamps (Figure 9).

4.3.7. Nonlinearity

Due to nonlinearity, some hyperspectral radiometers, measuring at different integration times, may show relative differences up to 4% (see Figure 12 in Section 4.1.4). According to recommendations, the non-linearity effects of good sensors should be correctable to less than 0.1%. The non-linearity correction (3) was applied to both calibration and measurement spectra, with residues expected to be less than 0.2%.

4.3.8. Spectral Stray Light

Spectral stray light of sensors is commonly not very relevant for measurements when the calibration and target source emissions have similar spectral composition. Value is estimated from Figure 14 in Section 4.1.5, and from [22,32].

4.3.9. Temperature

For array spectroradiometers with silicon detectors, the present estimate of standard uncertainty due to temperature variability (±1.5 °C) in the spectral region from 400 nm to 700 nm is around 0.1% and will increase up to 0.6% for longer wavelengths (950 nm) [13].

4.3.10. Temporal Instability of Radiation Source

The short-term instability of the source is relevant for the indoor measurements as they were not made simultaneously by all the participants. Thus, the time needed for intercomparison measurements, including power cycling the source between the two days of indoor experiment, has to be considered. This uncertainty component was estimated using the uncertainty in setting the lamp current and its effect on lamp emission. The drift of the irradiance values (at 500 nm) measured during the calibration campaign was ~0.1%, (2.2).

4.3.11. Stray Light in Laboratory

Sources of stray light are associated with the stray light in the laboratory during the indoor experiment. This component has been estimated in previous experiments made in the Tartu Observatory.

4.3.12. Type A Uncertainty of Repeated Measurements

For Type A uncertainty of time series of indoor measurements, white noise of measured series can be often expected. The analysis has indicated that sometimes the measurements are not completely independent and the autocorrelation of time series has been accounted for. If there is autocorrelation in the time series, the effective number of independent measurements n_e has to be considered instead of actual number of points n_t in the series [40]

$$n_e \approx n_t \frac{1 - r_1}{1 + r_1}, \tag{7}$$

where r_1 is the lag-1 autocorrelation of the time series.

5. Discussion and Conclusions

The LCE-2 exercise consisted of three sub-tasks: SI-traceable radiometric calibration of participating radiometers just before the intercomparison; laboratory intercomparison of stable lamp sources in controlled environment; outdoor intercomparison of natural radiation sources over terrestrial water surface. Altogether, 44 radiometric sensors from 11 institutions were involved: 16 RAMSES, 2 OCR-3000, 4 HyperOCR, 4 WISP-3, 1 SeaPRISM and 1 SR-3500 radiance sensors, and 10 RAMSES, 1 OCR-3000, 2 HyperOCR, 2 WISP-3, and 1 S R-3500 irradiance sensors. Additionally, the majority of sensors involved in LCE-2 were recalibrated at TO a year later (for FICE-AAOT) giving estimate for their long-term stability. More than 80% of the sensors changed during one year less than ±1%.

Agreement between the radiometers is mostly affected by the calibration state of sensor. For example, factory calibrations made at different times can cause differences exceeding ±10%. Former calibrations in different laboratories from several years ago can cause differences around ±3%. Different calculation schemes (corrections for non-linearity, stray light or for OLCI band values) can cause differences about ±1 ... 2% each factor. The best agreement of 0.5 ... 0.8% between participants has been achieved when measurements were carried out just after calibration and for data handling unified procedures have been used including application of nonlinearity correction and the same algorithm for calculation of OLCI band values.

Dependence of the calibration coefficients on temperature can also cause significant deviation from SI-traceable result, especially in the near-infrared spectral region. For maximum temperature difference of about 20 °C between calibration and later measurements (typically between 0 °C and 40 °C) a responsivity change more than 10% is possible [2,13]. For laboratory measurements in controlled environment the temperature effect is expected to be within (0.1 ... 0.5)%.

Effect of stray light correction evaluated for indoor measurements in the range (400 ... 700) nm has been less than 0.5%. Though, outside the range of (400 ... 700) nm the relative uncertainty may increase substantially if correction is not applied.

Maximum value of the nonlinearity effect due to integration times determined from calibration spectra of TriOS RAMSES radiometers for a group of 15 radiometers was in the range of (1.5...4)%. At the same time, variability between the instruments due to this effect if not corrected, remained within ±1% due to the systematic nature of the nonlinear behavior affecting all the instruments in similar manner. During laboratory measurements the non-linearity correction was applied to both calibration and measurement spectra, with residues expected to be less than 0.2%.

Field of view and cosine responsivity effects can significantly depend on the limits of error set by specifications of radiometers, and on results of individual tests showing how large is deviation from the specified values. In the laboratory, the cosine responsivity error of the sensor during calibration was close to the error during the intercomparison measurements due to similar illumination geometry, and therefore, the resulting systematic error is insignificant.

Through the indoor experiment, when conditions for later measurements and conditions specified for calibration were quite similar, high effectiveness of the SI-traceable radiometric calibration has been demonstrated, a large group of different type radiometers operated by different scientists achieved satisfactory consistency between results showing low standard deviations between radiance (27 in total) or irradiance (15 in total) results (s < 1%). This is provided when some unification of measurement and data processing was settled: alignment of sensors, structure of collected data, application of unified wavelength bands and non-linearity corrections. Nevertheless, variability between sensors may be insufficient for complete quantification of uncertainties in measurement. For example, standard deviation of nonlinearity estimates versus the mean effect (Figure 13) demonstrates that differences are not able to reveal full size of systematic errors common for all the instruments. Therefore, all radiometers still should be individually tested for all significant systematic effects which may affect the results as this is the only way to get full estimate of the effects degrading traceability to the SI scale.

Furthermore, in the frame of the outdoor experiment when conditions specified for calibration and in field are much more different from each other the variability between freshly calibrated individual sensors did increase substantially, demonstrating the limitation of typical OC field measurements with sensors having SI-traceable radiometric calibration. Including laboratory intercomparison to the LCE-2 exercise has clearly shown that further reduction of the uncertainty of radiometric calibration of sensors will not improve the agreement between field results significantly. Much more relevant for achieving better SI-traceability of field measurements are improved specifications of radiometers, additional characterization of individual sensors accounting for specific field conditions, and unified data handling which will be considered in [1].

Author Contributions: Conceptualization, V.V., J.K., I.A., R.V., K.A., K.R., C.D., T.C.; Data curation, V.V., I.A.; Formal analysis, V.V., J.K., I.A.; Investigation, J.K., I.A., R.V., K.A., K.R., A.A., M.B., H.B., M.C., D.D., G.D., B.D., T.D., C.G., K.K., M.L., B.P., G.T., R.D., S.W.; Methodology, V.V., J.K., I.A., K.A., K.R.; Project administration, R.V.; Software, I.A.; Validation, V.V., J.K., I.A.; Visualization, V.V., J.K., I.A., K.R., M.L.; Writing—original draft, V.V., J.K., I.A., R.V., K.A.; Writing—review & editing, V.V., J.K., I.A., R.V., K.A., K.R., A.A., M.B., H.B., M.C., D.D., G.D., B.D., T.D., C.G., K.K., M.L., B.P., G.T., R.D., S.W; A.B., C.D., T.C.

Funding: This work was funded by European Space Agency project Fiducial Reference Measurements for Satellite Ocean Color (FRM4SOC), contract no. 4000117454/16/I-Sbo.

Acknowledgments: Support from numerous scientists, experts, and administrative personnel contributing to the project are gratefully acknowledged. The authors express special gratitude to Giuseppe Zibordi (JRC of the EC), Anu Reinart (UT), Tiia Lillemaa (UT), Mari Allik (UT), Claire Greenwell (NPL).

References

1. Zibordi, G.; Ruddick, K.; Ansko, I.; Gerald, M.; Kratzer, S.; Icely, J.; Reinart, A. In situ determination of the remote sensing reflectance: An inter-comparison. *Ocean Sci.* **2012**, *8*, 567–586. [CrossRef]
2. Antoine, D.; Schroeder, T.; Slivkoff, M.; Klonowski, W.; Doblin, M.; Lovell, J.; Boadle, D.; Baker, B.; Botha, E.; Robinson, C.; et al. *IMOS Radiometry Task Team.* 2017. Available online: http://imos.org.au/fileadmin/user_upload/shared/IMOS%20General/documents/Task_Teams/IMOS-RTT-final-report-submission-30June2017.pdf (accessed on 7 May 2019).
3. Vabson, V.; Kuusk, J.; Ansko, I.; Vendt, R.; Alikas, K.; Ruddick, K.; Ansper, A.; Bresciani, M.; Burmester, H.; Costa, M.; D'Alimonte, D.; et al. Field intercomparison of radiometers used for satellite validation in the 400–900 nm range. *Remote Sens.* **2019**, in press.
4. Kuusk, J.; Ansko, I.; Vabson, V.; Ligi, M.; Vendt, R. *Protocols and Procedures to Verify the Performance of Fiducial Reference Measurement (FRM) Field Ocean Colour Radiometers (OCR) Used for Satellite Validation*; Tartu Observatory: Tõravere, Estonia, 2017; Available online: https://frm4soc.org/wp-content/uploads/filebase/FRM4SOC-D130_TR5_v1.2_TO_signed.pdf (accessed on 7 May 2019).
5. International Vocabulary of Metrology—Basic and General Concepts and Associated Terms (VIM), JCGM 200:2008, 3rd Edition, 2008 Version with Minor Corrections, JCGM, 2012. Available online: https://www.bipm.org/utils/common/documents/jcgm/JCGM_200_2012.pdf (accessed on 7 May 2019).
6. Johnson, B.C.; Yoon, H.; Rice, J.P.; Parr, A.C. Chapter 1.2—Principles of Optical Radiometry and Measurement Uncertainty. In *Experimental Methods in the Physical Sciences*; Zibordi, G., Donlon, C.J., Parr, A.C., Eds.; Optical Radiometry for Ocean Climate Measurements; Academic Press: Cambridge, MA, USA, 2014; Volume 47, pp. 13–67.
7. Ocean Optics Protocols for Satellite Ocean Color Sensor Validation: Instrument Specifications, Characterization, and Calibration. 2003. Available online: https://www.oceanbestpractices.net/bitstream/handle/11329/479/protocols_ver4_volii.pdf?sequence=1&isAllowed=y (accessed on 7 May 2019).
8. Salim, S.G.R.; Fox, N.P.; Hartree, W.S.; Woolliams, E.R.; Sun, T.; Grattan, K.T.V. Stray light correction for diode-array-based spectrometers using a monochromator. *Appl. Opt.* **2011**, *50*, 5130–5138. [CrossRef] [PubMed]
9. Hooker, S.B.; Firestone, E.R.; McLean, S.; Sherman, J.; Small, M.; Lazin, G.; Zibordi, G.; Brown, J.W.; McClain, C.R. *The Seventh SeaWiFS Intercalibration Round-Robin Experiment (SIRREX-7)*; TM-2003-206892; NASA Goddard Space Flight Center: Greenbelt, MD, USA, 2002; Volume 17.
10. Salim, S.G.R.; Woolliams, E.R.; Fox, N.P. Calibration of a Photodiode Array Spectrometer Against the Copper Point. *Int. J. Thermophys.* **2014**, *35*, 504–515. [CrossRef]
11. Ylianttila, L.; Visuri, R.; Huurto, L.; Jokela, K. Evaluation of a Single-monochromator Diode Array Spectroradiometer for Sunbed UV-radiation Measurements. *Photochem. Photobiol.* **2005**, *81*, 333–341. [CrossRef]
12. Seckmeyer, G. *Instruments to Measure Solar Ultraviolet Radiation Part 4: Array Spectroradiometers*; WMO/TD No. 1538; World Meteorological Organization: Geneva, Switzerland, 2010; 44p.
13. Zibordi, G.; Talone, M.; Jankowski, L. Response to Temperature of a Class of In Situ Hyperspectral Radiometers. *J. Atmos. Ocean. Technol.* **2017**, *34*, 1795–1805. [CrossRef]
14. Kuusk, J. Dark Signal Temperature Dependence Correction Method for Miniature Spectrometer Modules. *J. Sensors* **2011**, 1–9. [CrossRef]
15. Salim, S.G.R.; Fox, N.P.; Theocharous, E.; Sun, T.; Grattan, K.T.V. Temperature and nonlinearity corrections for a photodiode array spectrometer used in the field. *Appl. Opt.* **2011**, *50*, 866–875. [CrossRef]
16. Price, L.L.A.; Hooke, R.J.; Khazova, M. Effects of ambient temperature on the performance of CCD array spectroradiometers and practical implications for field measurements. *J. Radiol. Prot.* **2014**, *34*, 655. [CrossRef] [PubMed]
17. Li, L.; Dai, C.; Wu, Z.; Wang, Y. Temperature and nonlinearity correction methods for commercial CCD array spectrometers used in field. In Proceedings of the AOPC 2017: Space Optics and Earth Imaging and Space Navigation, Beijing, China, 4–6 June 2017; Volume 10463, p. 104631K.
18. Kostkowski, H. *Reliable Spectroradiometry*; Spectroradiometry Consulting: La Plata, MD, USA, 1997.
19. Zong, Y.; Brown, S.W.; Johnson, B.C.; Lykke, K.R.; Ohno, Y. Simple spectral stray light correction method for array spectroradiometers. *Appl. Opt.* **2006**, *45*, 1111–1119. [CrossRef]

20. Zong, Y.; Brown, S.W.; Meister, G.; Barnes, R.A.; Lykke, K.R. Characterization and correction of stray light in optical instruments. *SPIE Remote Sens.* **2007**, *6744*, 1–11. [CrossRef]

21. Nevas, S.; Wübbeler, G.; Sperling, A.; Elster, C.; Teuber, A. Simultaneous correction of bandpass and stray-light effects in array spectroradiometer data. *Metrologia* **2012**, *49*, S43. [CrossRef]

22. Talone, M.; Zibordi, G.; Ansko, I.; Banks, A.C.; Kuusk, J. Stray light effects in above-water remote-sensing reflectance from hyperspectral radiometers. *Appl. Opt.* **2016**, *55*, 3966–3977. [CrossRef]

23. Talone, M.; Zibordi, G. Non-linear response of a class of hyper-spectral radiometers. *Metrologia* **2018**, *55*, 747. [CrossRef]

24. Talone, M.; Zibordi, G. Polarimetric characteristics of a class of hyperspectral radiometers. *Appl. Opt.* **2016**, *55*, 10092–10104. [CrossRef] [PubMed]

25. Mekaoui, S.; Zibordi, G. Cosine error for a class of hyperspectral irradiance sensors. *Metrologia* **2013**, *50*, 187. [CrossRef]

26. Cordero, R.R.; Seckmeyer, G.; Labbe, F. Cosine error influence on ground-based spectral UV irradiance measurements. *Metrologia* **2008**, *45*, 406. [CrossRef]

27. Zibordi, G.; Mélin, F.; Berthon, J.-F.; Holben, B.; Slutsker, I.; Giles, D.; D'Alimonte, D.; Vandemark, D.; Feng, H.; Schuster, G.; et al. AERONET-OC: A Network for the Validation of Ocean Color Primary Products. *J. Atmos. Ocean. Technol.* **2009**, *26*, 1634–1651. [CrossRef]

28. Zibordi, G.; Voss, K.J. Chapter 3.1—In situ Optical Radiometry in the Visible and Near Infrared. In *Optical Radiometry for Ocean Climate Measurements*; Zibordi, G., Donlon, C.J., Parr, A.C., Eds.; Experimental Methods in the Physical Sciences; Academic Press: Cambridge, MA, USA, 2014; Volume 47, pp. 247–304.

29. Spectral Response Function Data. Available online: https://sentinel.esa.int/web/sentinel/technical-guides/sentinel-3-olci/olci-instrument/spectral-response-function-data (accessed on 3 September 2018).

30. Müller, J.W. Possible Advantages of a Robust Evaluation of Comparisons. *J. Res. Natl. Inst. Stand. Technol.* **2000**, *105*, 551–555. [CrossRef]

31. Evaluation of Measurement Data—Guide to the Expression of Uncertainty in Measurement (GUM), JCGM 100, First Edition, September 2008. Available online: http://www.bipm.org/utils/common/documents/jcgm/JCGM_100_2008_E.pdf (accessed on 7 May 2019).

32. Vabson, V.; Ansko, I.; Alikas, K.; Kuusk, J.; Vendt, R.; Reinat, A. *Improving Comparability of Radiometric In Situ Measurements with Sentinel-3A/OLCI Data*; EUMETSAT: Darmstadt, Germany, 2017.

33. Gergely, M.; Zibordi, G. Assessment of AERONET-OC L WN uncertainties. *Metrologia* **2014**, *51*, 40. [CrossRef]

34. IOCCG, International Network for Sensor Inter-comparison and Uncertainty Assessment for Ocean Color Radiometry (INSITU-OCR) White Paper. 2012. Available online: http://www.ioccg.org/groups/INSITU-OCR_White-Paper.pdf (accessed on 7 February 2017).

35. Ruddick, K.G.; De Cauwer, V.; Park, Y.-J.; Moore, G. Seaborne measurements of near infrared water-leaving reflectance: The similarity spectrum for turbid waters. *Limnol. Oceanogr.* **2006**, *51*, 1167–1179. [CrossRef]

36. Alikas, K.; Ansko, I.; Vabson, V.; Ansper, A.; Kangro, K.; Randla, M.; Uudenberg, K.; Ligi, M.; Kuusk, J.; Randoja, R.; et al. *Validation of Sentinel-3A/OLCI Data over Estonian Inland Waters*; Plymouth, UK, 2017. Available online: https://www.researchgate.net/publication/321275461_Validation_of_Sentinel-3AOLCI_data_over_Estonian_inland_waters (accessed on 7 May 2019).

37. Ruddick, K. *Technical Report TR-2 "A Review of Commonly Used Fiducial Reference Measurement (FRM) Ocean Colour Radiometers (OCR) Used for Satellite OCR Validation"*. 2018. Available online: https://frm4soc.org/wp-content/uploads/filebase/FRM4SOC-TR2_TO_signedESA.pdf (accessed on 7 May 2019).

38. Voss, K.J.; da Costa, L.B. Polarization properties of FEL lamps as applied to radiometric calibration. *Appl. Opt.* **2016**, *55*, 8829–8832. [CrossRef]

39. Bernhard, G.; Seckmeyer, G. Uncertainty of measurements of spectral solar UV irradiance. *J. Geophys. Res. Atmos.* **1999**, *104*, 14321–14345. [CrossRef]

40. Santer, B.D.; Wigley, T.M.L.; Boyle, J.S.; Gaffen, D.J.; Hnilo, J.J.; Nychka, D.; Parker, D.E.; Taylor, K.E. Statistical significance of trends and trend differences in layer-average atmospheric temperature time series. *J. Geophys. Res. Atmos.* **2000**, *105*, 7337–7356. [CrossRef]

Article

Field Intercomparison of Radiometers Used for Satellite Validation in the 400–900 nm Range

Viktor Vabson [1,*], Joel Kuusk [1], Ilmar Ansko [1], Riho Vendt [1], Krista Alikas [1], Kevin Ruddick [2], Ave Ansper [1], Mariano Bresciani [3], Henning Burmester [4], Maycira Costa [5], Davide D'Alimonte [6], Giorgio Dall'Olmo [7,8], Bahaiddin Damiri [9], Tilman Dinter [10], Claudia Giardino [3], Kersti Kangro [1], Martin Ligi [1], Birgot Paavel [11], Gavin Tilstone [7], Ronnie Van Dommelen [12], Sonja Wiegmann [10], Astrid Bracher [10], Craig Donlon [13] and Tânia Casal [13]

[1]   Tartu Observatory, University of Tartu, 61602 Tõravere, Estonia; joel.kuusk@ut.ee (J.K.); ilmar.ansko@ut.ee (I.A.); riho.vendt@ut.ee (R.V.); krista.alikas@ut.ee (K.A.); ave.ansper@ut.ee (A.A.); kersti.kangro@ut.ee (K.K.); martin.ligi@ut.ee (M.L.)
[2]   Royal Belgian Institute of Natural Sciences, 1000 Brussels, Belgium; kruddick@naturalsciences.be
[3]   National Research Council of Italy, 21020 Ispra, Italy; bresciani.m@irea.cnr.it (M.B.); giardino.c@irea.cnr.it (C.G.)
[4]   Helmholtz-Zentrum Geesthacht, Institute for Coastal Research, 21502 Geesthacht, Germany; henning.burmester@hzg.de
[5]   Geography Department at the University of Victoria, Victoria, BC V8P 5C2, Canada; maycira@uvic.ca
[6]   Center for Marine and Environmental Research CIMA, University of Algarve, 8005-139 Faro, Portugal; davide.dalimonte@gmail.com
[7]   Plymouth Marine Laboratory, Plymouth PL1 3DH, UK; gdal@pml.ac.uk (G.D.); GHTI@pml.ac.uk (G.T.)
[8]   National Centre for Earth Observation, Plymouth PL1 3DH, UK
[9]   Cimel Electronique S.A.S, 75011 Paris, France; bahaiddin.damiri@univ-lille1.fr
[10] Alfred Wegener Institute Helmholtz Center for Polar and Marine Research, D-27570 Bremerhaven, Germany; Tilman.Dinter@awi.de (T.D.); Sonja.Wiegmann@awi.de (S.W.); Astrid.Bracher@awi.de (A.B.)
[11] Estonian Marine Institute, University of Tartu, 12618 Tallinn, Estonia; birgot.paavel@ut.ee
[12] Satlantic, Sea Bird Scientific, Bellevue, WA 98005, USA; rvandommelen@seabird.com
[13] European Space Agency, 2201 AZ Noordwijk, The Netherlands; craig.donlon@esa.int (C.D.); tania.casal@esa.int (T.C.)
*   Correspondence: viktor.vabson@ut.ee; Tel.: +372-737-4552

Received: 26 March 2019; Accepted: 8 May 2019; Published: 11 May 2019

Abstract: An intercomparison of radiance and irradiance ocean color radiometers (the second laboratory comparison exercise—LCE-2) was organized within the frame of the European Space Agency funded project Fiducial Reference Measurements for Satellite Ocean Color (FRM4SOC) May 8–13, 2017 at Tartu Observatory, Estonia. LCE-2 consisted of three sub-tasks: (1) SI-traceable radiometric calibration of all the participating radiance and irradiance radiometers at the Tartu Observatory just before the comparisons; (2) indoor, laboratory intercomparison using stable radiance and irradiance sources in a controlled environment; (3) outdoor, field intercomparison of natural radiation sources over a natural water surface. The aim of the experiment was to provide a link in the chain of traceability from field measurements of water reflectance to the uniform SI-traceable calibration, and after calibration to verify whether different instruments measuring the same object provide results consistent within the expected uncertainty limits. This paper describes the third phase of LCE-2: The results of the field experiment. The calibration of radiometers and laboratory comparison experiment are presented in a related paper of the same journal issue. Compared to the laboratory comparison, the field intercomparison has demonstrated substantially larger variability between freshly calibrated sensors, because the targets and environmental conditions during radiometric calibration were different, both spectrally and spatially. Major differences were found for radiance sensors measuring a sunlit water target at viewing zenith angle of 139° because of the different fields of view. Major differences were found for irradiance sensors because of imperfect cosine response of diffusers. Variability between individual radiometers did depend significantly also on the type of the sensor and on the specific measurement target. Uniform SI traceable radiometric calibration ensuring fairly good consistency for indoor, laboratory measurements is insufficient for outdoor, field measurements, mainly due to the different angular variability of illumination. More stringent specifications and individual testing of radiometers for all relevant systematic effects (temperature, nonlinearity, spectral stray light, etc.) are needed to reduce biases between instruments and better quantify measurement uncertainties.

Keywords: ocean color radiometers; radiometric calibration; field intercomparison measurement; agreement between sensors; measurement uncertainty

---

1. Introduction

The FRM4SOC project aimed to support the consistency of the ground-based validation measurements for "ocean color (OC)", or water reflectance, with the SI units, and thus, contribute to higher quality and accuracy of Sentinel-2 Multispectral Instrument (MSI) and Sentinel-3 Ocean and Land Color Instrument (OLCI) products. For that, the second laboratory comparison exercise (LCE-2) comparison experiment was organized in the frame of the FRM4SOC project. A stepwise approach was chosen for the LCE-2: At first, calibration of sensors, secondly; indoor, laboratory comparisons using various levels of radiance or irradiance performed in stable conditions similar to those during radiometric calibration; and as a third, outdoor, field measurements of natural radiation sources in an environment significantly different from laboratory conditions. This paper only describes the field experiment, whilst the radiometric calibration and indoor exercise are covered in a related paper of the same journal issue [1].

Intercomparison of data produced by a number of independent radiometric sensors measuring simultaneously the same object allows assessment of the consistency of different results and their estimated uncertainties depending on the type of the sensor, the spectral composition, intensity and angular variability of the measured radiation, environmental temperature, and the particular method used for collecting and handling the measurement data [2,3]. This information can serve also for further elaboration of uncertainty estimation. Compared to the indoor experiment [1], much larger variability between radiometric sensors is expected in the outdoor experiment, due to much larger differences in target signal and environmental temperature with respect to the radiometric calibration conditions.

The analysis of field measurements is more complicated than for the indoor case. The main differences in field and laboratory measurements of LCE-2, causing a substantial increase of the field measurements uncertainty, are shown in Figure 1. The spectral composition and intensity of radiation from the target being measured (sky, water) are significantly different from the incandescent source used as the radiometric calibration standard. The angular distribution of downwelling irradiance also varies from the nearly collimated radiation source used during radiometric calibration. Ambient temperature in the field can differ from the stable laboratory temperature during the radiometric calibration by more than ±15 °C. The stray light effect may be an order of magnitude larger, due to different shapes of the calibration and field spectra. Strong autocorrelation in recorded time series data implies that statistical analysis of intercomparison results should be suitably rearranged.

Due to non-ideal performance of radiometers (temperature dependence, deviation from ideal cosine response for irradiance sensors, nonlinearity, spectral stray light, etc.), all the differences between conditions during radiometric calibration and field measurements can contribute to the bias between radiometers and increase the measurement uncertainty. The known measurement errors should be corrected and the unknown or residual errors have to be assessed and accounted for in the uncertainty budget. Unfortunately, the information needed for these corrections is often available only through highly time- and resource-consuming tests of individual radiometers, and it is often necessary to make such corrections based on the characterization of an instrument from the same family.

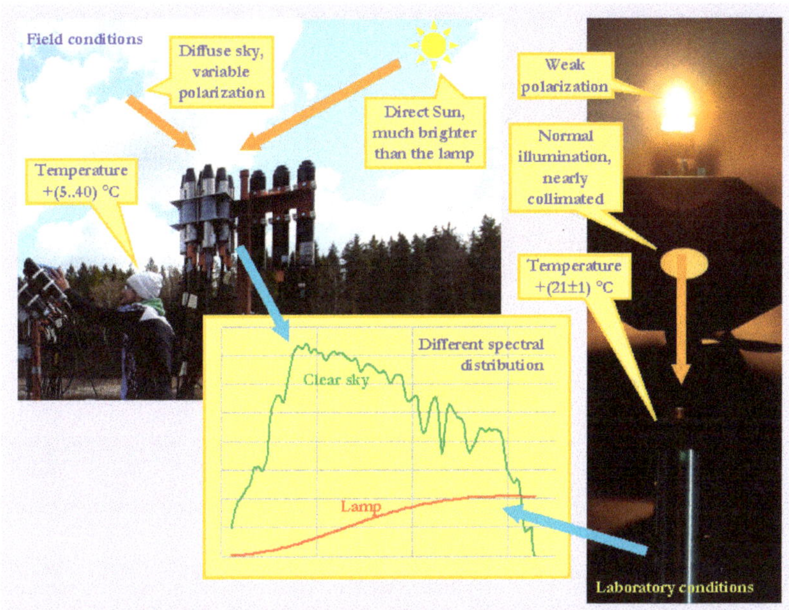

Figure 1. Main differences between the field and laboratory measurements of the second laboratory comparison exercise (LCE-2) causing a substantial increase in uncertainty of the field measurements.

This study aims to evaluate the effectiveness of SI-traceable radiometric calibration for consistency of OC field measurements, presents LCE-2 data processing results, and discusses techniques and procedures for improving traceability of OC field measurements.

2. Material and Methods

2.1. Participants of the LCE-2

In total 11 institutes or companies were involved in the LCE-2, see Table 1. Altogether 44 radiometric sensors from five different manufacturers were involved, as shown in Table 2.

Table 1. Institutes and instruments participating in the LCE-2 intercomparison.

Participant	Country	L-Radiance; E-Irradiance Sensor
Tartu Observatory (pilot)	Estonia	RAMSES (2 L, 1 E) WISP-3 (2 L, 1 E)
Alfred Wegener Institute	Germany	RAMSES (2 L, 2 E)
Royal Belgian Institute of Natural Sciences	Belgium	RAMSES (7 L, 4 E)
National Research Council of Italy	Italy	SR-3500 (1 L, 1 E) WISP-3 (2 L, 1 E)
University of Algarve	Portugal	RAMSES (2 L, 1 E)
University of Victoria	Canada	OCR-3000 (OCR-3000 is the predecessor of HyperOCR) (2 L, 1 E)
Satlantic; Sea Bird Scientific	Canada	HyperOCR (2 L, 1 E)
Plymouth Marine Laboratory	UK	HyperOCR (2 L, 1 E)
Helmholtz-Zentrum Geesthacht	Germany	RAMSES (2 L, 1 E)
University of Tartu	Estonia	RAMSES (1 L, 1 E)
Cimel Electronique S.A.S	France	SeaPRISM (1 L)

Table 2. Technical parameters of the participating radiometers.

Parameter	RAMSES	HyperOCR	WISP-3	SR-3500	SeaPRISM
Field of View (L/E)	7°/cos	6°(According to the manufacturer, the HyperOCR radiance sensors 444 and 445 have 6° FOV) or 23°/cos	3°/cos	5°/cos	1.2°/NA
Manual integration time	yes	yes	no	yes	no
Adaptive integration time	yes	yes	yes	yes	yes
Min. integration time, ms	4	4	0.1	7.5	NA
Max. integration time, ms	4096	4096	NA	1000	NA
Min. sampling interval, s	5	5	10	2	NA
Internal shutter	no	yes	no	yes	yes
Number of channels	256	256	2048	1024	12
Wavelength range, nm	320 … 1050	320 … 1050	200 … 880	350 … 2500	400 … 1020
Wavelength step, nm	3.3	3.3	0.4	1.2/3.8/2.4	NA
Spectral resolution, nm	10	10	3	3/8/6	10

2.2. Venue and Measurement Setup

The outdoor exercise took place at Lake Kääriku, Estonia, 58°0′5″N, 26°23′55″E on 11–12.05.2017. Lake Kääriku is a small eutrophic lake with 0.2 km^2 surface area. Maximum depth is 5.9 m, with an average of 2.6 m. The water color is greenish-yellow with measured transparency (Secchi disk depth) of 2.6 m. The average chlorophyll content Chl_a = 7.3 mg m^{-3}, total suspended matter content TSM = 3.9 g m^{-3}, absorption of the colored dissolved organic matter a_{CDOM}(442 nm) = 1.7 m^{-1}, diffuse attenuation coefficient of downwelling irradiance K_d(PAR) = 1.3 m^{-1}. The bottom is muddy. Lake Kääriku has a 50 m long pier and a diving platform on the southern coast. The diving platform has two levels. During LCE-2 the upper level was used for the instruments, computers and instrument operators were located on the lower level and the pier below the tower (Figure 2).

Figure 2. Pier and diving platform at the southern coast of Lake Kääriku.

The instruments were located roughly 7.5 m above the water surface. Depth of water around the diving platform was 2.6 m to 3.6 m and the bottom was not visible to observers. The closest trees were about 65 m south of the platform, the treetops are less than 20° above the horizon when viewed from the upper level of the platform. Purpose-built frames were used for mounting and aligning the participating radiometers (Figures 3 and 4). The irradiance sensors were mounted in a fixed frame ensuring the levelling of the cosine collectors. The front surfaces of all the cosine collectors were set at the same height so that the illumination conditions were equal and the instruments were not shadowing each other.

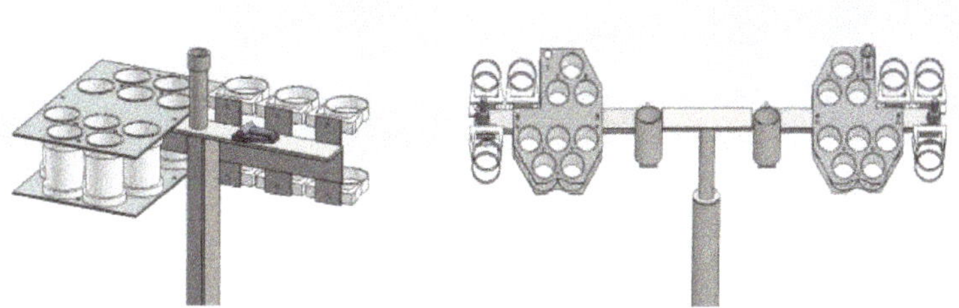

Figure 3. 3D CAD (computer-aided design) drawings of the frames for mounting irradiance (left) and radiance (right) sensors during the outdoor experiment.

Figure 4. All the radiance and irradiance radiometers were mounted in common frames during the LCE-2 outdoor experiment. Left frame—irradiance sensors; right frame—radiance sensors.

2.3. Environmental Conditions and Selection of Casts

The environmental conditions during the outdoor experiment were not ideal, mainly due to the presence of scattered cumulus clouds. The aerosol content was low, average daily aerosol optical depth at 500 nm (AOD500) was 0.077 on May 11 and 0.071 on May 12 (measured at Tõravere AERONET station, 30 km north of Lake Kääriku [4]). The air temperature was rather low, between 5 °C and 9 °C; water temperature was around 11 °C. Wind speed was mainly between 0.5 m s^{-1} and 4 m s^{-1} with occasional gusts of up to 7 m s^{-1}.

The outdoor measurements were performed in 5-minute casts, an exception of 25-minute irradiance cast no. 14. The beginning and end times of casts were announced and during the casts all the participants recorded the radiance and irradiance data at their usual fieldwork data acquisition rate. 30 casts were recorded in total, but only seven of them were included in the intercomparison. The selection of casts was based on the time series of 550 nm spectral band. The coordinating laboratory received the 550 nm time series data for 16 radiance and 10 irradiance sensors. Only the casts with the most stable signal and least missing data were selected for further analysis. All the selected casts were measured on May 12—the second day of the outdoor experiment. The all-sky camera images captured in the middle of the selected casts can be seen in Figure 5.

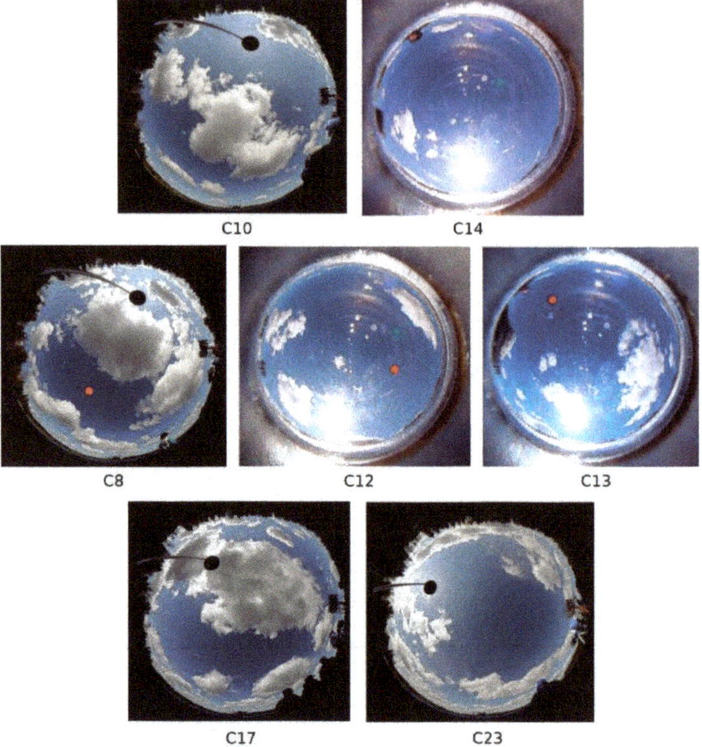

Figure 5. All-sky camera images captured in the middle of the casts used in the intercomparison analysis. Irradiance—C10, C12, C13, C14; blue sky radiance—C8, C12, C13; water radiance—C17, C23. Red dots in C8, C12, C13 indicate approximate view direction of the radiance sensors.

The casts used in the analysis of LCE-2 intercomparison are listed in Table 3. Four casts (C10, C12, C13, and C14) were chosen for irradiance, all recorded with direct sunlight, although with some clouds in the sky away from the sun. Five casts were chosen for radiance: Three casts (C8, C12, and C13) recorded with blue sky as a target, one (C17) measurement of the water surface in cloud shadow, and one (C23) measurement of sunlit water. Measurement C17 is made at a zenith angle suggested in the protocols for above-water radiometry, while measurement C23 is made at a slightly more oblique angle. These measurements are made for azimuth angles 107° and 143° with respect to the sun, in order to avoid sunglint and direct shadow from the platform. The 550 nm time series of one irradiance (RAMSES SAM_8329) and one radiance (RAMSES SAM_81B0) sensor for all the radiance and irradiance casts used for intercomparison are plotted in Figure 6. The initial cast start and stop times were adjusted based on Figure 6 to exclude the intervals with high temporal variability. Photographs of the radiance targets can be seen in Figure 7. Approximate field-of-view (FOV) footprints for WISP-3 (3°), RAMSES (7°), and HyperOCR (23°) are shown in Figure 7 as well. The images were taken with a handheld Nikon D40X digital single-lens reflex (DSLR) camera equipped with a Nikkor 18–200 mm zoom lens. According to the Exchangeable image file format (EXIF) meta-info of the images, the lens was completely zoomed out to 18 mm for C8, C12, C13, and C23. Considering the parameters of the lens and the camera, the horizontal FOV of these images is 67°. The lens was zoomed to 32 mm for C17 which corresponds to 41° horizontal FOV of the image. As the camera was not fixed to the frame in line with the radiometers, its collinearity with the radiometers is uncertain and the actual FOV-s of the radiometers may slightly differ from circles, shown in Figure 7.

Table 3. Casts used in the analysis.

Cast	Target	Time (UTC)	SZA	SAA	Relative VAA from Sun	VZA	Wind speed
C8	L_d (blue sky)	07:46:00–07:49:25	48°	131°	162°	43°	NA
C10	E_d	08:07:00–08:12:00	46°	137°	NA	NA	NA
C12	E_d, L_d (blue sky)	08:50:00–08:55:00	43°	151°	90°	43°	NA
C13	E_d, L_d (blue sky)	09:00:00–09:03:05	42°	154°	134°	58°	NA
C14	E_d	09:22:30–09:47:30	41°	162°	NA	NA	NA
C17	L_u (shadow)	10:30:00–10:35:00	40°	187°	107°	139°	2 m s^{-1}
C23	L_u (sunlit)	11:56:00–12:01:00	44°	217°	143°	130°	1 m s^{-1}

UTC—coordinated universal time; NA—not applicable; SZA—solar zenith angle; L_d—downwelling sky radiance; SAA—solar azimuth angle; L_u—total upwelling water radiance; VAA—view azimuth angle; E_d—downwelling irradiance; VZA—view zenith angle.

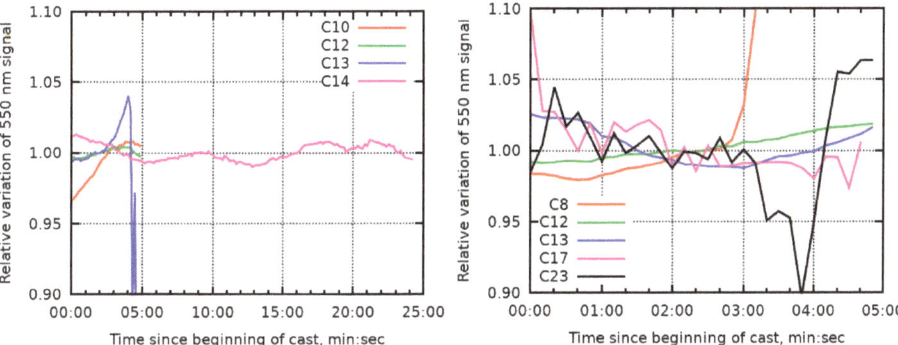

Figure 6. Relative variation of 550 nm signal of one RAMSES sensor during irradiance (**left**) and radiance (**right**; C8, C12, C13 blue sky; C17 water in cloud shadow; C23 sunlit water) casts selected for intercomparison analysis.

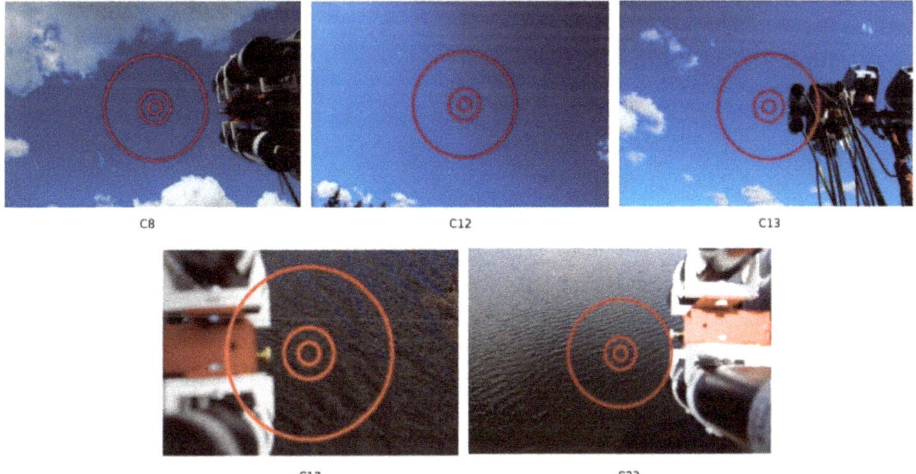

Figure 7. Photographs of radiance targets used in the intercomparison analysis. The circles denote approximate FOV of WISP-3 (smallest), RAMSES, and HyperOCR (largest).

2.4. Outdoor Experiment of the LCE-2

The initially planned outdoor intercomparison [5] accounted for two phases: (1) Direct intercomparison of the downwelling irradiance E_d, the downwelling sky radiance L_d, and the total upwelling water radiance L_u; (2) intercomparison of the remote sensing reflectance R_{rs} and the water-leaving radiance L_w derived from simultaneously measured E_d, L_d, and L_u. The radiance sensors were mounted on the frame in two groups which could be moved independently in the zenith direction. Additionally, the relative zenith angle between the two groups could be fixed, and both groups tilted together. The selected setting was to fix the relative azimuth angle between the two groups of sensors to 0° and move simultaneously all the radiance sensors in the azimuth direction. The design of the radiance frame allowed mounting the L_u radiometers to one group and L_d radiometers to another group for measuring L_w and R_{rs} in a typical 3-radiometer above-water configuration [6].

On the first day of the outdoor measurements, seven casts of simultaneous E_d, L_d, and L_u measurements at typical above-water 3-radiometer configuration were recorded. However, none of the casts was considered suitable for the analysis, due to cumulus clouds causing rather unsteady illumination conditions. On the second day of the outdoor experiment, priority was given to the phase (I) measurements and all the radiance sensors were simultaneously measuring either L_u or L_d.

2.5. Data Processing

In total, data for 40 out of 44 radiometers were reported back to the pilot. For the rest, the pilot carried out the data handling using the provided raw files. The data processing details are described in Sections 3.1 and 3.2 of the related paper [1]. The outdoor data processing chain contained the following steps:

- Separation of the raw data files, based on the casts' start and stop timestamps;
- Subtraction of the dark signal;
- Division by radiometric responsivity;
- Interpolation/convolution of spectra into the OLCI bands.

2.6. Consensus Value Used for the Analysis

The group median was used as the consensus value. Compared to the indoor measurements, outdoor variability between radiance sensors on average was about twice larger, and for irradiance sensors more than five times larger. Two irradiance and one radiance sensor were not accounted for in the variability estimate, because they had extremely large deviations from the group median.

2.7. Accuracy of Sensor Adjustment

The collinearity of groups of radiance sensors on the left and right frame was set by visual observation from the side of the frame and was better than ±1°. Due to the flexibility of the plastic clamps used to fix the HyperOCR radiometers, slight deflection from collinearity of HyperOCR and RAMSES sensors within the groups was noticed during the experiment (visually much larger than misalignment between the groups). Using Figure 8, the angle between HyperOCR and RAMSES sensors was measured to be 1.3°, the HyperOCR sensors were pointing lower than the RAMSES instruments. Image taken from the other side of the frame revealed that the HyperOCR sensors in the other group were pointing about 1.1° higher than the RAMSES instruments. The left and right radiance frames were visually aligned by the topmost RAMSES instruments, thus, the maximum angle between the HyperOCR instruments on the frames could have been about 2.5°. Although this is ten times smaller than the FOV of a standard HyperOCR instrument, it can have a significant impact when measuring spatially heterogeneous targets.

Figure 8. The angle between red lines marking the directions of HyperOCR and RAMSES sensors was measured to be 1.3° from this image.

3. Results

3.1. Results of Outdoor Comparison

The consensus spectra for the irradiance and radiance targets are presented in Figure 9. The difference between the casts of radiance sensors measuring the sky and water is evident. Radiation from the water with blue sky gave the smallest signal.

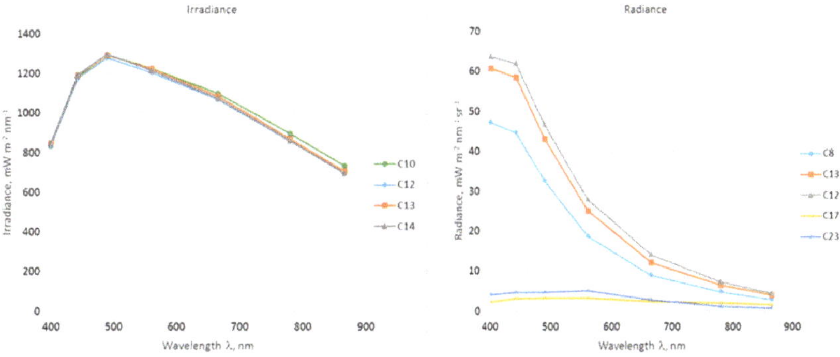

Figure 9. Irradiance and radiance consensus values in the outdoor experiment. C8, C10, C12, C13, C14—blue sky (**radiance**) or direct sunshine (**irradiance**); C17—water in cloud shadow; C23—sunlit water.

The measurement results for the field casts are presented in Figures 10 and 11 as the deviation from the consensus value. The different behavior of RAMSES and HyperOCR sensor groups became evident. For the irradiance measurements, the deviation of HyperOCR sensors from the consensus value was very small, and the group of RAMSES sensors caused the increase of mean variability, see Figure 10. Conversely, the variability of the radiance sensors during the indoor and outdoor exercises was almost at the same level for the RAMSES group, and the increase of the outdoor variability was caused largely by the HyperOCR sensors, see Figure 10.

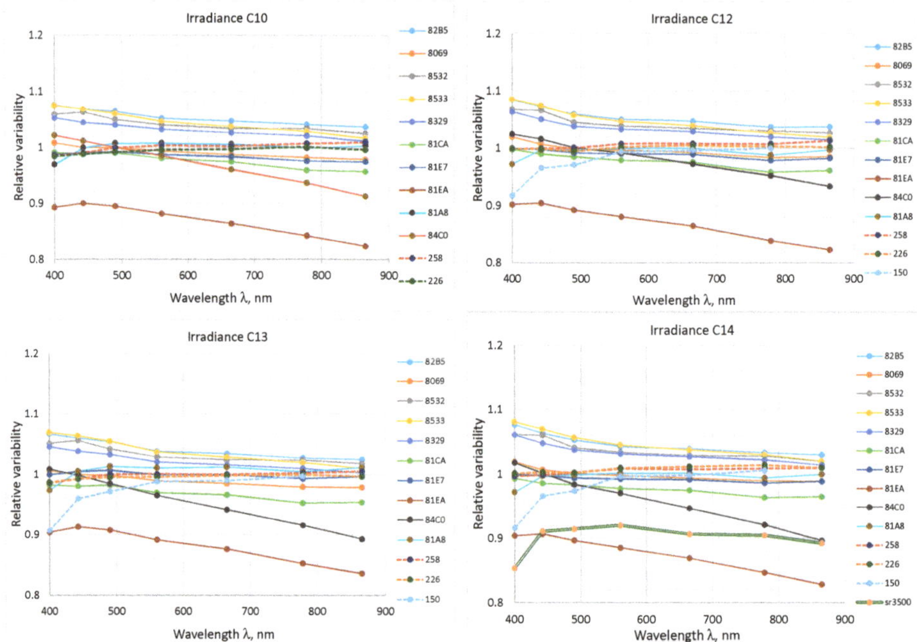

Figure 10. Irradiance sensors compared to the consensus value. Solid lines—RAMSES sensors; dashed lines—HyperOCR sensors; double line—SR-3500.

All the irradiance casts in Figure 10 were measured with direct sunshine and no big difference between casts can be observed for the consensus irradiance spectra (Figure 9). The group of HyperOCR sensors, shown in Figure 10 with dashed lines, are more consistent with the consensus value than the sensors of the RAMSES group shown with solid lines. Remarkable is much higher variability across sensors of the RAMSES group. Interestingly, the intra-sensor variability of irradiance is almost wavelength-independent, except at 400 nm.

The comparison of different radiance sensors (Figure 11) did show a very good agreement to within 1.2% across the full spectrum for all RAMSES sensors for casts C12 and C13—the most homogeneous blue sky targets. Higher variability between all sensors, and particularly the HyperOCR radiance sensors, is seen for the obliquely viewed water target C23 (Figure 11). This is probably caused by spatial heterogeneity of the target (C23 in Figure 7), and by slight bias from collinearity of the sensors (Figure 8). This assumption is supported by the fact that radiometers 151, 222, and 444 which are below the consensus value in Figure 11 were mounted on the left frame and radiometers 152, 223, and 445 which all remain above the consensus value in Figure 11 were mounted on the right frame. The water-viewing measurement C17 has better spatial heterogeneity and is more representative, due to more suitable zenith angle normally used for water reflectance measurements because the angular variability of the Fresnel reflection coefficient for 41° angle of incidence (cast C17) is smaller than for 50° (cast C23), and hence gives less spatial variability of skylight reflection.

In Figure 11, the SeaPRISM shows fairly good agreement with the consensus value of LCE-2, while SR-3500 is through all casts biased to somewhat smaller values. WISP-3 sensors show above an average scattering of results, partly because their alignment to the frame in line with the other radiometers was difficult, due to the ergonomic shape of these handheld instruments and lack of suitable reference surfaces for alignment. It is not possible to conclude which sensor(s) showed best agreement with SI, due to lack of a well-characterized SI-traceable reference radiometer involved simultaneously in the comparison.

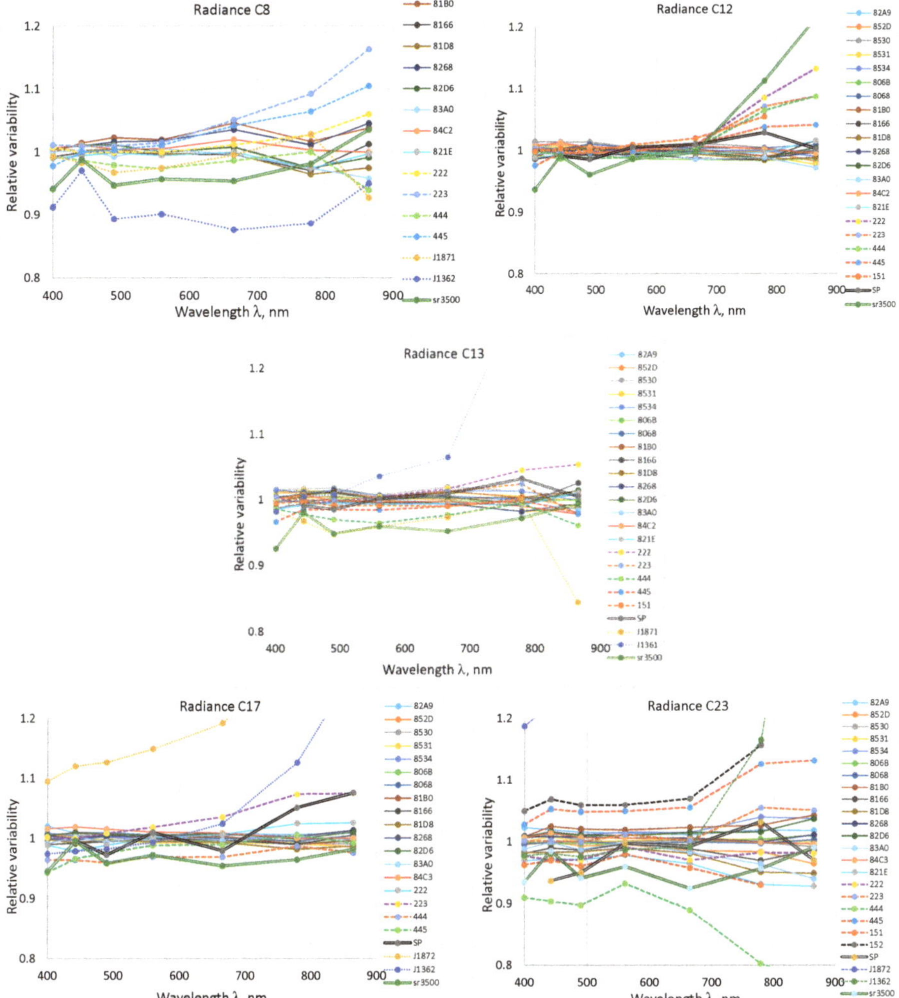

Figure 11. Radiance sensors compared to the consensus value in the outdoor experiment. C8, C12, C13—blue sky; C17—water in cloud shadow at 139° VZA; C23—sunlit water at 130° VZA. Solid lines—RAMSES sensors; dashed lines—HyperOCR sensors; double lines—SeaPRISM (SP) and SR-3500; dotted lines—WISP-3.

The variability of irradiance and radiance results in the LCE-2 in comparison with differences between sensors, due to calibration state before the experiment is summarized in Figure 12. All standard deviations of laboratory measurements are smaller than 1%. Standard deviations of the field results are substantially higher (1–5)%, but still much smaller than variability, due to calibration state of sensors before the experiment (5–10)%, i.e., the calibration that each participant would have used if the radiometers were not freshly calibrated just before the start of the LCE-2 intercomparison exercise. It must be noted, however, that some instruments had not been used for fieldwork in recent years, thus, the previous calibration coefficients were several years old.

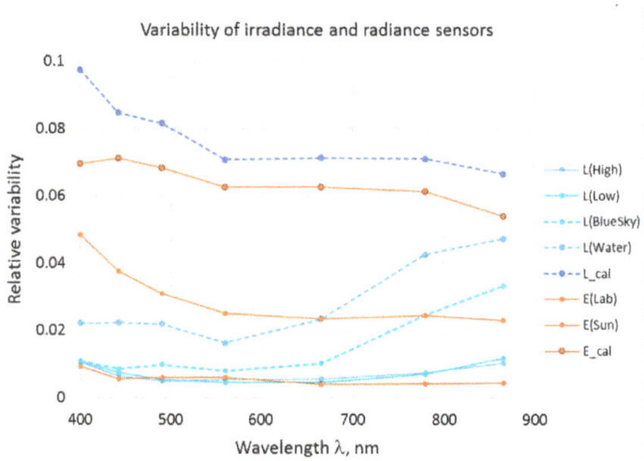

Figure 12. Variability between irradiance and radiance sensors. E_cal and L_cal—due to calibration state; E(Lab), L(Low) and L(High)—variability in laboratory intercomparison; E(Sun), L(BlueSky) and L(Water) variability in the field.

3.2. Measurements after the End of LCE-2 Comparison

Large variability between irradiance sensors of the RAMSES group during the outdoor exercise cannot be fully explained by poor stability of sensors, or by factors, such as temperature dependence (which is rather similar for the whole RAMSES group [7]), nonlinearity (which would be stronger for wavelengths with high digital counts), and stray light (which would show more spectral features). Most likely, the main reason for differences between RAMSES and between HyperOCR irradiance sensors comes from different properties of the entrance optics (angular response). The results of [8] for six RAMSES irradiance sensors suggest a cosine error within ±2% for sun zenith angles lower than 50° when radiometric calibration is conducted at 20° tilted sensor with respect to the incident irradiance. For the "conventional" calibration procedure at normal illumination somewhat larger cosine error may be expected. Therefore, after the end of LCE-2, in January 2019 the in-air cosine response error of five RAMSES irradiance sensors was measured, see Figure 13. One new sensor number 8598 measured was not involved in LCE-2.

Dependence of the cosine error on the zenith angle varies from radiometer to radiometer significantly with values ranging from −16% up to +9% at ±65°. Deviation from the ideal cosine response is irregular and does not always show a monotonic increase with the incidence angle. This is in agreement with the results of [8]. For one sensor, 8329, significant asymmetry is evident. The best of the characterized sensors, 81A8, has demonstrated in the outdoor experiment irradiance results very close to the consensus value (Figure 10), whereas the sensor 81EA with the largest cosine error, at the same time, had a deviation from consensus value about −10% to −15%, depending on wavelength.

Following the 20° "offsetting" calibration method suggested in Reference [8], the comparison data of Figure 10 were recalculated for two sensors by using the cosine response characterization results. Effect of calibration with tilted to 20° with respect to the incident irradiance sensor is shown in Figure 14. Improvement is evident for both sensors, but for 81EA the residual error is still large.

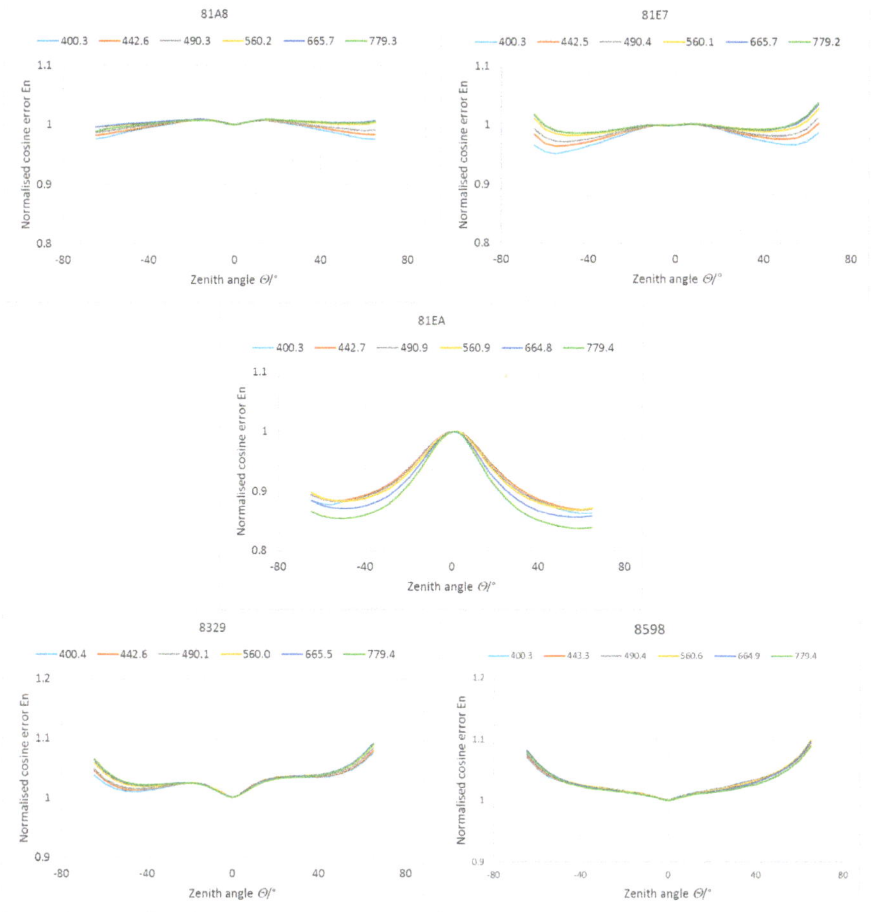

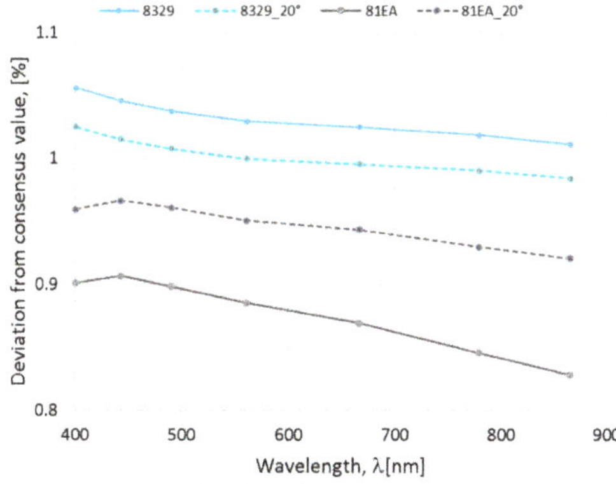

Figure 13. Normalized cosine response error of five RAMSES sensors.

Figure 14. Effect of calibration with tilted to 20° with respect to the incident irradiance sensor.

The manufacturer's specification of the HyperOCR [9] states that the cosine root mean square (RMS) error is within 3% at 0–60°, and within 10% at 60°–85° incidence angles. For RAMSES [10], accuracy is stated to be better than 6–10% depending on spectral range. The respective specification in Reference [11] is: For E_d measurement, the response to a collimated source should vary as $\cos\theta$ within less than 2% for angles $0° < \theta < 65°$ and 10% for angles $65° < \theta < 90°$. For easier comparison of different sensors the deviation from ideal cosine response was quantified as the integral of azimuth-independent absolute values of the cosine error for θ in the 0° to 85° interval, the index f_c in Reference [8] or cosine error f_2 in Reference [12], see Figure 15.

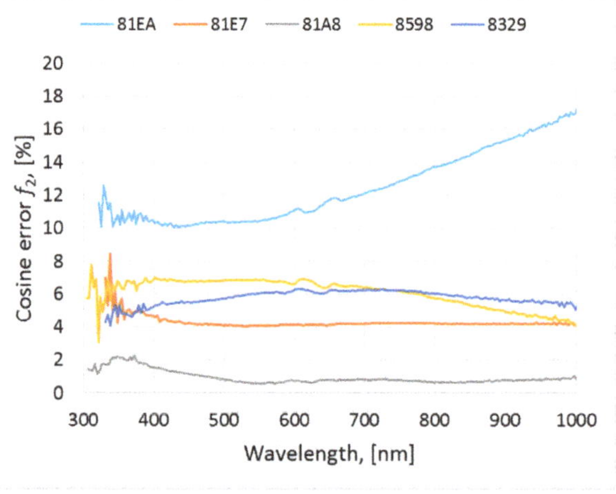

Figure 15. Integrated cosine error of the five RAMSES radiometers.

Increased variability between the RAMSES sensors in comparison with HyperOCR sensors presented in Figure 10 can be reasonably explained by a too tolerant specification of the cosine error, as departures from $\cos\theta$ imply analogous errors in E_d in the case of direct sunlight [11]. Although the majority of the RAMSES sensors meet the present specification, differences revealed during the field measurements may render the specification unsatisfactory for the users, unless laboratory characterization data and an indication of the angular variation of the downwelling radiance field, e.g., direct/diffuse ratio, is available to correct for imperfect cosine response.

Thus, rather large cosine errors of RAMSES irradiance sensors can be considered to be the main reason for the differences between irradiance sensors during the LCE-2 outdoor measurements.

4. Uncertainty Budgets of Outdoor Comparisons

An uncertainty analysis according to Reference [13,14] is undertaken for the outdoor measurements to understand the contribution of different factors to the observed variability between sensors. The outdoor downwelling irradiance uncertainty estimates are presented in Table 4; Table 5 corresponds to the blue sky radiance, and Table 6 to the radiance of sunlit water. All the uncertainty estimations in Tables 4–6 are based on experimental variability data of TriOS RAMSES sensors and information from References [2,6,15–19]. For the other radiometer models that took part in the intercomparison very little publicly available information can be found regarding various instrument characteristics that influence the measurement results [20]. In addition, the RAMSES was the only sensor model that was represented in sufficiently large number for statistical analysis.

Table 4. Relative uncertainty budget for the downwelling irradiance (in percent), based on the spread of individual sensors measuring the same target during the outdoor comparison. Data highlighted in green are not used for combined and expanded uncertainties. Last row: Relative experimental variability of sensors evaluated from the results of field comparisons.

	400 nm	442.5 nm	490 nm	560 nm	665 nm	778.8 nm	865 nm
Certificate	0.88	0.68	0.65	0.62	0.59	0.62	0.56
Interpolation	0.5	0.3	0.3	0.3	0.3	0.3	0.3
Instability (sensor)	0.05	0.03	0.04	0.03	0.04	0.03	0.02
Polarization	0.1	0.1	0.1	0.1	0.1	0.1	0.2
Nonlinearity	0.4	0.3	0.3	0.3	0.3	0.3	0.2
Stray light	0.9	0.7	0.3	0.3	0.7	0.9	1.0
Temperature	0.4	0.2	0.2	0.2	0.2	0.4	0.8
Cosine error	4.8	3.7	3	2.4	2.2	2.2	2
Signal, type A	0.01	0.01	0.01	0.01	0.01	0.02	0.02
Combined ($k = 1$)	4.9	3.8	3.1	2.5	2.3	2.4	2.3
Expanded ($k = 2$)	9.8	7.6	6.2	5	4.6	4.8	4.6
Variability ($k = 2$)	9.7	7.6	6.2	5	4.7	4.9	4.6

Table 5. Relative uncertainty budget for the radiance of blue sky (in percent), based on the spread of individual sensors pointing to the same target during the outdoor comparison. Data highlighted in green are not used for combined and expanded uncertainties. Last row: Relative experimental variability of sensors evaluated from the results of field comparisons.

	400 nm	442.5 nm	490 nm	560 nm	665 nm	778.8 nm	865 nm
Certificate	1.2	0.78	0.76	0.73	0.71	0.73	1.35
Interpolation	0.5	0.3	0.3	0.3	0.3	0.3	0.3
Instability (sensor)	0.04	0.03	0.02	0.01	0.01	0.02	0.01
Polarization	0.1	0.1	0.2	0.2	0.4	0.4	0.4
Nonlinearity	0.4	0.4	0.5	0.5	0.5	0.6	0.6
Stray light	0.8	0.6	0.2	0.2	0.5	0.9	1
Temperature	0.4	0.2	0.2	0.2	0.2	0.4	0.8
Alignment, FOV	0.3	0.4	0.6	0.6	0.5	2	2.9
Signal, type A	0.01	0.01	0.01	0.01	0.02	0.11	0.2
Combined ($k = 1$)	1.1	0.9	0.9	0.9	1	2.4	3.3
Expanded ($k = 2$)	2.2	1.8	1.8	1.8	2	4.8	6.6
Variability ($k = 2$)	2.2	1.8	2	1.6	2	4.8	6.6

Table 6. Relative uncertainty budget for the radiance of sunlit water (in percent), based on the spread of individual sensors pointing to the same target during the outdoor comparison. Data highlighted in green are not used for combined and expanded uncertainties. Last row: Relative experimental variability of sensors evaluated from the results of field comparisons.

	400 nm	442.5 nm	490 nm	560 nm	665 nm	778.8 nm	865 nm
Certificate	1.2	0.78	0.76	0.73	0.71	0.73	1.35
Interpolation	0.6	0.3	0.3	0.3	0.3	0.3	0.3
Instability (sensor)	0.04	0.03	0.02	0.01	0.01	0.02	0.01
Polarization	0.2	0.2	0.2	0.2	0.2	0.2	0.2
Nonlinearity	0.7	0.8	0.9	1	1.1	1.2	1.3
Stray light	0.9	0.7	0.3	0.3	0.7	0.9	1
Temperature	0.4	0.2	0.2	0.2	0.2	0.4	0.8
Alignment, FOV	1.7	1.8	1.8	1.6	1.8	4	4.3
Signal, type A	0.04	0.07	0.11	0.11	0.21	0.55	0.72
Combined ($k = 1$)	2.2	2.1	2.1	1.9	2.3	4.2	4.6
Expanded ($k = 2$)	4.4	4.2	4.2	3.8	4.6	8.4	9.2
Variability ($k = 2$)	4.4	4.4	4.4	3.2	4.6	8.6	9.4

In general the uncertainty is calculated from the contributions originating from: (1) The spectral responsivity of the radiometer, including data from the calibration certificate; (2) interpolation of the spectral responsivity values to the designated wavelengths and/or spectral bands; (3) temporal instability of the radiometer; (4) contribution caused by polarization sensitivity; (5) non-linearity effects; (6) effect of spectral stray light; (7) temperature effects; (8) error of cosine collector; (9) type A component of recorded signal; (10) alignment and FOV effects.

The calibration uncertainty is most relevant for traceability to the SI units. The remaining uncertainty sources in Tables 4–6 describe variability between the sensors while overlooking possible systematic effects which can influence all the instruments in a similar way. Moreover, there was no fully characterized reference instrument involved during the LCE-2 outdoor exercise. Thus, the uncertainty analysis presented here is not sufficient to link the measurements to the SI units.

For the RAMSES group, the variability of radiance sensors during indoor and outdoor exercises (Figure 11, except C8 and C23) was close. Therefore, variability due to significant influence factor—temperature, and respective estimate used in uncertainty budget, can be considered practically the same as rather large systematic change is likely similar for all sensors [7]. For example, during outdoor measurements, temperature was rather stable varying from 5 °C to 9 °C, a range fairly comparable with variation of temperature during indoor exercise from 21 °C to 24 °C. As the construction of radiance and irradiance sensors (except the input optics) is similar, the similar estimate is likely suitable also for the temperature caused variability between irradiance sensors.

Some increase in variability may be expected, due to nonlinearity and spectral stray light of outdoor results. Major differences in combined uncertainty estimates for outdoor measurements are likely caused by different FOV of the sensors (including deviation from cosine response for irradiance instruments), and due to temporal variation and nonuniformity of the targets.

4.1. Calibration Certificate

The calibration certificates of the radiometers provide calibration points following the individual wavelength scale of the radiometer. During the relatively short time needed for LCE-2 measurements, this uncertainty component normally is not contributing to the variability between radiometric sensors freshly calibrated at the same laboratory using the same calibration standards. Therefore, this component is presented only for reference and is not included in the combined and expanded uncertainties. At the same time, for the full uncertainty of SI traceable results, the radiometric calibration uncertainty shall always be accounted for.

4.2. Interpolation

Interpolation of radiometer's data is needed due to differences between individual wavelength scales of the radiometers. Therefore, measured values were transferred for comparison to a common scale basis (a selection of Sentinel-3/OLCI bands). The uncertainty contribution associated with the interpolation of spectra is estimated using different interpolation algorithms. The weights used for binning hyperspectral data to OLCI bands depend on the wavelength scale and exact pixel positions of the hyperspectral sensor. Interpolation component includes interpolation, as well as wavelength scale uncertainty contributions. Figure 16 shows the change of the OLCI band values of a measured spectrum as a function of the wavelength scale error of a radiometer, as determined for a single RAMSES radiance sensor for the casts C8, C12, C17, and C23. The precision of the wavelength scale of the RAMSES instrument is stated by the manufacturer as 0.3 nm. For ±0.3 nm shift of the scale, the changes of the OLCI band values for the different spectra remain less than ±0.5% except for the 400 nm spectral band where the radiance changes rapidly with wavelength and the effect of shifting the wavelength scale is stronger.

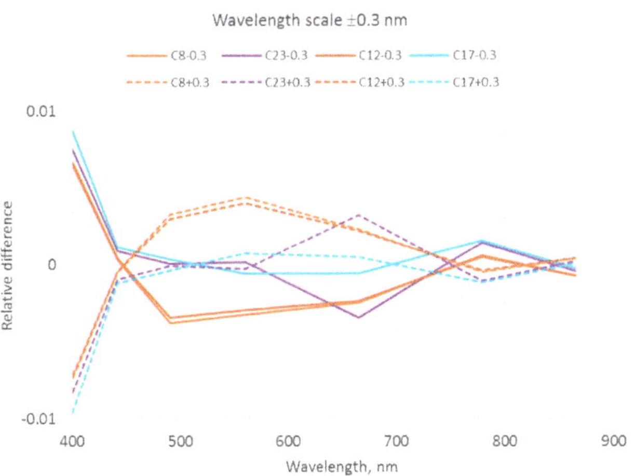

Figure 16. Relative variability due to wavelength error of ±0.3 nm of a radiance sensor.

4.3. Temporal Instability of Sensor

Instability of the radiometric responsivity can be estimated from data of repeated radiometric calibrations. For LCE-2, the instruments were calibrated just before the comparisons and only short-term instability relevant for the time needed for the measurements has to be considered. The values are derived from the data collected in calibration sessions of LCE-2 and FICE-AAOT a year later, see 4.1.1 and Figure 10 in Reference [1]. The instability over two weeks was interpolated from the yearly variability data assuming only smooth drift (no mechanical shocks, no abrupt changes). Besides the instability of the sensors, data in Figure 10 of [1] include other uncertainty components related to the calibration setup (e.g., alignment, short-term lamp instability, etc).

4.4. Polarization

For the outdoor radiance measurements, the uncertainty contribution caused by polarization sensitivity is estimated using worst-case data in Reference [17]. Evaluation of the polarization effect for the outdoor irradiance measurement is difficult as the degree of linear polarization (DoLP) depends on various factors, such as wavelength, solar zenith angle (SZA), aerosol optical depth (AOD), amount and location of clouds, etc. In addition, the DoLP can strongly vary over the hemisphere, being due to Rayleigh scattering the largest at 90° from the Sun, and for the direct solar flux decreasing to zero. However, according to Reference [17] the polarization sensitivity of RAMSES irradiance sensors is rather small, hence, regardless of the DoLP value of downwelling irradiance the contribution of polarization effect in the uncertainty budget is also small. Uncertainty component of solar irradiance associated with polarization is estimated to be less than 0.25%.

4.5. Nonlinearity

The nonlinearity of the participating radiometers was evaluated by varying the integration time during the calibration. As an automatically adjusted optimal integration time is typically used in the field conditions, the class-specific method for the RAMSES instruments was developed and validated using the indoor results, see Equation (6) in Reference [1]. Variability between sensors due to nonlinearity was evaluated by applying this equation to the different casts of the field spectra of five RAMSES sensors.

4.6. Spectral Stray Light

Figure 17 presents the impact of the stray light in outdoor measurements. The effect is much stronger than in indoor experiment, due to significantly different spectral shape of the target and calibration signals. The general impact of the stray light correction is similar for RAMSES and HyperOCR radiometers. Variability between sensors and between different measurements targets for HyperOCR radiometers increases significantly in the NIR spectral region. This is probably related to the uncertainty associated with the stray light correction procedure and is not characteristic to the actual impact of spectral stray light. The spectral stray light matrices of HyperOCR sensors used in the analysis had a, somewhat, higher noise level compared to the matrices of the RAMSES instruments. Data in Tables 5 and 6 is estimated from Figure 17, and from References [21,22].

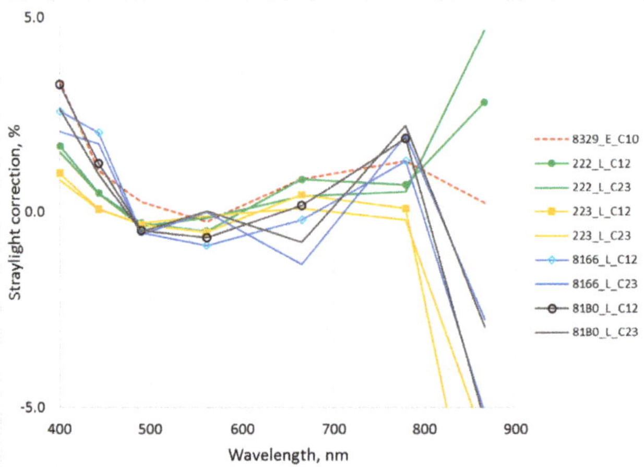

Figure 17. Stray light effects in the outdoor experiment. One RAMSES 8329 irradiance sensor—dashed line; two RAMSES and two HyperOCR radiance sensors: Solid lines with markers—blue sky (C12), and solid lines without markers—sunlit water (C23).

4.7. Temperature

For array spectroradiometers with silicon detectors, the present estimate for standard uncertainty, due to temperature variability ($\pm1.5\ ^{\circ}$C) in the spectral region from 400 nm to 700 nm is around 0.1% and will increase up to 0.6% for longer wavelengths (950 nm) [7]. In the case of outdoor measurements, the temperature differences between sensors quite likely were in the range of ($\pm2\ ^{\circ}$C), so temperature contribution is slightly larger than for indoor experiment. But outside air temperature between 5 °C and 9 °C was significantly lower than calibration temperature contributing to systematic biases common to all the instruments and not accounted in Tables 4–6.

4.8. Cosine Error

The irradiance sensors are calibrated using normal illumination, but during outdoor solar irradiance measurements the radiation arriving from hemisphere has to be measured with the angular dependence of responsivity ideally corresponding to the cosine of incidence angle. Typical class-specific values of uncertainty related to the deviation of cosine response are derived from Reference [8]. Measurements carried out after LCE-2 at TO (Figures 13 and 14) have shown that RAMSES sensors may have rather large cosine errors around ±10%. This may be a likely reason for excessive differences from +7% up to −16% evident for irradiance sensors during the LCE-2 outdoor measurements (Figure 10).

4.9. Type A Uncertainty of Repeated Measurements

The type A uncertainty was estimated from the ratio of two RAMSES radiometers. While there is strong autocorrelation in individual time series, due to the unstable nature of natural illumination, there was almost no correlation between individual ratios during one cast, and the effective number of measurements was close to the actual number of data points in the time series. The effective number of measurements was calculated by using the lag-1 autocorrelation coefficient, as shown in Reference [23].

The instability of the target signal during the outdoor measurements was significantly larger compared to the indoor experiment, however, all the instruments measured simultaneously and the impact of source temporal variability affected all the radiometers in a similar manner without causing differences between the sensors. This was verified by separately analyzing some shorter and more stable sections of the selected casts, no reduction of variability between the sensors was observed. Thus, the uncertainty due to the temporal variability of the target is not included in Tables 4–6.

4.10. Alignment and Field-of-View

During the outdoor radiance measurement, the spatial and temporal non-uniformity of the target can substantially contribute to the uncertainty, due to different FOV-s of the radiance sensors, and due to misalignment (Figures 7 and 8).

5. Discussion and Conclusions

For irradiance, the difference in cosine response is the main source of differences between different sensor groups revealed during the field experiment. For radiance, the angular response (different Field of Views) and spatial nonuniformity of the targets provides the main difference between different sensor groups. In the case of a spatially heterogeneous target (sky with scattered clouds, water at oblique viewing angle) the large differences of FOV of different sensors will likely cause significant discrepancies between sensors. Without reliable data or individual testing of the input properties of all involved sensors interpretation of measurement results may be strongly hindered. For field measurements the variability between radiance sensors was about two times larger than during indoor exercise, this can among others be explained by larger effects of outside influence factors like temperature, stray light and nonlinearity which all have not been corrected during the field experiment.

Dependence of the calibration coefficients on temperature can cause significant deviation from SI-traceable result. For maximum temperature difference of about 20 °C between calibration and later measurements (typically between 0 °C and 40 °C) a responsivity change more than 10% will be possible [3,7]. The calibration procedure may be improved if specified conditions will cover all situations possible during the use of a calibrated instrument. For example, if it is known that the radiometer has a linear response with temperature [7], the responsivity of the radiometer can be adequately evaluated when calibration is performed at three different temperatures covering the possible range of temperature variations during the later use.

Variability between irradiance sensors was about five times larger than that observed during indoor exercise. A large variability between sensors during outdoor exercise cannot be explained by poor stability of sensors, as stability check in lab conditions, a year later has shown smaller changes than during outdoor measurements some days after calibration. Variability cannot be fully explained by factors, such as temperature, nonlinearity, and stray light either as one could expect a smaller difference between radiance and irradiance sensors in this case. Most likely, the different behavior of RAMSES and HyperOCR sensors is largely due to a different construction of input optics of these sensors and hence imperfect cosine response. This hypothesis is supported by the angular response characterization of 5 RAMSES irradiance sensors and comparing the integral cosine error values in Figures 13–15 to the deviations from consensus value in the outdoor experiment, shown in Figure 10.

The different behavior of RAMSES and HyperOCR sensor groups was clearly revealed during the LCE-2 exercise. For the RAMSES group, the variability of radiance sensors during indoor and outdoor exercises was very similar, and larger variability for outdoor measurements was mostly caused by HyperOCR and WISP-3 sensors. For irradiance measurements, the deviation of HyperOCR sensors from the consensus value of the group was very small, and an increase in variability was caused mostly by the group of RAMSES sensors.

The indoor experiment has demonstrated great effectiveness of the radiometric calibration at the same laboratory just before intercomparison measurements [1] for obtaining consistent results. Nevertheless, a sufficient individual characterization of radiometers by testing them for all significant systematic effects beside regular radiometric calibration, is the shortest way to enable reduction of biases in outdoor intercomparisons, and thus smaller variability between measurements from different instruments, and more realistic and complete quantification of uncertainties in measurement.

Lessons Learned for the Design of Future Intercomparisons

In order to foster the interpretation of results, the following suggestions are proposed for the future outdoor intercomparison campaigns. The number of involved radiometers should be around ten for each radiometer type to strengthen the statistical representativeness of the analysis. Consistent calibration of the responsivity of all involved radiometers just before the campaign is indispensable. The 20° "offsetting" calibration method suggested in Reference [8] should be also tested by a comparison measurement. Calibration history of each radiometer should be available to detect long-term instabilities. Together with radiometric calibration, the angular response of all individual radiance and irradiance sensors should be measured if such information is not available from previous characterizations. Before radiometric calibration, all instruments involved should be tested or be characterized for temperature, nonlinearity, spectral stray light and wavelength scale effects. As these tests may be rather time-consuming they should be performed well before the radiometric calibration. Spectral responsivity should be calibrated at different ambient temperatures relevant for the campaign. Nonlinearity and wavelength correction coefficients should also be available. The usefulness of individual characterization of the spectral stray light should be further proven by thorough field tests using an independent validation method based on a reference instrument less affected by stray light.

During the outdoor campaign measurements, well-synchronized data acquisition for all instruments is strongly advised. Start timer should be aligned better than within ±1 s; setting exactly the same sampling interval for all sensors is indispensable. Data processing algorithms should be well defined and agreed between the participants. For that, sufficient calibration and test information should be available for each sensor in order to be able to apply likewise all needed corrections. Instruments' temperature should be recorded whenever possible. Using a well-characterized additional reference instrument is highly recommended, as well as using an aligned photo- or video camera to record the measurement scene during outdoor experiments simultaneously with radiometric sensors.

Metrological specifications of all involved radiometers should be based on suitable international standards whenever possible. Minimum requirements should be agreed between the participants, instruments involved should be tested to give evidence that all these minimum requirements are met.

Author Contributions: Conceptualization, V.V., J.K., I.A., R.V., K.A., K.R., C.D., T.C.; Data curation, V.V., J.K.; Formal analysis, V.V., I.A.; Investigation, J.K., I.A., R.V., K.A., K.R., A.A., M.B., H.B., M.C., D.D., G.D., B.D., T.D., C.G., K.K., M.L., B.P., G.T., R.V.D., S.W.; Methodology, V.V., J.K., I.A., R.V., K.A., K.R., M.C., K.K.; Project administration, R.V.; Software, I.A.; Validation, V.V., J.K., I.A.; Visualization, V.V., J.K., I.A., K.R., M.L.; Writing—original draft, V.V., J.K., I.A., R.V., K.A.; Writing—review and editing, V.V., J.K., I.A., R.V., K.A., K.R., A.A., M.B., H.B., M.C., D.D., G.D., B.D., T.D., C.G., K.K., M.L., B.P., G.T., R.V.D., S.W., A.B., C.D., T.C.

Funding: This work was funded by European Space Agency project Fiducial Reference Measurements for Satellite Ocean Color (FRM4SOC), contract No. 4000117454/16/I-Sbo.

Acknowledgments: Support from numerous scientists, experts and administrative personnel contributing to the project are gratefully acknowledged. The authors express special gratitude to Giuseppe Zibordi (JRC of the EC), Anu Reinart (UT), Tiia Lillemaa (UT), Viljo Allik (UT), Claire Greenwell (NPL).

Conflicts of Interest: The authors declare no conflict of interest. Most authors collecting experimental data are customers of the respective instrument manufacturers. Two authors (R.V.D., B.D.) are employees of radiometer manufacturers. Two authors (J.K., K.R.) are developing a new radiometer for commercialization in 2021. Study data analysis and conclusions focus on aspects which are common to all radiometers and results are anonymized to avoid any risk of bias. This study does not constitute an endorsement of any of the products tested by any of the authors or their organizations.

References

1. Vabson, V.; Kuusk, J.; Ansko, I.; Vendt, R.; Alikas, K.; Ruddick, K.; Ansper, A.; Bresciani, M.; Burmester, H.; Costa, M.; D'Alimonte, D.; et al. Laboratory intercomparison of radiometers used for satellite validation in the 400–900 nm range. *Remote Sens.* **2019**, *11*, 1101. [CrossRef]
2. Zibordi, G.; Ruddick, K.; Ansko, I.; Moore, G.; Kratzer, S.; Icely, J.; Reinart, A. In situ determination of the remote sensing reflectance: An inter-comparison. *Ocean Sci.* **2012**, *8*, 567–586. [CrossRef]
3. Antoine, D.; Schroeder, T.; Slivkoff, M.; Klonowski, W.; Doblin, M.; Lovell, J.; Boadle, D.; Baker, B.; Botha, E.; Robinson, C.; et al. IMOS Radiometry Task Team. Final Report. 2017. Available online: http://imos.org.au/fileadmin/user_upload/shared/IMOS%20General/documents/Task_Teams/IMOS-RTT-final-report-submission-30June2017.pdf (accessed on 7 May 2019).
4. Aerosol Robotic Network (AERONET) Homepage. Available online: https://aeronet.gsfc.nasa.gov/ (accessed on 9 August 2018).
5. Kuusk, J.; Ansko, I.; Vabson, V.; Ligi, M.; Vendt, R. *Protocols and Procedures to Verify the Performance of Fiducial Reference Measurement (FRM) Field Ocean Colour Radiometers (OCR) used for Satellite Validation*; Technical Report TR-5; Tartu Observatory: Tõravere, Estonia, 2017.
6. Ruddick, K.G.; de Cauwer, V.; Park, Y.-J.; Moore, G. Seaborne measurements of near infrared water-leaving reflectance: The similarity spectrum for turbid waters. *Limnol. Oceanogr.* **2006**, *51*, 1167–1179. [CrossRef]
7. Zibordi, G.; Talone, M.; Jankowski, L. Response to Temperature of a Class of In Situ Hyperspectral Radiometers. *J. Atmos. Ocean. Technol.* **2017**, *34*, 1795–1805. [CrossRef]
8. Mekaoui, S.; Zibordi, G. Cosine error for a class of hyperspectral irradiance sensors. *Metrologia* **2013**, *50*, 187–199. [CrossRef]
9. Sea-Bird Scientific, "Specifications for HyperOCR Radiometer". Available online: https://www.seabird.com/hyperspectral-radiometers/hyperocr-radiometer/family?productCategoryId=54627869935 (accessed on 31 January 2019).
10. TriOS, "RAMSES Technische Spezifikationen", TriOS Mess- und Datentechnik. Available online: https://www.trios.de/ramses.html (accessed on 31 January 2019).
11. Mueller, J.L.; Fargion, G.S.; McClain, C.R. (Eds.) *Ocean Optics Protocols for Satellite Ocean Color Sensor Validation: Instrument Specifications, Characterization, and Calibration*; NASA/TM-2003-21621/Rev-Vol II; NASA: Washington, DC, USA, 2003.
12. Pulli, T.; Kärhä, P.; Ikonen, E. A method for optimizing the cosine response of solar UV diffusers. *J. Geophys. Res. Atmos.* **2013**, *118*, 7897–7904. [CrossRef]
13. JCGM. *Evaluation of Measurement Data—Guide to the Expression of Uncertainty in Measurement (GUM)*, 1st ed.; JCGM 100, GUM 1995 with Minor Corrections; JCGM: Sèvres, France, September 2008; Available online: http://www.bipm.org/utils/common/documents/jcgm/JCGM_100_2008_E.pdf (accessed on 31 January 2019).
14. EUMETSAT. *Requirements for Copernicus Ocean Colour Vicarious Calibration Infrastructure*; Annex: Example of Quantified Uncertainty Budget; EUMETSAT: Darmstadt, Germany, 2017.
15. Zibordi, G.; Voss, K.J. Chapter 3.1—In situ Optical Radiometry in the Visible and Near Infrared. In *Optical Radiometry for Ocean Climate Measurements*; GZibordi, I., Donlon, C., Parr, A., Eds.; Academic Press: Cambridge, MA, USA, 2014; Volume 47, pp. 247–304.
16. Gergely, M.; Zibordi, G. Assessment of AERONET-OC L_{WN} uncertainties. *Metrologia* **2014**, *51*, 40. [CrossRef]
17. Talone, M.; Zibordi, G. Polarimetric characteristics of a class of hyperspectral radiometers. *Appl. Opt.* **2016**, *55*, 10092–10104. [CrossRef] [PubMed]
18. IOCCG. International Network for Sensor Inter-Comparison and Uncertainty Assessment for Ocean Color Radiometry (INSITU-OCR) White Paper. 2012. Available online: http://www.ioccg.org/groups/INSITU-OCR_White-Paper.pdf (accessed on 7 February 2017).

19. Alikas, K.; Ansko, I.; Vabson, V.; Ansper, A. Validation of Sentinel-3A/OLCI data over Estonian inland waters. In Proceedings of the AMT4 Sentinel FRM Workshop, Plymouth, UK, 20–21 June 2017.

20. Ruddick, K. A Review of Commonly Used Fiducial Reference Measurement (FRM) Ocean Colour Radiometers (OCR) Used for Satellite OCR Validation. Technical Report TR-2. 2018. Available online: https://frm4soc.org/wp-content/uploads/filebase/FRM4SOC-TR2_TO_signedESA.pdf (accessed on 7 May 2019).

21. Talone, M.; Zibordi, G.; Ansko, I.; Banks, A.C.; Kuusk, J. Stray light effects in above-water remote-sensing reflectance from hyperspectral radiometers. *Appl. Opt.* **2016**, *55*, 3966–3977. [CrossRef] [PubMed]

22. Vabson, V.; Ansko, I.; Alikas, K.; Kuusk, J.; Vendt, R.; Reinat, A. Improving comparability of radiometric in situ measurements with Sentinel-3A/OLCI data. In Proceedings of the S3VT Sentinel-3 Validation Team meeting, Eumetsat, Darmstadt, Germany, 13–15 March 2018.

23. Santer, B.D.; Wigley, T.M.; Boyle, J.S.; Gaffen, D.J.; Hnilo, J.J.; Nychka, D.; Parker, D.E.; Taylor, K.E. Statistical significance of trends and trend differences in layer-average atmospheric temperature time series. *J. Geophys. Res. Atmos.* **2000**, *105*, 7337–7356. [CrossRef]

Article

Comparison of Above-Water Seabird and TriOS Radiometers along an Atlantic Meridional Transect

Krista Alikas [1,*], Viktor Vabson [1], Ilmar Ansko [1], Gavin H. Tilstone [2], Giorgio Dall'Olmo [2], Francesco Nencioli [2], Riho Vendt [1], Craig Donlon [3] and Tania Casal [3]

[1] Department of Remote Sensing, Tartu University, Tartu Observatory, Observatooriumi 1, 61602 Tartu, Estonia; viktor.vabson@ut.ee (V.V.); ilmar.ansko@ut.ee (I.A.); riho.vendt@ut.ee (R.V.)

[2] Plymouth Marine Laboratory, Prospect Place, The Hoe, Plymouth PL1 3DH, UK; ghti@pml.ac.uk (G.H.T.); gdal@pml.ac.uk (G.D.); fne@pml.ac.uk (F.N.)

[3] European Space Agency, 2201 AZ Noordwijk, The Netherlands; craig.donlon@esa.int (C.D.); tania.casal@esa.int (T.C.)

* Correspondence: alikas@ut.ee

Received: 14 April 2020; Accepted: 18 May 2020; Published: 22 May 2020

Abstract: The Fiducial Reference Measurements for Satellite Ocean Color (FRM4SOC) project has carried out a range of activities to evaluate and improve the state-of-the-art in ocean color radiometry. This paper described the results from a ship-based intercomparison conducted on the Atlantic Meridional Transect 27 from 23rd September to 5th November 2017. Two different radiometric systems, TriOS-Radiation Measurement Sensor with Enhanced Spectral resolution (RAMSES) and Seabird-Hyperspectral Surface Acquisition System (HyperSAS), were compared and operated side-by-side over a wide range of Atlantic provinces and environmental conditions. Both systems were calibrated for traceability to SI (Système international) units at the same optical laboratory under uniform conditions before and after the field campaign. The in situ results and their accompanying uncertainties were evaluated using the same data handling protocols. The field data revealed variability in the responsivity between TRiOS and Seabird sensors, which is dependent on the ambient environmental and illumination conditions. The straylight effects for individual sensors were mostly within ±3%. A near infra-red (NIR) similarity correction changed the water-leaving reflectance (ρ_w) and water-leaving radiance (L_w) spectra significantly, bringing also a convergence in outliers. For improving the estimates of in situ uncertainty, it is recommended that additional characterization of radiometers and environmental ancillary measurements are undertaken. In general, the comparison of radiometric systems showed agreement within the evaluated uncertainty limits. Consistency of in situ results with the available Sentinel-3A Ocean and Land Color Instrument (OLCI) data in the range from (400 . . . 560) nm was also satisfactory (−8% < Mean Percentage Difference (MPD) < 15%) and showed good agreement in terms of the shape of the spectra and absolute values.

Keywords: ocean color; remote sensing; radiometry; TriOS RAMSES; Seabird HyperSAS; measurement uncertainty; validation; Sentinel-3 OLCI; Copernicus

---

1. Introduction

The European Commission provides daily global ocean color data via the Ocean and Land Color Instrument (OLCI) on board Sentinel-3 (S-3) satellite in the context of the Copernicus program. The first Sentinel-3 satellite was launched in 2016. The Sentinel-3 mission will continue for at least two decades through the sequential launch of a cluster of satellites. These will provide data to Europe's Copernicus environmental program to support monitoring, services, decision and policymaking, and climate change studies.

Based on the requirements of the Global Climate Observing System (GCOS), there is less than 5% uncertainty level expectation for water-leaving radiance (L_w) data contributing to climate studies. To reduce the uncertainties in the satellite products, System Vicarious Calibration (SVC) approach has been undertaken using field data to calibrate the combined system of satellite instrument and the processing algorithm [1,2]. SVC has been operationally used for previous, e.g., MEdium Resolution Imaging Spectrometer (MERIS) on board Environmental Satellite (ENVISAT) and on ongoing missions to meet ocean color mission requirements in open waters. For Sentinel-3 data, the SVC gains have now been applied to OLCI data on Sentinel-3A but not to Sentinel-3B yet. For S-3 OLCI radiometric products, Sentinel-3 mission requirements [3] foresee 5% uncertainty for bands (490, 510, 560) nm and 5–10% uncertainty for bands (400, 412, 442) nm depending on water types. To qualify as Fiducial Reference Measurements (FRM), quantification of the uncertainties in the Earth observation data is required. This can only be done by quantifying the uncertainties in the field data used to validate and through rigorous intercomparison exercises to assess differences in radiometer systems used for such validation.

The Fiducial Reference Measurements for Satellite Ocean Color (FRM4SOC) project aimed to evaluate and improve state of the art in ocean color radiometry through review of commonly used radiometers [4], SI (Système international) traceable calibrations [5], protocols for the downwelling irradiance [6] and water-leaving radiance [7], uncertainty evaluation at different stages of the traceability chain [5,8], and through series of radiometric comparisons [9–12]. These included:

- a comparison of radiance and irradiance sources used for calibration of radiometers National Physics Laboratory-UK(NPL, UK) [9];
- an indoor comparison of uniformly calibrated radiometers measuring stable radiance and irradiance sources where the illumination conditions and measurement geometry were strictly controlled and close to ideal [10];
- an outdoor comparison over a Case 2 water body with the radiometers installed on the fixed platform (Lake Kääriku, Estonia). The illumination conditions during this experiment were variable due to the weather, while the measurement geometry resembled as closely as possible to the realistic field conditions [11];
- a further outdoor comparison with the same instruments a year later on a fixed platform (the Aqua Alta Oceanographic Tower—AAOT) under near-ideal environmental conditions [12];
- a shipborne campaign on the Atlantic Meridional Transect 27 (AMT27), (the current study).

The indoor exercise consisted of calibration of the instruments at the same laboratory and demonstrated satisfactory consistency between sensors with a standard deviation within ±1% [10]. The field comparisons had substantially larger variability between the same sensors, implying to the respective increase of the uncertainty of the field results. At Lake Kääriku, Estonia, there was large variability between recently calibrated sensors due to high spectral and spatial variability in the targets and environmental conditions. At the AAOT in the Adriatic Sea off Venice, there was a <5% difference in normalized water-leaving radiance of both TriOS- Radiation Measurement Sensor with Enhanced Spectral resolution (RAMSES) and Seabird- Hyperspectral Surface Acquisition System (HyperSAS) sensors compared to Aerosol Robotic Network for Ocean Color (AERONET-OC) SeaWiFS Photometer Revision for Incident Surface Measurements (SeaPRiSM) [12]. The reasons for the increase from 1 to 5% are likely to be due to different measurement targets during field measurements and calibration, both spectrally and spatially; less stable ambient temperatures during the field campaigns, which can vary, compared to the calibration temperature, by more than ±15 °C.

In this study, we analyzed the difference between a TriOS-RAMSES and Seabird-HyperSAS systems that were used in the indoor laboratory intercomparison and the two field intercomparisons [10–12] on the 27th Atlantic Meridional Transect (AMT27) cruise, which crossed a range of ocean provinces and different environmental conditions. The in situ data was also compared to S-3A OLCI radiometric products.

The objectives of this work were: (1) to analyze above-water in situ radiometric data measured using two different systems, both of which used three radiometers each, under variable environmental conditions, in the context of the previous intercomparisons [10–12]; (2) to specify the uncertainties for both systems, and to evaluate the consistency of measured in situ data; (3) to evaluate the consistency of satellite data with the in situ results, accounting for estimated in situ uncertainties.

2. Materials and Methods

2.1. Study Site

The AMT27 field campaign took place from 23rd September to 5th November 2017 from Southampton, UK to South Georgia and the Falkland Islands on the UK-Natrual Environment Research Council (NERC) ship Royal Research Ship (*RRS*) *Discovery*. The AMT is a multidisciplinary research program, which undertakes biological, chemical, and physical oceanographic research during an annual voyage between the UK and destinations in the South Atlantic. The program has been running for 20 years and was established in 1995, in collaboration with National Aeronautics and Space Administration (NASA), as an independent platform to validate Sea-Viewing Wide Field-of-View Sensor (SeaWiFS) ocean color data. The transect covered several ocean provinces where key physical and biogeochemical variables, such as chlorophyll, primary production, nutrients, temperature, salinity, and oxygen, were measured. The stations sampled were principally in Case 1 waters [13,14]; in the North and South Atlantic Gyres, but also the productive waters of the Celtic Sea, Patagonian Shelf, and Equatorial upwelling zone were visited, which, therefore, offered a wide range of variability in which to conduct field intercomparisons. The measurement stations are listed in Table 1; Sentinel-3A OLCI quality controlled match-ups were available for station id-s 22, 32, 46, 48, 56.

Table 1. Overview of measurement conditions during the midday station during AMT27.

No	Station Id	Date	Latitude (Degree)	Longitude (Degree)	Sun Zenith Angle (Degree)	Wind Speed (W, m·s^{-1})	Temperature (t, °C)
1	1	24.09.2017	48.9	−7.6	52.37	1.48	16.2
2	3	25.09.2017	46.7	−12.0	51.52	7.23	17.3
3	6	27.09.2017	42.2	−18.8	46.31	2.24	19.3
4	8	28.09.2017	39.4	−22.7	45.31	5.94	23.0
5	10	30.09.2017	35.1	−26.3	38.87	1.69	24.3
6	12	01.10.2017	31.8	−27.2	35.84	5.69	23.5
7	16	03.10.2017	25.7	−28.7	30.52	7.15	24.5
8	18	04.10.2017	22.3	−29.5	28.58	1.69	25.7
9	20	05.10.2017	18.8	−29.7	26.4	5.47	26.6
10	22	06.10.2017	15.5	−28.8	23.21	4.31	27.8
11	24	07.10.2017	12.8	−28.2	20.4	8.43	28.0
12	26	08.10.2017	9.9	−27.4	18.38	6.89	28.3
13	28	09.10.2017	6.9	−26.7	15.41	5.13	27.6
14	32	11.10.2017	1.5	−25.4	10.42	6.34	26.0
15	34	12.10.2017	−1.8	−25.0	8.23	8.44	25.9
16	36	13.10.2017	−4.6	−25.0	6.07	10.74	25.7
17	38	14.10.2017	−7.1	−25.0	5.84	6.8	25.5
18	40	15.10.2017	−10.5	−25.1	5.93	6.63	25.1
19	42	16.10.2017	−13.7	−25.1	7.85	8.12	23.8
20	43	17.10.2017	−16.0	−25.1	8.22	8.07	22.9
21	46	19.10.2017	−21.8	−25.1	13	8.76	21.7
22	48	20.10.2017	−25.1	−25.0	15.56	6.26	21.2
23	50	21.10.2017	−27.9	−25.2	17.78	3.88	20.7
24	52	22.10.2017	−31.3	−26.2	21.22	8.74	19.4
25	54	23.10.2017	−33.9	−27.1	26.05	6.17	17.3
26	56	24.10.2017	−37.0	−28.3	28.62	8.12	15.8

Table 1. *Cont.*

No	Station Id	Date	Latitude (Degree)	Longitude (Degree)	Sun Zenith Angle (Degree)	Wind Speed (W, m·s^{-1})	Temperature (t, °C)
27	59	26.10.2017	−42.1	−30.4	34.22	7.92	10.0
28	61	27.10.2017	−45.2	−31.7	36.15	16.25	8.8
29	62	28.10.2017	−47.1	−32.6	54.53	8.03	6.4
30	64	29.10.2017	−50.4	−34.2	40.23	11.63	1.6
31	66	30.10.2017	−52.9	−35.7	43.25	9.25	0.9
32	67	01.11.2017	−53.7	−38.1	60.54	19.71	2.0

* AMT27—Atlantic Meridional Transect 27.

2.2. In Situ Above-Water Radiometric Data

Stations were sampled daily at 12:00 local time to ensure coincident in situ measurements within 1 h of the S-3 overpass. Radiometric measurements were performed with two sets of above water hyperspectral radiometers, both consisting of three separate sensors to measure radiance from the water surface $L_u(\lambda)$, radiance from the sky $L_d(\lambda)$, and downwelling solar irradiance $E_d(\lambda)$. Plymouth Marine Laboratory (PML) used three Seabird (formerly Satlantic) HyperOCR sensors, while the University of Tartu (TO) used three TriOS RAMSES sensors. All radiance and irradiance sensors of both radiometric systems were SI-traceably calibrated at the Tartu Observatory following procedures entirely described in [10]. To comply with FRM, the sensors were calibrated frequently; in this case, three times: in April 2017 before the second SI-traceable Laboratory inter-comparison experiment (LCE-2) campaign, just after the AMT-27 campaign in January 2018, and in June 2018 before AAOT. All of these sensors were involved also in the LCE-2 intercomparisons, and during indoor measurements [10] demonstrated differences of less than ±1% both for radiance and irradiance results. However, during the outdoor exercise under cloudy conditions, the PML irradiance sensors did show up to 6% higher values in E_d at 400 nm, and radiance sensors did show up to about 10% higher values in red and IR parts of spectrum than the respective TO sensors [11]. Technical parameters [15,16] of the applied radiometers are given in Table 2.

Table 2. Technical parameters of the radiometers.

Parameter	RAMSES	HyperOCR
Field of View (L/E)	7°/cos	6°/cos
Adaptive integration time	Yes	yes
Min. integration time, ms	4	4
Max. integration time, ms	4096	4096
Min. sampling interval, s	1	0.5
Recording dark signal	Opaque pixels	Internal shutter
Number of channels	256	256
Wavelength range, nm	320 … 1050	320 … 1050
Wavelength step, nm	3.3	3.3
Spectral resolution, nm	10	10

The sensors are based on the Carl Zeiss Monolithic Miniature Spectrometer (MMS), incorporating a 256-channel silicon photodiode array. The SI-traceable radiometric calibration covered the wavelength range of (350 … 900) nm. During the field measurements, the integration time was automatically adjusted to correspond to the measured light intensity. The data acquisition system consisted of power supplies, RS232 multiplexers, and logging computers. The instruments were mounted on a common steel frame, which was constructed to perform measurements under identical zenith and viewing angles (Figure 1). The radiance sensors ($L_d(\lambda)$) and ($L_u(\lambda)$) were positioned at the very front of the ship (Figure 1A), with an un-obscured view of the ocean and sky (Figure 1C), side by side at the same height at 40° and 120° angles from zenith (Figure 1B), respectively. The colinearity of the radiance sensors in the frame was set by visual observation from the side of the frame and was estimated to be within ±1°. The downwelling irradiance sensors were positioned on the same steel frame, higher from other sensors to avoid any ship shadows. A fixed mounting frame of the irradiance sensors ensured equal height and leveling of the cosine collectors.

2.3. In Situ Data Processing

For both systems, the radiometric raw data were logged through a laptop, which was set up in the meteorological laboratory, some 50 m away from the setup of the radiometers on the meteorological platform at the bow of the ship (Figure 1). The HyperOCR instruments produced proprietary binary data files as standard output from the manufacturer's software. The TriOS RAMSES spectrometers were operated by software developed in TO; spectra were stored in the American Standard Code for Information Interchange (ASCII) datafiles. The three HyperOCR spectrometers were individually measured in burst mode (i.e., continuously), while the three RAMSES devices performed synchronized measurement every 10 s. Altogether, five million spectra were collected by PML and 200,000 by TO. Additionally, ancillary meteorological and positional data were provided by the AMT crew in the form of Network Common Data Form (NetCDF) files. Particularly, position latitude and longitude, ship speed and direction, and wind speed and air temperature were used in this study.

A number of computer programs were created for this particular dataset to process the data. The algorithms were programmed directly without using external libraries. Third-party software was used to visualize the results and extract data from NetCDF files. Due to a large amount of spectra, a database was created first for all instruments, containing the filenames and positions within the data files, which was ordered by timestamps. Then, the database was used to dynamically extract the individual spectra from the raw data files without creating unnecessary copies of big data for the HyperOCR spectra, and the closest shutter measurement was subtracted from each field spectrum. In the case of TRiOS-RAMSES instruments, the average signal over the opaque pixels was used as a dark reference. In the case of HyperOCR, the L_u spectrum was derived first and then the closest (within ±3 s) L_d and E_d spectrum in order to form the consistent data triplet. The spectra were derived according to the cast start and stop timestamps, calibration and all necessary corrections/filtering applied, after which the output quantities were calculated, and the corresponding uncertainties were evaluated. The results were stored as ASCII data files, which are convenient to use for post-processing or spreadsheet programs. The hyperspectral data was convoluted into 19 OLCI channels, from 400 nm to 885 nm, based on channel definitions from [17].

The corrections and filtering criteria were sequentially applied via command line parameters during the data processing. The steps included in the processing were: straylight correction, NIR similarity correction, clear sky, and overcast screening. The straylight removal algorithm was based on [18]. The Line Spread Functions (LSF) have been previously measured at TO. The straylight algorithm was applied separately to the raw calibration signals and to all the raw field spectra, individually for the six participating radiometers. NIR similarity correction was based on Formulas (3) and (4). The clear/overcast condition was based solely on the E_d threshold level and was only used to assess the cosine error of the E_d sensors (Section 3.2).

The uncertainty of the results was evaluated according to the Guide to the Expression of Uncertainty in Measurement (GUM) [19]. For each input quantity, a relative standard uncertainty was estimated. The relative combined standard uncertainty of output quantity was calculated by combining relative standard uncertainties of all input estimates by using formula (12) of the Joint Committee for Guides in Metrology (JCGM) [19]. Radiometric calibration of the irradiance and radiance sensors and respective uncertainty budgets are described in [5]. The properties of the measured and derived quantities (water-leaving reflectance ρ_w, downwelling sky irradiance E_d, water-leaving radiance L_w) and evaluation of related uncertainties can be found in [11,20]. Measurement models for the evaluation of the uncertainty of ρ_w and L_w are defined by the Formulas (1), (2), and (5). Type B uncertainty of results measured by the E_d sensor included the calibration uncertainty and the aging, contribution due to the non-cosine response and temperature effects. The calibration uncertainty included the following components: alignment, repeatability, temporal stability of the calibration source, repeatability of the calibration and dark signals, thermal effects in the laboratory, polarization sensitivity. The absolute calibration of the source was excluded from the uncertainty budget of ρ_w because all the radiometers were calibrated against the same lamp. Type A uncertainty for ρ_w, E_d, and L_w was calculated as the standard deviation of the station average, taking into account the effective degree of freedom based on the lag1 autocorrelation of the time series [21]. The autocorrelation coefficient varied from 0 to 1, depending on the station. In the figures, the expanded uncertainties ($k = 2$) are shown. E_n numbers [22] were used to assess the agreement between the results from two radiometric systems.

The water-leaving reflectance spectra were calculated from the synchronized triplets measured with HyperSAS and TriOS-RAMSES hyperspectral radiometers following MERIS-Regional Validation of MERIS Chlorophyll Products in North Sea coastal waters (REVAMP) and the National Aeronautics and Space Administration (NASA) protocols [23,24]. The water-leaving reflectance $\lfloor \rho_w \rfloor_N$ was calculated as

$$\lfloor \rho_w \rfloor_N = \pi \frac{L_u(\lambda) - \rho(W)L_d(\lambda)}{E_d(\lambda)} \tag{1}$$

where $R_{rs}(\lambda)$ is the remote sensing reflectance, $L_u(\lambda)$ is the upwelling radiance from the sea, $L_d(\lambda)$ is the downwelling radiance from the sky, $E_d(\lambda)$ is the downwelling irradiance, and $\rho(W)$ is the sea surface reflectance as a function of wind speed (W, m·s^{-1}), calculated as

$$\rho(W) = 0.0256 + 0.00039W + 0.000034W^2 \tag{2}$$

The NIR similarity correction of the water-leaving reflectance spectra was based on [25,26]. First, the additive correction term for every individual spectrum was found as

$$\varepsilon = \frac{\alpha_{1,2} \cdot \rho'_w(\lambda_2) - \rho'_w(\lambda_1)}{\alpha_{1,2} - 1} \tag{3}$$

The constant parameter $\alpha_{1,2}$ of the NIR similarity correction [25] is determined in [26] and depends on the choice of wavelengths λ_1 and λ_2; $\alpha_{1,2} = 2.35$ for the $\lambda_1 = 720$ nm and $\lambda_2 = 780$ nm. The NIR similarity-corrected water-leaving reflectance, $\rho_w(\lambda)$, was then calculated as:

$$\rho_w(\lambda) = \rho'_w(\lambda) - \varepsilon \tag{4}$$

The NIR-corrected water leaving radiance was calculated from corrected $R_{rs}(\lambda)$ as

$$L_w = R_{rs}(\lambda)E_d(\lambda) \tag{5}$$

For the output quantities (ρ_w, E_d, and L_w), the median over the station results was calculated, and only the spectra within $\pm10\%$ in respect of the median were included in final averaging.

The evaluation of the agreement between the results from two radiometric systems, E_n numbers were calculated following [22] as

$$E_n = \frac{x_1 - x_2}{\sqrt{U_1^2 + U_2^2}} \tag{6}$$

where x_1 and x_2 are the independent results subject to comparison; U_1 and U_2 are the expanded uncertainties of these results with $k = 2$, respectively. The agreement between the compared values was considered satisfactory if $|E_n| \leq 1$ and non-satisfactory if $|E_n| > 1$.

2.4. Sentinel-3A OLCI Data

Sentinel-3A OLCI full resolution level-2 data was downloaded from EUMETSAT (https://eoportal.eumetsat.int/) from the same day with in situ measurements. OLCI values from the 3 × 3 pixel Region of Interest (ROI) centered at the coordinates of the in situ stations were extracted for further analyses.

The recommended set of flags (CLOUD, CLOUD_AMBIGUOUS, CLOUD_MARGIN, INVALID, COSMETIC, SATURATED, SUSPECT, HISOLZEN, HIGHGLINT, SNOW_ICE, WHITECAPS, ANNOT_ABSO_D, ANNOT_MIXR1, ANNOT_TAU06, RWNEG_O2, RWNEG_O3, RWNEG_O4, RWNEG_O5, RWNEG_O6, RWNEG_O7, RWNEG_O8) was applied on the data to eliminate possible invalid values. Additionally, quality control was performed by checking the OLCI zenith angle < 60° and Sun zenith < 70°. Then, the mean (μ) and standard deviation (σ) were calculated within the ROI.

Further, pixel outliers were removed if $\rho_w(\lambda) < (\mu - 1.5\sigma)$ or $\rho_w(\lambda) > (\mu + 1.5\sigma)$. After removing the outliers, the Coefficient of Variation (CV=σ/μ) was calculated for the full ROI, and match-ups with CV > 0.2 at 560 nm were discarded. After these filtering steps, five qualified match-ups were left corresponding to the in situ stations with station id's 22, 32, 46, 48, and 56, respectively, from Table 1.

For the match-ups, the Mean Absolute Percentage Difference (MAPD) to investigate dispersion and MPD to investigate bias were calculated between OLCI and in situ data:

$$MAPD_\lambda = \sum_{i=1}^{n} 100 \left| \frac{\rho_w(\lambda)_{insitu,i} - \rho_w(\lambda)_{olci,i}}{\rho_w(\lambda)_{insitu,i}} \right| \tag{7}$$

$$MPD_\lambda = \sum_{i=1}^{n} 100 \left(\frac{\rho_w(\lambda)_{insitu,i} - \rho_w(\lambda)_{olci,i}}{\rho_w(\lambda)_{insitu,i}} \right) \tag{8}$$

where $\rho_w(\lambda)_{insitu,i}$ and $\rho_w(\lambda)_{olci,i}$ are, respectively, in situ and OLCI-derived values for the band λ and match-up i.

The filtering of satellite and in situ data was done following the EUMETSAT "Recommendations for Sentinel-3 OLCI ocean color product validations in comparison with in situ measurements—Match-up Protocols" [27].

2.5. Measurement of Chlorophyll-a

Surface water samples were collected using 20 L Niskin bottles and between 1 and 6 L of seawater were filtered onto 0.7 µm GF/F filters. Discrete water samples were collected along the transects from an underway flow-through optical system and from Niskin bottle rosettes deployed with a Seabird Conductivity-Temperature-Depth sampling device. The water samples were filtered onto Whatman GF/F filters (nominal pore size of 0.7 µm), transferred to Cryovials and stored immediately in liquid nitrogen. High Performance Liquid Chromatography (HPLC) was then used to determine Total Chl *a* following the methods given in [28]. A WET Labs hyperspectral absorption-attenuation instrument (AC-S) was coupled to the ship's clean flow-through system, which continually pumps seawater from a nominal depth of 5 to 7 m beneath the ship's hull. The underway spectrophotometric method of determining Chl *a* is given in detail in [29]. Chl *a* concentrations were estimated from the absorption-attenuation instruments, using the absorption coefficient of particulate matter ($a_p(\lambda)$) data at 650, 676 and 715 nm [29].

3. Results

Radiometric measurements were conducted at 32 stations (Figure 2), covering the Solar Zenith Angle (SZA) range of (6 … 60)° and ambient temperature range of (1 … 30) °C. Chlorophyll-a (Chl *a*) concentrations varied from (0.05 … 1.0) mg·m^{-3}. They were highest on the UK shelf at 48°N where they reached 0.8 mg·m^{-3} and also in the South Atlantic from 33°S to 49°S where they were between 0.7 mg·m^{-3} and >0.9 mg·m^{-3} (Figure 3). For the results given in Figure 4 data from all stations were included without screening. In the following text and figures, The Seabird HyperOCR from Plymouth

Marine Laboratory and TriOS RAMSES data from Tartu Observatory are denoted as "PML" and "TO", respectively.

The uncorrected E_d and ρ_w spectra for all 32 stations are shown in Figure 4 together with the expanded uncertainties. The scatter in E_d at noon was caused by the high Solar Zenith Angle and by the sky conditions, which varied from overcast to clear. The water type was Case 1 during the whole campaign with slightly modified short-wavelength reflectance near the coastline in the beginning and end of the voyage, as characterized by the reflectance spectra (Figure 4, right).

The initial ρ_w spectra (Figure 4) showed two potential outliers from stations 67 and 50, respectively, the lowest and highest ρ_w spectra in Figure 4. While ρ_w from station 67 could be explained by poor measurement conditions (Table 1), the measurement conditions were optimal in station 50. Besides, as both radiometric systems resulted in the closely similar shape and absolute values of ρ_w at these stations, these spectra were included in further analyses in order to show the resulting processing steps.

The uncertainty lower limits (3% for E_d and 5% for ρ_w, $k = 2$) were determined by type B instrumental components and could be reached in the near-ideal measurement conditions.

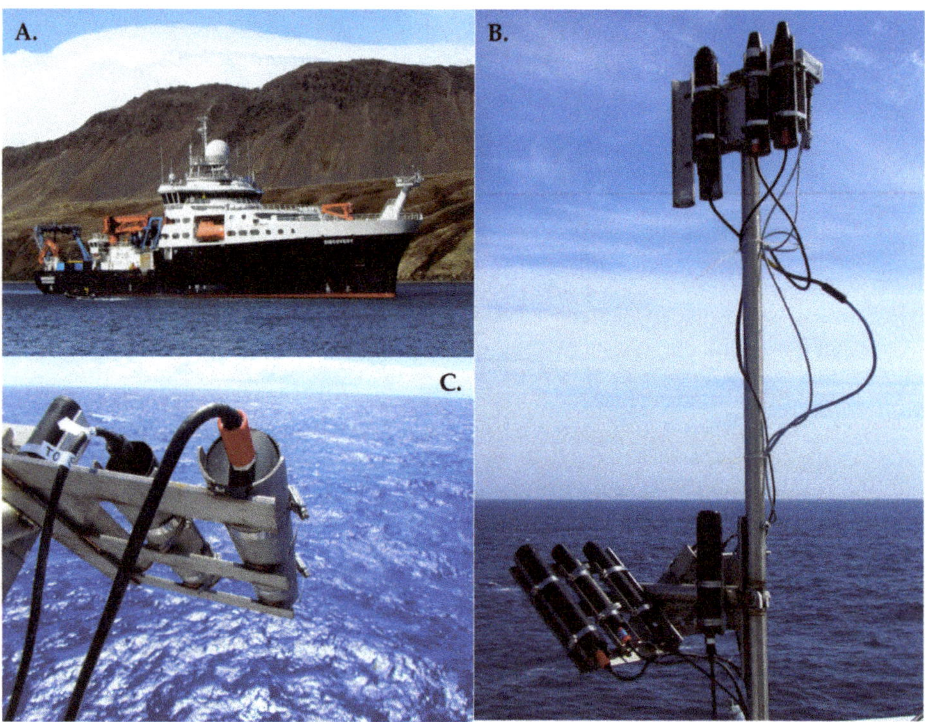

Figure 1. Radiometer set up on *RRS Discovery*; (**A**) showing the location of the meteorological mast on the bow of the ship where the radiometers were located; (**B**) Purpose-built frame for locating the sensors, side by side; (**C**) View of upwelling radiance sensors.

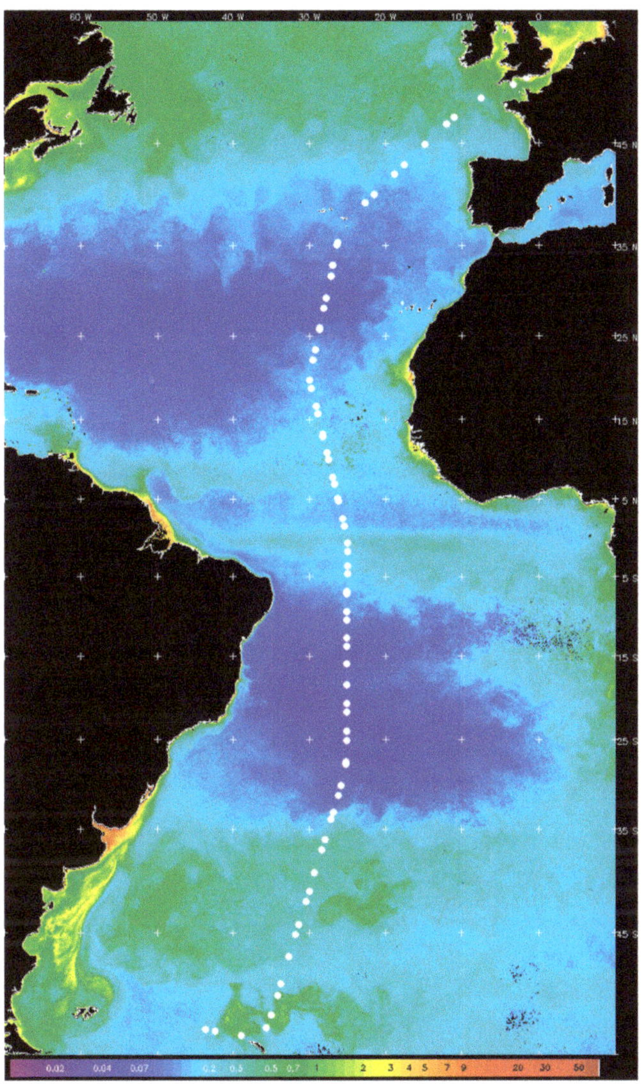

Figure 2. AMT27 campaign sampling stations superimposed on a NASA Visible Infrared Imager Radiometer Suite (VIIRS) Chl a composite satellite image from 19th September to 09th November 2017. The Chl a concentration scale is given below the image.

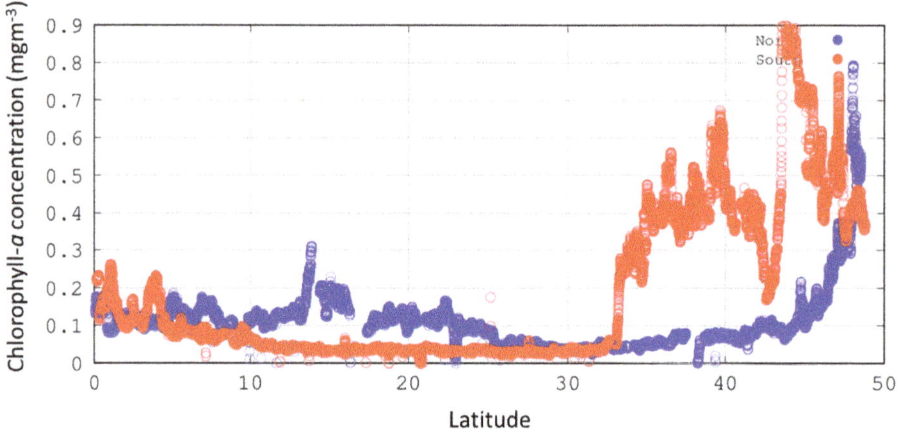

Figure 3. Chl *a* concentration derived from WetLABs absorption-attenuation measurements (ac-S) along AMT27. Blue points are for Northern latitudes; Red points are for Southern latitudes.

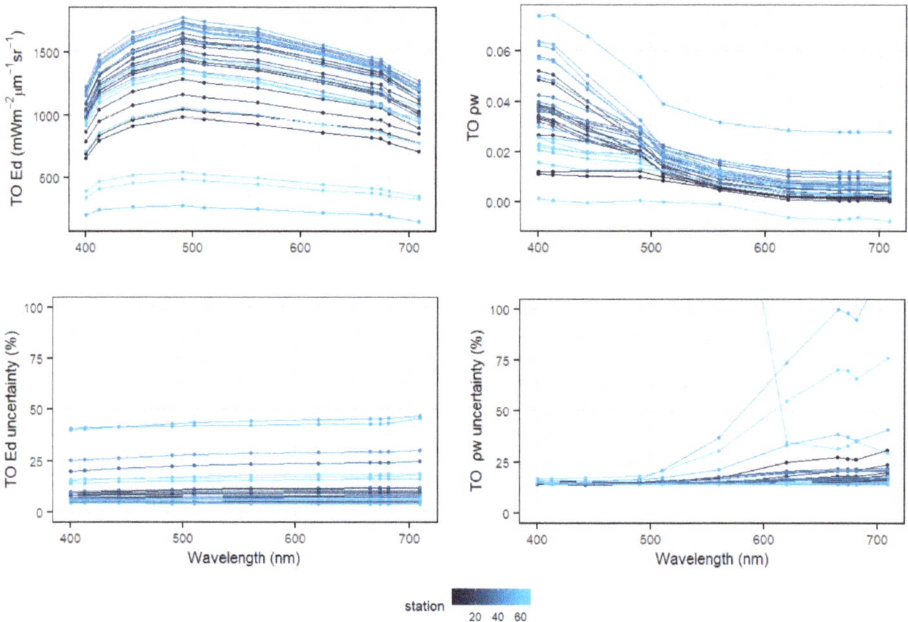

Figure 4. Upper panel: E_d (left) and ρ_w (right) spectra, re-calculated for OLCI (Ocean and Land Color Instrument) channels. Lower panel: expanded uncertainties of downwelling solar irradiance (E_d) and water-leaving reflectance (ρ_w). The color denotes the station ID, as listed in Table 1.

3.1. Environmental Effects

The difference in results of downwelling irradiance between PML and TO, as a function of ambient temperature and as a function of Solar Zenith Angle, is shown in Figure 5. The ratios of E_d(PML)/E_d(TO) were averaged over all OLCI band values and included all measurement stations.

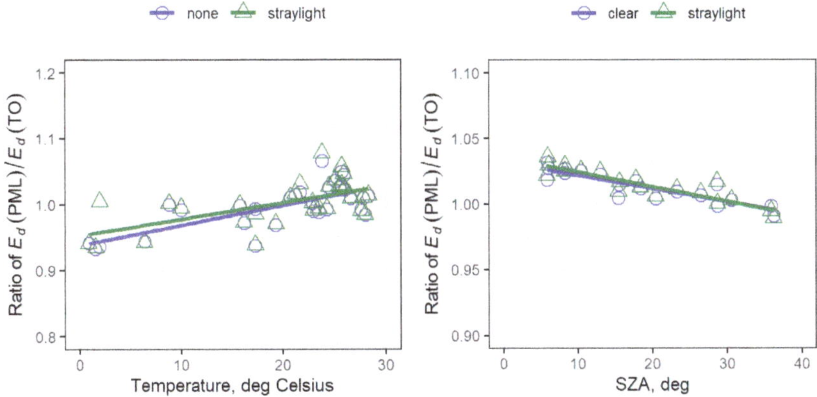

Figure 5. The difference in downwelling irradiance between PML (Plymouth Marine Laboratory) and TO (Tartu Observatory) as a function of ambient temperature (left panel) and function of Solar Zenith Angle (right panel). Blue color circles represent data without any correction and with green triangles after straylight correction.

3.2. Comparison Between the In Situ Radiometric Systems

A comparison between the PML and TO radiometric systems is shown in Figure 6. Results are given as ratios (PML/TO) for $\rho_w(\lambda)$, $L_w(\lambda)$, and $E_d(\lambda)$, with and without the NIR similarity correction (not applicable in the case of E_d). The outliers for $\rho_w(\lambda)$, $L_w(\lambda)$ ratios, without any correction (left panel on Figure 6) and with straylight correction (middle panel in Figure 6), originated from station 67. After NIR similarity correction, the consistency between TO and PML radiometric data was poorer, although there were no distinct outliers (Figure 6).

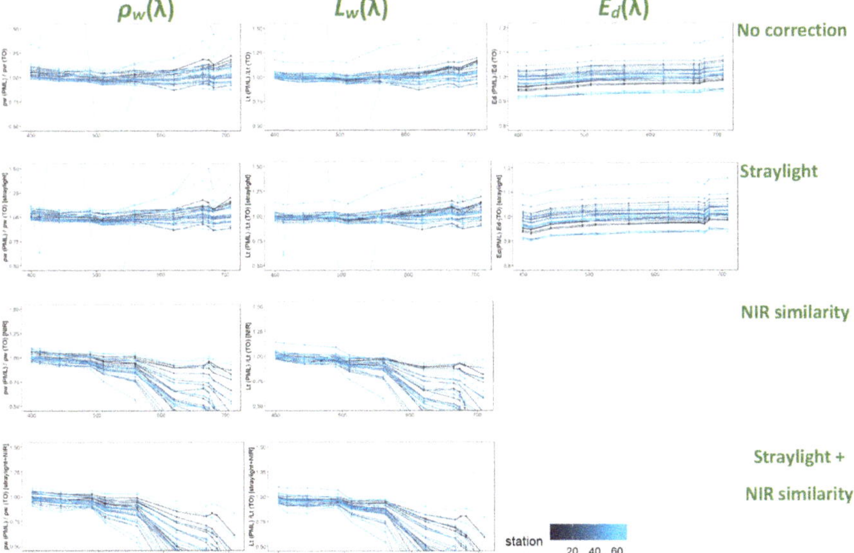

Figure 6. The ratio between PML and TO water-leaving reflectance (ρ_w) (left), water-leaving radiance (L_w) (middle), and downwelling solar irradiance (E_d) (right) in terms of different corrections (rows from top to bottom): no correction; staylight; near infra-red (NIR) similarity; straylight and NIR similarity.

The straylight correction was applied to all data. An example of the straylight and NIR similarity correction effects on ρ_w is shown in Figure 7.

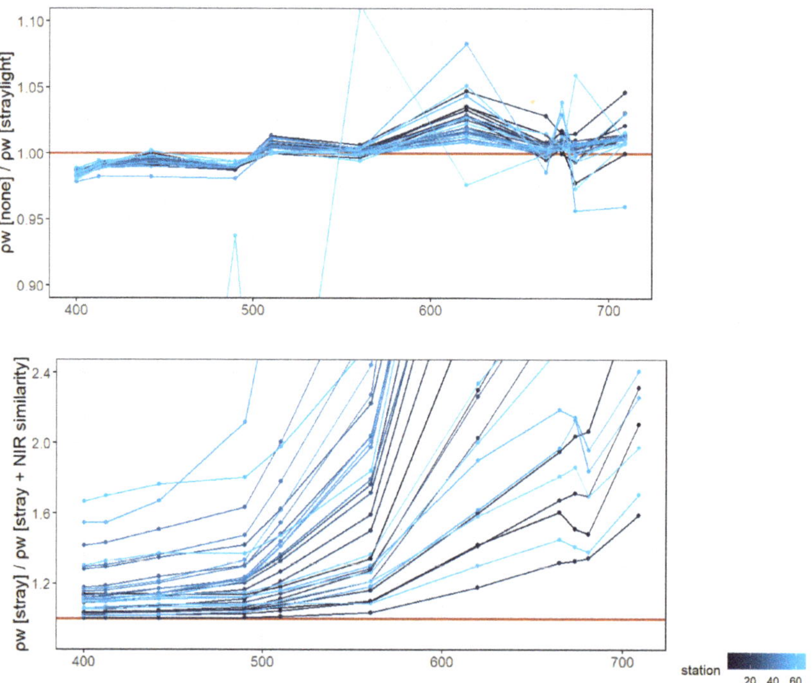

Figure 7. The ratio of uncorrected to straylight corrected ρ_w (above) and straylight corrected to straylight and NIR similarity corrected ρ_w (below) over all stations. Brown line is set to 1 to show the direction of the correction.

The NIR correction could substantially decrease the ρ_w values (Figure 7), but also retained the spectral shape (Figure A1).

In order to compare E_d measured by the RAMSES and HyperOCR instruments, only clear sky conditions were used (Figure 8, Figure A2). First, individual spectra were screened for the irradiance threshold of 1200 mW/m^2 at 560 nm. The resulting dataset was further reduced by rejecting all stations with an expanded uncertainty of E_d more than 5%. As a result, 10 stations were included in the comparison of E_d (Figure 8) from the whole dataset (Figure A3). The average E_d ratio between PML and TO over all 32 stations, together with the expanded uncertainty, is shown in Figure 9. The median of E_n numbers of individual station ratios is shown in Figure 10 as well.

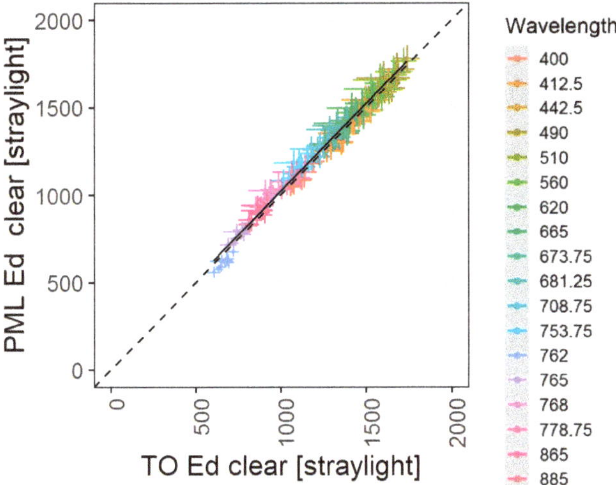

Figure 8. Comparison between TO and PML E_d (mW·m^{-2}·μm^{-1}) and respective uncertainties over all OLCI's band. The comparison in each band separately can be found in Figure A2 for clear conditions and in Figure A3 for all conditions.

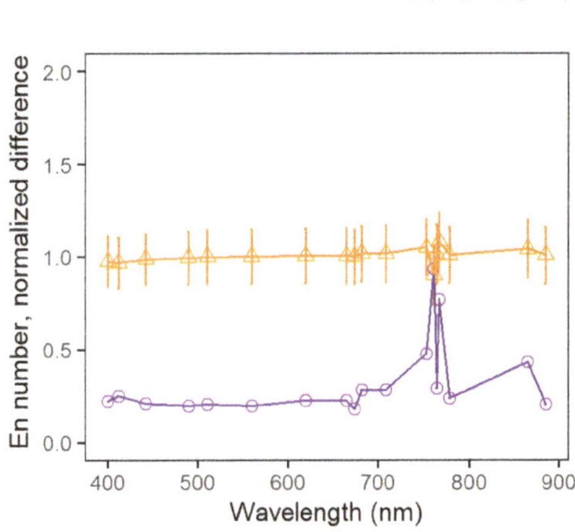

Figure 9. Mean ratio of E_d between TO and PML (triangles) and agreement between the results from two radiometric systems, expressed as median E_n number of the ratio (circles) over all stations.

The agreement between different radiometric systems is shown in Figure 10. The yellow line is the mean ratio (PML/TO) for $\rho_w(\lambda)$ together with uncertainty bars, and the purple line is the respective median of E_n number.

Comparison of $\rho_w(\lambda)$ values measured by PML and TO, after correcting for straylight only or after applying both straylight and NIR similarity correction, is shown in Figure 11. For straylight correction only (Figure A4), the agreement between PML and TO was good overall wavelengths, whereas for straylight and NIR similarity correction (Figure A5), the agreement became weaker from 510 nm wavelength.

Agreement between L_w values as measured by PML and TO, after correcting for straylight only or after applying straylight and NIR similarity corrections, is shown in Figures 12 and A6.

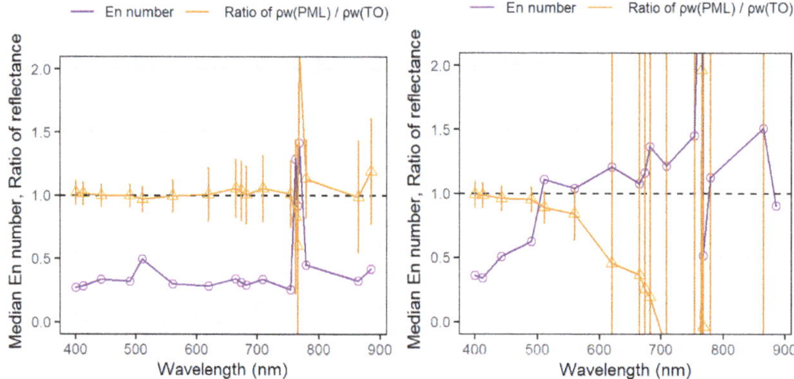

Figure 10. Comparison PML/TO averaged results for ρ_w, before (left) and after (right) NIR similarity corrections. Mean ratio with uncertainty bars—yellow line (triangles); Median of respective E_n values—purple line (circles).

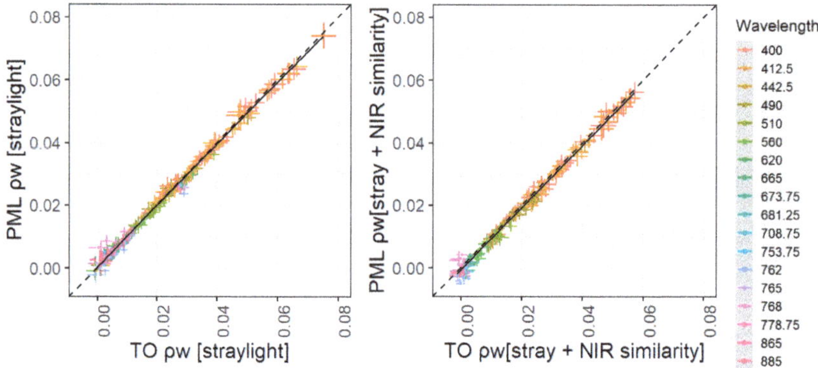

Figure 11. Comparison between TO and PML ρ_w after straylight (left) and straylight + NIR similarity correction over all OLCI's band. The comparison at each band can be found in Figure A4 for straylight correction and in Figure A5 for straylight + NIR similarity correction.

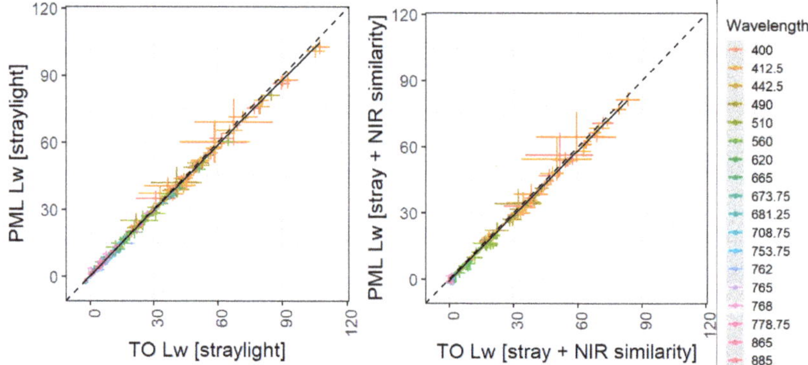

Figure 12. Comparison between TO and PML Lw after straylight (left) and straylight + NIR similarity correction over all OLCI's band. The comparison in each band separately can be found in Figure A6 for straylight correction.

3.3. Consistency between In Situ and OLCI Radiometric Data

After applying the filtering criteria on OLCI's data as described in Section 2.3, five match-ups were obtained. Comparison of in situ and OLCI-derived reflectances for all five match-ups is shown in Figure 13 in the case of different corrections of the in situ results. The corresponding scatterplots are shown in Figure 14 over all wavelengths and in Figure A7 for the wavelengths up to 560 nm separately. The statistics describing the agreement between in situ and satellite measurements are concluded in Table 3. The OLCI to in situ ratios of ρ_w together with uncertainty limits are shown in Figure 15.

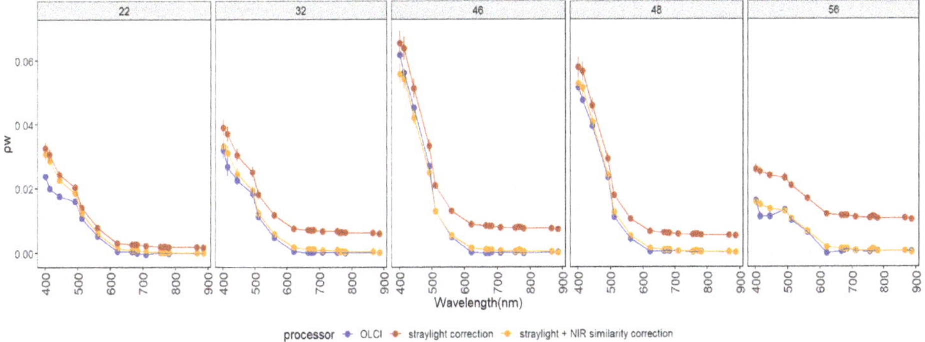

Figure 13. Comparison of OLCI ρ_w against TO in situ measured ρ_w after straylight correction and after straylight + NIR similarity correction in five match-up stations. The error bars denote respective uncertainties for in situ data and standard deviation inside the Region of Interest (ROI) for OLCI. The general information for each station can be found in Table 1.

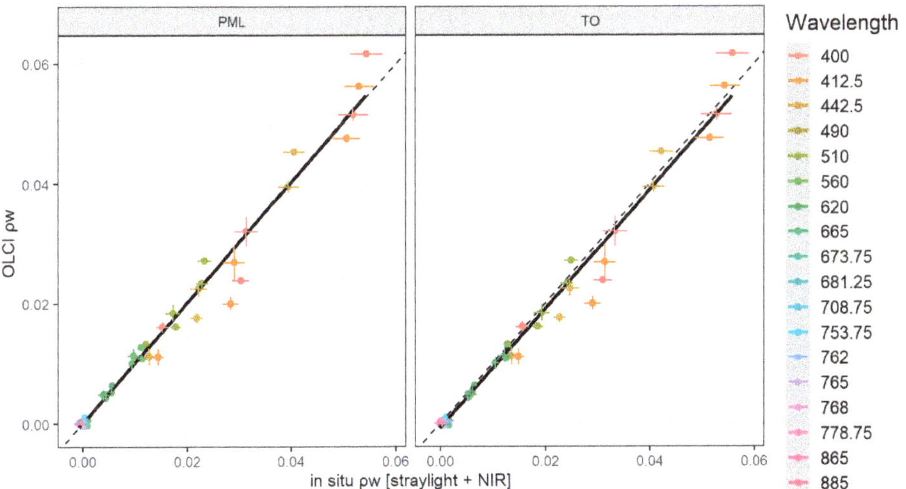

Figure 14. Correlation between OLCI-derived and in situ-measured ρ_w processed with straylight and NIR similarity correction over five match-ups over all wavelengths. The error bars denote respective uncertainties for in situ data and standard deviation inside the ROI for OLCI.

Table 3. The Mean Absolute Percentage Difference (MAPD) and Mean Percentage Difference (MPD) as calculated between Ocean Land Colour Instrument (OLCI) and in situ ρ_w data measured by TO or PML over five match-up stations. Straylight and NIR similarity correction was applied to in situ data. The mean in situ uncertainty for both radiometers is shown in the right column.

	MAPD (%)		MPD (%)		Mean Uncertainty of In Situ ρ_w (%)	
Band	TO	PML	TO	PML	TO	PML
400	9	9	3	0	6	6
412.5	16	14	15	12	6	6
442.5	12	9	9	3	6	6
490	7	9	2	−5	6	6
510	7	8	7	−6	7	7
560	11	10	11	−8	10	12
620	84	178	84	−53	30	>50
665	77	117	77	109	40	>50
673.75	61	102	61	31	42	>50
681.25	62	123	62	−9	41	>50
708.75	84	1411	84	−903	>50	>50
753.75	60	156	60	156	>50	>50
778.75	110	143	62	143	>50	>50
885	145	116	−72	116	>50	>50

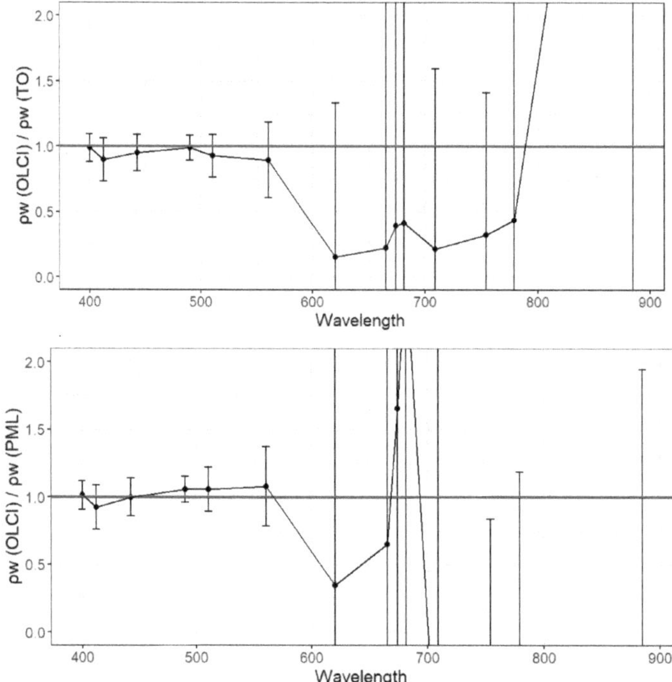

Figure 15. The ratio between mean OLCI ρ_w against in situ ρ_w (TO above and PML below) with straylight and NIR similarity corrections over four match-up stations (22, 32, 48, 56). The error bars denote the uncertainty for the ratio calculated from in situ and OLCI's data.

4. Discussion

4.1. Data Filtering Procedure

The aim of the paper was to compare the performance of the two different radiometric systems over different environmental conditions and water types. The variable environmental conditions (e.g., −53.65 < latitude (°) < 48.93; −38.05 < longitude (°) < −7.62; 5.84 < sun zenith angle (°) < 60.54; 1.48 < wind speed (m·s^{-1}) < 19.71; 0.9 < temperature (°C) < 28.3) allowed the comparison of radiometric data slightly outside of the strict rules applied to produce validation datasets for satellites. This is important in order to show the reliability of the in situ measurements. It is likely that the recommended optimal conditions in satellite validation protocols are associated with the measurement limits of the radiometers. These do not always correspond to the expectations of the users nor the realistic measurement conditions. Therefore, to study the behavior of existing radiometers close to (or even beyond) the specification limits is important, in order to plan the next-generation systems.

Outliers in the datasets are typically caused by: (1) instrumental errors; (2) unsuitable measurement conditions or natural variations outside the acceptable limits; (3) methodical errors in sampling and statistical treatment of results [11]. In our dataset, two potential outlier stations were present (Figure 4, Figure 6, Figure 7). Because two different measurement systems resulted in nearly identical results over these stations, the instrumental and sampling errors could be excluded. Measurement conditions fell into the recommended limits in terms of SZA. Measurement procedures, including most aspects of sampling, were the same for both systems in all stations. Therefore, there was no clear reason for removing these stations from the dataset, and, instead, the behavior of data was analyzed during the processing. Nevertheless, in order to avoid biased conclusions, we acknowledge the need to study the potential outliers and remove them if justified. The uncertainty also needs to be considered as a criteria for removal of the outliers and further research and analysis is required on this. Moreover, during the previous phases of the FRM4SOC project, the investigation of the outliers (not caused by the instrumental errors) helped to reveal errors in the recommended measurement and processing procedure and in the instrument characterization [9,11]. This would not be possible when the outliers have been removed from the datasets without identifying the causes for them. As a result, no data were screened out from the AMT27 dataset, except in the case of clear-sky E_d comparison (Figure 8). However, the in situ ρ_w data of the five match-ups used for S-3 OLCI validation agreed well with the main group of stations and had mean uncertainty (6 ... 7)% for (400 ... 510) nm, which increased towards longer wavelengths (Table 3).

4.2. Comparison of Radiometric Measurement Systems on a Moving Vessel

The intercomparison of simultaneous radiometric measurements on the AMT27 field campaign allowed us to assess the consistency of data and investigate the uncertainties on a moving vessel. An intercomparison exercise of optical systems at the stable, AAOT during near-ideal conditions showed spectrally averaged values of relative differences comprised between −1% and +6% and spectrally averaged values of absolute differences around 6% for the above-water systems and 9% for buoy-based systems [30]. As expected for a moving vessel, we found the differences to be slightly higher, depending both on the corrections and also on varying environmental conditions. Comparison of the in-water and above-water results of radiometric measurements [31] gave the best agreement at the 490 nm band, especially in Case 1 waters during cloud-free conditions, highlighting again that the agreement was better under ideal measurement conditions with minimal impact from the environmental parameters. Due to the responsivity of the radiometers and the influence of the environmental conditions on the radiance signal, the reflectance at wavelengths > 650 nm was weak and noisy, which is characteristic of these Case 1 oligotrophic waters [32]. The results above 650 nm were included to show the effects of straylight and NIR similarity corrections on the in situ radiometric data, especially in the context of comparisons with S-3A OLCI data. The signal at 760 nm was further affected by oxygen absorption, which was clearly sharper than the spectral resolution of the TRiOS and

Seabird radiometers. The behavior of the corrected signal around this narrow spectral band showed an improvement using straylight correction and also revealed a possible shift in the wavelength scales of the radiometers (the wavelength scales of radiometers were not individually tested during this study). A large increase in the uncertainties above 650 nm was expected and did not limit the use of these radiometers for Case 2 waters, where reflectance in the red and NIR was substantially higher [11]. Nevertheless, even with the weak signal in Case 1 waters, the comparison showed agreement within the evaluated uncertainty limits.

Simulations from [26] showed that the NIR similarity spectrum correction applied in this study, was valid for waters and was $\rho_w(780) > 0.0001$. All $\rho_w(780)$ values, including the type-A uncertainty, were above the threshold except one data point measured at very high wind speed (19.71 m·s^{-1}, station id 67 from Table 1). The mean values, both for TO and PML systems were $\rho_w(780) > 0.004$ before the NIR correction. Without the NIR similarity correction, the values for the ρ_w were too high due to sunglint. Therefore, the NIR correction was included in the standard data processing scheme, and its effect were evaluated. The results showed that for clear waters, with a very low radiometric signal and high noise in the NIR, the NIR similarity correction removed this and gave reasonable results. Although various corrections exist for Case 1 waters (e.g., [33,34]), due to the low signal/high noise in the NIR, it is difficult to choose a specific correction as the noise is higher than the signal strength, and these parameters are evaluated separately in the processing steps. This would require a reference value to eliminate the effect of wind and then to estimate the most accurate correction factor for each above-water measurement made. Historically, the scientific community has been encouraged to develop improved and universal correction methods for waters with very low signal in the NIR. To do this properly, however, it requires the development of a new instrument with improved signal-to-noise performance in the NIR spectral range and independent reference methods that are not dependent on the wind that causes air-water surface effects and artifacts.

4.3. Environmental Effects

The field data revealed variability in the responsivity between TRiOS and Seabird sensors calibrated at the same laboratory, which depended on the ambient and illumination conditions. The variability in responsivity of both sensors was likely to be much larger compared to the change of the signal ratio but was masked by the similar behavior of sensors. The differences in Figure 2 varied from approximately −5% to +5% over the temperature range from 1 °C to 30 °C, which was a substantially smaller change than expected temperature effect determined for other TriOS sensors [35]. A slight dependence on the straylight correction was also evident. The irradiance sensors showed the best agreement at 21 °C, which corresponded to the calibration temperature. Thus, the characterization of thermal effects in the full temperature range of field measurements would improve the traceability of results to SI considerably. In this study, due to the lack of characterization data, the temperature correction was not applied.

Differences in downwelling irradiance between PML and TO as a function of SZA showed that the variation was in agreement with known or expected errors of the cosine collectors of the sensors, which were within ±3% [11]. The ratio of L_w-s showed no correlation with SZA or temperature, while the ratio of $\rho_w(\lambda)$ showed the opposite pattern compared to the ratio for E_d-s, as expected. By comparison with the temperature and angle effects, straylight effects were negligible and given in Figure 5 for reference only.

4.4. Comparison of Water-Leaving Reflectance and Radiance Spectra

Agreement between ρ_w values measured by TO and PML over all stations and after correcting for straylight and NIR showed good agreement without any bias over all wavelengths. In general, the TO measured values were slightly higher than PML data. The best agreement was for bands (400 ... 442.5) nm ($R^2 = 0.99$, Figure A5), it was slightly lower for bands (490 ... 560) nm (R^2 decrease 0.96 to 0.82, respectively, Figure A5), and rapidly decreased towards longer wavelengths, where the signal was small and negligible. The agreement between TO and PML for L_w after applying two corrections was good at all wavelengths (Figure A6), although TO measured values were slightly higher than PML (Figure 12).

The NIR similarity correction increased both the bias and the scatter between the ρ_w and L_w ratios, but this could be expected as the L_w signal levels in the Case 1 waters above 550 nm as a result of this correction normally would be extremely low (Figure 11, ρ_w and L_w after straylight and NIR similarity correction).

The straylight effects for individual sensors, as well as for the derived quantities, were mostly within ±3%, which was consistent with previous results [10,11,36]. The effect of straylight correction on the ρ_w is shown in the upper panel of Figure 7. The straylight correction was applied to all of the data, regardless of the other corrections. The NIR similarity correction [25,26], on the other hand, changed the ρ_w and L_w spectra significantly and converged the outliers closer to 1 (Figures 6 and 7, lower panel). The NIR similarity correction removed the shift in ρ_w (and L_w) spectra, which was probably caused by high tilt, waves, and skyglint. While the agreement between the two in situ radiometric systems could be satisfactory or even better without the NIR similarity correction (Figures 6 and 7), distortion of the spectra was evident (Figure 13), and the comparison with satellite data showed significant bias without the NIR correction (Figure 13).

4.5. Comparison of In Situ and OLCI Radiometric Data

Agreement between OLCI and in situ ρ_w values, after both straylight and NIR similarity corrections were applied, was very good in terms of the shape of the spectra and absolute values (Figure 13). Although a similar shape in the ρ_w spectrum was obtained using straylight correction only, it was significantly larger than the OLCI-derived ρ_w spectrum, especially at longer wavelengths (Figure 13).

Based on the five match-ups available, there was good agreement between OLCI and in situ derived ρ_w values for bands (400 ... 510) nm and respective uncertainties less than 7%. For bands 620 nm onwards, the signal was very weak, and the relative uncertainties of the in situ data increased (>30%); the signal to noise ratio of the OLCI's five cameras was lower than for blue bands [37], which made the validation of satellite radiometric products over Case 1 waters in this range challenging.

The comparison between OLCI and in situ ρ_w for both PML and TO from 400 nm to 560 nm showed comparable results (Figure 14, Table 3), though at 412.5 nm, the dispersion was up to 16%, and bias was 15%. This was mainly caused by OLCI ρ_w at stations 22 (Figure A7) and 56 (Figure 13), which had a steep decrease from 400 nm to 412.5 nm, and was not seen in the field measurements. The reason could be due to poor characterization of the aerosol load and type in the S-3A OLCI atmospheric correction [38,39]. Five match-ups, however are too few to make robust conclusions on the accuracy of S-3A OLCI. In this study, they were used as a preliminary analysis of the potential difference between using two different above-water systems for the validation of S-3A OLCI. For a comprehensive accuracy assessment, more match-up data are required.

The estimation of the uncertainty of the ratio of OLCI and in situ ρ_w was up to 40% (Figure 15). The combined uncertainty for the ratio of ρ_w (OLCI) and ρ_w (in situ) (Figure 15) was within (10 ... 16)% for the (400 ... 510) nm, 39% for 560 nm, and above 100% for longer wavelengths, where the signal was very small (Figure 15). The contribution from in situ uncertainty was (6 ... 7)% for bands (400 ... 510) nm, about 11% for 560 nm, and 30% for longer wavelengths. The difference in the mean uncertainty between the TO and PML ρ_w was within 1% for bands up to 510 nm and increasing towards longer wavelengths.

S-3 mission requirements [3] specify a 5% uncertainty for bands (490, 510, 560) nm in Case 1 waters and (5 ... 10)% uncertainty for bands (400, 412, 442) nm depending on the water type. As per pixel uncertainty estimates on OLCI Level 2 data that do not include the uncertainty estimate from Level-1B products [40], it was currently not possible to validate the OLCI products against in situ data using uncertainty estimates for OLCI. The uncertainties for the match-ups were ~6% for the bands

(400 ... 510) nm, and based on the limited number of match-ups available, there was a difference of up to 16% between in situ and OLCI data.

5. Conclusions

The results of the AMT27 field intercomparison are a major international step in assessing differences in commonly used above-water radiometers under different environmental conditions on a moving ship. The AMT campaign was also a significant step, in the framework of both the FRM4SOC project and the international community, in enhancing the confidence in different sources of radiometric data used for satellite ocean color validation.

In general, the agreement between the two in situ systems during the field campaign was satisfactory, with up to a 5% difference over visible wavelengths before corrections applied to ρ_w. Over the wavelength range from (400 ... 510) nm, the relative mean uncertainty of in situ data was close to the S-3 mission requirements of 5%, but with an increase in wavelength beyond 500 nm, the relative uncertainty also increased mainly due to unstable targets, highly variable environmental conditions, and the low signal at red bands in oligotrophic waters.

The consistency between the satellite and in situ data for both sets of radiometers was similar. From 412 nm to 560 nm, the Seabird system showed $-8\% < \text{MPD} < 12\%$ difference and TriOS $2\% < \text{MPD} < 15\%$ difference compared to OLCI data. The consistency between the in situ and OLCI radiometric data was good, and the respective uncertainties were less than 7%.

SI traceable calibration of radiometers before or after field campaigns is very important to ensure traceability. Calibration alone is not sufficient however, and to trace where the variability between radiometric systems comes from, it is necessary to characterize a number of other parameters, including temperature dependence, nonlinearity and spectral straylight. In parallel to correction for instrument bias, and to improve measurement uncertainties, a detailed characterization of environmental conditions during deployment is required. The effect of different processing corrections applied to different radiometric sensors, including NIR similarity correction, correction for straylight, non-linearity, also needs further testing using a simultaneous independent reference instrument that is less affected by these systematic effects.

Author Contributions: Conceptualization, K.A., V.V., G.H.T., G.D., R.V., C.D. and T.C.; methodology, K.A., V.V., G.H.T. and G.D.; software, I.A.; formal analysis, K.A., I.A., V.V., G.H.T., G.D. and F.N.; investigation, K.A., V.V., I.A., G.H.T., and G.D.; resources, G.H.T. and R.V.; data curation, K.A, G.D. and F.N.; writing—original draft preparation, K.A.; writing—review and editing, G.H.T., G.D., V.V., I.A. and R.V.; visualization, K.A., G.H.T. and F.N.; project administration, R.V.; funding acquisition, G.H.T., R.V., C.D. and T.C. All authors have read and agreed to the published version of the manuscript.

Funding: This work was funded by European Space Agency project Fiducial Reference Measurements for Satellite Ocean Color (FRM4SOC), contract No. 4000117454/16/I-Sbo. G.H.T., G.D. and F.N. were also funded by AMT4SentinelFRM from the European Space Agency (contract No. ESRIN/RFQ/3-14457/16/I-BG) and UK Natural Environment Research Council (NERC) National Capability funding to Plymouth Marine Laboratory for the Atlantic Meridional Transect. The Atlantic Meridional Transect is funded by the UK Natural Environment Research Council through its National Capability Long-term Single Center Science Program, Climate Linked Atlantic Sector Science (grant number NE/R015953/1).

Acknowledgments: The authors would like to thank the captain and crew of *RRS Discovery* for AMT27. We thank Silvia Pardo of the Natural Environment Research Council Earth Observation Data Acquisition and Analysis Service (NEODAAS) for satellite imagery during the campaign and for producing Figure 2. This study contributes to the international IMBeR project, and 344 is the contribution number of the AMT program.

Conflicts of Interest: The authors declare no conflict of interest.

Appendix A

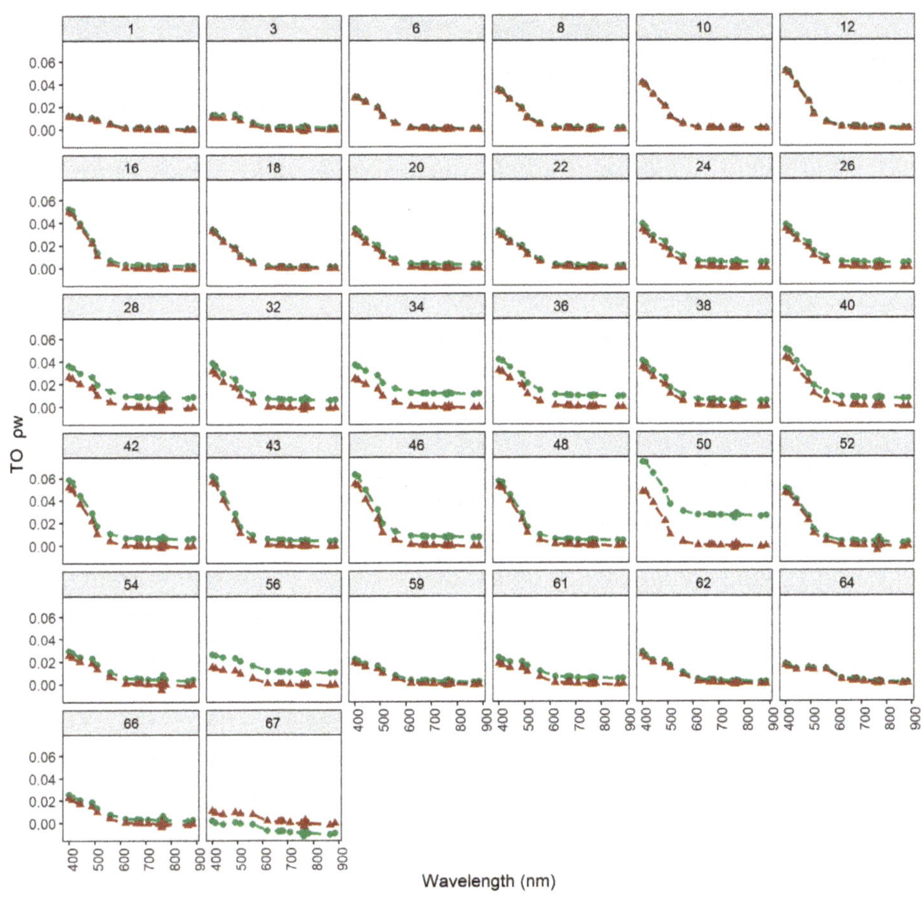

Figure A1. $\rho_w(\lambda)$ spectra at each measurement station corrected for straylight (green circles) and after the straylight and NIR similarity correction (brown triangles). More data from each station can be found in Table 1.

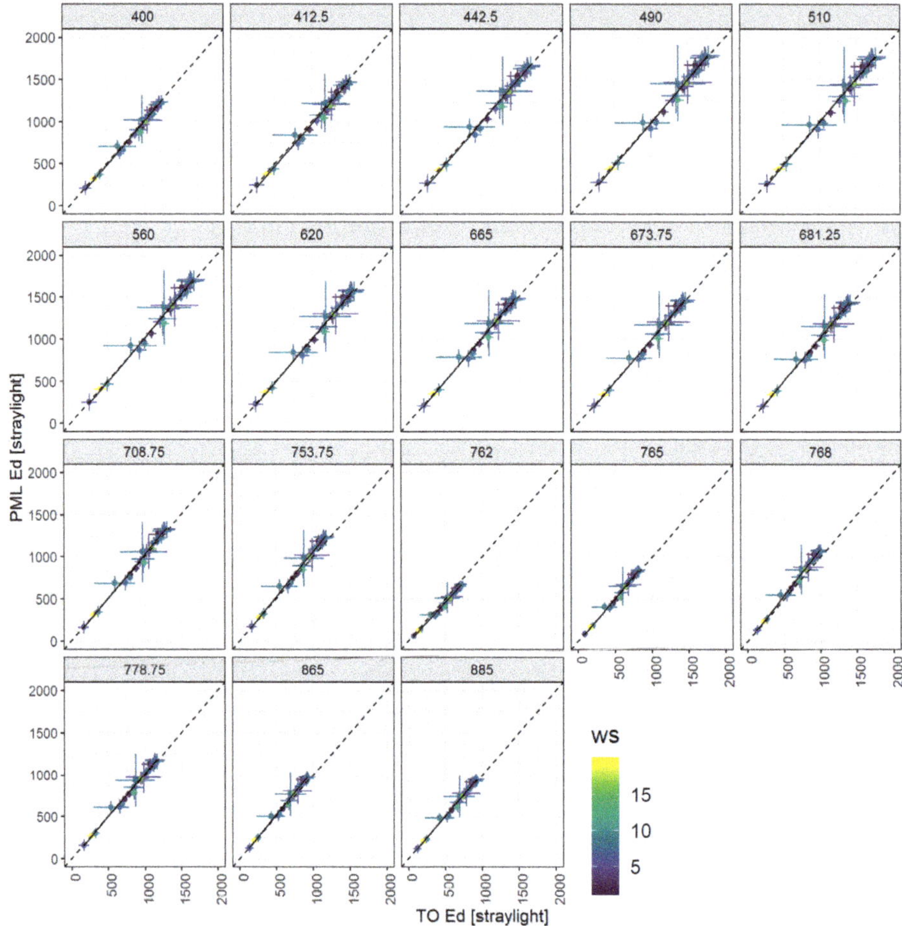

Figure A2. Comparison between TO and PML $E_d(\lambda)$ after straylight correction at OLCI's bands.

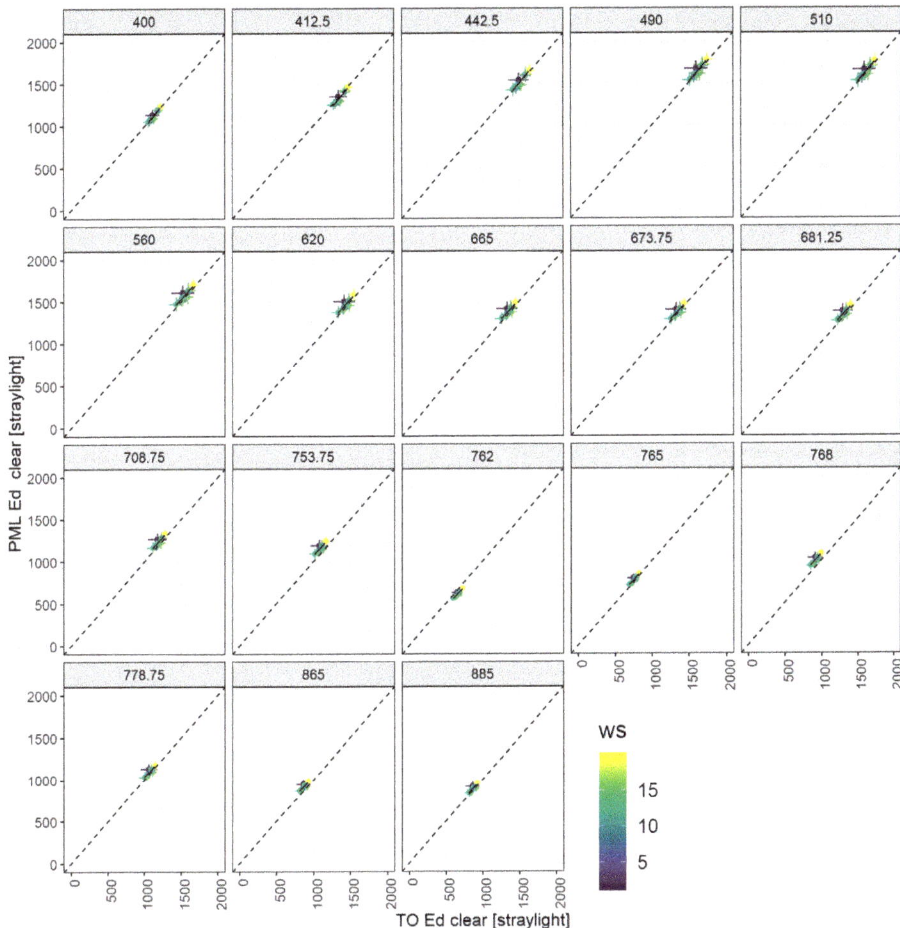

Figure A3. Comparison between TO and PML $E_d(\lambda)$ after straylight correction and NIR similarity correction at OLCI's bands.

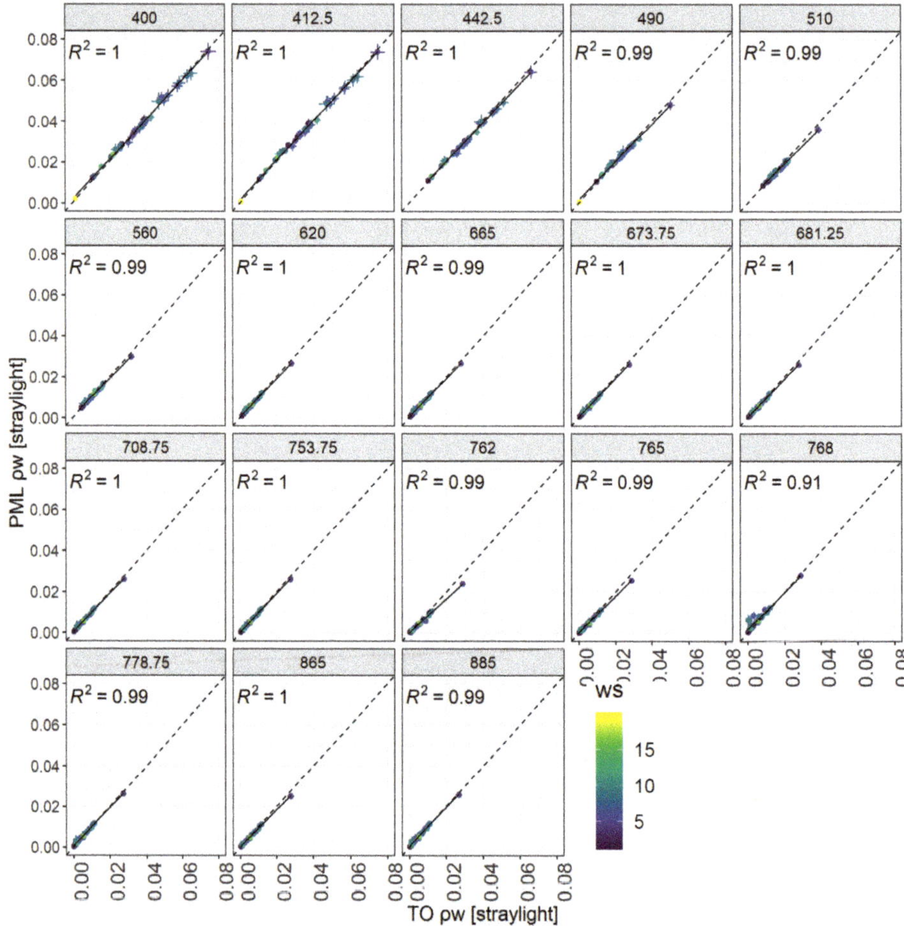

Figure A4. Comparison between TO and PML $\rho_w(\lambda)$ after straylight correction at OLCI's bands.

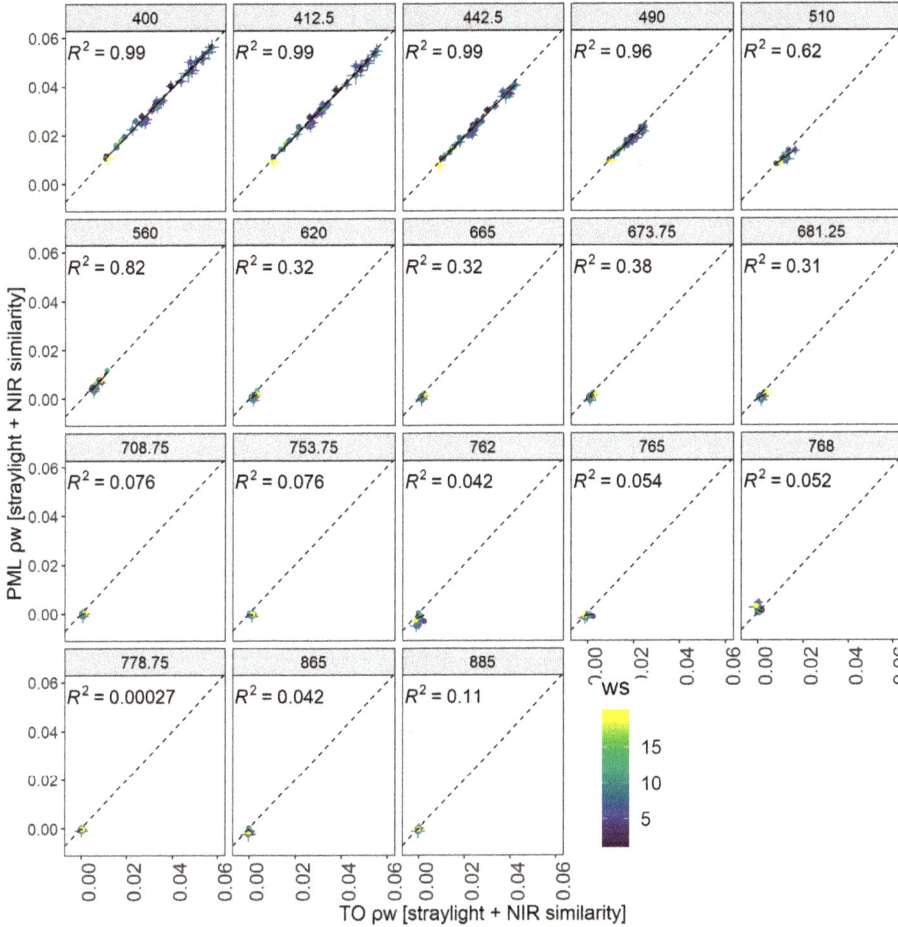

Figure A5. Comparison between TO and PML $\rho_w(\lambda)$ after straylight correction and NIR similarity correction at OLCI's bands.

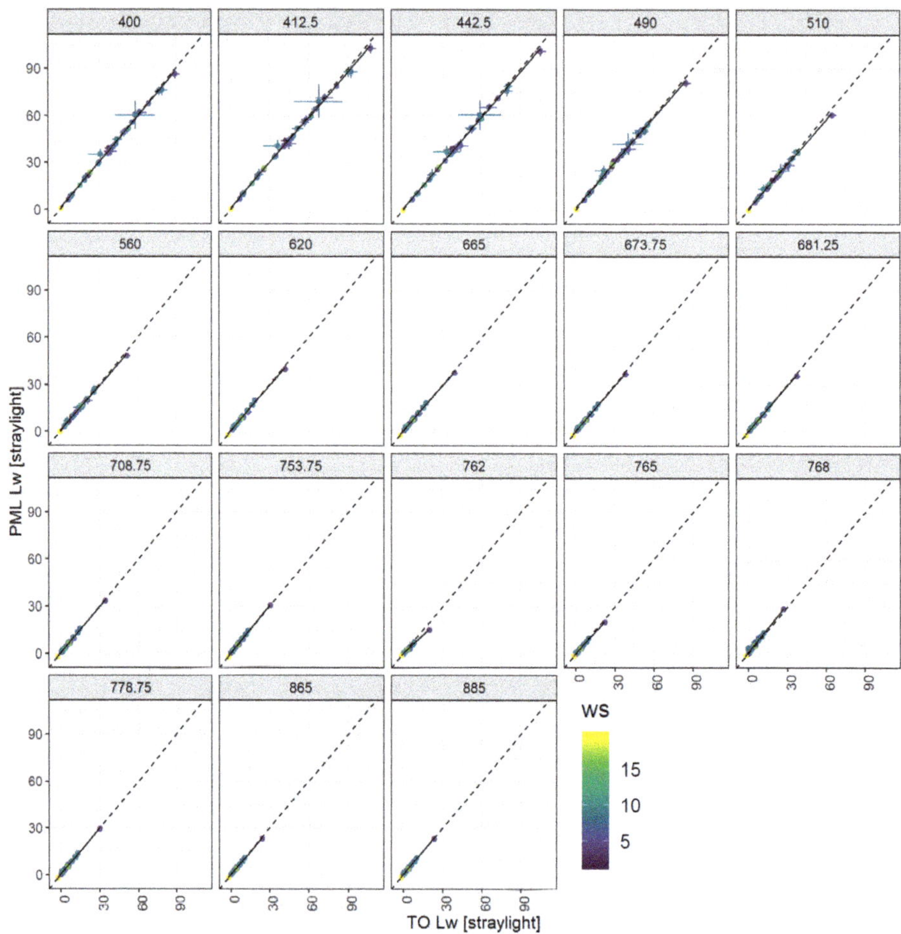

Figure A6. Comparison between TO and PML $L_w(\lambda)$ after straylight correction at OLCI's bands.

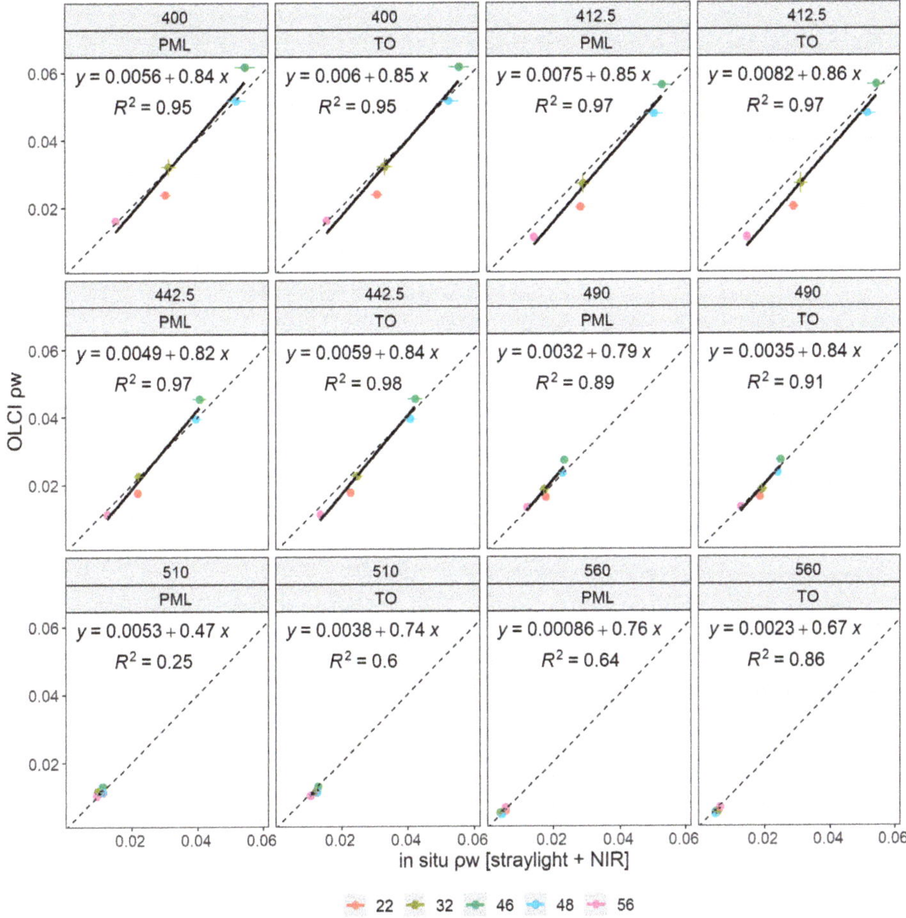

Figure A7. Scatterplots for the validation of OLCI radiometric data for bands 400 to 560 nm using PML's and TO's in situ data for five data match-ups.

References

1. Bailey, S.W.; Hooker, S.B.; Antoine, D.; Franz, B.A.; Werdell, P.J. Sources and assumptions for the vicarious calibration of ocean color satellite observations. *Appl. Opt.* **2008**, *47*, 2035–2045. [CrossRef]
2. Zibordi, G.; Mélin, F.; Voss, K.J.; Johnson, B.C.; Franz, B.A.; Kwiatkowska, E.; Huot, J.P.; Wang, M.; Antoine, D. System vicarious calibration for ocean color climate change applications: Requirements for in situ data. *Remote Sens. Environ.* **2015**, *159*, 361–369. [CrossRef]
3. ESA. *Sentinel-3 Mission Requirements Document (MRD), EOPSMO/1151/MD-md*; ESA: Paris, France, 2007.
4. Ruddick, K. Technical Report TR-2 „A Review of Commonly used Fiducial Reference Measurement (FRM) Ocean Colour Radiometers (OCR) used for Satellite OCR Validation" 2018. Available online: https: //frm4soc.org/wp-content/uploads/filebase/FRM4SOC-TR2_TO_signedESA.pdf (accessed on 21 May 2020).
5. Kuusk, J.; Ansko, I.; Vabson, V.; Ligi, M.; Vendt, R. *Protocols and Procedures to Verify the Performance of Fiducial Reference Measurement (FRM) Field Ocean Colour Radiometers (OCR) Used for Satellite Validation*; Technical Report TR-5; Tartu Observatory: Tõravere, Estonia, 2017.

6. Ruddick, K.G.; Voss, K.; Boss, E.; Castagna, A.; Frouin, R.; Gilerson, A.; Hieronymi, M.; Johnson, B.C.; Kuusk, J.; Lee, Z.; et al. A Review of Protocols for Fiducial Reference Measurements of Downwelling Irradiance for the Validation of Satellite Remote Sensing Data over Water. *Remote Sens.* **2019**, *11*, 1742. [CrossRef]
7. Ruddick, K.G.; Voss, K.; Boss, E.; Castagna, A.; Frouin, R.; Gilerson, A.; Hieronymi, M.; Johnson, B.C.; Kuusk, J.; Lee, Z.; et al. A Review of Protocols for Fiducial Reference Measurements of Water-Leaving Radiance for Validation of Satellite Remote-Sensing Data over Water. *Remote Sens.* **2019**, *11*, 2198. [CrossRef]
8. Białek, A.; Douglas, S.; Kuusk, J.; Ansko, I.; Vabson, V.; Vendt, R. Example of Monte Carlo Method Uncertainty Evaluation for Above-Water Ocean Colour Radiometry. *Remote Sens.* **2020**, *12*, 780. [CrossRef]
9. Białek, A.; Goodman, T.; Woolliams, E.; Brachmann, J.F.S.; Schwarzmaier, T.; Kuusk, J.; Ansko, I.; Vabson, V.; Lau, I.C.; MacLellan, C.; et al. Results from Verification of Reference Irradiance and Radiance Sources Laboratory Calibration Experiment Campaign. *Remote Sens.* **2020**, in press.
10. Vabson, V.; Kuusk, J.; Ansko, I.; Vendt, R.; Alikas, K.; Ruddick, K.; Ansper, A.; Bresciani, M.; Burmester, H.; Costa, M.; et al. Laboratory Intercomparison of Radiometers Used for Satellite Validation in the 400–900 nm Range. *Remote Sens.* **2019**, *11*, 1101. [CrossRef]
11. Vabson, V.; Kuusk, J.; Ansko, I.; Vendt, R.; Alikas, K.; Ruddick, K.; Ansper, A.; Bresciani, M.; Burmester, H.; Costa, M.; et al. Field Intercomparison of Radiometers Used for Satellite Validation in the 400–900 nm Range. *Remote Sens.* **2019**, *11*, 1129. [CrossRef]
12. Tilstone, G.; Dall'Olmo, G.; Hieronymi, M.; Ruddick, K.; Beck, M.; Ligi, M.; Costa, M.; D'Alimonte, D.; Vellucci, V.; Vansteenwegen, D.; et al. Field intercomparison of radiometer measurements for ocean colour validation. *Remote Sens.* **2020**, *12*, 1587. [CrossRef]
13. Morel, A.; Prieur, L. Analysis of variations in ocean color1. *Limnol. Oceanogr.* **1977**, *22*, 709–722. [CrossRef]
14. Organelli, E.; Dall'Olmo, G.; Brewin, R.J.W.; Tarran, G.A.; Boss, E.; Bricaud, A. The open-ocean missing backscattering is in the structural complexity of particles. *Nat. Commun.* **2018**, *9*, 1–11. [CrossRef]
15. Sea-Bird Scientific. Specifications for HyperOCR Radiometer. Available online: https://www.seabird.com/hyperspectral-radiometers/hyperocr-radiometer/family?productCategoryId=54627869935 (accessed on 31 January 2019).
16. TriOS. RAMSES Technische Spezifikationen. TriOS Mess- und Datentechnik. Available online: https://www.trios.de/ramses.html (accessed on 31 January 2019).
17. Spectral Response Function Data. Available online: https://sentinel.esa.int/web/sentinel/technical-guides/sentinel-3-olci/olci-instrument/spectral-response-function-data (accessed on 3 September 2018).
18. Slaper, H.; Reinen, H.A.J.M.; Blumthaler, M.; Huber, M.; Kuik, F. Comparing ground-level spectrally resolved solar UV measurements using various instruments: A technique resolving effects of wavelength shift and slit width. *Geophys. Res. Lett.* **1995**, *22*, 2721–2724. [CrossRef]
19. *JCGM 100, Evaluation of Measurement Data—Guide to the Expression of Uncertainty in Measurement (GUM)*, 1st ed.; JCGM: September 2008. Available online: http://www.bipm.org/utils/common/documents/jcgm/JCGM_100_2008_E.pdf. (accessed on 20 March 2020).
20. Alikas, K.; Ansko, I.; Vabson, V.; Ansper, A.; Kangro, K.; Uudeberg, K.; Ligi, M. Consistency of Radiometric Satellite Data over Lakes and Coastal Waters with Local Field Measurements. *Remote Sens.* **2020**, *12*, 616. [CrossRef]
21. Santer, B.D.; Wigley, T.M.L.; Boyle, J.S.; Gaffen, D.J.; Hnilo, J.J.; Nychka, D.; Parker, D.E.; Taylor, K.E. Statistical significance of trends and trend differences in layer-average atmospheric temperature time series. *J. Geophys. Res. Atmos.* **2000**, *105*, 7337–7356. [CrossRef]
22. ISO/IEC 17043:2010. *Conformity Assessment—General Requirements for Proficiency Testing*; ISO: Geneva, Switzerland, 2010.
23. Tilstone, G.H.; Moore, G.F.; Sørensen, K.; Doerfer, R.; Røttgers, R.; Ruddick, K.D.; Pasterkamp, R.; Jørgensen, P.V. *Regional Validation of MERIS Chlorophyll Products in North Sea Coastal Waters: REVAMP Protocols*; Presented at the ENVISAT Validation Workshop; ESA: Frascatti, Italy, 2004.
24. Mueller, J.L.; Davis, C.; Arnone, R.; Frouin, R.; Carder, K.; Lee, Z.P.; Steward, R.G.; Hooker, S.; Mobley, C.D.; McLean, S. Above-Water Radiance and Remote Sensing Reflectance Measurement and Analysis Protocols. In *Ocean Optics Protocols for Satellite Ocean Color Sensor Validation, Revision 2*; NASA/TM-2000-209966; Goddard Space Flight Space Center: Greenbelt, MD, USA, 2000.

25. Ruddick, K.; de Cauwer, V.; van Mol, B. Use of the near infrared similarity reflectance spectrum for the quality control of remote sensing data. *Remote Sens. Coast. Ocean. Environ.* **2005**, *5885*, 588501. [CrossRef]
26. Ruddick, K.G.; de Cauwer, V.; Park, Y.-J.; Moore, G. Seaborne measurements of near infrared water-leaving reflectance: The similarity spectrum for turbid waters. *Limnol. Oceanogr.* **2006**, *51*, 1167–1179. [CrossRef]
27. EUMETSAT. *Recommendations for Sentinel-3 OLCI Ocean Colour Product Validations in Comparison with in situ Measurements—Matchup Protocols*; User Guide EUM/SEN3/DOC/19/1092968; EUMETSAT: Darmstadt, Germany, 2019.
28. Barlow, R.G.; Cummings, D.; Gibb, S.W. Improved resolution of mono- and divinyl chlorophylls a and b and zeaxanthin and lutein in phytoplankton extracts using reverse phase C-8 HPLC. *Mar. Ecol. Prog. Ser.* **1997**, *161*, 303–307. [CrossRef]
29. Dall'Olmo, G.; Westberry, T.K.; Behrenfeld, M.J.; Boss, E.; Slade, W.H. Significant contribution of large particles to optical backscattering in the open ocean. *Biogeosciences* **2009**, *6*, 947–967. [CrossRef]
30. Zibordi, G.; Ruddick, K.; Ansko, I.; Moore, G.; Kratzer, S.; Icely, J.; Reinart, A. In situ determination of the remote sensing reflectance: An inter-comparison. *Ocean Sci.* **2012**, *8*, 567–586. [CrossRef]
31. Kowalczuk, P.; Durako, M.J.; Cooper, W.J.; Wells, D.; Souza, J.J. Comparison of radiometric quantities measured in water, above water and derived from seaWiFS imagery in the South Atlantic Bight, North Carolina, USA. *Cont. Shelf Res.* **2006**, *26*, 2433–2453. [CrossRef]
32. Brewin, R.J.W.; Dall'Olmo, G.; Pardo, S.; van Dongen-Vogels, V.; Boss, E.S. Underway spectrophotometry along the Atlantic Meridional Transect reveals high performance in satellite chlorophyll retrievals. *Remote Sens. Environ.* **2016**, *183*, 82–97. [CrossRef]
33. Lee, Z.; Ahn, Y.-H.; Mobley, C.; Arnone, R. Removal of surface-reflected light for the measurement of remote-sensing reflectance from an above-surface platform. *Opt. Express* **2010**, *18*, 26313–26324. [CrossRef] [PubMed]
34. Hooker, S.B.; Lazin, G.; Zibordi, G.; McLean, S. An Evaluation of Above- and In-Water Methods for Determining Water-Leaving Radiances. *J. Atmos. Ocean. Technol.* **2002**, *19*, 486–515. [CrossRef]
35. Zibordi, G.; Talone, M.; Jankowski, L. Response to Temperature of a Class of In Situ Hyperspectral Radiometers. *J. Atmos. Ocean. Technol.* **2017**, *34*, 1795–1805. [CrossRef]
36. Talone, M.; Zibordi, G.; Ansko, I.; Banks, A.C.; Kuusk, J. Stray light effects in above-water remote-sensing reflectance from hyperspectral radiometers. *Appl. Opt.* **2016**, *55*, 3966–3977. [CrossRef] [PubMed]
37. EUMETSAT. *Sentinel-3 OLCI Marine User Handbook, v1H e-Signed*; EUMETSAT: Darmstadt, Germany, 2018.
38. Li, J.; Jamet, C.; Zhu, J.; Han, B.; Li, T.; Yang, A.; Guo, K.; Jia, D. Error Budget in the Validation of Radiometric Products Derived from OLCI around the China Sea from Open Ocean to Coastal Waters Compared with MODIS and VIIRS. *Remote Sens.* **2019**, *11*, 2400. [CrossRef]
39. Zibordi, G.; Mélin, F.; Berthon, J.-F. A Regional Assessment of OLCI Data Products. *IEEE Geosci. Remote Sens. Lett.* **2018**, *15*, 1490–1494. [CrossRef]
40. EUMETSAT Mission Management. *Sentinel-3 Product Notice—OLCI Level-2 Ocean Colour, v1*; EUMETSAT: Darmstadt, Germany, 2019.

Article

Field Intercomparison of Radiometer Measurements for Ocean Colour Validation

Gavin Tilstone [1,*], Giorgio Dall'Olmo [1,2], Martin Hieronymi [3], Kevin Ruddick [4], Matthew Beck [4], Martin Ligi [5], Maycira Costa [6], Davide D'Alimonte [7], Vincenzo Vellucci [8], Dieter Vansteenwegen [9], Astrid Bracher [10], Sonja Wiegmann [10], Joel Kuusk [5], Viktor Vabson [5], Ilmar Ansko [5], Riho Vendt [5], Craig Donlon [11] and Tânia Casal [11]

[1]   Plymouth Marine Laboratory, Earth Observation Science and Applications, Plymouth PL1 3DH, UK; gdal@pml.ac.uk
[2]   National Centre for Earth Observations, Plymouth PL1 3DH, UK
[3]   Institute of Coastal Research, Helmholtz-Zentrum Geesthacht (HZG), 21502 Geesthacht, Germany; martin.hieronymi@hzg.de
[4]   Royal Belgian Institute of Natural Sciences, 29 Rue Vautierstraat, 1000 Brussels, Belgium; kruddick@naturalsciences.be (K.R.); mbeck@naturalsciences.be (M.B.)
[5]   Tartu Observatory, University of Tartu, 61602 Tõravere, Estonia; martin.ligi@ut.ee (M.L.); joel.kuusk@ut.ee (J.K.); viktor.vabson@ut.ee (V.V.); ilmar.ansko@ut.ee (I.A.); riho.vendt@ut.ee (R.V.)
[6]   Geography Department at the University of Victoria, Victoria, BC V8P 5C2, Canada; maycira@uvic.ca
[7]   Center for Marine and Environmental Research CIMA, University of Algarve, 8005-139 Faro, Portugal; davide.dalimonte@aequora.org
[8]   Sorbonne Université, CNRS, Institut de la Mer de Villefranche, IMEV, F-06230 Villefranche-sur-Mer, France; enzo@imev-mer.fr
[9]   Flanders Marine Institute (VLIZ), Wandelaarkaai 7, 8400 Ostend, Belgium; dieter.vansteenwegen@vliz.be
[10] Alfred Wegener Institute Helmholtz Centre for Polar and Marine Research, Department of Climate Sciences, D-27570 Bremerhaven, Germany; Astrid.Bracher@awi.de (A.B.); Sonja.Wiegmann@awi.de (S.W.)
[11] European Space Agency, 2201 AZ Noordwijk, The Netherlands; craig.donlon@esa.int (C.D.); tania.casal@esa.int (T.C.)
*   Correspondence: ghti@pml.ac.uk; Tel.: +44-1752-633-406

Received: 31 March 2020; Accepted: 11 May 2020; Published: 16 May 2020

Abstract: A field intercomparison was conducted at the Acqua Alta Oceanographic Tower (AAOT) in the northern Adriatic Sea, from 9 to 19 July 2018 to assess differences in the accuracy of in- and above-water radiometer measurements used for the validation of ocean colour products. Ten measurement systems were compared. Prior to the intercomparison, the absolute radiometric calibration of all sensors was carried out using the same standards and methods at the same reference laboratory. Measurements were performed under clear sky conditions, relatively low sun zenith angles, moderately low sea state and on the same deployment platform and frame (except in-water systems). The weighted average of five above-water measurements was used as baseline reference for comparisons. For downwelling irradiance (E_d), there was generally good agreement between sensors with differences of <6% for most of the sensors over the spectral range 400 nm–665 nm. One sensor exhibited a systematic bias, of up to 11%, due to poor cosine response. For sky radiance (L_{sky}) the spectrally averaged difference between optical systems was <2.5% with a root mean square error (RMS) <0.01 mWm^{-2} nm^{-1} sr^{-1}. For total above-water upwelling radiance (L_t), the difference was <3.5% with an RMS <0.009 mWm^{-2} nm^{-1} sr^{-1}. For remote-sensing reflectance (R_{rs}), the differences between above-water TriOS RAMSES were <3.5% and <2.5% at 443 and 560 nm, respectively, and were <7.5% for some systems at 665 nm. Seabird-Hyperspectral Surface Acquisition System (HyperSAS) sensors were on average within 3.5% at 443 nm, 1% at 560 nm, and 3% at 665 nm. The differences between the weighted mean of the above-water and in-water systems was <15.8% across visible bands. A sensitivity analysis showed that E_d accounted for the largest fraction of the variance in R_{rs}, which suggests that minimizing the errors arising from this measurement is the most important variable in reducing the inter-group differences in R_{rs}. The differences may also be due, in part, to using five of the above-water systems as a reference. To avoid this, in situ normalized water-leaving radiance (L_{wn}) was therefore compared to AERONET-OC SeaPRiSM L_{wn} as an alternative reference measurement. For the TriOS-RAMSES and Seabird-HyperSAS sensors the differences were similar across the visible spectra with 4.7% and 4.9%, respectively. The difference between SeaPRiSM L_{wn} and two in-water systems at blue, green and red bands was 11.8%. This was partly due to temporal and spatial differences in sampling between the in-water and above-water systems and possibly due to uncertainties in instrument self-shading for one of the in-water measurements.

Keywords: fiducial reference measurements; remote sensing reflectance; ocean colour radiometers; TriOS RAMSES; Seabird HyperSAS; field intercomparison; AERONET-OC; Acqua Alta Oceanographic Tower

---

1. Introduction

Fiducial reference measurements (FRM) are an important component of satellite missions for the validation of remote sensing products and are used to ensure that the most accurate data are distributed to the user community. FRMs are distinct from other in situ measurements in that they use protocols recommended by international ocean colour organizations and space agencies, are traceable to SI (Système international) units, are referenced to inter-comparison exercises and have a full uncertainty budget [1] to provide independent, high quality validation measurements for the duration of a satellite mission [2,3]. To underpin the validation of satellite ocean colour radiometry, it is therefore essential that radiometers used to collect FRMs are inter-compared to assess data consistency and characterise the potential differences between instruments and methods. In the absence of such intercomparisons, the traceability chain and uncertainty budget are not validated. The use of a wide range of instruments, methods and laboratory practices may only add to the uncertainty of satellite ocean colour products.

The primary data product in satellite ocean colour is remote-sensing reflectance, R_{rs}. Radiometric field measurements used to derive this parameter are generally obtained from in-water and above-water optical measurements. These include above-water radiometry, underwater profiling, underwater measurements at fixed depths or combined above and underwater measurements from floating systems. Within the numerous measurement systems that exist, differences in calibration sources, methods and data processing schemes lead to the greatest variation between them [1,4]. To minimise these, best practices on each of the steps used to generate radiometric FRM have been established previously and in this project [5–8].

Since the launch of the Sea-Viewing Wide Field-of-View Sensor (SeaWiFS) in 1997, a growing body of literature on intercomparisons between radiometers has developed, which were conducted to constrain the differences between in-water and satellite ocean colour parameters [9–17]. The National Aeronautics and Space Administration (NASA) SeaWiFS Intercalibration Round Robin Experiments (SIRREX) 1–8 focused on sensor calibration; specifically spectral radiance response using plaques and portable field sources for monitoring the stability of sensors [9–15]. The experiments established protocols for these, which reduced the uncertainties in measurements from 8% to 1%. During SIRREX 5, comparisons between in-water radiometers were conducted at Little Seneca Lake in Maryland, USA and comparisons of above-water radiometers were conducted in the laboratory [11]. Differences in in-water apparent optical properties were found to be related to the stability of the platform and illumination geometry. For the laboratory comparisons, systematic differences between radiometers were associated with the derivation of the downwelling irradiance from the reflected plaque radiance and smaller differences due to isolated problems with the radiometers. These were addressed in SIRREX 8 through the publication of new laboratory methods for characterising irradiance sensors [15]. Then followed the NASA Sensor Intercomparison and Merger for Biological and Interdisciplinary

Studies (SIMBIOS) Radiometric Intercomparison (SIMRIC) -1 and -2 [12,13]. The purpose of these experiments was to establish a common radiometric scale among the facilities that calibrate in situ radiometers used for ocean colour related research. The final result was an updated document on calibration procedures and protocols [9]. Using these protocols during SIMRIC-2, the SeaWiFS Transfer Radiometer (SXR-II) measured the calibration radiances at six wavelengths from 411 nm to 777 nm, which was compared against measurements from 10 other laboratories [13]. The agreement between laboratories was within the combined uncertainties for all but two laboratories and the errors for these laboratories were traced back to the sensor calibrations. Following the launch of the Medium Resolution Imaging Spectrometer (MERIS) in 2002, the MERIS and Advanced Along-Track Scanning Radiometer (AATSR) Validation Team (MAVT) conducted a series of intercomparison exercises (PlymCal 1-3 [14]) to compare above- and in- water radiometry (Bio-spherical, PR650, SATLANTIC, SIMBADA, SPMR, TACCS and TRIOS). The expected calibration of radiometric sensors is that they are within ±1 to 2% of each other and during these intercomparisons this was achieved, except for one sensor that exhibited degradation of the cosine collector. The MERIS Validation Team (MVT) then conducted a field campaign at a coastal site off South West Portugal and determined the accuracy of atmospheric and in-water measurements using a hyperspectral, SATLANTIC buoy radiometer with a tethered irradiance chain (TACCS), as well as band-pass filter radiometer of the same design [15]. The overall error for the TACCS system in these waters was 5% where the in-water attenuation coefficient (K_d) was known and 7% where K_d was modelled and extrapolated from the surface to depth. Under the Assessment of In Situ Radiometric Capabilities for Coastal Water Remote Sensing Applications (ARC) MERIS MVT intercomparison of above-water radiometers (SeaPRISM and RAMSES) and in-water radiometers (WiSPER-Wire-Stabilized Profiling Environmental Radiometer and TACCS) was performed under near ideal deployment conditions at the Acqua Alta Oceanographic Tower (AAOT) in the northern Adriatic Sea [17]. For this intercomparison, all sensors were inter-calibrated through absolute radiometric calibration with the same standards and methods. The spectral water-leaving radiance (L_w), as well as E_d and R_{rs} were compared. The relative difference in R_{rs} was between −1% and +6%. The spectrally averaged values of absolute difference were 6% for the above-water systems and 9% for the in-water systems. The good agreement between sensors was achievable because of the stability of the deployment platform used. The first in situ radiometer intercomparison exercise in support of the Ocean and Land Colour Instrument (OLCI) on-board the Sentinel-3 satellite was conducted at a lake in Estonia in May 2017 under non-homogeneous environmental conditions [4]. It highlighted that there was a large variability between recently calibrated sensors, due to high spectral and spatial variability in the targets and environmental conditions. For the radiance sensors tested, variation in the fields of view (FOV) contributed to the differences whereas for the irradiance sensors, this arose from imperfect cosine response. Following the success of ARC MERIS MVT, and because the environmental conditions are nearly ideal during summer, the AAOT was chosen to undertake the intercomparison reported here. The main difference over the previous intercomparisons at the AAOT was that: (1) more measurement systems were compared (10 in this study, five in [17]); (2) Seabird HyperSAS and C-OPS were included in this study, whereas in [17] two TACCS systems were included; (3) in this study the above-water sensors were located side by side on purpose-built frames, which ensured that all above-water optical systems pointed at the same patch of water or sky; (4) in [17] the measurements were referenced to in-water WiSPER, whereas in this study they were referenced to the weighted mean of RAMSES and HyperSAS systems and SeaPRISM. The difference from [4], is that in summer the AAOT experiences near-ideal homogeneous conditions for conducting such intercomparisons. The main aim of the AAOT intercomparison was to quantify differences in radiometric quantities determined using a range of above-water and in-water radiometric systems (including both different instruments and processing protocols). Specifically, we evaluated the differences among:

1. Hyperspectral (five above-water TriOS-RAMSES, two Seabird-HyperSAS, one Pan-and-Tilt System with TriOS-RAMSES sensors (PANTHYR), one in-water TriOS-RAMSES system) and multispectral (one in-water Biospherical-C-OPS) sensors.

2. In-water and above-water measurement systems.

2. Materials and Methods

2.1. Determination of Water-Leaving Radiance: Above-Water

Above-water methods generally rely on measurements of (i.) total radiance from above the sea $L_t(\theta, \theta_0, \Delta\phi, \lambda)$, that includes water-leaving radiance as well as sky and sun glint contributions; and (ii.) the sky radiance that would be specularly reflected towards the L_t sensor if the sea surface was flat $L_{sky}(\theta', \theta_0, \Delta\phi, \lambda)$. The measurement geometry is defined by the sea viewing zenith angle θ, the sky viewing zenith angle θ', and the relative azimuth angle between the sun (ϕ_0) and sensors (ϕ) $\Delta\phi = \phi_0 - \phi$ [18–21]. Illumination is largely defined by the sun zenith angle θ_0, and to a lesser extent atmospheric properties (assuming no clouds). The water-leaving radiance was computed by removing sky glint effects from L_t as follows:

$$L_w(\theta, \theta_0, \Delta\phi, \lambda) = L_t(\theta, \theta_0, \Delta\phi, \lambda) - \rho(\theta, \theta_0, \Delta\phi, U_{10})L_{sky}(\theta', \theta_0, \Delta\phi, \lambda), \tag{1}$$

where $\rho(\theta, \theta_0, \Delta\phi, U_{10})$ is an estimate of the sea surface reflectance typically expressed as a function of the sun-sensor geometry and of the wind speed 10 m above the sea surface, U_{10} [18].

2.2. Determination of Water-Leaving Radiance: In-Water

In-water methods to estimate water-leaving radiance require the measurement of the nadir upwelling radiance, $L_u(z, \lambda)$, or upwelling irradiance, $E_u(z, \lambda)$, as a continuous profile in the water column or at several fixed depths. The first type of measurement is generally performed with free-fall profilers deployed from a ship [22] or autonomous profiling floats [23]. The second type of measurement is generally performed with optical moorings [24], surface buoys [12] or when instruments have to be lowered manually in the water column [25] from a variety of platforms. Each method has its pros and cons, however all of them have a few principles in common to improve the accuracy of estimating L_w. These include a sufficient number of vertical or temporal measurements to reduce the effect of wave focusing, minimize the effect of instrument self-shading, platform shading and reflection, so that the measurements can also be made as close as possible to the surface and at nadir view. L_u is then used to extrapolate the radiometric quantities to just below the water surface (0^-) since this cannot be directly measured due to wave perturbations. Although linear extrapolation of log transformed radiometric quantities is the most commonly used technique, this is not always an ideal approach depending on the type of measurements, environmental conditions and wavelength. Finally, the $L_u(0^-, \lambda)$ is projected above water to obtain the water-leaving radiance:

$$L_w = L_u(0^-, \lambda) \cdot \frac{(1 - \rho')}{n^2}, \tag{2}$$

where ρ' is the water-air interface Fresnel reflection coefficient and n is the refractive index of seawater.

2.3. Determination of Remote-Sensing Reflectance and Normalized Water-Leaving Radiance

The remote-sensing reflectance $R_{rs}(\theta, \theta_0, \Delta\phi, \lambda)$ is defined as the ratio of the L_w to the above-water downwelling irradiance $E_d(0^+, \lambda)$ and was computed as:

$$R_{rs}(\theta, \theta_0, \Delta\phi, \lambda) = \frac{L_w(\theta, \theta_0, \Delta\phi, \lambda)}{E_d(0^+, \lambda)}. \tag{3}$$

The exact normalized water-leaving radiance was computed as per [26]:

$$L_{wn}(\lambda) = R_{rs}(\theta, \theta_0, \Delta\phi, \lambda)BRDF(\theta, \theta_0, \Delta\phi, \lambda, chl)F_0(\lambda), \tag{4}$$

where $F_0(\lambda)$ is the extraterrestrial solar irradiance [27] and the bidirectional reflectance distribution function (BRDF) factor $BRDF(\theta, \theta_0, \Delta\phi, \lambda, chl)$ normalizes R_{rs} to a standard geometry ($\theta = \theta_0 = 0$):

$$BRDF(\theta, \theta_0, \Delta\phi, \lambda, chl) = \frac{\Re(U_{10})}{\Re(\theta, U_{10})} \frac{f_0(\lambda, chl)}{Q_0(\lambda, chl)} \left[\frac{f(\theta_0, \lambda, chl)}{Q(\theta, \theta_0, \Delta\phi, \lambda, chl)} \right]^{-1}. \tag{5}$$

In (5), $\Re(\theta, U_{10})$ accounts for the reflection-transmission properties of the air-sea interface during the measurement. The term $\frac{\Re(U_{10})}{\Re(\theta, U_{10})}$ and Q remove from R_{rs} [28] the variability due to viewing angle. The term $\frac{f(\theta_0, \lambda, chl)}{Q(\theta, \theta_0, \Delta\phi, \lambda, chl)}$ describes the bi-directionality of the upwelling light field from the ocean, and the corresponding term $\frac{f_0(\lambda, chl)}{Q_0(\lambda, chl)}$ [28] normalizes it to the standard geometry. Look-up-tables of $\frac{f}{Q}$ and $\Re$ values were computed by [28] for selected wavelengths and were obtained from ftp://oceane.obs-vlfr.fr/pub/gentili/DISTRIB_fQ_with_Raman.tar.gz. The values of chl required for such a correction were obtained from daily averaged total chlorophyll-a high-performance liquid chromatography (HPLC) measurements [29].

2.4. Simulation of Ocean and Land Colour Instrument (OLCI) Bands

Hyperspectral measurements of L_t, L_{sky}, E_d and R_{rs} were converted into equivalent OLCI bands by applying the OLCI spectral response functions [30] as in the following example for L_t:

$$L_t(\lambda_{i,OLCI}) = \frac{\int S_{i,OLCI}(\lambda) L_t(\lambda) d\lambda}{\int S_{i,OLCI}(\lambda) d\lambda}, \tag{6}$$

where $L_t(\lambda_{i,OLCI})$ and $S_{i,OLCI}(\lambda)$ are L_t and the OLCI spectral response function (SRF) for the ith OLCI channel, respectively. For the multispectral measurements the R_{rs} was shifted to the OLCI bands following [31], and similarly the E_d was shifted using a solar irradiance model (see Section 2.9.1).

2.5. The Field Intercomparison

The field intercomparison was conducted at the Acqua Alta Oceanographic Tower (AAOT) which is located in the Gulf of Venice, Italy, in the northern Adriatic Sea at 45.31°N, 12.51°E during July 2018. The AAOT is a purpose-built steel tower with a platform containing an instrument house to facilitate the measurement of ocean properties under stable conditions such as clear skies, low wind speed and calm sea state (Figure 1). The platform has a long history of optical measurements to support and validate both NASA and European Space Agency (ESA) ocean colour missions [32–34]. An autonomous optical measurement system was developed at the tower in 2002, the data from which are widely used and accessed by the ocean colour community for satellite validation via the AERONET-OC network [35,36]. The ocean circulation in the north western Adriatic region where the tower is located, is mainly influenced by the coastal southward flow of the North Adriatic current and a North Adriatic (cyclonic) gyre in autumn [37,38]. The site is also influenced by discharge from northern Adriatic rivers: Piave, Livenza and Tagliamento [36]. The water type at the tower can vary depending on wind and swell conditions from clear open sea (for 60% of the time [35]) to turbid coastal. The atmospheric aerosol type is mostly continental and determined by atmospheric input from the Po valley, although occasionally this changes to maritime type aerosols [34].

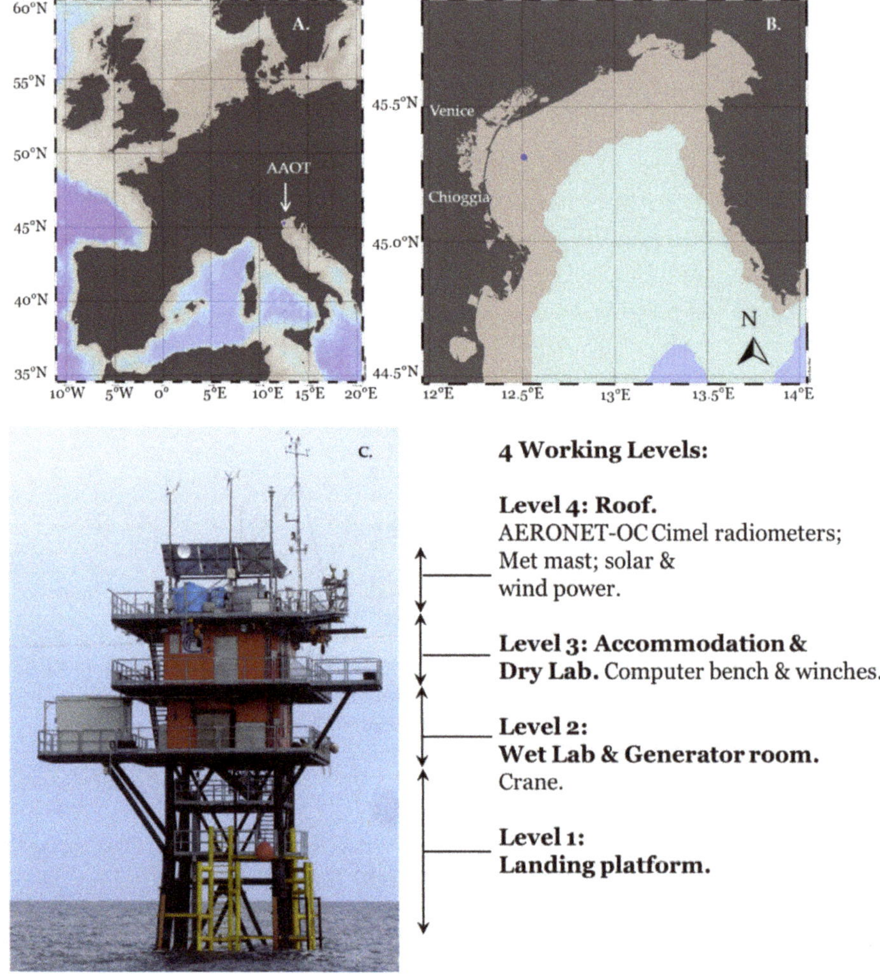

Figure 1. (**A**) Map showing the location of the Acqua Alta Oceanographic Tower (AAOT) in Europe and (**B**) in the Northern Adriatic Sea; (**C**) schematic of the AAOT. For the intercomparison, the radiance sensors were located on the deployment platform on level 3, on a 6 m pole that raised them above the solar panels on level 4. A telescopic (Fireco) mast for the irradiance sensors was located in the eastern corner of level 4.

2.6. Participants and Data Submission

In total, 10 institutes participated in the intercomparison enabling the comparison of 11 measurement systems comprising 31 radiometers (Table 1). To rule out any differences arising from absolute radiometric calibration, all of the sensors used during the campaign were calibrated at the University of Tartu (UT), under the same conditions, within ~1 month of the campaign. The sensors were then shipped directly to Venice prior to setting up the campaign on 9 and 10 July 2018. Each participant was asked to submit their data 'blind', so that the overall results were not seen by participants prior to submission. Processed L_{sky}, L_t, E_d and R_{rs} data with application of OLCI's spectral response function to obtain wavelengths corresponding to the OLCI channels (400, 412, 443, 490, 510, 560, 620,

665, 674, 681 nm) were submitted along with a UTC timestamp, the make, model, serial number of the instrument and integration time setting used during the acquisition.

Table 1. Field intercomparison measurement systems, sensors and institutes. All sensors are hyperspectral except the bio-spherical which is multispectral.

	Method (Identifier)	**Radiometers**	**Reference**	**Institute**
1	Above-water (RAMSES-A)	TriOS-RAMSES	[39]	University of Algarve, Portugal
2	Above-water (RAMSES-B)	TriOS-RAMSES	[40]	University of Tartu, Estonia
3	Above-water (RAMSES-C)	TriOS-RAMSES	[41]	Helmholtz-Zentrum Geesthacht, Germany
4	Above-water (RAMSES-D)	TriOS-RAMSES	[42]	Alfred Wegener Institute, Germany
5	Above-water (RAMSES-E)	TriOS-RAMSES	[43,44]	Royal Belgian Institute of Natural Sciences
6	Above-water (HyperSAS-A)	Seabird	[45]	Plymouth Marine Laboratory, United Kingdom
7	Above-water (HyperSAS-B)	Seabird	[46]	University of Victoria, Canada
8	Above-water (PANTHYR)	TriOS-RAMSES + pan and tilt	[47]	Flanders Marine Institute, Belgium
9	Above-water (SeaPRISM)	SeaPRISM	[11]	Joint Research Centre, Italy
10	In-water C-OPS (in-water A)	Biospherical microradiometers	[48]	Institut de la Mer de Villefranche, France
11	In-water TriOS (in-water B)	TriOS-RAMSES	[49]	Alfred Wegener Institute, Germany

2.7. Radiometer Set-Up and Experimental Design

All above-water radiometers except the PANTHYR system were located on the same purpose-built frames (Figure 2). The radiance sensors were located on the western corner of the AAOT and irradiance sensors and PANTHYR system were located at the eastern corner. For the radiance sensors, the frame was constructed to position the sensors side by side and at the same height (Figure 2A). The frame was fabricated from aluminium at a height of 12.3 m from the sea surface. All L_{sky} and L_t sensors had the same identical viewing zenith angles of $\theta = 40°$ and $\theta' = 140°$, respectively. A sundial was located mid-way down the mast of the frame with a vertical bar to turn it to the correct $\Delta\phi$ (Figure 2B,C). The deployment frame was adjusted for each measurement sequence so that $\Delta\phi = 135°$ or $\Delta\phi = 90°$, which are typically used to reduce sun glint [18]. The radiance mast was positioned at the same level as the SeaPRISM system (Figure 2B,C). The base of the mast was attached to a foldable knuckle joint so that the frame could be lowered, allowing for daily cleaning and servicing of the sensors. For irradiance measurements, a telescopic (Fireco) mast was used to minimize interference from the tower super-structure and other overhead equipment which was installed at a height of 18.9 m above the sea surface (Figure 1C, Figure 2E,F).

Measurements were made at 20 min intervals, from 08:00 to 13:00 UTC, over a discrete measurement period of 5 min (called "cast"), with all instruments having a synchronized start time so that the data collected were directly comparable. In-water C-OPS measurements were also coordinated to these times, though with a temporal delay that is inherent in the practicalities of the deployment. The PANTHYR above-water system is automated to measure every 20 min and was not synchronised to the other (manually-triggered) above-water measurements. In-water TriOS measurements were made immediately after the above-water casts, taking around six minutes for the downcast measurements. From all casts, the median, mean and standard deviation at each OLCI wavelength were calculated.

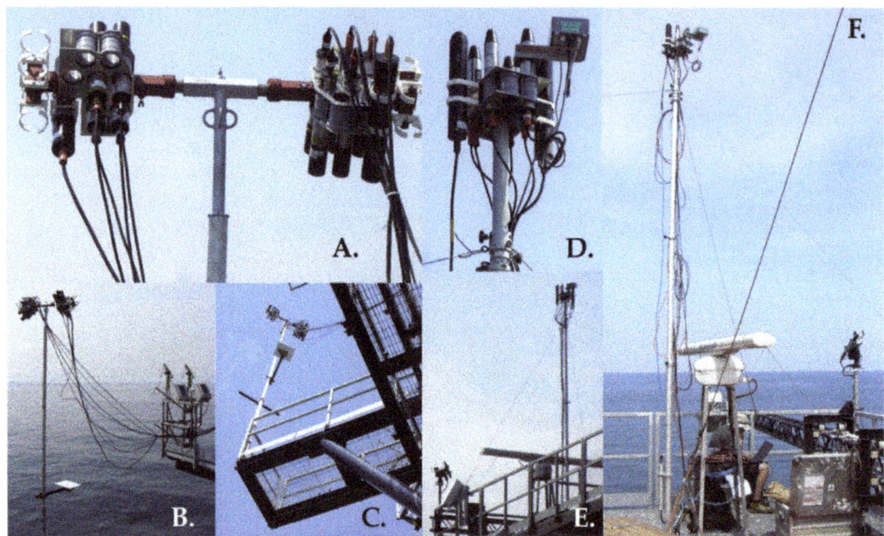

Figure 2. Photographs of the radiance sensors showing (**A**) the mounting for L_{sky} and L_t radiometers, (**B**) location of sensors next to the AERONET-OC SeaPRISM, (**C**) location of the sensors on level 3 of the AAOT, (**D**) location of the irradiance sensors on the mounting block, (**E**) telescopic mast with irradiance sensors at the eastern corner of the AAOT, (**F**) proximity of the telescopic mast with irradiance sensors and the PANTHYR system just above the railings below.

2.8. Above-Water Measurement Methods

All above-water systems measure E_d, L_t and L_{sky} which were interpolated to a spectral resolution of 1 nm. For each cast, the spectral response function for OLCI was applied to obtain the data at OLCI bands and the median, standard deviation and mean were calculated. R_{rs} was then computed using Equation (3). The HyperSAS and most RAMSES instrument systems used the ρ' factor from [18] and the specific values for 90° and 135° azimuth viewing angle with respect to the sun plane or a variation on this theme (RAMSES-C), except for RAMSES-E (Table 2).

2.8.1. TriOS-RAMSES

For above-water measurements RAMSES-A, -B, -C, -D and -E, three TriOS radiometers (TriOS Mess- und Datentechnik GmbH, Germany) were deployed by each institute; two RAMSES ARC-VIS hyperspectral radiance sensors for measuring L_{sky} and L_t respectively, and one RAMSES ACC-VIS irradiance sensor for measuring E_d. Measurements were made over the spectral range of 350–950 nm, with a resolution of approximately 10 nm, sampling approximately every 3.3 nm, with a spectral accuracy of 0.3 nm. The nominal full angle field-of-view (FOV) of the radiance sensors is 7°. The sensors are based on the Carl Zeiss Monolithic Miniature Spectrometer (MMS 1) incorporating a 256-channel silicon photodiode array. Integration time varies from 4 ms to 8 s and is automatically adjusted based on measured light intensity to prevent saturation of the sensors. The data stream from all three instruments is integrated by an IPS-104 power supply and interface unit and logged on a PC via a RS232 connection. A two-axis tilt sensor is incorporated inside the downwelling irradiance sensor in some models. The basic measurement method used was developed by [43] based on the generic Method 1 described in the Ocean Optics Protocols [50]. For the deployment and processing of data, all institutes followed published satellite validation protocols [5].

2.8.2. TriOS Data Processing

For RAMSES sensors, data were acquired every 10 s for the duration of each 5 min cast (except RAMSES-C and -D which used burst mode) using TriOS' proprietary MSDA XE software (except RAMSES-B who used their own code) and calibrated using the coefficients determined before the campaign by UT. Dark values were removed by the software's "dynamic offset" function, which makes use of blocked photodiode array channels inside the radiometer to determine a background response signal in the absence of any measurable light. Using MSDA XE, the data output interval is 2.5 nm.

For RAMSES-A, the number of data records collected for each cast and radiometric sensor was 30. A bi-directional phase function f/Q correction [28] for wavelengths within 412 nm and 665 nm was also applied to account for the viewing and illumination geometry. For this the $\mathfrak{R}$-tables from Gordon [26] were applied where the probability distribution of surface slope follows that of Ebuchi and Kizu [51].

Table 2. Differences between laboratories in the processing of data from E_d, L_t, L_{sky} to R_{rs}. Year (E_d, L_{sky}, L_t) is the year of manufacture of E_d, L_{sky}, $L_t/L_u/E_u$ sensors; N are the number of replicates used for processing each cast; QC flag are quality control flags used; FOV is the radiance field of view; ρ' is the Fresnel reflectance factor used to process the data. For in-water-B, the number (N) reported for L_t is actually N of $L_u(z)$.

Sensor Type	Year (E_d, L_{sky}, L_t)	N E_d	N L_{sky}	N L_t	QC Flag	FOV	ρ'
RAMSES-A	2015, 2015, 2015	3–30	3–30	3–30	Visual QC	7°	[18]
RAMSES-B	2004, 2006, 2010	3–30	3–30	3–30	Visual QC	7°	[18]
RAMSES-C	2006, 2006, 2006	117–140	116–140	102–140	5 min scans	7°	[18,41] *
RAMSES-D	2007, 2006 **, 2011	123–141	4–90	4–54	$L_t < 1.5\%$; $L_{sky} <$ 0.5% of min.	7°	[18]
RAMSES-E	2008, 2001, 2001	1st 5 QC	1st 5 QC	1st 5 QC	1st 5 scans ***	7°	[43,44]
HyperSAS-A	2006, 2006, 2006	280–345	284–398	93–198	5 min scans	6°	[18,45] †
HyperSAS-B	2004, 2004, 2004	~130	~86	~86	lower 20%	6°	[18,46]
PANTHYR	2016, 2016 #	2*3	2*3	11	See [45]	7°	[18,47]
In-water A	2010, N/A, 2010	3–4	N/A	3–4	Visual QC	N/A	[18]
In-water B	2007, N/A, 2010	150–200	N/A	‡	~86	7°	[52]

* Using Mobley [18] and wave height correction of Hieronymi [41]; ** RAMSES-C Lsky sensor was used by lab RAMSES-D; *** The first five scans are taken as long as: (1) Inclination from the vertical does not exceed 5°; (2) E_d, L_{sky} or L_t at 550 nm does not differ by more than 25% from either neighbouring scan; (3) the spectra are not incomplete or discontinuous; # One sensor used for both L_{sky} and L_t; † Mean of 750–800 nm also removed; ‡ For L_u, average from 2 m–8 m was extrapolated to the surface (using 30–40 measurements). N/A means not applicable or measured.

For RAMSES-B, data were collected using bespoke Python software. The irradiance sensor had GPS time and location and tilt and heading devices located next to the sensor was a fish eye camera (see images from this in Figures A2 and A3). No corrections were applied but spectra with missing or saturated values were removed from the database.

RAMSES-C measurements were conducted in "burst mode" over the common 5 min casts which typically gave between 100 and 140 spectra. All spectra were used for averaging and determination of standard deviation; no flagging was applied (visual quality control confirmed expected natural variability for clear sky conditions). Both radiance spectra, L_{sky} and L_t, were interpolated to the wavelengths of E_d. The ρ factor of Mobley [18] with roughness-considerations of Hieronymi [41] was used. The observed wave height was used to estimate the actual sea surface roughness [41]. If the observed significant wave height, Hs, was smaller than 0.5 m, "wind speed" was reduced by 30% and the rounded values were used to select ρ from Mobley's look-up-tables (LUT) [18]. The usage of Mobley's ρ follows the rationale and comparisons of Zibordi [52] for this setup and conditions. According to the institute's protocol however, different sea surface reflectance factors are used to estimate uncertainties in the determination of R_{rs}. These reflectance factors are taken from the LUT of [18,41,44,53] depending on sun- and sensor-viewing geometry and wind speed, with one additional factor calculated from L_t/L_{sky} at 750 nm under the assumption that at this wavelength, water-leaving radiance is negligible for this site.

RAMSES-D followed protocol [54]. L_t and E_d were recorded continuously throughout the day and the data for the casts were extracted for the five-minute measurement period. Two E_d sensors were used; an 81EA sensor (referred to as Sensor 1) and an 81E7 sensor (referred to as Sensor 2) for both the above water (RAMSES-D) and in-water (in-water B) measurements. No L_{sky} sensor was available, so the L_{sky} data from RAMSES-C were used. For each five-minute cast, the E_d data were matched and extrapolated to the same time resolution of L_t and L_{sky} data. Following the NASA and International Ocean Colour Coordinating Group (IOCCG) protocols [55,56], L_t and L_{sky} measurements were corrected for variations in E_d using the median E_d. In order to minimize fluctuations within one cast, only L_t and L_{sky}, which were equal or less than 1.5% and 0.5% of the minimum of L_t and L_{sky} with respect to the median E_d, were considered (Table 2). After convoluting the final L_t, L_{sky} and E_d spectra to the original TriOS sensor resolution, the OLCI spectral response function were applied.

For RAMSES-E, full details of the data processing are described in [44] and associated appendices. In brief, once the data were exported from the TriOS software, in-house Python scripts were used to implement several quality checks (QC), where a spectral scan was discarded if it met any of the following criteria: (a) inclination of the irradiance sensor exceeds 5° from the vertical, (b) E_d, L_{sky} or L_t at 550 nm differ by more than 25% from either neighboring scan, (c) $L_{sky}/E_d > 0.05$ sr^{-1} at 750 nm (indicating clouds either in front of the sun or in the sky-viewing direction), or (d) the scan spectra is incomplete or discontinuous (occasional instrument malfunction). Once all scans for a given cast were processed through QC, only the first five scans (relative to the start time of the station) that had complete spectra for all three of E_d, L_{sky} and L_t were used for further processing. From these data, "uncorrected" water-leaving radiance reflectance, $R'_w(\theta, \theta_0, \Delta\phi, \lambda)$, was calculated for each wavelength and for each of the five scans using Equations (1) and (3) above (with the distinction that $R'_w(\theta, \theta_0, \Delta\phi, \lambda)$ is equal to $R_{rs}(\theta, \theta_0, \Delta\phi, \lambda)$ multiplied by a factor of π). A simple quadratic function of wind speed for ρ was used as approximation of the LUT of [18]. Minimization of perturbations due to wave effects was achieved through the turbid water near-infrared (NIR) similarity correction (Equation (8) in [43]). This was applied to $R'_w(\theta, \theta_0, \Delta\phi, \lambda)$ by determining the departure from the NIR similarity spectrum with:

$$\varepsilon = \frac{\alpha_{1,2} \cdot R'_w(\lambda_2) - R'_w(\lambda_1)}{\alpha_{1,2} - 1}, \tag{7}$$

where wavelengths λ_1 and λ_2 are chosen in the NIR, and the constant $\alpha_{1,2}$ is set according to Equation (7) from [43] and Table 2 of [44]. For this exercise, λ_1 and λ_2 were set to 780 nm and 870 nm respectively, generating a value of $\alpha_{1,2} = 1.912$. It is noted that this approach is similar to that proposed by [57], although relying on different wavelengths and values of sea surface reflectance. The NIR similarity-corrected water-leaving reflectance, $R_w(\lambda)$, is then calculated as:

$$R_w(\lambda) = R'_w(\lambda) - \varepsilon. \tag{8}$$

A final pass of QC checks is performed on this NIR-corrected $R_w(\lambda)$ data, resulting in the entire station being discarded if the coefficient of variation (CV, standard deviation divided by the mean) of the five scans is >10% at 780 nm.

2.8.3. Seabird-HyperSAS

The measurement system consists of three hyperspectral Seabird (Washington, DC, USA; formerly SATLANTIC) spectro-radiometers, two measuring radiance and one measuring downwelling irradiance. The sensors measure over the wavelength range 350–900 nm with a spectral sampling of approximately 3.3 nm and a spectral width of about 10 nm. Integration time can vary from 4 ms to 8 s and was automatically adjusted to the measured light intensity. The data stream from all three instruments is integrated by an interface unit and logged on a PC via a RS232 connection. The radiance sensors have a FOV of 6°. Both HyperSAS-A and -B were first dark corrected in the same way; each instrument is equipped with a shutter that closes periodically to record dark values. The E_d, L_t and L_{sky} data were first dark corrected by interpolating the dark value data in time, to match the light measurements for each sensor. Then dark values were subtracted from the light measurements at each wavelength. The E_d, L_t and L_{sky} data for both instrument systems were then interpolated to a common set of wavelengths (every 2 nm from 352–796 nm).

2.8.4. Seabird HyperSAS Data Processing

For HyperSAS-A, data processing follows "Method 1" of [53]. In brief, data were first extracted from the raw instrument files and the pre-campaign calibration coefficients were applied. Given that the optical conditions at the AAOT can often be considered Case 1 waters, besides the standard processing described for the other above-water sensors (no NIR correction), HyperSAS-A also implemented a processing method specific for open-ocean conditions (NIR correction). For NIR correction, $R_{rs}(750)$ was subtracted from each R_{rs} spectrum. The effect of NIR and no NIR correction were compared. For HyperSAS-B, data were processed using the lowest 20% values to minimize contamination by sky and sun glint.

2.8.5. The Pan-and-Tilt Hyperspectral Radiometer System (PANTHYR)

The PANTHYR is a new system designed for autonomous hyperspectral water reflectance measurements and described in detail in [47]. The instrument consists of two TriOS-RAMSES hyperspectral radiometers, mounted on a FLIR PTU-D48E pan-and-tilt pointing system, controlled by a single-board-computer and associated custom-designed electronics which provide power, pointing instructions, and data archiving and transmission. The TriOS radiometer specifications are the same as those outlined in Section 2.8.1 above. The instrument is capable of full pan ($\pm174°$) and tilt ($+90°/-30°$) movement. The radiance sensor is fixed at an angle of 40° to the irradiance sensor, giving a zenith angle range for the irradiance sensor of 180° (downwelling irradiance measurement) to 60° (parked) and a zenith angle range for the radiance sensor of 140° (sky radiance measurement) through 40° (water radiance measurement) to 20° (parked). The PANTHYR system performs automated measurements every 20 min from sunrise until sunset. Each cycle consists of measurements with a 90°, 135°, 225°, and/or 270° relative azimuth to the sun. In general, and depending on the installation location, platform geometry, time of day (sun location), and associated platform shading of the water target, only one or two (or sometimes zero) of these azimuth angles are appropriate for measurement of water reflectance; other azimuth angles will be contaminated by platform shading or even direct obstruction of the water target as defined from the instrument FOV. A selection of acceptable azimuth angles is made a priori, based on expert judgement. For each measurement cycle, the system performs a sub-cycle for each of the configured relative azimuth angles. Based on the AERONET-OC protocol [8,22], but with repetition of the E_d and L_{sky} replicates, each azimuthal measurement sub-cycle consists of 2×3 replicate scans each of E_d and L_{sky}, and 11 replicate scans of L_t, where "scan" refers to acquisition of a single instantaneous spectrum. Firstly, the irradiance sensor is pointed upward, with the radiance sensor offset by 40°, and three replicates of E_d followed by three replicates of L_{sky} are measured. The radiance sensor is then moved to a 40° downward viewing angle to make 11 replicate L_t scans. The irradiance and radiance sensors are then repositioned to make three more replicate scans of both L_{sky} and E_d. The PANTHYR system was deployed on the east side of the top deck of the platform (Figure 2E), as opposed to the west side where the other above-water systems including AERONET-OC were located (Figure 2B). The irradiance sensor collector was 2 m above the top deck floor and, hence, about 14 m above sea level as opposed to being located on the telescopic mast with the other irradiance sensors in the exercise and hence at 18.9 m above sea level.

2.8.6. PANTHYR Data Processing

L_{sky} and L_t scans with >25% difference between neighbouring scans at 550 nm were removed as well as any scans with incomplete spectra. E_d scans were removed using the same criteria after normalizing E_d by $\cos(\theta')$, where θ' is the sun zenith angle. The data are further processed if a sufficient number of scans passes the quality control criteria: for L_t this is 9 of the possible 11 scans, and for E_d and L_{sky} this is 5 of the possible 6 scans. The remaining E_d and L_{sky} measurements are then grouped and mean-averaged. For each L_t scan, L_w (Equation (1)) is computed by removing sky-glint radiance using the look-up table (LUT) given in [18]. Wind speed was retrieved from ancillary data files in this intercomparison, but can alternatively be set to a user-defined default value if wind speed data are unavailable. The data in the LUT are linearly interpolated to the current observation geometry and wind speed. The L_w scans are then converted into "uncorrected" water-leaving radiance reflectance $R'_w(\theta, \theta_0, \Delta\phi, \lambda)$ scans and NIR similarity spectrum correction is applied to remove any white error from inadequate sky-glint correction, following the "RAMSES-E" TriOS Data Processing sub-section in Section 2.8.1 above. The final quality control to retain or reject the NIR corrected spectra, $R_w(\theta, \theta_0, \Delta\phi, \lambda)$, is performed according to Ruddick et al. [44]. Measurements were rejected when L_{sky}/E_d >0.05 sr^{-1} at 750 nm (indicating clouds either in front of the sun or in the sky-viewing direction), or when the coefficient of variation (CV, standard deviation divided by the mean) of the $R_w(\theta, \theta_0, \Delta\phi, \lambda)$ scans was >10% at 780 nm.

2.8.7. SeaPRISM AERONET-OC

The SeaWiFS Photometer Revision for Incident Surface Measurements (SeaPRISM) is a modified CE-318 sun-photometer (CIMEL, Paris, France) that has the capability to perform autonomous above-water measurements. Measurements are made with a FOV of 1.2° every 30 min in order to determine L_w at a number of narrow spectral bands with centre-wavelengths of 412, 441, 488, 530, 551, 667 nm [32,35,36]. These measurements are: (1) the direct sun irradiance $E_s(\Theta_0, \Phi_0, \lambda)$ acquired to determine the aerosol optical thickness $\tau_a(\lambda)$ used for the theoretical computation of E_d (0$^+$,λ), and (2) a sequence of 11 sea-radiance measurements for determining $L_t(\theta, \Delta\phi, \lambda)$ and of three sky radiance measurements for determining $L_{sky}(\theta, \Delta\phi, \lambda)$. These sequences are serially repeated for each λ with $\Delta\phi$ = 90°, θ = 40° and θ' = 140°. The larger number of sea measurements, when compared to sky measurements, are required because of the higher environmental variability (mostly produced by wave perturbations) affecting the sea measurements during clear skies. Quality flags are applied at the different processing levels to remove poor data. Quality flags include checking for cloud contamination, high variance of multiple sea- and sky-radiance measurements, elevated differences between pre- and post- deployment calibrations of the SeaPRISM system, and spectral inconsistency of the normalized water-leaving radiance L_{wn} [35]. The data are made available through the AERONET-OC web site (https://aeronet.gsfc.nasa.gov/new_web/ocean_color.html) version 3 to processing levels 1.5 and 2. At the time of the field campaign, only level 1.5, real time cloud screened data were available, which were therefore used to compare against the other measurement systems. The difference between version 3 level 1.5 and 2 is in the application of post-deployment calibration and further QC checks. On 13 July 2018, three coincident measurements were available; on 14 July four were available and on 17 July two were available. Of these, four were available at 21 min past the hour, when measurements from the above-water system were taken from 20–25 min past the hour. Five measurements were available at 49 min past the hour, when measurements from the above-water system were made between 40–45 min past the hour. The SeaPRiSM bands are centered at 412, 441, 488, 530, 551 and 667 nm. From R_{rs} above-water hyperspectral data, using a spectrally flat window of 10 nm with ±5 nm centered at the SeaPRISM bands, the average, median and standard deviation were computed and converted to L_{wn} using the BRDF function described in Section 2.2 (except for PANTHYR).

2.9. In-Water Methods

2.9.1. Compact Optical Profiling System (C-OPS)

C-OPS (Biospherical Instruments Inc., USA) was designed specifically to operate in shallow coastal waters and from a wide range of deployment platforms [58,59]. The light sensors are mounted into a frame using a kite-shaped back plane with a hydrobaric chamber mounted along the top of the profiler with a set of floats immediately below it (Figure 3A,B). This allows the sensor to be vertically buoyant in the water column whilst ensuring that both light sensors are kept level (Figure 3B). For this intercomparison exercise, the instrument was deployed at approximately 30 m distance from the stern of Research Vessel (*RV*) *Litus* (Figure 3A) near coincident with the above-water measurements made on the AAOT. The surface reference sensor was mounted on a custom support on the foredeck of the *RV Litus*, and verticality was determined with a level when the ship was in port. For each cast, at least three consecutive profiles of upwelling nadir irradiance, $E_u(t,z,\lambda)$, were acquired between surface and approximately 13 m depth, at the same time as the measurement of surface downwelling irradiance, $E_d(t,0^+,\lambda)$. Both radiometers collected data at 20 Hz and included a pitch and roll sensor. A dark correction and a tare of depth were performed at least twice a day (at the beginning of the morning and afternoon casts). A second degree local polynomial fit function was used to interpolate and extrapolate $E_u(t,z,\lambda)$ and $E_d(t,0^+,\lambda)$ in order to derive the upwelling irradiance just beneath the surface, E_u, and the surface irradiance at the beginning of the cast $E_d(t_0,0^+,\lambda)$, respectively. Data with an absolute tilt $>10°$ for $E_u(t,z,\lambda)$ and $>20°$ for $E_d(t,0^+,\lambda)$ were filtered out from the analysis. The fitted upwelling irradiance profile was corrected with a factor, $f_t(t)$, to account for possible variations in the surface irradiance:

$$f_t(t) = \frac{E_d(t,0^+,\lambda)}{E_d(t_0,0^+\lambda)}. \tag{9}$$

The E_u is corrected for radiometer self-shading following [48], where the absorption coefficient is estimated following [48] and initialized with the in situ total chlorophyll-a (TChl *a*). TChl *a* used in the calculations was derived from $K_d(443)$ [60] because the HPLC data were only analysed after the date of submission of the radiometry data. No correction was applied for the shading from the profiler, however the E_u sensor was deployed in a way that minimize this effect (i.e., with the E_u sensor side of the profiler oriented toward the sun). The water-leaving radiance, $L_W(\lambda)$, was calculated from the upwelling irradiance just below the sea surface as:

$$L_w(\lambda) = E_u(t_0,0^-,\lambda)\cdot\frac{1}{Q(\theta',TChla)}\cdot\frac{(1-\rho')}{n^2}, \tag{10}$$

where θ' is the solar zenith angle (at t_0), ρ' is the water–air interface Fresnel reflection coefficient (depending on θ' and sea roughness) and n is the refractive index of seawater for a flat surface and the $Q(\theta,TChla)$ factor is log-linearly interpolated from LUTs as provided in [28]. The ρ' is 0.043 and the refractive index of seawater, n, is 1.34. A ρ' of 0.02 is generally applied for a flat sea and uniform sky radiance and $\theta' < 30°$ and increases to 0.03 for θ' at $40°$. The values also increase with sea roughness. A ρ' value of 0.043 is reported for a wind speed of 15 m s^{-1} ($\theta' = 30°$). The choice of 0.043 was made when the operational data processing was set to take into account average conditions at the deployment site for sea roughness and θ' and was not modified as the difference in latitude with the AAOT is relatively low. Assuming 2% instead of 4.3% would have a limited impact on the comparison for R_{rs}. Finally the remote-sensing reflectance was calculated using (Equation (3)) and shifted to OLCI central bands, when not coincident, following [31]. Specifically, C-OPS bands ($\lambda_{1/2,COPS}$) at 395, 555/565, 625, 665/683 and 683 nm were shifted to 400, 560, 620, 674 and 681 nm, i.e., one wavelength is used when the wavelength difference is ≤5 nm, two wavelengths are used when difference is >5 nm or two measurements with ≤5 nm difference are available. Similarly, the $E_d(t_0,0^+\lambda)$ was shifted to OLCI bands following:

$$E_d\left(t_0,0^+,\lambda_{OLCI}\right) = \frac{E_d(t,0^+,\lambda_{1,COPS})}{E_{d_{th}}(t_0,0^+,\lambda_{1,COPS})}\cdot E_{d_{th}}\left(t_0,0^+,\lambda_{OLCI}\right); \; \Delta\lambda_1 \leq 5 \, nm \tag{11}$$

or

$$E_d(t_0, 0^+, \lambda_{OLCI}) =$$

$$\frac{(\lambda_{OLCI}-\lambda_{1,COPS})\frac{E_d\left(t,0^+,\lambda_{1,COPS}\right)}{E_{d_{th}}\left(t_0,0^+,\lambda_{1,COPS}\right)}\cdot E_{d_{th}}(t_0,0^+\lambda_{OLCI})+\left(\lambda_{2,COPS}-\lambda_{OLCI}\right)\frac{E_d\left(t,0^+,\lambda_{2,COPS}\right)}{E_{d_{th}}\left(t_0,0^+,\lambda_{2,COPS}\right)}\cdot E_{d_{th}}(t_0,0^+\lambda_{OLCI})}{\lambda_2-\lambda_1}; \Delta\lambda > 5\ nm \quad (12)$$

$$or\ \Delta\lambda_1\ and\ \Delta\lambda_2 \leq 5\ nm$$

where the $E_{d_{th}}(t_0, 0^+,)$ denotes the theoretical surface irradiance for a clear sky and standard atmosphere computed from [26,61] using a spectrally flat window of 10 nm with ±5 nm centered on the C-OPS and OLCI bands. The same correction scheme was applied for the comparison with the Sea-PRISM AERONET-OC with band shift from 443, 490, 532, 555, and 665 nm to 441, 488, 530, 551, and 667 nm.

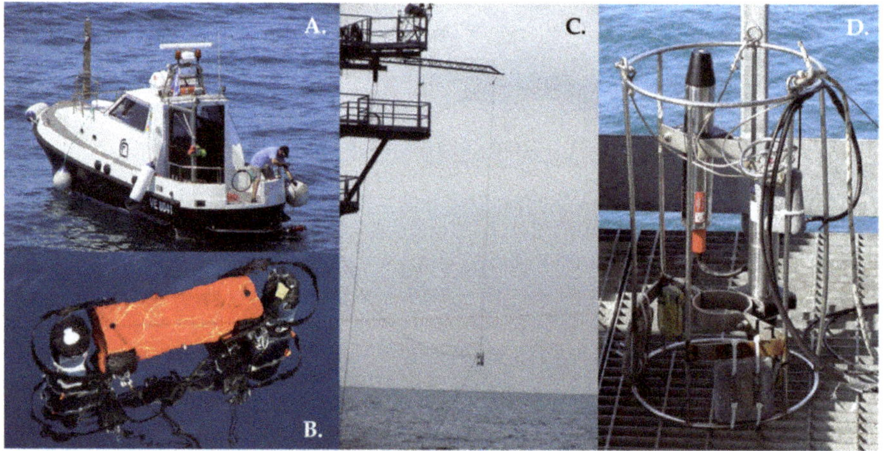

Figure 3. In-water sensors (**A**) C-OPS being deployed from *RV Litus*, (**B**) positioning of C-OPS in-water, (**C**) in-water TriOS deployment from an extendable boom on the AAOT, (**D**) TriOS in-water irradiance sensor in metal deployment frame.

2.9.2. In-Water TriOS-RAMSES

Hyperspectral TriOS-RAMSES radiometers, (Figure 3D) measured profiles of upwelling radiance, L_u, and downwelling irradiance, E_d, following the methods outlined in [49,54]. All measurements were collected with sensor-specific automatically adjusted integration times (between 4 ms and 8 s). The radiance and irradiance sensors were deployed from an extendable boom to 12 m off the south western corner of the AAOT (Figure 3C). The height of the boom was 12 m above sea surface, and is designed to reduce shadow and scatter from the tower. The E_d sensor was equipped with an inclination and a pressure sensor. For this study, we only used the depth and inclination information from this sensor. During the intercomparison, the in-water inclination in either dimension was <6° [54]. For all casts, the instruments were first lowered to just below the surface, at approximately 0.5 m, for 2 min to adapt them to the ambient water temperature. The frame was then lowered to approximately 14 m, with stops every 1 m for a period of 30 s each, to obtain representative average values at each depth. Data were directly extracted from the calibrated instrument files applying the pre-campaign calibration coefficients and factory supplied immersion factors from the last factory calibration (2016) to obtain in water calibrations. Following the NASA and IOCCG protocols [55,56], $L_u(t,z,\lambda)$ data were corrected for incident sunlight (e.g., changing due to varying cloud cover) using simultaneously obtained downwelling irradiance $E_{d+}(\lambda)$ measured above the water surface with another hyperspectral RAMSES irradiance sensor (either RAMSES-D E_d Sensor 1 or Sensor 2) which was located on the telescopic mast on level 4 of the AAOT (Figure 2D–F). As surface waves strongly affect measurements in the upper few meters, measurements made at depth were used and extrapolated to the sea surface [28], since they are more reliable. Similar to Stramski et al. [62] a depth interval was defined ($z' = 2$ m to 8 m) at which the instrument was stopped, so that average light fluctuations at a series of discrete depths could then be used to calculate the vertical attenuation coefficients for upwelling radiance, (i.e., $K_u(\lambda, z')$). Using $K_u(\lambda, z')$, the subsurface radiance $L_u(t_0, 0^-, \lambda)$ was extrapolated from the profile of $L_u(\lambda, z)$. For the calculation of R_{rs}, $L_u(0^-, \lambda)$ was multiplied by a coefficient of 0.5425, which accounts for the reflection and refraction effects at the air-sea interface, as in [62]. Then R_{rs} was calculated using the median above-water downwelling irradiance median of $E_d(t, \lambda)$:

$$R_{rs}(\lambda) = \frac{0.5425\, L_u(\lambda, 0)}{E_d(t_0, \lambda)} \tag{13}$$

The water-leaving reflectance $\lfloor \rho_w \rfloor_N$ was then calculated multiplying $R_{rs}(\lambda)$ (at nadir) by a factor of π. L_{wn} was determined following IOCCG Protocols [57], using $F_0(\lambda)$ from [63]:

$$L_{wn}(\lambda) = R_{rs}(\lambda)\, F_0(\lambda) \tag{14}$$

2.10. Environmental Conditions and Selection of Casts

Wind speed data was measured as part of the meteorological platform on the AAOT. Only casts with wind speeds <5 m s^{-1} and with clear skies and no clouds, characterised from the standard deviation in E_d within which there is a flat signal (Figure 4A–I), were considered in the intercomparison. Using these criteria, 13 casts were valid from 13 July, 15 casts from 14 July and 7 casts on 17 July (Figure 4).

2.11. Inherent Optical Properties and Biogeochemical Concentrations

An AC-9 absorption meter (Wetlabs, USA) with a 25 cm pathlength was used to measure particulate (a_P) and coloured dissolved organic material (a_{CDOM}) absorption coefficients as well as particulate attenuation (c_P) and scattering (b_P) coefficients every hr. Waters samples were collected from the base of the tower using a stainless steel bucket which was deployed to ~2–3 m depth and then raised by hand to the surface. One litre of the water was filtered through 0.2 μm nucleopore filters using an all-glass Sartorius filtration system. Discrete seawater samples of both filtered and unfiltered seawater were then used to measure a_{CDOM} and a_P, respectively. The absorption of purified water (milliQ) was measured after every 10 measurements and used as a blank to correct for the absorption of water (a_w).

Pigment composition was analysed on triplicate samples by HPLC following the method of [64] and adjusted following [49]. In brief, samples were measured using a Waters 600 controller (Waters GmbH, Eschborn, Germany) combined with a Waters 2998 photodiode array detector and a Waters 717plus auto sampler. Details of the solvent and solvent gradient used are given in Table 1 in [49]. As an internal standard, 100 μL canthaxanthin (Roth) was added to each sample. Identification and quantification of the different pigments were carried out using the program EMPOWER by Waters. The pigment data were quality controlled according to [65]. The total chlorophyll *a* concentration (TChl *a*) was derived from the sum of monovinyl-chlorophyll *a*, chlorophyllide *a* and divinyl-chlorophyll *a* concentrations, although the latter two pigments were not present in these samples.

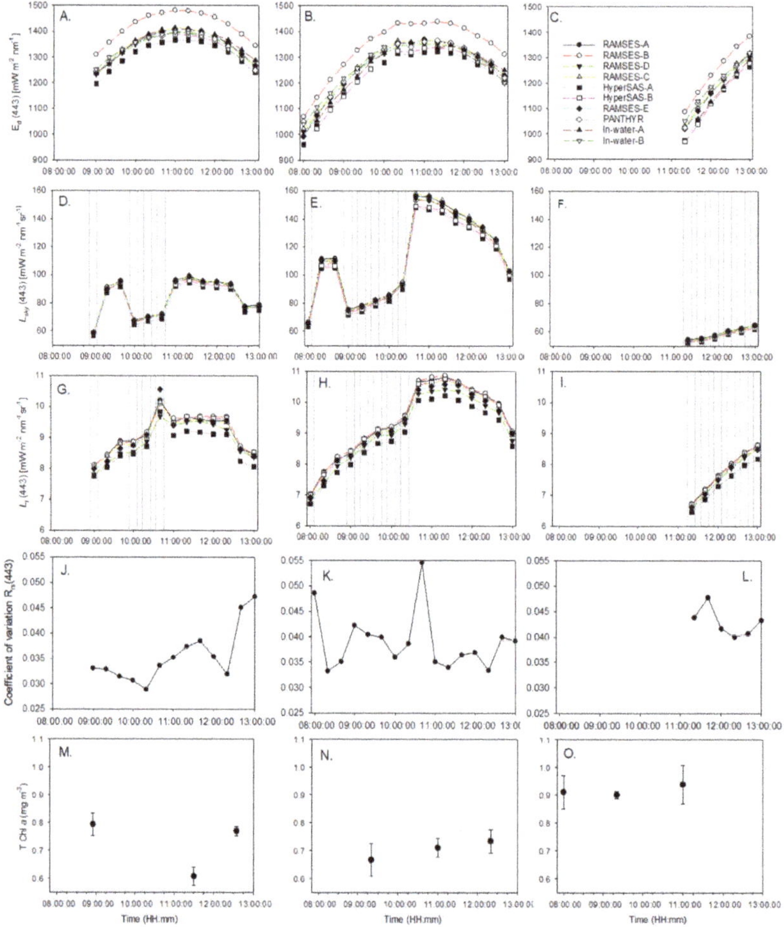

Figure 4. Variation in measurements used for the intercomparison for (**A**) $E_d(443)$ on 13 July 2018, (**B**) 14 July 2018, (**C**) 17 July 2018; $L_{sky}(443)$ on (**D**) 13 July 2018, (**E**) 14 July 2018, (**F**) 17 July 2018; $L_t(443)$ on (**G**) 13 July 2018, (**H**) 14 July 2018, (**I**) 17 July 2018; (**J**) coefficient of variation in $R_{rs}(443)$ on 13 July 2018, (**K**) 14 July 2018, (**L**) 17 July 2018 and TChl a on (**M**) 13 July 2018, (**N**) 14 July 2018, (**O**) 17 July 2018. Only above water sensor results are shown. L_{sky} and L_t were measured at 90 and 135° relative azimuth. Grey shaded bars represent measurements taken at 135° relative azimuth; the un-shaded area are measurements made at 90° relative azimuth.

2.12. Statistical Analyses

For all above-water systems E_d, L_t, L_{sky} and R_{rs} were acquired over a 5 min period for each cast. After each institute's quality control procedure was applied (Table 2), mean, median and standard deviation values were then submitted. These were compared to the weighted mean of above-water systems that were submitted by the 'blind' submission date, and subsequently used as a reference. The mean of 3 × TriOS-RAMSES (RAMSES-A, -B and -C) systems was calculated, then the mean of two Seabird-HyperSAS systems (HyperSAS-A, HyperSAS-B) was calculated and from these, the weighted mean was calculated. Since HyperSAS-B were not available for all casts 1, 2, 3, 4, 5, 6, 12, 14 and 20, only HyperSAS-A data for these casts were used. In-water systems were excluded from the computation of reference values to allow a direct comparison with above-water systems and because of the lower number of comparable radiometric products.

The following statistical metrics were then computed against the reference data:

$$\text{RPD} = 100\frac{1}{N}\sum_{n=1}^{N}\frac{Rc(n) - Rr(n)}{Rr(n)}. \tag{15}$$

RPD is the relative percentage difference, where N is the number of measurements, $Rc(n)$ is the institute measurement or method and $Rr(n)$ is the reference measurement.

$$\text{RMS} = \sqrt{\frac{1}{N}\sum_{n=1}^{N}[Rc(n) - Rr(n)]^2}. \tag{16}$$

RMS is the root mean square difference.

2.13. Sources of Variability in R_{rs}

Finally, we investigated which of the input terms in Equation (3) (i.e., L_t, L_{sky}, E_d, ρ') contributed most of the inter-group variability of R_{rs}. To this aim, we first removed the variability in R_{rs} that was due to variations in environmental conditions by calculating anomalies (with respect to the median value of all measurements) of R_{rs} and of L_t, L_{sky}, E_d and ρ'. We then used the standard law of propagation [2] to compute the combined variance in the anomaly of R_{rs} as follows:

$$u^2_{R_{rs}} = \sum_i \left(\frac{\partial R_{rs}}{\partial x_i}u_{x_i}\right)^2 + 2\sum_i\sum_j \frac{\partial R_{rs}}{\partial x_i}\frac{\partial R_{rs}}{\partial x_i}u_{x_i}u_{x_j}r(x_i, x_j), \tag{17}$$

where $u^2_{R_{rs}}$ is the variance in the anomaly of R_{rs}; x_i is the anomaly of the i-th input term of (Equation (3)); u_{x_i} is the robust standard deviation in the anomaly of the i-th input term; and $r(x_i, x_j)$ is the correlation coefficient between the anomalies of input terms x_i and x_j. The term $\left(\frac{\partial R_{rs}}{\partial x_i}u_{x_i}\right)^2$ is the variance in the anomaly of R_{rs} due to the variance in the anomaly of the i-th input term. The term $2\sum_i\sum_j \frac{\partial R_{rs}}{\partial x_i}\frac{\partial R_{rs}}{\partial x_i}u_{x_i}u_{x_j}r(x_i, x_j)$ are adjustments for the correlations among the input terms. We then computed the fractional contribution from the variance of each input term (as well as from the adjustment for the correlations) to the variance in the anomaly in R_{rs} by calculating the ratios of each $\left(\frac{\partial R_{rs}}{\partial x_i}u_{x_i}\right)^2$ term and of the adjustment for the correlations to $u^2_{R_{rs}}$.

3. Results

3.1. Data Submission

SeaPRISM is a permanent fixture at the AAOT. Of the other nine institutes that participated in the field intercomparison, eight submitted data 'blind' by the submission deadline of 15 August 2018. One institute submitted their final data sets (PANTHYR and RAMSES-E) after the first results had been circulated. Of the original eight, two institutes re-processed their data. For RAMSES-D and in-water B, the irradiance Sensor 1 was found to have an angular response significantly deviating from cosine [1,4]. Data for RAMSES-D and in-water B were, therefore, re-processed using irradiance Sensor 2. The in-water B radiance sensor 2 still exhibited large deviations from the reference measurements, which was due to errors in the data processing. The final corrected in-water B data set was submitted on 3 January 2020. For HyperSAS-A the original data were submitted using a 'Case 1 water-type' processor which were later re-processed using a 'case 2 type' processor. RAMSES-B, RAMSES-C, RAMSES-D and HyperSAS-A submitted $N = 35$ casts. Due to problems with sensor logging at the beginning of the campaign, RAMSES-A submitted $N = 34$ and HyperSAS-B submitted $N = 27$ casts. In-water-B submitted $N = 28$ casts due to power outage on 13 July and on 14 July the cable for the upwelling radiance sensor broke, which led to omission of 5 casts. Due to time constraints on deployment and retrieval of the in-water-A, this institute submitted N = 28 casts.

3.2. Inherent Optical Properties (IOPs) and Biogeochemical Concentrations

Median (±median absolute deviation) and range of IOPs and HPLC TChl *a* during the campaign are given in Table 3. $a_P(440)$, $c_p(440)$, $a_{CDOM}(412)$, TChl *a* were slightly lower than values reported during the ARC MERIS intercomparison during July 2010 at the AAOT [17]. Notably $a_P(440)$ and $a_{CDOM}(440)$ were similar indicating riverine influence from neighbouring Northern Adriatic Rivers. The TChl *a* concentrations were typical for this site and time of year (Table 3). The standard deviation for the HPLC TChl *a* triplicates ranged from 1% to 9% (median and mean 5%, stdev <3%). Diel variability in TChl *a* was evident on 13 July and on 14 July (Figure 4M–O) and varied by 23% and 9%, respectively.

Table 3. Median and median absolute deviation of IOPs (absorption coefficients of particulate material (a_P), coloured dissolved organic material (a_{CDOM}) and water (a_w); scattering coefficient of water (b_w); attenuation coefficient of particulate material (C_P)) and HPLC TChl *a* during the AAOT intercomparison.

Quantity [Units]	Median ± abs Dev	Min–Max Range
$a_P(440)$ [m^{-1}]	0.079 ± 0.014	0.063–0.092
$a_{CDOM}(412)$ [m^{-1}]	0.112 ± 0.010	0.107–0.131
$a_{CDOM}(440)$ [m^{-1}]	0.080 ± 0.009	0.070–0.091
$c_P(440)$ [m^{-1}]	0.929 ± 0.166	0.856–1.229
$a_w(440)$ [m^{-1}]	0.0063 ± 0.001	N/A
$b_w(440)$ [m^{-1}]	0.004 ± 0.001	N/A
TChl *a* [mg m^{-3}]	0.77 ± 0.12	0.61–0.94

3.3. Intercomparison of $E_d(\lambda)$, $L_{sky}(\lambda)$, $L_t(\lambda)$, $R_{rs}(\lambda)$ and $L_{wn}(\lambda)$

The variability in $E_d(443)$, $L_{sky}(443)$ and $L_t(443)$ for the days and casts used in the intercomparison are shown in Figure 4. The criteria for selection of data used in the intercomparison was a smooth $E_d(443)$ signal, indicating the absence of clouds (see also Appendix A for variability of these parameters on cloudy days and Figures A2 and A3 for fish eye images on 13 and 14 July to show the cloud-free sky conditions). Although the conditions were clear, there were some fairly large variations in $L_{sky}(443)$. On 13 July the variation in $L_{sky}(443)$ was 40% and on 14 July it was 57%, with the largest changes at 09:20 and 10:40, respectively (Figure 4D–F). The variations in $L_{sky}(443)$ were partly due to temporal changes in sky conditions and partly due to using either 90° or 135° viewing angles (shaded areas in Figure 4D–J indicate 135° viewing angles). $L_{sky}(443)$ was always higher for both RAMSES and HyperSAS sensors (see Figure A4) with viewing angles of 90°. For example, on 14 July the viewing angle at 10:20 was 135° and was changed to 90° at 10:40. Changes in $L_t(443)$ were more uniform, indicating lower variability of in-water conditions except at 10:20 on 13 July and 10:40 on 14 July, when the changes were 20% and 35% of early morning values, respectively. Again $L_t(443)$ were consistently higher using viewing angles of 90° compared to 135° (Figure A4). The variation in $L_t(443)$ at 10:40 on 14 July co-varied with $L_{sky}(443)$, but at 10:20 on 13 July $L_t(443)$ and $L_{sky}(443)$ diverged even though the sky conditions were similar (Figure A2), possibly due to a sea surface microlayer slick. On 13 July from 12:20 to 12:40 there was a decrease in $L_t(443)$, due to a change in the viewing angle from 90° to the right to 90° to the left. Prior to 17 July there was a storm and rain on 15 and 16 July, which may change the atmospheric aerosol type and in-water conditions on 17 July, compared to those on 13 and 14 July. To further assess whether the variation in $L_{sky}(443)$ on 13 and 14 July reduced the quality of the radiometric data used in the intercomparison, across-group coefficient of variation in $R_{rs}(443)$ from the above-water systems on 13, 14 and 17 July are also presented in Figure 4J–L. The change in viewing angle from 90° to 135° and co-varying temporal changes in E_d, L_t and L_{sky} cancel out when computing R_{rs} (e.g., Figures A5 and A6). The $R_{rs}(443)$ coefficient of variation, therefore, represents temporal changes in both in-water constituents and the bi-directionality of the light field. This varied from 0.029 to 0.055, with the highest value recorded on 13 July at 10:40 when there was a decrease in L_{sky} and increase in L_t. Despite the relatively large variations in $L_{sky}(443)$, the coefficient of variation in $R_{rs}(443)$ was <5.5% across above-water systems, indicating that the observed variability in $L_{sky}(443)$ was not introducing large variations in R_{rs}. Comparison of variations in $E_d(443)$, $L_t(443)$ and $L_{sky}(443)$ during cloudy conditions on 15 and 16 July 2018 are given in Figure A1. These data were not used for the intercomparison, but illustrate that under non-ideal conditions there are larger variations in both $E_d(443)$, $L_t(443)$ and $L_{sky}(443)$.

The intercomparison results of E_d and the corresponding residuals are shown in Figure 5 and the statistics are given in Table 4. There was good agreement between the TriOS-RAMSES above-water systems, with an RMS <0.03 mWm^{-2}nm^{-1} across the visible spectrum and with most sensors having an RPD <5% (with some exceptions; e.g., RAMSES-A >5% at 665 nm; RAMSES-B > 6.5% at 443 nm; Table 4). There was a tendency for RAMSES irradiance sensors to over-estimate at 400 nm relative to the weighted mean, which was highest for RAMSES-B (Figure 5). RAMSES-A tended to under-estimate in red and green channels. There was a systematic bias in the RAMSES-D and in-water-B Sensor 1 E_d data, compared to the weighted mean, whereby all channels were underestimated by >5% varying from ±7.9% at 443 nm to ±10.6% at 665 nm. This bias was due to poor cosine response of the sensor. To correct for this, E_d from the RAMSES-D sensor 2 were processed for both RAMSES-D and in-water B data, which significantly reduced the differences against the weighted mean (Figure 6, Table 4). Due to the problems discovered with RAMSES-D and in-water B Sensor 1 data, only the corrected data using Sensor 2 are therefore plotted in Figure 5. The two Seabird-HyperSAS irradiance sensors exhibited an RMS of <0.01 mWm^{-2}nm^{-1} across the visible spectrum which was higher in the blue. Both in-water systems measured E_d above-water, but tended to underestimate E_d. The in-water A system was within <3% at 443, 560 and 665 nm (Table 4), although it exhibited high scatter at 400 nm (Figure 5).

Table 4. Relative percentage difference (RPD) and root mean square (RMS) in mWm^{-2}nm^{-1} for spectral values of E_d at OLCI bands 443, 560 and 665 nm. N is the number of measurements. For RAMSES-D and in-water B, S1 is Sensor 1 and S2 is Sensor 2.

Sensor Type	N	RMS 443	RPD 443	RMS 560	RPD 560	RMS 665	RPD 665
RAMSES-A	34	0.007	−1.55	0.023	−4.88	0.027	−5.74
RAMSES-B	35	0.029	6.64	0.014	3.01	0.017	3.76
RAMSES-C	35	0.006	1.13	0.005	−0.92	0.004	−0.28
RAMSES-D S1	35	0.034	−7.92	0.046	−9.66	0.048	−10.58
RAMSES-D S2	35	0.004	0.35	0.009	−1.64	0.007	−1.03
RAMSES-E	35	0.004	0.73	0.005	−1.03	0.003	−0.24
PANTHYR	30	0.007	0.85	0.011	2.18	0.018	4.07
HyperSAS-A	35	0.011	−2.30	0.008	1.61	0.004	0.78
HyperSAS-B	27	0.009	−1.79	0.005	−0.19	0.006	0.65
In-water A	28	0.005	0.03	0.014	−2.90	0.005	0.03
In-water B S1	28	0.032	−6.64	0.044	−9.04	0.046	−9.55
In-water B S2	28	0.010	1.21	0.011	−0.82	0.010	−0.22

There was also a temporal difference between casts for the above- and in-water systems. Some outliers are observed for the PANTHYR system in Figure 5, with respect to other systems. This is thought to be due to variation of sky conditions, including sun zenith angle, between the time of the automated PANTHYR measurements and the other measurements, which were synchronized at a different time. It could also be due to the PANTHYR irradiance sensor being located on the railings rather than the mast so it may be affected by the surrounding AAOT infrastructure. The outliers in the box plots are from the first measurement taken on 14 July (Cast 14), when $E_d(443)$ was varying most quickly in time (Figure 4A).

The comparison between L_{sky} measurements from the above-water systems (except PANTHYR which was viewing at different azimuth) and the corresponding residuals are presented in Figure 7 and the statistics relative to the weighted mean of the above-water systems in Table 5. There was very good agreement between the TriOS-RAMSES with an RMS <0.011 mWm^{-2}nm^{-1}sr^{-1} and RPD <2.5%

across all channels. The Seabird-HyperSAS sensors exhibited similar results with <2.5% difference at 443, 560 and 665 nm, though HyperSAS-A at 400 nm was −5.3%. For HyperSAS-B, there was an underestimate in the green and red channels of −2.5 and −1.9%, respectively, and a slight overestimate in the 400 nm channel.

L_t measurements and the corresponding residuals are shown in Figure 8, and the statistics relative to the weighted mean of the above-water systems are given in Table 6.

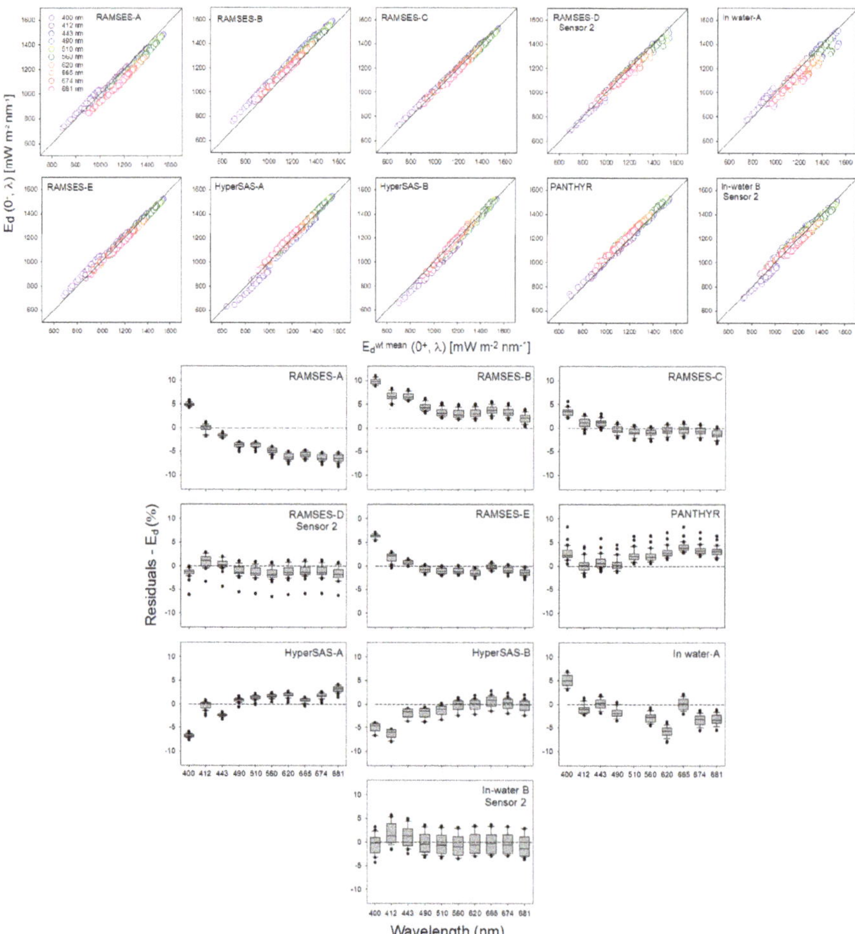

Figure 5. Top Panel: Scatter plots of E_d for each cast from the different above- and in-water systems versus weighted mean E_d ($E_d^{wt\ mean}$) from above-water systems (RAMSES-A, -B, -C, HyperSAS-A, -B). The different coloured points correspond to the different Ocean and Land Colour Instrument (OLCI) bands given in the RAMSES-A sub-plot. **Bottom Panel:** residuals of E_d from each cast expressed as percent residuals for the different above- and in-water systems. The residuals at each wavelength are calculated for each system as $[(E_d - E_d^{wt\ mean})/E_d^{wt\ mean}]*100$. The weighted mean from above-water systems (RAMSES-A, -B, -C, HyperSAS-A, -B), is given as the dotted line. The boundary of the box closest to zero indicates the 25th percentile, the solid line within the box is the median, the dashed line is the mean and the boundary of the box farthest from zero indicates the 75th percentile. The error bars above and below the box indicate the 90th and 10th percentiles and the points beyond the error bars are outliers. For RAMSES-D and in-water B data from Sensor 2 are shown.

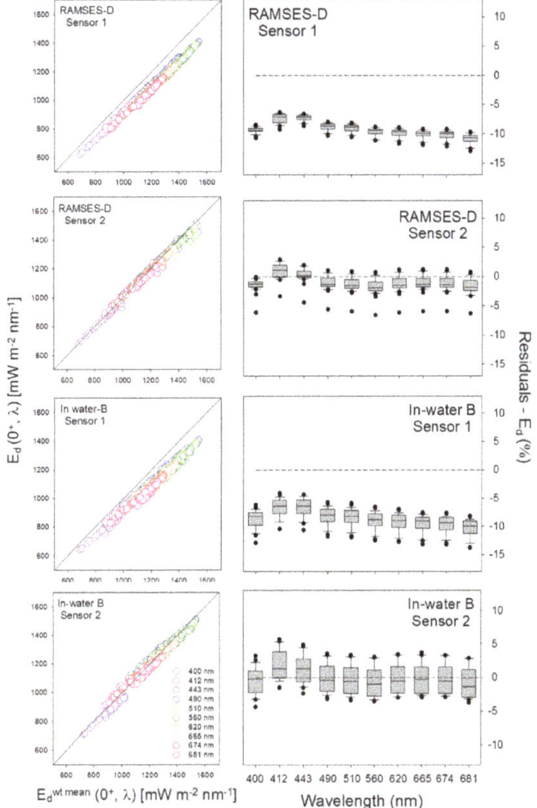

Figure 6. Scatter plots and residuals of E_d for RAMSES-D and in-water B for Sensor 1 and Sensor 2.

Table 5. RPD and RMS in mWm^{-2}nm^{-1}sr^{-1} for spectral values of L_{sky} at OLCI bands 443, 560 and 665 nm, used to quantify differences between systems and methods. N is the number of measurements.

Sensor Type	N	RMS 443	RPD 443	RMS 560	RPD 560	RMS 665	RPD 665
RAMSES-A	34	0.011	2.42	0.003	0.19	0.003	0.38
RAMSES-B	35	0.003	0.65	0.003	−0.26	0.004	−0.38
RAMSES-C	35	0.011	2.37	0.004	0.41	0.005	0.93
RAMSES-D	35	0.004	0.46	0.004	−0.49	0.005	−0.41
RAMSES-E	35	0.008	1.78	0.007	1.23	0.008	1.27
HyperSAS-A	35	0.011	−2.47	0.007	1.43	0.0041	−1.31
HyperSAS-B	27	0.005	−0.82	0.012	−2.49	0.010	−1.88

Table 6. RPD and RMS in mWm^{-2}nm^{-1}sr^{-1} for spectral values of L_t at OLCI bands 443, 560 and 665 nm, used to quantify differences between systems and methods. N is the number of measurements.

Sensor Type	N	RMS 443	RPD 443	RMS 560	RPD 560	RMS 665	RPD 665
RAMSES-A	34	0.007	1.57	0.004	−0.83	0.003	0.31
RAMSES-B	35	0.009	3.00	0.002	0.97	0.008	2.97
RAMSES-C	35	0.005	0.86	0.006	−1.21	0.003	−0.12
RAMSES-D	35	0.006	−0.57	0.005	−0.33	0.015	−2.66
RAMSES-E	35	0.005	0.39	0.007	−0.56	0.006	−0.37
HyperSAS-A	35	0.015	−2.35	0.006	1.23	0.005	−0.8
HyperSAS-B	27	0.006	1.26	0.003	−0.24	0.004	0.06

Again, there was good agreement between TriOS-RAMSES and Seabird-HyperSAS sensors, with RMS values generally <0.015 mWm^{-2}nm^{-1}sr^{-1} and RPD values were <3.0% at 443, 560 and 665 nm.

Similar to L_{sky}, at 400 nm, HyperSAS-A L_t exhibited a consistent underestimate and RAMSES-A, -B, -C and −E overestimated at 400 nm (Figure 8). RAMSES-D had a low but consistent off-set from the weighted mean. For L_{sky} and L_t the pattern between TriOS-RAMSES and Seabird-HyperSAS was similar, but reduced compared to E_d, with a slight over-estimate in blue channels for RAMSES and under-estimate for HyperSAS. RAMSES-E showed a higher variability in L_t compared to the other RAMSES systems, which was due to outliers on the first measurement of July 13 (Cast 1) and second measurement of July 14 (Cast 15), which correspond to low values $L_{sky}(443)$ and $L_t(443)$ or large variations in $L_{sky}(443)$ (Figure 3D–H).

Scatter plots for R_{rs} and the accompanying residuals relative to the weighted mean are given in Figure 9 and the accompanying statistics are given in Table 7. Spectral comparison of R_{rs} at OLCI bands for selected Casts on 13, 14 and 17 July are also given in Figures A5–A7. The TriOS-RAMSES and one Seabird-HyperSAS system tended to slightly overestimate R_{rs} by <5% in the blue, <3% in the green, but >5% in the red (Figure 9). TriOS-RAMSES systems generally had a similar RMS in blue, green and red bands (<0.02 sr^{-1}, <0.01 sr^{-1}, <0.027 sr^{-1}, respectively) to Seabird-HyperSAS systems (<0.04 sr^{-1}, <0.01 sr^{-1}, <0.042 sr^{-1}, respectively; Table 7). For RAMSES-D, the underestimate in E_d using Sensor 1 caused an overestimate in R_{rs} of between 9% in the blue to 10% in the red. This was reduced to 2% and 4% respectively when Sensor 2 was used to compute R_{rs}. HyperSAS-A tended to underestimate R_{rs}, where the RPD at 443 nm was −1.5% and at 665 nm was −4%. The PANTHYR system showed consistent precision with an RMS <0.032 sr^{-1} at the blue and <0.026 sr^{-1} at green bands, with differences from the weighted mean of R_{rs} of <5.6% at 443 nm, <5% at 560 nm, and 13% at 665 nm (Table 7). The in-water systems underestimated R_{rs}, and they exhibited a high scatter and bias compared to the weighted mean of the five above-water systems (see also magnitude and shape of In-water spectra in Figures A5–A7). For in-water A, the difference was <10% across visible bands. For in-water B Sensor 2, the RPD and RMS were −17% and 0.1 sr^{-1} at 443 nm and 12.3% and 0.065 sr^{-1} at 560 nm, respectively. These differences may in part be due to comparisons against a weighted mean from some of the above-water systems. For in-water B, the effect of using an E_d sensor with poor cosine response on R_{rs} (Sensor 1), gave higher values which fortuitously agreed better with the weighted mean of the above-water systems in blue and green bands (RPD of 10% at 443 nm and 4% at 560 nm). For R_{rs} red bands, both in-water B Sensor 1 and 2 the difference was much higher (~30%).

For L_{wn}, SeaPRiSM was used as an independent reference for both above-water and in-water systems rather than the weighted mean from three of the TriOS-RAMSES and two of the Seabird HyperSAS systems. There were, however, only nine near-coincident casts with SeaPRiSM during 13, 14 and 17 July 2018 and this was reduced to six casts for HyperSAS-B and in water-B. PANTHYR was not compared with SeaPRiSM, due to the differences in relative azimuth angles and the BRDF correction used for RAMSES and HyperSAS systems. TriOS-RAMSES systems tended to underestimate L_{wn}, which was generally <8.0% at 441 nm with an RMS <0.052 mWm^{-2}nm^{-1}sr^{-1}, <6.0% at 551 nm and an RMS <0.031 mWm^{-2}nm^{-1}sr^{-1} and <9.5% at 667 nm and RMS <0.057 mWm^{-2}nm^{-1}sr^{-1} (Figure 10, Table 8). The HyperSAS systems also underestimated L_{wn} compared to SeaPRISM which were between −1.4% and −5.5% at 441 nm, −4.0% and −7.5% at 551 nm and <5.0% at 667 nm.

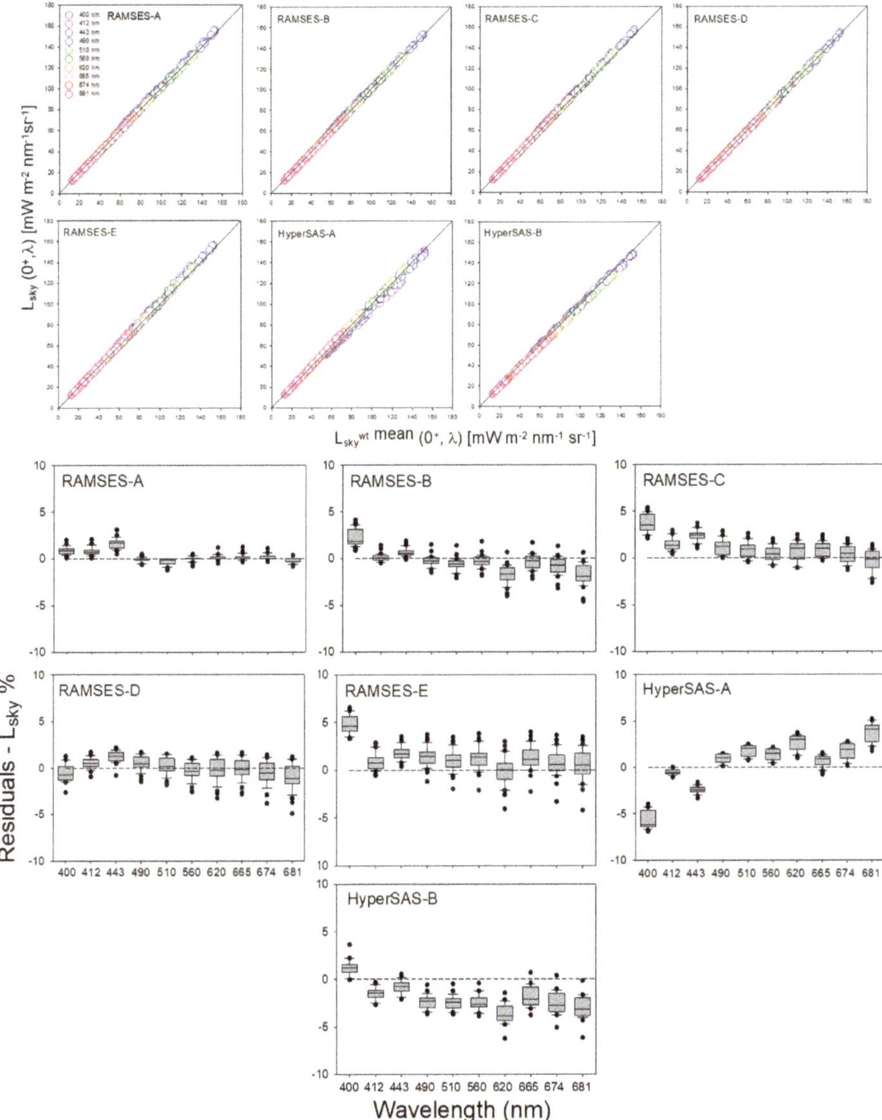

Figure 7. Top Panel: Scatter plots of L_{sky} at 90° and 135° relative azimuth from the different above-water systems versus weighted mean L_{sky} ($L_{sky}^{wt\ mean}$) from above-water systems (RAMSES-A, -B, -C, HyperSAS-A, -B). The different coloured points correspond to the different OLCI bands given in the RAMSES-A subplot. **Bottom Panel:** Residuals of L_{sky} expressed as a percent for the different above–water systems. The residuals at each wavelength are calculated from each system [(L_{sky} − $L_{sky}^{wt\ mean}$)/$L_{sky}^{wt\ mean}$] ∗ 100.

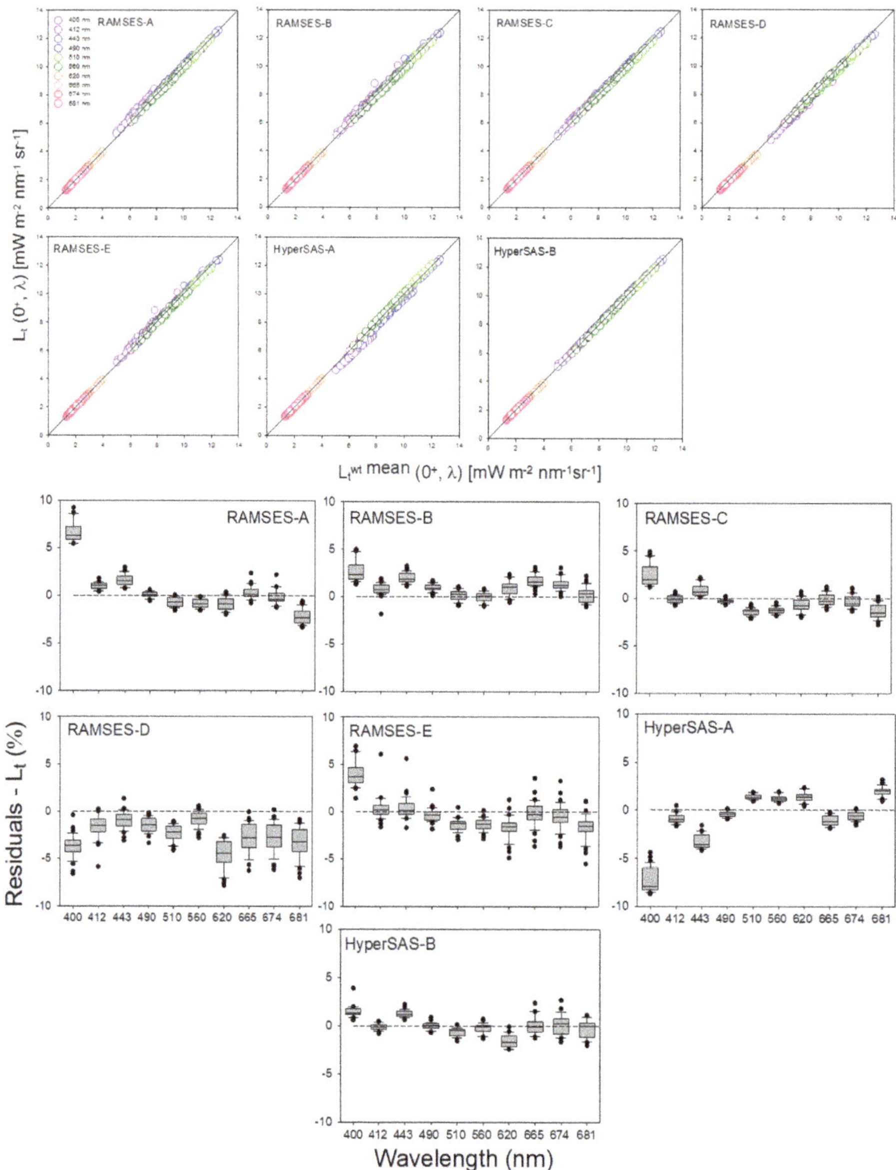

Figure 8. Top Panel: Scatter plots of L_t at 90° and 135° relative azimuth from the different above-water systems versus weighted mean L_t ($L_t^{wt\ mean}$) from above-water systems (RAMSES-A, -B, -C, HyperSAS-A, -B). The different coloured points correspond to the different OLCI bands given in the RAMSES-A subplot. **Bottom Panel:** Percent residuals of L_t for the different above- and in-water systems. The residuals at each wavelength are calculated from each system as $[(L_t - L_t^{wt\ mean})/L_t^{wt\ mean}] \times 100$.

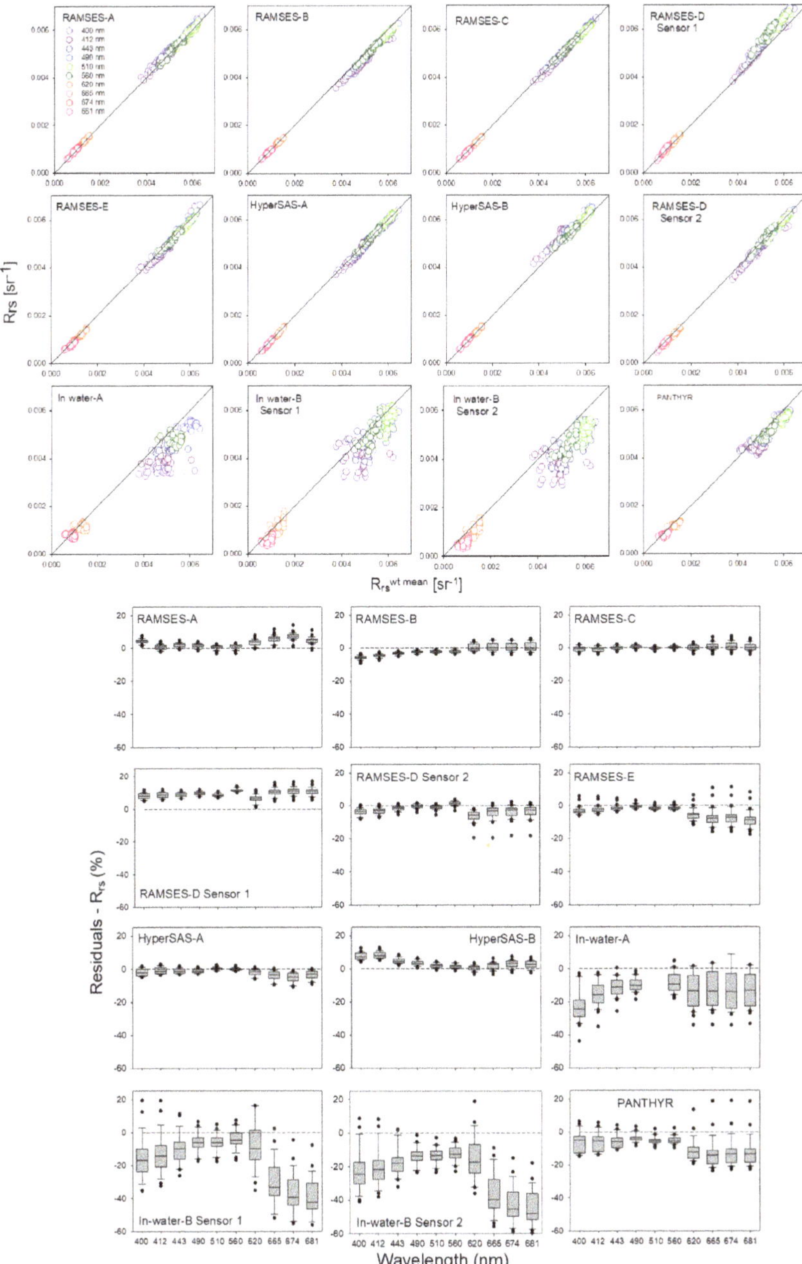

Figure 9. Top Panel: Scatter plots of R_{rs} from the different above- and in-water systems versus weighted mean R_{rs} ($R_{rs}^{wt\ mean}$) from above-water systems (RAMSES-A, -B, -C, HyperSAS-A, -B). For RAMSES-D and in-water B, S1 is Sensor 1 and S2 is Sensor 2. **Bottom Panel:** Percent residuals of R_{rs} for the different above- and in-water systems. The residuals at each wavelength are calculated from each system as $[(R_{rs} - R_{rs}^{wt\ mean})/R_{rs}^{wt\ mean}] * 100$.

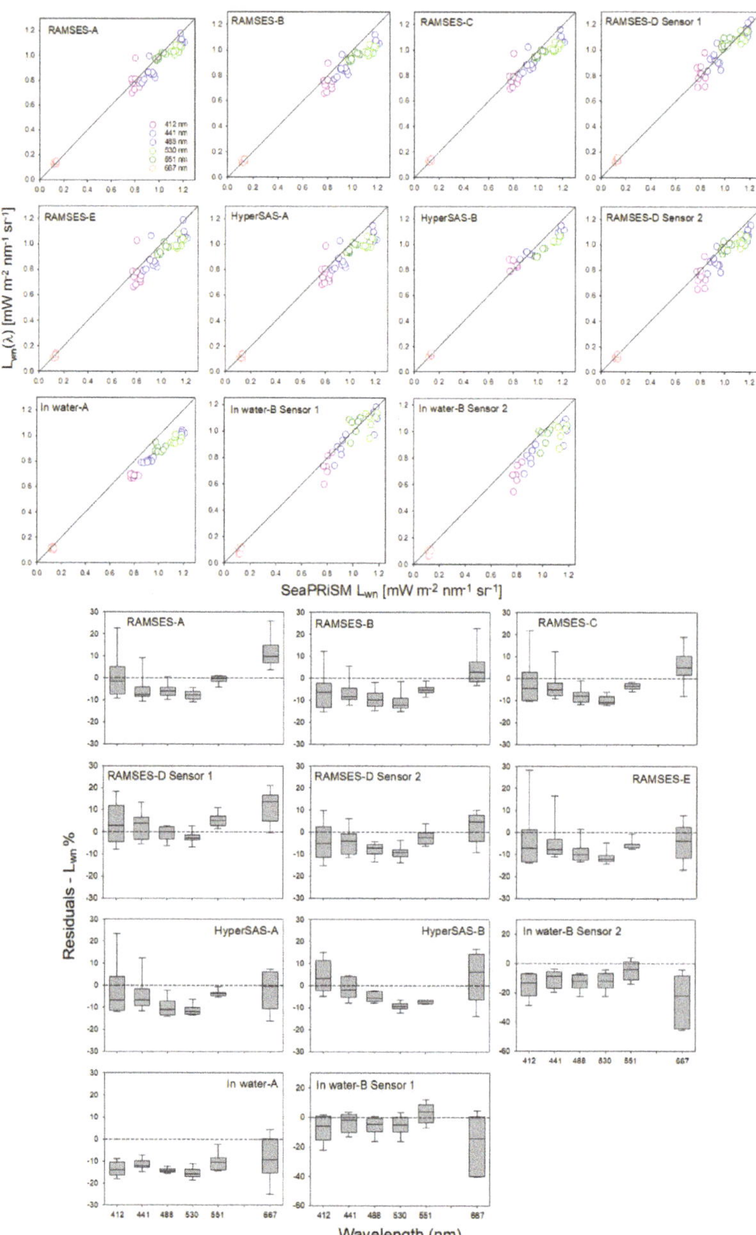

Figure 10. Top Panel: Scatter plots of $L_{wn}(\lambda)$ from the different above- and in-water systems versus AERONET-OC SeaPRiSM $L_{wn}(\lambda)$. The different coloured points correspond to the different SeaPRiSM bands given in the panel for RAMSES-A. For RAMSES-D and in-water B, S1 is Sensor 1 and S2 is Sensor 2. **Bottom Panel:** Percent residuals of L_{wn} for the different above- and in-water systems. The residuals at each wavelength are calculated from each system as $[(L_{wn} - \text{SeaPRiSM } L_{wn})/\text{SeaPRiSM } L_{wn}] * 100$. Note the change of scale for in-water B Sensor 1 and 2.

Table 7. RPD and RMS in sr^{-1} for spectral values of R_{rs} at OLCI bands 443, 560 and 665 nm to quantify differences in R_{rs} between systems and methods. Data sets are compared against the weighted mean R_{rs} from above-water systems (RAMSES-A, -B, -C, HyperSAS-A, -B). N is the number of measurements. For RAMSES-D and in-water B, S1 is E_d Sensor 1 and S2 is E_d Sensor 2.

Sensor Type	N	RMS 443	RPD 443	RMS 560	RPD 560	RMS 665	RPD 665
RAMSES-A	34	0.010	1.89	0.008	0.98	0.026	5.55
RAMSES-B	35	0.016	−3.39	0.011	−2.23	0.011	0.39
RAMSES-C	35	0.006	0.01	0.005	0.19	0.004	0.75
RAMSES-D S1	35	0.038	8.79	0.049	11.61	0.045	10.30
RAMSES-D S2	35	0.010	−1.63	0.008	1.34	0.027	−4.01
RAMSES-E	35	0.010	−1.46	0.004	−1.42	0.023	−7.42
HyperSAS-A	35	0.032	−1.15	0.008	−0.01	0.042	−4.02
HyperSAS-B	27	0.023	5.04	0.009	1.49	0.015	2.14
PANTHYR	29	0.032	−5.58	0.026	−5.00	0.077	−13.22
In-water A	27	0.065	−12.07	0.051	−8.20	0.097	−8.85
In-water B S1	28	0.066	−10.09	0.034	−4.39	0.19	−29.98
In-water B S2	28	0.096	−17.05	0.065	−12.30	0.229	−36.59

Table 8. RPD and RMS for spectral values of $L_{wn}(\lambda)$ at SeaPRiSM bands 441, 551 and 667 nm, used to quantify differences between systems and methods compared to AERONET-OC $L_{wn}(\lambda)$. For RAMSES-D and in-water B, S1 is Sensor 1, S2 is Sensor 2.

Sensor Type	N	RMS 441	RPD 441	RMS 551	RPD 551	RMS 667	RPD 667
RAMSES-A	9	0.046	−6.32	0.009	−0.83	0.057	9.45
RAMSES-B	9	0.051	−7.89	0.031	−5.82	0.046	2.55
RAMSES-C	9	0.011	−3.49	0.004	−3.69	0.005	4.73
RAMSES-D S1	9	0.037	1.22	0.027	5.10	0.064	9.00
RAMSES-D S2	9	0.048	−6.00	0.020	−2.59	0.052	−0.10
RAMSES-E	9	0.052	−5.65	0.030	−5.62	0.055	−5.90
HyperSAS-A	9	0.048	−5.52	0.021	−3.94	0.064	−4.81
HyperSAS-B	6	0.025	−1.39	0.041	−7.37	0.058	3.99
In-water A	8	0.063	−11.51	0.051	−10.25	0.071	−9.06
In-water B S1	6	0.039	−3.38	0.036	−3.18	0.164	−17.44
In-water B S2	6	0.069	−10.49	0.043	−4.71	0.200	−24.67

In-water A exhibited a −10.2% difference across 441, 555 and 667 nm bands. In-water B Sensor 2 exhibited smaller differences at blue and green bands with an RMS 0.11 to 0.23 $mWm^{-2}nm^{-1}sr^{-1}$ and RPD of −10% and −5% at 441 nm, respectively. At 667 nm, the RPD for in-water B was more than double that of in-water A (Table 8). For in-water B using E_d Sensor 1, L_{wn} was higher and more accurate than Sensor 2. This was probably caused by the extrapolation of L_u from in-water profiles to above surface which compensated the error in E_d due to non-normal cosine response for Sensor 1.

4. Discussion

Using a stable measurement platform, under near-ideal illumination and environmental conditions to compare a range of above-water optical measurement systems, there was generally <6% difference in E_d among sensors with an RMS <0.03 mW m^{-2} nm^{-1}. This was also the case for the sole instrument deployed on a ship (in-water A), for which the unavoidable tilt can introduce additional uncertainty. In a previous intercomparison at the AAOT, [17] reported similar differences between two TriOS-RAMSES sensors and the WiSPER system, which were <5.5% at blue, green and red wavebands. In this study, compared to the weighted mean, TriOS-RAMSES tended to slightly overestimate, and Seabird-HyperSAS slightly underestimated E_d (Table 4), also reported by [4]. These differences were always greater at 400 nm for both types of E_d sensors (Figure 5). The weighted mean does not represent the true value of E_d, so we can only conclude that the RAMSES and HyperSAS sensor types are different, particularly at 400 nm. This wavelength is also where the highest uncertainty is expected for calibration coefficients, since at 400 nm the calibration lamp signal is at its lowest. The greatest difference was for RAMSES-D, in which Sensor 1 exhibited a systematic bias (Figure 6), due to a poor cosine response of the E_d sensor (see also sensor 81EA data in Figures 13 and 15 of [4]). By contrast, RAMSES-D Sensor 2 had an appropriate cosine response (sensor 81E7 in [4]), and performed similarly to the other E_d sensors. In the absence of this field inter-comparison the poor cosine response may have been used unchecked for satellite validation. The <6% difference observed for the other sensors may arise from smaller differences in cosine response among or between sensor types [4], as well as from the temperature effects of individual and specific sensors. Even within a single sensor type, cosine response may be quite different from unit to unit, as shown by [66]. For the in-water systems, the differences were also low and generally <2.5%. Similar to the study of [17], the sensors were pre-calibrated at the same laboratory [1] and potential biases due to differences in the calibration coefficients from different sources were, therefore, removed. This contributed to the small differences in E_d between sensor types and methods, both above- and in-water.

For radiance measurements, the differences in above-water L_{sky} over visible bands compared to the weighted mean, except at 400 nm, were <2.5% with an RMS <0.01 mWm^{-2}nm^{-1}sr^{-1} (Table 5). The differences between RAMSES and HyperSAS sensors were similar at blue bands and higher for HyperSAS in green and red bands. The differences within RAMSES sensors were generally small, with one group showing a slightly higher deviation. This may in part arise from the processing methods used and specifically the number of replicates processed per cast (Table 2), especially in view of the high variability in L_{sky} (Figure 4). This is further discussed in Section 4.1.5. For L_t, the differences for both above-water sensor types were <3.5% with an RMS <0.009 mWm^{-2}nm^{-1}sr^{-1}. The differences in L_t in the blue and green for RAMSES and HyperSAS were similar, but were lower for HyperSAS in the red where the signal is lower (Table 6, Figure 8).

Of the radiometric quantities measured, E_d showed the largest variation between sensor types and methods, and the differences in L_{sky} and L_t between above-water sensor types were smaller. In this study, the differences between sensors were smaller than [4] who made the measurements under heterogenous (partially cloudy) conditions. The factors below contribute to the differences found.

4.1. Sources of Uncertainty

4.1.1. Effects of Sensor Absolute Calibration

All sensors were calibrated at UT in June 2018 under the same laboratory conditions, using the same calibration standards and by the same operator prior to the field intercomparison. Standard uncertainty of the calibration coefficients were of the order of 1% for irradiance and radiance over the whole spectrum. Potential biases due to differences in the calibration coefficients from different sources were therefore removed. According to [1], the long term stability of the calibrations was good with 80% of the sensors experiencing a change of <1% over one year. The largest differences in E_d were at 400 nm especially for two of the RAMSES sensors (RAMSES-B, RAMSES-E) and the two HyperSAS sensors. RAMSES-B and -C, HyperSAS-A and -B E_d sensors were the oldest used for the intercomparison (Table 2). Both RAMSES and HyperSAS have redesigned the geometry of the sensor head over time possibly suggesting that the deviation at blue bands may in part be due to the age and geometry design of the cosine collector. These effects were not visible over one year of calibration [1]. For L_{sky} and L_t there was a similar trend at 400 nm, but the magnitude of the difference was much lower for both RAMSES and HyperSAS sensors. These effects need to be carefully tracked through full and regular sensor characterisation, especially for the E_d sensors.

4.1.2. Differences in Cosine Response

The largest difference in measured E_d was found in the RAMSES-D E_d Sensor 1, in which a poor cosine response caused a high bias in the measurements. As highlighted in [4], the RAMSES-D Sensor

1 had large cosine error of 10–12% in the visible channels, which resulted a negative bias of ~12% at 400 nm and 7–10% in bands >490 nm (Figure 6). The angular dependence of responsivity of the irradiance instrument should correspond to the cosine of incidence angle, but for the RAMSES-D Sensor 1 this was not the case (see also Figures 13 and 15 in [4]). This caused the overestimate and offset in R_{rs} over all spectral bands for RAMSES-D (Figure 9). In addition, [4] highlighted that the manufacturer's specification of the Seabird-HyperSAS [67] is that the cosine RMS error is <3% at 0–60°, and within 10% at 60–85° incidence angles and that for TriOS-RAMSES [68], the accuracy is between 6–10% depending on spectral range. It is noted, however, that the manufacturer specifications are vague and laboratory measurements on individual radiometers can show quite different behaviour [66]. For the Biospherical E_d sensor the cosine response was measured in February 2014 by the manufacturer and the average cosine error was <2% at 0–60° and within 9% at 60–85°. Deviations from these may be one of the main sources of error contributing to the differences in the field measurements. The cosine response is likely to be the principal cause of the differences between and among sensor types.

4.1.3. Differences in Field of View (FOV) of Radiance Sensors

For all TriOS-RAMSES radiance sensors, the manufacturers state that the FOV is 7° (Table 2). For the Seabird HyperSAS sensors used in this study, FOV is 6°. Theoretically there should be small differences due to the FOV between TriOS-RAMSES and Seabird HyperSAS-A radiance sensors. Instrument-specific differences between the sensors may however, contribute to the differences in L_{sky} observed. This is illustrated in Figure 7 of [4], especially when sky conditions are heterogeneous, with partial clouds, in the viewing direction. In this study, instrument specific differences may be more difficult to distinguish, since the sky conditions were cloud-free and stable (Figures A2 and A3). In addition, the above-water systems were mounted on the same frame so in theory they were viewing the same area of sky, although differences may arise from instrument-specific FOV or alignment differences of the sensors [4]. To assess the difference due to FOV, firstly L_{sky} spectra for each cast from one of the TriOS-RAMSES systems (RAMSES-C; FOV 7°) are compared with those from Seabird-HyperSAS-A (FOV 6°; Figure 11).

In general, and especially for casts with $L_{sky}(443)$ >100 mWm^{-2}nm^{-1}sr^{-1}, Seabird-HyperSAS-A (Figure 11A) was slightly lower than RAMSES-C (Figure 11B), suggesting possible differences in FOV. To verify this trend, the ratio between $L_{sky}(443)$ HyperSAS-A/$L_{sky}(443)$ RAMSES for each cast is plotted for each RAMSES sensor (Figure 11C) and then for HyperSAS-A and –B against the mean of the RAMSES sensors (Figure 11D). In theory, the ratio between $L_{sky}(443)$ HyperSAS-A or -B/$L_{sky}(443)$ RAMSES should be close to 1, since HyperSAS have a similar FOV (6°) compared to RAMSES (7°). Compared to individual RAMSES sensors and the mean of the RAMSES sensors, HyperSAS-A and -B $L_{sky}(443)$ are consistently lower. Compared to the RAMSES mean, HyperSAS-A is consistently lower than HyperSAS-B, further suggesting that the small differences in FOV between HyperSAS and RAMSES have an impact on the observed differences. The influence of the number of replicates used to compute median L_{sky} values (Table 2) and the integration time used during these measurements will also contribute to these differences. In theory, this effect should also be seen in the L_t radiance measurements. In Figure 11E–H the same L_t data as for L_{sky} are presented. The difference between HyperSAS-A and –B and RAMSES is less apparent for L_t than it is for L_{sky} (Figure 11A–D). The ratio between $L_t(443)$ HyperSAS/$L_t(443)$ mean RAMSES is clearly lower for HyperSAS-A compared to HyperSAS-B, further suggesting that differences may be due to these small variations in FOV. This is specific to the conditions during this intercomparison, under clear skies and on a stable platform. Figure 7/C23 of [4] shows variation of sky reflection for different FOV sensors. In addition, Figure 6 of [7] shows that the reflectivity of the sea surface varies strongly and non-linearly over an angular range of 23° around the 40° incidence angle, suggesting that a smaller FOV is preferable for these measurements. Under non-homogeneous sky and sea conditions and on moving vessels, the different FOV could generate greater differences in both L_{sky} and L_t. This warrants further investigation.

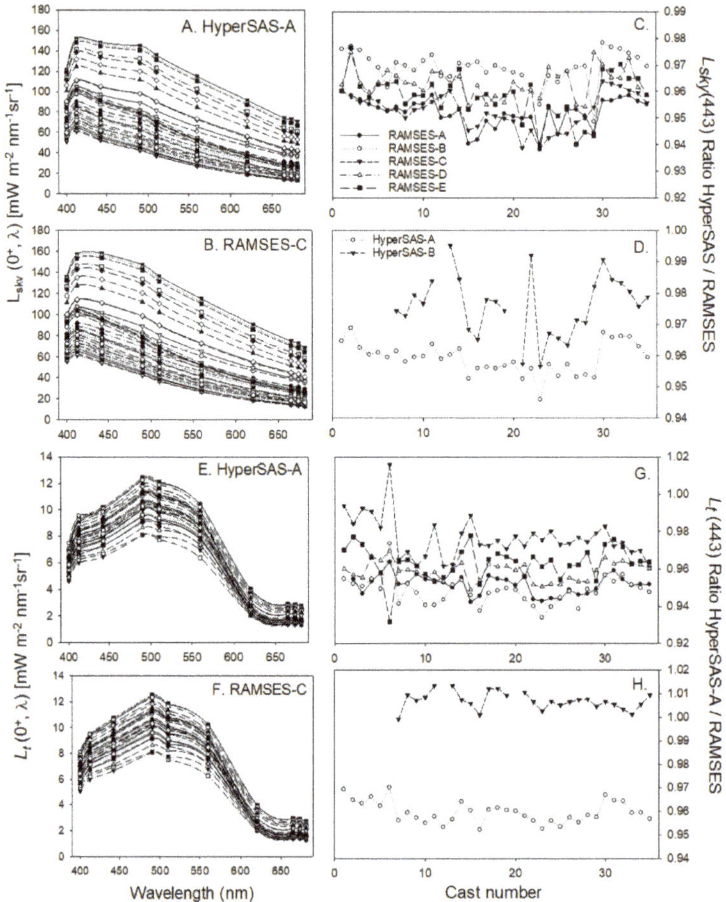

Figure 11. L_{sky} spectra of all casts for (**A**) HyperSAS-A, (**B**) RAMSES-C, and for each cast over time as the ratio of (**C**) L_{sky}(443) for HyperSAS-A/each RAMSES system and (**D**) HyperSAS-A and -B/mean of RAMSES systems (RAMSES-A, -B, -C, -D, -E). L_t spectra of all casts for (**E**) HyperSAS-A, (**F**) RAMSES-C, and for each cast over time as the ratio of (**G**) L_t(443) for HyperSAS-A/each RAMSES system and (**H**) HyperSAS-A and -B/mean of RAMSES systems (RAMSES-A, -B, -C, -D, -E).

4.1.4. Temperature Effects

Variations in temperature can affect the performance of the radiometer photo-diode array which can have a significant effect on the uncertainty of the instrument [69]. For TriOS-RAMSES sensors, temperature coefficients vary from -0.04×10^{-2} °C^{-1} at 400 nm to $+0.33 \times 10^{-2}$ °C^{-1} at 800 nm [69]. For biospherical microradiometers, typical temperature coefficients of -3.65×10^{-4} °C^{-1} have been reported [48]. Temperature can affect the dark and light counts of an instrument, across spectral regions differently. In this intercomparison, the calibration temperature (at the University of Tartu) was 20 °C, whereas the air temperature at the AAOT from 13–17 July 2018 varied from 23 °C to 26 °C. Due to heating of the metal super-structure of the AAOT and the sensor body, the internal temperature of the photo-diode array may have been considerably higher than 26 °C. The internal temperature of HyperSAS-B was far higher (~40 °C). Vabson et al. [4] identified that differences in calibration and ambient temperature during field intercomparisons may contribute to the bias in the results. Due to the small difference between the calibration and ambient temperature at the AAOT, theoretically the bias should be smaller, however the internal temperature of each sensor type may respond differently under the same air temperature. The Seabird-HyperSAS sensors are manufactured from plastic with a black finish and have a larger volume, whereas the TriOS-RAMSES are fabricated from stainless steel with a metallic finish and have a smaller volume. The biospherical surface reference sensor is manufactured from plastic with a white painted finish. Although temperature biases should be similar to all sensors with the same internal spectrometer, the design characteristics of TriOS-RAMSES versus the Seabird-HyperSAS may alter the internal temperature of each instrument type with respect to the ambient temperature, which could result in biases between instrument types. Due to the black finish, the HyperSAS sensors used in this study may have a higher internal temperature, which may account for some of the differences seen. In addition the RAMSES power consumption is smaller compared to the HyperSAS sensors, which may contribute to varying the internal temperature of the instrument. Theoretically, the differences should increase throughout the day from the morning casts into the afternoon as the sensors heat up over the course of the day. This needs to be carefully characterized and verified in future intercomparisons.

4.1.5. Differences Due to Data Processing

The differences among sensor systems may arise from the diverse methods of processing and quality control in data implemented between institutes. The procedure for data processing includes quality control of measured data, time binning, spectral interpolation and applying appropriate Fresnel reflectance factors, ρ'. The main differences in data processors between systems are summarised in Table 2. Of the TriOS-RAMSES processors, RAMSES-A and -B used the same number of replicates for E_d, L_{sky} and L_t and ρ' to process R_{rs} (Table 2). RAMSES-E used a different number of replicates for E_d, L_{sky} and L_t, a different ρ' to process R_{rs}, and added NIR correction to the processing. For RAMSES-C and RAMSES-D Sensor 2 there were large differences in the number of replicate E_d, L_{sky} and L_t used, though RAMSES-D used RAMSES-C L_{sky} measurements. To assess the effect of differences between processing chains, we assessed two steps in the processing for one Cast (Cast 7). Firstly, a subset of TriOS-RAMSES data (RAMSES-B, -C, -E) were run through one processing chain (that of RAMSES-C) to assess the differences in R_{rs} due to processing methods (Figure 12). The differences are only significant at red bands and increase from 620 to 685 nm. Across blue to green bands, the differences are minimal. Using the individual processors, the difference over visible bands for RAMSES-B, and -E compared to RAMSES-C were 3.45%. By comparison using the RAMSES-C processor for all three datasets reduced the difference to 1.31%. For TriOS-RAMSES systems, differences in processors therefore only accounted for ~2% in the blue and green, but up to 8% in the red. Secondly, for Cast 7 we evaluated the differences in processing due to the ρ' value used by each institute against using a single ρ' factor (Figure 12). Using the same ρ' value, the difference was reduced to ~1% at red bands. The difference between using a single ρ' value was, therefore, as important as using a common processor, but the effect was only significant for red bands. Although a common processor has been advocated for use in the future, this study suggests that the use of common ρ' values for RAMSES systems is the important aspect for reducing differences between institutes. The use of the same ρ' for the in-water A system would also be beneficial in reducing differences in R_{rs} of about 1.7%. Further work should focus on deriving ρ' values from in-water and above-water measurements.

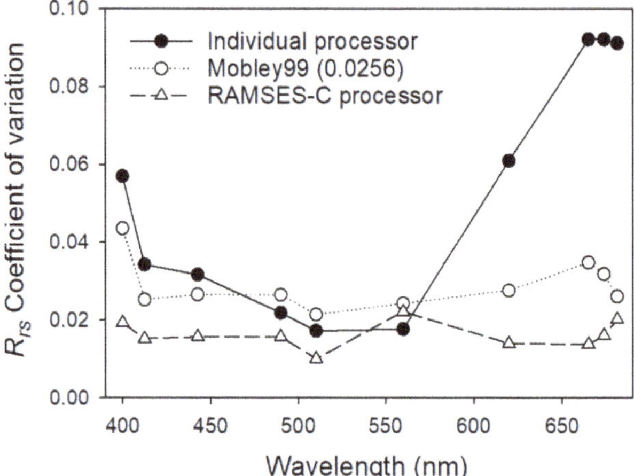

Figure 12. Coefficient of variation in R_{rs} spectra from for Cast 7 for RAMSES-B, -C, -E processed using individual processors (solid circle), the same ρ' factor (open circle) and the same (RAMSES-C's) processor (open triangles).

4.1.6. Differences between Case 1 and Case 2 Water-Type Processors

NIR reflectance is expected to be close to zero in waters with low particle scattering [10]. There has been much discussion in the literature around this topic, and the assumption for Case 1 waters where particle scattering is considered low and application to Case 2 waters where it can be significant. An offset from zero in the NIR has been attributed mainly to residual surface water effects (spray, sun glint, whitecaps, and sky radiance including scattered cloud reflected on waves). Any offset observed in the NIR that is not due to particle scattering is expected to be spectrally neutral and can be compensated for by subtracting this signal from the $R_{rs}(\lambda)$. For high particle scattering, the shape of $R_{rs}(\lambda)$ should reflect the spectral dependence of the reciprocal of water absorption [42]. If this is not the case, high particle scattering cannot account for the NIR offset and may thus be subtracted. In Figure 13, the effect of not including and including NIR correction on HyperSAS-A data are compared for three casts (1, 4, 20). When NIR correction is included, the shape of the HyperSAS-A spectra at OLCI bands are closer to the in-water A spectra (Figure 13A–C). When the NIR correction is not implemented the shape of the HyperSAS-A spectra are closer to the mean of the RAMSES spectra (Figure 13D–F). We have to consider which of the $R_{rs}(\lambda)$ spectral shapes are correct; the above-water or in-water? As an independent measurement, we have referenced each system to SeaPRiSM, although this is also an above-water system. Zibordi et al. [17] showed however, that SeaPRISM L_{wn} are within −0.1% of in-water WiSPER measurements. Although we have no measurements from WiSPER in this study, in-water A compare well with SeaPRISM with a slight under-estimate of up to 10% at 441, 551 and 667 nm (Figure 10; Table 8). This possibly suggests that NIR correction for these waters may be necessary. The difference between applying or not the NIR correction on HyperSAS-A data compared to SeaPRiSM is given in Figure 13G,H. HyperSAS-A with NIR correction resulted in a consistent under-estimate in L_{wn} over visible bands, although the scatter was small. For HyperSAS-A with no NIR correction, the data were closer to the 1:1, although the scatter increased. This suggests that no NIR correction, at least for HyperSAS-A, is recommended.

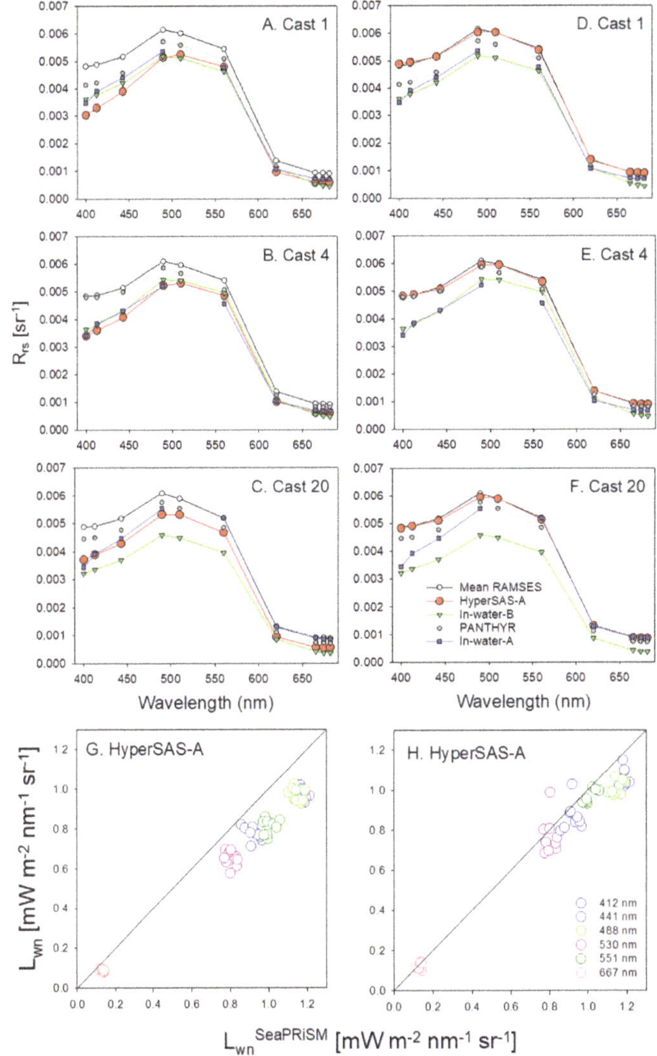

Figure 13. Variation in L_{wn} spectra at OLCI bands using near-infrared (NIR) correction in HyperSAS-A processing on (**A**) Cast 1, (**B**) Cast 4, (**C**) Cast 20, and no NIR correction in HyperSAS-A processing on (**D**) Cast 1, (**E**) Cast 4 (**F**) Cast 20. Scatter plots of L_{wn} for HyperSAS-A versus AERONET-OC SeaPRiSM L_{wn} using (**G**) NIR correction and (**H**) not applying NIR correction.

4.1.7. Other Effects

A number of other effects that may have caused significant differences between the sensors have already been considered in detail by [4]. Of these, the impact of the stray light in field measurements is expected to have the greatest effect, which was higher in the blue (<3.5%), and smaller in green and red (<1%). Data interpolation effects during processing can sometimes cause discrepancies. For example the interpolation of the radiometric measurements to OLCI bands is reported to contribute <0.5% of the variance between sensors except at 400 nm where the variance is expected to be larger. This depends however on whether the interpolation is done on the irradiance and radiance parameters that are used to compute reflectance or on the reflectance data directly. If on the latter, the error can be 5% [70]. No large difference between multi- and hyperspectral E_d data were observed (Figure 4), however the uncertainty related to the band shift needs to be assessed further. Polarization, the effect of light exhibiting different properties in different directions, is considered to be smaller still (<0.25%) [4].

4.2. Differences in $R_{rs}(\lambda)$ and $L_{wn}(\lambda)$

For R_{rs} the mean absolute differences among TriOS-RAMSES systems were 2.5% at 443 nm, 2.0% at 560 nm and 8% at 665 nm. In a previous study, [17] compared R_{rs} from two TriOS-RAMSES sensors against a WiSPER at the AAOT in 2010 and found that the RPD was <8% at 443 nm, <4% at 555 nm and <11% at 665 nm. The WiSPER was not available to us as an independent reference. However, the above-water sensors that we deployed were located side by side on the same measurement frame with the ability to turn them away from the sun and shade, as opposed to being fixed on the railings of the AAOT in the [17] intercomparison. For Seabird-HyperSAS the differences in R_{rs} were 3.1%, 0.75% and −3.1%, respectively. Only two HyperSAS systems were compared as opposed to five for TriOS-RAMSES, so the lower variability for HyperSAS is expected. The PANTHYR system showed consistent precision, with differences from the weighted mean of R_{rs} of <2.5% in the blue, <3.0% in the green, and <5.5% in the red. Since the PANTHYR was pointed at a different water area than the other sensors during the intercomparison, this result is promising. The differences in R_{rs} were higher between above- and in-water methods, though interpretation of this is compounded by the fact that the reference measurement was from above-water systems only. For the in-water systems the RPD were within 11% at 443 nm, 7.5% at 560 nm and up to 17% at 665 nm, respectively. A large part of the difference comes from the application of BRDF correction to account for the angular response of variation of upwelling light measured by the above-water systems. In addition, the in-water A and B casts did not match exactly with the above-water casts in either location or time, and fewer casts were made in-water.

For the comparison against SeaPRISM data, we eliminated these potential biases by performing BRDF correction to all above-water systems, to compute L_{wn} (Figure 13). The SeaPRISM has a long established legacy as a FRM and for satellite validation and therefore provides high quality data to compare to each system. All above water systems showed a similar pattern with a slightly lower L_{wn} in the blue and green compared to SeaPRISM. The TriOS-RAMSES systems showed a slightly lower difference in L_{wn}, which was generally <8%, <6% and <9.5% at 441, 551 and 667 nm, respectively. The differences were probably caused by a combination of imperfect correction of sky glint, propagation of differences in E_d and radiometer calibration and characterization. For HyperSAS, the difference compared to SeaPRiSM were <6%, <8% and <5% at 441, 551 and 667 nm, respectively. For in-water A and in-water-B, there were differences of 10 and 13% respectively, compared to SeaPRiSM across the 441, 551 and 667 nm bands. Processing of the in-water data requires instrument self-shading correction [71,72]. An in-water correction for instrument self-shading of L_u and shading by the deployment cage was not performed for in water-B, which is likely to account for a significant proportion of the error, especially in the red bands. In addition for in water-B, the influence of scattered light from the deployment frame, significant influence of the deployment cable on the light field, or the influence of the AAOT super-structure on the in-water light field and extrapolation of $L_u(z, \lambda)$ to L_{wn}, were not accounted for. If these effects were corrected for, undoubtedly the comparison with SeaPRiSM would have improved. Moreover, the use of a Case 1 water model for deriving L_u from irradiance measurements in complex waters is likely to introduce further uncertainty [60]. Further uncertainties for in-water L_{wn} data may result from extrapolating the measured spectra at deeper depths to the surface in order to obtain the subsurface radiance (in-water B) or from the approximation of converting the subsurface upwelling radiance (L_u) to L_{wn}. A proportion of the difference between the above-water and in-water system and SeaPRiSM will be due to differences in the exact time of casts and interpolation over them.

No uncertainty budget was calculated for each individual sensor or system during this intercomparison. Vabson et al. [4] computed the relative uncertainty of the same above-water sensors under field conditions at an Estonian Lake. The relative uncertainty for irradiance sensors varied from 9.7% to 4.7% from blue to red bands. For L_{sky} and L_t, the uncertainty was ~2% and 4%, respectively across visible spectral bands. This study and that of [4] are similar. Both use the same sets of radiometers, and the same procedures to calibrate them. Differences arise from the environmental conditions especially temperature and the repeatability of the measured signal. Both studies do not however, characterise stray light and non-linearity which may be different under the different environmental conditions experienced. The uncertainty budget of differences between radiometers during this study is, therefore, likely to be similar to the budget given in [4]. Zibordi et al. [17] included an uncertainty budget for each measurement system and compared relative uncertainties for each against the reference system. They found that the differences between methods and systems could be explained by the combined uncertainties determined for the systems compared [17].

4.3. Propagation of Errors in $E_d(\lambda)$, $L_{sky}(\lambda)$ and $L_t(\lambda)$ to $R_{rs}(\lambda)$

We evaluated which of the individual inputs terms in Equation (3) contributed the greatest fraction of the variance in R_{rs} (Figure 14). From 400 nm to 610 nm, E_d accounted for the largest fraction variance in R_{rs}. At 443 nm, both E_d and L_t accounted for a similar fraction of the variance in R_{rs}.

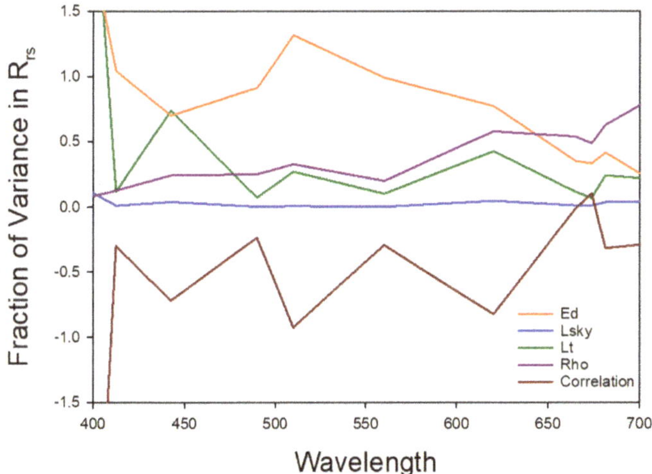

Figure 14. Fraction of variance in R_{rs} due to the different input terms of $E_d(\lambda)$, $L_{sky}(\lambda)$, $L_t(\lambda)$, $\rho'(\lambda)$ and the correlation among them.

Wavelengths higher than 610 nm, the contribution of ρ' (given as Rho in Figure 14) became dominant. L_{sky} consistently exhibited the lowest contribution to the variance in R_{rs} which was <3% over the visible spectrum.

The adjustment for the correlations among the input terms also contributed a relatively large fraction of the variance in R_{rs}. The sign of the adjustment was negative indicating that the correlation terms decrease the variance in R_{rs}. Overall this analysis showed that minimizing the errors arising from the measurement of E_d is the most important variable for reducing the inter-group differences in R_{rs}.

Recommendations: This field intercomparison illustrated that the differences in E_d, L_{sky} and L_t were low and generally less than the target 5%, although there were some anomalies above this for individual sensors and bands. The exercise was also pivotal in highlighting some errors in protocols and anomalies arising from some sensors. The difference in E_d between systems was the highest, which was greater for RAMSES (<7%) compared to HyperSAS (<2.5%) irradiance sensors. For L_{sky} the differences were low and similar at blue bands for RAMSES and HyperSAS but higher in green and red bands for HyperSAS. For L_t, differences in the blue and green for RAMSES and HyperSAS were similar, but lower for HyperSAS in the red. The differences in E_d are likely due to differences in cosine response of individual sensors. Differences in L_{sky} and L_t are possibly due to a combination of differences in viewing geometry, angular response and temperature effects between the sensors plus processing methods used. In this study, we could not isolate all of these factors to evaluate the magnitude of their effect individually. Recommendations for future intercomparisons are as follows:

- For both above- and in-water systems, the cosine collector of the E_d sensor needs to be carefully characterised to ensure the most accurate measurements are made.
- We found that above-water Fresnel reflectance factor ρ' caused a high variability between processing chains which was greater than other differences between processors, as demonstrated by using a single community processor. Future studies should assess further differences between above and in-water systems and the resulting ρ' under a range of environmental conditions and on moving vessels.
- The experimental design should be carefully considered in order to balance between representative sensor types of different above-water, in-water, and new technological systems whilst capturing a broad international range of participants that are active in satellite ocean colour validation.
- This intercomparison focused mainly on differences within and between TriOS-RAMSES systems. Differences within RAMSES systems were low. Future intercomparisons should include a wider range of sensors and systems to capture a further cross-section of the community, rather than just RAMSES systems.
- A more detailed characterisation of stray light, cosine response, linearity, temperature response and polarization sensitivity of individual instruments should be made to assess the contribution of each of these factors to the overall measurement uncertainty. Once these have been assessed, it is recommended to compute a full uncertainty budget as demonstrated in [4,17,73], to evaluate relative differences in uncertainty between instruments.
- Differences between sensors with varying FOV should be further investigated under non-homogeneous sky and sea conditions. In particular the use of a large FOV may be suboptimal when viewing the sea surface which has strong angular variability at the viewing nadir angle of 40°.
- Further intercomparisons of this nature are required from other types of platforms, such as on moving ships as in [74], and under non-ideal environmental conditions such as high sea states and partially cloudy skies when the errors between sensors are expected to increase.

5. Conclusions

A field intercomparison was conducted at the Acqua Alta Oceanographic Tower (AAOT) in the northern Adriatic Sea, from 9 to 19 July 2018, to assess combined differences in the accuracy of measurements collected using a range of in- and above-water optical systems. Prior to the intercomparison, the absolute radiometric calibration of all sensors was carried out using the same standards and methods at the same reference laboratory (University of Tartu) and the same operator. Measurements were performed at the AAOT under near-ideal conditions, on the same deployment platform and frame, under clear sky conditions, relatively low sun zenith angles and moderately low sea state (<5 m s^{-1}). For E_d, there was generally good agreement with differences of <7% between institutes with an RMS of <0.03 mWm^{-2}nm^{-1}, except for one E_d sensor which exhibited a systematic bias in the data due to poor cosine response. The difference in E_d was greater for RAMSES than for HyperSAS sensors. For L_{sky} and L_t the differences between systems and institutes were consistently lower. For L_{sky}, the differences were <2.5% with an RMS <0.01 mWm^{-2} nm^{-1}sr^{-1}, and RAMSES and HyperSAS sensors were similar at blue bands, but HyperSAS was higher in green and red bands. For L_t, the differences for both above-water sensor types were <3.5% with an RMS <0.009 mWm^{-2}nm^{-1}sr^{-1}. RAMSES and HyperSAS were similar at blue and green bands and HyperSAS was lower in the red. For R_{rs}, the differences among TriOS-RAMSES systems varied from 0.01% to −7.5% at visible bands, whereas for HyperSAS the differences were −0.01% to 5.0%. For in-water A the difference in R_{rs} was <10%. For the in-water B system the differences were greater and varied from −12.3% to 36.6%, although this may be largely constrained by using a weighted mean based on above-water measurements. L_{wn} was therefore computed to compare all sensors to SeaPRISM AERONET-OC as an independent reference measurement. The above-water TriOS-RAMSES had an average difference of <4.7% at 441, 551 and 667 nm compared to SeaPRISM. For Seabird-HyperSAS the mean difference over these bands was 4.9%, for in-water A 10.3% and for in-water B 13.3%. Differences between the in-water and above-water systems arise from differences in spatial and temporal sampling and extrapolating the in-water data from depth to the subsurface. The differences between above-water systems mainly arose from differences in E_d cosine response and FOV between L_{sky} and to a lesser extent L_t sensors, and the Fresnel reflectance value used and whether or not an NIR correction was applied at the data processing stage.

Author Contributions: Conceptualization was derived by C.D., T.C., G.T., G.D. and R.V.; methodology was derived and implemented by G.T. and G.D.; software was developed by G.D., M.H., I.A., M.L., D.D., V.V. (Vincenzo Vellucci), D.V., K.R., and A.B.; validation was performed by G.T. and G.D.; formal analysis was conducted by G.T. and G.D.; investigation was performed by G.T., G.D., M.H., M.L., M.C., D.D., V.V. (Viktor Vabson), D.V., K.R., M.B., A.B., V.V. (Vincenzo Vellucci), S.W., J.K., and I.A.; resources to support the research were provided by C.D., T.C., R.V., and G.T.; writing—original draft preparation was undertaken by G.T.; writing—review and editing, was done by G.T., G.D., M.H., M.L., M.C., D.D., V.V. (Viktor Vabson), K.R., M.B., A.B., Vnzo.V., S.W., J.K., and I.A.; visualization (data, tables, and figures) was done by G.T., G.D., M.H., M.L., M.C., D.D., V.V. (Viktor Vabson), K.R., M.B., A.B., V.V. (Vincenzo Vellucci), S.W., J.K., and I.A.; supervision was from G.T., and G.D.; project administration was from R.V.; funding acquisition to support the research were sought by C.D., T.C., R.V., and G.T. All authors have read and agreed to the published version of the manuscript.

Funding: This research was funded by European Space Agency contract Fiducial Reference Measurements for Satellite Ocean Colour (FRM4SOC) Contract number; ESA/AO/1-8500/15/I-SBo. M.B. was also funded by the European Space Agency PRODEX/HYPERNET-OC project. D.V was also funded by BELSPO (Belgian Science Policy Office) in the framework of the STEREO III program project HYPERMAQ (SR/00/335).

Acknowledgments: We are grateful to Giuseppe Zibordi for his support in the concept of FRM4SOC, in facilitating the set-up at the AAOT and for use of AERONET-OC SeaPRiSM data. We are also grateful to the crew of *RV Litus* and to Maurio Bastianini and Angela Pomaro for logistics on the AAOT. I personally thank P. R. H. Tilstone who was a lifetime inspiration to my science, who passed away on the first day of the campaign.

Conflicts of Interest: The authors declare no conflict of interest.

Appendix A

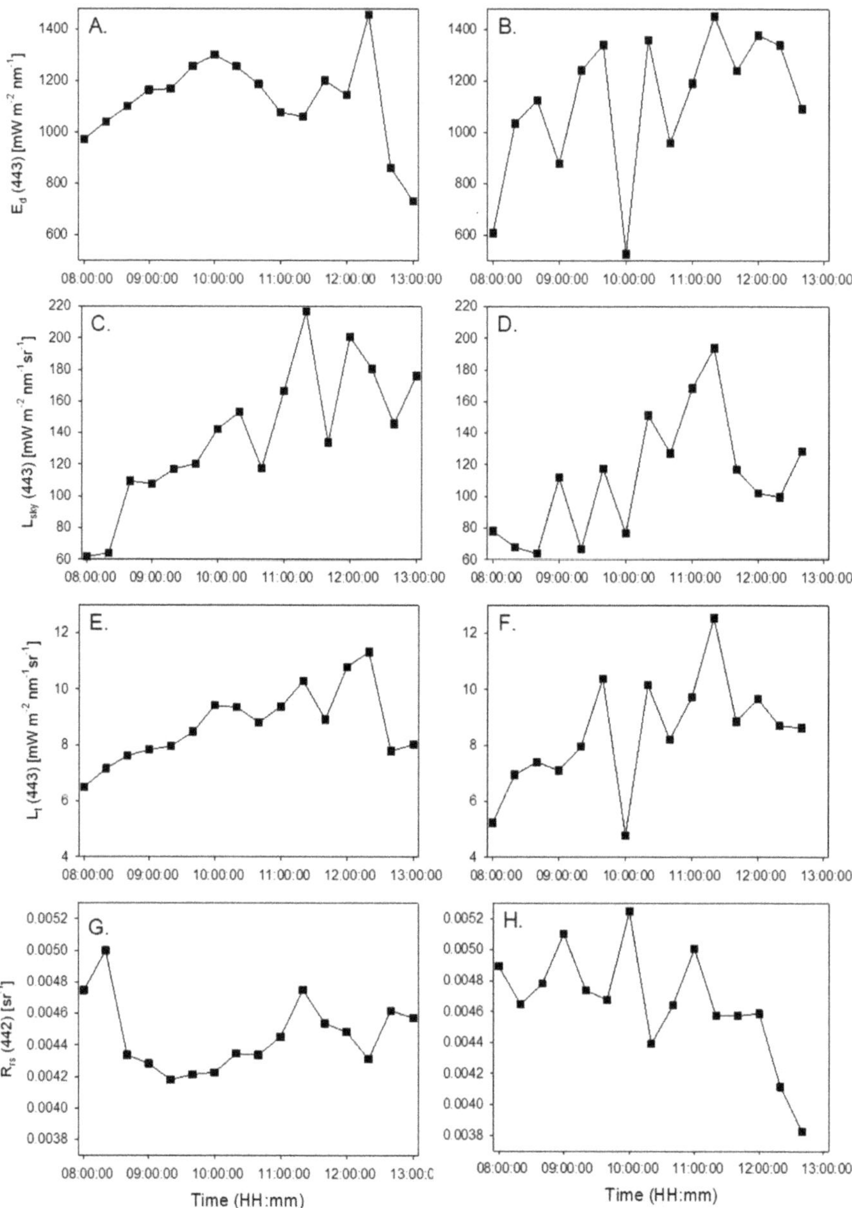

Figure A1. Variation in measurements that were NOT USED in the intercomparison for $E_d(442)$ on (**A**) 15 July 2018, (**B**) 16 July 2018; $L_{sky}(442)$ on (**C**) 15 July 2018, (**D**) 16 July 2018; $L_t(442)$ on (**E**) 15 July 2018, (**F**) 16 July 2018; $R_{rs}(442)$ on (**G**) 15 July 2018, (**H**) 16 July 2018. This illustrates the variation in these parameters under the influence of cloud.

Appendix B

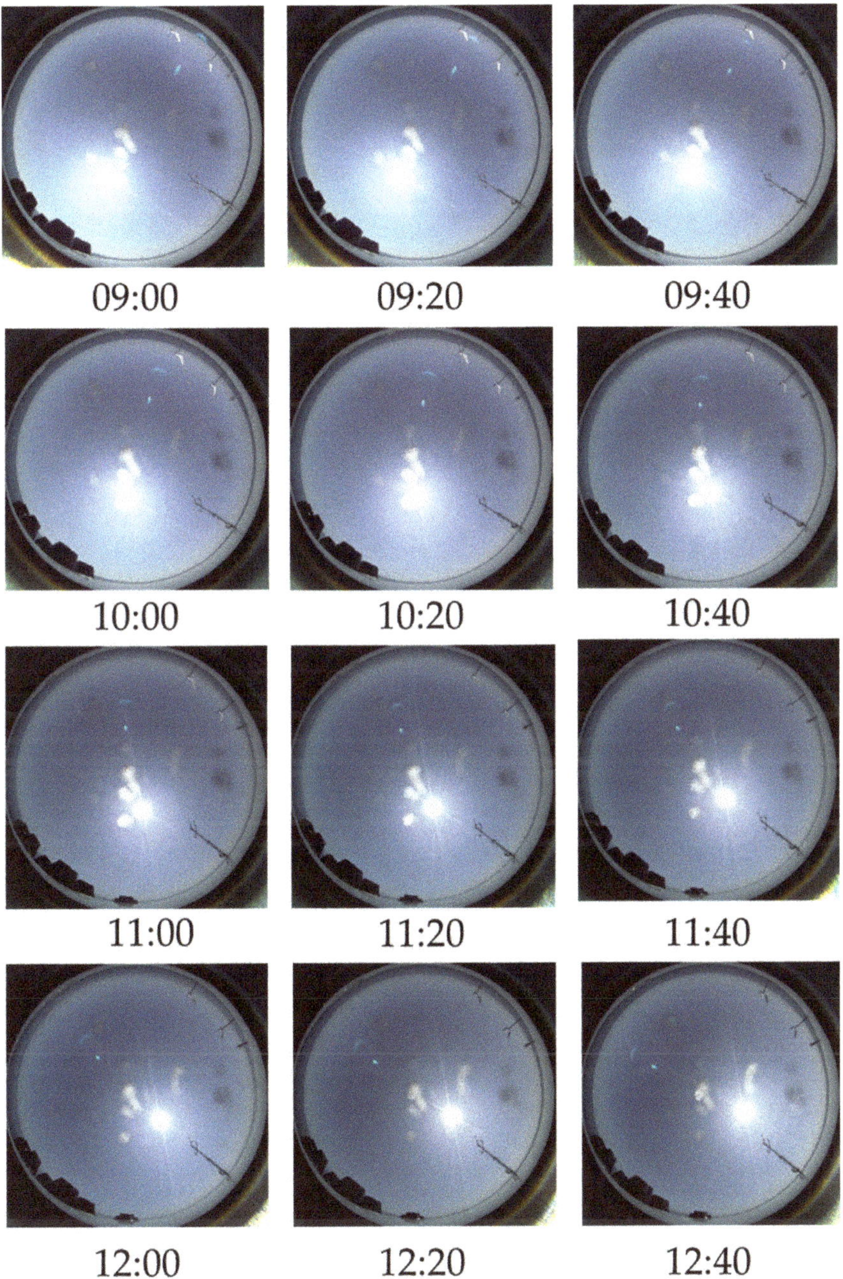

Figure A2. Fish eye images of sky conditions at the start of each Cast on 13 July.

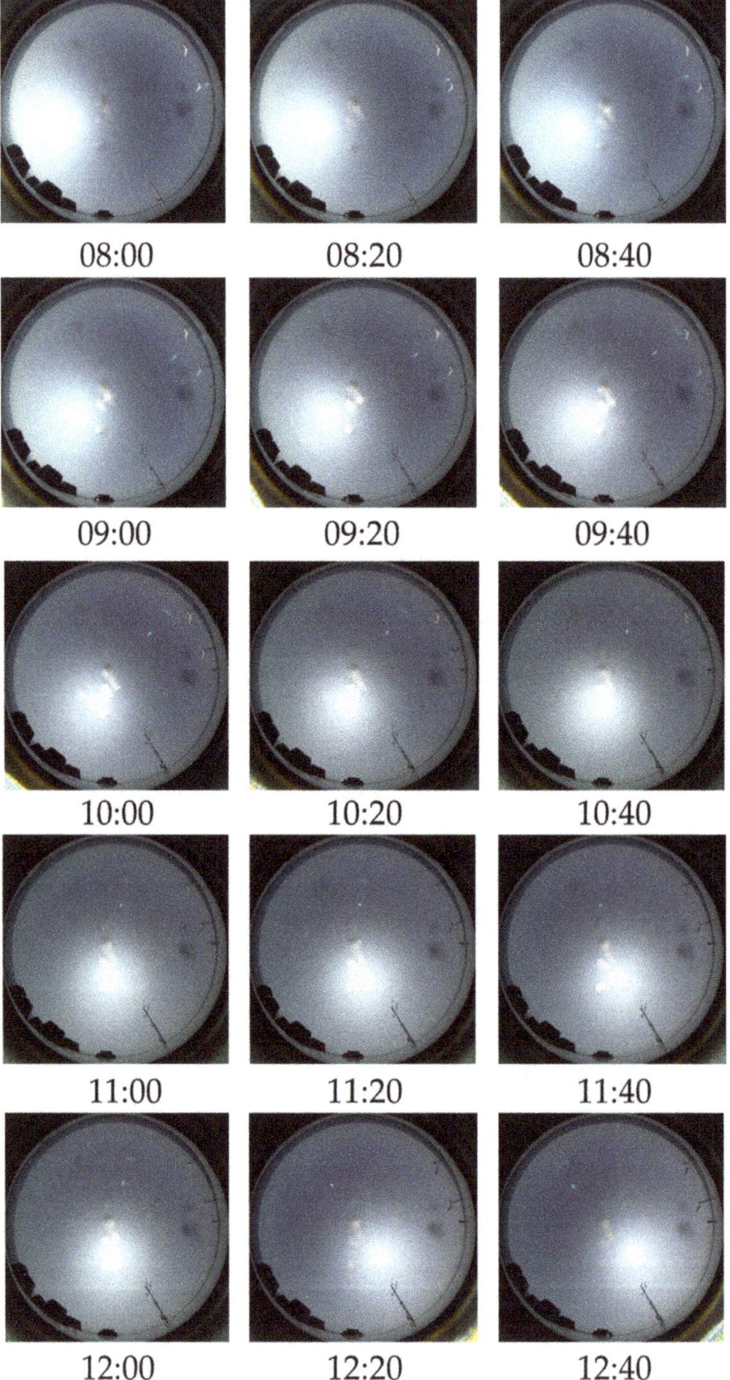

Figure A3. Fish eye images of sky conditions at the start of each Cast on 14 July.

Appendix C

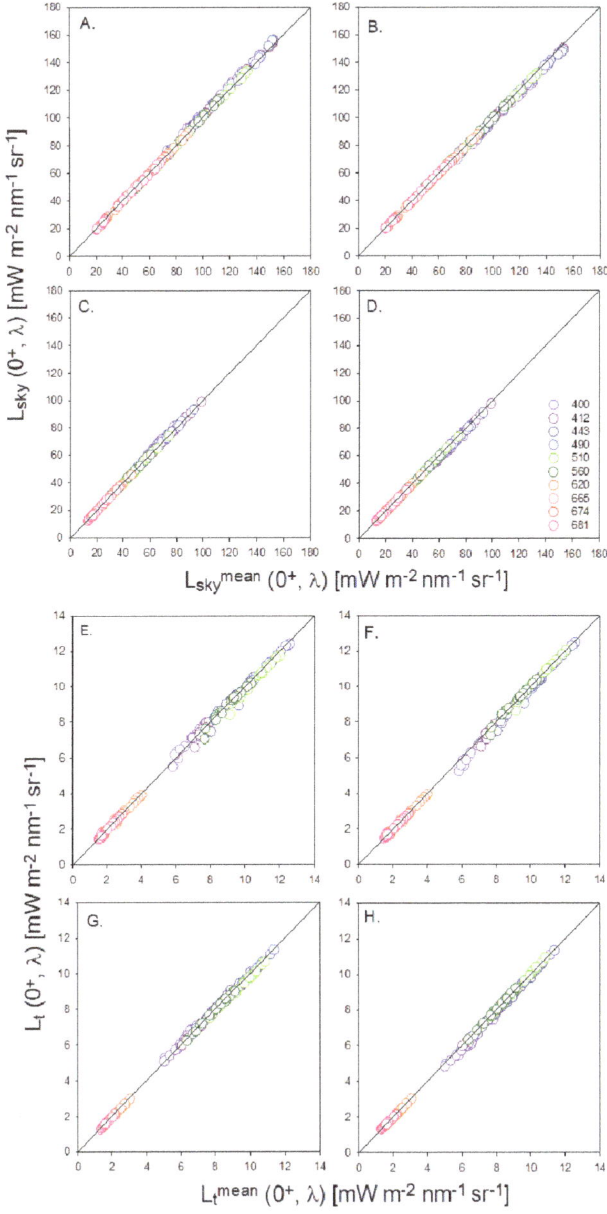

Figure A4. Scatter plots of L_{sky} and L_t at 90 and 135° relative azimuth from the different above- water systems versus weighted mean L_{sky} from above-water systems (RAMSES-A, -B, -C, HyperSAS-A, -B) for (**A**) mean L_{sky} for RAMSES at 90°, (**B**) mean L_{sky} for HyperSAS at 90°, (**C**) mean L_{sky} for RAMSES at 135°, (**D**) mean L_{sky} for HyperSAS at 135°, (**E**) mean L_t for RAMSES at 90°, (**F**) mean L_t for HyperSAS at 90°, (**G**) mean L_t for RAMSES at 135°, (**H**) mean L_t for HyperSAS at 135°. The different coloured points correspond to the different OLCI bands given in (**D**).

Appendix D

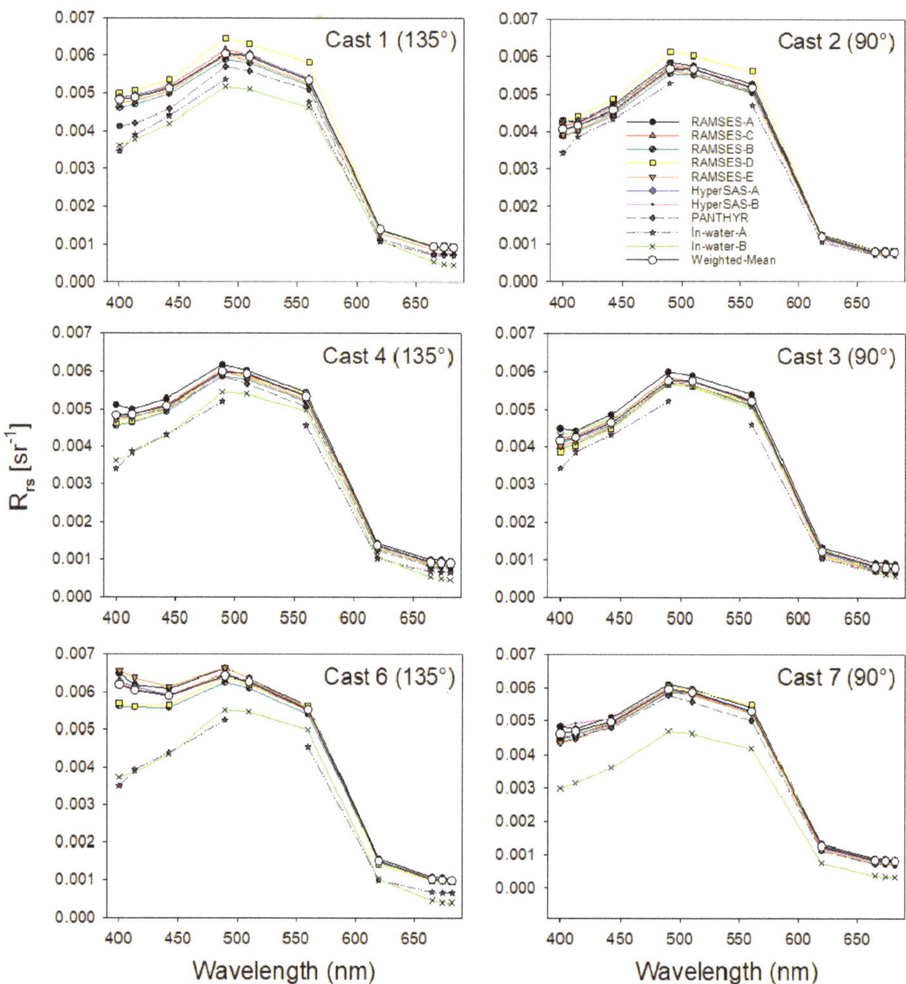

Figure A5. Variation in R_{rs} spectra of all instrument systems at OLCI bands for Casts with adjacent viewing angles at 135° and 90° on 13 July 2018.

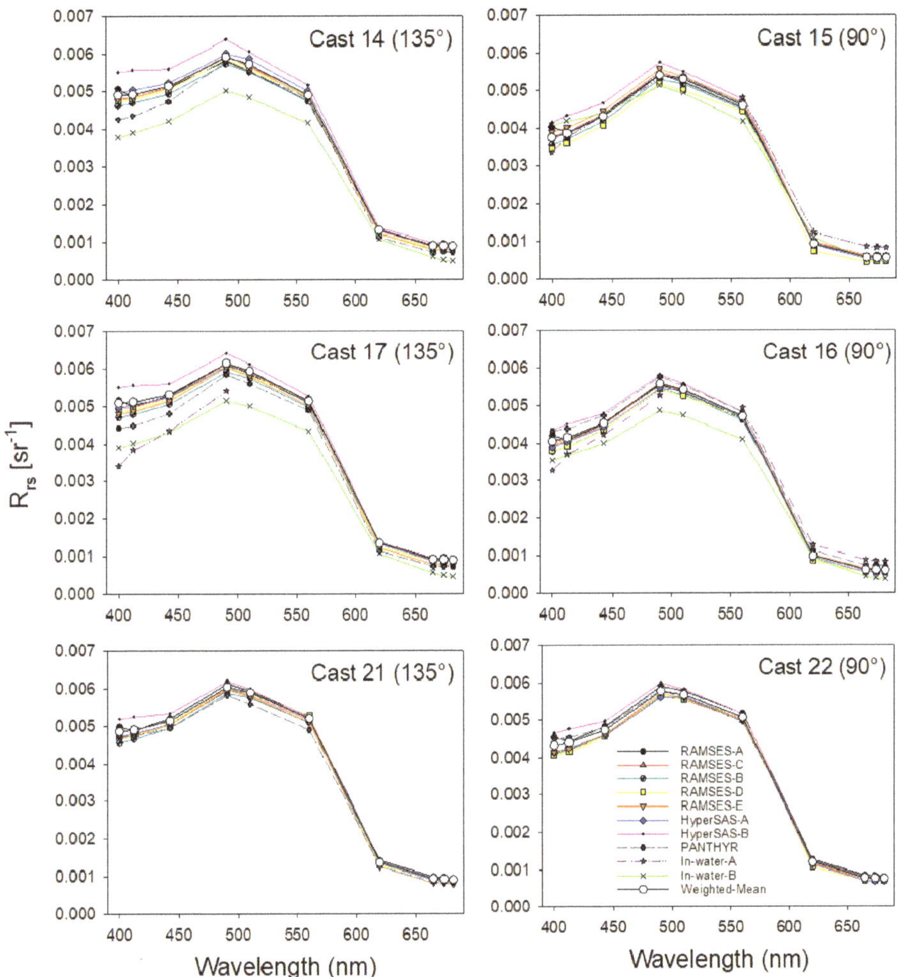

Figure A6. Variation in R_{rs} spectra of all instrument systems at OLCI bands for Casts with adjacent viewing angles at 135° and 90° on 14 July 2018.

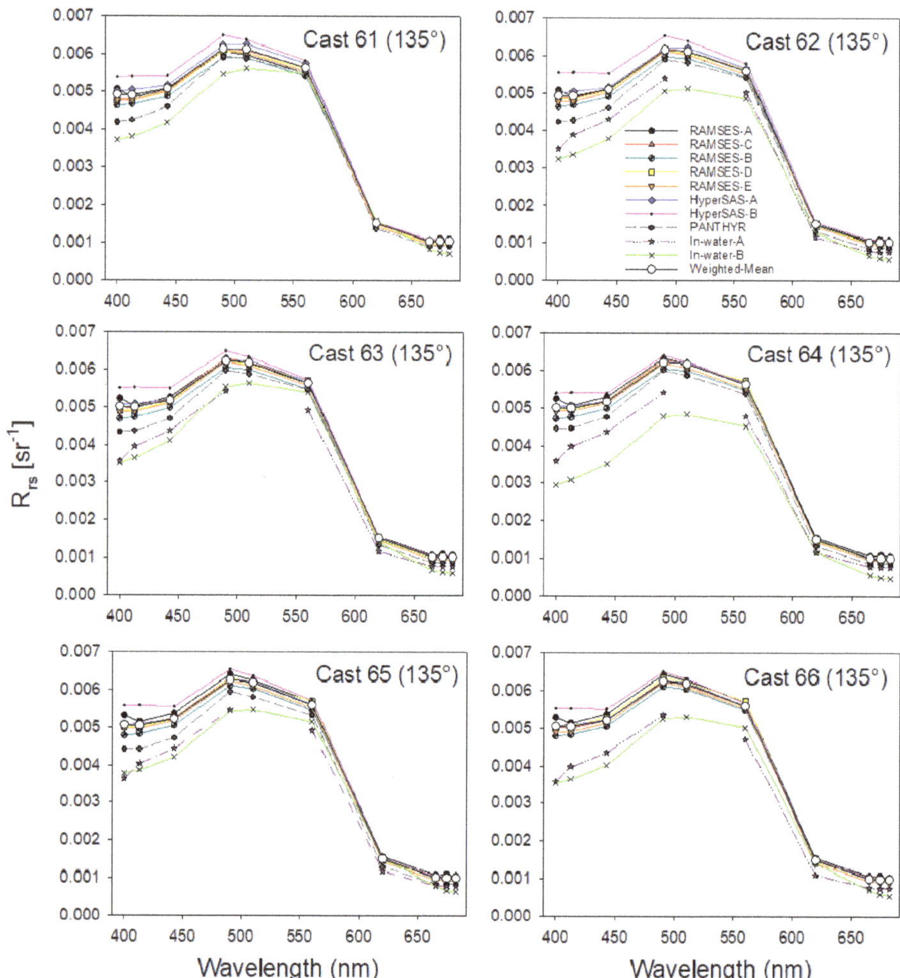

Figure A7. Variation in R_{rs} spectra of all instrument systems at OLCI bands on 14 July 2018. All Casts had a viewing angle of 135°.

References

1. Vabson, V.; Kuusk, J.; Ansko, I.; Vendt, R.; Alikas, K.; Ruddick, K.; Ansper, A.; Bresciani, M.; Burmester, H.; Costa, M.; et al. Laboratory intercomparison of radiometers used for satellite validation in the 400–900 nm range. *Remote Sens.* **2019**, *11*, 1101. [CrossRef]
2. JCGM. *Guide to the Expression of Uncertainty in Measurement—JCGM 100:2008 (GUM 1995 with Minor Corrections—Evaluation of Measurement Data*; Joint Committee for Guides in Metrology, 2008; pp. 1–134.
3. Zibordi, G.; Donlon, C.J. In Situ Measurement Strategies. In *Experimental Methods in the Physical Sciences*; Zibordi, G., Donlon, C.J., Parr, A.C., Eds.; Academic Press: Cambridge, MA, USA, 2014; Chapter 5; Volume 47, pp. 527–529.
4. Vabson, V.; Kuusk, J.; Ansko, I.; Vendt, R.; Alikas, K.; Ansper, A.; Bresciani, M.; Burmeister, H.; Costa, M.; D'Alimonte, D.; et al. Field intercomparison of radiometers used for satellite validation in the 400–900 nm range. *Remote Sens.* **2019**, *11*, 1129. [CrossRef]

5. Mueller, J.L. Above-water radiance and remote sensing reflectance measurements and analysis protocols. In *Ocean Optics Protocols for Satellite Ocean Color Sensor Validation*; Fargion, G.S., Mueller, J.L., Eds.; National Aeronautical and Space Administration: Washington, DC, USA, 2000.

6. Tilstone, G.H.; Moore, G.F.; Sorensen, K.; Doerffer, R.; Rottgers, R.; Ruddick, K.G.; Jorgensen, P.V.; Pasterkamp, R. Regional Validation of MERIS Chlorophyll Products in North Sea Coastal Waters: REVAMP Protocols. In *ENVISAT Validation Workshop*; European Space Agency: Frascatti, Italy, 2004; Available online: http://envisat.esa.int/workshops/mavt_2003/ (accessed on 13 May 2020).

7. Ruddick, K.G.; Voss, K.; Boss, E.; Castagna, A.; Frouin, R.; Gilerson, A.; Hieronymi, M.; Johnson, B.C.; Kuusk, J.; Lee, Z.; et al. A Review of Protocols for Fiducial Reference Measurements of Water-leaving Radiance for the Validation of Satellite Remote Sensing Data over Water. *Remote Sens.* **2019**, *11*, 2198. [CrossRef]

8. Ruddick, K.G.; Voss, K.; Banks, A.C.; Boss, E.; Castagna, A.; Frouin, R.; Hieronymi, M.; Jamet, C.; Johnson, B.C.; Kuusk, J.; et al. A Review of Protocols for Fiducial Reference Measurements of Downwelling Irradiance for the Validation of Satellite Remote Sensing Data over Water. *Remote Sens.* **2019**, *11*, 1742. [CrossRef]

9. Hooker, S.B.; Lazin, G.; Zibordi, G.; McClean, S. An evaluation of above- and in-water methods for determining water leaving radiances. *J. Atmos. Ocean. Tech.* **2002**, *19*, 486–515. [CrossRef]

10. Hooker, S.B.; McLean, S.; Sherman, J.; Small, M.; Lazin, G.; Zibordi, G.; Brown, J.W. *The Seventh SeaWiFS Intercalibration Round-Robin Experiment (SIRREX-7), March 1999. NASA Tech. Memo*; NASA/TM-2002-206892/VOL17; NASA Goddard Space Flight Center: Prince George's County, MD, USA, 2002; p. 78.

11. Johnson, B.C.; Yoon, H.W.; Bruce, S.S.; Shaw, P.-S.; Thompson, A.; Hooker, S.B.; Eplee, R.E.; Barnes, R.A., Jr.; Maritorena, S.; Mueller, J.L. *The fifth SeaWiFS Intercalibration Round Robin Experiment (SIRREX-5), July 1996; NASA Tech Memo. 1999-206892*; Hooker, S.B., Firestone, E.R., Eds.; NASA Goddard Space Flight Center: Prince George's County, MD, USA, 1999; Volume 7, p. 75.

12. Meister, G.; Abel, P.; McClain, C.; Barnes, R.; Fargion, G.; Cooper, J.; Davis, C.; Korwan, D.; Godin, M.; Maffione, R. *The First SIMBIOS Radiometric Intercomparison (SIMRIC-1), April–September 2001*; National Aeronautics and Space Administration, Goddard Space Flight: Greenbelt, MD, USA, 2002.

13. Meister, G.; Abel, P.; Carder, K.; Chapin, A.; Clark, D.; Cooper, J.; Davis, C.; English, D.; Fargion, G.; Feinholtz, M.; et al. *The Second SIMBIOS Radiometric Intercomparison (SIMRIC-2), March–November 2002*; NASA Technical Memorandum; National Aeronautics and Space Administration, Goddard Space Flight Center: Prince George's County, MD, USA, 2003; Volume 210006, pp. 1–65.

14. Tilstone, G.H.; Moore, G.F.; Sorensen, K.; Doerffer, R.; Rottgers, R.; Ruddick, K.G.; Pasterkamp, R. Protocols for the validation of MERIS products in Case 2 waters. In Proceedings of the ENVISAT MAVT Conference, Frascatti, Italy, 20–24 October 2003.

15. Moore, G.F.; Icely, J.D.; Kratzer, S. Field Inter-comparison and validation of in-water radiometer and sun photometers for MERIS validation. In Proceedings of the ESA Living Planet Symposium, Bergen, Norway, 28 June–2 July 2010.

16. Zibordi, G.; D'Alimonte, D.; van der Linde, D.; Berthon, J.-F.; Hooker, S.B.; Mueller, J.-L.; Lazin, G.; McLean, S. *The Eighth SeaWiFS Intercalibration Round-Robin Exercise (SIRREX-8), September–December 2001. NASA Tech Memo. 2002-206892*; Hooker, S.B., Firesetone, E.R., Eds.; NASA Goddard Space: Prince George's County, MD, USA, 2002; Volume 20, p. 39.

17. Zibordi, G.; Ruddick, K.; Ansko, I.; Moore, G.; Kratzer, S.; Icely, J.; Reinart, A. In situ determination of the remote sensing reflectance: An intercomparison. *Ocean Sci.* **2012**, *8*, 567–586. [CrossRef]

18. Mobley, C.D. Estimation of the remote-sensing reflectance from above-surface measurements. *Appl. Opt.* **1999**, *38*, 7442–7455. [CrossRef]

19. Deschamps, P.-Y.; Fougnie, B.; Frouin, R.; Lecomte, P.; Verwaerde, C. SIMBAD: A field radiometer for satellite ocean-color validation. *Appl. Opt.* **2004**, *43*, 4055–4069. [CrossRef]

20. Zibordi, G.; Melin, F.; Hooker, S.B.; D'Alimonte, D.; Holben, B. An autonomous above-water system for the validation of ocean color radiance data. *IEEE Trans. Geosc. Rem. Sens.* **2004**, *42*, 401–415. [CrossRef]

21. Hooker, S.B.; Lazin, G. *The SeaBOARR-99 Field Campaign*; NASA: Greenbelt, MD, USA, 2000; p. 46.

22. Waters, K.J.; Smith, R.C.; Lewis, M.R. Avoiding ship-induced light-field perturbation in the determination of oceanic optical properties. *Oceanography* **1990**, *3*, 18–21. [CrossRef]

23. Leymarie, E.; Penkerc'h, C.; Vellucci, V.; Lerebourg, C.; Antoine, D.; Boss, E.; Lewis, M.R.; d'Ortenzio, F.; Claustre, H. ProVal: A new autonomous profiling float for high quality radiometric measurements. *Front. Mar. Sci.* **2018**, *5*, 437. [CrossRef]

24. Clark, D.K.; Yarbrough, M.A.; Feinholz, M.; Flora, S.; Broenkow, W.; Kim, Y.S.; Johnson, B.C.; Brown, S.W.; Yuen, M.; Mueller, J.L. *MOBY, a Radiometric Buoy for Performance Monitoring and Vicarious Calibration of Satellite Ocean Color Sensors: Measurement and Data Analysis Protocols*; National Aeronautics and Space Administration, Goddard Space Flight: Greenbelt, MD, USA, 2003; Chapter 2.

25. Morel, A. Résultats expérimentaux concernant la pénétration de la lumière du jour dans les eaux Méditerranéennes. *Cah. Océanograph.* **1965**, *17*, 177–184.

26. Gordon, H.R. Normalized water-leaving radiance: Revisiting the influence of surface roughness. *Appl. Opt.* **2005**, *44*, 241–248. [CrossRef] [PubMed]

27. Thuillier, G.; Floyd, L.; Woods, T.N.; Cebula, R.; Hilsenrath, E.; Herse, M.; Labs, D. Solar irradiance reference spectra for two solar active levels. *Adv. Space Res.* **2004**, *34*, 256–261. [CrossRef]

28. Morel, A.; Antoine, D.; Gentili, B. Bidirectional reflectance of oceanic waters: Accounting for Raman emission and varying particle phase function. *Appl. Opt.* **2002**, *41*, 6289–6306. [CrossRef]

29. Berthon, J.F.; Zibordi, Z. Bio-optical relationships for the northern Adriatic Sea. *Int. J. Remote. Sens.* **2004**, *25*, 1527–1532. [CrossRef]

30. OLCI spectral response functions. In *Technical Guide*; European Space Agency, ESRIN: Frascatti, Italy, 2016.

31. Melin, F.; Scleps, G. Band shifting for ocean color multi-spectral reflectance data. *Opt. Express* **2015**, *23*, 2262–2279. [CrossRef]

32. Zibordi, G.; Melin, F.; Berthon, J.-F. Comparison of SeaWiFS, MODIS and MERIS radiometric products at a coastal site. *Geophys. Res. Lett.* **2006**, *33*, L06617. [CrossRef]

33. Zibordi, G.; Berthon, J.-F.; M'elin, F.; D'Alimonte, D.; Kaitala, S. Validation of satellite ocean color primary products at optically complex coastal sites: Northern Adriatic Sea, Northern Baltic Proper and Gulf of Finland. *Remote Sens. Environ.* **2009**, *113*, 2574–2591. [CrossRef]

34. Zibordi, G.; D'Alimonte, D.; Berthon, J.-F. An evaluation of depth resolution requirements for optical profiling in coastal waters. *J. Atmos. Ocean. Tech.* **2004**, *21*, 1059–1073. [CrossRef]

35. Zibordi, G.; Mélin, F.; Berthon, J.; Holben, B.; Slutsker, I.; Giles, D.; D'Alimonte, D.; Vandemark, D.; Feng, H.; Schuster, G. AERONET-OC: A Network for the Validation of Ocean Color Primary Products. *J. Atmos. Ocean. Tech.* **2009**, *26*, 1634–1651. [CrossRef]

36. Zibordi, G.; Hooker, S.B.; Berthon, J.-F.; D'Alimonte, D. Autonomous above–water radiance measurement from an offshore platform: A. field assessment. *J. Atmos. Ocean. Tech.* **2002**, *19*, 808–819. [CrossRef]

37. Artegiani, A.; Bregant, D.; Paschini, E.; Pinardi, N.; Raicich, F.; Russo, A. The Adriatic Sea general circulation. Part I: Air-sea interactions and water mass structure. *J. Phys. Oceanogr.* **1997**, *27*, 1492–1514. [CrossRef]

38. Zavatarelli, M.; Pinardi, N.; Kourafalou, V.H.; Maggiore, A. Diagnostic and prognostic model studies of the Adriatic Sea general circulation: Seasonal variability. *J. Geophys. Res.* **2002**, *107*, 3004. [CrossRef]

39. D'Alimonte, D.; Zibordi, G.; Kajiyama, T. Effects of integration time on in-water radiometric profiles. *Opt. Express* **2018**, *26*, 5908–5939. [CrossRef]

40. Uudeberg, K.; Ansko, I.; Põru, G.; Ansper, A.; Reinart, A. Using Optical Water Types to Monitor Changes in Optically Complex Inland and Coastal Waters. *Remote Sens.* **2019**, *11*, 2297. [CrossRef]

41. Hieronymi, M. Polarized reflectance and transmittance distribution functions of the ocean surface. *Opt. Express* **2016**, *24*, A1045–A1068. [CrossRef]

42. Theis, A. Validation of MERIS, MODIS and SeaWiFS Level-2 Products with Ground Based in-situ Measurements in Atlantic Case 1 Waters. Mater's Thesis, University of Bremen, Bremen, Germany, 2009; pp. 1–91. Available online: https://epic.awi.de/21447/1/The2009a.pdf (accessed on 13 May 2020).

43. Ruddick, K.; Cauwer, V.D.; van Mol, B. Use of the near infrared similarity spectrum for the quality control of remote sensing data. In Proceedings of the SPIE International conference on Remote Sensing of the Coastal Oceanic Environment, San Diego, CA, USA, 23 August 2005; Volume 5885, p. 588501.

44. Ruddick, K.; De Cauwer, V.; Park, Y.-J. Seaborne measurements of near infra-red water leaving reflectance: The similarity spectrum for turbid waters. *Limnol. Oceanogr.* **2006**, *51*, 1167–1179. [CrossRef]

45. Brewin, R.J.W.; Dall'Olmo, G.; Pardo, S.; van Dongen-Vogel, V.; Boss, E.S. Underway spectrophotometry along the Atlantic Meridional Transect reveals remarkable performance in satellite chlorophyll data. *Remote Sens. Environ.* **2016**, *183*, 82–97. [CrossRef]

46. Carswell, T.; Costa, M.; Young, E.; Komick, N.; Gower, J.; Sweeting, R. Evaluation of MODIS-Aqua Atmospheric Correction and Chlorophyll Products of Western North American Coastal Waters Based on 13 Years of Data. *Remote Sens.* **2017**, *9*, 1063. [CrossRef]

47. Vansteenwegen, D.; Ruddick, K.; Cattrijsse, A.; Vanhellemont, Q.; Beck, M. The Pan-and-Tilt Hyperspectral Radiometer System (PANTHYR) for Autonomous Satellite Validation Measurements—Prototype Design and Testing. *Remote Sens.* **2019**, *11*, 1360. [CrossRef]

48. Morrow, J.H.; Hooker, S.B.; Booth, C.R.; Bernhard, G.; Lind, R.N.; Brown, J.W. *Advances in Measuring the Apparent Optical Properties (AOPs) of Optically Complex Waters*; NASA/TM-2010-215856; NASA: Greenbelt, MD, USA, 2010.

49. Taylor, B.B.; Torrecilla, E.; Bernhardt, A.; Taylor, M.H.; Peeken, I.; Röttgers, R.; Piera, J.; Bracher, A. Bio-optical provinces in the eastern Atlantic Ocean and their biogeographical relevance. *Biogeosciences* **2011**, *8*, 3609–3629. [CrossRef]

50. Mueller, J.L.; Giulietta, S.; Fargion, C.; McClain, R. *Ocean Optics Protocols For Satellite Ocean Color Sensor Validation*; Revision 5; Biogeochemical and Bio-Optical Measurements and Data Analysis Protocols National Aeronautical and Space Administration: Washington, DC, USA, 2003; Volume V.

51. Ebuchi, N.; Kizu, S. Probability Distribution of Surface Wave Slope Derived Using Sun Glitter Images from Geostationary Meteorological Satellite and Surface Vector Winds from Scatterometers. *J. Oceanogr.* **2002**, *58*, 477–486. [CrossRef]

52. Zibordi, G. Experimental evaluation of theoretical sea surface reflectance factors relevant to above-water radiometry. *Opt. Express* **2016**, *24*, A446. [CrossRef] [PubMed]

53. Mobley, C.D. Polarized reflectance and transmittance properties of windblown sea surfaces. *Appl. Opt.* **2015**, *54*, 4828–4849. [CrossRef]

54. Bracher, A.; Taylor, M.; Taylor, B.B.; Dinter, T.; Röttgers, R.; Steinmetz, F. Using empirical orthogonal functions derived from remote sensing reflectance for the prediction of phytoplankton pigment concentrations. *Ocean Sci.* **2015**, *11*, 139–158. [CrossRef]

55. Mueller, J.L.; Fargion, G.S.; McClain, C.R. *Ocean Optics Protocols for Satellite Ocean Color Sensor Validation*; Revision 4; Radiometric Measurements and Data Analysis Protocols; NASA Goddard Space Flight Center: Greenbelt, MD, USA, 2003; Volume III.

56. IOCCG Protocol Series; Protocols for Satellite Ocean Colour Data Validation: In situ Optical Radiometry. In *IOCCG Ocean Optics and Biogeochemistry Protocols for Satellite Ocean Colour Sensor Validation*; Zibordi, G.; Voss, K.J.; Johnson, B.C.; Mueller, J.L. (Eds.) IOCCG: Dartmouth, NS, Canada, 2019; Volume 3.0.

57. Gould, R.W.; Arnone, R.A.; Sydor, M. Absorption, scattering, and remote sensing reflectance relationships in coastal waters: Testing a new inversion algorithm. *J. Coastal Res.* **2001**, *17*, 328–341.

58. Hooker, S.B.; Zibordi, G.; Berthon, J.-F.; Brown, J.W. Above-Water Radiometry in shallow coastal waters. *Appl. Opt.* **2004**, *43*, 4254–4268. [CrossRef]

59. Hooker, S.B.; Morrow, J.H.; Matsuoka, A. Apparent optical properties of the Canadian Beaufort Sea—Part 2: The 1% and 1 cm perspective in deriving and validating AOP data products. *Biogeosciences* **2013**, *10*, 4511–4527. [CrossRef]

60. Morel, A.; Maritorena, S. Bio-optical properties of oceanic waters: A reappraisal. *J. Geophys. Res.* **2001**, *106*, 7163–7180. [CrossRef]

61. Gregg, W.W.; Carder, K.L. A simple spectral solar irradiance model for cloudless maritime atmospheres. *Limnol. Oceanogr.* **1990**, *35*, 1657–1675. [CrossRef]

62. Stramski, D.; Reynolds, R.A.; Babin, M.; Kaczmarek, S.; Lewis, M.R.; Röttgers, R.; Sciandra, A.; Stramska, M.; Twardowski, M.S.; Franz, B.A.; et al. Relationships between the surfaceconcentration of particulate organic carbon and optical properties in the eastern South Pacific and eastern Atlantic Oceans. *Biogeosciences* **2008**, *5*, 171–201. [CrossRef]

63. Chance, J.; Kurucz, R.L. An improved high-resolution solar reference spectrum for earth's atmosphere measurements in the ultraviolet, visible, and near infrared. *J. Quant. Spectrosc. Radiat. Transf.* **2010**, *111*, 1289–1295. [CrossRef]

64. Barlow, R.G.; Cummings, D.G.; Gibb, S.W. Improved resolution of mono-and divinyl chlorophylls a and b and zeaxanthin and lutein in phytoplankton extracts using reverse phase C-8 HPLC. *Mar. Ecol. Prog. Ser.* **1997**, *161*, 303–307. [CrossRef]

65. Aiken, J.; Pradhan, Y.; Barlow, R.; Lavender, S.; Poulton, A.; Holligan, P.; Hardman-Mountford, N. Phytoplankton pigments and functional types in the Atlantic Ocean: A decadal assessment, 1995–2005. *Deep-Sea Res. Part II Top. Stud. Oceanogr.* **2009**, *56*, 899–917. [CrossRef]
66. Mekaoui, S.; Zibordi, G. Cosine error for a class of hyperspectral irradiance sensors. *Metrologia* **2013**, *50*, 187. [CrossRef]
67. Sea-Bird Scientific. Specifications for HyperOCR Radiometer. 2019. Available online: https://www.seabird. com/hyperspectral-radiometers/hyperocr-radiometer/family?productCategoryId=54627869935 (accessed on 13 May 2020).
68. TriOS. RAMSES Technische Spezifikationen, TriOS Mess- und Datentechnik. 2019. Available online: https://www.trios.de/ramses.html (accessed on 13 May 2020).
69. Zibordi, G.; Talone, M.; Jankowski, L. Response to Temperature of a Class of In Situ Hyperspectral Radiometers. *J. Atmos. Ocean. Technol.* **2017**, *34*, 1795–1805. [CrossRef]
70. Burggraaff, O. Biases from incorrect reflectance convolution. *Opt. Express* **2020**, *28*, 13801. [CrossRef]
71. Gordon, H.; Ding, K. Self-shading of in-water optical measurements. *Limnol. Oceanogr.* **1992**, *37*, 491–500.
72. Talone, M.; Zibordi, G. Spectral assessment of deployment platform perturbations on above-water radiometry. *Opt. Express* **2019**, *27*, A878–A889. [CrossRef]
73. Białek, A.; Douglas, S.; Kuusk, J.; Ansko, I.; Vabson, V.; Vendt, R.; Casal, T. Example of Monte Carlo Method Uncertainty Evaluation for Above-Water Ocean Colour Radiometry. *Remote Sens.* **2020**, *12*, 780. [CrossRef]
74. Alikas, K.; Vabson, V.; Ansko, I.; Tilstone, G.H.; Dall'Olmo, G.; Vendt, R.; Donlon, C.; Casal, T. Comparison of above-water Seabird and TriOS radiometers along an Atlantic Meridional Transect. *Remote Sens.* **2020**. (in revision).

remote sensing

Article

Example of Monte Carlo Method Uncertainty Evaluation for Above-Water Ocean Colour Radiometry

Agnieszka Białek [1,*], Sarah Douglas [1], Joel Kuusk [2], Ilmar Ansko [2], Viktor Vabson [2], Riho Vendt [2] and Tânia Casal [3]

[1] National Physical Laboratory, Teddington TW11 0LW, UK; sarah.douglas@npl.co.uk
[2] Tartu Observatory, University of Tartu, 61602 Tõravere, Estonia; joel.kuusk@ut.ee (J.K.);
     ilmar.ansko@ut.ee (I.A.); viktor.vabson@ut.ee (V.V.); riho.vendt@ut.ee (R.V.)
[3] European Space Agency, 2201 AZ Noordwijk, The Netherlands; tania.casal@esa.int
*   Correspondence: agnieszka.bialek@npl.co.uk; Tel.: +44-208-943-6716

Received: 27 December 2019; Accepted: 25 February 2020; Published: 29 February 2020

Abstract: We describe a method to evaluate an uncertainly budget for the in situ Ocean Colour Radiometric measurements. A Monte Carlo approach is chosen to propagate the measurement uncertainty inputs through the measurements model. The measurement model is designed to address instrument characteristics and uncertainty associated with them. We present the results for a particular example when the radiometers were fully characterised and then use the same data to show a case when such characterisation is missing. This, depending on the measurement and the wavelength, can increase the uncertainty value significantly; for example, the downwelling irradiance at 442.5 nm with fully characterised instruments can reach uncertainty values of 1%, but for the instruments without such characterisation, that value could increase to almost 7%. The uncertainty values presented in this paper are not final, as some of the environmental contributors were not fully evaluated. The main conclusion of this work are the significance of thoughtful instrument characterisation and correction for the most significant uncertainty contributions in order to achieve a lower measurements uncertainty value.

Keywords: ocean colour; downwelling irradiance; water-leaving radiance; satellite validation; Fiducial Reference Measurements

---

1. Introduction

One of the main successes of the Coastal Zone Color Scanner (CZCS) [1] were the first meaningful ocean colour satellite measurements, particularly considering it was a proof of concept instrument. At the same time, ocean colour measurements have proven to be among the most challenging of all Earth observation measurements. This is due to the small fraction of the ocean related signal that is present in the top of atmosphere satellite readings. An enormous amount of research and measurements went into supporting missions like the Sea-viewing Wide Field-of-view Sensor (SeaWIFS) [2] via the Seventh SeaWiFS Intercalibration Round-Robin Experiment (SIRREX) [3] and the Second Intercomparison and Merger for Interdisciplinary Ocean Studies (SIMBIOS) [4] programs. The second produced further support for many more sensors including for example the Moderate Resolution Imaging Spectrometer (MODIS) [5] and the Medium Resolution Imaging Spectrometer (MERIS) [6]. All that work was summarised in the Ocean Optics Protocols for Satellite Ocean Color Sensor Validation; of particular interest for this study are volumes I to III and VI [7–10] and a number of publications [11–14]. All these work stress a particular need for in situ validation and vicarious calibration in ocean colour radiometry (OCR). With the advances in technology and the use of hyperspectral sensors in situ, some aspects of this work need to be repeated. Particularly since significant differences (up to 25%) are found when compared to theoretical modelling and conventional multispectral instruments [15]. It seems that the well known and well characterised multispectral instruments that are described in the NASA protocols are not utilised as much nowadays, as the world focuses on hyperspectal instruments, which provide greater spectral information. However, generally, they are less well characterised but require a wider array of characterisation tests (e.g., spectral stray light). Similarly with the new generation of Ocean Colour sensors like the Sentinel 3 series [16] and planned PACE [17] mission there is a need for even better quality measurements. The old Ocean Colour accuracy requirements, set at 5%, are now being pushed to achieve a level of 3%. In order to achieve this very ambitious target we need to better understand the individual uncertainty components that effect ocean colour measurements as well as their contribution to the final products.

The Fiducial Reference Measurements for Satellite Ocean Colour (FRM4SOC) project, with funding from ESA, aimed to provide support for evaluating and improving the state of the art in satellite ocean colour validation through a series of comparisons under the auspices of the Committee on Earth Observation Satellites (CEOS). The project makes a fundamental contribution to the European system for monitoring the Earth (Copernicus) by ensuring high quality ground-based Fiducial Reference Measurements (FRM) for ocean colour radiometry for use in validation of ocean colour products from missions like Sentinel-3 Ocean Colour and Land Imager (OLCI) [16] and Sentinel-2 Multi Spectral Imager (MSI) [18]. The main aim of FRM4SOC is to establish and maintain SI traceability of ground-based FRM for OCR. Specifically the project develops, documents, implements and reports OCR measurement procedures and protocols. FRM4SOC project activities addressed the SI traceability, comparability and uncertainty evaluation of measurements at several stages of the processing chain. These included comparison of radiance and irradiance sources, that are used for radiometric calibration of the in situ measurement radiometers. Additionally, an indoor comparison of the radiometers took place in controlled laboratory conditions with stable radiance and irradiance sources [19]. The same instruments were taken out of the laboratory to outdoor conditions where they were compared again at two different locations. The first comparison was in Estonia, for measurements of a lake, where the environmental conditions were unfortunately not optimal for above-water radiometry measurements [20]. Nevertheless, this did show the importance of instrument characterisation and the effects due to changeable measurement conditions. The uncertainty budget for the indoor and outdoor exercises were evaluated [20] as relative uncertainty and there was also an attempt to address the variability of the sensors used during the comparisons. The second field comparison was at sea, at the Acqua Alta Oceanographic Tower (AAOT) and used the same instruments. The comparison took place in 2018 and, in this case, the environmental conditions were good.

This paper presents uncertainty budgets for FRM4SOC FRM OCR systems used to validate satellite OCR products. The aim here is to provide an uncertainty budget for in situ ocean colour products. The data used in this evaluation was taken from the second of the above field intercomparison exercises. The instruments selected for this analysis are RAMSES Hyperspectral Radiometers that belong to the University of Tartu. These instruments, in addition to radiometric calibration, had the most optical characterisation tests performed [20]. The characterisation test results were made available for further analysis and allowed correction coefficients to be applied to the data to allow for differences between ideal and non-ideal instrument performance, such as cosine response of a diffuser or detector spectral stray light. In particular, these data enabled the development of a few case studies for uncertainty evaluation with and without corrections applied. The radiometers were used for above-water measurements. Since the FRM4SOC project is intended to be useful for validation of ocean colour satellite products, the hyperspectral results were convoluted with an OLCI spectral band response and thus the results are reported for the Sentinel-3 spectral bands.

The first comparison of OCR measurement at AAOT was performed in 2010 [21] and showed a higher than expected variability of results, often higher than quoted measurements uncertainties. This exercise clearly indicated the need for more frequent comparisons and a more robust measurement uncertainty evaluation. The evaluation of in situ uncertainty is rarely reported, except for the measurement network such as AERONET-OC [22,23]. However, the AERONET instruments are multispectral rather than hyperspectral, which are the most commonly used during validation cruises. Hyperspectral instruments tend to have more instrument characteristics that might affect instrument performance including stray light [24,25] and detector linearity [26], in addition to temperature response [27,28] and cosine diffuser response [29] that are common for both radiometer types. These features of hyperspectral instruments are incorporated into the in situ uncertainty budget presented here. The final products of that evaluation are downwelling irradiance and water-leaving radiance at given viewing configurations, which are all in situ derived quantities. The next step of the in situ data processing leads to remote sensing reflectance or fully normalised water-leaving radiance. This step is not included here, as this is mostly related to the modelling rather than pure measurement errors. We hope to address the next step in future activities, after a number of various models have been used for in situ data modelling, which will allow these methods to be compared and uncertainty results evaluated. The uncertainties are evaluated using the Guide to the expression of Uncertainty in Measurement (GUM) [30] methodology, in particular the Monte Carlo Method (MCM) [31] presented in Supplement 1 to the GUM.

2. Methods

The main purpose of this study is to demonstrate a method of how to determine the uncertainty in downwelling irradiance and water-leaving radiance. Specifically, we investigate two scenarios in which the biases are corrected (the ideal case), and the biases are not corrected but treated as an uncertainty contributor (the non-ideal case). In addition, we present how the non-ideal case shows an under-estimation of measurement uncertainty because the biases are not corrected and the errors due to a lack of that correction are not accounted for. The required data for this activity include downwelling irradiance, downwelling radiance and upwelling radiance as well as all correction factors, the Fresnel reflectance of the water surface, and the fraction of diffuse to direct radiation at the time of measurement. This section provides a description of the lab and field sites and how each parameter was measured, along with how an estimation of uncertainty in the measurement was deduced. Towards the end of the section, the band integration step is described in more detail as is the error correlation. Other effects that have not been taken into account are also described.

2.1. Instruments Used

Three RAMSES instruments (ACC-VIS measuring irradiance and two ARC-VIS measuring radiance) manufactured by TriOS Mess- und Datentechnik GmbH, Rastede, Germany, and supplied by the optical radiometry laboratory of the University of Tartu were used in this activity. The technical parameters are summarised in Table 1.

Table 1. Technical parameters of the instruments used.

Parameter	RAMSES Instruments
Field of View	7°/cos
Manual integration time	yes
Adaptive integration time	yes
Min. integration time, ms	4
Max. integration time, ms	4096
Min. sampling interval, s	5
Internal shutter	no
Number of channels	256
Wavelength range, nm	320–1050
Spectral sampling interval step, nm	3.3
Spectral resolution, nm	10

2.2. Laboratory Calibration

Radiometric calibration, non-linearity, spectral stray light and the cosine correction measurements were performed at a room temperature of (21.5 ± 1.5) °C in a cleanroom equivalent to ISO 14644 Class 8 located at the optical radiometry laboratory of Tartu Observatory, Estonia between 2 and 7 May 2017.

2.3. Field Site AERONET-OC Station

The characterisation of the instruments was carried out as part of the comparison exercise at Lake Kääriku, Estonia. However, due to variable illumination conditions caused by cumulus clouds, few casts were deemed usable and of these, none measured irradiance, upwelling radiance and downwelling radiance simultaneously, hence the decision to use the same instruments at a later campaign site at AAOT near Venice (45.314°N, 12.508°E).

The Aerosol Robotic Network (AERONET) is a system of ground-based sun photometers which support atmospheric studies by providing a readily accessible public domain database of atmospheric properties [32]. The sub network dedicated to Ocean Colour validation is called AERONET-OC [22] and in addition to atmospheric measurements provides water-leaving radiance measurements. The aerosol optical depth, water vapour content and total column ozone values used in this study were measured by the AAOT AERONET-OC station. This is located on the same tower used as the radiance and irradiance measurements (see Figure 1).

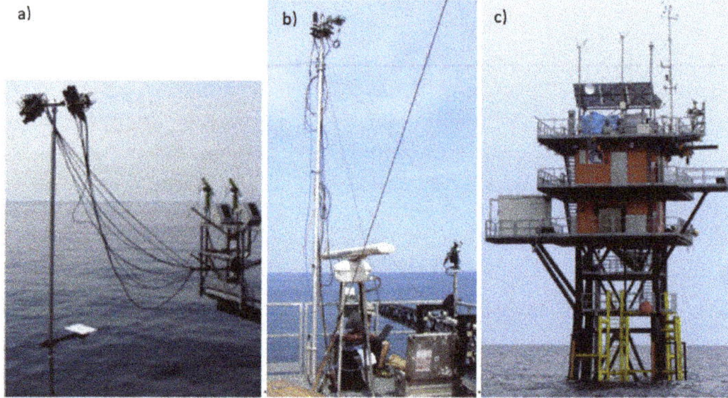

Figure 1. (**a**) Upwelling and downwelling radiance sensors mounted on a purpose-built mast; (**b**) Irradiance sensor mounted on a 6.6 m mast to minimise interference from overhead equipment on the tower; (**c**) The Acqua Alta Oceanographic Tower (AAOT). The radiance sensors were located on the middle platform on the right of the photo; the irradiance sensors were located on the top level. This tower is the site of the AAOT AERONET station which can be seen in this image on the highest level.

2.4. Field Measurements

Two sets of observations of irradiance and radiance were taken at the AAOT, one on the 13 July 2018 between 11:00 and 11:04 ('cast 1') and another on the 14 July 2018 between 11:40 and 11:44 local time ('cast 2'). At these times, downwelling irradiance, downwelling radiance and upwelling radiance were all measured simultaneously. Measurements were performed at the AAOT under near ideal conditions, on the same deployment platform and frame (see Figure 1), under clear sky conditions, sun zenith angles of approximately 24° and moderately low sea state with wind speed of 3.1 m s^{-1} and 0.5 m s^{-1} for each cast. The average chlorophyll content was Chl = 0.77 mg m^{-3} and absorption of the coloured dissolved organic matter was a CDOM (442 nm) = 0.12 m^{-1}.

2.5. Uncertainty Evaluation Methodology

The uncertainties are calculated for the two in situ measurement products: downwelling irradiance, E_d, and water-leaving radiance, L_w that are convoluted to S3 OLCI spectral bands as the final product of interest. The same in situ input data can be used for validation of other satellite sensors as they come from hyperspectral instruments, thus derived radiance and irradiance values can be convoluted with any spectral bands of interest. The wavelength dependence is addressed but omitted in the equations below for better readability. The measurement function for downwelling irradiance is

$$E_d(\theta_s) = f_{dirr}E_{OLCI}(\theta_s)f_{cos} + (1 - f_{dirr})E_{OLCI}(\theta_s)f_{cos_h} + 0. \tag{1}$$

This measurement equation for downwelling irradiance, E_d, at given Sun zenith angle, θ_s is split into two components, one for direct solar irradiance and the other for diffuse sky irradiance. The first term includes the direct-to-total-fraction of irradiance, f_{dirr}, the cosine response for direct irradiance, f_{cos}, and the total measured irradiance, E_{OLCI}, which has already been convolved to OLCI bands and has had various correction factors applied to it. These corrections are further explained in Section 2.5.4. The second term contains, f_{cos_h}, rather than f_{cos}. f_{cos_h} is the cosine response correction for the full hemispherical diffuse irradiance. The fraction of direct to total irradiance, f_{dirr}, is also shifted to provide the diffuse to total fraction instead. The term 0 is used as a placeholder for any currently undefined model error.

The measurement function for water-leaving radiance is

$$L_w(\theta, \Delta\phi, \theta_s) = L_{OLCI,u}(\theta, \Delta\phi, \theta_s) - \rho(\theta, \Delta\phi, \theta_s, W)L_{OLCI,d}(\theta', \Delta\phi, \theta_s) + 0. \tag{2}$$

For water-leaving radiance, L_w, the measurement setup consists of two radiometers, one pointing upwards towards the sky with the zenith angle, $\theta' = 140°$ and the other downwards towards the water, $\theta = 40°$. They are both at the same azimuth angle i.e., the difference between the sun and the sensor ($\Delta\phi = 90°$ or $\Delta\phi = 135°$). The upward-facing instrument measures the downwelling radiance from the sky (marked here with superscript d for downwelling) while the down-facing instrument measures the upwelling radiance from the water (upwelling, u). In the measurement equation, $L_{OLCI,u}$ is defined as the upwelling radiance from the water which has had the same corrections factors applied as the downwelling irradiance with the addition of a polarisation correction. Polarisation effects can be assumed to be negligible in irradiance sensors due to their cosine diffuser. The measured values have also been convoluted to OLCI bands. $L_{OLCI,d}$, is the equivalent for downwelling radiance. The upwelling radiance includes both light from below the surface (water-leaving radiance) and light which is reflected from the surface of the water. The reflectance of the water is characterised by the Fresnel reflectance, ρ that is a function of the sensor viewing geometry ($\theta, \Delta\phi$), solar zenith angle, θ_s and wind speed, W.

Figures 2 and 3 illustrate the error contributions for the measurement equations for downwelling irradiance and water-leaving radiance respectively. These diagrams were first designed in the Horizon 2020 FIDUCEO project [33] to show the sources of uncertainty from their origin through to the measurement equation. The outer labels describe the effects which cause the corresponding uncertainty.

The colour coding of the error contributions presented in Figures 2 and 3 can be treated as the classification of the errors due to instrument, environment and applied modeling, although that classification in not always straightforward. All yellow highlighted boxes in both figures represent a class of instruments related contributors related to instruments signal, S, absolute radiometric calibration, (c_{cal}), and instruments characteristics such as temperature, (c_T), detector linearity, (c_{lin}), spectral stray light, (c_{stray}) etc. In addition in Figure 2 pinkish boxes address the errors related to the cosine diffuser angular response. The turquoise box represents the spectral band convolution step, thus this can be classified as modelling class as uses the satellite spectral response function. For the irradiance case the green box represent the fraction of the direct to total irradiance and this is an example where the environmental and modeling errors contributions are tangled together, as some input to the model uses environmental condition such as actual value of AOD (see Section 2.5.10 for more details). Similar situation applies to the red box in Figure 3, for water-leaving radiance to estimate Fresnel reflectance values, this is again a combination of environmental errors such as wind speed and modeling errors that are used to derive this value. Blue box, in both figures, with +0 term represents a placeholder for any other environmental effects that are not fully accounted for in the current version of the uncertainty budget, this can be the structure shadings effect.

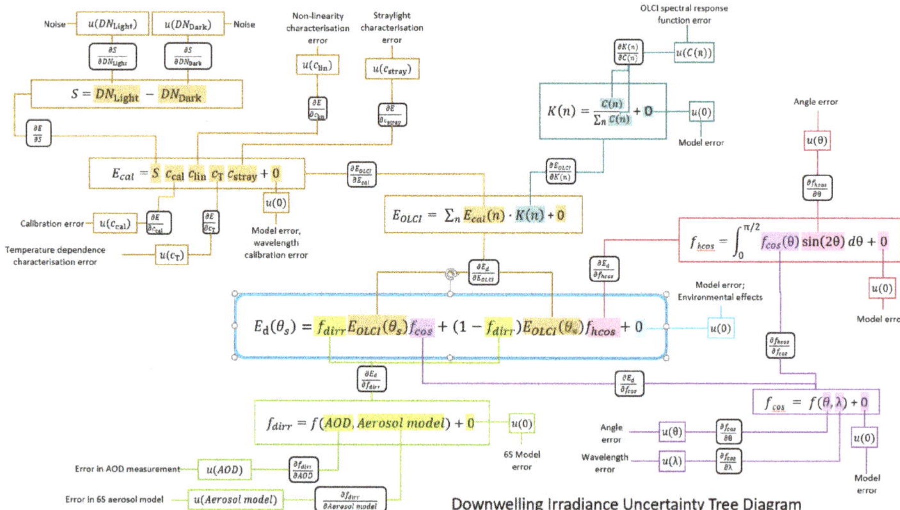

Figure 2. Uncertainty tree diagram for downwelling irradiance.

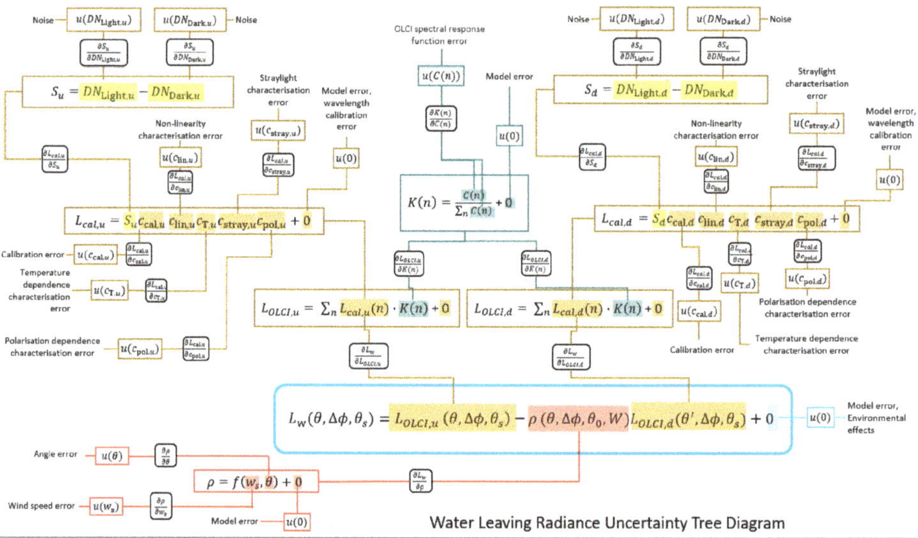

Figure 3. Uncertainty tree diagram for water-leaving radiance.

2.5.1. Practical Implementation of GUM/MCM

GUM [30] provides a framework for how to determine and express the uncertainty of the measured value of a given quantity which is being measured. The analytical method known as the Law of Propagation of Uncertainties is given by:

$$u^2(y) = \sum_{(i=1)}^{N} c_i^2 u^2(x_i) + 2 \sum_{i=1}^{N-1} \sum_{j=i}^{N} c_i c_j u(x_i, x_j), \tag{3}$$

where $u^2(y)$ is the combined variance, $u(y)$ is the combined standard uncertainty, which is the positive square root of the combined variance, c_i are the sensitivity coefficients of the measurand to the input quantities, $\dfrac{\partial f}{\partial x}$, $u(x_i)$ are the uncertainties of the input quantities, $u(x_i, x_j)$ are the covariance between input quantities. The covariance between input quantities may be written as,

$$u(x_i, x_j) = r(x_i, x_j) u(x_i) u(x_j), \tag{4}$$

where $r(x_i, x_j)$ is the error-correlation coefficient between x_i and x_j (note: $r(x_i, x_j) = r(x_j, x_i)$)

The analytical method can become difficult to apply on complex functions with many correlated input parameters where the calculation of sensitivity coefficients is not straightforward. Monte Carlo Methods (MCM) for uncertainty estimation are recognised, accepted and summarised in the GUM supplement [31]. MCM is a numerical method that requires a distinct probability distribution function (PDF) for each of the input components; if input components are correlated, then the joint PDF and the measurement equation are required. The MCM will then run a large number of numerical calculations of the measurement equation, with each iteration using a random choice of each of the inputs from the available range defined by the relevant PDF. The large number of output values calculated using different input values at each iteration provides the uncertainty of the output value with its PDF.

Both methods are allowed in the GUM, with the main document [30] focussing on the use of analytical methods to evaluate the propagation of uncertainty and the MCM (detailed in Supplement 1) [31] using a numerical approach. To evaluate the measurement uncertainty, regardless of the chosen method, several steps need to be taken:

1. A measurement function must be defined.
2. All sources of uncertainty must be identified.
3. Then according to the Law of Propagation of Uncertainty (see Equation (3)) the sensitivity coefficients must be calculated. Nominally, sensitivity coefficients are a partial derivative of a measurement function and a given contributor. For complex functions, it might be impossible to calculate these coefficients analytically. This step is not needed for Monte Carlo approach.
4. Using the analytical method, all inputs with non-Gaussian PDFs must be converted to that shape. To do this, divisors are specified for typical shapes. For example, to convert a rectangular PDF to its equivalent in normal distribution, the uncertainty value is divided by $\sqrt{3}$. This step is not needed when the MCM is used, as the original PDF can be propagated through the model.
5. If any of the input quantities are correlated with each other, an error covariance matrix is necessary (second part of Equation (3)). To combine uncertainties, Equation (3) is used for the analytical approach. In the MCM approach, the measurement function is run many times, each time using randomly selected inputs from previously defined PDFs of the input quantities.
6. The results of the analytical approach by default have a Gaussian distribution and when quoted as the output of Equation (3) called the standard uncertainty, which means it has a coverage factor $k = 1$, or 1σ defined as one standard deviation from the mean assuming a normal distribution function. This information expresses the confidence level at 68% that the estimated value is within its quoted uncertainty. The result in the MCM is a PDF of the output quantity and depending on its shape the uncertainty is defined as the standard deviation of this distribution (Gaussian PDF) or by the upper and lower limits that define 68% coverage of that distribution.

7. For some applications higher confidence is necessary, and then the standard uncertainty is expanded to $k = 2$ for 95% coverage or $k = 3$ for 99% coverage. This is called the expanded uncertainty.

Practical implementation of these steps can vary depending on application. To propagate uncertainty for the measurands of interest for this project (E_d and L_w) we propose the following approach:

1. Measurement functions are defined based on the tree diagrams including all inputs (see Figure 2 and Figure 3). The inputs are defined as quantities that can have an influence on the measurand.
2. All inputs have their standard uncertainty identified in terms of magnitude (value) and PDF shape.
3. The MCM method is chosen so we run Equations (1) and (2) a large number of times (10^4 in this case).
4. We handle the correlation between some input quantities (for example, the absolute radiometric calibration coefficients of the different instruments) by treating them as systematic contributions, thus the draws from that distribution are not randomised.
5. The final uncertainty value is derived from the resultant PDF.
6. All uncertainties are reported with a $k = 1$ coverage factor.

2.5.2. Error Correlation

A common assumption made during evaluation of measurement uncertainty is the independence of the input components which simplifies Equation (3) to the first term only and leads to the well-known sum of squares expression. However, this might lead to under- or over-estimation of the final uncertainty values if not true.

It might be difficult to evaluate partial error correlation of various components. A practical method to address this is splitting the uncertainty contributors into two categories a) uncorrelated (random) and b) fully correlated (systematic) for which the correlation coefficient is equal to 1.

When applying the MCM using a distribution for a parameter with a particular mean and standard deviation, each draw will be randomly selected. In the case of uncorrelated inputs, all the draws together will make up a PDF with the correct mean and standard deviation. However, it is possible to allow for correlated distributions by selecting identical draws for two or more distributions. It is important to consider this in the MCM since this leads to the correct propagation of systematic errors created by the correlated nature of two parameters.

An obvious example of a systematic error (leading to correlations) is uncertainty in absolute radiometric calibration. For example, in the above-water radiometry situation considered here, the two radiometers that measure radiance are both calibrated against the same radiance source and although the calibration uncertainty has a small contribution from the random errors related to the noise during the measurements, it is dominated by the systematic components related to the radiometric scale realisation; hence the radiometric calibrations are highly correlated. Here, the PDFs representing the radiometric calibration of each instrument should be correlated, meaning the same draws which make up the PDF would be taken for each.

As another example, consider the calculation of the final OCR product for a given OLCI band, in which the values from several pixels are used and convoluted with the OLCI band spectral response function (SRF) to provide the final band integrated value. To avoid underestimation of uncertainty in the integrated value it is necessary to consider if a given uncertainty contributor has a systematic effect on all pixels used for one band.

Each input quantity of the measurement equations is considered whether it is most suitable having a distribution which is assigned as uncorrelated with other distributions or whether the distribution should be correlated with another distribution. For example, for the seven distributions representing the temperature correction in downwelling irradiance (one for each of the seven OLCI bands of interest), each one is correlated with all the others.

2.5.3. Bias Correction

The true value of a measurement can never be exactly known; only an estimate can be made which is as good as the instrument and method used. Therefore, an error (bias) will always exist between the measured and best estimate value.

Another assumption, this time made by the creators of the GUM is that when all identified biases are corrected, only a residual uncertainty in that correction is propagated. In practice, especially for any field measurements, these corrections are not applied and often are unknown.

2.5.4. Data Processing Steps

The processing follows the structure of the measurement equations in Figures 2 and 3. Firstly, all parameters are assigned a PDF and a decision is made over whether any correlation should be assigned for this parameter. Following this, the correction factors (calibration, non-linearity, temperature and stray light and polarisation for radiance only) are applied to the signal. The next step involves a band integration which is performed to convolve the hyperspectral instrument wavelengths with seven OLCI bands. The final step is to calculate the downwelling irradiance and water-leaving radiance respectively using the defined and calculated parameters in the final measurement equations (Equations (1) and (2)). Each PDF is assigned 10,000 draws for this MCM process.

2.5.5. Instrument Signal (S)

The instrument signal is defined as the digital numbers when the sensor is exposed to light conditions (DN_{light}) subtracted by the digital numbers in dark conditions (DN_{dark}). The RAMSES radiometers do not have a mechanical internal shutter. Instead, black-painted pixels on the photodiode array are used to derive the dark signal and electronical drifts [24]. The values obtained by NPL for further analysis were already converted to radiometric values. Nevertheless, their statistics are used for assessment of the signal uncertainty.

The main steps in defining the PDF for the signal are firstly interpolating the signal to align with OLCI wavelengths and secondly deciding how best to approach the low number of repeated measurements (15 repeats measurements for cast 1 and 12 for cast 2). Here, we consider both procedures.

The three instruments used in this study have a spectral range from 320 to 1050 nm, however for this exercise we are interested in seven OLCI bands (400, 442.5, 490, 560, 665, 778.8, 865 nm) since this procedure is intended to inform comparisons of ground in situ OCR data with satellite sensors. Therefore, the signal is extracted from only the wavelengths overlapping with the OLCI SRF of the seven OLCI bands and then linearly interpolated to match the 200 wavelengths of the OLCI SRF [34] (see Figure 4). There are 200 OLCI SRF values for each band which means the aforementioned signal interpolation to OLCI wavelengths produces 200 values for each repeated measurement. In the next step, the mean of the repeated measurements is calculated for each of the 200 wavelengths, producing 200 signal values per band. These mean values are assigned as the mean of 200 Gaussian distributions which represents the PDF of the signal at each of these wavelengths.

Later in the processing, once all parameters have been defined, the band integration step, described in Section 2.5.14, convolves these values with the OLCI bands to produce one value per band for E_{OLCI}.

The readings taken for DN_{light} and DN_{dark} only covered a 5-min window, meaning only 15 repeat measurements were taken for cast 1 and 12 repeats for cast 2. This is a very small number of repeated measurements and is far from enough to get a representative mean and standard deviation of the measurements. Since there is not a smooth distribution, it is difficult to decide whether to use the mean, median or mode of the repeated measurements. In this study, the typical standard uncertainty of the mean formula is used Equation (5). Additionally, the mean of the repeated measurements is chosen for the PDF.

$$u^2_{light,mean} = \left(\frac{s_{light}}{\sqrt{N}}\right)^2,$$ (5)

where u is the uncertainty, N is the number of repeat measurements and s_{light} is the standard deviation of light values.

Signal

wvl	Repeat 1	Repeat 2	Repeat 3	...
382.93	0.172	0.164	0.170	
386.28	0.170	0.163	0.169	
389.62	0.177	0.170	0.175	
392.97	0.186	0.179	0.184	
396.32	0.213	0.205	0.211	
399.67	0.260	0.250	0.257	
403.01	0.299	0.289	0.296	
406.36	0.324	0.312	0.321	
409.71	0.347	0.336	0.344	
413.06	0.374	0.362	0.371	
416.41	0.399	0.386	0.396	
419.77	0.421	0.408	0.417	
423.12	0.438	0.425	0.434	
426.47	0.446	0.433	0.442	
429.82	0.460	0.447	0.456	
433.18	0.494	0.480	0.489	
436.53	0.535	0.521	0.530	

219

SRF

wvl	SRF
387.75	5.7E-08
387.86	1.1E-07
387.98	2.1E-07
388.10	3.9E-07
388.22	7.1E-07
388.34	1.3E-06
388.46	2.3E-06
388.57	4.1E-06
388.69	7.2E-06
388.81	1.2E-05
388.93	2.1E-05
389.05	3.6E-05
389.17	5.9E-05
389.28	9.6E-05
389.40	1.6E-04
389.52	2.5E-04
389.64	3.9E-04
...	...
410.94	7.1E-07
411.06	3.8E-07
411.18	2.0E-07
411.30	1.0E-07

200

Signal interpolated to SRF wvls

Repeat 1	Repeat 2	Repeat 3	...
0.173	0.166	0.166	
0.174	0.167	0.167	
0.174	0.167	0.167	
0.174	0.167	0.167	
0.174	0.167	0.167	
0.175	0.168	0.168	
0.175	0.168	0.168	
0.175	0.168	0.168	
0.175	0.168	0.168	
0.176	0.169	0.169	
0.176	0.169	0.169	
0.176	0.169	0.169	
0.176	0.169	0.169	
0.177	0.170	0.170	
0.177	0.170	0.170	
0.177	0.170	0.170	
0.177	0.170	0.170	
...	...	...	
0.357	0.345	0.345	
0.358	0.346	0.346	
0.359	0.347	0.347	
0.360	0.348	0.348	

200

Mean of repeats

Mean
0.169
0.169
0.169
0.169
0.170
0.170
0.170
0.170
0.171
0.171
0.171
0.171
0.172
0.172
0.172
0.172
0.173
...
0.349
0.350
0.351
0.352

200

Figure 4. Description aid for the process of interpolating the signal values to the SRF values and finding the mean of the repeats. Note that this is for one band only; each band will have its own set of tables. The blue number below each table shows how many rows that table has.

2.5.6. Calibration Coefficients (c_{cal})

The calibration of each instrument was performed in the optical radiometry laboratory at University of Tartu. The calibration coefficients and associated uncertainty values are incorporated into the Monte Carlo analysis by appropriate PDF. The instrument readings are automatically adjusted for the calibration coefficients, thus in the MCM, the PDF is taken to be a Gaussian distribution with mean equal to 1. The standard deviation is equal to the standard uncertainty associated with the calibration. These values are presented in Table A1 in the Appendix A.

2.5.7. Non-Linearity Correction (c_{lin})

The integration time used when capturing the measured spectra can lead to non-linearity effects in the results. The maximum value of non-linearity effects was determined in the indoor calibration for the two radiance instruments, but not for the irradiance instrument [20]. As the principal aim of this study is to outline the method and only secondary provide results, the radiance instrument corrections are used for the irradiance instrument. The non-linearity correction values are provided for all instrument wavelengths. Similarly, to the instrument signal, in order to align with the OLCI bands, the only non-linearity values used are those whose wavelengths overlap with the SRFs of each of the seven OLCI bands. These are then interpolated to match the 200 OLCI SRF wavelengths. The interpolation method chosen was linear interpolation due to the smoothness of the non-linearity correction curve (see Figure 5).

Each of the 200 resultant non-linearity correction values is assigned a PDF, which is a rectangular distribution with a mean value of 1 and half-width equal to the linearity correction value.

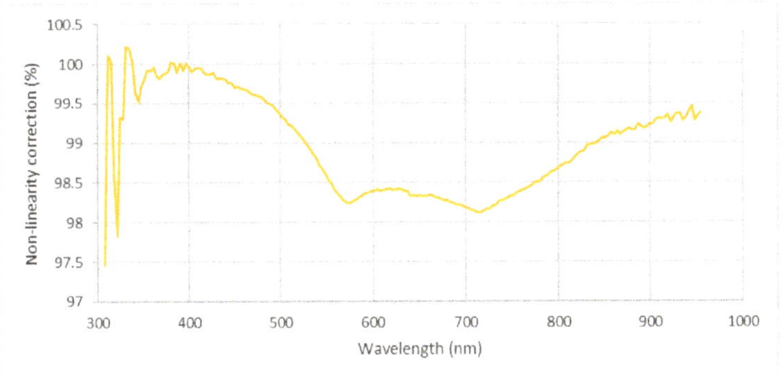

Figure 5. The non-linearity correction curve derived at the optical radiometry laboratory of Tartu Observatory.

Note that the detector might and often does exhibit non-linearity due to the radiance flux levels as well as integration time [26], however tests showing these effects were not used in this study.

2.5.8. Temperature Correction (c_T)

The variation of the instrumental calibration coefficients due to temperature is based on a previous evaluation [27] presented in Figure 6. The variability in % at the seven central wavelengths of the bands of interest (400, 442.5, 490, 560, 665, 778.8, 865 nm) were selected. The seven PDFs which represent the temperature correction at each band were defined to be Gaussian distributions with a mean value of 1 and a standard deviation equal to the aforementioned variability. The standard deviation values are presented in Table A1 in the Appendix A.

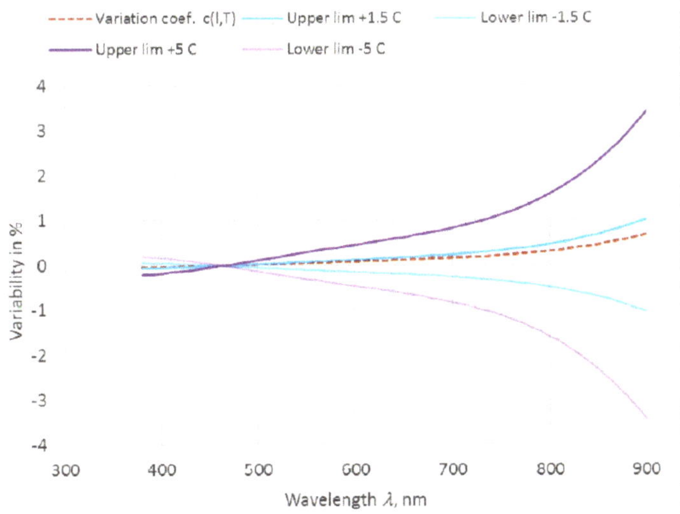

Figure 6. Variation in calibration coefficients as a function of wavelength for a range of temperatures for a RAMSES instrument, based on [27].

2.5.9. Stray Light Correction (c_{stray})

Scattering or reflections in the radiometer optics cause light from one part of the spectrum to fall on pixels associated with light from another part of the spectrum. This effect is known as spectral stray light and is common in hyperspectral instruments and must be corrected for.

This study was intentionally planned so that all instruments would be well characterised, whereas typical campaigns have much less information about the instrument performance, including the stray light characterisation. Hence, we consider two scenarios for the stray light. The first case is for an ideal situation in which the stray light is corrected for based on the performed characterisation and we assign an uncertainty to this method as described in point (i) below; the second, non-ideal, case does not correct for stray light and instead we demonstrate the scale of uncertainty which will result from this in point (ii).

(i) The stray light characterisation provides correction values for each of the instrument wavelengths (see Figure 7). The correction values obtained are quite erratic. Due to the correction values being low in magnitude, it is possible that the erratic behaviour could be due to noise from the stray light measurement.

The stray light correction values used were calculated by convoluting the stray light values with the OLCI SRF [34]. The selected values are presented in Figure 7 as the green cross series.

For the ideal case, the PDFs assigned for stray light for each band is a Gaussian distribution with a mean equal to the stray light correction acquired from the polynomial. The stray light characterisation does have an uncertainty associated with it, but this is unknown. A value of 5% is chosen which accommodates for the true uncertainty and is negligible in comparison to the assigned uncertainty of other variables [20], meaning that the quantity will have little or no effect on the further results.

(ii) For the non-ideal case, no correction is applied. The assigned distribution is a Gaussian distribution with mean equal to 1 and standard deviation equal to the fitted polynomial stray light correction (see Figure 7 and Table 2) at the seven OLCI bands of interest.

The total uncertainty from this method will also need to take into account the difference the lack of correction makes in the value of the downwelling irradiance, E_d and water-leaving radiance, L_w.

Section 3 gives an overview of the impact of both of the above scenarios.

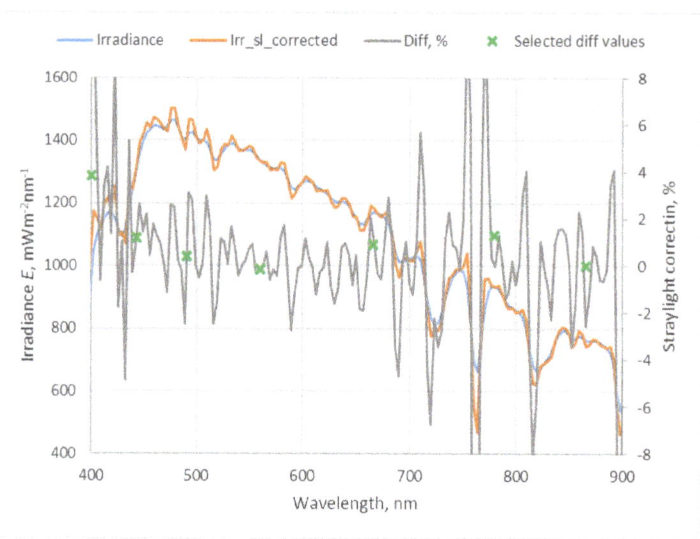

Figure 7. The stray light correction of the irradiance measurements and the percentage difference between non-corrected and corrected. The selected values are calculated using a weighted interpolation technique.

Table 2. Stray light correction values applied to the three radiometers in the ideal case and used as the standard deviation of a Gaussian distribution in the non-ideal case.

Wavelength (nm)		400	442.5	490	560	665	778.8	865
	Downwelling irradiance	3.82	1.17	0.38	−0.17	0.89	1.31	0.03
Stray light correction	Upwelling radiance	2.40	0.84	−0.01	0.30	−1.01	−0.21	−15.20
(%)	Downwelling radiance	2.77	1.92	−0.48	−0.81	0.27	1.45	−7.02

2.5.10. The Fraction of Direct to Total Irradiance $\left(f_{dirr}\right)$

The fraction of direct to total irradiance is applicable to downwelling irradiance measurements only and can be estimated using measurements of the aerosol optical depth, water vapour content and total column ozone in a radiative transfer model. This takes into account the atmospheric transmission of radiation for the conditions specified. In this study, an AERONET-OC [22] station at the observation site provided the atmospheric conditions at the time of data acquisition, and then the radiative transfer model, 6S [35], was used to estimate the direct to total ratio. The atmospheric parameters for both casts were AOD at 500 nm 0.112 and 0.297, precipitable water in 2.83 cm and 3.13 cm, and ozone 330.1 DU and 329.5 DU. The uncertainty components of f_{dirr} consist of i) the accuracy of the 6S radiative transfer model, ii) the uncertainty of the inputs to 6S and iii) an error related to the designated atmosphere type.

(i) The 6S does not provide an estimate of its own accuracy, however a comparison of 6S with a highly accurate Monte Carlo radiative transfer yielded a maximum observed relative difference between the two methods of 0.79 % for a maritime atmosphere [36]. This value is used as an estimate of 6S uncertainty and is shown as "6S model accuracy" row in Table 3.

(ii) There are several inputs of the AERONET data to 6S, namely aerosol optical depth (AOD) at 550 nm, precipitable water and total column ozone. The only of the three variables which causes a change in f_{dirr} is the AOD at 550 nm. Therefore, the only variable which we need to consider the uncertainty of is AOD at 550 nm which has an uncertainty of 0.01 according to [32]. The sensitivity of the resultant f_{dirr} in 6S with a change of ± 0.01 in AOD has been calculated. The difference observed makes up the corresponding uncertainty values shown in the rows labelled "AOD 550 nm" in Table 3.

(iii) The AAOT site is eight nautical miles from the coast of Venice and as such the atmosphere is between continental and maritime and should possibly be considered coastal. However, the 6S model has several defined atmospheres for the user to select which does not include coastal, hence for this study the 'maritime' option will be used and the error in this assumption will be estimated by comparing the results of both 'maritime' and 'continental'. The differences between the two atmospheric types varies across wavelengths, hence the uncertainty due to this will also be wavelength dependent (see the row labelled "Atmosphere type assumptions" Table 3).

Assuming that each contributor is independent, using the Law of Propagation of Uncertainties, we can calculate the uncertainty of f_{dirr} (see Table 3). This uncertainty is applied as half the width of a rectangular distribution centred on a mean value of 1. Note that, here we assume Table 3 shows the uncertainty in terms of the width of a rectangular PDF, whereas in further processing this uncertainty is used as the standard deviation of the Gaussian PDF (see Figure 10 where the values used in MCM are quoted). This means that the values for f_{dirr} are not equal since the standard deviation of a rectangular distribution is smaller than half the width of a rectangular distribution.

Table 3. Uncertainty components for the model, input data (AOD 550 nm) and assumptions relating to the atmosphere specified for each band of interest. All values, except model accuracy, are written in terms of the uncertainty applied to the output, f_{dirr}.

	400	442.5	490	560	665	778.8	865
6S model accuracy			0.79%				
AOD 550 nm (cast 1)	0.0077	0.0081	0.0085	0.0083	0.0082	0.0077	0.0077
AOD 550 nm (cast 2)	0.0063	0.0066	0.0070	0.0069	0.0069	0.0068	0.0067
Atmosphere type assumptions	0.008	0.001	0.010	0.023	0.041	0.058	0.067
fdirr (cast 1)	0.66	0.73	0.78	0.83	0.86	0.88	0.89
fdirr (cast 2)	0.54	0.60	0.64	0.68	0.72	0.74	0.75
Uncertainty fdirr (cast 1)	0.012 (1.82%)	0.010 (1.36%)	0.014 (1.83%)	0.025 (3.08%)	0.042 (4.92%)	0.059 (6.65%)	0.068 (7.63%)
Uncertainty fdirr (cast 2)	0.011 (2.01%)	0.008 (1.36%)	0.013 (2.03%)	0.025 (3.63%)	0.042 (5.86%)	0.058 (7.92%)	0.068 (9.05%)

2.5.11. Cosine Response (f_{cos} and f_{cos_h})

These two components are applicable to downwelling irradiance measurements only. To capture irradiance over a full hemisphere, instruments are equipped with cosine diffusers which collect radiances from the hemisphere and transmit them onto the radiometer. The ideal diffuser will transmit light in proportion with the cosine of the incident angle. However, instruments always differ from the theoretical ideal, hence the need for this to be characterised and corrected for, meaning the residual elements of uncertainty depend on the correction method.

Downwelling irradiance is made up of two components: direct sun irradiance and diffuse sky irradiance. These components are not affected the same way by a non-perfect cosine response and so are separated in the measurement equation. The cosine response term, f_{cos}, incorporates only the direct solar component thus relates to SZA during the measurement and the error in cosine response diffuser for this particular angle. Whereas, f_{cos_h}, the cosine response over the full hemisphere, integrates the diffuse light component across the hemisphere, and integrates the deviation from the perfect diffuser across the whole hemisphere as well.

In this study, the cosine response has been fully characterised with four repeated measurements and is propagated as the ideal case below (i). However, often the cosine response is unknown and thus not corrected for. This section demonstrates the impact of an additional scenario (ii), in which the only information known about the cosine response is the manufacturer's quote of the uncertainty due to the cosine response. Here we choose the value of 3% as this is similar to the cosine response error in this study for all bands and is also a typical value quoted by manufacturers for angles under 60° [37].

The ideal case is having the cosine response errors for a range of solar zenith angles and wavelengths. In this study, the diffuser was characterised in Tartu Observatory before the field comparison in May 2017 for 45 angles across the hemisphere and seven wavelengths. Figure 8 shows the results of the laboratory test for the instruments used in this study. The average of the four repeated measurements was linearly interpolated to the solar zenith angle and the central wavelength of each of the seven bands of interest. The cosine response correction is treated as a Gaussian distribution centred on the interpolated value. The standard deviation of this distribution is assigned from the standard deviation of the four repeated measurements.

The non-ideal case examines the typical scenario in which a manufacturer has quoted the cosine response to be equal to 3% with no additional information. Typically, this is assigned as the standard deviation of a Gaussian PDF of mean 1 (i.e., no correction is applied, just an uncertainty). However, the cosine response is a systematic error and should shift the mean value of E_d as in the ideal scenario (i), but this will not happen if parameterised as a PDF of mean 1 hence the mean value is incorrect. This should be accounted for by applying an additional measure of uncertainty based on the impact not accounting for the bias has on E_d.

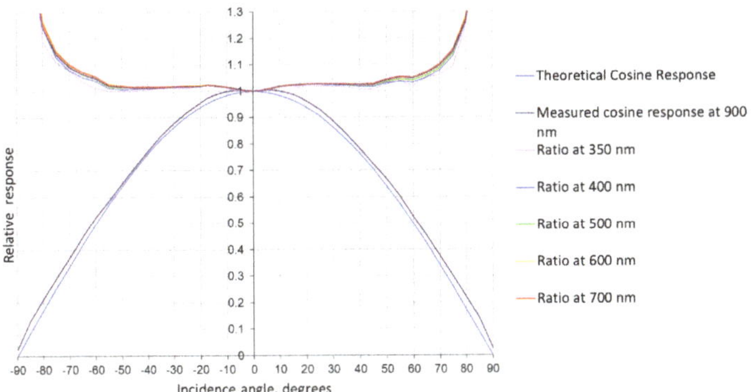

Figure 8. Diffuses cosine response tests results. Series marked as ratio at particular wavelength are the ratios of the measured instrument response to the theoretical cosine response for a given angle and normalised to $0°$

The diffuse component requires all cosine responses to be integrated over the hemisphere. Assuming an isotropic sky radiance distribution, this is calculated using:

$$f_{cos_h} = \int_0^{2\pi} f_{cos}(\theta)sin(2\theta)d\theta, \tag{6}$$

where f_{cos_h} is the integrated cosine response over the full hemisphere, and $f_{cos}(\theta)$ is the cosine response for a given illumination angle. We again look at two similar scenarios, the ideal case (i) demonstrates a scenario in which all cosine response errors are known, and the non-ideal (ii) demonstrates a situation where the only information known regarding the response is that it is within 3% for angles $< 60°$ and 10% for angles $> 60°$. No values were provided by the manufacturer in this study, but these values are typical for manufacturers (e.g., Sea-Bird Scientific HyperOCR radiometer [37]).

The isotropic sky is a simplification and clear-sky radiance distributions are not isotropic and normally show larger radiances for large zenith angles (e.g., [38]) and band circumsolar brightening (aureole) [39]. For measurements presented here with SZA approximatetly $24°$ the aureole effect is minimised by the small errors in cosine response of the diffuser for small incidence angles (see Figure 8). For the horizon brightening the $sin(2\theta)$ factor in Equation (6) would however minimize their contribution to f_{cos_h}. The diffuse component of the downwelling irradiance for clear skies is about 40-30 % for short wavelengths and decreases at longer wavelengths (see Table 3 for the actual values observed during the field measurements), further reducing the impact of f_{cos_h} on the E_d uncertainty evaluation. Obviously, the situation is different for hazier skies, or for high values of the solar zenith angle, which anyway are not recommended conditions for the satellite validation activities.

For the ideal case, the cosine response values are corrected for by taking an average over the four repeated measurements for f_{cos} at each angle and each wavelength. Then these can be integrated over all angles and interpolated to the wavelength of interest. The PDF assigned for the diffuse component is a Gaussian distribution with mean equal to these integrated cosine values and standard deviation originating from the standard deviation of the four repeated measurements of the cosine response.

The non-ideal case does not correct for the cosine response, therefore the mean of the Gaussian distribution is equal to 1 and the standard deviation relates to 3% for angles $\leqslant 60°$ and 10% for angles $> 60°$ (which are typical values for manufacturers (e.g., Sea-Bird Scientific HyperOCR radiometer [37]). However, the cosine response is a systematic error which should be corrected for, hence an additional measure of uncertainty must be applied to take into account the lack of completing a correction. This is equal to the difference in the mean values of the resultant E_d calculated with and without a correction applied (i.e., the difference between the resultant E_d from the ideal and non-ideal cases).

2.5.12. Polarisation Correction (c_{pol})

Polarisation effects can be assumed to be negligible in irradiance sensors due to their use of a cosine diffuser that acts as a scrambler. However, polarisation is an important correction for radiance sensors.

The sensitivity of the instrument to the polarisation of light was not determined during the outdoor inter-comparisons and so instead the worst-case data from [40] was used. This data was used as a measure of uncertainty of the values, rather than used as a correction. Hence, the mean of the polarisation distribution is equal to 1 and standard deviation derived from [40] (see Table A1 in the Appendix A for the values of polarisation used).

2.5.13. Fresnel Reflectance (ρ)

The Fresnel reflectance values for each cast were calculated at the AAOT field site based on approach proposed by Ruddick [41,42] that goes back to Mobley [43] and related Hydrolight simulation, the correction takes into account the wave height by using the wind speed as input factor. The values are 0.0258 and 0.0271 for cast 1 and 2 respectively. Just one value was used across all wavelengths for each cast, which is not ideal. The uncertainty of the reflectivity has previously been determined by [43] at the AAOT site for wavelengths 412, 443, 488, 551 and 667 nm. These values have been interpolated to the wavelengths of interest in this study and are presented in Table A1 in the Appendix A.

2.5.14. Band integration

In order to compare the ground-based in situ ocean colour data with the satellite sensor, OLCI, the hyperspectral instrument must be convolved with OLCI's SRF [34]. For this, we use the following equations.

The weight factors for each wavelength are:

$$K(n) = \frac{c(n)}{\sum_n c(n)},\tag{7}$$

where n is the interpolated wavelength, $c(n)$ is the SRF of a band, and $K(n)$ is the normalized weight coefficient for the n'th wavelength. The summation of the $c(n)$ term is the summation of the SRF values over the 200 wavelengths in one band. The irradiance/radiance value for the corresponding OLCI channel is calculated as:

$$I_{OLCI} = \sum_n I_{cal}(n)K(n),\tag{8}$$

where $I_{cal}(n)$ denotes the irradiance/radiance which has had the correction factors applied (E_{cal} or L_{cal}) at the n'th wavelength.

The two inputs to this band integration are $I_{cal}(n)$ and $c(n)$. The former is calculated from applying the correction factors to irradiance/radiance and interpolating the values to match the 200 SRF wavelengths (see Section 2.5.5). The latter is composed of the SRF of OLCI published by ESA [34]. This has a Gaussian PDF of mean equal to the SRF and standard deviation calculated from guideline SRF relative accuracy for OLCI [44]. The accuracy has been interpolated to match each of the 200 wavelengths in the OLCI SRF. Each of the wavelengths for each band is assigned a PDF for $C(n)$.

For the first wavelength of the first band, a PDF for $c(n)$ will first be assigned. Then a summation over all 200 wavelengths in that band will occur $\sum_n c(n)$ which can be used to determine $K(n)$. This is then multiplied by the first of the 200 values of $I_{cal}(n)$. The result is stored waiting for the result of the next iteration to be added. This procedure occurs for each of the 200 wavelengths in the first band at which point the value of I_{OLCI} for the first band has been determined.

This process is then repeated for each of the seven OLCI bands yielding the seven values for I_{OLCI} which can then be applied to the measurement equations to find the downwelling irradiance or water-leaving radiance.

2.5.15. Error Correlation Summary

Below we summarise all uncertainty contributors that have correlated errors. Please note that the correlation can be between different levels. For example: correlation can exist between all pixels in a given band, or between two different instruments. We do not include correlation between bands in this study.

The parameters which are considered to have correlated errors are listed below, together with the approach used to allow for these correlations in the uncertainty evaluation:

c_{cal} There is a different set of PDFs and draws for each OLCI band, but within the band the 200 contributing wavelengths are correlated with each other. For the radiance case, the same PDFs and draws for each band are used for both instruments (i.e., the instruments are correlated with each other).

c_{temp} There is one set of draws for all OLCI bands, but each band has an individual PDF (i.e., the mean and standard deviation for each band differs but the draws are identical).

c_{lin} There is a different set of PDFs and draws for each OLCI band; but within the band the 200 contributing wavelengths are correlated with each other. For the radiance case, different PDFs and draws are used for each instrument.

c_{stray} Different set of PDFs draws for each OLCI band, but within the band the 200 contributing wavelengths are correlated. For radiance case different PDFs and draws are used for each instrument.

2.5.16. Effects Not Accounted for

So far in this paper, we have not mentioned environmental effects, such as those caused by waves, shadows and sun glint, which may cause systematic variations in the signal which propagate through to the resultant radiance. No additional information was available about the field campaign and so we neglect these terms here. In a number of cases, interpolation has been used to estimate a variable for the wavelengths of interest. The use of interpolation does need to be accommodated for in the MCM with an uncertainty. However, here the uncertainty will be negligible as compared to the other factors, hence it is omitted from our method.

3. Results

This section presents the outputs of the uncertainty analysis for the ideal and non-ideal cases, as well as a corrected case where an extra correction is applied to show the true resultant uncertainty when not corrected. The MCM for downwelling irradiance and water-leaving radiance was run over two casts and results are presented for the seven OLCI bands of interest (400, 442.5, 490, 560, 665, 778.8, 865 nm).

3.1. Downwelling Irradiance

The ideal scenario relates to the case in which the cosine correction and stray light were well characterised and were corrected for. As previously mentioned, the MCM requires each parameter in the measurement equations to be assigned a PDF based on collected data, or best knowledge, etc. These distributions as shown in Figure 9 were then propagated through the measurement equation to find the downwelling irradiance, E_d.

Figure 10 presents the summary values of the MCM simulation for cast 1. The value of several of the variables (S, c_{lin}, E_{cal}, $K(n)$) is chosen for just one wavelength out of 200. Most of these variables vary over the 200 wavelengths. Here we present the 99 value as an example for reference. In the calculations, all 200 values are propagated.

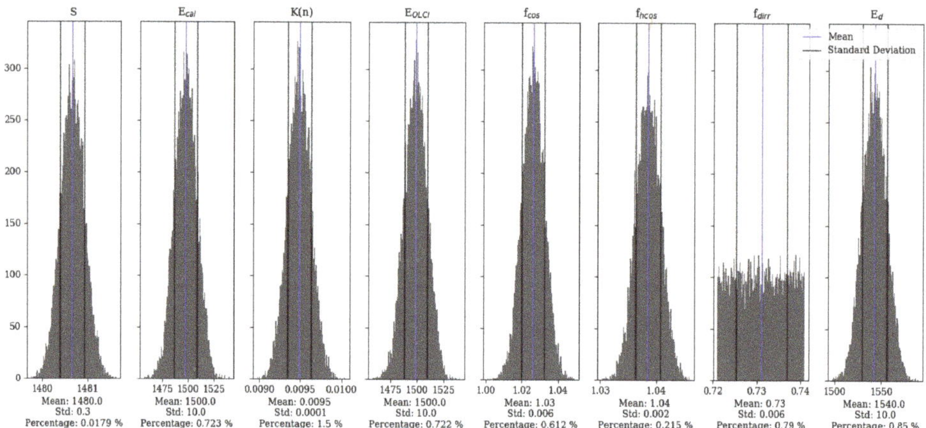

Figure 9. Histograms of several variables in the measurement process for the ideal case of cast 1, band 442.5 nm. The mean, standard deviation and standard deviation as a percentage of the mean (coefficient of variation CV) are labelled below the histograms.

a)

	S	c_{cal}	c_{lin}	c_{temp}	c_{stray}	E_{cal}	K(n)	E_{OLCI}	f_{cos}	f_{hcos}	f_{dirr}	E_d
400	1070.0	1.0	1.0	1.0	1.04	1110.0	0.0084	1120.0	1.03	1.04	0.66	1150.0
442.5	1480.0	1.0	1.0	1.0	1.01	1500.0	0.0095	1500.0	1.03	1.04	0.73	1540.0
490	1590.0	1.0	1.0	1.0	1.0	1600.0	0.0094	1600.0	1.03	1.04	0.78	1650.0
560	1510.0	1.0	1.0	1.0	0.998	1510.0	0.0094	1510.0	1.03	1.05	0.83	1550.0
665	1320.0	1.0	1.0	1.0	1.01	1330.0	0.0094	1330.0	1.03	1.05	0.86	1370.0
778.8	1060.0	1.0	1.0	1.0	1.01	1070.0	0.008	1060.0	1.03	1.05	0.88	1090.0
865	873.0	1.0	1.0	1.0	1.0	873.0	0.0072	876.0	1.02	1.04	0.89	899.0

	S	c_{cal}	c_{lin}	c_{temp}	c_{stray}	E_{cal}	K(n)	E_{OLCI}	f_{cos}	f_{hcos}	f_{dirr}	E_d
400	0.0228	0.873	0.0171	0.203	0.19	0.913	2.2	0.913	0.619	0.312	1.0	1.0
442.5	0.0179	0.683	0.133	0.203	0.057	0.723	1.5	0.722	0.612	0.215	0.79	0.85
490	0.017	0.647	0.292	0.203	0.019	0.741	1.2	0.741	0.596	0.257	1.1	0.873
560	0.0142	0.619	0.902	0.305	0.0083	1.13	0.9	1.13	0.585	0.194	1.8	1.23
665	0.0147	0.599	0.957	0.711	0.044	1.35	0.78	1.35	0.554	0.188	2.8	1.43
778.8	0.0158	0.618	0.878	1.32	0.065	1.69	0.74	1.69	0.542	0.199	3.8	1.76
865	0.0246	0.557	0.508	2.74	0.0012	2.86	0.73	2.86	0.536	0.174	4.4	2.9

b)

	S	c_{cal}	c_{lin}	c_{temp}	c_{stray}	E_{cal}	K(n)	E_{OLCI}	f_{cos}	f_{hcos}	f_{dirr}	E_d
400	1070.0	1.0	1.0	1.0	1.0	1070.0	0.0084	1080.0	1.0	1.0	0.66	1080.0
442.5	1480.0	1.0	1.0	1.0	1.0	1480.0	0.0095	1480.0	1.0	1.0	0.73	1480.0
490	1590.0	1.0	1.0	1.0	1.0	1590.0	0.0094	1590.0	1.0	1.0	0.78	1590.0
560	1510.0	1.0	1.0	1.0	1.0	1510.0	0.0094	1510.0	1.0	1.0	0.83	1510.0
665	1320.0	1.0	1.0	1.0	1.0	1320.0	0.0094	1320.0	1.0	1.0	0.86	1320.0
778.8	1060.0	1.0	1.0	1.0	1.0	1060.0	0.008	1050.0	1.0	1.0	0.88	1050.0
865	873.0	1.0	1.0	1.0	1.0	872.0	0.0072	875.0	1.0	1.0	0.89	876.0

	S	c_{cal}	c_{lin}	c_{temp}	c_{stray}	E_{cal}	K(n)	E_{OLCI}	f_{cos}	f_{hcos}	f_{dirr}	E_d
400	0.0228	0.873	0.0171	0.203	3.9	3.96	2.3	3.96	3.01	1.34	1.1	4.44
442.5	0.018	0.683	0.133	0.203	1.2	1.38	1.5	1.38	2.99	1.35	0.78	2.62
490	0.0165	0.647	0.292	0.203	0.38	0.839	1.1	0.84	2.99	1.33	1.1	2.51
560	0.0141	0.619	0.902	0.305	0.16	1.14	0.91	1.14	2.99	1.36	1.8	2.73
665	0.015	0.599	0.957	0.711	0.89	1.62	0.79	1.62	2.98	1.33	2.8	3.04
778.8	0.016	0.618	0.878	1.32	1.3	2.13	0.74	2.13	3.0	1.32	3.8	3.41
865	0.025	0.557	0.508	2.74	0.025	2.86	0.74	2.86	2.96	1.34	4.4	3.9

Figure 10. Irradiance MCM outputs. (**a**) Ideal case results. The mean (top panel) and standard uncertainty as a percentage of the mean (second panel) of each variable. This is shown for all OLCI bands of interest and relates to cast 1. The colour code intensity indicates the magnitude, (**b**) Non-Ideal case results. The mean (third panel) and standard uncertainty as a percentage of the mean (lower panel) of each variable.

The quoted number of significant figures exceeds the number which are conventionally quoted because the values are intended to be compared with values in the other figures and tables of the report. The final values which are quoted in Section 4 are in line with standard significant figure reporting.

The results for the cases in which neither stray light nor the cosine error are corrected for look different (the two bottom panels in Figure 10) they were each assigned a mean value of 1 and an uncertainty relating to typical manufacturers' estimates (cosine) and the correction itself (stray light). It is possible to see the effects of stray light and the cosine error separately since each parameter exists in separate parts of the measurement equation. For example, it is clear that the mean of c_{stray} is set to 1 and the standard deviation is much higher in the non-ideal case than ideal. Additionally, the changes in parameters can be tracked through up to E_{OLCI}. Beyond this, the correction is mixed with the cosine correction which can be seen to dominate the standard deviation of the majority of bands of E_d.

The uncertainty presented in the lower panel of Figure 10 is an under-representation of the true uncertainty that is applicable when corrections are not applied. This is because we have taken into account the uncertainty in the measurement but have not accounted for the lack of correction.

Hence, we add the bias which is the difference between the ideal mean value and the non-ideal mean value (see Equation (9)) and add this to the standard deviation of the non-ideal case.

$$bias = mean(L_{w,ideal}) - mean(L_{w,non-ideal}), \tag{9}$$

$$u_{correctednon-ideal} = u_{non-ideal} + |bias|, \tag{10}$$

The result is illustrated in Figure 11 as the "Corrected non-ideal" case. This exhibits a much wider standard deviation, which is the true standard deviation which should be used but cannot be calculated without the ideal characterisation.

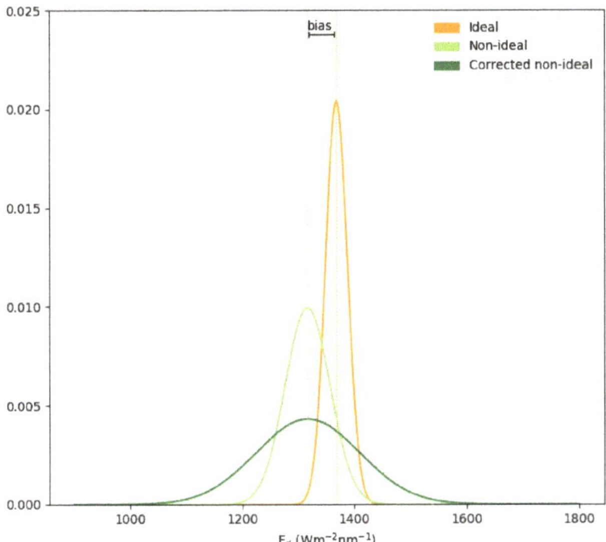

Figure 11. Representations of the histograms of E_d for the ideal, non-ideal and corrected case showing the differences in the standard deviation. This is for cast 1 and the band with central wavelength of 778.8 nm.

Table 4 provides an overview of the values of interest in this correction. The mean values of the PDF function obtained in the MCM processing for downwelling irradiance are shown for the ideal and non ideal case in irradiance units (column 2 and 3 in the Table 4). Column 4 contains the bias values, thus the difference between the two means calculated according to the Equation (9). The columns 5 to 7 present relative standard uncertainties for the three different scenarios.

Table 4. The mean and standard uncertainty as a percentage of the mean of the downwelling irradiance, E_d [mWm^{-2}nm^{-1}] presented for the ideal and non-ideal cases.

Band (nm)	Mean E_d			Standard Uncertainty $u(E_d)$ [%]		
	Ideal	Non-ideal	Bias of E_d	Ideal	Non-ideal	Corrected non-ideal
400	1150.0	1080.0	73.9	1.0	4.44	11.30
442.5	1540.0	1480.0	60.8	0.85	2.62	6.72
490	1650.0	1590.0	53.8	0.87	2.51	5.89
560	1550.0	1510.0	43.5	1.23	2.73	5.61
665	1370.0	1320.0	51.9	1.43	3.04	6.99
778.8	1090.0	1050.0	44.6	1.76	3.41	7.669
865	899.9	876.0	23.6	2.9	3.9	6.6

3.2. Water-Leaving Radiance

The same approach as for the downwelling irradiance case is used to present, in Figure 12, the water-leaving radiance results. The value of several of the variables (S_u, S_d, $c_{lin,u}$, $c_{lin,d}$, $c_{stray,u}$, $c_{stray,d}$, $L_{cal,u}$, $L_{cal,d}$, $K(n)$) presented in Figure 12 is chosen for just one wavelength out of 200. Most of these variables vary over the 200 wavelengths. Here we present the 99 value as an example for reference. In the calculations, all 200 values are propagated.

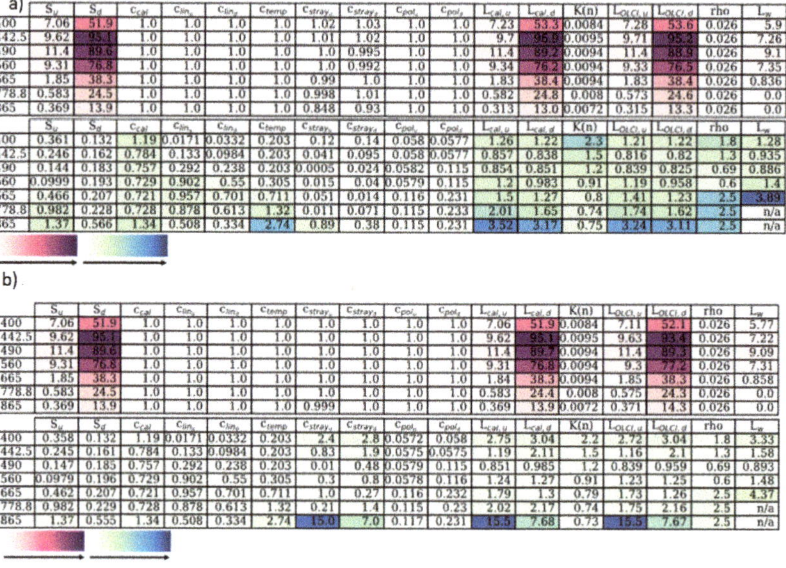

Figure 12. Water-leaving radiance MCM outputs. (**a**) Ideal case results. The mean (top panel) and standard uncertainty as a percentage of the mean (second panel) of each variable. This is shown for all OLCI bands of interest and relates to cast 1. The colour code intensity indicates the magnitude, (**b**) Non-Ideal case results. The mean (third panel) and standard uncertainty as a percentage of the mean (lower panel) of each variable.

The ideal scenario relates to the case in which stray light was well characterised and corrected for as is explained thoroughly in Section 2.5.9. For the non-ideal scenario stray light is not corrected for and was assigned a mean value of 1 (i.e., no correction applied) and an uncertainty relating to the previously used correction values. From Figure 12 it is clear that the mean of the stray light correction factors is 1 and the standard deviation is larger as compared to the ideal case. These changes can be seen to affect downstream parameters in the measurement equation.

The 778.8 nm and 865 nm bands resulted in a negative water-leaving radiance. A negative value here is not theoretically possible, so is likely due to an error in the measurement model. However, we do not expect to observe any water-leaving radiance measurable with the used radiometers at AAOT for these wavelengths. We quote the water leaving radiance as $0\,\mathrm{mWm}^{-2}\mathrm{sr}^{-1}\mathrm{nm}^{-1}$ with no meaningful uncertainty.

The uncertainty for non-ideal case is an under-representation of the true uncertainty that is applicable when corrections are not applied. This is because we have taken into account the uncertainty in the measurement but have not accounted for the lack of correction. Hence, we add the difference between the ideal mean value and the non-ideal mean value and add this to the standard deviation of the non-ideal case. This exhibits a much wider standard deviation, which is the true standard deviation which should be used but cannot be calculated without the ideal characterisation. Table 5 provides an overview of the values of interest in this correction. The mean values of the PDF function obtained in the MCM processing for water-leaving radiance are shown for the ideal and non ideal case in radiance units (column 2 and 3 in the Table 5). Column 4 contains the bias values, thus the difference between the two means calculated according to the Equation (9). The columns 5 to 7 present relative standard uncertainties for the three different scenarios.

Table 5. The mean and standard uncertainty as a percentage of the mean of water-leaving radiance L_w $[\mathrm{mWm}^{-2}\mathrm{sr}^{-1}\mathrm{nm}^{-1}]$ presented for the ideal and non-ideal cases.

	Mean L_w			Standard Uncertainty	$u(L_w)[\%]$	
Band (nm)	Ideal	Non-Ideal	Bias of L_w	Ideal	Non-Ideal	Corrected Non-Ideal
400	5.9	5.77	0.134	1.28	3.33	5.65
442.5	7.26	7.22	0.0353	0.94	1.58	2.07
490	9.1	9.09	0.0097	0.886	0.893	1.00
560	7.35	7.31	0.0442	1.40	1.48	2.08
665	0.836	0.858	0.0218	3.89	4.37	6.91

4. Discussion

This paper presents an evaluation of an uncertainty budget for above-water OCR measurements performed as part of FRM4SOC field inter-comparison exercise which demonstrates the importance of correcting for instrumental biases. The data from one participant (University of Tartu) were used here for further investigations. These radiometers were chosen because the radiometers, in addition to radiometric calibration, had a set of optical characterisations completed. This enabled an investigation of different scenarios, where we correct the measurand for the known instruments' characteristic (we call this the ideal case). Since having this set of optical characterisations is rather rare, we present a so-called non-ideal case, where the biases caused by instrument characteristics are not corrected for. The most important case is, however, the corrected non-ideal scenario where we evaluate the real uncertainty related to the uncorrected bias case.

We presented the measurement equations using uncertainty tree diagrams that in graphical form show all the relationships between different uncertainty contributors. The uncertainty was evaluated by assigning PDFs to each contributor and propagating this through the measurement equations using the MCM method.

To test the stability of the model, the MCM was run 15 times for both the irradiance and radiance ideal cases. The standard deviation of the standard uncertainty of E_d and L_w are shown in Table 6.

Table 6. Stability of E_d and L_w found by conducting 15 runs of the Monte Carlo Method with 10,000 draws. The values shown are the standard deviation of the standard uncertainty of E_d and L_w as a percentage of the mean of the standard deviations for each OLCI band.

Variability in Estimates	400	442.5	490	560	665	778.8	865
$u(E_d)[\%]$	0.34	0.59	0.40	0.61	0.41	0.26	0.17
$u(L_v)[\%]$	0.05	0.06	0.04	0.14	0.16	0.24	0.11

The variability in the uncertainty of E_d and L_w allow an estimate of the number of significant figures to be quoted. The final standard uncertainty values are quoted to two significant figures as shown in Table 7. It is important to note that components related to environmental conditions such as structure shading, sun and sky glint, wave focusing etc are not included in this uncertainty budget, therefore the true uncertainty value is likely to be higher.

Table 7. Final standard uncertainty values for corrected non-ideal case expressed in percentage.

	$u(E_d)[\%]$		$u(L_w)[\%]$	
Wavelength	Cast 1	Cast 2	Cast 1	Cast 2
400	11.3	11.3	5.7	7.3
442.5	6.7	6.6	2.1	2.3
490	5.9	5.8	1.0	1.1
560	5.6	5.6	2.1	2.3
665	7.0	6.9	6.9	8.3
778.8	7.7	7.6	–	22.4
865	6.6	6.6	–	152.0

The fully evaluated uncertainty values as presented in Table 7 are potentially higher than expected, commonly agreed or used, especially for the downwelling irradiance, are caused by the biases due to non-perfect cosine diffuser response and stray light added to the uncertainty estimate. We evaluate the biases as the measurement uncertainty during AAOT field inter-comparisons for one participant's radiometers as the results provided for the comparison pilot were not corrected for any of the instruments' characteristics. When these biases are treated not as errors but as uncertainty components, firstly the final product mean value is incorrect since no correction takes place. Secondly, the uncertainty associated with that value is underestimated (see Figure 11). This is because the uncertainty in the measurement has been accounted for (typical manufacturers' estimates are used for the cosine error and the correction itself for stray light) but not the lack of correction. Hence the need to find the difference between the ideal mean value and the non-ideal mean value and add this to the standard uncertainty of the non-ideal case, producing the results of the corrected non-ideal scenario.

We chose to correct only the cosine diffuser and stray light characterisations to present the three scenarios (ideal, non-ideal and corrected non-ideal). However, in practice, any parameter which is the dominant source of uncertainty and can be feasibly corrected should be corrected for. Once that parameter has been corrected for, the next most dominant should also be considered, and so on. The limiting factor in this step-wise procedure is the absolute calibration which cannot be corrected for. Hence any parameters with an uncertainty less than that of the absolute calibration does not need to be corrected for. For example, looking at Figures 10 and 12 that show the magnitude of uncertainty contributors, one could recommend that any component with uncertainty greater than c_{cal} should be corrected for. This would include three of the bands of both temperature and linearity for E_d and similar for L_w.

In this study, the environmental uncertainty is not considered, but this may be the limiting factor since it is likely to be larger than the absolute calibration uncertainty. An evaluation of how to correctly estimate environmental uncertainty for a range of conditions is yet to be completed.

We presented the MCM to evaluate uncertainty as we found it simpler to apply especially for the hyperspectral data that require band integration to fit the satellite bands. However, the equation for L_w (see Equation (A1)) leads itself to being a good example to show the traditional Law of Propagation of Uncertainties defined in GUM [30]. Appendix B presents a comparison of MCM to correctly applied analytical method and a case where the so-called sum of squares does not work.

5. Conclusions

This study has demonstrated how to conduct an uncertainty analysis for *in situ* radiometers of Ocean Colour measurements. The results of the three scenarios (ideal, non-ideal and corrected non-ideal) in Tables 4 and 5 highlight the importance and benefits of carrying out instrument characterisations before campaigns and performing instrument corrections in addition to absolute radiometric calibration. It is recommended that the sources of uncertainty which are likely to dominate over the absolute calibration uncertainty (or other more dominant uncertainty contributors which cannot be corrected for) should be characterised before campaigns so that these can be corrected for. This will produce results with reduced uncertainties as demonstrated in the ideal scenario (Tables 4 and 5). The most likely parameters which will need prior characterisations are stray light, cosine, temperature and non-linearity corrections. Following these guidelines will support compliance with the requirements of *in situ* Ocean Colour measurements for use in satellite product validation.

Author Contributions: Conceptualization, A.B.; methodology, A.B.; software, S.D.; validation, A.B., S.D.; data curation, J.K., I.A., V.V.; writing—original draft preparation, A.B., S.D.; writing—review and editing, A.B., J.K., V.V., R.V. and T.C.; visualization, S.D.; project administration, R.V. and T.C. All authors have read and agreed to the published version of the manuscript.

Funding: This work was funded by European Space Agency project Fiducial Reference Measurements for Satellite Ocean Colour (FRM4SOC), contract no. 4000117454/16/I-Sbo.

Conflicts of Interest: The authors declare no conflict of interest.

Appendix A. Uncertainty Values

Table A1. The mean and uncertainty values used for each parameter for the radiance simulation.

Wavelength (nm)		400	442.5	490	560	665	778.8	865
Calibration	Mean	1	1	1	1	1	1	1
	Std (%)	1.2	0.78	0.76	0.73	0.71	0.73	1.35
Temperature	Mean	1	1	1	1	1	1	1
	Std (%)	0.2	0.2	0.2	0.3	0.7	1.3	2.7
Polarisation,	Mean	1	1	1	1	1	1	1
upwelling	Std (%)	0.1	0.1	0.1	0.1	0.2	0.2	0.2
Polarisation,	Mean	1	1	1	1	1	1	1
downwelling	Std (%)	0.1	0.1	0.2	0.2	0.4	0.4	0.4
Fresnel reflectance,	Mean	0.0258	0.0258	0.0258	0.0258	0.0258	0.0258	0.0258
cast 1	Std (%)	1.8	1.3	0.7	0.6	2.5	2.5	2.5
Fresnel reflectance,	Mean	0.02714	0.02714	0.02714	0.02714	0.02714	0.02714	0.02714
cast 2	Std (%)	1.8	1.3	0.7	0.6	2.5	2.5	2.5

Appendix B. Analytical Method Example

Sum of squares of percentage uncertainties is commonly used to evaluate uncertainty budgets and is derived from the Law of Propagation of Uncertainties but applicable solely in the cases where the measurement equation contains only multiplication or division and there is no correlation between terms. Thus, clearly that form of the sum of squares cannot be used in the evaluation of uncertainty related to L_w, because there is a subtraction in a measurement equation and correlation exists between $L_{OLCI,u}$ and $L_{OLCI,d}$.

Starting from a simplified measurement equation (see full equation in Equation (2))

$$L_w = L_{OLCI,u} - \rho L_{OLCI,d}. \tag{A1}$$

To use the Law of Propagation of Uncertainties we merge Equations (3) and (4) together and substitute the terms of Equation (A1) to give the combined variance of L_w.

$$
\begin{aligned}
u^2(L_w) = {} & \left(\frac{\partial L_w}{\partial L_{OLCI,u}}\right)^2 u^2(L_{OLCI,u}) + \left(\frac{\partial L_w}{\partial \rho}\right)^2 u^2(\rho) + \left(\frac{\partial L_w}{\partial L_{OLCI,d}}\right)^2 u^2(L_{OLCI,d}) + \\
& + 2\left(\frac{\partial L_w}{\partial L_{OLCI,u}}\right)\left(\frac{\partial L_w}{\partial L_{OLCI,d}}\right) r(L_{OLCI,u}, L_{OLCI,d}) u(L_{OLCI,u}) u(L_{OLCI,d}),
\end{aligned}
\tag{A2}
$$

where $\left(\frac{\partial L_w}{\partial L_{OLCI,u}}\right)$, $\left(\frac{\partial L_w}{\partial \rho}\right)$ and $\left(\frac{\partial L_w}{\partial L_{OLCI,d}}\right)$ are the sensitivity coefficients formed from the partial derivatives of the measurement function L_w with respect to input quantities ($L_{OLCI,u}$, ρ and $L_{OLCI,d}$). $u(L_{OLCI,u})$, $u(L_{OLCI,d})$ and $u(\rho)$ are the standard uncertainties of each input quantity respectively and $r(L_{OLCI,u}, L_{OLCI,d})$ is the correlation coefficient. Here, we use the Pearson correlation coefficient that measures the linear relationship between two datasets. The coefficients vary between -1 and $+1$ with 0 implying no correlation. Perfectly correlated datasets have coefficients of $+1$ or -1. The sign of the result describes whether the relationship is positive (as one increases, the other increases) or negative (as one increases, the other decreases).

The sensitivity coefficients are:

$$
\begin{aligned}
\frac{\partial L_w}{\partial L_{OLCI,u}} &= 1 \\
\frac{\partial L_w}{\partial \rho} &= -L_{OLCI,d} \\
\frac{\partial L_w}{\partial L_{OLCI,d}} &= -\rho
\end{aligned}
\tag{A3}
$$

We use the uncertainty in L_u, L_d and ρ form the results of the non-ideal case (see Figure 12 lower panel b columns $L_{OLCI,u}$, $L_{OLCI,d}$ and ρ) and error correlation between L_d and L_u from Table A2 to calculate the uncertainty in L_w using the analytical GUM method.

Please note that the uncertainty terms referred to in the equation above are absolute, thus expressed in the units of the input values, not as percentages. Note that the second term of Equation (3) is only non-zero when two parameters are correlated. Figure A1 presents the error correction between the three inputs components for two selected wavelengths and for both cases ideal and non-ideal. There is no correlation between Fresnel reflectance and radiance measurements, thus its omission in Equation (A2).

The sum of squares is calculated as

$$u(L_w) = \sqrt{u^2(L_{OLCI,u}) + u^2(\rho) + u^2(L_{OLCI,d})}, \tag{A4}$$

where all standard uncertainty inputs are expressed in percentages.

As the values in Table A3 the uncertainty calculated through each of the correctly applied GUM methods for upwelling radiance (MCM and analytical), L_w, is the same. Generally, both the analytical approach and MCM should agree with each other, if all correlation and model non-linearities are correctly addressed in the analytical model, and such a comparison should be used as a validation of the other. In this example, we did not completely derive the analytical GUM uncertainty budget because we used pre-calculated values of three input components to the measurement equation that were derived from the MCM. It was intended to present an example where the simple sum of squares does not work as it is not fit-for-purpose, but the properly applied analytical approach and MCM do provide the same results.

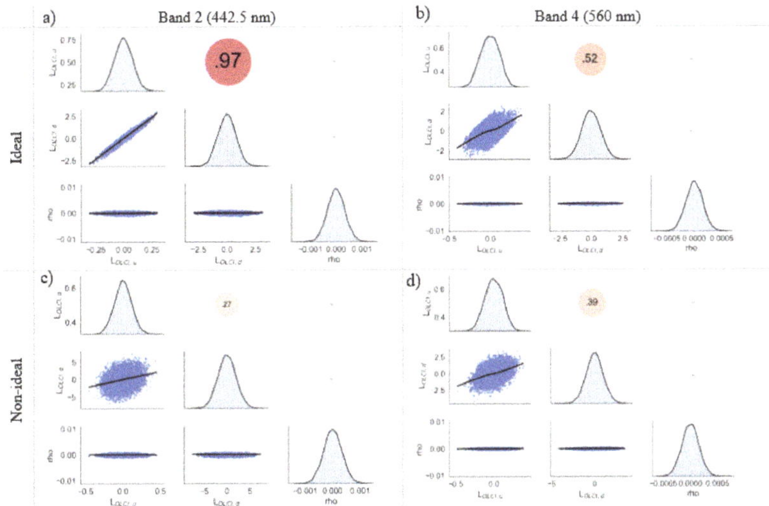

Figure A1. Error correlation between $L_{OLCI,u}$, $L_{OLCI,d}$ and ρ in the water-leaving radiance measurement equation observed in the Monte Carlo analysis for cast 1 only.

Table A2. Pearson correlation coefficients between the $L_{OLCI,u}$ and $L_{OLCI,d}$ errors for all seven bands for cast 1 only.

	Correlation Coefficients	
Band (nm)	**Ideal**	**Non-Ideal**
400	0.99	0.18
442.5	0.97	0.27
490	0.89	0.78
560	0.52	0.39
665	0.60	0.48
778.8	0.80	0.59
865	0.93	0.08

Table A3. Uncertainty evaluation methods comparison results. The first four columns are the uncertainty of the input components, followed by the uncertainty of the water-leaving evaluated using the MCM. All uncertainties here are presented as percentages. The analytical result calculates L_w according to Equation (A2) and sum of squares using Equation (A4). The 778.8 and 865 nm bands are not included due to them having a water-leaving radiance of 0 and hence no meaningful uncertainty.

Band (nm)	$u(L_{OLCI,u})$	$u(L_{OLCI,d})$	$u(\rho)$	MCM $u(L_w)$	Analytical $u(L_w)$	Sum of Squares $u(L_w)$
400	2.72	3.04	1.8	3.33	3.33	4.46
442.5	1.16	2.10	1.3	1.58	1.58	3.73
490	0.84	0.96	0.7	0.89	0.89	1.45
560	1.23	1.25	0.6	1.48	1.47	1.85
665	1.73	1.26	2.5	4.37	4.39	3.3

References

1. Ball Aerospace Division. *Development of the Coastal Zone Color Scanner for Nimbus-7, vol 2: Test and Performance Data, Final Report*; NASA contract NAS5-20900 Rep. F78-11; Ball Aerospace Division: Muncie, IN, USA, 1979.

2. Hooker, S.B.; Esaias, W.E.; Feldman, G.C.; Gregg, W.W.; Mc Clain, C.R. *An Overview of SeaWiFS and Ocean Colour in SeaWiFS Technical Report Series*; NASA Tech. Memo 104566; NASA Goddard Space Flight Centre: Greenbelt, MD, USA, 1992.

3. Hooker, S.B.; McLean, S.; Sherman, J.; Small, M.; Lazin, G.; Zibordi, G.; Brown, J. *The Seventh SeaWiFS Intercalibration Round-Robin Experiment (SIRREX-7)*; NASA Tech. Memo 2002-206892' NASA Goddard SpaceFlight Center: Greenbelt, MD, USA, 2002; Volume 17.

4. McClain, C.R. *SIMBIOS Background*; NASA Tech. Memo TM-1999-208645,1–2; NTRS: Springfield, VA, USA, 1998.

5. Salomonson, V.V.; Barnes, W.L.; Maymon, P.W.; Montgomery, H.E.; Ostrow, H. MODIS: Advanced facility instrument for studies of the Earth as a system. *IEEE Trans. Geosci. Remote Sens.* **1989**, *27*, 145–153. doi:10.1109/36.20292. [CrossRef]

6. Rast, M.; Bezy, J.L.; Bruzzi, S. The ESA Medium Resolution Imaging Spectrometer MERIS a review of the instrument and its mission. *Int. J. Remote. Sens.* **1999**, *20*, 1681–1702, doi:10.1080/014311699212416. [CrossRef]

7. Mueller, J.L.; Austin, R.; Morel, A.; Fargion, G.; McClain, C. *Ocean Optics Protocols for Satellite Ocean Color Sensor Validation, Revision 4, Volume I: Introduction, Background and Conventions*; Technical Report NASA/TM-2003-21621/Rev-Vol I; NASA Goddard Space Flight Center: Greenbelt, MD, USA, 2003.

8. Mueller, J.L.; Pietras, C.; Hooker, S.B.; Austin, R.; Miller, M.; Knobelspiesse, K.; Frouin, R.; Holben, B.; Voss, K. *Ocean Optics Protocols for Satellite Ocean Color Sensor Validation, Revision 4, Volume II: Instrument Specifications, Characterization and Calibration*; Technical Report NASA/TM-2003-21621/Rev-Vol II; NASA Goddard Space Flight Center: Greenbelt, MD, USA, 2003.

9. Mueller, J.L.; Morel, A.; Frouin, R.; Davis, C.; Arnone, R.; Carder, K.; Lee, Z.P.; Steward, R.G.; Hooker, S.; Mobley, C.D.; et al. *Ocean Optics Protocols for Satellite Ocean Color Sensor Validation, Revision 4, Volume III: Radiometric Measurements and Data Analysis Protocols*; Technical Report NASA/TM-2003-21621/Rev-Vol III; NASA Goddard Space Flight Center: Greenbelt, MD, USA, 2003.

10. Mueller, J.L.; Clark, D.K.; Kuwahara, V.S.; Lazin, G.; Brown, S.W.; Fargion, G.S.; Yarbrough, M.A.; Feinholz, M.; Flora, S.; Broenkow, W.; et al. *Ocean Optics Protocols for Satellite Ocean Color Sensor Validation, Revision 4, Volume VI: Special Topics in Ocean Optics Protocols and Appendices*; Technical Report NASA/TM-2003-211621/Rev4-Vol.VI; NASA Goddard Space Flight Center: Greenbelt, MD, USA, 2003.

11. Gordon, H.R.; Clark, D.K. Clear water radiances for atmospheric correction of coastal zone color scanner imagery. *Appl. Opt.* **1981**, *20*, 4175–4180, doi:10.1364/AO.20.004175. [CrossRef] [PubMed]

12. Gordon, H. In-Orbit Calibration Strategy for Ocean Color Sensors. *Remote. Sens. Environ.* **1998**, *63*, 265–278. doi:10.1016/S0034-4257(97)00163-6. [CrossRef]

13. McClain, C.R.; Esaias, W.E.; Barnes, W.; Guenther, B.; Endres, D.; Hooker, S.; Mitchell, G.; Barnes, R. *Calibration and Validation Plan for SeaWiFS*; NASA Tech. Memo. 104566; Hooker, S.B., Firestone, E.R., Eds.; NASA Goddard Space Flight Center: Greenbelt, MD, USA, 1992; Volumw 3, pp. 503–514.

14. Franz, B.A.; Bailey, S.W.; Werdell, P.J.; McClain, C.R. Sensor-independent approach to the vicarious calibration of satellite ocean color radiometry. *Appl. Opt.* **2007**, *46*, 5068–5082. doi:10.1364/AO.46.005068. [CrossRef]

15. Chang, G.C.; Dickey, T.D.; Mobley, C.D.; Boss, E.; Pegau, W.S. Toward closure of upwelling radiance in coastal waters. *Appl. Opt.* **2003**, *42*, 1574–1582. doi:10.1364/AO.42.001574. [CrossRef]

16. Donlon, C.; Berruti, B.; Buongiorno, A.; Ferreira, M.-H.; Féménias, P.; Frerick, J.; Goryl, P.; Klein, U.; Laur, H.; Mavrocordatos, C.; et al. The Global Monitoring for Environment and Security (GMES) Sentinel-3 mission. *Remote. Sens. Environ.* **2012**, *120*, 37–57, doi:10.1016/j.rse.2011.07.024. [CrossRef]

17. Cetinić, I.; McClain, C.; Werdell, P. *PACE Technical Report Series, Volume 6: Data Product Requirements and Error Budgets*; Nasa tech. memo; NASA Goddard Space Flight Space Center: Greenbelt, MD, USA, 2018.

18. Drusch, M.; Bello, U.D.; Carlier, S.; Colin, O.; Fernandez, V.; Gascon, F.; Hoersch, B.; Isola, C.; Laberinti, P.; Meygret, A.; et al. Sentinel-2: ESA's Optical High-Resolution Mission for GMES Operational Services. *Remote. Sens. Environ.* **2012**, *120*, 25–36, doi:10.1016/j.rse.2011.11.026. [CrossRef]

19. Vabson, V.; Kuusk, J.; Ansko, I.; Vendt, R.; Alikas, K.; Ruddick, K.; Ansper, A.; Bresciani, M.; Burmester, H.; Costa, M.; et al. Laboratory Intercomparison of Radiometers Used for Satellite Validation in the 400–900 nm Range. *Remote. Sens.* **2019**, *11*, 1101, doi:10.3390/rs11091101. [CrossRef]

20. Vabson, V.; Kuusk, J.; Ansko, I.; Vendt, R.; Alikas, K.; Ruddick, K.; Ansper, A.; Bresciani, .; Burmester, H.; Costa, M.; et al. Field Intercomparison of Radiometers Used for Satellite Validation in the 400–900 nm Range. *Remote. Sens.* **2019**, *11*, 1129, doi:10.3390/rs11091129. [CrossRef]

21. Zibordi, G.; Ruddick, K.; Ansko, I.; Moore, G.; Kratzer, S.; Icely, J.; Reinart, A. In situ determination of the remote sensing reflectance: An inter-comparison. *Ocean. Sci.* **2012**, *8*, 567–586, doi:10.5194/os-8-567-2012. [CrossRef]

22. Zibordi, G.; Mélin, F.; Berthon, J.-F. AERONET-OC: A Network for the Validation of Ocean Color Primary Products. *J. Atmos. Ocean. Technol.* **2009**, *26*, 1634–1651, doi:10.1175/2009JTECHO654.1. [CrossRef]

23. Gergely, M.; Zibordi, G. Assessment of AERONET-OC L-WN uncertainties. *Metrologia* **2014**, *51*, 40–47, doi:10.1088/0026-1394/51/1/40. [CrossRef]

24. Talone, M.; Zibordi, G.; Ansko, I.; Banks, A.C.; Kuusk, J. Stray light effects in above-water remote-sensing reflectance from hyperspectral radiometers. *Appl. Opt.* **2016**, *55*, 3966–3977, doi:10.1364/AO.55.003966. [CrossRef]

25. Kuusk, J.; Ansko, I.; Bialek, A.; Vendt, R.; Fox, N. Implication of Illumination Beam Geometry on Stray Light and Bandpass Characteristics of Diode Array Spectrometer. *IEEE J. Sel. Top. Appl. Earth Obs. Remote. Sens.* **2018**, pp. 1–8, doi:10.1109/JSTARS.2018.2841772. [CrossRef]

26. Talone, M.; Zibordi, G. Non-linear response of a class of hyper-spectral radiometers. *Metrologia* **2018**, *55*, 747–758, doi:10.1088/1681-7575/aadd7f. [CrossRef]

27. Zibordi, G.; Talone, M.; Jankowski, L. Response to Temperature of a Class of In Situ Hyperspectral Radiometers. *J. Atmos. Ocean. Technol.* **2017**, *34*, 1795–1805, doi:10.1175/JTECH-D-17-0048.1. [CrossRef]

28. Hueni, A.; Bialek, A. Cause, Effect, and Correction of Field Spectroradiometer Interchannel Radiometric Steps. *IEEE J. Sel. Top. Appl. Earth Obs. Remote. Sens.* **2017**, *10*, 1542–1551, doi:10.1109/JSTARS.2016.2625043. [CrossRef]

29. Mekaoui, S.; Zibordi, G. Cosine error for a class of hyperspectral irradiance sensors. *Metrologia* **2013**, *50*, 187–199, doi:10.1088/0026-1394/50/3/187. [CrossRef]

30. JCGM100. Evaluation of Measurement Data - Guide to the Expression of Uncertainty in Measurement. Guidance Document, BIPM, 2008. Available online: https://www.bipm.org/utils/common/documents/jcgm/JCGM100-2008-E.pdf (accessed on 5 February 2020).

31. JCGM101. Evaluation of Measurement Data - Supplement 1 to the "Guide to the Expression of Uncertainty in Measurement" - Propagation of Distributions Using a Monte Carlo Method. Guidance Document, BIPM, 2008. Aviaiable online: https://www.bipm.org/utils/common/documents/jcgm/JCGM-101-2008-E.pdf (accessed on 5 February 2020).

32. Holben, B.N.; Eck, T.F.; Slutsker, I.; Tanré, D.; Buis, J.P.; Setzer, A.; Vermote, E.; Reagan, J.A.; Kaufman, Y.J.; Nakajima, T.; et al. AERONET—A Federated Instrument Network and Data Archive for Aerosol Characterization. *Remote Sens. Environ.* **1998**, *66*, 1 – 16. doi:10.1016/S0034-4257(98)00031-5. [CrossRef]

33. FIDUCEO FIDelity and Uncertainty in Climate Data Records from Eart Obervations. Available online: https://www.fiduceo.eu/ (accessed on 5 February 2020).

34. Spectral Response Function Data. Available online: https://sentinel.esa.int/web/sentinel/technical-guides/sentinel-3-olci/olci-instrument/spectral-response-function-data (accessed on 5 February 2020).

35. Vermote, E.F.; Tanre, D.; Deuze, J.L.; Herman, M.; Morcette, J.. Second Simulation of the Satellite Signal in the Solar Spectrum, 6S: an overview. *IEEE Trans. Geosci. Remote. Sens.* **1997**, *35*, 675–686, doi:10.1109/36.581987. [CrossRef]

36. Kotchenova, S.Y.; Vermote, E.F. Validation of a vector version of the 6S radiative transfer code for atmospheric correction of satellite data. Part II. Homogeneous Lambertian and anisotropic surfaces. *Appl. Opt.* **2007**, *46*, 4455–4464. doi:10.1364/AO.46.004455. [CrossRef] [PubMed]

37. Sea-Bird Scientific HyperOCR Radiometer. Available online: https://www.seabird.com/hyperspectral-radiometers/hyperocr-radiometer/family?productCategoryId=54627869935 (accessed on 5 February 2020).

38. Zibordi, G.; Voss, K.J. Geometrical and spectral distribution of sky radiance: Comparison between simulations and field measurements. *Remote. Sens. Environ.* **1989**, *27*, 343–358, doi:10.1016/0034-4257(89)90094-1. [CrossRef]

39. Coombes, C.; Harrison, A. Calibration of a three-component angular distribution model of sky radiance. *Atmosphere-Ocean* **1988**, *26*, 183–192, doi:10.1080/07055900.1988.9649298. [CrossRef]

40. Talone, M.; Zibordi, G. Polarimetric characteristics of a class of hyperspectral radiometers. *Appl. Opt.* **2016**, *55*, 10092–10104, doi:10.1364/AO.55.010092. [CrossRef] [PubMed]

41. Ruddick, K.; Cauwer, V.D.; Mol, B.V. Use of the near infrared similarity reflectance spectrum for the quality control of remote sensing data. In *Remote Sensing of the Coastal Oceanic Environment*; Frouin, R.J., Babin, M., Sathyendranath, S., Eds.; International Society for Optics and Photonics: San Francisco, CA, USA, 2005; Volume 5885, pp. 1–12, doi:10.1117/12.615152. [CrossRef]

42. Ruddick, K.G.; De Cauwer, V.; Park, Y.J.; Moore, G. Seaborne measurements of near infrared water-leaving reflectance: The similarity spectrum for turbid waters. *Limnol. Oceanogr.* **2006**, *51*, 1167–1179, doi:10.4319/lo.2006.51.2.1167. [CrossRef]

43. Mobley, C.D. Estimation of the remote-sensing reflectance from above-surface measurements. *Appl. Opt.* **1999**, *38*, 7442–7455, doi:10.1364/AO.38.007442. [CrossRef]

44. Sentinel-3 OLCI-A Spectral Response Functions. Technical Note, ESA. S3-TN-ESA-OL-660. Available online: https://sentinel.esa.int/documents/247904/2700436/Sentinel-3-OLCI-B-spectral-response-functions (accessed on 5 February 2020).

Article

ROSACE: A Proposed European Design for the Copernicus Ocean Colour System Vicarious Calibration Infrastructure

David Antoine [1,2,*], Vincenzo Vellucci [3], Andrew C. Banks [4], Philippe Bardey [5], Marine Bretagnon [6], Véronique Bruniquel [6], Alexis Deru [6], Odile Hembise Fanton d'Andon [6], Christophe Lerebourg [6], Antoine Mangin [6], Didier Crozel [7], Stéphane Victori [7], Alkiviadis Kalampokis [4], Aristomenis P. Karageorgis [8], George Petihakis [4], Stella Psarra [4], Melek Golbol [3], Edouard Leymarie [1], Agnieszka Bialek [9], Nigel Fox [9], Samuel Hunt [9], Joel Kuusk [10], Kaspars Laizans [10] and Maria Kanakidou [11]

[1] Laboratoire d'Océanographie de Villefranche LOV, CNRS, Sorbonne Université, F-06230 Villefranche-sur-Mer, France; leymarie@obs-vlfr.fr

[2] Remote Sensing and Satellite Research Group, School of Earth and Planetary Sciences, Curtin University, Perth, WA 6845, Australia

[3] Institut de la Mer de Villefranche IMEV, CNRS, Sorbonne Université, F-06230 Villefranche-sur-Mer, France; enzo@imev-mer.fr (V.V.); melek.golbol@imev-mer.fr (M.G.)

[4] Hellenic Centre for Marine Research, HCMR, Institute of Oceanography, Hersonissos, 71500 Crete, Greece; andyb@hcmr.gr (A.C.B.); alkiviadis.kalampokis@hcmr.gr (A.K.); gpetihakis@hcmr.gr (G.P.); spsarra@hcmr.gr (S.P.)

[5] ACRI-IN, Sophia Antipolis, 06904 Valbonne, France; prb@acri.fr

[6] ACRI-ST, Sophia Antipolis, 06904 Valbonne, France; marine.bretagnon@acri-st.fr (M.B.); veronique.bruniquel@acri-st.fr (V.B.); alexis.deru@acri-st.fr (A.D.); oha@acri-st.fr (O.H.F.d.); christophe.lerebourg@acri-st.fr (C.L.); antoine.mangin@acri-st.fr (A.M.)

[7] CIMEL Électronique, 75011 Paris, France; didier.crozel@cimel.fr (D.C.); s-victori@cimel.fr (S.V.)

[8] Hellenic Centre for Marine Research, HCMR, Institute of Oceanography, 19013 Anavyssos, Greece; ak@hcmr.gr

[9] National Physical Laboratory, Teddington TW11 0LW, UK; agnieszka.bialek@npl.co.uk (A.B.); nigel.fox@npl.co.uk (N.F.); sam.hunt@npl.co.uk (S.H.)

[10] Tartu Observatory, University of Tartu, 61602 Tõravere, Estonia; joel.kuusk@ut.ee (J.K.); kaspars.laizans@ut.ee (K.L.)

[11] Department of Chemistry, University of Crete, 71003 Heraklion, Greece; mariak@uoc.gr

* Correspondence: david.antoine@obs-vlfr.fr or david.antoine@curtin.edu.au; Tel.: +61-8-9266-3572

Received: 15 April 2020; Accepted: 6 May 2020; Published: 12 May 2020

Abstract: The European Copernicus programme ensures long-term delivery of high-quality, global satellite ocean colour radiometry (OCR) observations from its Sentinel-3 (S3) satellite series carrying the ocean and land colour instrument (OLCI). In particular, the S3/OLCI provides marine water leaving reflectance and derived products to the Copernicus marine environment monitoring service, CMEMS, for which data quality is of paramount importance. This is why OCR system vicarious calibration (OC-SVC), which allows uncertainties of these products to stay within required specifications, is crucial. The European organisation for the exploitation of meteorological satellites (EUMETSAT) operates the S3/OLCI marine ground segment, and envisions having an SVC infrastructure deployed and operated for the long-term. This paper describes a design for such an SVC infrastructure, named radiometry for ocean colour satellites calibration and community engagement (ROSACE), which has been submitted to Copernicus by a consortium made of three European research institutions, a National Metrology Institute, and two small- to medium-sized enterprises (SMEs). ROSACE proposes a 2-site infrastructure deployed in the Eastern and Western Mediterranean Seas, capable of delivering up to about 80 high quality matchups per year for OC-SVC of the S3/OLCI missions.

Remote Sens. **2020**, *12*, 1535

Keywords: satellite ocean colour; system vicarious calibration; fiducial reference measurements; radiometry; SI-traceability; uncertainty budget; Mediterranean Sea; BOUSSOLE; MSEA

1. Introduction

As of 2020, a number of low-earth-orbit satellites together provide systematic coverage of ocean colour radiometry (OCR) observations over the world's oceans and coastal zones. Two programmes among this constellation are operational, which means that they are expected to maintain delivery of their observations and products over the long-term, in order to sustain a variety of uses, from, e.g., science of the long-term, climate-driven trends of the marine ecosystem, to services to government and industry users, e.g., water quality monitoring. These two programmes are the joint polar satellite system (JPSS) of the US National Oceanographic and Atmospheric Administration (NOAA), and the European Copernicus programme. The JPSS satellites carry the visible infrared imaging radiometer suite (VIIRS), delivering OCR observations in eight spectral bands in the visible and near infrared (VisNIR) spectral region. The Copernicus Sentinel-3 (S3) satellites carry the ocean and land colour instrument (OLCI), delivering OCR observations in twenty spectral bands in the VisNIR.

For these and any other OCR missions to deliver products of the desired accuracy, a system vicarious calibration (SVC) programme has to run over their entire lifetime [1], which is also a requirement for other types of Earth observation missions that intend to address climate change related questions [2]. Ocean colour SVC (OC-SVC) consists of adjusting the prelaunch mission spectral calibration coefficients of the onboard radiometers through comparison of the top-of-atmosphere (TOA) radiance measured by the mission, to the same quantity derived from using high-quality radiometry data collected in the field and radiative transfer computations to propagate the bottom-of-atmosphere field measurements to TOA [3]. SVC therefore requires a sustained post-launch field programme to collect the necessary data, with the aim of maintaining the level of uncertainty of the derived satellite products within predefined requirements [4–7]. This need for SVC has been demonstrated since the early times of the satellite OCR era [8]. Modern satellite OCR observations must indeed provide the water-leaving radiance in the blue part of the electromagnetic spectrum with a < 5% accuracy over meso- to oligotrophic waters [9,10], which translates as an uncertainty of about 0.002 in reflectance at 443 nm [11] when reflectance is modelled from chlorophyll following [12]. Since the marine signal is generally < 10% of the TOA radiance measured by the spaceborne sensor, achieving this goal requires that the instruments be calibrated to better than 1% uncertainty. This cannot be reached using only onboard calibration devices such as sun diffusers, hence the need for OC-SVC. It is worth remembering that OC-SVC is not an absolute calibration of the sensor. It is an adjustment of the overall response of the sensor plus the atmospheric correction algorithm [1,8]. The goal is to absorb residual uncertainties in order to obtain final products, e.g., the normalized water-leaving reflectance or radiance with the desired accuracy.

Until now, two field programmes have provided space agencies with OC-SVC-quality field observations, both using a moored optical buoy. The marine optical buoy (MOBY [13,14]) has been in operation since 1995 off the island of Lanaï in the Hawaiian archipelago. It continuously delivers hyperspectral reflectance for SVC of the NASA and NOAA instruments (sea-viewing wide field-of-view sensor, SeaWiFS, moderate resolution imaging spectroradiometer, MODIS and VIIRS), and has also been used by international missions, e.g., the European Space Agency (ESA) medium resolution imaging spectrometer (ENVISAT/MERIS), and the Japan Aerospace Exploration Agency (JAXA) global change observation mission-climate, second generation global imager (GCOM-C/SGLI). Another SVC programme was set up to support European OCR missions, at that time the ESA ENVISAT/MERIS. This programme is named BOUSSOLE ("BOUée pour l'acquiSition d'une Série Optique à Long termE" [15,16]) and has been operating continuously since 2003 at an open ocean site in the Northwestern Mediterranean Sea. It is currently used for SVC of the Copernicus Sentinel-3A and -3B OLCI instruments and for the GCOM-C/SGLI. Data from both programmes have also been used for the SVC of satellite sensors not specifically designed for ocean colour applications such as the thematic mapper (TM) on Landsat 5 and 7, the operational land imager (OLI) on Landsat 8, and the multispectral instrument (MSI) on Sentinel-2A&B [17–19].

The European Organisation for the Exploitation of Meteorological Satellites (EUMETSAT) is in charge of the S3/OLCI marine ground segment as part of Copernicus operations. As such, they envision having a fully independent European in situ OC-SVC infrastructure deployed and operated for the long-term, as has been recommended for the support of Copernicus ocean colour missions [20]. In order to properly set up this significant investment, they first commissioned a study in 2017 to summarize requirements for such an infrastructure [21]. They then called for ideas for what this European OC-SVC infrastructure could be through an invitation to tender (ITT) in 2018, from which they commissioned two studies that worked in parallel during 2019 and delivered their proposed preliminary designs by December of the same year.

This paper summarises the main characteristics of one of the proposed OC-SVC infrastructures, named ROSACE, which stands for "Radiometry for Ocean colour Satellites Calibration and community Engagement". The high-level rationale for the proposed solution is presented first. The characteristics of the two sites that compose the field segment are then detailed along with the deployment platform, followed by the overall strategy and equipment for the optical system and its calibration. The methodology used to assess the preliminary uncertainty budget of the proposed system is then summarised and results presented. Finally the organisation of the ground segment and the infrastructure operations are outlined. In addition, an autonomous platform capable of hosting the optical system is presented as an option to improve the overall capacity of the infrastructure.

2. High-Level Rationale for the Proposed ROSACE OC-SVC Infrastructure

The high-level rationale underlying the ROSACE preliminary design (Figure 1) is to:

1. Take advantage as much as possible of existing European expertise in OC-SVC that has been developed in the past two decades under European (ESA in particular) and national funding, including field operations, SI-traceability and evaluation of uncertainties, associated data quality assurance and quality control and processing and generation of satellite matchups and OC-SVC gains.
2. Reinforce the above and, in addition, build a new European capability in the domain of OC-SVC field radiometry, in order to ensure long-term sovereignty and stability of the Copernicus SVC infrastructure. This includes two sites, collaboration with a national metrology institute (NMI), and new technology developments. The two field sites are in the Ligurian Sea (BOUSSOLE heritage) and the Cretan Sea (MSEA). They were selected by the ESA "Fiducial Reference Measurements For Satellite Ocean Colour" (FRM4SOC [22,23]) study as the two of the best European locations among those currently under evaluation, in part because significant logistical capabilities adapted to maintaining an OC-SVC infrastructure already exist at these sites.
3. Rely on strong national support in conjunction with the main Copernicus funding to ensure that the ROSACE OC-SVC infrastructure is backed by long-term sustainable staff, capability and logistical resources.
4. Be ready to incorporate additional partnerships in order to improve redundancy inside the OC-SVC infrastructure, to enlarge the database onto which the uncertainty budget is built, to improve, if needed, the methodological baseline used to compute the SVC gains, and to increase the matchup capacity of the OC-SVC infrastructure.

In this logic, collaboration with a possible third site in the Eastern Indian Ocean (EIO, off Western Australia) is envisaged, which has also been identified by the FRM4SOC study (see also [24]). This is an option for future extension of the infrastructure, only possible if the EIO site would have the technical capability. Additionally, included in the preliminary design is the possibility to increase the capacity of the infrastructure to deliver in situ data by deploying autonomous profiling floats [25].

5. Ensure close communication with international bodies that work on establishing OC-SVC requirements and fostering international collaboration (e.g., the International Ocean Colour Coordinating Group, IOCCG and the Committee on Earth Observation Satellites Ocean Colour Virtual Constellation, CEOS OCR-VC).

6. Maintain development activities that are vital to allow improvement in data acquisition and processing procedures and ultimately improve the data quality of the OC-SVC infrastructure, and also to better secure national support for this European infrastructure.

The rationale supporting ROSACE is rather simple. It consists of leveraging the two-decade-long experience as well as previous European investments in order to be in a position to deliver an operational system meeting the Copernicus requirements in the time frame anticipated by EUMETSAT, i.e., likely in line with the launch of the Sentinel-3 C and D units and beyond (from 2023/2024).

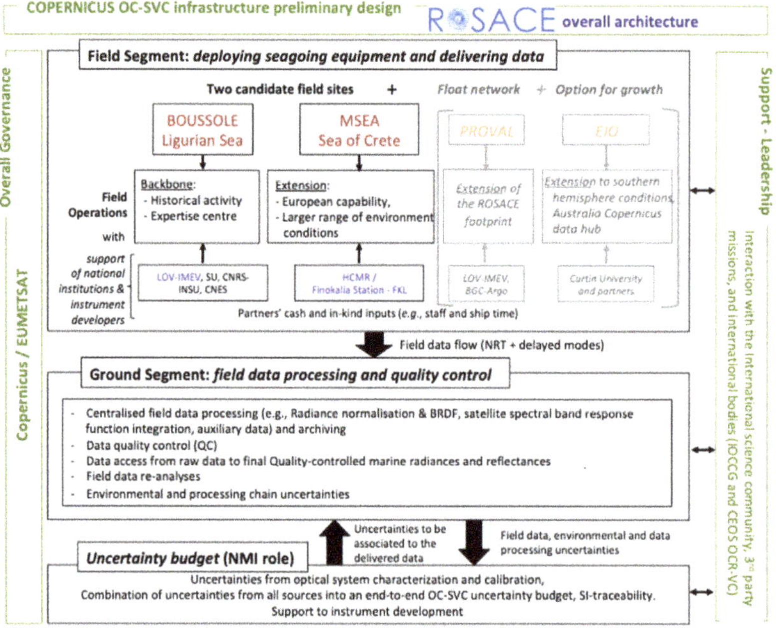

Figure 1. Overall architecture proposed for radiometry for ocean colour satellites calibration and community engagement (ROSACE).

That is why ROSACE capitalises on the single existing European OC-SVC infrastructure, i.e., the BOUSSOLE site and project, as the backbone that supports the development of a more European-integrated system, including a second European site in the best suited location of the European Seas for OC-SVC, i.e., the MSEA site. This strategy appears as the most robust approach to achieve, in a safely managed way, the full compliance with the stringent requirements of an operational programme like Copernicus/Sentinels, including lowering the uncertainties in OC-SVC.

ROSACE also incorporates innovative solutions, in the development of specifically designed European instruments and calibration devices that will fully match the OC-SVC requirements and fulfils the need for modularity, and the possibility to integrate a network of autonomous profiling floats to complement the fixed sites.

The success of such a long-term service for the Copernicus programme also depends on adequate competence transfers and periodic training. This is one of the roles assigned to universities and public research institutions involved in ROSACE, i.e., to transfer knowledge, and it constitutes a fundamental task for building a long-term infrastructure that will operate across generations.

3. The Field Segment

3.1. Metrology Rationale

At the present time, SVC is the only means to achieve the uncertainties needed for remotely sensed OCR observations. It is well established that the SVC reference system (radiometry measurements) needs to be robustly tied to the International System of Units (SI) and be collected in meso- to oligotrophic waters, ideally from a location with stable oceanographic conditions [6,8]. Provided that there are enough reliable observations of the site from space for any given sensor to allow statistically reliable calibration, it could be argued that the SVC requirements can be achieved by a single infrastructure, in a single location. This would also require that the instrumentation, maintenance and recalibration, is unequivocal and guaranteed to be operational in this manner for the foreseeable future. In the case of climate monitoring and the Copernicus series of sensors this timescale must span many decades. This guarantee must stand irrespective of funding and unexpected events, either natural or manmade. The 2020 global pandemic [26] is an example where large numbers of assets had to be recovered from sea because their servicing or emergency recovery were no longer feasible during lockdown measures taken by many governments (e.g., [27]). Equipment or site redundancy and autonomous platforms can help in such circumstances.

It soon becomes clear that implementation of a single site infrastructure is of high risk and in engineering terms considered a single critical point of failure. Even if multiple spares are ready and waiting to be deployed, a single (non-independently checked) calibration route provides a risk that no metrology institute would safely rely upon for any measurement. Primary radiometric scales and ancillary measurements are regularly checked across international borders through formal comparison with peers to ensure consistency and avoidance of any potential errors, (standards/procedures/typographical). This comparison process is a fundamental requirement to ensure international consistency of measurements and trade.

It thus becomes apparent that a minimum requirement to ensure long term consistency and reliability of a reference measurement system for OC-SVC is two independently calibrated systems of similar performance and in different geographical locations. It is also clear that providing the two independent measurement systems always agree (within their uncertainties) then this strategy of two measurement systems can probably be relied upon. However, as anyone making comparative measurements knows, if the two references disagree then you are left with the question of which one do you believe?

For this reason it is normal practise, for any reference, e.g., a standard light source (FEL lamp) or detector, to consist of a group of a minimum of three entities, to provide redundancy and the likelihood of at least two being consistent and a probabilistic indicator of closeness to the truth. The recent redefinition of the kilogram for example was only considered reliable and acceptable once there were three independent measurements made that were consistent within their uncertainties (also below a prescribed minimum level) and that at least two of the measurements had to be from a completely different traceability route. A robust long-term reference for OC-SVC should follow the same minimal metrological rigour, i.e., at least three independent measurement infrastructures and at least two independent traceability paths. The ROSACE proposal of two linked European sites, when added to the existing North American MOBY infrastructure, meets both these metrological requirements.

There are many other benefits, such as a redundancy from unforeseen system failures, increased number of matchup opportunities, randomisation of environmental effects, etc. but fundamentally, a global system of just two sites would be metrologically and operationally high risk. Following the same metrological rationale, ROSACE will allow "built in" systematic round robin intercalibration exercises to validate (1) the calibration procedures between the two field sites and a third independent calibration system operated by the Tartu Observatory and (2) the consistency of measurements and uncertainties between the two sites with an independent transfer profiling radiometer (see also Section 7.1).

3.2. Practical Considerations, Specific Role of Each Site

Other elements, of a more practical type, have also been considered in selecting two sites to develop the ROSACE OC-SVC infrastructure. Essentially, what we proposed was not a simple doubling of identical (twin) sites. It rather takes advantage of specific aspects present at the two locations in order to leverage their complementarity. These elements are:

- Geophysical properties are similar yet different enough at the two sites, so that they are complementary, not simply redundant (see the sites descriptions in Sections 3.3 and 3.4).
- Expertise in deploying and maintaining large oceanographic buoys is similar at the two sites, i.e., the Laboratoire d'Océanographie de Villefranche from the Institut de la Mer de Villefranche (LOV-IMEV) and the Hellenic Centre for Marine Research (HCMR). However, a longer and more developed experience with optics, radiometry and OC-SVC exists at LOV-IMEV.
- BOUSSOLE buoys and instrumentation exist at LOV-IMEV, ready to be used (until improved versions become available) in continuation of the present effort and to host the new optical system for testing, and ensure continuity of the time series.

Therefore, the logic is to use the BOUSSOLE site not only for operational delivery of OC-SVC data, which certainly remains a main objective, but also as a development and test site, where:

- Improvements in the buoy structure can be tested before new versions are built and operationally deployed at MSEA and BOUSSOLE.
- The new optical system can be deployed in parallel to the one currently used (SeaBird HyperOCR radiometers). This is not to qualify the new radiometer system, which will be of superior radiometric quality to that of the current one, but rather is an opportunity to ensure continuity of the time series between MERIS and OLCI observations.
- Testing of new equipment or further improvement of the optical system, can be carried out on one site before being transitioned to permanent upgrades at both sites.

3.3. The BOUSSOLE Site

3.3.1. Location and General Characteristics

The BOUSSOLE site is located at 32 nmi (59 km) offshore from Nice (France) in the NW Mediterranean Sea (43°22′ N, 7°54′ E; Figure 2). Water depth at the site is about 2440 m. This site has been already described in detail in [28], which readers are referred to. Here we provide a summary of the features that are most relevant to OC-SVC. Time series, histograms and climatological values of a number of parameters can be found in [29].

3.3.2. Weather and General Hydrology

Wind speeds at BOUSSOLE are generally moderate (Figure 3a). Over the past 15 years, only 3% of recorded wind speeds were greater than 15 m s^{-1}, and 16% of records were above 10 m s^{-1}. These higher wind speeds, and the associated larger waves are concentrated in the 5 months from November to March. Dominant wind directions are from the west to southwest and from the northeast sectors (Figure 3b). These are channelled into these two main directions on the one hand by the general atmospheric circulation of the region, and, on the other hand, by the topography formed by the Alps and Corsica.

Significant wave heights at BOUSSOLE were mostly less than 2 m, although values as high as 5 m could be occasionally recorded (Figure 3c). The wave period (not shown) was around 4–6 s, which is typical of the Mediterranean Sea.

This site was selected in particular because currents are extremely low. This peculiarity is due to the position close to the centre of the cyclonic circulation that characterizes the Ligurian Sea. The northern branch of this circulation is the Ligurian current, forming a jet flowing close to the shore in the NE to SW direction and creating a front whose position is seasonally varying, closer to the shore in winter than in summer. The southern branch is a SW to NE current flowing north of Corsica and the eastern part of the circulation is simply imposed by the geometry of the basin. The 17 years of BOUSSOLE deployments have confirmed the low-current character of the site (high currents would have been identified as strong buoy tilts). Stronger currents have been very occasionally revealed through observing the buoy plunging unusually deep. This situation occurred only four times during the 17 years of deployment and, for each instance, the event lasted generally less than a day.

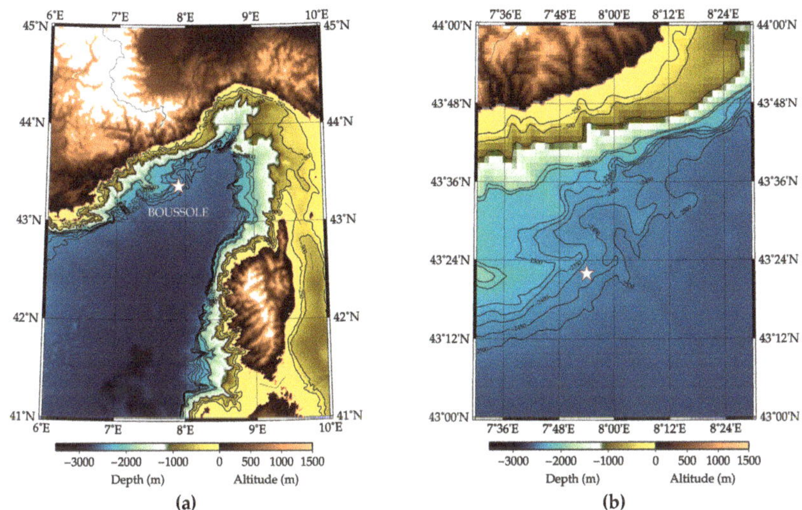

Figure 2. (a) The BOUSSOLE site location in the NW Mediterranean Sea (white star) including bathymetry of the area and (b) zoom on the site itself showing more details of the bathymetry.

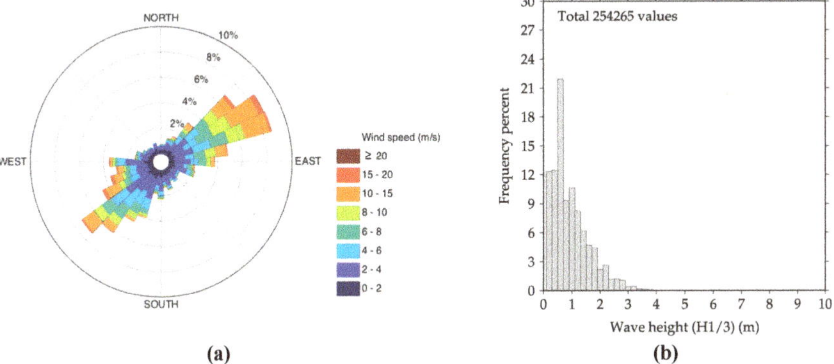

Figure 3. (a) Wind speed and direction and (b) significant wave height, from data collected by the Côte d'Azur meteorological buoy (Météo-France) located 2 nmi from BOUSSOLE.

The temperature and salinity conditions at BOUSSOLE are illustrated in Figure 4a. The minimum sea surface temperature (SST) was about 12.7 °C (associated with a salinity of 38.4 psu), which is a constant value reached in the coldest winters when the water mass was fully mixed down to the bottom. This deep mixing contributed to the formation of the bottom waters of the Western Mediterranean Sea.

3.3.3. Phytoplankton Chlorophyll-a, Water Type and Inherent Optical Properties

The seasonality of the physical forcing drove the seasonal changes of nutrients and phytoplankton, as illustrated here by the surface nitrate and total chlorophyll-*a* concentrations ([TChl-*a*], Figure 4b). Oligotrophic conditions prevailed during the summer with undetectable nitrate levels and [TChl-*a*] lower than 0.1 mg m^{-3} (with a minima around 0.05 mg m^{-3}). The higher concentrations are on average in the range 0.8-1.7 mg m^{-3} (with a maxima up to about 5 mg m^{-3}), during the early spring bloom (February to March or April) when surface waters are nitrate replete. Moderate [TChl-*a*], between 0.1 and 0.2 mg m^{-3}, characterize most of the other periods of the year. Interannual variability of physical forcing, specifically the depth of the winter convection, determines the magnitude of the phytoplankton spring bloom [30].

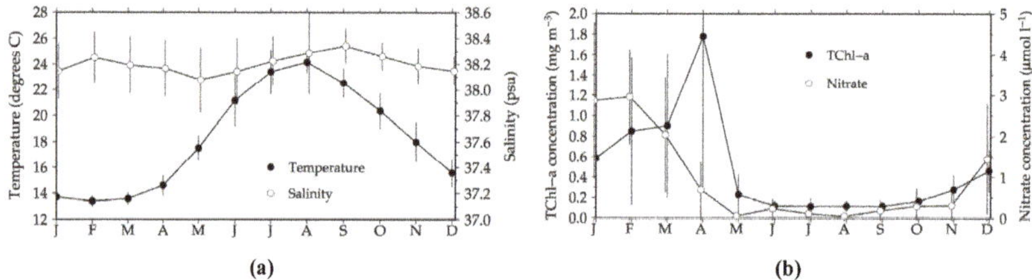

(a)   (b)

Figure 4. (**a**) Average seasonal cycles of sea-surface temperature (black symbols) and sea-surface salinity (open symbols) at BOUSSOLE (depth < 10 m), from the buoy data collected from 2003 to 2019. (**b**) Average seasonal cycles of sea-surface [TChl-*a*] (black symbols) and nitrate concentrations (open symbols). Vertical bars are standard deviations of the displayed monthly means (slightly shifted between the two parameters for clarity).

An important consequence of the above characteristics, in particular water depth, circulation, and distance from shore, is that waters at the BOUSSOLE site are permanently of the Case-1 category (following [31]). This assertion is quantitatively evaluated by plotting the irradiance reflectance at 560 nm, determined from the buoy measurements, as a function of [TChl-*a*] (Figure 5), and superimposing on top of the data a theoretical upper limit of this reflectance for Case-1 waters [32]. With the exception of a few outliers, all data points are below the curve, demonstrating that waters permanently belong to the Case-1 type at the BOUSSOLE site.

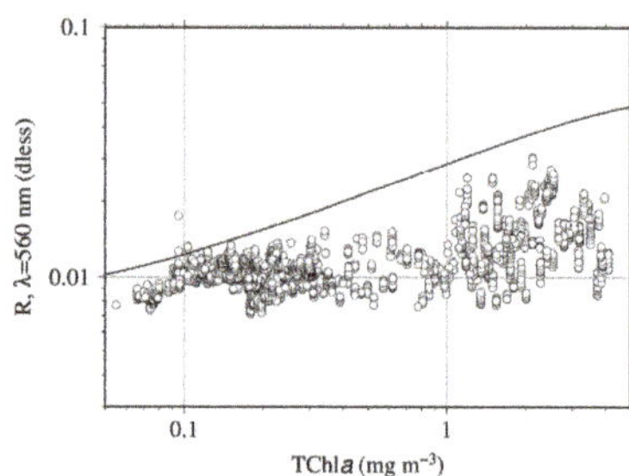

Figure 5. Irradiance reflectance at 560 nm, R(560), as a function of [TChl-*a*]. The points are from 3 years of clear-sky quality-checked buoy measurements taken within one hour of solar noon. The curve is the upper limit for Case-1 waters [32] (adapted from [28]).

Inherent optical properties (IOPs) measured at the BOUSSOLE site include the particulate absorption coefficient (a_p, from samples collected during monthly cruises), the particulate backscattering coefficient, b_{bp}, and the particulate beam attenuation coefficient, c_p. Average seasonal cycles are displayed in Figure 6. These parameters also clearly illustrate the seasonal variability at BOUSSOLE, consistent with the description already provided for [TChl-*a*]. This is expected for typical Case-1 waters.

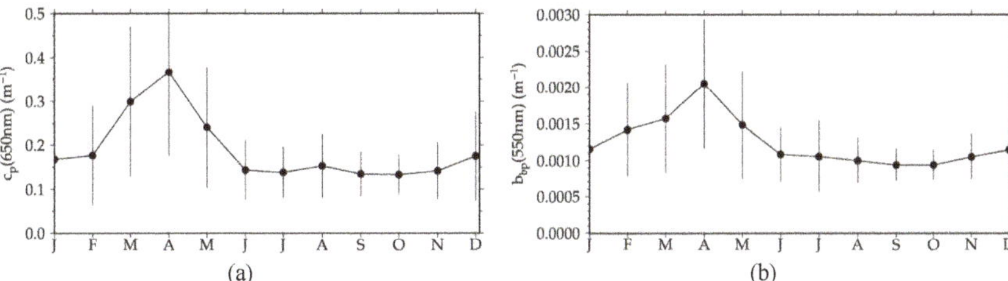

Figure 6. (a) Average seasonal cycles at BOUSSOLE of the particulate beam attenuation coefficient at 650 nm and (b) the particulate backscattering coefficient at 550 nm, from the buoy data collected from 2003 to 2019 (depth < 10 m).

3.3.4. Remote-Sensing Reflectance

Underwater downward and upward irradiances, upwelling nadir radiances and above-water downward irradiances have been measured continuously at BOUSSOLE since September 2003, initially with 7-band multispectral radiometers (Satlantic 200-series; wavelengths 412, 443, 490, 510, 560, 670 and 683 nm) and, from 2007, also using hyperspectral radiometers (Satlantic HyperOCR series; 350–850 nm with a 3 nm resolution). The multispectral instruments have been decommissioned at the end of 2017.

Apparent optical properties (AOPs) are derived from these measurements, such as the diffuse attenuation coefficient for downward irradiance or the remote sensing reflectance [16,28]. The latter is shown in Figure 7, for λ = 443 nm, as an example illustrating the seasonal signal as well as the quasi-systematic observations through the 14 years displayed here.

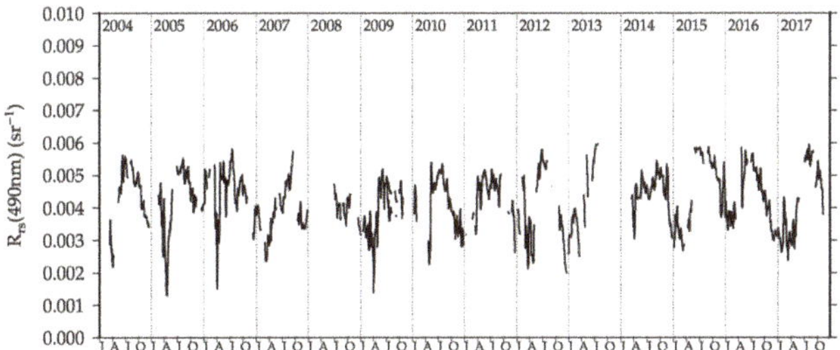

Figure 7. 2004–2017 time series of the remote sensing reflectance at 443 nm at BOUSSOLE.

3.3.5. Atmospheric Parameters

BOUSSOLE benefits from the general conditions prevailing in the Mediterranean Sea, which is known for being a rather clear-sky area. Seasonal cycles derived from the MODIS cloud fraction products are displayed in Figure 8a and they clearly illustrate these characteristics.

The aerosol optical depth (AOD) and aerosol Angstrom exponent have been derived both from measurements at a coastal AERONET site (Cap Ferrat; 43°41′ N, 7°20′ E; automatic CIMEL CE318 sun photometer) and from measurements at the BOUSSOLE site offshore, using a hand-held CIMEL CE317 sun photometer. These measurements reveal that AODs are actually extremely similar at both sites (Figure 8b), with seasonal values from around 0.05 to 0.1 at 865 nm. The Angstrom exponents, which are determined by the aerosol types and their size distribution, are however markedly different, with values steadily around 1.4 at the Cap Ferrat station and around 0.6 at the BOUSSOLE site. This is also expected, because coastal aerosols include a larger proportion of small particles of continental origin than marine aerosols, and therefore have a larger Angstrom exponent.

The total ozone content over BOUSSOLE varied from about 290 to 400 Dobson units (DU, not shown), with maxima generally in winter and minima at the end of summer. Average values in summer were around 340 DU. These values were derived from data of the NASA AIRS (Atmospheric Infrared Sounder), also indicating a decline of concentrations since around 2011.

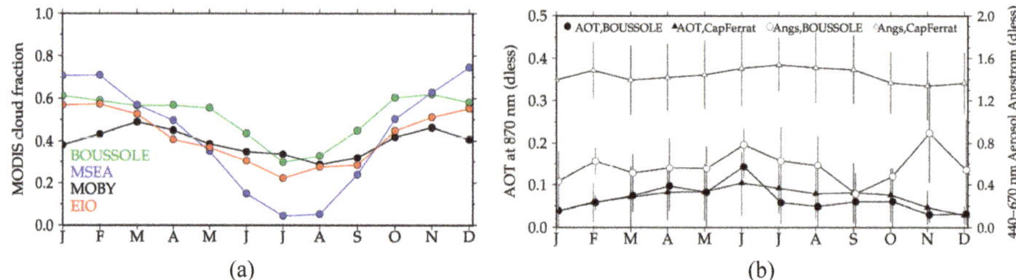

(a)   (b)

Figure 8. (a) Average seasonal cycles of the cloud fraction (moderate resolution imaging spectroradiometer (MODIS) product [33], average from 2012 to 2015) for the four sites indicated. The curve for the southern hemisphere Eastern Indian Ocean (EIO) site is artificially shifted by 6 months. (b) Average seasonal cycles of the aerosol optical thickness at 865 nm (black symbols) and of the aerosol Angstrom exponent (open symbols), both at the Cap Ferrat coastal AERONET station (triangles) and at the offshore BOUSSOLE site (circles).

3.3.6. Spatial Homogeneity

In situ measurements have a horizontal sampling scale on the order of tens of metres, whereas satellite-derived quantities have sampling scales of hundreds of metres to 1 km. Therefore, when comparing parameters derived from both approaches, an important aspect to consider is the spatial homogeneity of the measurement site.

Spatial surveys have, therefore, been conducted during BOUSSOLE monthly cruises [34], by following a grid pattern of about one square nautical mile centred on the buoy site, during which along-track fluorescence measurements were performed. Changes of this small-scale spatial variability of the chlorophyll-a fluorescence are illustrated in Figure 9 (see also Figure 4 in [28]). They show that the variability is, as expected, low during the oligotrophic summer (around ±10%). The horizontal gradients can reach larger values during the spring bloom and during fall, with values from −70% to +35% (not shown).

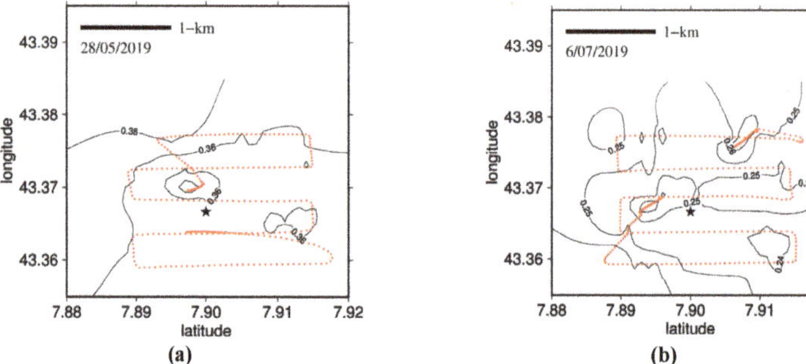

(a)   (b)

Figure 9. Example of grid surveys performed (**a**) in May and (**b**) in July 2019. The red dotted line shows the ship track, and the black curves are the contours of the chlorophyll-*a* fluorescence (values are indicated on the contour lines).

A ±10% change of chlorophyll when chlorophyll was initially 0.1 mg m-3 (oligotrophic conditions at BOUSSOLE) translates into changes in reflectance at 443 nm of the order of 0.0015 (3.5%).

3.3.7. Summary of the BOUSSOLE Site Characteristics Relevant to OC-SVC

The above sections have shown that BOUSSOLE is a fully characterised site in all aspects that are needed to define relevance for OC-SVC. Extremely few examples such as this exist globally, where hydrology, IOPs, AOPs, biogeochemical quantities (e.g., [TChl-a]) and atmospheric properties have been continuously sampled for nearly two decades.

The site's main geophysical characteristics are:

- Low currents.
- Moderate wind speed/wave height.
- Case-1 waters throughout the year.
- [TChl-a] concentration $< 0.1–0.2$ mg m^{-3} in summer and fall.
- Characterised kilometre-scale spatial variability.
- High occurrence of clear skies.
- Low atmospheric aerosol load throughout the year.

As for the logistics:

- Well-established oceanographic institute (>130 years) close to the site (LOV-IMEV, Villefranche-sur-Mer, France).
- Well-established expertise in marine optics (>60 years).
- Well-trained permanent staff.
- Ships and other necessary equipment all available.
- Proven capacity to manage the BOUSSOLE platform.
- Proven record (>16 years) of uninterrupted acquisition of OC-SVC-quality observations.
- Field site identified on marine charts within an area identified for scientific work.
- Meteorological buoy managed by the French weather forecast agency 2 nmi from BOUSSOLE.

3.4. The MSEA Site

3.4.1. Location and General Characteristics

The MSEA site (Cretan Passage, South Sea of Crete; Figure 10) has been suggested by the global evaluation work of Zibordi and Melin [24], Zibordi et al. [20] and the FRM4SOC international OC-SVC workshop report [22] as the best region in the European Seas for OC-SVC. A more detailed analysis of available in situ oceanographic, optical, atmospheric and satellite data demonstrated very similar conditions in the Cretan Sea (north of Crete Island) and that a field infrastructure meeting the requirements for Copernicus OC-SVC can be deployed there.

When combined with logistical considerations, the best site in the Cretan Sea is in the vicinity of the HCMR operational physical and biogeochemical monitoring buoy (E1-M3A) [35]. This location, referred to as MSEA similarly to [24], is at 35°44′ N, 25°20′ E approximately 26 nmi (48 km) north of the HCMR headquarters in Crete. Specifically, it is 10 nmi east of the E1-M3A buoy to optimize the position with respect to Sentinel-3A and B overpasses.

The site has open ocean characteristics representative of a wider area of the Eastern Mediterranean [36], and a water depth of 1400–1500 m. HCMR has operated the E1-M3A buoy in this locale for over 20 years within the framework of the Monitoring, Forecasting and Information System for the Greek Seas (POSEIDON) network [37] and as part of the European contribution to the global ocean observing system (GOOS) ocean sites operational network [38]. From this buoy, a vast array of oceanographic and biogeochemical data has been, and continues to be, gathered for the area. Augmenting the data from the buoy itself are monthly sampling site visits, FerryBox data from the

Athens-Crete ferry that passes to the west of the buoy, multiple deployment Argo profiler data of the local area and data from an AUV/glider that is autonomously and continuously operating on transects past the buoy [35].

The proximity of the 6000 m^2 HCMR-Crete research centre to the buoy provides the extensive support facilities, in terms of laboratories, human resources and research vessel support (62 m R/V Aegaeo; 24m R/V Philia), that are necessary for OC-SVC buoy operations. Furthermore, the waters surrounding the island of Crete, in the Eastern Mediterranean, are characterized by very low suspended particle concentrations, and correspond to a "no bloom" trophic regime according to [39]. The Eastern Mediterranean, including the Cretan Sea, is generally characterized by oligotrophy throughout the year making it very similar in this respect to the MOBY site near Hawaii [14]. The following sections further detail the site's characteristics that support its prime suitability for Copernicus OC-SVC.

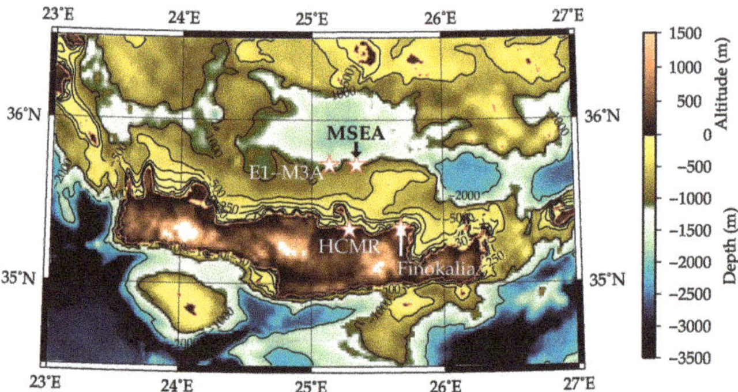

Figure 10. The proposed location of the MSEA site offshore from the Island of Crete, including bathymetry of the area.

3.4.2. Weather and General Hydrology

In general, weather and circulation conditions at the MSEA site were calm and stable with very few extremes. Wind speeds at MSEA were generally lower and varied less than at the BOUSSOLE site. Over the 11-year period analysed (2007–2018) very few days were recorded with an established wind speed above 15 m s^{-1}. The average wind speed at the site was only 5.3 m s^{-1} and the majority were within the range 2–8 m s^{-1}. High wind speeds (> 15 m s^{-1}) and the associated large swells were rare and might occur in the winter season from November to February. Dominant winds were from the W–NW (Figure 11a), and were channelled into this main direction by the general atmospheric circulation of the region. The Meltemi/Etesian N or NW winds are a repeating summer event for the Cretan Sea, occurring mainly in August. They are rarely very intense and average around 6 m s^{-1}. Air temperature and pressure 11 year averages were 20.09 °C (min. 2 °C; max. 32 °C; standard deviation (SD) 4.63 °C) and 1015.20 hPa (min. 995 hPa; max. 1035 hPa; SD 5.73 hPa).

Surface current data were collected from the MSEA site using an acoustic Doppler current profiler (ADCP) installed at 1 m over the last 11 years. The currents were extremely low, i.e., in the entire Cretan Sea the flows away from the coast were on the order of 0.1–0.3 m $^{-1}$, in contrast to the Levantine Basin, where the meso-scale circulation structures at the near-surface layers were characterized by higher velocities reaching 0.4–0.5 m $^{-1}$ [40]. Thus the majority of the currents at the MSEA site were of this low order with a measured average over the last 11 years of 0.23 m $^{-1}$. The dominant currents were from the NNW towards the SSE (Figure 11b) and this was dependent on the position of the semi-permanent slow-moving anticyclonic circulation feature in the Cretan Sea [41].

Wave direction generally follows the prevailing wind direction with a slight offset further west and further north. Significant wave heights also followed the wind strength of the area with nearly all waves under 3 m since 2007 and on average less than 1 m (Figure 11c). High swell and high waves (4–7 m) were extremely rare and occurred sporadically in the winter season from November to February. The wave period for the majority of the 11 years analysed falls within the range of 2–6 s with an average of 3.96 s.

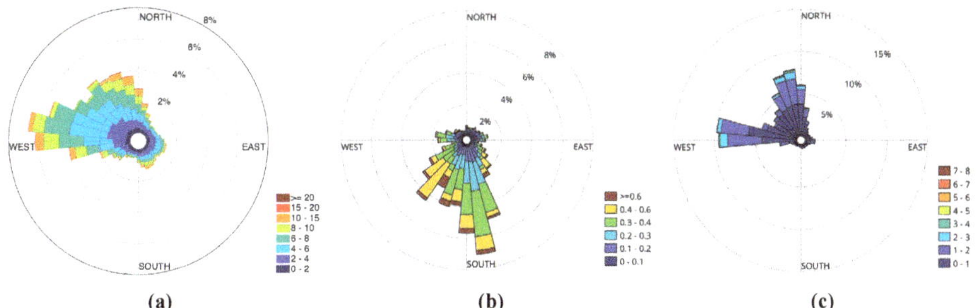

(a)   (b)   (c)

Figure 11. (a) Wind speed (m s^{-1}) and direction, (b) current speed (m s^{-1}) and direction and (c) significant wave height (m) and direction at the MSEA site (2007–2018). Data from the E1-M3A site.

Average surface water temperature and salinity levels at the MSEA site were 21.35 °C and 39.23 psu respectively. The minimum surface temperature was 15 °C (associated with a salinity of 38.6 psu), which was a constant value reached in winter when the water mass was fully mixed down to the bottom. This deep mixing contributes to the formation of the bottom waters of the Eastern Mediterranean Sea. During summer the maximum surface temperature was 29 °C (associated with a salinity of 39.65 psu). Assimilation of these data together with Ferrybox and glider measurements around MSEA into the POSEIDON models have also allowed a better description of the hydrodynamics of the site and the wider southern Aegean area [42].

3.4.3. Phytoplankton Chlorophyll-*a*, Water Type and Inherent Optical Properties

In the Cretan Sea, low concentration [TChl-*a*] maxima tended to occur during late winter to early spring, i.e., late February to March, associated with a very limited phytoplankton spring "bloom". Monthly in situ sampling of essential biogeochemical variables since 2010 and the deployment of a sediment trap at the site contribute to the study of ecosystem functioning and carbon export potential at MSEA [35].

Phytoplankton surface values of [TChl-*a*] in the offshore Cretan Sea waters varied between <0.05 and 0.2 mg m^{-3}, during the stratified and mixing periods, with values < 0.1 mg m^{-3} being dominant throughout the year. Subsurface maxima (around 20 m) may rarely reach 1 mg m^{-3} during the spring bloom while deep chlorophyll maxima (below 75 m) are consistently formed for most of the year [43] and annual primary production values yield < 25 g C m^{-2} [44]. This very low level of surface chlorophyll was confirmed through the analysis of bottle samples collected from the location of the MSEA site since the end of 2012. The HPLC data were considered the most accurate and the triangles in Figure 12 confirmed that chlorophyll concentrations rarely rose above 0.2 mg m^{-3} and values < 0.1 mg m^{-3} dominate.

Based on productivity and light attenuation data, Ignatiades [44] found that the waters of the Cretan Sea have a deep blue colour and an average value of the spectral attenuation coefficient (K$_d$), at 480 nm, of 0.040 m^{-1}. During the EU-funded CINCS project (Pelagic-Benthic Coupling in the oligotrophic Cretan Sea), five oceanographic cruises were conducted in the central Cretan Sea, focusing on the marine sector from the coast off Heraklion and up to 1700 m depth. Transmissometry profiles (at 660 nm) and bottle data from these revealed the presence of very faint nepheloid layers and very low suspended particulate matter concentrations (SPM <1.5 mg m^{-3}) [45].

In terms of hyperspectral data of IOPs and AOPs a few unpublished profiles exist for the MSEA site. These were derived mostly from the EU FP7 project PERSEUS research cruises in the area in 2013.

The first results from the analysis of the complete data set revealed that the Aegean, including the Cretan Sea and MSEA site, has similar backscatter properties to the rest of the Eastern Mediterranean [46], with b_{bp} at low levels across the visible spectrum (0.001–0.003 m^{-1}). The CDOM concentrations were also found to be insignificant with values of around 3–5 ppb. The data therefore shows that MSEA has oligotrophic clear oceanic waters that fit into Jerlov's definition of the most transparent deep blue waters (Case-1) [47].

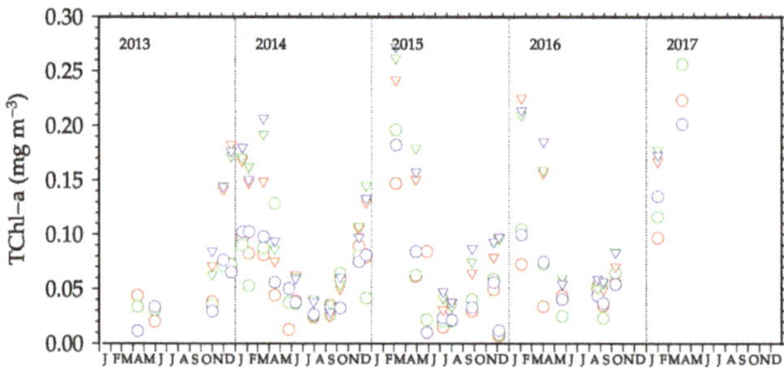

Figure 12. Total chlorophyll-*a* at various depths in the surface layer at the MSEA site (red: 2 m; green: 10 m; blue: 20 m). Circles are for fluorometric [TChl-*a*] determinations and triangles for HPLC.

3.4.4. Remote-Sensing Reflectance

HCMR runs a marine optics suite that has been taking profiles of underwater downward irradiance, upwelling nadir radiance and above-water downward irradiance, as well as IOPs, in the Eastern Mediterranean from various research cruises since 2012 [48–51]. Recently, a move towards fiducial reference measurements for Sentinel-3/OLCI validation from this optical suite has been made by HCMR, with high quality Rrs satellite matchup analysis carried out using the radiometry profiles taken in the waters around Crete from the PERLE-2 oceanographic cruise (February-March 2019) [52]. Furthermore, in the immediate area surrounding E1-M3A and MSEA a ProVal float was deployed between 26/09/2019 and 17/10/2019 and took potential OC-SVC quality radiance and irradiance profiles (see Section 8.1). Figure 13 shows an example of one of the upwelling radiance profiles down to 150 m depth and the average derived surface Rrs spectra from all the ProVal profiles. These results compare very well with the validation spectra and radiometry profiles of [53], all with low diffuse attenuation coefficients (Kd and K_u) and high penetration of light across the visible spectrum, indicative of the oligotrophic and clear transparent waters at MSEA.

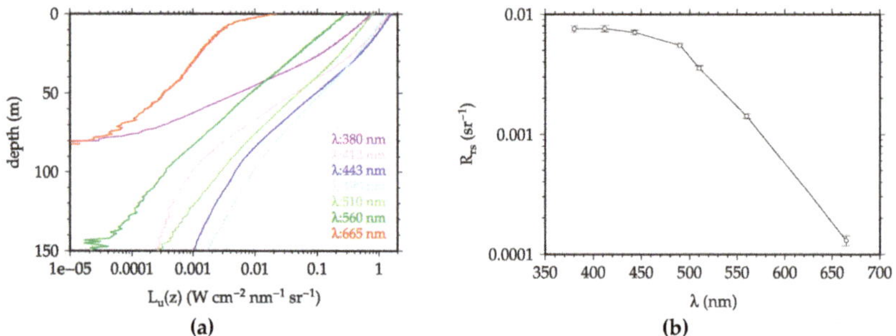

Figure 13. (**a**) A typical vertical profile of the spectral upwelling nadir radiance at MSEA (ProVal data, 27/09/2019), (**b**) Average remote sensing reflectance spectrum at MSEA derived from 35 spectra derived from ProVal data collected in 2019 (note the very small standard deviation).

3.4.5. Atmospheric Parameters

Figure 8a (Section 3.3.4) emphasizes that outside of the winter months of November to February, where all sites suffer from overcast skies, MSEA had significantly more clear skies than all other sites. This is a significant advantage for MSEA as an OC-SVC site because atmospheric clarity is one of the key factors in enabling enough relevant in situ to satellite matchup data.

An AERONET station has been running at HCMR-Crete since 2003 [53]. As already mentioned this is on the northern coast of Crete approximately 26 nmi directly south from the proposed MSEA buoy site and therefore considered to be representative of the atmosphere above the site. In general the AERONET data indicate a typical clear marine atmosphere averaging 0.11 AOD at 870 nm with very occasional maxima above 0.4–0.5, which probably coincide with infrequent Saharan dust events that the entire Mediterranean basin experience [54]. Water vapour values are generally low, with maxima in summer (August) where clear skies and high levels of solar insolation lead to higher evaporation from the sea surface and the warmer air above the sea being able to hold more moisture. The value of the Angstrom exponent is also an indicator of aerosol particle size [55]. Values of $\alpha \leq 1$ indicate size distributions dominated by coarse mode aerosols (radii ≥ 0.5 µm) that are typically associated with marine aerosols, and values of $\alpha \geq 2$ indicate size distributions dominated by fine mode aerosols (radii ≤ 0.5 µm) that are usually associated with urban pollution and biomass burning [56–58]. For MSEA the average Angstrom exponent over the period 2003–2018 was 1.11 ranging between almost 0 (likely clouds or dust events) and 2 (probably due to the coastal location of the AERONET station and the fact that coastal aerosols include a larger proportion of small particles of continental origin than marine aerosols). Nevertheless, these aerosol figures were indicative that MSEA had a clean Mediterranean maritime environment where there was little to no urban pollution or biomass burning and the dominant aerosol was water vapour.

In 2017, the HCMR-Crete AERONET station was moved further east along the coast to Finokalia (35°20′ N, 25°40′ E) to be colocated with the rest of the international atmospheric monitoring equipment of this European supersite [59]. The Environmental Chemical Processes Laboratory (ECPL) of the University of Crete (UoC) has operated the Finokalia station since 1993, participating in the most important international research networks of atmospheric research such as ACTRIS [60], ICOS [61], GAW [62] and EMEP [63]. Ozone monitoring at Finokalia started by the end of 1997 [64–67] and continuous PM10 observations in 2004 [68], while several intensive campaigns have taken place at the site during the past 25 years. Thus, Finokalia observatory is now well characterized and documented as representative of the Eastern Mediterranean background and MSEA site atmosphere. In addition to AERONET and a fully equipped scientific weather station, its great range of atmospheric measurement equipment allows a detailed characterization and monitoring of the atmosphere over the MSEA site. For example, the aerosol scattering coefficient from Finokalia (Figure 14) shows an annual cycle with maximum values observed during summer, minima during winter, while during spring secondary maxima were observed, which were attributed to the infrequent dust transport towards the Eastern Mediterranean. The absorption coefficient presents minimum values during winter, a local maximum was observed in spring followed by a local minimum in June. Maximum values were observed during summer. The same trend was observed for elemental carbon (EC) concentrations, suggesting that EC determines the levels in the area [69].

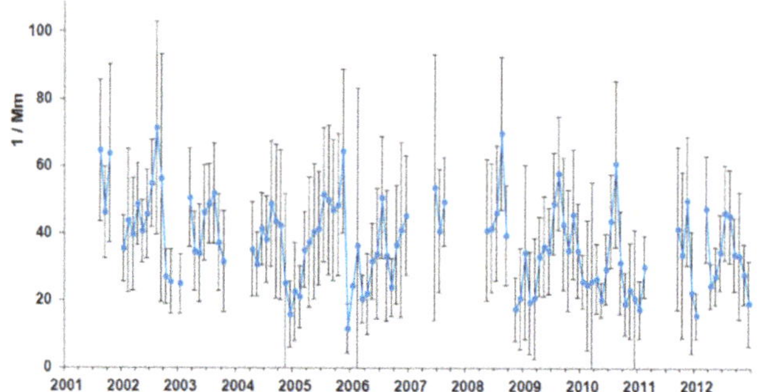

Figure 14. Aerosol scattering based on wet nephelometer measurements at Finokalia indicate a declining trend since 2001. Standard deviation shown as light grey bars. Data from [70].

In situ Finokalia and satellite observations show that dust events over the MSEA (Crete) site are infrequent and short-lived [69,71]. The frequency of desert dust events from 2000 to 2016 over the whole Mediterranean can be seen from [72,73]. On average over the MSEA site there were 1–3 dust events per year based on the data between 2000 and 2016. The results of these analyses for the Eastern Mediterranean were supported by the AOD data from the HCMR-Crete AERONET station. Furthermore, dust transport to this part of the Mediterranean region is rather a phenomenon of episodic nature with each event lasting no more than two days [69]. At the Finokalia station the dust events are well captured and identified from a variety of instruments and methods and therefore each event can be identified and characterized with high temporal resolution, which will facilitate their removal from the OC-SVC process. In addition to the ground-based observations at Finokalia, it is quite easy to identify the dust events from the satellite OCR observations themselves [74], and then to discard contaminated observations and not use them in the OC-SVC process.

Similarly, the air quality is very good over the MSEA site. Significant pollution from the oxides of nitrogen gases and other urban and industrial emissions that could adversely affect the clarity of the atmosphere for optical measurements were not evident at MSEA from satellite measurements, model simulations, or in situ measurements at Finokalia. Seasonal means varied between 0.20 ± 0.16 ppbv in winter and 0.56 ± 0.60 ppbv in summer, which agreed with the low tropospheric columns of NO_2 seen by satellites (e.g., TROPOMI on Sentinel-5P) over the MSEA local area [75–77].

3.4.6. Spatial Homogeneity

Figure 15 indicates stable levels of top of the atmosphere radiance as measured by S3A/OLCI over the area of the MSEA site, both temporally and spatially. Through the 3-year analysis of the S3A/OLCI time series and across the area of interest around the MSEA site, a consistent median of 0.13 W m^{-2} sr^{-1} was found with an SD of only 0.03–0.045 W m^{-2} sr^{-1}. Furthermore, the combined spatio-temporal analysis also confirmed these stable levels with the standard deviation varying between similar levels of 0.03 and 0.04 W m^{-2} sr^{-1}.

Evidence for high spatial homogeneity has also been obtained from surface [TChl-*a*] estimations for the wider area around Crete Island, with a low spatial and temporal variability during an annual cycle [44,78].

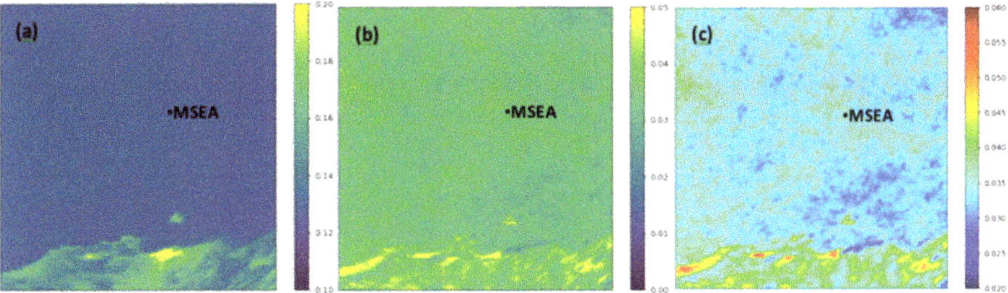

Figure 15. Spatial and temporal variability of top-of-atmosphere radiance (L_{toa}) from S3A/ ocean and land colour instrument (OLCI; April 2016–May 019) over the MSEA site: (**a**) median of L_{toa} (W m^{-2} sr^{-1}) for entire time series; (**b**) standard deviation (SD) of L_{toa} time series; (**c**) combined temporal and spatial (5 pixels × 5 pixels area) SD. (Locational coordinates (same for all panels): top left corner = 24.65°E, 36.30°N; top right = 25.85°E, 36.10°N; bottom right = 25.60°E, 35.20°N; bottom left = 24.40°E, 35.40°N).

3.4.7. Summary of the MSEA Site Characteristics Relevant to OCR-SVC

MSEA represents the most appropriate and stable open ocean calibration target in European Seas in terms of atmospheric clarity and in situ measurement conditions. This translates into the possibility of achieving a greater number of good matchups with the Copernicus OCR satellite sensors than at any other site in Europe. Furthermore, oligotrophic conditions like those at MOBY and MSEA facilitate very precise tracking of the calibration of in situ optical instruments, which is fundamental to accurate satellite ocean colour system vicarious calibration.

3.5. The Deployment Structure

3.5.1. General Architecture and Design

To the best of our knowledge there are no commercially available off-the-shelf solutions for deployment structures that fully answer the requirements that an OC-SVC infrastructure must meet. These requirements, as a minimum, include:

- Minimizing shading on underwater instruments.
- Maximizing stability (low tilt, no vertical movements), under the specific weather conditions encountered at the deployment site (in terms of currents, tides, wind and wave characteristics).
- Being deployable on a deep-water site.
- Giving easy access to divers for handling and cleaning instruments.
- Keeping above-water radiometers far enough from the sea surface to minimize sea spray.
- Installing in-water radiometers close enough to the sea surface to enable a proper extrapolation of measured quantities to the "0-" level and far enough from the sea surface to minimize possible impacts from storms and occasional yachting activity.

The BOUSSOLE mooring plus buoy [79] (Figure 16) was conceived precisely to meet these requirements. It was also designed to cope with the wave heights, wave periods and currents specific to BOUSSOLE, which a tethered buoy solution could not cope with. Readers are referred to [80] for a full description of the existing design and deployment procedures.

This design and construction have proven to be robust and efficient through (as of today) a 17-year uninterrupted deployment time series, providing adapted hosting conditions for high-quality radiometry measurements. Therefore, the reasons why it is recommended to continue using it in the frame of ROSACE are:

- The concept has been theoretically evaluated, then tested on a reduced-scale model, and then deployed successfully in real conditions.

- The concept has been further validated by external experts [80,81].
- The concept has been demonstrated to be successful. Maintenance procedures effectiveness has been proven over the long-term.
- Options for further improvement of the structure's behaviour at sea (reducing tilt) have been identified (see the next section).

In addition, analyses of the buoy tilt under actual current conditions at BOUSSOLE combined with records of currents at MSEA also showed that this concept is suitable for deployment on both sites. The same superstructure design would therefore be used at both sites, with the only difference being the length of the mooring line cable, allowing HCMR to capitalize on more than fifteen years of experience in the use and optimization of this European OC-SVC infrastructure. This is an additional justification of reusing the BOUSSOLE concept.

Furthermore, a strategy to further improve the appropriateness of the current BOUSSOLE structure design has been created for ROSACE, so as to increase the capacity to deliver OC-SVC matchups.

3.5.2. Upgrade of the BOUSSOLE Buoy

The design of the BOUSSOLE buoy has undergone some preliminary modifications for ROSACE in order to optimize stability and reduce tilt. Such improvements will increase the percentage of observations for which the uncertainty meets the requirements for OC-SVC matchups. The structural variables to be modified to increase the buoy stability are:

1. The overall buoyancy, which is primarily determined by the volume of the main buoyancy sphere (about 95% of the total buoyancy).
2. The distance between the centre of buoyancy of the entire buoy and the connection point to the mooring cable.
3. The platform plus payload mass.
4. The distance between the centre of gravity (determined by the distribution of masses) and the connection point of the buoy to the mooring cable.

Improving stability and decreasing tilt is obtained by increasing either (1) or (2) or both and by decreasing either (3) or (4) or both.

By design, the BOUSSOLE buoy tilt is highly correlated with oceanic currents [81,82], through:

$$\text{tilt} = \tan^{-1}\!\left(\alpha\, v^2\right) \tag{1}$$

where v is the current speed (m s^{-1}), and α is a constant specific to the buoy design, its volumes and masses.

This equation has been used to derive surface currents from the entire record of the tilt at BOUSSOLE, because currents are not available there from direct measurements. This calculation showed that current speeds are similar at BOUSSOLE compared to MSEA, where they are directly measured (Figure 17a). This result supports the deployment of a BOUSSOLE-type buoy at MSEA.

Furthermore, using either the inferred (BOUSSOLE) or measured (MSEA) currents, we have calculated the tilt distribution that would be obtained on both sites with the improved design, i.e., where the distance between the centre of buoyancy and the connection point to the mooring cable is increased (which essentially ends up with the α constant in Equation (1) being lowered as compared to the current BOUSSOLE design). Results are displayed in Figure 17b, and show a modal buoy tilt of 2° and about 70% of data with a tilt < 5° at BOUSSOLE. The numbers for MSEA are even better with a modal value of 0.5° and 83% of data with a tilt < 5° (Figure 17c).

This exercise will be refined during a final design phase and confirmed through reduced-scale model tests in water tanks (as was done for the initial BOUSSOLE design).

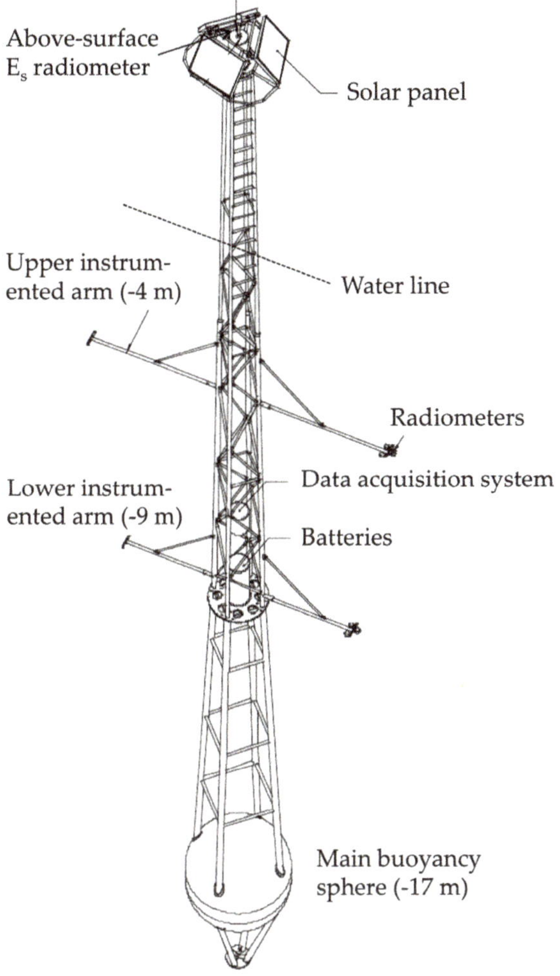

Figure 16. Three-dimensional model of the BOUSSOLE buoy.

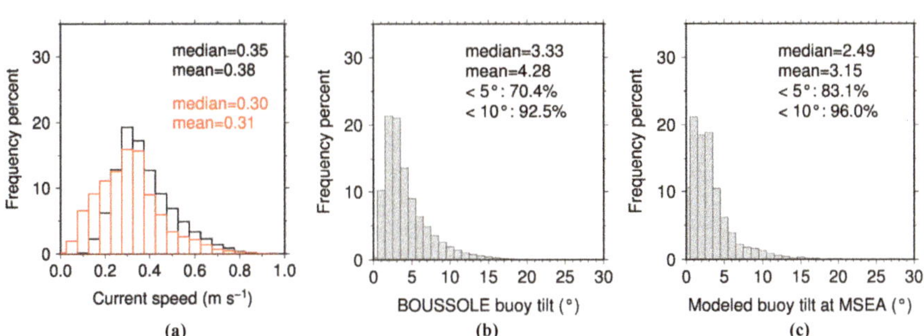

Figure 17. Distribution of (**a**) the current speed at BOUSSOLE (modelled; black) and MSEA (measured; red), the modelled buoy tilt at (**b**) BOUSSOLE, and (**c**) MSEA with the improved buoy design.

3.6. Matchup Potential of the Two Sites

Two complementary studies were performed with the aim of evaluating the matchup potential of both the BOUSSOLE and MSEA sites. By "matchup potential" we mean the potential number of collocations between a satellite overpass (here S3A) and a field measurement that both respect current criteria used for the determination of OC-SVC gains.

This potential was evaluated for the MSEA site for the first time here, and is somewhat known for BOUSSOLE, which is already used for the SVC of S3A&B/OLCI. We nevertheless included BOUSSOLE in this study because the improvements in the superstructure that we proposed for ROSACE have a great potential to improve the number of days where the field data are qualified for OC-SVC. The matchup potential of BOUSSOLE would be consequently increased as compared to its current status.

The first approach is similar to that of Zibordi and Mélin [24], except it used the S3A orbit characteristics and products, instead of SeaWiFS products. This approach gives a potential number of SVC-qualified matchups at both sites assuming that field data of the appropriate quality are available at the time of each of the selected satellite overpasses.

The second approach uses real field observations, i.e., eleven years of data collected at BOUSSOLE from 2007 to 2017, combined with typical overpass predictions of S3A. It allows analysing the impact on the number of matchups of various criteria related to the quality of the field data at both sites (such as the buoy tilt or the sun zenith angle, SZA).

3.6.1. Potential for OC-SVC Matchups: Satellite Approach

For this approach, all S3A overpasses for 2016, 2017 and 2019 were considered as potential matchups, regardless of the field measurement conditions (e.g., buoy tilt, wind). The year 2018 was not included due to technical issues delaying the access to S3A data. It should be noted that for the sake of comparison the MOBY site was also included in the exercise.

Satellite data might be used for OC-SVC when (see [3,4]):

- The SZA is lower than 70°.
- There is no sun glint or saturation of the image.
- There is no cloud contamination in the image or whitecaps.
- The viewing zenith angle (VZA) is lower than 56°.

Two additional criteria were also considered:

- Aerosol optical thickness (AOT) < 0.15 at 865 nm (low enough so that atmospheric correction has a chance to perform well).
- [TChl-*a*] concentration < 0.2 mg m^{-3} (meso- to oligo-trophic waters).

To estimate the most sensitive parameter at MSEA and at BOUSSOLE, these criteria were considered sequentially (e.g., one by one) and then globally when all were applied. The annual number of matchups satisfying the conditions was compared to the number of overpasses. To do that, the total number of matchups was divided by the number of months covered by this study (i.e., all months covered in 2016 + 2017 + 2019) and then multiplied by 12, to get an annual estimation. Results are summarized in Table 1. Note that the VZA criterion was not included because it did not eliminate any data.

Considering all OC-SVC requirements, adding MSEA to BOUSSOLE allows increasing the number of matchups by a factor of 3 as compared to a BOUSSOLE-only scenario. Therefore, 44 potential matchups per year might be expected when both sites are combined.

Both sites appeared to be strongly affected by the criterion on the AOT, which eliminated 60% of matchups on average for MSEA and BOUSSOLE. This was actually inconsistent with the field measurements of AOT at BOUSSOLE and MSEA, which showed average values at 865 nm generally <0.1, with very few situations where it was >0.15. This clearly indicates an overestimation of AOT by OLCI. Therefore, the AOT criterion has to be considered with caution.

Table 1. Impact of each selection criterion when taken individually, and when combined together (last column) with the threshold values indicated in the text. The percentage reductions (rounded to integers) are calculated from the number of matchups after excluding the glint risk (so $N = 123$ for BOUSSOLE, 103 for MSEA and 81 for MOBY). GLO corresponds to the OC4ME algorithm and Med to the MedOC4ME algorithm.

	N Overpass	SZA	Glint	Cloud	AOT	[TChl-*a*]		All Criteria	
						GLO	Med	GLO	Med
BOUSSOLE									
N matchup	149	134	123	80	59	45	74	12	20
% reduction		10	17	46	60	70	50	92	87
MSEA									
N matchup	144	144	103	95	57	88	95	32	32
% reduction		0	28	34	60	39	34	78	78
MOBY									
N matchup	111	111	81	66	58	74		31	
% reduction		0	27	40	48	33		72	

The second most influencing criterion was the total chlorophyll-*a* concentration. It was somewhat less critical for MSEA, with 39% of matchups eliminated compared to BOUSSOLE with 70%. This estimation was obtained with the global chlorophyll algorithm (OC4Me, [82]), however, which was known to significantly overestimate the chlorophyll concentration in the Mediterranean Sea, in particular for low-chlorophyll waters (e.g., [83]). Performing the same exercise using the MedOC4ME regional algorithm [84] greatly decreased the impact of the [TChl-*a*] criterion, ending up with only 50% of discarded matchups at BOUSSOLE and 34% at MSEA (i.e., comparable to MOBY where the percentage was 31%), for a total of 52 potential matchups for the whole infrastructure. These numbers were confirmed by the study using the BOUSSOLE field measurement (see the subsequent section). The MSEA site was also clearly shown as the one where clear skies most often occurred.

3.6.2. Potential for OC-SVC Matchups: Field Data Approach

For this study, we used the actual 2017 S3A overpasses over BOUSSOLE and MSEA, and we matched these times with the field observations at BOUSSOLE for each and every one of the eleven years from 2007 to 2017. This means that for the MSEA site we assumed that a buoy would have been deployed there and would have provided the same in situ time series as BOUSSOLE, except for the [TChl-*a*] and buoy tilt. The total chlorophyll-*a* was replaced by the MODIS chlorophyll-*a* at MSEA for the period 2007–2017. The tilt was computed from the measured current speeds at MSEA using the tilt vs. current equation. The same method was applied for both the current and improved buoy designs. Performing the analysis in this way, we did not take into account the slight shift that occurs year after year in the overpass times, yet this was not critical inasmuch as we used 11 years of field observations, which allowed the calculation of a relevant average of how many field observations would qualify for a matchup.

The parameters that were considered to determine whether a matchup pair (overpass and field observation) is suitable for OC-SVC were the sun zenith angle at the time of the overpass, the satellite view angle, whether there is a glint risk, the buoy tilt, a clear sky index (corresponding to the cloud elimination with the satellite approach), the wind speed and the total chlorophyll-*a* concentration. The last four parameters were taken from the field measurement closest to the satellite overpass (maximum allowed time difference is 3 h). Note that, by virtue of the buoy design, the wind speed and tilt criteria were somewhat redundant because the wind speed largely determines the surface currents at BOUSSOLE and MSEA. For BOUSSOLE, the clear-sky index was calculated as the absolute value of one minus the ratio of the measured above-water downward irradiance to its value modelled for a clear sky [85]. Therefore, values lower than, e.g., 0.1, mean that downward irradiance is within 10%

of its theoretical clear sky value. For MSEA, this index was calculated using AERONET data and a radiative transfer model.

The following thresholds were used [3,4]:

- No glint risk.
- SZA at the time of the satellite overpass < 70°.
- VZA < 56°.
- Clear-sky index < 0.1 at BOUSSOLE and MSEA, as in [3].
- [TChl-a] < 0.2 mg m^{-3}.
- Buoy tilt < 5°.
- Wind speed < 7.5 m s^{-1}.

These thresholds are the ones currently used for OC-SVC or recommended here for selecting data with the lowest uncertainties. For both sites, we evaluated the matchup numbers using both the observed buoy tilt and the modelled buoy tilt that the proposed design adaptations would provide (see Section 3.5.2). Results are displayed in Figure 18 and in the following we only refer to results obtained with the modelled tilt for brevity.

With such a selection, the average number of matchups was 29 at BOUSSOLE and 42 at MSEA (Table 2). These numbers were obtained with the actual BOUSSOLE time series, during which the buoy collected data on average during 83% of the time (was 90%–100% in many years, and down to a minimum of 58% in 2008 when a ship collision occurred). If we assume a fully operational system working 100% of the time, the number of matchups increases to 36 at BOUSSOLE and 52 at MSEA.

Therefore, a conservative estimate of the matchup potential for the ROSACE infrastructure was of about 70 matchups every year and about 90 matchups for a fully operational infrastructure.

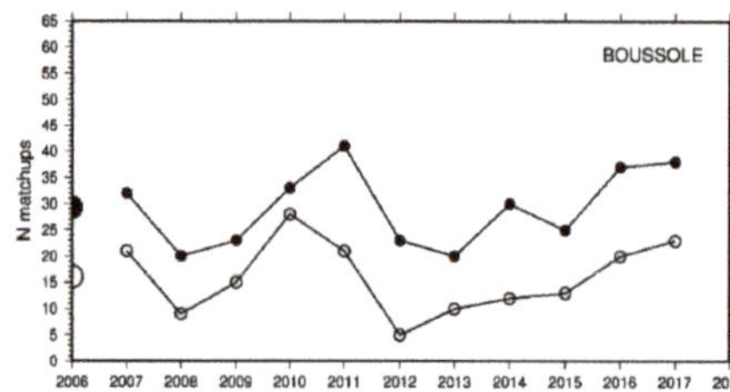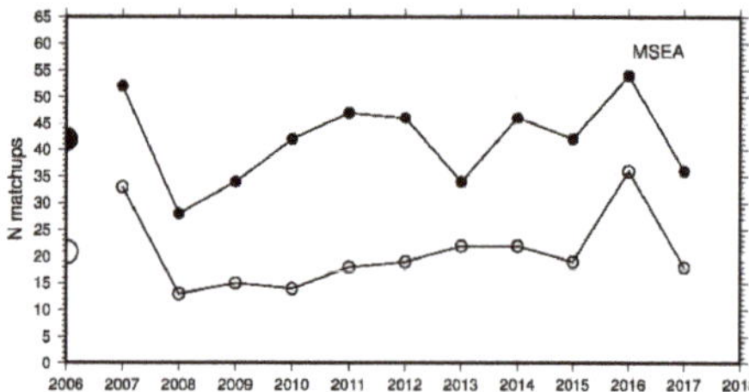

Figure 18. Potential number of valid system vicarious calibration (SVC) matchups at BOUSSOLE and MSEA. The open symbols are for the current buoy design and tilt values. The dark symbols are for tilt values that would be observed with the improved buoy design. The larger symbols on the left side of each panel show the average values for each time series.

Table 2. Statistics of matchups reduction following quality criteria.

| | | Satellite Measurements | | Field Measurements | | | | All | |
	Collocation	Glint	SZA	Clear-sky	[TChl-a]	Buoy Tilt	Wind Speed	83% rate	100% rate
BOUSSOLE									
N matchups	175	133	120	55	88	89	96	29	**36**
% reduction		24	10	58	34	34	28	78	73
MSEA									
N matchups	168	115	115	52	109	98	84	42	**52**
% reduction		30	0	54	6	14	28	63	55

Elimination Criteria

Table 2 shows the impact of each criterion when taken individually. Accounting for the "glint risk" eliminated about 24% of all satellite overpasses at BOUSSOLE, and 32% at MSEA. This difference is a direct effect of MSEA being at a lower latitude, making it slightly more prone to sun glint than BOUSSOLE. This criterion determines a fixed starting point in terms of the number of potentially usable matchups, before other criteria are taken into account.

Table 2 also shows that the observation and sun geometries were not critical. All matchups at MSEA were below the thresholds for the SZA and VZA at the time of the satellite overpass. Only 10% of matchups were eliminated at BOUSSOLE for the SZA being larger than 70° and none at MSEA.

The parameter with the strongest impact on the number of matchups was the clear-sky index. Note, however, that it was set to a rather low value here so the estimate is conservative. The [TChl-*a*], wind speed and tilt criteria had roughly the same impact at BOUSSOLE. For MSEA, [TChl-*a*] was not a significant criterion at that site as it was low throughout the year. The analysis was not repeated for Sentinel-3B because similar, actually probably identical, numbers can be anticipated inasmuch as the orbit characteristics are globally the same as those for S3A.

It should be remembered that this exercise uses thresholds on various parameters for the selection of in situ data. This is not what is eventually recommended for the future Copernicus OC-SVC infrastructure, where the selection should be driven by the uncertainty associated with each measurement. It is foreseeable that data not included here, on the sole basis of how they compare to thresholds on selected parameters, could actually be selected if their uncertainty was accounted for (i.e., because it was revealed to be low). The rather stringent criteria (thresholds) we have used here make the opposite situation very unlikely (i.e., where data selected here would eventually be revealed as not appropriate). In the end, the number of matchups is likely to be equivalent to what we found here, i.e., between 70 and 90 each year for the combined ROSACE infrastructure (MSEA + BOUSSOLE).

4. The Optical System and Calibration Strategy and Equipment

4.1. Technical Requirements for the Field Radiometers

The technical requirements for a field radiometer dedicated to providing measurements in support of OC-SVC were defined in [21]. These can be summarised as:

- The radiometers have to measure the underwater nadir radiance, L_u, the above-water downward irradiance, E_s, and the underwater downward irradiance, E_d. The latter is not strictly speaking used for OC-SVC, yet is needed to allow the determination of the diffuse attenuation coefficient for downward irradiance, K_d, as a proxy for absorption, itself useful for self-shading assessment.
- Spectral coverage should be sufficient to cover both the in-band and full-band spectral characteristics of the satellite sensor, specifically 380–900 nm for S3/OLCI, S2/MSI and ideally from 340 nm for the future NASA Plankton, Aerosol, Cloud and ocean Ecosystem (PACE) mission. The measurements should be hyperspectral, i.e., sampling the full spectral range at a resolution better than 3 nm, with a sampling interval of about 1 nm. The spectral calibration and its stability must allow maintaining the radiometric accuracy of the retrieved products (typically 0.2 nm for each channel of the field spectrometer).
- Stray light shall be characterised for each radiometer, so that appropriate correction can be applied to measurements.
- Radiometric calibration of the radiometers (in air) must be held to 1%–2% uncertainty in the VIS domain (above 400 nm) and traceable to SI units.
- For underwater radiance radiometers, the half-angle field of view (FOV) should be < 10°, although this requirement can be relaxed in open ocean waters as far as the instrument does not see the deployment platform.
- The immersion factors should be experimentally determined for each radiometer.

- The temperature response of the detectors shall be determined, and the internal temperature of the instrument measured continuously so that temperature effects are known, quantified and can be corrected.
- The polarisation sensitivity of the radiometers must be less than 1% and fully characterised.
- Dark signals shall be measured and corrected.
- The linearity of the detectors must be characterised and corrected to have an uncertainty of less than 0.1%.
- The cosine response of the surface irradiance radiometers shall be characterised, so that appropriate correction can be applied to measurements.

These requirements cannot be met with commercial, off-the-shelf instrumentation [86], which is why ROSACE includes development of new European radiometers specifically dedicated to OC-SVC. Besides the need to match stringent technical requirements, this development also conforms to the long-term vision of ROSACE, which implies that we assume and maintain control of the instrumentation we use. European expertise in developing high-quality optical and radiometric instrumentation for EO and in related domains is at the highest level and its use is strongly recommended.

4.2. Predesign of the Optical System

A spectrometer-based solution is proposed to meet the requirements previously summarised. In order to properly cover the spectral range and to avoid low signal-to-noise ratio (SNR) in a given part of the spectrum when the other part is optimised, two spectrometers are included, one for the blue part of the spectrum and the other one for the longer wavelengths, similarly to what was developed for MOBY [14]. After an evaluation of various manufacturer designs, preliminary specifications for radiance and irradiance sensors that meet the OC-SVC needs have been defined and are provided in Table 3.

The radiance instrument predesign is shown in Figure 19a. It uses a Gershun tube design coupled to a multimode fibre located between the collector and the spectrometer. Three components follow on the optical path: a motorized beam shutter allowing precise measurement of the dark signal, a short pass filter with cut-on wavelength at 900 nm to block NIR illumination (especially for calibration with a high power FEL lamp), and a dichroic beam splitter that separates the red and blue parts of the spectrum. Each of the two beams is coupled into a multimode fibre that is connected to each spectrometer.

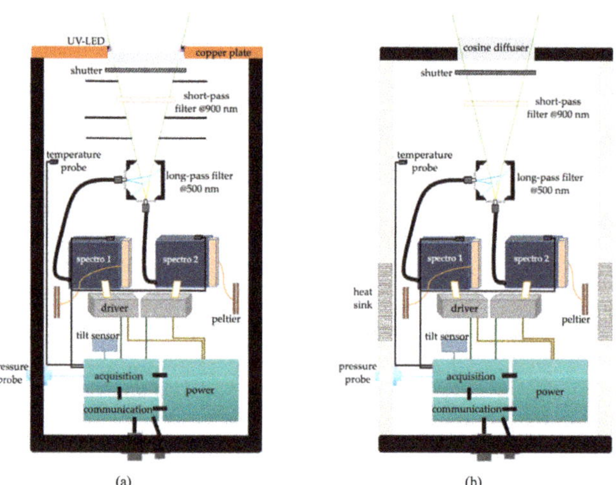

Figure 19. Preliminary design of (**a**) radiance radiometer and (**b**) irradiance radiometer.

The irradiance instrument (Figure 19b) shares most of the predesign of the radiance instrument, apart from being equipped with a flat cosine diffuser. The above-surface radiometer will also be equipped with a heat sink (dissipater) to avoid excessive heating of the instrument body when exposed to direct sun.

More detailed characteristics of the radiance instrument predesign are as follows:

- The optics is made of fused silica (HPFS 7980 or 8655) with minimum 92% internal transmission in the whole spectral range (340–900 nm). The standard MgF_2 coating as a broadband antireflection coating will be used.
- The Gershun tube includes four levels of black anodized aluminium baffles, for FOV definition and geometric stray light reduction. The half FOV is set at 7° in water.
- The shutter is mounted in a translation/rotation stage. It is made of black anodized aluminium on both sides.
- To avoid light scattering into pixels from the high intensity of the incandescent calibration source in the NIR domain, a short-pass filter is inserted in the optical path with a cut-off wavelength centred at 900 nm.
- The dichroic beam splitter is a broadband dichroic mirror with wavelength cutting at 500 nm. All beam splitters have little dependence on polarization at 45° with a cone half angle of 7°.

Table 3. Preliminary specification of the radiometers based on commercially available spectrometer modules.

Characteristics	Preliminary Specifications
Field of view	Radiance half angle: 7° (9° in air) Irradiance: cosine response in fused silica diffuser
Detectors	2048 × 1 CMOS/2048 × 20 Back-thinned CCD
Entrance slit	10 × 750 µm
Pixel size	14 × 200 µm/14 × 14(×64) µm
Bandwidth range	340–1100 nm
Spectral sampling	0.3 nm/pixel
Spectral accuracy	0.2 nm
Spectral resolution	1 nm
Stray light	<0.03% @ ± 40 nm from peak
Temperature shift	<0.02 nm/°C
Acquisition module	16 bit ADC
Integration time	>10 ms
Frame rate	Typical 6 Hz- Programmable (remote, depends on spectrometer driver and processing)
Power requirements	<5 W without cooling (to be confirmed with real measurement)
Housing	Black anodized aluminium
Size	300 mm long (not including connector) 100 mm diameter
Weight	<2 kg
Operating temperature	−10 to +50 °C

The fibre optics are kept as short as possible, and include a loop to ensure good depolarization. They have a multilayer, armed black protection to ensure no stray light from outside of the fibre. They are linked via a coaxial connector (SMA905 type), have a broadband anti reflection hard coating and are solarisation resistant. The SNR is dependent on the signal (S), on the dark signal (D) and on the reading noise (read N_p). The uncertainty is given by the same parameters and the number of pixels used for the dark (N_d). N_p and t_{int} are the number of pixels used for the signal and the integration time respectively. The equation for the SNR is therefore:

$$\frac{S}{N} = \frac{S\, t_{int}\, N_p}{\sqrt{(S+D)\, t_{int}\, N_p + read\, N_p}} \tag{2}$$

and the relative uncertainty is given by:

$$\frac{\Delta S}{S} = \frac{1}{S\, t_{int}\, N_p} \left(N + \frac{\sqrt{D\, t_{int}\, N_d} + read\, N_d}{N_d} \right) \tag{3}$$

With this approach, we neglected the noise introduced by analog-to-digital conversion (ADC). The ADC is chosen carefully enough not to degrade the signal to noise ratio. Typical SNR values are provided in Table 4, for an integration time of 2 s (accumulation of 20 measurements each at 0.1 s of integration time), and a pixel binning leading to a 1 nm spectral resolution. The slit size was 10×750 μm.

Table 4. Signal to noise ratio at 2 s acquisition duration for average (L_{typ}) and maximum (L_{max}) levels of radiance recorded at BOUSSOLE and associated uncertainties.

2 s	SNR		Uncertainty (%)	
	L_{typ}	L_{max}	L_{typ}	L_{max}
683	511	1797	0.211	0.057
665	503	1772	0.214	0.058
560	2327	7564	0.044	0.013
510	2216	7214	0.046	0.014
490	2170	7069	0.047	0.014
443	2059	6717	0.050	0.015
412	1982	6474	0.052	0.016

4.3. Predesign of the Radiometers and Data Acquisition and Transmission System

Each radiometer (Figure 20) was equipped with the following components:

- The light collecting optics.
- Two spectrometers and their drivers.
- A depth sensor.
- Three temperature/relative humidity sensors.
- One-stage Peltier systems to thermally stabilize the spectrometers.
- Radiators in contact with the structure for heat dissipation (above-water instrument only).
- A 2-axis tilt sensor.
- A driver for the motorized shutter.
- An acquisition and communication card.
- A power supply card.
- Standard marine connectors.

The diameter of the instrument housing was minimized to reduce self-shading. Protection against biofouling will be provided by the use of UV diodes. Power requirements were less than 5 W peak without the thermoelectric cooling.

The central acquisition system will use an embedded Linux system, which will have the capability to drive the acquisition scenario (automatic and manual modes), to process any kind of data from all the instruments (data compression without loss, data reduction by analysis). The communication between most of the components will use serial protocols. The system will be able to check the state of every component, to start and shut it down if necessary, and to reboot the complete buoy system. There was no risk of loss of data due to the use of redundant flash memory (4 TB × 2) with and/or without connectors. The embedded software will be configurable and upgradable through remote control.

The transmission system will be separated from the central acquisition system to avoid any electromagnetic perturbations. It will provide a secured remote access of all components through a Wi-Fi connection (useful for maintenance from a ship). It will have a GPS system and be able to communicate via satellite bidirectional communication (typically iridium). Priority will be given to near real-time (NRT) transmission of diagnostic data about the platform behaviour, and essential subsets of radiometry measurement. The entire full resolution data set will be downloadable during buoy servicing.

Power will be supplied by three 75 W solar panels (similar to what BOUSSOLE currently uses). One of the defining parameters is how much energy will be needed for the spectrometer cooling. The produced energy will go through a charge regulation system to protect the batteries. High-performance LiFePO$_4$ batteries will be used, allowing a high number of charge cycles.

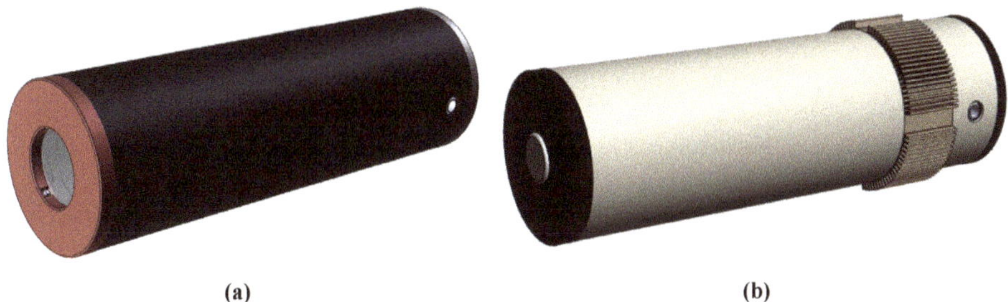

(a)   (b)

Figure 20. Computer assisted predesign of (**a**) the radiance instrument with the front optics, copper plate, ultraviolet (UV) light-emitting diode (LED) and the depth sensor and (**b**) the above-surface irradiance sensor with the cosine collector and the thermal cooling interface (radiator).

Adapted versions of the radiometers will be developed for use on a free-fall profiling system to be deployed from ships during servicing cruises to the mooring sites. This profiling system will be conceived as a transfer radiometer, also allowing inter-comparisons to be performed between the infrastructure sites and other similar OC-SVC sites, e.g., MOBY.

4.4. Absolute and Relative Calibration and Characterization: Strategy and Equipment

4.4.1. General Architecture

The characterization and calibration strategy must allow the OC-SVC uncertainty requirements to be met for the field instrumentation. For this to be accomplished, many steps of absolute and relative calibration and characterisation of the field radiometers are required. This is why the ROSACE infrastructure involves an NMI, i.e., the United Kingdom National Physical Laboratory (NPL), which will be responsible for SI traceability—providing absolute calibration standards and overall supervision and realisation of the full characterisation and calibration process.

For absolute radiometric calibration of irradiance sensors at the field site facilities (LOV-IMEV and HCMR), we proposed to use FEL lamps as irradiance standards. FEL is the American National Standards Institute (ANSI) designation for a standard 1000 W quartz tungsten halogen lamp. These lamps are commonly used as calibration sources and will be routinely calibrated at NPL using the spectral radiance and irradiance primary spectroradiometer (SRIPS) facility [87].

For the radiance sensors we proposed a new absolute radiometric calibration system based on the spectroscopically tuneable absolute radiometric calibration and characterisation optical ground support equipment (STAR-CC-OGSE; Figures 21 and 22) facility that is currently under development at NPL. STAR-CC-OGSE will be used for the initial characterization and calibration of the new radiometers by NPL. The use of the STAR-CC-OGSE system allows very low uncertainty absolute radiometric calibrations for the optical system of ROSACE, including detector nonlinearity, polarisation sensitivity and stray light.

Then for the routine radiance sensor calibrations we proposed to build a smaller integrating sphere-based system (SMART-CC; Figure 23) that will be based at the field site facilities (LOV-IMEV and HCMR) and used more regularly.

This laboratory-based equipment will be complemented by a hand-held, field deployable, radiometric stability monitoring device (IN-SITU-SC), to be deployed during monthly servicing of the buoy and "in air" intercalibration facilities for comparisons between radiometers against common natural or standard targets for pre- and post-deployment verifications in realistic conditions.

The main characteristics of these innovative elements are provided in the three subsequent sections.

4.4.2. NMI Primary Absolute Radiometric Calibration and Characterisation: The STAR-CC-OGSE System

The STAR-CC-OGSE is a versatile facility for the radiometric calibration and characterisation of various sensors. This system is comprised of an integrating sphere fed by a tuneable Ti-Sapphire based laser (including harmonic generation) to produce a monochromatic radiance source. It allows monochromatic continuous tunability from 260 to 2700 nm (although more commonly available in more limited spectral ranges to suit customer applications). In addition, the same sphere includes a set of calibrated and uncalibrated lamps to produce a broadband (white light) radiation source extending over the same spectral extent. The sphere output is monitored by a set of monitoring photodiodes. The radiance of this system at chosen wavelengths can be calibrated with an uncertainty of better than 0.5% via a silicon trap detector that is itself calibrated against the NPL cryogenic radiometer [88]. The laser radiation is despeckled, depolarised and intensity stabilised.

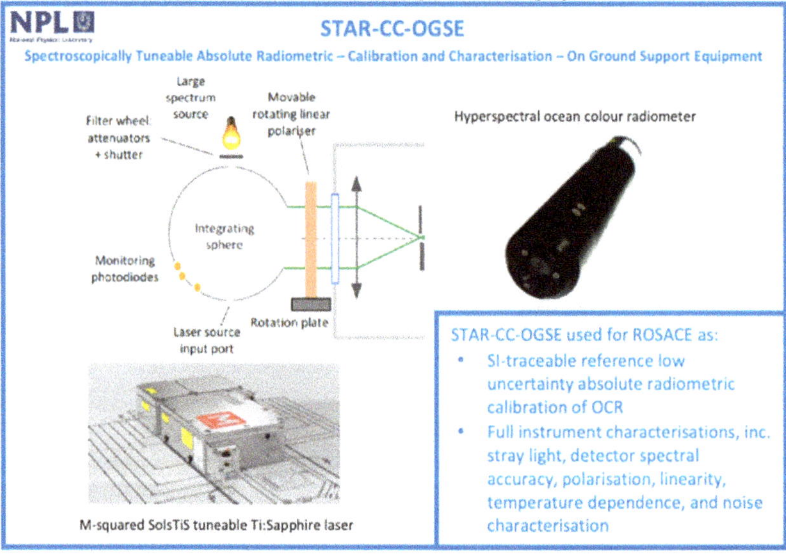

Figure 21. The main elements of the National Physical Laboratory (NPL) spectroscopically tuneable absolute radiometric calibration and characterisation optical ground support equipment (STAR-CC-OGSE) system.

Figure 22. Computer assisted design of the STAR-OGSE-CC (excluding the laser).

The STAR-CC-OGSE comprises of two illumination systems each able to deliver broadband and monochromatic radiation:

- A large aperture SI-traceable calibrated integrating sphere source primarily for radiometric calibration and characterisation.
- A collimated beam source, equipped with an interchangeable, position fine-tuneable feature mask for optical performance characterisation. The features of the mask can be readily customised to suit various applications.

The main radiometric specifications of the STAR-CC-OGSE are listed in Table 5.

Table 5. Main radiometric specifications of the STAR-OGSE-CC.

Radiometric source aperture size	Max. 200 mm diameter
Monochromatic spectral range	260–2700 nm
Broadband spectral range	250–2700 nm (equivalent to 3000 K blackbody).
Monochromatic typical radiance (radiometric sphere)	Over full spectral range 0.05 W m^{-2} sr^{-1} (laser bandwidth)$^{-1}$ to 5 W m^{-2} sr^{-1} (laser bandwidth)$^{-1}$ Corresponds to input laser power of 10 mW to 1 W
Broadband typical radiance	Variable in steps (ten times 10%) spectrally invariant. Max. > 3000 W m^{-2} sr^{-1} μm^{-1}
Radiometric accuracy	<0.5% (in air/vacuum)
Radiance spatial uniformity	Typically <0.15% PV (application dependent)
Radiance temporal stability	<0.1% duration of a measurement
Monochromatic source line width	<0.2 pm
Monochromatic source tuning step size	<1–5 pm
Monochromatic source wavelength calibration	<0.2 pm (PV)
Calibrated TVAC-compatible radiance monitor	<0.5% (k = 1) [TBC]
Collimator focal length and F/#	1000 mm and F/5 (max collimated beam size 200 mm diameter)

4.4.3. Site-Based Routine Absolute Radiometric Calibration: The SMART-CC System

This system will be developed by NPL and be permanently placed at LOV-IMEV and HCMR's ROSACE calibration laboratories to allow for routine calibration and stability checks without the need for the instruments to be shipped. Using an integrating sphere-based system will reduce time and repeatability issues related to the traditional FEL and panel calibrations by minimising alignment efforts. The SMART-CC (Figure 23) will be an integrating sphere-based system allowing for absolute radiometric calibration, instrument stability checks and stray light and spectral calibration checks using a few key wavelengths. It can be seen as a mini version of the STAR-CC-OGSE without the tuneable laser option. This system will be calibrated at NPL on the SRIPS facility and will provide a broadband calibration source. It will have an accompanying calibrated radiometer to check for stability variance. In addition, diode lasers will be fed into the sphere providing a monochromatic source for selected wavelengths. Thus, this system will allow detection of any changes in spectral accuracy or stray light for the selected wavelengths. Any inconsistency detected using the SMART-CC system will have to be fully investigated by the STAR-CC-OGSE.

Some practical aspects of using the SMART-CC system are:

- Radiometer mounting: custom designed self-centring mounts to ensure the same position to within a few microns between calibrations.
- Light sources: the broadband source will be a quartz-tungsten-halogen (QTH) lamp. This may simply be a QTH bulb positioned at one of the sphere ports and powered with a suitable, stable power supply. For the greatest stability it is best to use a lamp in a housing with a feedback loop provided by an internal, thermoelectrically cooled photodiode. In addition, to minimize the use of the sphere surface for coupling, quasi-monochromatic sources such as temperature controlled single mode lasers, will feed into the sphere via optical fibre.
- Room stray light: to ensure external stray light does not affect the calibration of the radiometers, the SMART-CC system will be enclosed to block the ambient light during measurements.
- Sphere monitoring: silica detectors are well known to be linear over several decades' incident optical power and stable over long periods of time. Use of a trap detector in the accompanying radiometer will allow identifying any drift in the SMART-CC system in excess of 0.2%.

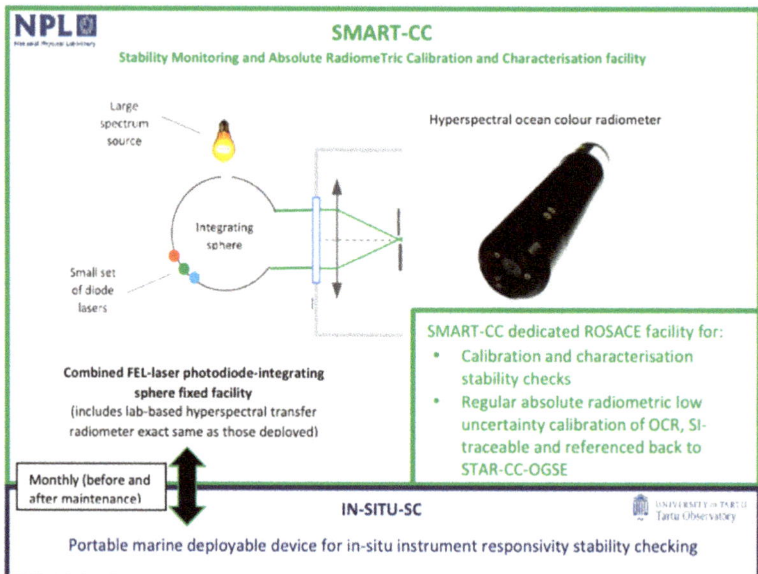

Figure 23. The main elements of the NPL SMART-CC system and interaction with the IN-SITU-SC.

4.4.4. Routine Relative Calibration in the Field: IN-SITU-SC

In order to follow possible changes in calibration and also biofouling contamination, a portable, field-deployable, relative calibration source will be developed for ROSACE. It will be used during buoy servicing to characterize the responsivity of complete instruments (including entrance optics) before and after cleaning. While the relative calibration source does not have to be SI-traceable, it must be well characterized and have good short-term stability (i.e., the duration of the cleaning and source measurements). Long-term stability of the source, i.e., stability across successive 6-month deployments of the radiometers, is not strictly required, but desirable and will be monitored with the SMART-CC.

A broadband emission in the VisNIR spectral range is needed and should not change with time or temperature. The radiant flux level and/or spectral shape will be easily switchable between pre-sets for validation of different instrument modes (e.g., upwelling radiance, downward irradiance). The instrument will be operable down to 20 m, be easily mounted onto underwater radiometers, and have an operating ambient temperature range of 10–30°. It will be battery powered and have slightly positive buoyancy so it is not lost in case of accidental release.

The solution proposed is based on LEDs, whose optical power and thermal control are needed to ensure constant intensity and spectral shape of the emitted light. The optical design concept suggested is presented in Figure 24a. The radiant flux from a white LED source is directed towards a diffusing window. The radiometer being tested measures the flux that is transmitted through the diffuser and is scattered forward towards the fore optics of the radiometer. An optical feedback photodiode (PD) measures the radiant flux that is backscattered from the diffuser. The photocurrent generated by the PD is used for adjusting the LED current.

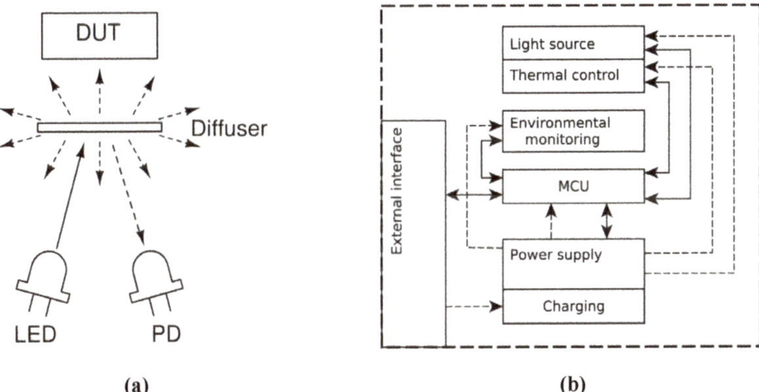

Figure 24. (**a**) General optical design concept of the IN-SITU-SC. DUT—optical entrance of the device under test, PD—feedback photodiode. Solid arrows denote the radiant flux from LED to the diffuser, dashed arrows mark the flux scattered from the diffuser. (**b**) Subsystem module diagram for the portable stability-monitoring device, with electrical connections shown with dashed lines, data connections with solid lines.

The optical power and thermal control of the light source, the user interface for switching modes, and battery management will be handled by a main control unit (MCU). The MCU will also take care of monitoring and logging of internal environmental parameters (temperature and humidity) as well as light source operating parameters (current, voltage and LED temperature). All the logs will be time stamped using an internal real time clock and stored in a non-volatile memory for retrieval via the external interface after the field deployment. The subsystem module diagram of the IN-SITU-SC is shown in Figure 24b.

The mechanical design must be robust enough to tolerate field operations and moderate overpressure, yet light enough to be able to float on water and not sink if accidentally released by the diver. In addition, the mechanical design must be corrosion-resistant in a marine environment and provide sufficiently good heat transfer between the thermoelectric cooler of the LED and the surrounding water. The mechanical design should ensure a precise and repeatable mechanical fit to the spectrometer fore optics while blocking all possible ambient radiation. The surroundings of the optical output window must allow air bubbles to escape easily and prevent them getting trapped between the IN-SITU-SC and the entrance of the radiometer being validated.

4.4.5. Summary of the Calibration Strategy

In summary, the main elements of the calibration strategy for ROSACE are:

- STAR-CC-OGSE, allowing very low uncertainty absolute radiometric calibration and full radiometric characterisation using the new NPL tuneable laser system, to be used on brand new radiometers and then as needed (for example if a radiometer had to be repaired).
- SMART-CC: low uncertainty (1%–1.5%) regular calibration stability checks and absolute radiometric recalibrations of radiance sensors via instrument swap-out using the NPL designed non-tuneable laser-integrating sphere-FEL based system. If the results of these regular calibration and stability tests are unsatisfactory, for example showing significant change in radiometer responsivity, then the instruments are sent to NPL for recalibration/recharacterisation using the STAR-CC-OGSE.
- FEL lamps: regular calibration stability checks and absolute radiometric calibration of irradiance sensor using standard 1000 W quartz tungsten halogen lamps. These standards are regularly recalibrated at NPL after maximum 50 h of use.
- IN-SITU-SC: a field deployable relative calibration source to check stability of the optical instrumentation in situ during monthly maintenance cruises.
- In-air reality checks: a custom intercalibration bench will be developed to allow relative comparison between radiometers before and after deployment in realistic conditions. This will be realized by acquiring data with radiometers pointing toward the sun or a common target (calibrated blue fabric) in a natural environment (in air) not perturbed by reflections and shadowing and with a spectrum closer to what the instruments experience when deployed. These verifications extend the strict laboratory conditions in which the calibration is realized in order to detect possible anomalies in the radiometer response.

4.5. Biofouling Mitigation

At least three types of biological perturbations have to be taken into account when making radiometry measurements from an ocean buoy: biofouling on instrument collectors and housings, large biological organism shading/reflections, and biological waste.

Biofouling is unavoidable at sea and can have a strong impact on optical measurements. Several mitigation devices have been used in the past and have proven to be instrumental in increasing the number of bio-optical observations not affected by growing bacteria and larger marine organisms, for example the use of copper sheets, plates or robotic shutters and pulses of ultraviolet radiation. A combination of the copper and UV approach will be used for ROSACE [89,90].

Birds occasionally rest on buoys, which can generate shadow or deposit biological waste. The possibility to adapt existing auditory or mechanical devices to the marine environment for keeping birds away from the mooring will be considered [91]. Comparing the relative temporal changes in the above-water E_s and in-water E_d is also an efficient way to detect such anomalies, when both diverge in their changes (implemented for the BOUSSOLE data processing).

Moorings also unavoidably attract marine life. While very large marine organisms like cetaceans only occasionally swim close to moorings, smaller size fishes habitually find shelter and food close to moorings and can potentially perturb radiometric measurement. They can generate "peak" shadows or reflections, which are easily removed by filtering the raw data (because data are recorded at a high frequency during the acquisition sequences that last at least one minute). The presence of those fishes does not modify the bulk optical properties of the water mass, however.

5. Uncertainty Budget

5.1. Methodology

The guide to the expression of uncertainty in measurement (GUM) [92] provides a framework for how to determine and express the uncertainty of the measured value of a given measurand (the quantity that is being measured). Within the GUM framework, uncertainty analysis begins with understanding the measurement function and is then performed by considering in turn each of the different input quantities to the measurement function. Each input quantity may be influenced by one or more error effects, which are described by an uncertainty distribution. These separate distributions may then be combined to determine the uncertainty of the measurand, using the analytical method called the law of propagation of uncertainties or the numerical method using a Monte Carlo method (MCM) simulation that is summarized in the GUM supplement [93]. An example of BOUSSOLE's in situ radiometric measurement uncertainty budget using MCM is described in [93].

In addition, we proposed to use a recently developed framework within the fidelity and uncertainty in climate data records from Earth observations (FIDUCEO) [94] project. This framework applies rigorous GUM-based metrological methods to satellite sensor fundamental climate data records (FCDRs) [95], where a schematic representation of the measurement function, called the uncertainty tree diagram, formed the basis of the analysis (Figure 25).

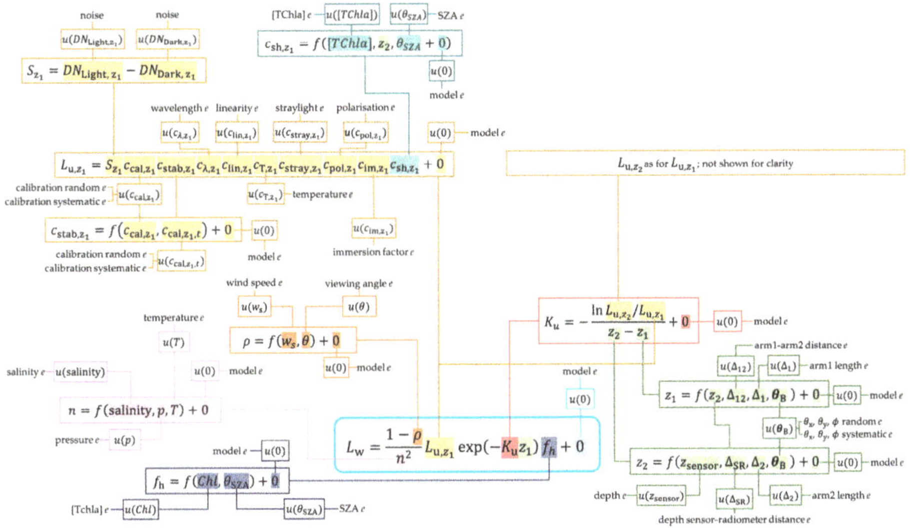

Figure 25. Water leaving radiance in situ measurement uncertainty tree diagram.

At the centre of this diagram is the measurement function—here with the measurand as water-leaving radiance, L_w, and input quantities such as underwater upwelling radiance at depth z_1, L_{u,z_1}, and the attenuation coefficient for the nadir upward radiance, K_u. From this function, branches spread out from each input quantity, which may themselves be determined by their own measurement functions (for example here K_u is determined from the upwelling radiance measurements acquired at two different depths), to their uncertainty. This uncertainty can be traced back through to its impact on the measurand by the sensitivity coefficients on each branch. Finally, the effects that cause each respective uncertainty are connected to the end of each branch.

Note that we should also consider the extent to which the measurement function describes the true physical state of the instrument, which is accounted for by including the term +0 at the end of the measurement function. This explicitly represents effects that are expected to have a zero mean and are not captured by the measurement function (i.e., there is an uncertainty associated with this quantity being zero).

Each of the effects identified at the end of each of the branches should then be understood, quantified and reported in an "Effects Table" that documents:

- The uncertainty associated with the given effect.
- The sensitivity coefficient required to propagate the uncertainties associated with that effect to uncertainty associated with the measurand.
- The correlation structure over spatial, temporal and spectral dimensions for errors from this effect.

Table 6 shows an adapted form of the FIDUCEO "Effects Table" for use in describing error effects in the OC-SVC process, with description of how it should be populated.

Table 6. Table for codifying the uncertainty due to an error effect and its correlation structure.

Table Descriptor		How This is Codified
Name of effect		A unique name for each source of uncertainty in a term of the measurement function
Affected term in measurement function		Name and standard symbol of affected term
Instruments in the series affected		Identifier of the specific instrument/deployment where this effect matters
Correlation type and form	Temporal within deployment Temporal between deployments Spectral (hyperspectral in-situ)	Forms of correlation described in detail in [96]
Correlation scale	Temporal within deployment Temporal between deployments Spectral (hyperspectral in-situ)	In units of spectral pixels, measurements or deployments in time—what is the scale of the correlation shape?
Channel/band	List of channels affected Error correlation coefficient matrix	OLCI channel names in standard form OLCI cross-channel correlation matrix
Uncertainty	PDF shape units magnitude	Functional form of estimated error distribution for the term Units in which PDF shape is expressed Value(s) or parameterisation estimating the PDF width
Sensitivity coefficient		Value, equation or parameterisation of sensitivity of measurand to term

A full description of all components presented in Figure 25 and examples of their effects tables are provided in Appendix A.

To propagate uncertainty for the measurands of interest (water-leaving radiance, downward irradiance, water-leaving reflectance and normalised water-leaving reflectance) we proposed the following approach.

Measurement functions are defined based on the tree diagrams, including all raw inputs (see Figure 25), and which are defined as quantities that can have an influence on the measurand values. These were used for all processing steps rather than intermediate quantities (for example K_u in the calculation of L_w). All raw inputs have their standard uncertainty identified in terms of magnitude (value) and probability distribution function (PDF). Due to the complexity of the measurement function it is challenging to derive all sensitivity coefficients analytically (for example $\frac{\partial L_w}{\partial Chl}$ or $\frac{\partial L_w}{\partial z_{sensor}}$ would be extremely difficult), thus we proposed using MCM to propagate the raw inputs uncertainties using a measurement function. We handled the partial correlation between some input quantities (for example, the absolute radiometric calibration coefficients of the different instruments) by decomposing these to a set of variables that are independent of each other (though these independent variables may themselves be correlated through time).

The final uncertainty value will be derived from the PDF of the output values generated through the MCM. All uncertainties will be reported with the $k = 1$ coverage factor. The uncertainties will be evaluated for water-leaving radiance, downward irradiance and water-leaving reflectance as an output of the field segment, and normalised water-leaving reflectance as the input to the gains calculation. They will be reported as one value for each of these outputs, and then split into three categories of random, fully correlated within one deployment and fully correlated within a satellite mission's lifetime.

The evaluation of the uncertainties of individual matchup and mission average OC-SVC gains will not be under the responsibility of the OC-SVC infrastructure. However, we recommended estimating the uncertainties of the mission averaged OC-SVC gains including different temporal correlation terms:

$$u^2(\overline{g}) = \frac{\sum_i u^2_{\text{rand}}(g_i)}{N} + \frac{\sum_j \left[\frac{1}{N_j} \sum_k u_{\text{d sys}}(g_k)\right]^2}{M} + \left[\frac{1}{N} \sum_i u_{\text{sys}}(g_i)\right]^2 \tag{4}$$

where $\overline{g}$ is the mission average gain, g_i is individual matchup gains, N is the total number of matchups and M is a number of deployments that have correlated inputs. N_j is the number of matchups in a given deployment, j, where g_k are the subset of individual matchups in that deployment.

5.2. Preliminary Uncertainty Budget

A demonstration data set was created from BOUSSOLE in situ hyperspectral data, by selecting data that are considered a priori suitable for the OC-SVC process based on environmental parameters such as the buoy tilt or the SZA. This data set covers the period from May 2016 to March 2017, during which 20 valid matchups were found for the S3A/OLCI overpasses.

The uncertainties were evaluated using the measurement equation presented in Figure 25 and the method described in [94]. The existing absolute calibration capabilities were used to estimate radiometric calibration uncertainties. In addition, instrument characteristics such as stray light and detector linearity were incorporated into the model and corrected for using the methods presented in [96]. Relative uncertainty for water-leaving radiance is shown in Figure 26 for twenty in situ measurements and the first ten S3A/OLCI spectral bands. The results indicate uncertainties generally below 3.5% for the spectral bands 1-7 covering the spectral range from 400 to 620 nm, then increasing to about 5% for longer wavelengths.

Table 7 presents the results of the same simulation run for the hypothetical new optical system with a new calibration facility for measurement number 15. The environmental conditions for this measurement seem to be close to perfect as the measured buoy tilt was $-1.3°$ and $1.2°$ for x and y axes, respectively, and [TChl-*a*] was 0.11 mg m^{-3}.

A significant reduction in uncertainty values could be anticipated for the shorter wavelengths where the instrument-related effects, which were small, had a larger relative impact on the overall uncertainty values. The reduction was smaller for the longer wavelengths for which the environmental conditions and modelling applied in the processing chain were the main relative contributors to the uncertainty values.

The proposed methodology addresses in detail the uncertainty propagation process, considering the temporal scale correlations present in mission averaged gain calculation. The practical realization of this method can be incorporated in the data processing chain and would enable evaluation of an uncertainty per measurement, always including appropriate ancillary data. This method therefore provides a concept of dynamic uncertainty per in situ measurement and matchup that we considered the most appropriate for both MSEA and BOUSSOLE as Copernicus OC-SVC.

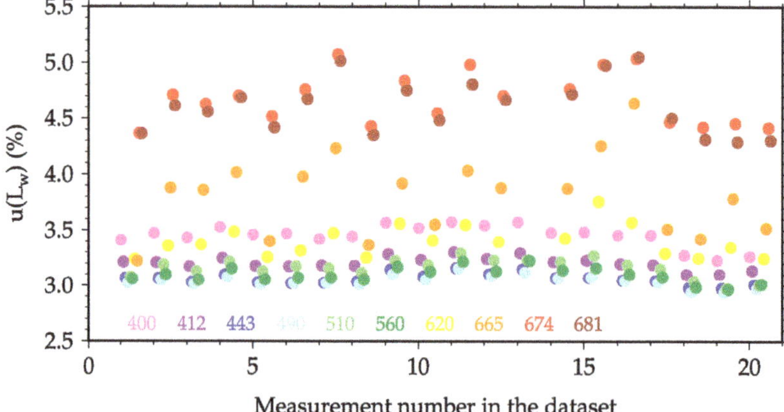

Figure 26. Relative uncertainties ($k = 1$) of the water-leaving radiance for 20 measurements (see text). The colour coding for the 10 spectral bands from 400 to 681 nm is indicated at the bottom of the figure. Symbols have been slightly shifted along the horizontal axis for the sake of clarity.

Table 7. Relative uncertainties (%, $k = 1$) in water-leaving radiance for the matchup number 15.

OLCI Band (nm)	400	412.5	442.5	490	510	560	620	665	673.75	681.25
and band (no.)	(1)	(2)	(3)	(4)	(5)	(6)	(7)	(8)	(9)	(10)
Current BOUSSOLE radiometers	3.5	3.2	3.1	3.1	3.3	3.2	3.8	4.3	4.9	4.9
Anticipated new ROSACE radiometers	2.8	2.4	2.3	2.3	2.6	2.5	3.1	3.8	4.6	4.6

6. The Ground Segment

6.1. Role and General Architecture

The ROSACE ground segment sits at the interface between the OC-SVC field component (field sites) and the ground segment of the satellite mission. It takes raw data from the field sites and performs all processing and quality control (QC) operations needed to generate the fully qualified data that the OC-SVC process requires. This section described the preliminary design of this ground segment, which will be located in, and operated from, ACRI-ST premises in Sophia Antipolis, France (Figure 27). The ground segment included near-real time, adjusted and delayed modes of data processing (NRT, AM and DM, respectively).

- The NRT processing was performed as soon as the data were received from the buoys, with products generated on the fly along with their quality flag and uncertainty estimates, following automatic and documented procedures (including QC. See Sections 6.4 and 6.5). An additional functionality will allow the site operator to change the QC flags through a semisupervised quality control interface. This information, i.e., the change of flag and its old and new value, will be incorporated into the NRT product.

- The adjusted mode (AM) products were generated by using data collected during monthly servicing cruises, when the instrumentation onboard the ROSACE buoys were cleaned, checked and maintained. During these operations, in situ sampling and radiometric profiles will be carried out. These data will be used for adjusting the instrument calibration for data collected during the previous month, and therefore to readjust the processing. Existing data products were thus reprocessed with these updated inputs.

- The delayed mode (DM) products will be generated through reanalysis of a longer time series of data, typically every 6 months at each rotation of instrumentation. This processing step allows better assessment of the overall consistency of the data (e.g., by examining seasonality, trends and comparison with climatology). The results of this reanalysis might lead to changes in the products' annotations.

Each of the data product types will by default contain the value of the parameter and its time of acquisition, its uncertainty and QC flags.

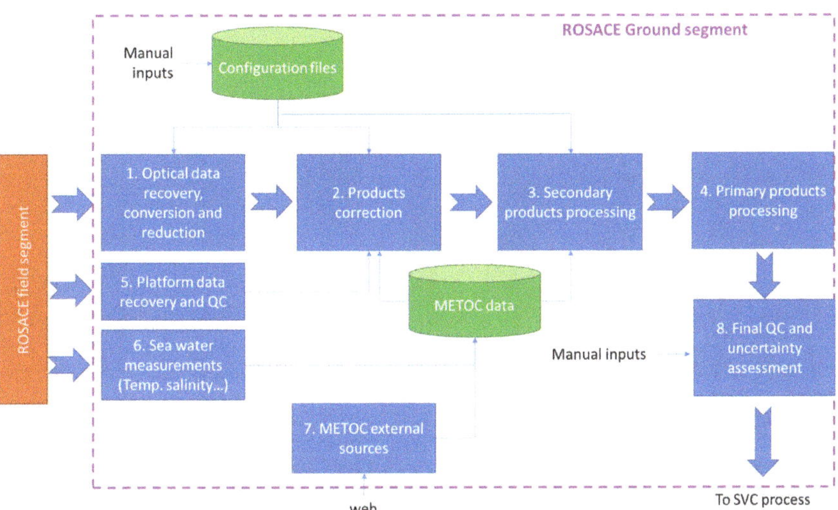

Figure 27. ROSACE data processing flow diagram.

6.2. Data Products and Levels

Raw data (RD) from the radiometric, optical and physical payloads were received in near-real time (NRT) from the buoys. They were converted into geophysical units with instrument-specific calibration coefficients, which are part of the auxiliary data (AD). These RD could be processed to derive statistically representative values at a lower frequency than the initial data, referred to as basic products (BP). For instance, a single representative value of $L_u(z)$ at a given depth was derived from the (60 × v) individual measurements available from a 1-min acquisition sequence of an instrument collecting data at frequency v (Hz).

As for the radiometry measurements, these BP will include the upwelling nadir radiance and the downward irradiance at the measurement depths, $L_u(z)$ and $E_d(z)$, and the above-water downward irradiance (E_s) at the native spectral resolution of the ROSACE radiometers. BP will also include diagnostic and ancillary data useful to monitor the status of the deployment platform and instruments in NRT.

Primary and secondary products (PP and SP, respectively) were then derived from the BP in conjunction with the data from instrument characterisation, modelling or other external sources (also included in the definition of AD). The SP was the same radiometric quantities corrected for instrument and environmental errors, data used for QC or by-products used to generate the PP. These included for example the diffuse attenuation coefficients for radiance and downward irradiance (K_L and K_d respectively, (m^{-1})).

The PP were the highest-level products, to be directly ingested into the SVC process generated both at full spectral resolution and for the spectral bands of any satellite instrument under consideration, through convolution with the spectral bands response functions of that sensor (here the OLCI instruments aboard the S3 satellite series):

- The spectral fully-normalised water-leaving radiance, $[L_w]_N$ ($W\ m^{-2}\ nm^{-1}\ sr^{-1}$).
- The remote-sensing reflectance, R_{rs} (sr^{-1}).

The equations involved in the derivation of PP and SP are available in [16,28] and were not detailed here.

These radiometric data products were stored along with other data in a single database that eventually included:

1. Optical data products, including the normalized water leaving radiance used for OC-SVC.
2. Platform data products, e.g., buoy tilt.
3. Metocean data products, e.g., wave conditions and wind speed.
4. Calibration and correction data, e.g., calibration gains.
5. Configuration management parameters, e.g., processor version.
6. Logbooks including operators' comments on specific situations and the local environmental background.

Data will be safeguarded using an archival system that includes a daily automatic backup of the in situ datasets. A 50 TB shared disk space is available at ACRI-ST, which is secured due to a Quantum iScalar 80 library of LTO5 drives. This system will be used to backup the entire ROSACE data and software, avoiding data losses.

6.3. Data Processing and Storage

The data processing in the ROSACE ground segment is illustrated in Figure 27 (note that for readability, archiving and storage are not shown). It will be a centralised entity, linked to local transmission centres that are dedicated to each of the OC-SVC field segments. In addition, it is intended to be almost fully automatic and triggered by data availability, with no human interaction other than the following exceptions:

- The configuration (e.g., of calibration or correction, depending on [TChl-*a*], etc.) may evolve from one rotation of the optical system or buoy to the other and then the corresponding parts of the database will need to be updated (manually) by maintaining the versioning in the database.
- The QC macro flags can be updated by an operator through a semisupervised quality control (SSQC— see Section 6.4.
- The data need to be reprocessed (back in time) each time there is an update in the configuration files.

6.4. Semisupervised Quality Control

As soon as the raw data are processed and available to the database, the operator can visualize them through a dashboard. The main capabilities of this tool are, in short:

- To check the product at each stage of its elaboration (from raw data to the final product).
- To check its consistency with respect to previous acquisitions and/or ancillary information from on-site campaigns or Metocean or satellite data.
- To visualize all related contextual information (e.g., sea state, platform behaviour) and auxiliary products that have been used to derive the consolidated product, its uncertainty and its quality annotation.

The dashboard will also allow the operator to:

- Annotate the product with a specific text comment that will be stored in the equivalent of a logbook.
- Raise a new quality flag on the product. This new flag will be added in parallel to the flag that was given automatically by the processing system (i.e., the initial flag is not replaced). This allows traceability of flagging.

Note that by 'operator' we mean a user who is an expert on the particular site and has been given relevant access rights to the system and database. Access to this dashboard will be allowed to any user after registration but will be limited to data consultation (no annotation) and, after a period of time, dedicated to product consolidation.

A similar SSQC system is operated for the biogeochemical (BGC-) Argo floats through a system available on-line (www.seasiderendezvous.eu [97]). In summary the ROSACE SSQC will have a similar functionality that will include:

- A hierarchical visualization web tool allowing any operator to navigate from a higher level product down to initial raw data while at the same time having access to all associated uncertainties and contextual information.
- The capability to annotate the data by adding a new QC flag value, and/or providing a comment in a textual form that will be stored in a log book.
- The capability to edit and consult the logbook.
- The capability to communicate with other operators, at other sites, through a simple chat system. This capability is also extended to communication between operators and the ROSACE ground segment manager.
- The capability to extract information in a simple file format (e.g., csv) to allow extra analysis.

6.5. Final QC

This final stage (stage 8—see Figure 27) represents a key component in the data product elaboration and evaluation. It consists of assigning confidence metrics and uncertainty to the final product that is delivered to the OC-SVC processing system.

Qualitative assessment:

All elementary flags that were evaluated at each stage of the process (stages 1 to 7—see Figure 27) will be combined to derive a global QC flag for each of the final products. These flags will follow the standard that has been adopted in the in situ thematic assembly centre (INSTAC) of the Copernicus marine service (i.e., (0) no QC applied, (1) good data, (2) probably good data, (3) bad data potentially correctable, (4) bad data and (5) value changed). The algorithm and rules that will be used to combine the elementary flags will be defined during the specification and design phase. Qualitative assessment is seen as a confidence rating attached to ROSACE products ensuring that all elements entering into the product derivation have been in line with acceptable ranges and expectations.

Quantitative assessment:

A model has been developed to provide estimates of the uncertainty (as an absolute value or percentage) of the final product used in the OC-SVC process [94]. This uncertainty model will be refined to be in line with the final data processing (Section 5.1). The final purpose is to quantitatively assess the data products and deliver three levels of data quality (Q1, Q2 and Q3) depending on their relative uncertainty, and to do this for each processing mode (NRT, AM and DM). In addition, a flag will identify products suitable for matchup analyses based on a combination of criteria (e.g., cloud filtering etc.). Table 8 shows an example of how the different phases and quality levels might combine according to relative uncertainty levels (u%). Different u% could be attributed to different wavelengths.

Both quantitative and qualitative evaluations will be attached to the final products delivered to the OC-SVC centre, allowing an informed selection of which set of measurements are eventually used for the gain computations.

Table 8. Examples of data quality levels as a function of the data uncertainty.

PHASE	Q1	Q2	Q3
NRT	u% < 3	$3 \leq u\% \leq 5$	u% > 5
DM	u% < 3	$3 \leq u\% \leq 5$	u% > 5
AM	u% < 3	$3 \leq u\% \leq 5$	u% > 5

7. Infrastructure Operations

7.1. Field Segment

The two field sites, BOUSSOLE and MSEA, will be coordinated by LOV-IMEV and HCMR, respectively. As such, they will be responsible for the calibration, deployment, monitoring, maintenance and recovery of the optical system radiometers (OSR) at the two sites. Each site will have three OSR sets available to rotate in the field. This will be with the support of NPL for the characterization of the OSR and recalibration of standards, of CIMEL for OSR and acquisition/transmission unit refurbishment and UT-TO and NPL for yearly round robin intercalibrations. LOV-IMEV and HCMR, with the support of ACRI-IN, will also be responsible for the deployment, monitoring, maintenance, recovery and refurbishment of the full mooring line, buoy upper superstructure and buoy lower superstructure, as well as the installation of the OSR on the platform.

LOV-IMEV and HCMR will conduct monthly cruises at the buoy site for maintenance, deployment of the IN-SITU-SC and of ship-deployed optical systems (free-fall profiling radiometers). They will ensure transfer of information to the ground segment for all aspects that might have an impact on data quality. They will transmit auxiliary and cruise data to the ground segment. LOV-IMEV and HCMR will also organize on demand cruises for extraordinary maintenance of the buoys' optical system and platform. LOV-IMEV and HCMR will also ensure transfer of information to the ground segment for all aspects that might have an impact on data quality and will transmit auxiliary and cruise data to the ground segment.

These activities at both sites will be supported by human and material resources committed by the national institutes as a significant in-kind support. For instance, these contributions have represented half of the total budget needed to run BOUSSOLE on average over 2003–2019. Funding of the other half was equally shared between ESA and the French space agency (CNES). A similar model is proposed for ROSACE.

Absolute calibration of the three OSR at each site will be performed before and after their deployment at sea with a 6-month rotation that is independent from the platform deployment (i.e., divers can swap out instruments in the field). Pre- and post-calibrations will be complemented by reality checks of the instrumentation, which consist of "in air" relative intercalibrations of OSR against the sky or a diffuser target. These verification and calibration processes will be the basis to decide whether or not the instrument needs on demand module replacements/repairs at CIMEL and recharacterisation at NPL.

Visual inspection of the OSR in the field will also be performed on a monthly basis by divers and technical staff. Refurbishment of the OSR will be performed at CIMEL after every three rotations in the field. The profiling system radiometers (PSR) will be calibrated at the same time as the OSR. A systematic refurbishment of the PSR every 2 years will also be performed at CIMEL. In-air intercalibration of the PSR will be integrated with the OSR in-air intercalibration.

An intercalibration round robin will be performed every year with a profiling transfer radiometer. This will consist of absolute calibration at LOV-IMEV, HCMR and UT-TO, and field deployment at BOUSSOLE and MSEA along with the PSR in the vicinity of the buoys. CIMEL and UT-TO will also act as a technology work group to maintain the state of the art of the optical system, whereas NPL will have the same role for absolute calibration standards and protocols. LOV-IMEV and HCMR will act as the technology work group for field segment operational activities in connection with the supervisor of platform maintenance.

7.2. Ground Segment

The operational ROSACE ground segment (GS) will be run by ACRI-ST. This GS will operate in both a NRT and an off-line mode. The NRT mode will process the field segment data as soon as they are available, and the off-line mode will reprocess the data back in time from the starting date when a new configuration has been applied.

Preventive maintenance, i.e., adaptation to information technology and operating system evolutions, will take place at a frequency of 6 months to one year during operations. Adaptive maintenance will also occur during the operations, and includes implementation of new or improved algorithms and QC procedures.

LOV-IMEV and HCMR will contribute to development of the data processing and quality control procedures in connection with the ground segment team.

7.3. Governance

A consortium of European Institutions and SMEs proposes the ROSACE infrastructure. With such an involvement of several entities, an appropriate governance structure has to be put in place, so that coordination of all activities is ensured, the overall schedule of the project is maintained, and possible evolutions of the infrastructure are adequately phased in.

A project office will oversee the activity, working closely with the five teams dedicated to the main technical functions of the infrastructure. These five teams will manage the optical system and calibration, the deployment platform, the field segment, the ground segment, and the SI-traceability, metrology aspects and uncertainty budget. The project office will be the main interface with EUMETSAT on scientific, administrative and financial matters. It will evaluate and validate the project deliverables and reports, coordinate the project on a day-to-day basis, monitor the overall progress and milestones, organise reviews and meetings, manage all administrative and financial matters.

A steering committee will be responsible for the strategic management and decision making in the project, which will include recommending possible directions/choices for the ROSACE project, evaluating the results achieved and making recommendations about strategic directions, performing risk assessment and suggesting strategies if issues arise, ensuring links with the scientific community and international groups (e.g., IOCCG) and anticipating evolutions in OC-SVC.

A technical advisory board composed of a team of international independent experts will support the development of the infrastructure, by providing recommendations, expertise and by reviewing the activity in an on-going manner.

The project office and steering committee will also promote the scientific exploitation of data collected by the ROSACE OC-SVC infrastructure via dedicated research projects, with attention to improvements in optical measurement protocols, uncertainty estimates, correction methodologies, QC procedures and the OC-SVC process.

Possible innovations will be discussed within the steering committee under EUMETSAT and Copernicus oversight. They will be tested for a sufficiently long time in parallel to operational standards in order to ensure continuity and a robust evaluation before integration into the operational activities.

8. The Profiling Float Network Option

8.1. Increasing the Matchup Capacity of the Infrastructure

Collecting radiometry measurements from autonomous profiling floats is gaining increasing attention in the oceanographic and Earth observation communities. Such floats, whose initial use was for measuring temperature and salinity over the global oceans (Argo programme [98]), now come with various payload configurations including an increasing number of instruments [99]. Some of these floats include radiometers, and their data have been used for satellite OCR validation purposes [100,101]. An advanced float including a specifically designed radiometer system has recently been proposed as an option that meets most of the requirements for the OC-SVC of the future NASA PACE mission [102].

Our approach here was not to use profiling floats as a central element of the OC-SVC infrastructure, but as an option to improve the matchup capacity of the infrastructure and eventually extend its footprint beyond the fixed sites. This extension allows confirmation of the OC-SVC gains and uncertainties, and will allow increasing the matchup capability during the commissioning phase of future elements of the Sentinel-3 constellation (e.g., S3C and S3D). About 50 matchups are needed to get stable OC-SVC gains [3]. A network of floats working in conjunction with the fixed sites could help reach this number faster. For this combination to be effective, the radiometry from the floats should, however, be fully consistent with the radiometry from the fixed sites. This is among the specifications that have been set for the development of the new ROSACE radiometer.

8.2. The ProVal Float and its ROSACE Upgrade

The development of a float dedicated to satellite OCR validation started in 2011 at LOV-IMEV. The float is named ProVal [25] and uses a two-arm configuration inspired by the BOUSSOLE mooring. It allows for two identical radiometers to be hosted on either side of the float. Both sensors measure the downward irradiance (E_d) and the upwelling nadir radiance (L_u). With this configuration the irradiance sensors are at 21 cm depth and the radiance sensors at 45 cm depth when the float is at surface, which is referred to as the "buoy mode". Platform shading issues are mitigated by always having one sensor outside of the float's shadow. This configuration also ensures data redundancy, which is helpful to monitor the relative behaviour of the instruments over time and for QC.

Vertical E_d and L_u profiles are acquired during the ascending phase with increasing depth resolution from parking depth to surface (e.g., every 1 m up to 60 m and every 10 cm up to the surface). When the float reaches the surface, one minute of data acquisition occurs in "buoy mode". Four instances of the remote sensing reflectance (R_{rs}) are generated from the collected data. Two are obtained by extrapolating to the surface the vertical E_d and L_u profiles of both pairs of radiometers, and two are generated from the data collected by both pairs of radiometers during the "buoy mode". The profiling sequence can be modified remotely, following trade-offs between the amount of data to be transmitted, the available energy and the data transmission costs. Rescheduling the float mission is also feasible remotely, including maintaining the float at surface for recovery.

The profiling mode described above would generate about 140 KB per profile, which would take 12 min to be sent back to the ground segment via satellite transmission. A single float will have enough energy to perform 300 such profiles.

First deployments of ProVal floats exhibited very good navigation behaviour in terms of tilt, vertical ascent speed and capacity to target a particular surfacing time. They also show a very good matchup probability with S3/OLCI of about 20% (Table 9), which expresses how many profiles are needed to generate a matchup. Note that the difference in matchups ratio between BOUSSOLE and MSEA may be related to the season. A July deployment at MSEA would likely end up with a higher matchup ratio. The matchup processing is described in [25] and mainly follows validation criteria.

Table 9. Matchups ratio for three deployments in the Mediterranean Sea with S3 OLCI.

Mission ID	Deployment Start Date	Area	Number of Profiles	Matchups Ratio
Lovapm006f	09/06/2017	BOUSSOLE	81	29%
Lovapm006h	11/06/2018	Ionian Sea	101	18%
Lovapm006i	26/09/2019	MSEA	35	23%

Modifications of the current ProVal platform will be necessary to meet the ROSACE requirements. These include:

- Implementation of the new fit-for-purpose CIMEL radiometer, which will require, in particular, an increased depth rating as compared to the version to be installed on the buoys.
- Development of on-board data processing in order to make NRT transmission possible despite the large amount of data generated by hyperspectral radiometers. A new acquisition board is under development for the next generation of ProVal floats, which will be used for the ROSACE floats. The full data set can be downloaded at float recovery.
- Inclusion of a rechargeable battery. Opening a float to swap batteries is time consuming and may compromise waterproofness of the float, so we will implement a rechargeable battery as is done now on gliders [103].

8.3. Minimum Configuration and Operations

One float will be available at each site to increase the sampling and matchup capability during operation of the Sentinel constellation, in particular during the satellite's critical phases, e.g., commissioning phase or programmed band shifts. These floats may also be used as a backup in case the main mooring is unavailable.

The human resources required to operate a float depends essentially on how often it is repositioned on site. At the BOUSSOLE site, a repositioning every two weeks will probably be sufficient but this frequency may have to be higher for MSEA because of the proximity of small islands. This will be adjusted according to the season and prevailing circulation patterns. During each recovery, the float will be cleaned and the sensors compared with the portable, field-deployable, relative calibration source. Every 150 profiles, the float will be brought back to the laboratory to recharge the batteries and recalibrate the sensors. Nonetheless the centralized ground segment and the simplicity of deployment/recovery of such a platform, make it flexible enough to envision different scenarios of operations to fit the evolving requirements of OC-SVC and Copernicus marine services.

9. Conclusions

A European solution was proposed for an infrastructure serving the OC-SVC needs of the Copernicus Sentinel-3 operational missions. It is named ROSACE, which stands for radiometry for ocean colour satellites calibration and community engagement, and has been described in this paper. Key features that can be highlighted in these concluding remarks are that ROSACE:

- Answers all OC-SVC requirements that were established in [21].
- Includes two sites (BOUSSOLE and MSEA) that are demonstrably suited to providing a large number of OC-SVC-quality matchups every year.
- Minimises risks because it builds on an existing quasi-operational capability (BOUSSOLE).
- Minimises the development time, because of the above point, plus it exploits existing expertise and elements that can still be used in the meantime before a new system becomes fully operational.
- Minimises costs for delivering the requirement of multiple sites.
- Maintains, optimises and expands current European expertise in order to provide a coordinated effort and sovereignty for the Copernicus OCR SVC infrastructure.

High-level rationales and features were presented in this paper, while details of existing elements forming the base of the proposed infrastructure were not repeated when already available in the literature, e.g., the full description of the BOUSSOLE buoy design and testing [80–82], the associated overall project structure and realisations [16,28], the details of published surveys indicating MSEA as the best location for OC-SVC in European waters [20,22–24], the data processing protocols [28], the approach to delivering per-measurement uncertainties [94] or the characteristics and preliminary achievements of the ProVal floats [25]. Further details about the proposed new elements, e.g., the fit-for-purpose hyperspectral radiometers or the MSEA site, will become available and published in the open literature when mature enough.

The emphasis was put here on the main items that demonstrate compliance of the proposed infrastructure with OC-SVC requirements described in [29]. This report includes 62 requirements, the compliance of ROSACE with all relevant ones (numbers 9–62) having been demonstrated in [29]. The corresponding compliance matrix was not replicated here. This full compliance was reached by combining outstanding expertise and knowledge in OC-SVC gained over the past 20 years at BOUSSOLE with a number of new developments. The growth from a long-term project run by essentially a single research group (BOUSSOLE) to an integrated and operational European infrastructure (ROSACE) is feasible through bringing together European expertise of partners who already have a history of collaboration, and through including new elements and technology.

As a matter of conclusion, we show the combined OC-SVC matchup capability of BOUSSOLE and MOBY (Figure 28), which is likely the best way of demonstrating the above claims about what ROSACE could deliver with BOUSSOLE and MSEA working together. The capability illustrated in Figure 28 for BOUSSOLE is already significant and indeed is used in support of the Copernicus Sentinels [104], although it is actually below the minimum capability of the proposed ROSACE infrastructure, which has the potential of delivering up to about 80 high-quality matchups every year from the combination of BOUSSOLE and MSEA.

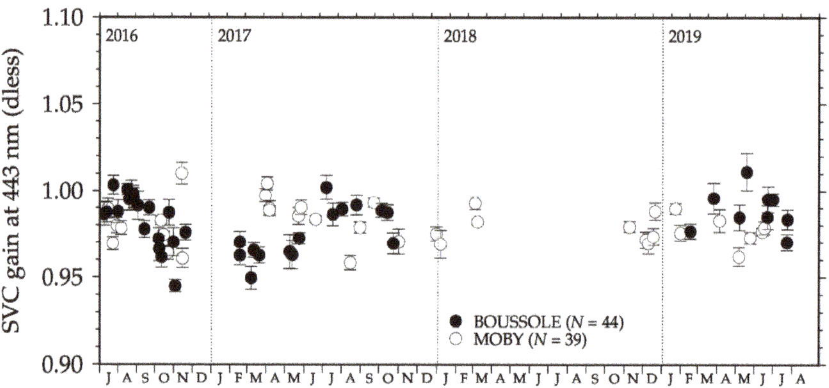

Figure 28. Ocean colour SVC (OC-SVC) gains for Sentinel 3A/OLCI obtained from either BOUSSOLE or MOBY. Dysfunction of both systems plus technical issues delaying the access to S3A data are responsible for the data gap in 2018.

Author Contributions: Conceptualization (alphabetic order used), A.B., A.C.B., A.M., A.P.K., D.A., D.C., G.P., J.K., K.L., N.F., O.H.F.d., P.B., S.P., S.V., V.V.; resources, A.B., A.D., A.K., A.C.B., A.M., A.P.K., D.A., E.L., G.P., J.K., M.K., M.B., M.G., S.P., S.V., V.V.; writing—original draft preparation, A.B., A.C.B., A.M., C.L., D.A., J.K., K.L., M.B., N.F., S.P., S.V., V.V.; writing—review and editing, ALL; project administration, A.B., J.K., P.B., S.P., S.V., V.V.; funding acquisition, A.B., A.C.B., A.P.K., D.A., D.C., J.K., N.F., O.H.F.d., S.P., S.V., V.V. All authors have read and agreed to the published version of the manuscript.

Funding: This work was funded by EUMETSAT, contract EUM/C0/18/4600002162/EJK issue 1A. BOUSSOLE is currently funded by the European Space Agency (ESA), contract 4000119096/17/I-BG, and by the Centre National d'Etudes Spatiales (CNES).

Acknowledgments: The proposed ROSACE infrastructure largely builds on the experience gained with developing and running the BOUSSOLE project. Many individuals contributed to the success of BOUSSOLE, who cannot all be named here. They are the ships' crews helping us with mooring operations or monthly servicing to the buoy site, technical staff and divers from the Villefranche observatory who contribute to operations at sea, to buoy preparation and maintenance, to calibration of instruments and to data processing and quality control. Our finance team have had to manage significant administrative tasks as well over the 20 years this project has now been operating. There is also the French Oceanographic Fleet for the management and provision of the research vessels supporting the BOUSSOLE cruises. They are all warmly thanked for their dedication to BOUSSOLE, and we are sure that their combined effort in supporting BOUSSOLE will be instrumental in making ROSACE a success in the event it would be selected as part of the Copernicus OCR SVC infrastructure. Data from the Cote d'Azur meteorological buoy were obtained from the HyMeX program, sponsored by Grants MISTRALS/HyMeX and Météo-France. Nutrient data were obtained from the DYnamique des Flux Atmosphériques en MEDiterranée (DYFAMED) project. We also acknowledge the Finokalia/University of Crete/PANACEA team (in particular G. Kouvarakis, N. Kalivitis and A. Bais), N. Spyridakis and the HCMR E1-M3A team for important inputs to the MSEA site characterisation. Lelia Proniewski (CIMEL Electronique) is kindly acknowledged for help with radiometers drawings. We thank Ewa Kwiatkowska and other EUMETSAT personnel (I. Cazzaniga, J. Chimot, D. Dessailly, F. Montagner and E. Obligis), as well as the independent review panel of the 2019 preliminary design study (K. N. Babu, P. Goryl, F. Jacq, B. C. Johnson, A. Reppucci, M. Wang and G. Zibordi), who all provided helpful comments on our proposal. The European Space Agency is also acknowledged for having funded the Fiducial Reference Measurements for Satellite Ocean Colour (FRM4SOC) project, which has delivered groundwork and foundations for the ROSACE proposal.

Conflicts of Interest: The authors declare no conflict of interest.

Appendix A

This appendix presents the uncertainty components displayed in Figure 25 in Table A1. A few examples of the effects tables are shown in Tables A2–A4. Note that the anticipated uncertainty values are presented for 490 nm only. Some of these uncertainties vary with wavelength and therefore might be higher or lower than what is provided here in Tables A2–A4.

Table A1. List of all symbols used in Figure 25.

Symbol	Description, and Units when Relevant
z_1, z_2	The two measurements depths on the buoy (m)
L_{u,z_1}, L_{u,z_2}	The upwelling radiance measured in water at a given depth (W m^{-2} nm^{-1} sr^{-1})
S_{z_1}, S_{z_2}	The dark-corrected signal per instrument, for the two measurement depths (W m^{-2} nm^{-1} sr^{-1})
Instrument-specific Quantities	
$DN_{\text{Light},z_1}\ DN_{\text{Light},z_2}$ $DN_{\text{Dark},z_1}\ DN_{\text{Dark},z_2}$	The median light and dark readings (counts)
$c_{\text{cal},z_1}, c_{\text{cal},z_2}$	Radiometric calibration coefficient
$c_{\text{stab},z_1}, c_{\text{stab},z_2}$	Radiometric stability evaluated post deployment
$c_{\lambda,z_1}, c_{\lambda,z_2}$	Spectral calibration actual central wavelength of each pixel and its accuracy
$c_{\text{lin},z_1}, c_{\text{lin},z_2}$	Detector linearity correction
c_{T,z_1}, c_{T,z_2}	Temperature correction
$c_{\text{stray},z_1}, c_{\text{stray},z_2}$	Spectral stray light correction
$c_{\text{pol},z_1}, c_{\text{pol},z_2}$	Polarisation sensitivity correction
$c_{\text{im},z_1}, c_{\text{im},z_2}$	Immersion factor
$c_{\text{sh},z_1}, c_{\text{sh},z_2}$	Shading correction
Derived Parameters	
K_{L_u}	The diffuse attenuation coefficient for upwelling radiance (m^{-1})
f_h	The Hydrolight-based [105] extrapolation correction (see Appendix A in [28]).
ρ	The Fresnel reflection coefficient for the water-air interface
n	The refractive index of seawater
Input to Various Models	
f	Generic term for a function
[TChl-*a*]	Total chlorophyll-*a* concentration (mg m^{-3})
θ_{SZA}	Solar Zenith Angle (degrees)
salinity	Seawater salinity (psu)
P	Atmospheric pressure (hPa)
T	Water Temperature (degrees C)
w_s	Wind speed (m s^{-1})
θ	Viewing angle (degrees)
Actual Instrument Depth Evaluation [*]	
z_{sensor}	Depth at the pressure sensor (m)
θ_B	The buoy/instrument tilt derived from 2-axis tilt sensors (degrees)
Δ_{SR}	Distance between the lower arm and the pressure sensor (m)
Δ_1, Δ_2	Lower and upper buoy arm length (m)
Δ_{12}	Distance between the arms (m)

[*] This is based on the current BOUSSOLE system, where the depth of instruments is derived from the measurement of a single pressure sensor installed on the buoy structure. The ROSACE instruments will each have a pressure sensor, reducing uncertainty of the depth evaluation.

The simple example of random thus uncorrelated uncertainty is the detector noise and dark current. The effects table for the detector noise is presented in Table A2.

Table A2. Effects table for detector noise.

Name of Effect		Noise in Light Counts [§]
Affected term in measurement function		DN_{Light}
Instruments in the series affected		All
Correlation type and form	Temporal within deployment	Random
	Temporal between deployments	Random
	Spectral (hyperspectral in-situ)	Random
Correlation scale	Temporal within deployment	0
	Temporal between deployments	0
	Spectral (hyperspectral in-situ)	0
Channels/bands	List of channels/bands affected	All
	Error correlation coefficient matrix	Identity – No correlation
Uncertainty	PDF shape	Gaussian
	Units	Counts
	Magnitude	Less than 0.1%
Sensitivity coefficient		$\dfrac{\partial f}{\partial DN_{Light1}}$, $\dfrac{\partial f}{\partial DN_{Light2}}$

[§] Same thing for the noise in dark counts, DN_{dark}.

A more complicated correlation structure is expected for instrument calibration, as shown in Table A3. The absolute radiometric uncertainty is combined from systematic and random effects. We will spit them, so that the systematic part of that uncertainty will stay fully correlated between deployments within the timescale related to the absolute radiometric standards recalibration (i.e., across calibrations) and the random part is correlated only within a deployment (i.e., between calibrations).

Table A3. Effects table for detector calibration.

Name of Effect		Detector Calibration Systematic Error	Detector Calibration Random Error 1, 2	Detector Calibration Stability Model Error 1, 2
Affected term in measurement function		$c_{cal,z_1}, c_{cal,z_2},$ $c_{cal,z_1,t}, c_{cal,z_2,t}$	$c_{cal,z_1}, c_{cal,z_2},$ $c_{cal,z_1,t}, c_{cal,z_2,t}$	$c_{stab,z_1}, c_{stab,z_2}$
Instruments in the series affected		All	All	All
Correlation type and form	Temporal within deployment	Rectangular Absolute	Rectangular Absolute	Rectangular Absolute
	Temporal between deployments	Rectangular Absolute	Random	Random
	Spectral (hyperspectral in-situ)	To be defined	To be defined	To be defined
Correlation scale	Temporal within deployment	$-\infty, +\infty$	$-\infty, +\infty$	$-\infty, +\infty$
	Temporal between deployments	a,b [§]	0	0
	Spectral (hyperspectral in-situ)	To be defined	To be defined	To be defined
Channels/bands	List of channels/bands affected	All	All	All
	Error correlation coefficient matrix	Identity – No correlation	Identity – No correlation	Identity – No correlation
Uncertainty	PDF shape	Gaussian	Gaussian	Gaussian
	Units	Radiance/Counts	Radiance/Counts	Radiance/Counts
	Magnitude	0.70%	0.25%	Less than 1%
Sensitivity coefficient		$\dfrac{\partial f}{\partial c_{cal\_s}}$	$\dfrac{\partial f}{\partial c_{cal\_r1}}$, $\dfrac{\partial f}{\partial c_{cal\_r2}}$	$\dfrac{\partial f}{\partial c_{stab1}}$, $\dfrac{\partial f}{\partial c_{stab2}}$,

[§] When items a and b depend on the recalibration schedule of the SMART-CC system.

Some of the effects need further investigation to fully understand their correlation structure. For example, modelling errors presented in Table A4.

Table A4. Effects tables for models' errors.

Name of Effect		Shading Correction Model Error	Hydrolight Correction Model Error	Refractive Index of Seawater Model Error	Fresnel Reflection Model Error
Affected term in measurement function		c_{sh}	f_h	n	ρ
Instruments in the series affected		All	All	All	All
Correlation type and form	Temporal within deployment	Random	Random	Rectangular Absolute	Rectangular Absolute
	Temporal between deployments	Random	Random	Rectangular Absolute	Rectangular Absolute
	Spectral (hyperspectral in-situ)	To be defined	To be defined	To be defined	To be defined
Correlation scale	Temporal within deployment	0	0	$-\infty,+\infty$	$-\infty,+\infty$
	Temporal between deployments	0	0	$-\infty,+\infty$	$-\infty,+\infty$
	Spectral (hyperspectral in-situ)	To be defined	To be defined	To be defined	To be defined
Channels/bands	List of channels/bands affected	All	All	All	All
	Error correlation coefficient matrix	Identity – No correlation	Identity – No correlation	Matrix of 1's – fully correlated	Matrix of 1's—fully correlated
Uncertainty	PDF shape	Gaussian	Gaussian	Gaussian	Gaussian
	Units	%	%	%	%
	Magnitude	2	0.5%	0.9% [§]	0.04% negligible
Sensitivity coefficient		$\frac{\partial f}{\partial c_{sh}}$	$\frac{\partial f}{\partial f_h}$	$\frac{\partial f}{\partial n}$	$\frac{\partial f}{\partial \rho}$

[§] When the refractive index is considered wavelength independent.

References and Note

1. Gordon, H.R. In-orbit calibration strategy for ocean color sensors. *Remote Sens. Environ.* **1998**, *63*, 265–278. [CrossRef]
2. Ohring, G.; Tansock, J.; Emery, W.; Butler, J.; Flynn, L.; Weng, F.; St. Germain, K.; Wielicki, B.; Cao, C.; Goldberg, M.; et al. Achieving satellite instrument calibration for climate change. *Eos Trans. AGU* **2007**, *86*, 136. [CrossRef]
3. Franz, B.A.; Bailey, S.W.; Werdell, P.J.; McClain, C.R. Sensor-independent approach to the vicarious calibration of satellite ocean color radiometry. *Appl. Opt.* **2007**, *46*, 5068–5082. [CrossRef] [PubMed]
4. Bailey, S.W.; Werdell, P.J. A multi-sensor approach for the on-orbit validation of ocean color satellite data products. *Remote Sens. Environ.* **2006**, *102*, 12–23. [CrossRef]
5. McClain, C.R.; Esaias, W.E.; Barnes, W.; Guenther, B.; Endres, D.; Hooker, S.B.; Mitchell, G.; Barnes, R. Calibration and Validation Plan for SeaWiFS. *NASA Tech. Memo.* **1992**, *3*, 104566.
6. McClain, C.R.; Hooker, S.B.; Feldman, G.C.; Bontempi, P. Satellite data for ocean biology, biogeochemistry and climate research. *Eos Trans. AGU* **2006**, *87*, 337–343. [CrossRef]
7. Zibordi, Z.; Mélin, F.; Voss, K.J.; Johnson, B.C.; Franz, B.A.; Kwiatkowska, E.; Huot, J.P.; Wang, M.; Antoine, D. System Vicarious Calibration for Ocean Color Climate Change Applications: Requirements for In Situ Data. *Remote Sens. Environ.* **2015**, *159*, 361–369. [CrossRef]
8. Gordon, H.R. Calibration requirements and methodology for remote sensors viewing the ocean in the visible. *Remote Sens. Environ.* **1987**, *22*, 103–126. [CrossRef]
9. Gordon, H.R. Atmospheric correction of ocean color imagery in the Earth observing system era. *J. Geophys. Res.* **1997**, *102*, 17081–17106. [CrossRef]
10. GCOS, Systematic Observation Requirements For Satellite-Based Data Products For Climate, 2011 update, Supplemental details to the satellite-based component of the "Implementation Plan for the Global Observing System for Climate in Support of the UNFCCC (2010 Update)", December 2011, GCOS-154, WMO-IOC. Available online: https://library.wmo.int/doc_num.php?explnum_id=3710 (accessed on 11 May 2020).
11. Antoine, D.; Morel, A. A multiple scattering algorithm for atmospheric correction of remotely-sensed ocean colour (MERIS instrument): Principle and implementation for atmospheres carrying various aerosols including absorbing ones. *Int. J. Remote Sens.* **1999**, *20*, 1875–1916. [CrossRef]
12. Morel, A.; Maritorena, S. Bio-optical properties of oceanic waters: A reappraisal. *J. Geophys. Res.* **2001**, *106*, 7763–7780. [CrossRef]
13. Clark, D.K.; Gordon, H.R.; Voss, K.J.; Ge, Y.; Broenkow, W.; Trees, C. Validation of atmospheric correction over the oceans. *J. Geophys. Res.* **1997**, *102D*, 17209–17217. [CrossRef]
14. Clark, D.K.; Yarbrough, M.A.; Feinholz, M.; Flora, S.; Broenkow, W.; Kim, Y.S.; Johnson, B.C.; Brown, S.W.; Yuen, M.; Mueller, J.L. MOBY, a radiometric buoy for performance monitoring and vicarious calibration of satellite ocean color sensors: Measurement and data analysis protocols. In *NASA Ocean Optics Protocols For Satellite Ocean Color Sensor Validation*; Chapter 2; Goddard Space Flight Space Center: Greenbelt, MD, USA, 2003; Volume 11, pp. 138–170.
15. The BOUSSOLE acronym, which is the French word for «compass», translates as «Buoy for the acquisition of a long-term optical time series».
16. Antoine, D.; Chami, M.; Claustre, H.; D'Ortenzio, F.; Morel, A.; Bécu, G.; Gentili, B.; Louis, F.; Ras, J.; Roussier, E.; et al. BOUSSOLE: A joint CNRS-INSU, ESA, CNES and NASA Ocean Color Calibration And Validation Activity. *NASA Tech. Memo.* **2006**, 214147.
17. Pahlevan, N.; Sarkar, S.; Franz, B.A.; Balasubramanian, S.V.; He, J. Sentinel-2 MultiSpectral Instrument (MSI) data processing for aquatic science applications: Demonstrations and validations. *Remote Sens. Environ.* **2017**, *201*, 47–56. [CrossRef]
18. Pahlevan, N.; Balasubramanian, S.; Sarkar, S.; Franz, B. Toward Long-Term Aquatic Science Products from Heritage Landsat Missions. *Remote Sens.* **2018**, *10*, 1337. [CrossRef]
19. Pahlevan, N.; Chittimalli, S.K.; Balasubramanian, S.V.; Vellucci, V. Sentinel-2/Landsat-8 product consistency and implications for monitoring aquatic systems. *Remote Sens. Environ.* **2019**, *220*, 19–29. [CrossRef]
20. Zibordi, G.; Mélin, F.; Talone, M. *System Vicarious Calibration for Copernicus Ocean Colour Missions: Requirements and Recommendations for a European Site*; Publications Office of the European Union: Brussels, Belgium, 2017. [CrossRef]

21. Mazeran, C.; Brockmann, C.; Ruddick, K.; Voss, K.J.; Zagolsky, F. Requirements for Copernicus Ocean Colour Vicarious Calibration Infrastructure; EUMETSAT, 2017; SOLVO/EUM/16/VCA/D8; EUMETSAT EUM/CO/16/4600001772/EJK. Available online: https://www.eumetsat.int/website/home/Data/ScienceActivities/ScienceStudies/CopernicusOceanColourVicariousCalibrationInfrastructure/RequirementsforCopernicusOceanColourVicariousCalibrationInfrastructure/index.html?lang=EN#SD (accessed on 11 May 2020).

22. Banks, A.C.; Vendt, R.; Alikas, K.; Bialek, A.; Kuusk, J.; Lerebourg, C.; Ruddick, K.; Tilstone, G.; Viktor Vabson, V.; Donlon, C.; et al. Fiducial Reference Measurements for Satellite Ocean Colour (FRM4SOC). *Remote Sens.* **2020**, *12*, 1322. [CrossRef]

23. FRM4SOC D-240, Proceedings of WKP-1 (PROC-1). Report of the International Workshop. 2017. Available online: https://frm4soc.org/wp-content/uploads/filebase/FRM4SOC-WKP1-D240-Workshop_Report_PROC-1_v1.1_signedESA.pdf (accessed on 13 January 2020).

24. Zibordi, G.; Mélin, F. An evaluation of marine regions relevant for ocean color system vicarious calibration. *Remote Sens. Environ.* **2017**, *19*, 122–136. [CrossRef]

25. Leymarie, E.; Penkerc'h, C.; Vellucci, V.; Lerebourg, C.; Antoine, D.; Boss, E.; Lewis, M.; D'Ortenzio, F.; Claustre, H. A new autonomous profiling float for high quality radiometric measurements. *Front. Mar. Sci.* **2018**, *5*, 437. [CrossRef]

26. Zhou, P.; Yang, X.-L.; Wang, X.-G.; Hu, B.; Zhang, L.; Zhang, W.; Si, H.-R.; Zhu, Y.; Li, B.; Huang, C.-L.; et al. A pneumonia outbreak associated with a new coronavirus of probable bat origin. *Nature* **2020**, *579*, 270–273. [CrossRef]

27. Maier, B.F.; Brockmann, D. Effective containment explains sub-exponential growth in confirmed cases of recent COVID-19 outbreak in Mainland China. *Science* **2020**. [CrossRef] [PubMed]

28. Antoine, D.; D'Ortenzio, F.; Hooker, S.B.; Bécu, G.; Gentili, B.; Tailliez, D.; Scott, A.J. Assessment of uncertainty in the ocean reflectance determined by three satellite ocean color sensors (MERIS, SeaWiFS and MODIS-A) at an offshore site in the Mediterranean Sea (BOUSSOLE project). *J. Geophys. Res.* **2008**, *113*, C07013. [CrossRef]

29. Vellucci, V.; Antoine, D.; Banks, A.C.; Bardey, P.; Bretagnon, M.; Bruniquel, V.; Deru, A.; Hembise Fanton d'Andon, O.; Lerebourg, C.; Mangin, A.; et al. *ROSACE, Radiometry for Ocean Colour SAtellites Calibration & Community Engagement, Preliminary Design Document (PDD-V4.1) EUMETSAT Contract EUM/CO/184600002162/EJK-Order No. 4500017110*. 2019. Available online: https://www.eumetsat.int/website/home/Data/ScienceActivities/ScienceStudies/CopernicusOceanColourVicariousCalibrationInfrastructure/PreliminaryDesignProjectPlanandCostingforCopernicusOceanColourVicariousCalibrationInfrastructure/index.html (accessed on 11 May 2020).

30. Mayot, N.; D'Ortenzio, F.; Taillandier, V.; Prieur, L.; Pasqueron de Fommervault, O.; Claustre, H.; Bosse, A.; Testor, P.; Conan, P. Impacts of the deep convection on the phytoplankton blooms in temperate seas: A multiplatform approach over a complete annual cycle (2012-2013 DEWEX experiment). *J. Geophys. Res.* **2017**, *122*. [CrossRef]

31. Morel, A.; Prieur, L. Analysis of variations in ocean color. *Limnol. Oceanogr.* **1977**, *22*, 709–722. [CrossRef]

32. Morel, A.; Bélanger, S. Improved Detection of turbid waters from Ocean Color information. *Remote Sens. Environ.* **2006**, *102*, 237–249. [CrossRef]

33. Platnick, S.; Hubanks, P.; Meyer, K.; King, M.D. *MODIS Atmosphere L3 Monthly Product (08_L3). NASA MODIS Adaptive Processing System*; Goddard Space Flight Center: Greenbelt, MD, USA, 2015. [CrossRef]

34. Golbol, M.; Vellucci, V.; Antoine, D. *BOUSSOLE Set of Cruises*. 2000. Available online: https://campagnes.flotteoceanographique.fr/series/1/ (accessed on 11 May 2020).

35. Petihakis, G.; Perivoliotis, L.; Korres, G.; Ballas, D.; Frangoulis, C.; Pagonis, P.; Ntoumas, M.; Pettas, M.; Chalkiopoulos, A.; Sotiropoulou, M.; et al. An integrated open-coastal biogeochemistry, ecosystem and biodiversity observatory of the eastern Mediterranean–the Cretan Sea component of the POSEIDON system. *Ocean Sci.* **2018**, *14*, 1223–1245. [CrossRef]

36. Henson, S.; Beaulieu, C.; Lampitt, R. Observing climate change trends in ocean biogeochemistry: When and where. *Glob. Chang. Biol.* **2016**, *22*, 1561–1571. [CrossRef]

37. POSEIDON: Monitoring, Forecasting and Information System for the Greek Seas. Available online: http://www.poseidon.hcmr.gr (accessed on 24 March 2020).

38. Global Ocean Observing System (GOOS) OceanSITES, A Worldwide System of Deepwater Reference Stations. Available online: http://www.oceansites.org/index.html (accessed on 24 March 2020).

39. D'Ortenzio, F.; Ribera d'Alcalà, M. On the trophic regimes of the Mediterranean Sea: A satellite analysis. *Biogeoscience* **2009**, *6*, 139–148. [CrossRef]

40. Menna, M.; Poulain, P.-M.; Zodiatis, G.; Gertman, I. On the surface circulation of the Levantine sub-basin derived from Lagrangian drifters and satellite altimetry data. *Deep-Sea Res. I* **2012**, *65*, 46–58. [CrossRef]

41. Theocharis, A.; Balopoulos, E.; Kioroglou, S.; Kontoyiannis, H.; Iona, A. A synthesis of the circulation and hydrography of the South Aegean Sea and the Straits of the Cretan Arc (March 1994–January 1995). *Prog. Oceanogr.* **1999**, *44*, 469–509. [CrossRef]

42. Korres, G.; Ntoumas, M.; Potiris, M.; Petihakis, G. Assimilating Ferry Box data into the Aegean Sea model. *J. Mar. Syst.* **2014**, *140*, 59–72. [CrossRef]

43. Psarra, S.; Tselepides, A.; Ignatiades, L. Primary productivity in the oligotrophic Cretan Sea (NE Mediterranean): Seasonal and interannual variability. *Prog. Oceanogr.* **2000**, *46*, 187–204. [CrossRef]

44. Ignatiades, L. The productive and optical status of the oligotrophic waters of the Southern Aegean Sea (Cretan Sea), Eastern Mediterranean. *J. Plank. Res.* **1998**, *20*, 985–995. [CrossRef]

45. Chronis, G.; Lykousis, V.; Georgopoulos, D.; Zervakis, V.; Stavrakakis, S.; Poulos, S. Suspended particulate matter and nepheloid layers over the southern margin of the Cretan Sea (N.E. Mediterranean): Seasonal distribution and dynamics. *Prog. Oceanogr.* **2000**, *46*, 163–185. [CrossRef]

46. Berthon, J.F.; Mélin, F.; Zibordi, G. Ocean Colour Remote Sensing of the Optically Complex European Seas. In *Remote Sensing of the European Seas*; Barale, V., Gade, M., Eds.; Springer: Dordrecht, Germany, 2008.

47. Jerlov, N.G. *Marine Optics*, 2nd ed.; Elsevier: New York, NY, USA, 1976.

48. Karageorgis, A.P.; Drakopoulos, P.G.; Psarra, S.; Pagou, K.; Krasakopoulou, E.; Banks, A.C.; Velaoras, D.; Spyridakis, N.; Papathanassiou, E. Particle characterization and composition in the NE Aegean Sea: Combining optical methods and biogeochemical parameters. *Cont. Shelf Res.* **2017**, *149*, 96–111. [CrossRef]

49. Drakopoulos, P.G.; Banks, A.C.; Kakagiannis, G.; Karageorgis, A.P.; Lagaria, A.; Papadopoulou, A.; Psarra, S.; Spyridakis, N.; Zervakis, V. Estimating chlorophyll concentrations in the optically complex waters of the North Aegean Sea from field and satellite ocean colour measurements. In Proceedings of the SPIE Third International Conference on Remote Sensing and Geoinformation of the Environment (RSCy2015), Paphos, Cyprus, 19 June 2015.

50. Banks, A.C.; Karageorgis, A.; Drakopoulos, P.G.; Psarra, S.; Zeri, C.; Pitta, E.; Papadopoulou, A.; Spyridakis, N. PERSEUS Marine Optics–Ocean Colour in the Aegean Sea. In Proceedings of the PERSEUS 2nd Scientific Workshop, Marrakesh, Morocco, 4 December 2014.

51. Karageorgis, A.P.; Drakopoulos, P.G.; Chaikalis, S.; Lagaria, A.; Spyridakis, N.; Psarra, S. The LEVECO project bio-optics experiment in the northwestern Levantine Sea: Preliminary results. In Proceedings of the SPIE 11174, Seventh International Conference on Remote Sensing and Geoinformation of the Environment (RSCy2019), Paphos, Cyprus, 27 June 2019. [CrossRef]

52. Banks, A.C.; Drakopoulos, P.G.; Chaikalis, S.; Spyridakis, N.; Karageorgis, A.P.; Psarra, S.; Taillandier, V.; D'Ortenzio, F.; Sofianos, S.; Durrieu de Madron, X. An in situ optical dataset for working towards fiducial reference measurements based satellite ocean colour validation in the Eastern Mediterranean. In Proceedings of the SPIE, Eighth International Conference on Remote Sensing and Geoinformation of the Environment (RSCy2020), Paphos, Cyprus, 16–18 March 2020.

53. NASA Aerosol Robotic Network (AERONET), FORTH_CRETE Station. Available online: https://aeronet.gsfc.nasa.gov/cgi-bin/data_display_aod_v3?site=FORTH_CRETE&nachal=2&level=3&place_code=10 (accessed on 22 March 2020).

54. Moulin, C.; Lambert, C.E.; Dayan, U.; Masson, V.; Ramonet, M.; Bousquet, P.; Legrand, M.; Balkanski, Y.J.; Guelle, W.; Marticorena, B.; et al. Satellite climatology of african dust transport in the Mediterranean atmosphere. *J. Geophys. Res.* **1998**, *103*, 13137–13144. [CrossRef]

55. Angstrom, A. On the atmospheric transmission of Sun radiation and on dust in the air. *Geogr. Ann.* **1929**, *11*, 156–166.

56. Schuster, G.L.; Dubovik, O.; Holben, B.N. Angstrom exponent and bimodal aerosol size distributions. *J. Geophys. Res.* **2006**, *111*. [CrossRef]

57. Eck, T.; Holben, B.N.; Reid, J.; Dubovik, O.; Smirnov, A.; O'Neill, N.; Slutsker, I.; Kinne, S. Wavelength dependence of the optical depth of biomass burning, urban, and desert dust aerosols. *J. Geophys. Res.* **1999**, *104*, 349. [CrossRef]

58. Westphal, D.; Toon, O. Simulations of microphysical, radiative, and dynamical processes in a continental-scale forest fire smoke plume. *J. Geophys. Res.* **1991**, *96*, 379–400. [CrossRef]
59. Finokalia Station-University of Crete (Greece). Available online: http://finokalia.chemistry.uoc.gr (accessed on 23 March 2020).
60. ACTRIS: The European Research Infrastructure for the observation of Aerosol, Clouds and Trace Gases. Available online: https://www.actris.eu (accessed on 23 March 2020).
61. ICOS: The Integrated Carbon Observation System, a European Research Infrastructure. Available online: https://www.icos-cp.eu (accessed on 23 March 2020).
62. World Meteorological Organisation Global Atmosphere Watch Programme (GAW). Available online: https://community.wmo.int/activity-areas/gaw (accessed on 23 March 2020).
63. EMEP (European Monitoring and Evaluation Programme), the Co-operative Programme for Monitoring and Evaluation of the Long Range Transmission of Air Pollutants in Europe. Available online: https://www.emep.int (accessed on 23 March 2020).
64. Kouvarakis, G.; Tsigaridis, K.; Kanakidou, M.; Mihalopoulos, N. Temporal variations of surface background ozone over Crete Island in the South-East Mediterranean. *J. Geophys. Res.* **2000**, *105*, 4399–4410. [CrossRef]
65. Kouvarakis, G.; Vrekoussis, M.; Mihalopoulos, N.; Kourtidis, K.; Rappenglueck, B.; Gerasopoulos, E.; Zerefos, C. Spatial and temporal variability of tropospheric ozone (O_3) in the boundary layer above the Aegean Sea (eastern Mediterranean). *J. Geophys. Res.* **2002**, *107*, 8137. [CrossRef]
66. Gerasopoulos, E.; Kouvarakis, G.; Vrekoussis, M.; Kanakidou, M.; Mihalopoulos, N. Ozone variability in the marine boundary layer of the Eastern Mediterranean based on 7-year observations. *J. Geophys. Res.* **2005**, *110*, D15309. [CrossRef]
67. Gerasopoulos, E.; Kouvarakis, G.; Vrekoussis, N.; Donoussis, C.; Mihalopoulos, N.; Kanakidou, M. Photochemical ozone production in the Eastern Mediterranean. *Atmos. Environ.* **2006**, *40*, 3057–3069. [CrossRef]
68. Kalivitis, N.; Gerasopoulos, E.; Vrekoussis, M.; Kouvarakis, G.; Kubilay, N.; Hatzianastassiou, N.; Vardavas, I.; Mihalopoulos, N. Dust transport over the eastern Mediterranean derived from Total Ozone Mapping Spectrometer, Aerosol Robotic Network, and surface measurements. *J. Geophys. Res.* **2007**, *112*, D03202. [CrossRef]
69. Kalivitis, N.; Bougiatioti, A.; Kouvarakis, G.; Mihalopoulos, N. Long term measurements of atmospheric aerosol optical properties in the Eastern Mediterranean. *Atmos. Res.* **2011**, *102*, 351–357. [CrossRef]
70. Pandolfi, M.; Alados-Arboledas, L.; Alastuey, A.; Andrade, M.; Angelov, C.; Artiñano, B.; Backman, J.; Baltensperger, U.; Bonasoni, P.; Bukowiecki, N.; et al. A European aerosol phenomenology–6: Scattering properties of atmospheric aerosol particles from 28 ACTRIS sites. *Atmos. Chem. Phys.* **2018**, *18*, 7877–7911. [CrossRef]
71. Antoine, D.; Nobileau, D. Recent increase of Saharan dust transport over the Mediterranean Sea, as revealed from ocean color satellite (SeaWiFS) observations. *J. Geophys. Res.* **2006**, *111*, D12214. [CrossRef]
72. Gkikas, A.; Hatzianastassiou, N.; Mihalopoulos, N. Aerosol events in the broader Mediterranean basin based on 7-year (2000–2007) MODIS C005 data. *Ann. Geophys.* **2009**, *27*, 3509–3522. [CrossRef]
73. Gkikas, A.; Basart, S.; Hatzianastassiou, N.; Marinou, E.; Amiridis, V.; Kazadzis, S.; Pey, J.; Querol, X.; Jorba, O.; Gassó, S.; et al. Mediterranean intense desert dust outbreaks and their vertical structure based on remote sensing data. *Atmos. Chem. Phys.* **2016**, *16*, 8609–8642. [CrossRef]
74. Nobileau, D.; Antoine, D. Detection of blue-absorbing aerosols using near infrared and visible (ocean color) remote sensing observations. *Remote Sens. Environ.* **2005**, *95*, 368–387. [CrossRef]
75. Im, U.; Kanakidou, M. Impacts of East Mediterranean megacity emissions on air quality. *Atmos. Chem. Phys.* **2012**, *12*, 6335–6355. [CrossRef]
76. Vrekoussis, M.; Liakakou, E.; Mihalopoulos, N.; Kanakidou, M.; Crutzen, P.J.; Lelieveld, J. Formation of HNO_3 and NO_3 in the anthropogenically influenced eastern Mediterranean marine boundary layer. *Geophys. Res. Lett.* **2006**, *33*, L05811. [CrossRef]
77. Vrekoussis, M.; Mihalopoulos, N.; Gerasopoulos, E.; Kanakidou, M.; Crutzen, P.J.; Lelieveld, J. Two-years of NO_3 radical observations in the boundary layer over the Eastern Mediterranean. *Atmos. Chem. Phys.* **2007**, *7*, 315–327. [CrossRef]

78. Kalaroni, S.; Tsiaras, K.; Petihakis, G.; Economou-Amilli, A.; Triantafyllou, G. Modelling the Mediterranean pelagic ecosystem using the POSEIDON ecological model. Part I: Nutrients and chlorophyll-a dynamics. *Deep-Sea Res. Part II* **2020**, *171*. [CrossRef]

79. Antoine, D.; Guevel, P.; Desté, J.-F.; Bécu, G.; Louis, F.; Scott, A.J.; Bardey, P. The «BOUSSOLE» buoy A new transparent-to-swell taut mooring dedicated to marine optics: Design, tests and performance at sea. *J. Atmos. Ocean. Technol.* **2008**, *25*, 968–989. [CrossRef]

80. LeBoulluec, M.; Aoustin, Y.; Bigourdan, B. Expertise du Flotteur Boussole, Rapport IFREMER TMSI/RED 02-028; Issy Les Moulineaux, France. 2002. Available online: http://www.obs-vlfr.fr/Boussole/html/technological/rapports/Ifremer-rapport-complet.pdf (accessed on 11 May 2020).

81. Hellan, Ø.; Leira, B.; Barrholm, R.; Erling Heggelund, S.; Lie, H. *Expert Evaluation of Boussole Buoy Design*; Marintek Rep. 700203.00:01; Marintek Rep.: Trondheim, Norway, 2002.

82. Morel, A.; Claustre, H.; Antoine, D.; Gentili, B. Natural variability of bio-optical properties in Case 1 waters: Attenuation and reflectance within the visible and near-UV spectral domains, as observed in South Pacific and Mediterranean waters. *Biogeosciences* **2007**, *4*, 913–925. [CrossRef]

83. Bricaud, A.; Bosc, E.; Antoine, D. Algal biomass and sea surface temperature in the Mediterranean basin: Intercomparison of data from various satellite sensors, and implications for primary production estimates. *Remote Sens. Environ.* **2002**, *81*, 163–178. [CrossRef]

84. Santoleri, R.; Volpe, G.; Marullo, S.; Nardelli, B.B. Open Waters Optical Remote Sensing of the Mediterranean Sea. In *Remote Sensing of the European Seas*; Springer: Berlin/Heidelberg, Germany, 2008; pp. 103–116.

85. Gregg, W.W.; Carder, K.L. A simple spectral solar irradiance model for cloudless maritime atmospheres. *Limnol. Oceanogr.* **1990**, *35*, 1657–1675. [CrossRef]

86. 'FRM4SOC D-70: Technical Report TR-2, A Review of Commonly Used Fiducial Reference Measurement (FRM) Ocean Colour Radiometers (OCR) used for Satellite Validation'. 2018. Available online: https://frm4soc.org/wp-content/uploads/filebase/FRM4SOC-TR2_TO_signedESA.pdf (accessed on 14 January 2020).

87. Woolliams, E.R.; Fox, N.P.; Cox, M.G.; Harris, P.M.; Harrison, N.J. The CCPR K1-a key comparison of spectral irradiance from 250 nm to 2500 nm: Measurements, analysis and results. *Metrologia* **2006**, *43*, S98–S104. [CrossRef]

88. Martin, J.E.; Fox, N.P.; Key, P.J. A Cryogenic Radiometer for Absolute Radiometric Measurements. *Metrologia* **1985**, *21*, 147. [CrossRef]

89. Manov, D.V.; Chang, G.C.; Dickey, T.D. Methods for reducing biofouling of moored optical sensors. *J. Atmos. Ocean. Technol.* **2004**, *21*, 958–968. [CrossRef]

90. Whelan, A.; Regan, F. Antifouling strategies for marine and riverine sensors. *J. Environ. Monit.* **2006**, *8*, 880–886. [CrossRef] [PubMed]

91. Bishop, J.; McKay, H.; Parrott, D.; Allan, J. *Review of International Research Literature Regarding the Effectiveness of Auditory Bird Scaring Techniques and Potential Alternatives*; Food and Rural Affairs: London, UK, 2003.

92. JCGM100. *Evaluation of Measurement Data-Guide to the Expression of Uncertainty in Measurement, Guidance Document*; BIPM: Saint-Cloud, France, 2008.

93. Białek, A.; Vellucci, V.; Gentili, B.; Antoine, D.; Gorroño, J.; Fox, N.; Underwood, C. Monte Carlo–based quantification of uncertainties in determining ocean remote sensing reflectance from underwater fixed-depth radiometry measurements. *J. Atmos. Ocean. Technol.* **2020**, *37*, 177–196. [CrossRef]

94. Fidelity and Uncertainty in Climate Data Records from Earth Observations (FIDUCEO). Available online: https://www.fiduceo.eu (accessed on 11 May 2020).

95. Mittaz, J.; Merchant, C.J.; Woolliams, E.R. Applying principles of metrology to historical Earth observations from satellites. *Metrologia* **2019**, *56*, 032002. [CrossRef]

96. Białek, A.; Douglas, S.; Kuusk, J.; Ansko, I.; Vabson, V.; Vendt, R.; Casal, T. Example of Monte Carlo Method Uncertainty Evaluation for Above-Water Ocean Colour Radiometry. *Remote Sens.* **2020**, *12*, 780. [CrossRef]

97. Obolensky, G.; Cancouet, R.; Mangin, A.; Poteau, A.; Schmechtig, C.; Thierry, V. *Argo Dataset Production: Real-time Data-management and Delayed-mode Qualified Dataset for O_2, Chlorophyll-a, Backscattering and NO_3*; AtlantOS Deliverable: Kiel, Germany, 2019.

98. Roemmich, D.; Johnson, G.C.; Riser, S.; Davis, R.; Gilson, J.; Owens, W.B.; Garzoli, S.L.; Schmid, C.; Ignaszewski, M. The Argo Program: Observing the global ocean with profiling floats. *Oceanography* **2009**, *22*, 34–43. [CrossRef]

99. Johnson, K.S.; Claustre, H. Bringing Biogeochemistry into the Argo Age. *Eos Trans. AGU* **2016**, *97*. [CrossRef]

100. Gerbi, G.; Boss, E.; Werdell, J.; Proctor, C.W.; Haentjens, N.; Lewis, M.R.; Brown, K.; Sorrentino, D.; Zaneveld, J.R.V.; Barnard, A.H.; et al. Validation of Ocean Color Remote.Sensing Reflectance Using Autonomous Floats. *J. Atmos. Ocean. Technol.* **2016**, *33*, 2331–2352. [CrossRef]

101. Wojtasiewicz, B.; Hardman-Mountford, N.J.; Antoine, D.; Dufois, F.; Slawinski, D.; Trull, T.W. Use of bio-optical profiling float data in validation of ocean colour satellite products in a remote ocean region. *Remote Sens. Environ.* **2018**, *209*, 275–290. [CrossRef]

102. Barnard, A.H.; Boss, E.; Van Dommelen, R.; Plache, B. A new paradigm for ocean color satellite calibration and validation: Accurate measurements of hyperspectral water leaving radiance from autonomous profiling floats (HYPERNAV). In Proceedings of the Ocean Optics Conference 2018, Dubrovnik, Croatia, 7–12 October 2018.

103. Pla, P.; Tricarico, R. Towards a low cost observing system based on low logistic SeaExplorer glider. *IEEE Underw. Technol.* **2015**, 1–3. [CrossRef]

104. EUMETSAT, Sentinel-3 Product Notice–OLCI Level-2 Ocean Colour, EUM/OPS-SEN3/TEN/19/1068317, S3.PN-OLCI-L2M.001. March 2019. Available online: https://sentinels.copernicus.eu/web/sentinel/technical-guides/sentinel-3-olci/data-quality-reports (accessed on 13 April 2020).

105. Mobley, D.D. *Light and Water: Radiative Transfer in Natural Waters*; Elsevier: Amsterdam, The Netherlands, 1994.

 *remote sensing* 

Article

European Radiometry Buoy and Infrastructure (EURYBIA): A Contribution to the Design of the European Copernicus Infrastructure for Ocean Colour System Vicarious Calibration

Gian Luigi Liberti [1,2,*], Davide D'Alimonte [3], Alcide di Sarra [4], Constant Mazeran [5], Kenneth Voss [6], Mark Yarbrough [7], Roberto Bozzano [8], Luigi Cavaleri [1,2], Simone Colella [1,2], Claudia Cesarini [1,2], Tamito Kajiyama [3], Daniela Meloni [4], Angela Pomaro [1,2], Gianluca Volpe [1,2], Chunxue Yang [1,2], Francis Zagolski [5] and Rosalia Santoleri [1,2]

[1]   Consiglio Nazionale delle Ricerche, Istituto di Scienze Marine (CNR-ISMAR), 00133 Rome, Italy; luigi.cavaleri@ismar.cnr.it (L.C.); simone.colella@cnr.it (S.C.); claudia.cesarini@artov.ismar.cnr.it (C.C.); angela.pomaro@ve.ismar.cnr.it (A.P.); gianluca.volpe@cnr.it (G.V.); chunxue.yang@artov.ismar.cnr.it (C.Y.); rosalia.santoleri@cnr.it (R.S.)

[2]   Consiglio Nazionale delle Ricerche, Istituto di Scienze Marine (CNR-ISMAR), 30122 Venice, Italy

[3]   Aequora, 1600-774 Lisbon, Portugal; davide.dalimonte@gmail.com (D.D.); tamito.kajiyama@gmail.com (T.K.)

[4]   Agenzia Nazionale per le Nuove Tecnologie, l'Energia e lo Sviluppo Economico Sostenibile (ENEA), 00123 Rome, Italy; alcide.disarra@enea.it (A.d.S.); daniela.meloni@enea.it (D.M.)

[5]   SOLVO, 06600 Antibes, France; constantmazeran@solvo.fr (C.M.); franciszagolski@solvo.fr (F.Z.)

[6]   Physics Department, University of Miami, Miami, FL 33143, USA; kvoss@miami.edu

[7]   San Jose State University Research Foundation, Moss Landing Marine Laboratories, MOBY Project, Honolulu, HI 96817, USA; yarbrough@mlml.calstate.edu

[8]   Consiglio Nazionale delle Ricerche, Istituto per lo studio degli impatti Antropici e Sostenibilità in ambiente marino (CNR-IAS), 16149 Genoa, Italy; roberto.bozzano@cnr.it

*   Correspondence: gianluigi.liberti@cnr.it; Tel.: +39-6-4993-4281

Received: 28 February 2020; Accepted: 2 April 2020; Published: 7 April 2020

Abstract: In the context of the Copernicus Program, EUMETSAT prioritizes the creation of an ocean color infrastructure for system vicarious calibration (OC-SVC). This work aims to reply to this need by proposing the European Radiometry Buoy and Infrastructure (EURYBIA). EURYBIA is designed as an autonomous European infrastructure operating within the Marine Optical Network (MarONet) established by University of Miami (Miami, FL, USA) based on the Marine Optical Buoy (MOBY) experience and NASA support. MarONet addresses SVC requirements in different sites, consistently and in a traceable way. The selected EURYBIA installation is close to the Lampedusa Island in the central Mediterranean Sea. This area is widely studied and hosts an Atmospheric and Oceanographic Observatory for long-term climate monitoring. The EURYBIA field segment comprises off-shore and on-shore infrastructures to manage the observation system and perform routine sensors calibrations. The ground segment includes the telemetry center for data communication and the processing center to compute data products and uncertainty budgets. The study shows that the overall uncertainty of EURYBIA SVC gains computed for the Sentinel-3 OLCI mission under EUMETSAT protocols is of about 0.05% in the blue-green wavelengths after a decade of measurements, similar to that of the reference site in Hawaii and in compliance with requirements for climate studies.

Keywords: ocean colour; system vicarious calibration; fiducial reference measurement; Lampedusa; Copernicus; MOBY; MarONet; radiometry; research infrastructure; uncertainty budget

1. Introduction

The European Copernicus Program for Earth monitoring (http://www.copernicus.eu/) aims to deliver remote sensing capabilities, field measurements, and data processing services that provide users with reliable and up-to-date information related to environmental, climate, and security issues. As part of the Copernicus space component, the European Commission is developing, operating, and planning a series of Sentinel missions for the next 15 years, together with ESA and EUMETSAT. Among these missions, the Sentinel-3 Ocean and Land Colour Instrument (S3 OLCI) is a key component for monitoring the biogeochemical status of the ocean. OLCI is a primary source of ocean color (OC) data for the Copernicus Marine Environment Monitoring Service (CMEMS). For instance, OLCI data are operationally used to advance our understanding of living marine environments and assimilated in the CMEMS models to forecast the evolution of marine ecosystems, together with the other OC third party missions. The Sentinel-2 mission, although mainly developed for land applications, already demonstrates a great potential for coastal applications like sediment and chlorophyll retrieval (e.g., reference [1]), and it is expected to become an operational complement of the CMEMS next phases. OC is also an important dataset for Copernicus Climate Change Service (C3S) to monitor the impact of climate change on marine ecosystems and the ocean carbon cycle [2].

OC science is making an increasing contribution to Ocean Observing systems thanks to continuous global monitoring of biogeochemical variables, to the point that OC missions have become fundamental to the success of the Copernicus Marine and Climate services. This has also raised expectations on the quality of data products. The current threshold requirement for calibration uncertainty of Copernicus satellite OC sensors is 0.5% in the blue-green spectral region, with a further goal of 0.3% [3–5]. OC System Vicarious Calibration (SVC) [4,6] complements pre-launch and on-board calibrations through highly accurate in-situ measurements of water-leaving radiances. These measurements are indeed the principal source for the vicarious calibration of space-born radiometric data. By reducing residual biases in water-leaving radiances, SVC is currently the only way to attain the target OC product uncertainties [7], satisfy OC mission requirements [8], and enable marine and climate data services.

Presently, OC-SVC relies on two sites—the Marine Optical BuoY (MOBY; https://www.mlml. calstate.edu/moby/) [9] and the BOUée pour l'acquiSition d'une Série Optique à Long termE (BOUSSOLE; http://www.obs-vlfr.fr/Boussole/) [10] —established in the last two decades with different instrumental designs, respectively, in the Pacific Ocean and in the Mediterranean Sea. None of these systems is, however, fully compliant with the operational requirements of the Copernicus Programme. The creation of an OC-SVC infrastructure dedicated to the Copernicus Programme is a key point to improve the performance of OC missions that meet the long-term and high-quality standards underpinning marine bio-geochemistry data products delivered by Climate and Marine Services [5]. EUMETSAT has identified a series of steps to achieve this goal. The first step accomplished in 2017 defined the requirements for the Copernicus OC-SVC Infrastructure [3]. These requirements represent the baseline for the present second step: preliminary design of the Copernicus OC-SVC Infrastructure. The overall objective of this paper is to contribute to this European effort by designing the OC-SVC infrastructure able to provide highly accurate, state-of-the-art Fiducial Reference Measurements (FRM) in the very tight development schedule (36 months from the start development) requested by Copernicus Programme. Our design is to commit to the highest radiometry standards and fast operational readiness. This work will then contribute to the third step: technical definition, specifications, and engineering design that are expected to start in 2020.

The main objective of this paper is to describe the proposed European Radiometry Buoy and Infrastructure (EURYBIA), named for the goddess of power over and mastery of the sea (Εὐρυβία: wide-force; who presided over external forces that influence the main such as the rise of the constellations, seasonal weather, and the power of the winds; Theoi Greek Mythology, www.theoi.com.) for SVC of current and future Copernicus missions (Sentinel-3 A, B, C, D, Sentinel-2 C, D), as well as of third party missions (e.g., NOAA/VIIRS, NASA/PACE). The paper will demonstrate that the proposed OC-SVC Infrastructure (Figure 1) meets these requirements in terms of core optical system and field deployment structure hosting it, data processing, and delivery. The deployment site selection will be justified on the basis of environmental conditions, logistic factors, and a complete estimate of the uncertainty budget of the products for the proposed OC-SVC infrastructure.

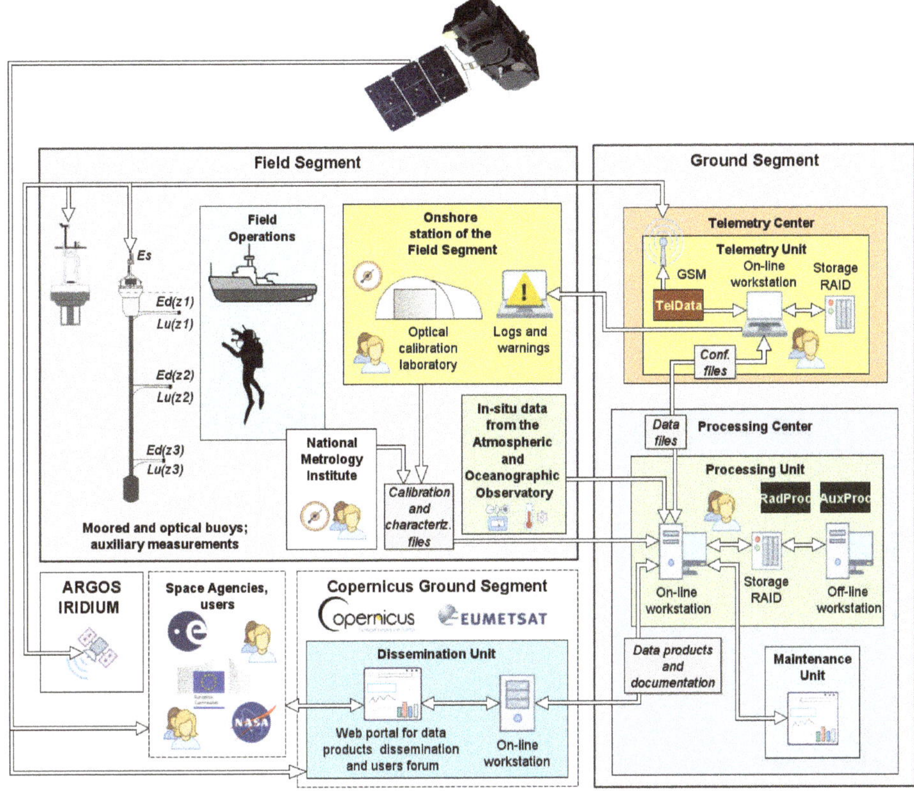

Figure 1. The overall concept of the European Radiometry Buoy and Infrastructure (EURYBIA), with the field segment (**left**) and ground segment (**right**).

Based on OC-SVC requirements, this paper is structured as follows. Section 2 describes the core EURYBIA optical system, the characterization and calibration procedures, as well as acquisition protocol. The optical buoy and the mooring buoy for the deployment of the optical system and ancillary equipment are presented in Section 3. Solutions featured by the Ground Segment for data telemetry, processing, quality control, dissemination, and the scheduling of maintenance operations are discussed in Section 4. Regarding the location of the site, previous studies [4,11–13] investigated the suitability of a set of candidate regions using, as reference metrics, the number of valid match-ups resulting from a set of single threshold criteria applied to a restricted set of variables. With respect to the previous studies, the OC-SCV requirements [3] encompass a larger number of variables and corresponding constraints taken into account by this paper. This environment assessment includes oceanographic and atmospheric properties and adjacency effect of the proposed selected sites and is summarized in Section 5. Section 6 summarizes the suitability of the site by developing a model to estimate the uncertainty on the vicarious calibration gains and by applying it to OLCI observations [14] and to the results of the characterization study. Final conclusions are addressed in Section 7.

2. The Optical System

This section provides a summary of the proposed optical system (2.1) and its characterization and calibration procedures (2.2). Additional information can be found in the existing literature and technical documents [15–18].

2.1. Optical System Description

The core instrumentation of EURYBIA is the optical system of the Marine Optical Network (MarONet), developed by the MOBY team on the basis of the long-term experience acquired in the last three decades. In order to fulfil spectral range and resolution requirements, the system actually consists of a common optical path from the radiation collectors to a device that directs the measured signals into two similar single spectrographs covering different spectral regions and named, on the basis of the range the "Blue" (BSG: 350–690 nm) and the "Red" (RSG: 550–900 nm). Figure 2 shows a scheme of the optical system.

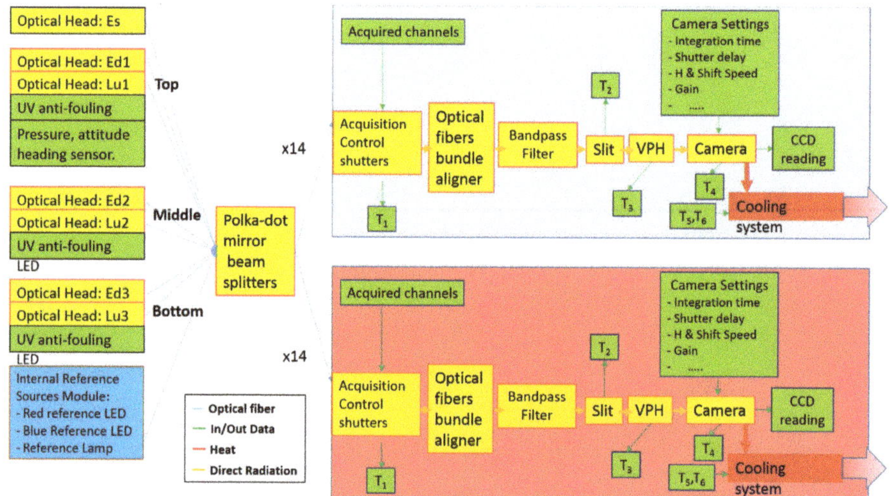

Figure 2. Scheme of the optical system.

The optical signal is collected by the optical heads deployed into the water at the end of each of the three arms (upwelling radiance and downward irradiance measurements, Lu and Ed, respectively) and on the top of the mast of the Optical Buoy (above-water irradiance measurement, Es) (see Section 3). The structure of the optical head depends on the radiometric variable (radiance vs. irradiance) to be acquired, as detailed by Zibordi et al. [12]. To mitigate bio-fouling effects on the optical windows, they are surrounded by a copper bezel and equipped with a UV anti-biofouling module consisting of a fiber optic coupled source of sterilizing UV light (285 nm). The exposure scheme will be programmable to be optimized for the actual site characteristics.

Optical fibers transfer the radiation from each collector to the bottom of the buoy where the two spectrometers are mounted. The signal is split with a polka-dot mirror and transmitted to the two spectrometers. In the following, we describe a single spectrometer. The optical fibers from each collector are connected to a shutter block consisting of 14 single shutters that can also host neutral density filters to adjust the signal level. These devices are also connected, by means of optical fibers, the reference lamps for system monitoring during deployment. Each shutter can be activated independently, allowing the sensor to be illuminated with all possible combinations of inputs.

Optical fibers are aligned on the entrance slit of the spectrometer to allow for the projection on the spectral dispersion system. The alignment order is determined to minimize stray light between different inputs. Band pass filters and short and long wave cutoff filters are mounted at the entrance of the spectral dispersion system to reduce the effects of radiation outside the respective spectral range of the BSG or SRG spectrometer.

The spectral dispersion system is a transmission Volume Phase Holographic spectrograph (VPH) distributed by Resonon Inc., Bozeman, MT 59715, USA. The VPH reduces the stray light by more than one order of magnitude with respect to the concave reflective gratings adopted in MOBY Heritage [19]. The VPH allows imaging all the environmental optical inputs at once: downwelling sky irradiance (Es), downwelling irradiance measurements from the three arms (Ed$_{Top}$, Ed$_{Mid}$, and Ed$_{Bot}$), and upwelling irradiance from the three arms (Lu$_{Top}$, Lu$_{Mid}$, and Lu$_{Bot}$).

The radiation sensor consists of a scientific Andor Ikon M934 camera, Oxford, UK. The camera is a back illuminated Charged Coupled Device (CCD) with a thicker active photosensitive area (i.e., deep depletion CCD) and fringe suppression to prevent etalon effects. The camera is retrofitted with wedged windows to minimize straylight effect (UV ghosting). Different base settings allow the definition of data acquisition characteristics and the sensor working temperature to optimize the camera performances. System synchronization capabilities are used to control the integrated C-mount shutter that is closed during readout to avoid vertical smear. The system is thermo-electrically cooled to reduce the dark current. For MarONet, the cooling temperature has been set to −60 °C as trade-off between noise reduction, power consumption, long term stability, and lowest drift. Cooling the CCD array provides stability to the camera system eliminating bright and dark pixels while also reducing thermal noise and dark current.

An internal reference system is incorporated into the optical system to monitor the stability of the radiometric detectors, the electronics, and internal optics. These measurements are critical for establishing confidence in the observations acquired during a deployment cycle. The reference system consists of one optical feedback-regulated incandescent lamp and two current regulated, temperature stabilized LEDs. The diodes emit at a wavelength centered at 465 nm (Blue LED) and 705 nm (Red LED), with a bandwidth of approximately 100 nm. The lamps are run with current controlled circuitry and the temperature of the lamp holder block is monitored. Each source output is channeled through a one-inch Spectralon sphere to split between the RSG and BSG. The diffuser is illuminated in sequence by the incandescent lamp, the blue LED, and the red LED. These lamps are observed at the end of each data acquisition set. The internal reference sources give no information on changes of optical properties of the single collectors (for example due to bio-fouling) since these components are not included in the optical path of the reference lamps signals.

The system is designed to allow for fully (also remotely) programmable measurement protocols. This functionality, which can be done remotely, will be exploited mostly in the implementation phase to optimize the characteristics of the measurement protocol scenarios needed for SVC. Table 1 summarizes the characteristics of the proposed optical system and compares them against the corresponding requirements.

A set of additional sensors are associated with the optical system to:

- monitor the status of the system: e.g., temperature of different sub-components, power supply, coolant flow, etc.;
- provide ancillary/auxiliary variables for data processing and/or data quality assessment: e.g., sea water temperature and salinity, top arm depth sensor, tilt along two horizontal axes, system heading.

2.2. System Characterization and Calibration

The main objectives of the system characterization are to

- allow the identification of interactions and dependencies between optical and electronic components;
- identify, evaluate, and correct any systematic disturbances;
- allow the design of the Level-0 to Level-1 data processing, estimate input variables, and verify the radiometric model (see Section 4);
- allow the estimate of the contribution of each component to the uncertainty budget associated with the radiometric measurement (see Section 6).

Table 1. Compliancy table of the proposed optical system with respect to the ocean color infrastructure for system vicarious calibration (OC-SVC) requirements [3]. T = Threshold, G = Goal.

Item	OC-SVC Requirement	MarONet Specs
Spectral Coverage	380–900 nm (T) 340–900 nm (G)	350–900 nm (B: 350–690 nm + R: 550–900 nm)
Spectral resolution	3 nm (T) 1 nm (G)	0.8–1 nm (FWHM)
Spectral sampling	<1 nm	0.6–0.9 nm
Spectral calibration	0.2 nm	0.01–0.02 nm
Stray light	*Characterization Monitoring Correction*	Spectral straylight < 5 orders of magnitude over the central signal. Cross track stray light negligible
Radiometric Calibration	1–2% (in air and in the VIS domain)	0.514% @410 nm
Radiometric Stability	1% (during deployment)	1%/deployment
Radiance detector Angular Response	<10° half angle FOV	1.73° full-angle FOV
Immersion Factor	*must be measured*	Estimated associated (k = 1) uncertainty ≈ 0.05%
Temperature Dependence	*shall be determined continuously*	Temperature dependence is obtained from characterization in thermal bath. Operationally, temperature of different system subcomponents is monitored continuously by means of a set of currently 6 (max 14) sensors.
Dark Signal	*shall be measured and corrected*	Acquisition protocol includes dark signal acquisition mode before and after each collector/reference lamps signal acquisition.
Polarization Sensitivity	<1%	<0.2%
Linearity	*Shall be characterized*	>99% (but not significant in the measurement range)
Noise Level	*Instrument noise shall be kept at levels that do not impact the total uncertainty of a measurement.*	CCD readout noise ~2.9 electrons → < 2 ADU → ~ 0.003% at full scale

This is achieved through a set of activities consisting of

- full system characterization after complete assembling;
- in-field monitoring;
- post-deployment characterization in case of identified anomaly.

A detailed description of the characterization activities carried out in the period 2014–2018 can be found in Voss et al. (2018) [15]. The complete system characterization should be done for each new unit when the optical system is built and fully integrated. Leaving the details of the characterization to dedicated existing or future literature, we list the following relevant issues that are planned for the system characterization:

- CCD tracks binning factor;

- full images (saturation checks);
- behavior of dark counts;
- integration time correction factor (shutter delay);
- linearity with optical flux (double aperture; variable radiance source);
- temperature sensitivity (water bath);
- wavelength calibration—atomic line emission sources (pre/post), Fraunhofer lines (field);
- spectral and cross track stray light;
- cosine response (Es);
- polarization sensitivity;
- immersion factors.

In addition to the complete system-level characterization when fully integrated, extensive partial characterizations are planned to be performed pre- and post-deployment (see Section 3). The aspects that need to be periodically characterized will depend on the results of the pre-operational phase to be carried out after the infrastructure deployment. For this purpose, a set of tests that can monitor the status of the system can be introduced in the measurement protocol and/or data processing.

The objective of the radiometric calibration is to obtain absolute values of radiometric measurements compliant with the International System of Units (SI) and directly traceable to National Metrology Institute (NMI) primary standards. This is obtained by absolutely calibrating each system before and after optical buoy deployment and by monitoring the radiometric calibration during each deployment [18].

3. EURYBIA Field Segment

Based on the MarONet system design [5], the EURYBIA Infrastructure consists of a twin buoy system: an optical buoy tethered to a mooring buoy anchored to the sea bottom (Figure 3). The first buoy carries the optical system to perform the radiometric measurements, whereas the second buoy hosts additional scientific meteorological and oceanographic instrumentation and collects ancillary data.

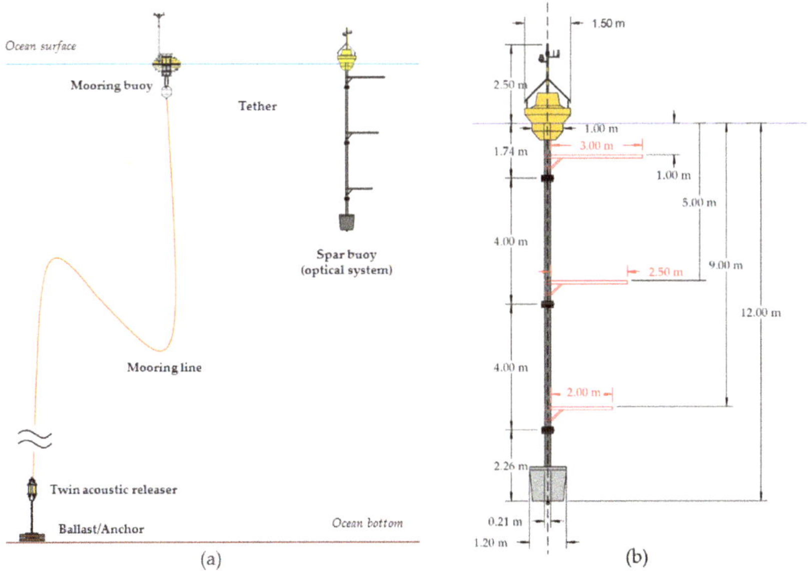

Figure 3. (**a**) Sketch (not to scale) of the main components of the EURYBIA deployment infrastructure; (**b**) scheme of the optical buoy.

Indeed, collecting measurements useful for OC-SVC implies strict constraints on the motion of the buoy that, on one hand, must follow the wavy surface to keep each optical collector at a constant depth and, on the other, must maintain tilting oscillations to within a few degrees in order not to bias the optical measurements.

In order to guarantee the operational measurements, we propose to design and build two identical infrastructures in order to enable rotation in the field, always having one system operational at sea and the other ashore for maintenance operations to make it ready for the next deployment. This design allows the system to be compliant with the critical requirements of the measurement traceability achievable only with frequent periodic on-site and on-land maintenance and calibration operations while ensuring the continuity of the long-term observations.

3.1. Optical Buoy

The optical buoy, hosting the radiometric system, consists of a medium-size spar buoy [9] whose actual dimensions, distribution of masses along the structure, and tether connection to the mooring buoy are properly designed to optimize the buoy stability due to local environmental conditions (i.e., reduced tilting with respect to the sea state conditions) and self-shading effects. Figure 3 shows a preliminary scheme of the optical buoy; the reported dimensions and depths of the three arms hosting the in-water optical collectors are based on MarONet design. The final design, optimized to the environmental conditions of the selected site, might require minor adjustments. Indeed, a spar buoy is characterized by a reduced section at the water surface, a deep draft, a long vertical hull, a disk on the deepest level, and a ballast. These characteristics, relevant to reduce heave (i.e., the buoy vertical motion) and tilting also allow the deployment of instruments along the long vertical hull. For the EURYBIA application, the buoy is also designed to follow the surface in order to keep the optical instruments at constant depths with respect to the sea surface, which is achieved by ensuring a substantial capacity of floating reaction.

The structure of the optical buoy can be divided into the following main parts, each one contributing to the overall structure buoyancy while housing instruments and electronic components essential for the system operational activity: a top floatation element, a hull, and a lower bay.

The top floatation element hosts a payload consisting of an Es irradiance cosine collector, antennas for the positioning and communication system, a radar reflector, a lantern, photovoltaic panels and batteries to power up all electronic systems, an industrial controller collecting data and operating the whole system through A/D converters and/or serial channels, a modem for the communication system, and controllers.

Along the hull, at different fixed depths, three arms support the in-water optical collectors—one for downwelling irradiance and one for upwelling radiance. The arms have different lengths, the longer being closest to the surface and the shorter at the deepest position in order to avoid any interference on upwelling radiance measurements. On the shallower arm, a pressure sensor is also installed in order to monitor the depth of the upper arm measurement.

Two industrial data loggers, connected through an Ethernet connection, are deployed on the optical spar buoy: one data logger collects the optical measurements, located underwater in the watertight lower bay, and a second data logger, which is installed in the watertight compartment on the main floatation element, collects all other data and monitoring parameters (i.e., orientation of the arms, temperatures inside the watertight compartment, depth of the shallower arm, trim of the buoy, current produced by the photovoltaic panels, voltage of the batteries, etc.). At the bottom of the buoy, an orientation module consisting of a pressure sensor and an attitude and heading reference system are installed in order to monitor the trim of the collector and to provide information for any compensation of the acquired irradiance. All data acquired on board are packed into a data cluster, timestamped, and geographically annotated by the upper controller, locally stored on the hard disk while constituting the stream of bytes to be transferred ashore with a communication (telemetry) system.

All the above-mentioned electronic systems and sensors are powered by the photovoltaic panels and batteries installed on the infrastructure itself, which are self-sustained for the entire period of activity, considering up to four acquisition cycles per day, also including the payload required by the necessary alert and monitoring services.

3.2. Mooring Buoy

A mooring buoy consisting in a standard disc-shaped surface buoy is anchored to the sea bottom and is tethered to the spar buoy to keep the optical buoy in place. It also hosts additional equipment to support the radiometric data collection, processing, quality assessment, and results interpretation.

The following parts of the payload of the mooring buoy are similar to the those installed in the optical buoy: antennas for the positioning and communication system, a radar reflector, a lantern, photovoltaic panels and batteries to power all electrical systems, an industrial controller collecting data and operating the whole system through A/D converters and/or serial channels, a modem for the communication, and controllers for the photovoltaic panels.

Additionally, the mooring buoy hosts an attitude and heading reference system to measure the motion of the buoy and instruments for collecting ancillary measurements above the sea surface and underwater. A meteorological station is required to process and interpret the radiometric observations. Optionally, above-water instrumentation may include radiometric measurements with the main objective of estimating the presence of clouds, and in-water standard bio-optical instrumentation could be installed to support main radiometric data processing and quality assessment.

As for the optical buoy, all above-mentioned electronic systems and sensors get their electrical power supply from a system of photovoltaic panels, charge regulators, and batteries.

4. Ground Segment

Ground segment (GS) tasks include telemetry operations, processing of the in-situ measurements to derive radiometric products and related uncertainties for the computation of SVC gains, dissemination of results, documentation of processing code, and scheduling of maintenance operations. The telemetry center and the processing center described in the next sub-sections are the two main infrastructural components that allow for GS activities, as detailed hereafter.

4.1. The Telemetry Center

The telemetry center executes direct and independent data transfers with the mooring and optical buoy by means of GSM communication. Its specific tasks are (1) uploading configuration files with instrumentation settings and acquisition protocols, (2) uploading firmware programs for the EURYBIA equipment, and software modules for the onboard controllers, (3) executing preliminary data processing and checks, (4) downloading in-situ measurements, and (5) verifying the status of buoys at sea and alerting in the case of out-of-order conditions. This latter assignment relies on a continuous (24/7) acquisition of monitoring parameters, even when radiometric and auxiliary measurements are not collected. Integrity checks are performed to detect transfer errors (checksum), missing values in data records, or out-of-range conditions (e.g., temperature, position and power supply). Redundant communication and storage resources, as well as functionalities to remotely supervise and restart telemetry operations, ensure continuous control of the buoys system at sea. In case of anomalies, including missing communication signal, alerts are immediately released to start recovery actions.

4.2. The Processing Centre

In-situ measurements are transferred from the telemetry center to the processing center to compute data products. The data processing is structured in levels to control the complexity of code, simplify its maintenance and evolution, facilitate the assessment of intermediate results, and allow for starting data reprocessing on demand from the specified processing level (Figure 4).

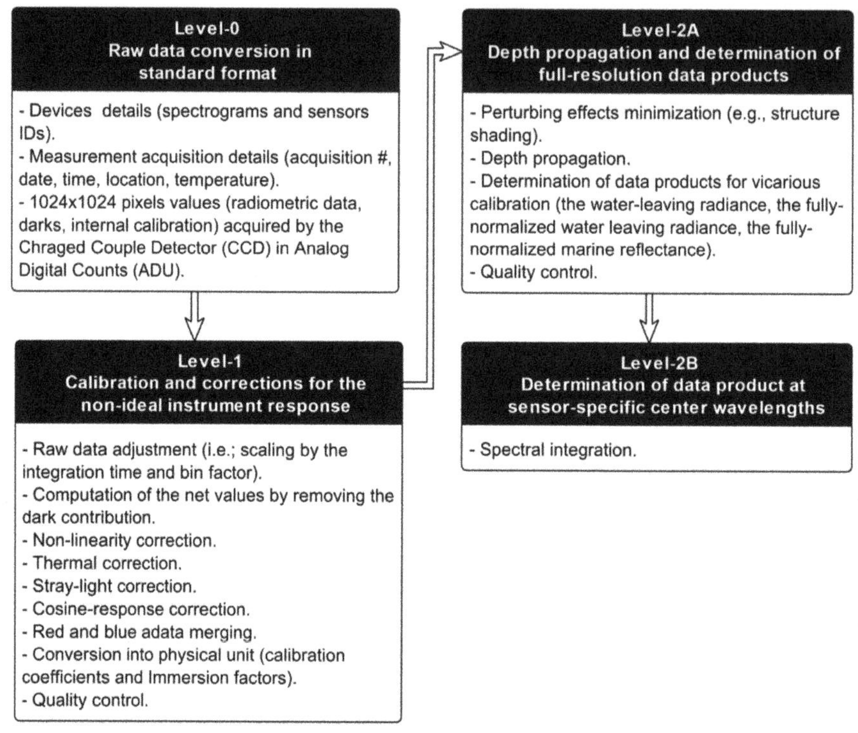

Figure 4. Schematics of processing levels.

The Level-0 processing accounts for the raw data conversion into a standard format. The Level-1 processing applies corrections for the non-ideal instrument response and computes calibrated values in physical units. The Level-2A proceeds to the minimization of environmental perturbations, performs depth propagation of in-water radiometric measurements and computes full-resolution data products such as the water-leaving radiance. The Level-2B finally derives data products at center wavelengths of specific space-borne sensors (spectral integration). Intermediate results and final products are archived as independent output files. Processing settings and metadata for the identification of input measurements and the version of the processing model are also recorded.

4.2.1. Level-0

The format of raw data produced by the radiometric system might change through time due to instrument updates or revisions of the EURYBIA data acquisition program. The Level-0 processing is then scoped to convert raw data into a standard format.

4.2.2. Level-1

Data recorded at the CCD of the Resonon spectrogram are expressed in Analog Digital Units (ADU). The Level-1 processing corrects these input data accounting for the non-ideal radiometer response and generates results in physical units. An additional task is a spectral calibration to relate CCD column-wise pixels to wavelengths.

The formalism adopted in this manuscript is derived from the "Protocols for Satellite Ocean Colour Data Validation: In-situ Optical Radiometry" report of the International Ocean-Colour Coordinating Group (IOCCG) [7]. Denoting with the symbol $\Im(\lambda)$ either $L(\lambda)$ or $E(\lambda)$ in physical units ($W{\cdot}cm^{-2}{\cdot}sr^{-1}{\cdot}nm^{-1}$ and $W{\cdot}cm^{-2}{\cdot}nm^{-1}$, respectively), the Level-1 conversion equation is

$$\Im(\lambda) = F_I(\lambda) \cdot F_C(\lambda) \cdot \aleph(\lambda) \cdot \widetilde{\Im}(\lambda), \tag{1}$$

where:

1. $\widetilde{\Im}(\lambda)$ represents net radiometric data obtained by subtracting the dark contribution from signal data after scaling the raw CCD readouts by the measurement integration time (sec) and the binning factor (unit of a pixel, pix)—$\widetilde{\Im}(\lambda)$ is thus in units of ADU·sec^{-1}·pix^{-1}, see for instance [20];

2. The term $\aleph(\lambda) = \aleph_{NL}(\lambda, \mathbf{X}_{NL}) \cdot \aleph_T(\lambda, \mathbf{X}_T) \cdot \aleph_{SL}(\lambda, \mathbf{X}_{SL}) \cdot \aleph_{CR}(\lambda, \mathbf{X}_{CR})$ accounts for the non-ideal instrument response due to non-linearity (NL) [21], temperature (T) [22], stray-light (SL) [23–25]; non-cosine response (CR) of the above-water irradiance sensor [26–28] (with $\mathbf{X}$ denoting input parameters for the specific correction; e.g., temperatures for T, sun elevation and tilt for CR, etc.);

3. $F_C(\lambda)$ indicates the in-air absolute calibration coefficient (i.e., the absolute responsivity) in units of $\left(\text{W·cm}^{-2}\text{·sr}^{-1}\text{·nm}^{-1}\right)/\left(\text{ADU·sec}^{-1}\text{·pix}^{-1}\right)$ and $\left(\text{W·cm}^{-2}\text{·nm}^{-1}\right)/\left(\text{ADU·sec}^{-1}\text{·pix}^{-1}\right)$ for the radiance and irradiance data, respectively;

4. $F_I(\lambda)$ is the immersion factor accounting for the change in responsivity of the sensor when immersed in water with respect to air (dimensionless).

Results obtained from measurements performed by the blue and red spectrographs are finally merged into a unique spectrum (for each sampling acquisition and radiometric sensor). This processing step requires as input the predefined cut-off wavelength at which the spectra are joined.

4.2.3. Level-2A

The first processing step is the minimization of shading effects due to the deployment structure and the radiometers housing by means of look-up tables (LUTs) produced with Monte Carlo (MC) simulations of the radiative transfer process for given boundary conditions [29–31]. The scheme to determine the shading correction factors executes, for the same environmental case, MC ray tracing with and without accounting for the presence of the optical buoy. The difference between the two results defines the shading correction coefficients [32].

The biofouling effect is evaluated by means of specific quality checks performed during the periodical cleaning of the radiometers. These monitoring results are used to evaluate changes in the instrument performance and compile information to assess the data products quality.

The sub-surface value of the upwelling radiance is derived from $L_u(z_i, \lambda)$ measurements performed at three fixed depths z_i ($i = 1, 2, 3$), with $z_1 > z_2 > z_3$ and assuming positive depth increasing towards the bottom [28,33]. Precisely, the value of the upwelling radiance just below the sea-surface $L_u(0^-, \lambda)$ is derived from the $L_u(z_1, \lambda)$ measurement at the shallowest depth z_1 as

$$L_u(0^-, \lambda) = L_u(z_1, \lambda)e^{K_L(0,1, \lambda) \cdot z_1}, \tag{2}$$

where $K_L(0, 1, \lambda)$ is the attenuation coefficient of the diffuse upwelling radiance for the depths from 0 to z_1—i.e., K_L^{01} [33]. The diffuse attenuation coefficient $K_L(i, j, \lambda)$ in the depth interval $[z_i, z_j]$, is defined as

$$K_L(i, j, \lambda) = -\frac{\ln\left[\dfrac{L_u(\lambda, z_j)}{L_u(\lambda, z_i)}\right]}{z_j - z_i}. \tag{3}$$

The value of $K_L(0, 1, \lambda)$ cannot be directly measured and is determined from $L_u(z_1, \lambda)$, $L_u(z_2, \lambda)$ and $L_u(z_3, \lambda)$ values [33]. To this end, the data processing accounts for the increasing fraction of light generated by inelastic radiative processes (Raman effect and chlorophyll fluorescence) from the blue region of the spectrum, where energy is abundant, to the red region, where the incoming light is rapidly attenuated (this implies a K_L tendency to decrease with depth for $\lambda > 575$ nm). MC simulations are employed to compute K_L lookup tables accounting for Raman scattering [33]. The effect of chlorophyll fluorescence is instead estimated from in-situ measurements in the $[0, z_1]$ interval performed with ancillary instrumentation of the moored buoy. The K_L^{01} estimate is finally modelled as a function of the K_L^{12} and K_L^{13} [33].

The water-leaving radiance [6,34,35], $L_w(\theta, \phi, \lambda_j)$, is computed as

$$L_w(\theta, \phi, \lambda_j) = L_u(0^-, \theta', \phi, \lambda_j) \frac{1 - \rho(\theta', \theta)}{\left[n(\lambda_j, T, S)\right]^2}, \tag{4}$$

where ϕ is the azimuth angle, θ is the zenith angle in air, θ' is the corresponding refracted nadir angle in water, $\rho(\theta, \theta')$ is the internal Fresnel reflectance, and $n(\lambda_j, T, S)$ is the refractive index of seawater, with T and S indicating the temperature and salinity, respectively [36].

The fully normalized water-leaving radiance, $L_{wN}(\lambda_i)$, is the radiance leaving the sea surface from the water body (i.e., excluding the contribution of incident light reflected above the sea surface) as if the sun was at the zenith, the Earth was at its mean distance from the sun and in the absence of any atmospheric loss [37]. $L_{wN}(\lambda_i)$ is determined as

$$L_{wN}(\lambda_i) = C_{f/Q}(\theta_0, \lambda_i, \text{Chl}a) \cdot F_0(\lambda_j) \cdot \frac{L_w(\lambda_i)}{E_s(\lambda_i)}, \tag{5}$$

where (1) $E_s(\lambda_i)$ is the incident solar irradiance; (2) $C_{f/Q}(\theta_0, \lambda_j, \text{Chl}a)$ is the BRDF correction term obtained from look-up-tables as a function of the solar zenith angle θ_0, the wavelength λ_j, and the Chlorophyll-a concentration $\text{Chl}a$ [38]; and 3) $F_0(\lambda_i) = \overline{F_0}(\lambda_i) \cdot \delta_0(\text{SDY})$ is the extraterrestrial solar irradiance $\overline{F_0}$ corrected for the Earth-sun distance $\delta_0(\text{SDY})$ indicating with SDY the Sequential Day of the Year.

The fully normalized marine reflectance, $\rho_{wN}(\lambda_i)$, is finally derived from $L_{wN}(\lambda_i)$ as

$$\rho_{wN}(\lambda_i) = \frac{L_{wN}(\lambda_i)}{F_0(\lambda_i)} \cdot \pi. \tag{6}$$

4.2.4. Level-2B

Full resolution data are determined in the 350–900 nm interval. Spectral integration is addressed to compute radiometric values for the channel identified by the central wavelength λ_i of selected space-borne sensors as

$$L_w(\lambda_i) = \frac{\int L_w(\lambda_j) \cdot SRF_i(\lambda_j) d\lambda_j}{\int SRF_i(\lambda_j) d\lambda_j}, \tag{7}$$

indicating with SRF_i the spectral response function of the $i-$th band of the space-borne sensor. Note that the numerical integral corresponding to Equation (7) is performed after interpolation of SRF discrete values to the central wavelengths of the in-situ measurement. The numerical integration scheme is then designed to minimize the uncertainty budget. Selected ocean color space sensors of interest include

1. Ocean and Land Color Instrument (OLCI); Sentinel 3 missions; Copernicus Program; EU.
2. Moderate Resolution Imaging Spectroradiometer (MODIS); Terra/Aqua; NASA; USA.
3. Visible Infrared Imaging Radiometer Suite (VIIRS); Suomi NPP, NOAA-20, JPSS-2, JPSS-3; NASA/NOAA; USA.
4. Ocean Color Instrument (OCI); Plankton, Aerosol, Cloud, ocean Ecosystem (PACE) program, NASA [39] (https://pace.gsfc.nasa.gov/, planned launch in 2022).

4.3. Quality Assurance and Quality Control

The determination of data products is complemented by the quality assurance and control (QA/QC) of results. This task is jointly based on objective parameters (e.g., coefficient of variation of CCD data at

Level-1, stability of pre- and post-deployment calibration coefficients, presence of optical stratifications limiting the capability to perform depth propagation of the upwelling radiance at Level-2), as well as the visual inspection of results (e.g., abnormal presence of spikes in Level-1 data) and complementary information (e.g., cloud-cover conditions). Additional QA/QC is performed checking the consistency of individual records with historical measurements once the time series permits (in analogy to the procedure adopted by the MOBY Team). For a final assessment, the stability of calibration lamps within their life cycle (i.e., determined based on traceability with respect to standards performed by the National Metrology Institute) is also verified.

QA/QC results are finally summarized through the following labels:

- GOOD indicates the highest quality data for SVC applications (reprocessed data only) or for validation and monitoring (these include near-real-time products release);
- QUESTIONABLE refers to data products derived upon applying significant corrections (e.g., excessive sensor tilting or instability of the instrument response); and
- BAD is applied to data that are unreliable and cannot be used even if the current set of correction schemes are applied (usually caused by unanticipated problems during data acquisition or clouds).

4.4. Data Products Dissemination

The data processing cycle accounts for the following three main release versions:

1. Near Real-Time (NRT) results are delivered once the following operations have been executed: visual data inspection, data correction and calibration, computation of data products such as the normalized water-leaving radiance, and preliminary data quality assurance and control (QA/QC. The timely dissemination of high-quality daily data products derived from in-situ measurements is scoped to release measurements results for the routine assessment and monitoring of orbiting OC remote sensing sensors.
2. Post-Deployment Calibration (PDC) results are computed by performing a data re-processing when the optical system has been recovered to shore and re-calibrated. Namely, pre- and post-deployment calibration coefficients are used to account for instruments responsivity change. Any instrument re-characterization for the non-ideal instrument response is also to be taken into consideration.
3. Finally, Traceable to Calibration Standards (TCS) of NMI supporting OC-SVC activities are determined by means of an additional data re-processing at the end of the operational cycle of calibration lamps (which is in the order of 50 working hours). At this stage, in fact, the lamps are sent to NMI for traceability. Data products are then recomputed when reference calibration sources are replaced (note that the source stability is continuously monitored at the calibration facilities to allow for prompt actions in the case of out-of-order variations).

All GS dissemination services are performed within the Copernicus infrastructure and in compliance with the Copernicus principles of offering free, full, and open access to data and processing models. An interactive data browser will be designed to allow end-users to look up field measurements and derived products for specified search conditions and generate a quick statistical summary of the selected data. Deliverables include details on the pre- and post-deployment calibration coefficients, as well as the stability of calibration sources. Additional services are (1) a web service for the documentation of processing modules and (2) a forum for registered users to provide general information, answer questions, and receive comments and suggestions.

4.5. Maintenance Operations

A web interface is finally dedicated to facilitate the maintenance of the ground segment (e.g., software revisions, hardware replacement) and the field segment (e.g., infrastructure turn-over, instrumentation calibration, cleaning, and replacement). Provided on-line services include the scheduling of maintenance operations and the posting of services logs.

5. Site Characterization

5.1. Location

5.1.1. Geography and Illumination

As discussed in Section 1, the area around the island of Lampedusa, south of the Strait of Sicily, is proposed as the EURYBIA candidate site for the OC-SVC infrastructure. Lampedusa, part of the small Pelagian archipelago, is a small island (about 20 km^2 surface area) in the open Mediterranean Sea, at 35.5° latitude North. About 6000 inhabitants live in Lampedusa and Linosa, the two major islands of the archipelago. The Lampedusa area has been extensively studied with respect to geomorphology, marine and terrestrial ecology, and many environmental parameters. The archipelago is located on the African tectonic plate, at the southern shoulder of a Plio-Quaternary rift zone that shows deep fault-controlled structural depressions oriented along the North-West to South East direction (e.g., the grabens of Pantelleria, Linosa, and Malta). The associated deep depressions, just North of Lampedusa, are more than 1500 m deep and separate the African continental shelf from that of Sicily. The rift zone is characterized by a moderate seismicity, mostly located in the Linosa graben.

Geologically, the island of Lampedusa is formed by Upper Miocene neritic limestones [40]. The island is about 10 km wide, with a tabular morphology and a surface sloping toward the south and south-east. The maximum altitude is about 130 m.

Calculations of the solar illumination and occurrence of sun glint conditions have been carried out. The illumination conditions at Lampedusa and MOBY are similar in summer, while in winter, the available solar energy at Lampedusa is a minimum of 70% of that available at MOBY. Higher solar zenith angles associated with higher latitude at Lampedusa than at MOBY are favorable for minimizing the influence of sun glint conditions.

5.1.2. Existing Infrastructures

Relevant logistical and scientific infrastructures, and several natural protection areas are present in and around Lampedusa (http://www.ampisolepelagie.it/). An airport and a primary and a secondary port exist on the island. Regular daily flights and ferry boats connect Lampedusa and Sicily throughout the year.

With respect to scientific infrastructures, Lampedusa hosts a climate observatory (http://www.lampedusa.enea.it), which has been operational since 1997 and is composed of an atmospheric observatory (35.52°N, 12.63°E) [41] and an oceanographic observatory (35.49°N, 12.47°E) [42]. A large number of atmospheric and oceanographic parameters are routinely measured at Lampedusa; these data, together with satellite observations and model results, have been used for the characterization of the Lampedusa area for OC-SVC. In addition, they offer relevant and reliable information for the definition of the design requirements of the buoys structure in order to meet the necessary stability for a higher number of quality measurements throughout the year. The Lampedusa Climate Observatory contributes to global measurement networks and to European environmental research infrastructures, such as the Integrated Carbon Observation System (ICOS) and the Aerosol, Clouds, and Trace Gases (ACTRIS). Lampedusa also contributes to the national network of the European Multidisciplinary Seafloor and water-column Observatory (EMSO), research infrastructure, with a number of oceanographic measurements. The operational instruments include sun photometers, also part of the Aerosol Robotic Network (AERONET), a ceilometer, an aerosol and water vapor lidar, radiometers and spectrometers for downwelling and upwelling radiation and atmospheric remote sensing, oceanographic sensors for temperature, salinity, chlorophyll, CDOM, backscattering, and underwater radiation.

As part of the Lampedusa Climate Observatory, instruments and capabilities in radiometers and spectrometers characterization and calibration (see e.g., references [43–45]) are available on site.

Scientific data acquired at the Lampedusa Climate Observatory have been extensively used, and more than 200 international scientific papers referring to these data have been published so far.

Many studies make advantage of these data for inter-comparison, validation, and combined analyses of satellite observations.

5.1.3. Bathymetry

Two regions with bottom depth larger than 300 m at 35.5248°N–12.7667°E and 35.7430°N–12.3579°E, LMP1 and LMP2 respectively, have been identified as possible sites for OC-SVC Infrastructure (Figure 5). These two sites are characterized by a negligible impact from fishing activities and ship traffic and are located at a reasonable distance from Lampedusa (less than 20 km), thus enabling technical operations and allowing for timely interventions, while avoiding any possible adjacency effect (see Section 5.4) and possible influence from emissions occurring on the island.

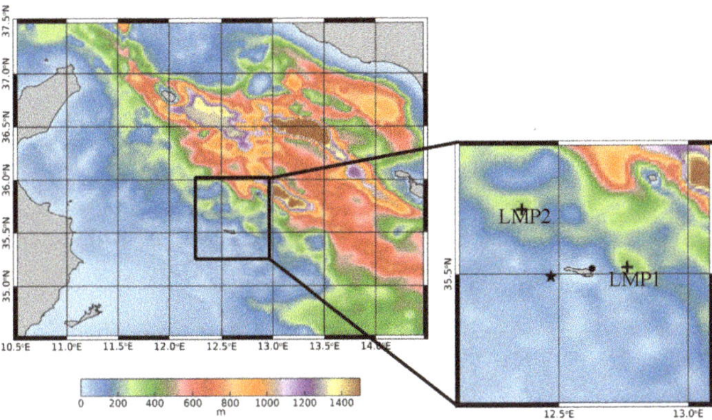

Figure 5. Bathymetry of the Sicily Channel with a zoom on the area around the Island of Lampedusa. Black crosses represent possible sites for the future installation (LMP1 and LMP2, respectively, at 35.5248°N–12.7667°E and 35.7430°N–12.3579°E). Black star and filled point show the position of Oceanography and Atmospheric Observatories, respectively.

The bottom effect has been analyzed by performing radiative transfer MC simulations [29,30]. The study acknowledges that the selected region of the Mediterranean Sea is characterized by a relatively high CDOM concentration in comparison with oceanic oligotrophic waters. Light penetration in the water column at the lower end of the spectral region is then significantly limited by the exponential increase of CDOM absorption as wavelength reduces. On the other hand, water absorption lessens the red and near-infrared light transmission. On this basis, 470 nm has been considered as a test wavelength for the bottom effect analysis (specific simulation parameters in Table 2). The bottom contribution has been estimated as percent difference of L_u^S and L_u^B values assuming a sandy (S) and a black (B) bottom:

$$\delta_{SB}(z_{btm}) = 100 \cdot \frac{L_u^S(z_{sns}) - L_u^B(z_{sns})}{L_u^B(z_{sns})} \tag{8}$$

where z_{btm} and z_{sns} indicate the bottom and the sensor depth, respectively. The value of z_{btm} has been varied between 50 and 200 m. The value adopted for z_{sns} is 14 m, which corresponds to the depth of the sensor that is mostly affected by bottom reflectance.

Table 2. Parameters employed for the radiative transfer simulations at $\lambda = 470$ nm.

Sky Radiance			IOPs			Bottom	
E_s [W m^{-2} sr^{-1} nm^{-1}]	Diff/dir	Sun zenith [deg.]	a [m^{-1}]	c [m^{-1}]	b_b/b	Reflectance [%]	Depth [m]
1.57	0.45	15	0.028	0.14	0.0183	0 (B) and 36.2 (S)	50 to 200

The Harrison–Coombes model has been employed for the sky radiance [46,47]. The values of total irradiance and the diffuse-to-direct irradiance ratio have been determined according to [48]. Inherent Optical Properties (IOPs) to model light transfer in the water column have been selected considering the lowest values of absorption, attenuation and backscattering fraction collected in the field [49]. It is highlighted that the minimum measured b_b/b value is 0.021. However, MC simulations have been executed considering a lower b_b/b value, namely employing the Volume Scattering Function of Petzold (i.e., $b_b/b = 0.0183$) [50]. The simulated bottom effect is hence expected to overestimate reality due to larger forward scattering. A Lambertian surface with the reflectance value of coral sand is considered for the bottom [51]. Raman scattering [33] has been neglected because it only accounts for a second-order contribution. Examples of optical profiles for a black and sandy bottom are presented in Figure 6 for water depths of 50 m and 200 m.

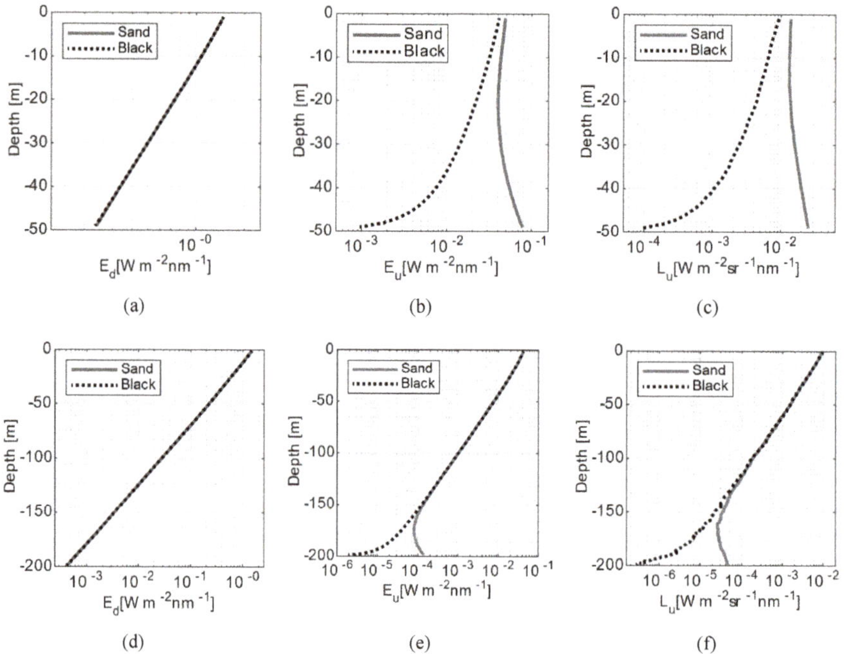

Figure 6. Examples of simulation results. (**a**–**f**) refer to a bottom depth of 50 and 200 m, respectively. E_d, E_u, and L_u profiles are from left to right column panels (E_d and E_u values presented only for completeness). Simulation results (Table 3 and Figure 7) indicate a negligible contribution to the upwelling radiance at 14 m below the sea surface for a water column of 200 m or deeper.

Results presented in Table 3 and Figure 7 suggest that the impact of bottom reflectance is negligible for a water column of 200 m or deeper, making the selected site with 300 m depth suitable considering the radiometric accuracy required for SVC (i.e., assuming the representativeness of the parameters employed for the MC simulations).

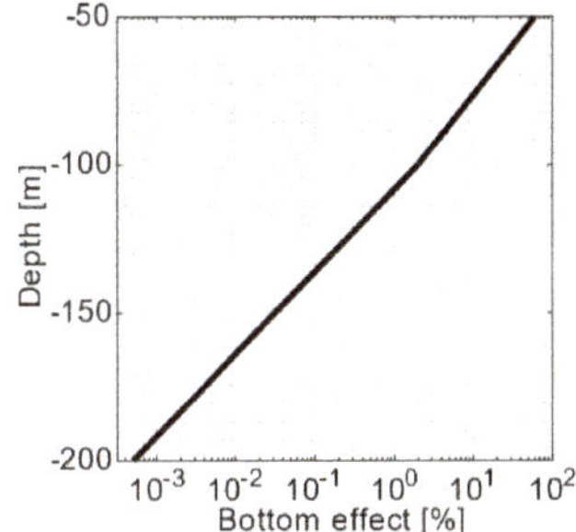

Figure 7. Summary of bottom effects simulations (see also Table 3).

Table 3. Estimated bottom contribution (percent) to the upwelling radiance at 14 m below the sea surface for different bottom depths.

| Bottom Depth [m] | L_u at 14 m Depth [W m^{-2} sr^{-1} nm^{-1}] | | δ [%] |
| | Black Bottom | Sand Bottom | |
	Reflectance 0%	Reflectance 39.4%	
−50	5.289×10^{-3}	1.362×10^{-2}	61.18
−100	6.911×10^{-3}	7.053×10^{-3}	2.01
−150	6.005×10^{-3}	6.007×10^{-3}	0.03
−200	5.895×10^{-3}	5.895×10^{-3}	0.00

5.2. Oceanographic Properties

5.2.1. Temperature and Salinity

The characterization of the ocean temperature and salinity in the Lampedusa area has been based on in-situ and satellite observations. Data from a conductivity temperature depth (CTD) at 18 m depth, and from a temperature and pressure sensor at a 1 m depth at the Lampedusa Oceanographic Observatory have been used for the reference period 2018–2019. The annual evolution of the temperature and salinity monthly mean values at 18 m depth is shown in Figure 8. By comparison, the range of temperature and salinity at the MOBY site, derived from Feinholz et al. [16], are also shown in Figure 8. The annual temperature cycle is larger at Lampedusa than at MOBY; conversely, the annual salinity changes are significantly smaller at Lampedusa. These variations constitute essential information for the determination of the water refractive index.

Temperature measurements at a 1 m depth show a slightly larger annual cycle than at 18 m, with similar temperatures in winter and somewhat larger temperatures in summer. The temperature daily cycle is generally within ± 3% of the average daily value.

The Sea Surface Temperature (SST) spatial and temporal variability in the region around Lampedusa has been studied based on the daily Level-4 (SST_MED_SST_L4_NRT_OBSERVATIONS_010_004) product from the Copernicus Marine Environment Monitoring Service (CMEMS) for the reference year 2013. The annual average temperature in the Lampedusa area is about 21 °C. Average day-to-day SST anomalies are generally smaller than 1.5% in the whole area around Lampedusa.

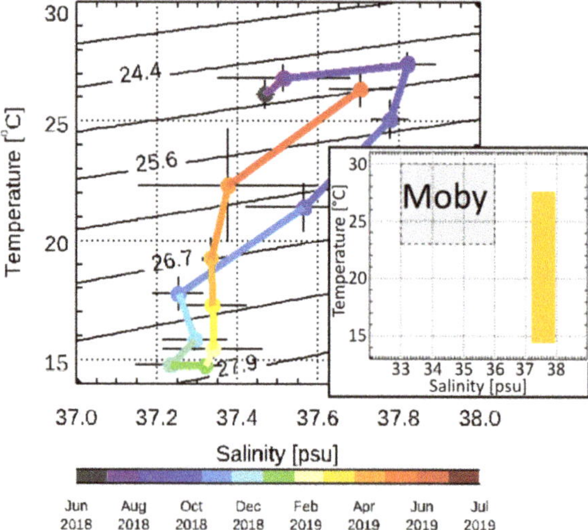

Figure 8. Time series of monthly average temperature and salinity, from the 18 m conductivity temperature depth (CTD), at Lampedusa during 2018–2019. The annual range for temperature and salinity at MOBY (grey rectangle) and at Lampedusa (yellow rectangle) is shown in the graph inlet.

5.2.2. Currents

Data for currents at different depths at the proposed sites have been extracted from the CMEMS Med Sea Physical Reanalysis dataset [52]. Daily data for the period 1987–2013 have been used. The information on currents at the Lampedusa area has been compared with current data for the MOBY site retrieved from the CMEMS GLORYS12V1 global reanalysis (http://resources.marine.copernicus.eu/documents/PUM/CMEMS-GLO-PUM-001-030.pdf). The reference period that has been considered is 1993–2013.

Figure 9 shows profiles of the median, 99th percentile, and maximum values of currents at LMP1, LMP2, and MOBY. For comparison, also values at the Boussole area (Ligurian Sea), also derived from the CMEMS Med Sea Physical Reanalysis dataset, are displayed.

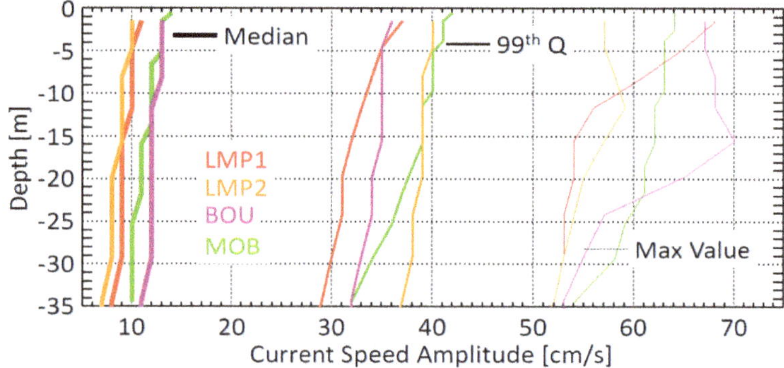

Figure 9. Profiles of the median, 99th percentile, and maximum values of currents at LMP1, LMP2, and MOBY. For comparison, also values in the Boussole area (Ligurian Sea).

Currents are generally weaker at the Lampedusa area than at MOBY and in the Ligurian sea, although higher current may occur in specific events. All sites show a marked seasonal variability for the currents, with median smaller values at Lampedusa in all seasons. Dominant currents are from the north-west [53] and are associated with the Atlantic–Tunisian current flowing on the Tunisian shelf.

5.2.3. Waves

Data on wave characteristics have been derived from the ECMWF ERA-5 dataset [54]. Data from the period 1979–2019 have been used in the analysis of wave characteristics for both the Lampedusa and MOBY areas. The median of the significant wave height at Lampedusa is about 0.9 m, while it is about 1.1 m in the MOBY area. The significant wave height shows a very limited spatial variability in the region around Lampedusa, much smaller than in the MOBY region. For Lampedusa, the significant wave height is expected to be > 1 m for about 40% of the time. Moreover, relatively high waves are mainly concentrated in winter. Waves from the North-North-West sector are the most frequent, with other smaller contributing sectors from East and South-East.

Significant differences in the wave period exist between the Lampedusa and the MOBY areas. Wave periods are generally shorter at Lampedusa than at MOBY, although a wider range of periods may be found in Lampedusa. The wave characteristics are expected to impact the stability of the buoy attitude; hence, the full wave climate characterization provided with this study aims to offer the backbone of the necessary requirements for the buoys structure design, which is essential for good quality measurements in the long-term.

5.2.4. Inherent Optical Properties

The Inherent Optical Properties (IOPs) were investigated using different the CMEMS datasets, for the January 2017–September 2019 period in the Mediterranean, and for the July 2018–November 2019 for the MOBY area; in-situ measurements of chlorophyll, CDOM and backscatter made at the Lampedusa Oceanographic Observatory at 4.5 m depth in the period June 2018–July 2019 [49].

Overall, a very good agreement between satellite and in-situ determinations of chlorophyll is found and confirms that the central Mediterranean is an oligotrophic region. The chlorophyll amount shows a clear annual cycle, with a maximum in winter and a minimum in summer. The annual average chlorophyll amount is <0.08 mg m^{-3} in both the Lampedusa and MOBY regions. The day-to-day variability, as retrieved from satellite observations, is <10% at both sites. The chlorophyll spatial variability is also small (<10%) in the Lampedusa area. Figure 10 shows the chlorophyll seasonal cumulative distribution as derived from in-situ observations. The chlorophyll amount exceeds 0.2 mg m^{-3} only during winter. In-situ continuous observations also show that CDOM and volume backscattering coefficients display an annual cycle, with maxima in winter. Maxima of CDOM and volume backscattering coefficient are 1.1 ppb and 6×10^{-4} m^{-1} sr^{-1}, respectively. Minima are around 0.5 ppb for CDOM and 2.5×10^{-4} m^{-1} sr^{-1} for the backscattering coefficient.

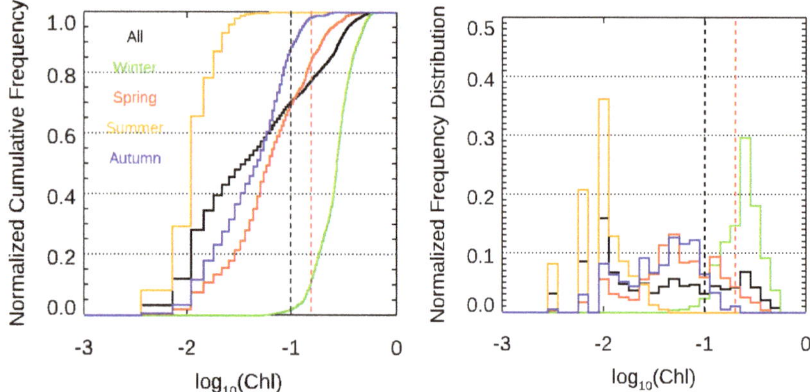

Figure 10. Normalized cumulative (**left**) and frequency (**right**) distribution of the log10-transformed chlorophyll concentration measured at a depth of 4.5 m. Different colors indicate seasons. The black dashed vertical line marks 0.1 mg m^{-3}, the red vertical dashed line marks 0.2 mg m^{-3}.

5.3. Atmospheric Properties

In this section, the main atmospheric properties that may affect SVC are presented and analyzed considering ground-based, satellite, and in-situ measurements.

5.3.1. Meteorology and Local Scale Effects

Because of its reduced surface extension and low and smooth topography, the meteorological conditions at Lampedusa are dominated by synoptic-scale phenomena, while local effects are negligible. Weather patterns display a significant seasonal cycle, with a larger variability during autumn and winter, and generally stable and dry conditions during spring and, particularly, in summer. Precipitation is concentrated in autumn and winter, with a maximum in October. Highest temperatures are registered in August (about 26 °C) and minima in January, February, and March (13.5–14 °C).

The annual cycle in weather variability is also evidenced by tropospheric temperature profiles, showing a temperature inversion from April to August, due to the differential warming of the air and sea surface. From September to March, the absence of temperature inversions is linked to more unstable conditions [55].

The comparison of the temporal evolution of meteorological parameters measured at the atmospheric observatory and at the oceanographic observatory confirms that the island of Lampedusa produces a very small perturbation to the atmospheric dynamics, and local effects are negligible.

5.3.2. Wind

Wind speed is linked to wave formation and with the sea surface roughening and foam, thus limiting the availability of optimal conditions for SVC.

Wind speed patterns are driven by the main synoptic structures, showing prevailing directions from NW, with limited interannual variability. The most frequent wind speed is between 4 m/s and 8 m/s (annual average of about 6 m/s); in about 67% of the cases the wind speed values are <8 m/s (Figure 11).

The wind speed and direction display a seasonal variability, also connected with seasonal changes of the mesoscale Mediterranean atmospheric circulation, with a larger variability during autumn and winter.

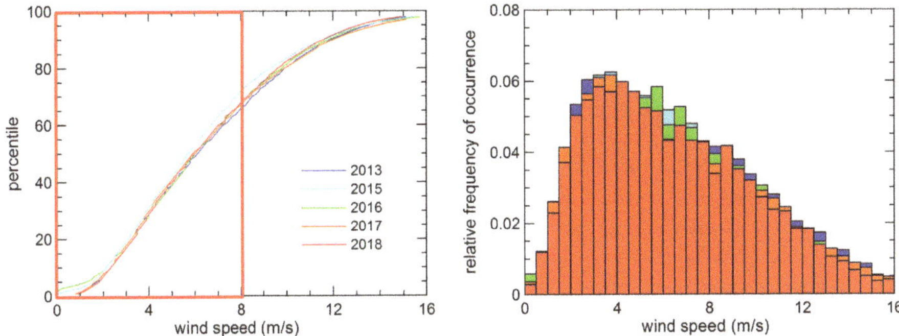

Figure 11. Cumulative distribution (**left**) and histogram of relative occurrence (**right**) of the wind speed at Lampedusa. The different curves/histograms are relative to years 2013, 2015, 2016, 2017, and 2018.

5.3.3. Cloud Cover

The Sicily channel is an area with relatively low cloud cover occurrence, a moderate seasonal cycle ranging from about 10% in summer to 40% in winter and a relatively low (<5%) interannual variability [56].

The detection of cloud-free periods at Lampedusa is based on ground-based measurements collected by all sky imagers and radiometers. A specific algorithm has been developed to identify cloud-free cases (cloud cover <2 oktas, sun unobstructed) from global and diffuse irradiance measurements made with a Multi Filter Rotating Shadowband Radiometer (MFRSR) [57]. The algorithm is based on the scheme proposed by Long and Ackerman [58], which has been modified to identify periods devoid of clouds also during cases with elevated aerosol amounts, in particular Saharan dust. The method is capable to identify cloud-free cases and has been tested against all-sky imager data. In the paper by [59] (Figure 4 of the cited reference) the frequency of occurrence of cloud-free conditions during the central part of the day (UTC time between 09:00 and 14:00), as derived from 15 years of MFRSR measurements, is shown. Cloud free conditions occur in more than 60% of the cases at Lampedusa during June, July, and August, with peaks of 80%. The frequency is <20% in February, November and December. On average, the area has a 37% overall probability of cloud free occurrence; the interannual variability is relatively low.

Cloud properties retrieved from the MODIS sensor on-board Aqua have been investigated over a region of 2° × 2° around Lampedusa analyzing Level-3 monthly mean cloud optical thickness, cloud fraction (CF), and cloud top pressure (CTP) from July 2002 to December 2016 (see Figure 10 [59]).

The annual evolution of cloud fraction, with summer minima, is consistent with that of the occurrence of cloud-free conditions. Moreover, covariance of CF and CTP is observed, indicating lowest clouds and smallest CF in July, when low cloud optical thickness values are also observed; this may be associated to the formation of thin low-level cloud at the top of the atmospheric inversion characterizing the marine boundary layer in summer [55]. The spatial distribution of the cloud cover occurrence is highly homogeneous around Lampedusa throughout the year.

Recently, an algorithm to derive cloud cover from measurements of diffuse irradiance at two wavelengths was developed. This algorithm is based on the method developed by Min et al. [60], which uses measurements of diffuse irradiances at 415 nm and 870 nm obtained with an MFRSR instrument. An improved version of the algorithm was implemented, taking into account the effect of aerosol and varying solar zenith angle and leading to an improved determination of cloud cover and to a more general applicability of the method. High time resolution cloud cover was obtained throughout the year 2016 and compared with ground-based sky imager observations, and cloud cover derived from MODIS satellite data. The analysis of those data has shown the absence of a daily cycle in cloud cover.

5.3.4. Aerosols

Aerosol transport patterns and optical properties over Lampedusa have been extensively studied using ground-based measurements and satellite data (e.g., references [57,61–63]). Measurements of the aerosol optical properties were started at Lampedusa in 1999 using an MFRSR instrument; a CIMEL sunphotometer contributing to AERONET was set up in 2000, although it worked with some discontinuity during the first years of operation. An extended dataset combining these two datasets has been implemented, and a correction scheme has been developed to take into account differences in the field of view of the two instruments [63].

Several studies show that, except for the cases directly influenced by Saharan dust transport events, Lampedusa is characterized by low aerosol optical depth (AOD). When such events are not considered in the statistics the annual mean aerosol optical depth at 870 nm is about 0.09 (as derived from the data by [61]). The distribution of the AOD daily average values at 870 nm for the period 1999–2018 (> 4500 days with measurements) is shown in Figure 12: about 75% of all AOD values at 870 nm are below 0.15, while larger AOD values are measured during sporadic events of Saharan dust transport that have a seasonal cycle and typically last for one or two days [57]. The distribution of Ångström exponent values, which is linked to the particles' dimension, is shown in Figure 12. Cases of pollution transport from the European continent, characterized by small particles, i.e., larger Ångström exponent values [61], are infrequent; values of Ångström exponent below 1 represents 78% of all values.

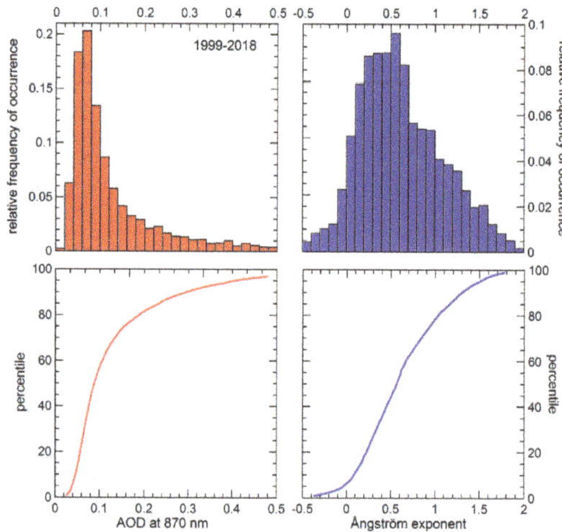

Figure 12. Histogram of relative occurrence (**upper**) and cumulative distribution (**lower**) of the values of AOD at 870 nm (**left**) and of the Ångström exponent calculated between 500 nm and 870 nm (**right**). Daily values (> 4500) measured between 1999 and 2018 are used.

Chemical analyses of in-situ aerosol samples [64] show that natural sources (sea salt, mineral dust, and biogenic emissions) give the largest contribution to PM10 all year round. Sea salt represents the largest absolute and relative contribution in all seasons, with a maximum (54%) in winter and minima (33%) in summer, while desert dust contributes by 17%–37% (largest in autumn).

5.3.5. Interfering Gases

Among all the atmospheric gases with absorption in the UV-visible-near infrared spectral intervals, water vapor, ozone, nitrogen dioxide, and sulphur dioxide have the largest absorption.

Total ozone column (TOC) has been measured at Lampedusa since 1998 by means of the Brewer spectrophotometer #123. TOC time series in the period 2000–2016 has a clear annual cycle with maxima in spring and minima in autumn, which is the result of a balance between transport and photochemical loss, whereby the former is dominant in winter, while photochemical loss dominates in summer, leading to a minimum in autumn. The TOC mean annual cycle at Lampedusa obtained from daily average has an amplitude of about 60 DU. The TOC diurnal variability is mainly caused by photochemical and transport processes occurring in the troposphere. At Lampedusa, diurnal variability is limited, below 2.5% under cloud-free conditions.

The total NO_2 column is not measured at Lampedusa, so data from the Ozone Monitoring Instrument satellite (OMI) onboard Aura satellite have been examined. The temporal evolution of NO_2 column density provided by OMI measurements have been derived in the period 2010–2018 from the NASA Giovanni web site (https://giovanni.gsfc.nasa.gov/giovanni/). Maxima are reached in summer and minima in winter, with an amplitude of the annual cycle of about 2×10^{15} cm^{-2}. The day-to-day column NO_2 variability is <17%.

Finally, in-situ measurements of O_3 and NO_2 mixing ratio show that their diurnal variability is negligible.

5.4. Adjacency Effect

The contribution of the diffuse radiance reflected by the presence of land to the radiance measured by the satellite for an ocean scene, called the adjacency effect (AE), perturbs the spectral satellite data, thus leading to uncertainties in derived primary products. The AE has to be minimized by considering a minimum distance from the coast (i.e., [12]).

The adjacency radiance contribution depends on the land surface radiative properties (spectral and geometric), on the extension and morphology. The adjacency effect field around Lampedusa island has been studied by [65] simulating OLCI measurements at the central wavelengths of four OLCI channels (λ = 490 nm, 555 nm, 670 nm, and 865 nm), accounting for typical surface and atmospheric conditions derived from satellite or remote sensing surface observations.

According to the results of this study, AE is avoided at distances larger than 15 km from the coast of the island.

Since land surface albedo has a key role in the estimation of the AE, spectrally detailed measurements of this quantity should be made at the same time and at the same wavelength as OC measurements [65]. The estimation of the AE will take advantage of surface albedo measurements carried out at the Atmospheric Observatory. While spectral downward irradiance has been measured by spectroradiometers since 2013, measurements of reflected spectral irradiance are about to be started.

Measurements of broadband solar albedo at the Atmospheric Observatory started in 2016. The annual distribution of the surface albedo monthly mean values presents maxima of 0.19–0.21 from July to January, and a decrease to 0.16–0.17 in April and May, when the occurrence of rain favors the growth of vegetation, decreasing the reflecting properties of the surface.

6. Uncertainty Budget

6.1. Uncertainty of the Radiometric Measurement

The estimated uncertainty budget (EUB) of the infrastructure is firstly assessed for the radiometric measurements. For each radiometric correction, the measurement equation is implemented in the ground segment to provide end-to-end uncertainties of every individual measurement (Figure 13). Uncertainty propagation is then theoretically achievable with this formalism (Type A uncertainty, defined in a statistical sense), using either the Guide to the Expression of Uncertainty in Measurement (GUM) first-order assumption [66] or Monte Carlo simulations [67].

For the quantified EUB of the optical system, we prefer, however, to rely on the actual uncertainties recently assessed by the MOBY team with the MarONet optical system [17,18]. This refers to Type B uncertainties, i.e., uncertainties not derived by a statistical analysis, but field experiences, calibration reports, and background knowledge on the system. It shall be noted that uncertainties of

MarONet are reduced compared to the previous MOBY system [68]. The EUB of the optical system is split into three main parts:

1. Uncertainty of water-leaving radiance (Lw(0+)) covering the optical system (Table 4), deployment structure and ground segment processing (Table 5);
2. Uncertainty of the downwelling irradiance (Ed) covering the optical system and environmental effects (Table 6);
3. Uncertainty of the resulting remote-sensing reflectance, Rrs (Table 7).

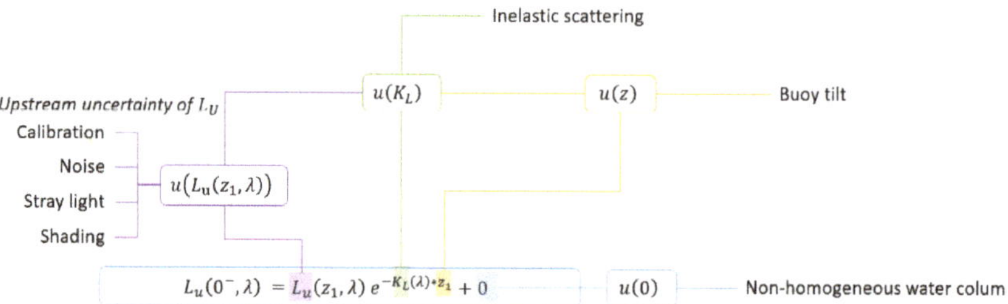

Figure 13. Example of measurement equation for uncertainty propagation: Lu depth propagation.

A basic principle in the uncertainty budget is that corrections are multiplicative. The formalism adopted in the study follows the protocols of the IOCCG report "Protocols for Satellite Ocean Color Data Validation: In-situ Optical Radiometry" [28]. Denoting with the symbol $\mathfrak{I}(\lambda)$ either $L(\lambda)$ or $E(\lambda)$ in physical units (W·cm^{-2}·sr^{-1}·nm^{-1} and W·cm^{-2}·nm^{-1}, respectively), the conversion equation for a given correction, $\aleph_i(\lambda)$, from measurement $\mathfrak{I}_{i-1}$ to measurement $\mathfrak{I}_i$, is

$$\mathfrak{I}_i(\lambda) = \aleph_i(\lambda)\cdot\mathfrak{I}_{i-1}(\lambda) \tag{9}$$

The term $\aleph_i(\lambda)$ accounts for the non-ideal instrument response of the instrument and implicitly depends on various input parameters (possibly other wavelengths, like for straylight correction). The final radiometry is thus given by the product of all corrections:

$$\mathfrak{I}(\lambda) = \prod_{i=1}^{N} \aleph_i(\lambda)\cdot\mathfrak{I}_0(\lambda) \tag{10}$$

where $\mathfrak{I}_0$ represents the net radiometric data obtained by subtracting the dark contribution from signal data after scaling the raw CCD readouts by the measurement integration time and the bin factor. Applications of the GUM gives

$$\frac{u^2(\mathfrak{I}(\lambda))}{\mathfrak{I}(\lambda)^2} = \sum_{i=1}^{N} \frac{u^2(\aleph_i(\lambda))}{\aleph_i(\lambda)^2} + \frac{u^2(\mathfrak{I}_0(\lambda))}{\mathfrak{I}_0(\lambda)^2} \tag{11}$$

This justifies why the quadratic uncertainty budget is additive, in relative unit, in the next tables. Furthermore, the tables tentatively distribute the uncertainties sources between their random and systematic parts. This has an important consequence on the SVC gains, further discussed in the next section.

Table 4. The estimated uncertainty budget (EUB) of the in-situ optical system (Lu), in %. See text for justification on the numbers.

Requirement from [3]	Uncertainty Source	Uncertainty Type (A/B)	412 nm		443 nm		490 nm		560 nm		674 nm	
			rand.	syst.	rand.	syst.	rand.	syst.	rand.	syst.	rand.	syst.
OC-VCAL-RD-14	Spectral resolution											
OC-VCAL-RD-15	Spectral calibration	B		0.38		0.29		0.22		0.14		0.06
OC-VCAL-RD-16	Stray-light	B		0.18		0.06		0.02		0.06		0.04
OC-VCAL-RD-17	Radiometric calibration	B		0.81		0.72		0.61		0.53		0.46
		B		0.42		0.46		0.51		0.53		0.49
		B	0.20		0.20		0.20		0.20		0.20	
		A	0.51		0.21		0.22		0.12		0.10	
		B		0.20		0.15		0.03		0.03		0.03
OC-VCAL-RD-18	Angular response											
OC-VCAL-RD-19	Immersion factor	B	0.05		0.05		0.05		0.05		0.05	
OC-VCAL-RD-20	Thermal stability	B	0.16		0.16		0.16		0.16		0.16	
OC-VCAL-RD-21	Dark current											
OC-VCAL-RD-22	Polarisation sensitivity	B	0.20		0.20		0.20		0.20		0.20	
OC-VCAL-RD-23	Non-linearity response											
OC-VCAL-RD-24	Noise characterisation	B	0.15		0.15		0.15		0.15		0.15	
		B	1.43		1.12		0.95		0.87		0.74	
OC-VCAL-RD-25	Environ. conditions (like-to-like rule)	B	1.65		1.37		1.69		1.82		1.25	
Total uncertainty on optical system Lu			2.27	1.02	1.82	0.92	1.99	0.82	2.05	0.77	1.50	0.68
Total uncertainty on optical system Lu (rand. + syst.)			**2.49**		**2.04**		**2.15**		**2.19**		**1.65**	

Table 5. EUB of the deployment structure (Lu in the field) and ground segment up to Lw(0+), in %. See text for justification on the numbers.

Requirement from [3]	Uncertainty Source	Uncertainty Type (A/B)	412 nm		443 nm		490 nm		560 nm		674 nm	
			rand.	syst.	rand.	syst.	rand.	syst.	rand.	syst.	rand.	syst.
Deployment structure (Lu)												
OC-VCAL-RD-26	Shading	A		0.50		0.50		0.60		1.25		4.00
OC-VCAL-RD-27	Tilting & BRDF	A	0.20		0.20		0.20		0.20		0.20	
	Environmental fluctuation	B	0.10		0.10		0.10		0.15		0.20	
OC-VCAL-RD-49	Bio-fouling	A	1.00		1.00		1.00		1.00		1.00	
Ground segment up to in-situ Lw(0+)												
OC-VCAL-RD-28	Depth-extrapolation	A		0.40		0.40		0.40		0.40		0.40
OC-VCAL-RD-29	Surface propagation	A		0.10		0.10		0.10		0.10		0.10
OC-VCAL-RD-30	Data reduction	A	2.00		2.00		2.00		3.00		3.00	
Total uncertainty on in-situ Lw(0+)			3.20	1.21	2.89	1.12	3.00	1.10	3.78	1.52	3.51	4.08
Total uncertainty on in-situ Lw(0+) (rand. + syst.)			**3.42**		**3.10**		**3.19**		**4.07**		**5.38**	

Table 6. EUB of in-situ Es, in %. See text for justification on the numbers.

Requirement from [3]	Uncertainty Source	Uncertainty Type (A/B)	412 nm		443 nm		490 nm		560 nm		674 nm	
			rand.	syst.	rand.	syst.	rand.	syst.	rand.	syst.	rand.	syst.
	Optical system and ground segment (Es)											
	Spectral calibration	B		0.34		0.25		0.19		0.12		0.06
	Cosine response	B		0.10		0.10		0.10		0.10		0.20
	Stray-light	B		1.07		0.35		0.11		0.05		0.27
OC-VCAL-RD-33	Radiometric calibration	B		0.52		0.46		0.43		0.39		0.34
		B		0.77		0.76		0.69		0.67		0.61
		B		0.49		0.49		0.49		0.49		0.49
		A	0.31		0.14		0.10		0.06		0.06	
		B		0.10		0.10		0.10		0.10		0.10
	Thermal stability	B	0.16		0.16		0.16		0.16		0.16	
	Noise characterisation	B	0.15		0.15		0.15		0.15		0.15	
		B	1.14		0.92		0.76		0.69		0.53	
	Environmental uncertainty	B	2.00		2.00		2.00		2.00		2.00	
	Total uncertainty on in-situ Es		2.33	1.54	2.22	1.11	2.15	0.98	2.13	0.93%	2.08	0.92
Total uncertainty on in-situ Es (rand. + syst.)			**2.80**		**2.48**		**2.37**		**2.32**		**2.27**	

Table 7. EUB of in-situ remote-sensing reflectance (Rrs), resulting from Tables 5 and 6, in %.

	412 nm		443 nm		490 nm		560 nm		674 nm	
	rand.	syst.	rand.	syst.	rand.	syst.	rand.	syst.	rand.	syst.
Total uncertainty on in-situ Rrs	3.96	1.96	3.64	1.58	3.69	1.47	4.33	1.79	4.08	4.18
Total uncertainty on in-situ Rrs (rand. + syst.)	**4.42%**		**3.97**		**3.98**		**4.69**		**5.84**	

Table 4 gives the EUB of the optical system (Lu), and Table 5 gives the EUB of the deployment structure (Lu in the field) and of the depth extrapolation and post-processing up to Lw(0+). Most of the terms are from Voss et al. 2017 and Johnson et al. 2017 [17,18]. Uncertainty due to spectral resolution is covered in the next section. The five sources of uncertainty related to radiometric calibration refer, respectively, to the sphere calibration, sphere drift, non-uniformity, calibration measurement uncertainty and spectral interpolation at the radiometer bands. The effect of angular responses is neglected thanks to very limited FOV ($1.73°$ full angle). The uncertainty of the immersion factor comes from Feinholz et al. 2017 [16]. Thermal stability refers to terms "instantaneous temperature" in Johnson et al. 2017 [18]. Dark current is implicitly taken into account in the temporal averaging and random component of noise calibration below. Uncertainty related to polarization comes from Voss et al. 2017 [17] and is very conservative, considering the use of a depolarizer. Noise characterization covers MOBY integration time and reproducibility of Johnson et al. 2017 [18]. The effect of environmental conditions on Lu is assessed with a very conservative assumption as the sum of uncertainties related to MOBY stability during deployment—system response, in-water internal calibration, and wavelength stability [68].

The quantified uncertainty of the depth propagation at MOBY of 0.4% [17] is kept for EURYBIA. Indeed, with chlorophyll concentration below 0.15 mg/m^3 most of the time relevant for SVC (only up to 0.2 in January), the attenuation coefficient at Lampedusa is below the maximum considered in Voss et al. 2017 [33]; real acquisitions at Lampedusa done with a Satlantic profiler (with OCR507 radiometers) in 2017 confirms values of K_L below 0.04 m^{-1} from 412 nm to 510 nm, of about 0.06 m^{-1} at 555 nm and 0.22 m^{-1} at 665 nm, fully compatible with the analysis of Voss et al. 2017 [33] at MOBY (see their Figure 6). A conservative uncertainty of 1% is assigned to the biofouling correction [68].

The total uncertainty of Lw(0+) ranges from 3% to 5.5%, with spectral variation essentially due to spectral emission of the calibration lamp and noise level.

EUB of solar irradiance (Es) is given in Table 6. The uncertainty induced by the cosine correction depends on the solar zenith angle and the wavelength. For relevant conditions used in the SVC (most of the time $\theta_0 < 60°$), the uncertainties range from below 0.1% in the blue and green bands to 0.2% in the red. Uncertainty of radiometric calibration accounts for these five sources respectively [18]: FEL lamp calibration, drift, bench effect, calibration measurement uncertainty, and spectral interpolation. The EUB includes also an estimated 2% environmental uncertainty, yielding a total uncertainty between 2% and 3% at all bands.

Table 7 eventually gives the uncertainty of remote-sensing reflectance (Rrs), through simple combination between Lw and Es. We emphasize that the requirement to conduct SVC with Lw rather than Rrs = Lw/Es is not agreed by all agencies; using Rrs may increase the total EUB and is here considered as a conservative assumption. The total uncertainty remains below 5% from 412 nm to 560 nm and below 6% in the red. These numbers come from various conservative assumptions and provide more an upper threshold than optimal estimates for the EURYBIA concept. Slightly smaller uncertainties can likely be expected during the detailed design of the infrastructure.

6.2. Uncertainty of the SVC Gains

The second level of the EUB refers to the SVC process, i.e., uncertainty of the vicarious gains. Although gain computation is not part of the SVC infrastructure, we consider that it is of highest importance to quantify the end-to-end uncertainty of the system. This computation is done for the identified locations around Lampedusa as well as for the MOBY location in Hawaii (MOBY), and using OLCI data and effective criteria defined by EUMETSAT, detailed below. Because both Lampedusa candidate locations yield very similar numbers, in the following we only report the analysis for location LMP1 (East of Lampedusa). The EUB contains two parts, both based on realistic assumption on the number of match-ups and radiometry at the calibration site.

1. Ground segment post-processing for in-situ and satellite data merging in the gain computation, i.e., match-ups, spectral integration (Table 8);

2. Gain computation, up to final mission average gains over a decade (Figure 14).

We cover two years of Sentinel-3 OLCI data (Full Resolution products) from July 2017 to July 2019. We apply the strict criteria used by EUMETSAT for SVC gain generation, assuming that the criteria related to the in-situ measurement should be met with an operational infrastructure:

1. 5×5 box centered on single pixels;
2. For each pixel, sensor zenith angle should be below $56°$ and the sun zenith angle below $70°$;
3. Excluding pixels contaminated by known disturbances, i.e., masked with at least one of the following flags: CLOUD, CLOUD_AMBIGUOUS, CLOUD_MARGIN, INVALID, COSMETIC, SATURATED, SUSPECT, HISOLZEN, HIGHGLINT, SNOW_ICE, WHITECAPS, ANNOT_ABSO_D, ANNOT_MIXR1, ANNOT_TAU06, RWNEG_O2, RWNEG_O3, RWNEG_O4, RWNEG_O5, RWNEG_O6, RWNEG_O7, RWNEG_O8;
4. All pixels within the 5×5 region of interest (ROI) should be valid;
5. Mean chlorophyll within the ROI should be below $0.2 \ \mathrm{mg/m^3}$;
6. Mean aerosol optical depth (AOD) for the channel at 865 nm should be below 0.15;
7. Coefficient of Variation $CV = \frac{\sigma^*}{\mu^*}$ should be below 0.15 for remote sensing reflectances at bands between 412 and 560 nm, where μ^* and σ^* are mean and standard deviation, respectively, computed after the removal of pixel outliers. A pixel is considered as an outlier for variable x_i if $|x_i - \mu| < 1.5 \ \sigma$ where μ is the mean and σ is the standard deviation of that variable over the 5×5 box.

In two years, the number of valid data is 36 at Lampedusa and 64 at Hawaii (Table 9). The difference is essentially explained by the threshold on AOD, and to a lesser extent on chlorophyll concentration, noticing that the number of match-ups after the flag and geometry screening only is exactly similar (70 match-up in two years for both sites). However, recent analyses suggest that OLCI may overestimate AOD [69], possibly reducing the number of cases for valid matchups at Lampedusa.

In the uncertainty of the ground segment post-processing, our analysis includes the spatial variability of the satellite observation (the first item of Table 8). This uncertainty is quantified for every match-up in each relative unit by the Coefficient of Variation (CV) and strongly grows toward red bands for clear waters. This term appears to be crucial in the EUB; neglecting it would drastically reduce the final uncertainty on the SVC gains in the NIR.

Table 8. EUB of the match-up process. See text for justification on the numbers.

Requirement from [3]	Uncertainty Source	Uncertainty Type (A/B)	412 nm		443 nm		490 nm		560 nm		674 nm	
			rand.	syst.	rand.	syst.	rand.	syst.	rand.	syst.	rand.	syst.
OC-VCAL-RD-35-36, 43-44-45	Environmental variability against satellite	A	4.00		3.50		3.60		9.40		56.10	
OC-VCAL-RD-41	Spectral integr. to satellite SRF	A		0.50		0.50		0.50		0.50		0.50
OC-VCAL-RD-42	Normalisation & BRDF corr.	A	1.00	1.00	1.00	1.00	1.00	1.00	1.00	1.00	1.00	1.00
Total uncertainty on targeted Rrs			5.71	2.26	5.15	1.93	5.25	1.85	10.40	2.11	56.26	4.33
Total uncertainty on targeted Rrs (rand. + syst.)			**6.14**		**5.50**		**5.57**		**10.61**		**56.42**	

Table 9. Number of Ocean and Land Color Instrument (OLCI) match-ups between July 2017 and July 2019 when applying System Vicarious Calibration (SVC) criteria of EUMETSAT.

Criteria	Lampedusa (EURYBIA)	Hawaii (MOBY)
Total number of acquisitions	159	148
100 % valid pixels (flags and geometry)	70 [44%]	71 [48%]
Chl $\leq$ 0.2 mg/m^3	134 [84%]	147 [100%]
AOD(865) $\leq$ 0.15	89 [56%]	105 [71%]
CV < 0.15 from 412 to 560 nm	95 [60%]	108 [73%]
Combined criteria	36 [23%]	64 [43%]

For a given band λ in the visible part of the solar spectrum, the SVC of OLCI consists first in computing individual gains, $g(\lambda)$, for every match-up and pixels [70]:

$$g(\lambda) = \frac{\rho_{path}(\lambda) + t(\lambda).\rho_w^t(\lambda)}{\rho_{gc}(\lambda)} \tag{12}$$

where $\rho_w^t(\lambda)$ is the in-situ marine reflectance (de-normalise for BRDF effect, i.e., in the observation and solar geometry of the actual pixel), $\rho_{path}(\lambda)$ and $t(\lambda)$ are the atmospheric path reflectance and diffuse total transmittance, respectively, accounting for the Rayleigh and aerosol components reflectance and $\rho_{gc}(\lambda)$ is the actual TOA reflectance pre-corrected for gaseous absorption and sun glint effect. Note that computation and application of gains directly on the TOA Level-1 radiometry would not change the present analysis. When dealing with standard atmospheric correction [37,71], the aerosol amount and type are identified by the near-infrared (NIR) bands only, assumed to be calibrated beforehand by other means [70]. Application of the GUM first-order assumption [66] gives the uncertainty of individual gain:

$$u(g)^2 = \left(\frac{t\rho_w^t}{\rho_{gc}}\right)^2 \left[\left(\frac{u(\rho_w^t)}{\rho_w^t}\right)^2 + \left(\frac{u(\rho_{path})}{\rho_{path}}\right)^2 \left(\frac{\rho_{path}}{t\rho_w^t}\right)^2 + \left(\frac{u(t)}{t}\right)^2 + \frac{u(\rho_{path}, t)}{t\rho_{path}} \frac{\rho_{path}}{t\rho_w^t} \right] \tag{13}$$

This equation implicitly discards the uncertainty of the satellite radiometry before SVC, i.e., $u(\rho_{gc})$, because it essentially depends on the satellite characteristics (radiometric noise) and not on the OC-SVC infrastructure nor aerosol or marine conditions. In term of covariances, we do not expect either any correlation with ρ_{gc}, insofar as the standard atmospheric correction does not detect aerosol from the VIS domain: $r(\rho_{gc}, \rho_{path})$=0 and $r(\rho_{gc}, t)$=0. On the contrary, covariance between scattering functions exists and is included through $u(\rho_{path}, t)$. The multiplicative factor $t\rho_w^t/\rho_{gc}$ decreases in absorbing waters and theoretically minimizes the uncertainty of g. However, complex waters may counterbalance this decrease by an increased uncertainty of the additive components in the square brackets—less reliable in-situ marine reflectance (temporal variability, radiometric measurement) and possibly more complex atmosphere. Also, a lower ratio gives more importance to the atmospheric uncertainty through the factor $\rho_{path}/t\rho_w^t$. The balance between all terms is important to get the realistic uncertainty estimate.

Eventually, when the uncertainties of individual gains are computed, the uncertainty on the mission average gains $\bar{g}$ is simply given by

$$u(\bar{g})^2 = \frac{1}{N^2} \sum_{match-ups} u(g) \tag{14}$$

This formula assumes a simple average of the individual gains. More evolved averaging techniques are also possible, such as the mean of the semi-interquartile range to remove outliers, or a weighted average. In the present context we have no access to the distribution of the gains, only to their uncertainties, so the arithmetic averaging is kept as a conservative estimate. The value of $u(\overline{g})$ is affected by the number of years considered in the time-series, here $Y = 2$. A more objective metric is the Relative Standard Error of the Mean (RSEM), introduced by Zibordi et al. 2015 [4] to study the stability requirements of SVC gains per decade. With an average SVC gain of the order of unity, RSEM, can be related to our computed $u(\overline{g})$ by simply scaling the number of match-ups obtained in Y years to a decade.

$$RSEM = u(\overline{g}) \,/\, \sqrt{\frac{10}{Y}} \tag{15}$$

The relative uncertainty of in-situ marine reflectance, $\dfrac{u(\rho_w^t)}{\rho_w^t}$, is assumed to be of 5% at all bands and all calibration sites (Lampedusa, Hawaii). This assumption, previously used in Zibordi et al. 2015 [4], is consistent with the overall quantification detailed in the previous sections. It has obviously a direct impact on the final uncertainties and could be refined in a later stage. The purpose of the present analysis is to simulate the gain uncertainties that would be achieved with OLCI over these locations assuming high-quality and daily radiometric measurements. All other terms involved in the computation of $u(g)$ are assessed by satellite data, notably the uncertainties of the atmospheric scattering functions, $u(\rho_{path})$ and $u(t)$. This allows conducting a harmonized uncertainty budget for all sites under consideration, while only Lampedusa is fully characterized by long-term AERONET measurements. The choice is made here to fully rely on OLCI aerosol products, as it would be done in real SVC operation. The actual uncertainties $u(\rho_{path})$ and $u(t)$ for a given match-up used in the SVC process may be due to many sources, such as the aerosol modelling in the atmospheric correction, the aerosol identification in the NIR, the propagation in the visible, and include both systematic and random components. Because the purpose of SVC is to remove the systematic effects, the present analysis determines $u(\rho_{path})$ and $u(t)$ by the local variation of ρ_{path} and t around the calibration site. This approach considers that heterogeneity in the atmospheric functions degrades the confidence in the gains. The method is implemented as follows:

1. Consider, for each potential match-up, a ROI of 5×5 OLCI pixels (full resolution), consistent with EUMETSAT protocols for SVC as defined previously;
2. Consider all OLCI metadata and ancillary data at pixel level (wind speed, pressure, geometry);
3. Consider OLCI aerosol optical thickness and Angstrom coefficient;
4. Identify, for each pixel, the best matching aerosol model in the OLCI LUTs;
5. Given geometry and ancillary data, compute ρ_{path} and t by the OLCI LUTs (Rayleigh + aerosol), as done in OLCI atmospheric correction;
6. Compute $u(\rho_{path})$ and $u(t)$ by their local standard-deviation over the ROI.

Final values of RSEM are given in Figure 14—the top-left for Lampedusa (LMP1 location) and the top-right for the Hawaii location. RSEM is of about 0.05% at all bands and both sites, with a peak closer to 0.08% in the Mediterranean at 490 nm. These numbers are consistent with the stability requirement over a decade in the blue-green bands but do not reach the requirement of 0.005% in the red. It is worth noting that this target is hardly reached by any existing SVC infrastructure. Indeed, the RSEM values computed by Zibordi et al. 2015 [4] based on real SVC gains with real in-situ measurements shows that only MOBY gains reach the required RSEM of 0.05% in the blue bands, and not in the red (their Figure 4, reprinted here in Figure 14, bottom-right). Despite the consideration of spatial variability in our uncertainty estimate (more than 50% at 674 nm, Table 8), this strongly suggests that the actual uncertainties of marine reflectance are, for many infrastructures, largely above 5%, especially in the red bands for clear waters.

Another analysis has been conducted in parallel following [4], by simplifying the uncertainty computation of individual gain by $u(g) = \left(t\rho_w^t/\rho_{gc}\right) * u(\rho_w^t)/\rho_w^t$, where the overline stands for a temporal average. At Lampedusa, this average factor is of 8.4% at 412 nm, 10.7% at 443 nm, 12.4% at 490%, 5.6% at 560 nm, and 1.2% at 674 nm. While neglecting the atmospheric uncertainties, this formulation allows easy computation of the uncertainty of gains, hence of RSEM, for testing various distributions of in-situ uncertainty sources between their random and systematic parts. Here two cases have been investigated:

1. A purely random case: the uncertainty sources are all assigned to the random component.
2. A mix between random and systematic sources (Figure 14, bottom-left, triangles): in Tables 4–8, an attempt is made to assign the uncertainty of various sources to the systematic component, only (i.e., without any random counterpart). Note that the global uncertainty of a given source is identical to the previous case (i.e., the quadratic sum of the random and systematic components equals the square random component of the previous case).

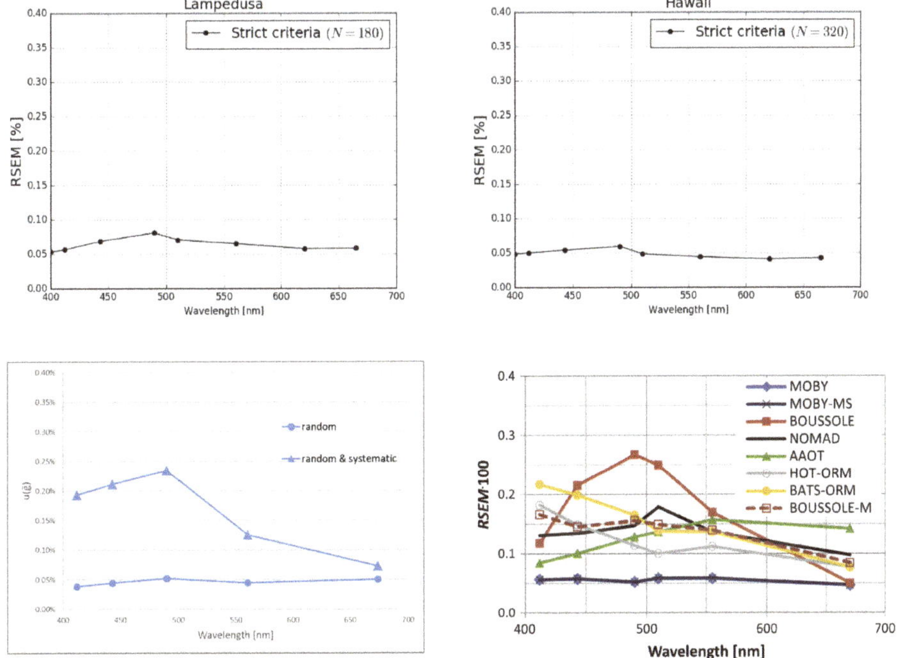

Figure 14. Relative Standard Error of the Mean (RSEM) of vicarious gains over a decade as a function of wavelength from various methods. Top: computation from GUM uncertainty propagation and OLCI data at Lampedusa (**left**) and Hawaii (**right**) assuming 5% uncertainty on in-situ Rrs with random source. Bottom left: simplified uncertainty of SVC gains from detailed EUB at Lampedusa (Tables 4–6) accounting for either purely random effects (circles) or both random and systematic effects (triangles; see previous tables for distribution). Bottom right: RSEM measured on real SVC gains computed at various site (MOBY in dark blue) for the SeaWiFS sensor (except BOUSSOLE-M: for MERIS) under a purely random assumption; reprinted from Zibordi et al. 2015 [4].

The first case is an ideal condition, that reduces the uncertainty of mission-average gains as a function of the square root of the number of match-ups. Such an assumption is generally retained in the SVC literature: Antoine et al 1999 [70] and Zibordi et al. 2015 [4] have estimated the final SVC uncertainty through the standard-error of the mean or the relative standard error of the mean of SVC gains, what amounts to dividing the standard-deviation of individual gains by the square root of number of match-ups. Under this purely random assumption, the EURYBIA assessment retrieves an uncertainty of about 0.05% over the full spectral range (Figure 14, bottom-left with circles) which is extremely similar to the assessment of Zibordi et al. 2015 [4] for MOBY (reprint in

Figure 14, bottom-right, blue curve for MOBY). The perfect consistency between these two assessments, made under completely different datasets and methods, is noteworthy—the EURYBIA computation relies on the uncertainty budget of the optical system (MarONet quantification) over Lampedusa conditions observed by OLCI, while the Zibordi et al. 2015 [4] estimate relies on the statistical analysis of real SVC gains computed at MOBY for the SeaWiFS sensor. These results demonstrate the relevance of EURYBIA, based on both the state-of-the-art optical system and favorable conditions, while other systems cannot meet such performance (Figure 14, bottom-right, other systems than MOBY).

This purely random assumption may, however, not be strictly true for various sources of uncertainty. For instance, it is likely that the radiometric calibration, stray light correction, depth extrapolation, have systematic uncertainties. The resulting uncertainty of average gains is then not driven by the number of match-ups, but by these systematic contributions which do not cancel out in the averaging process. Uncertainties can now reach nearly 0.25% in the blue and 0.1% in the red (Figure 14, bottom-left with triangles). Importantly, although these numbers show some similarity with the curves of Zibordi et al. 2015 [4] for non-MOBY systems, they are not comparable in the sense that the relatively poor performance of these latter systems still benefits from the ideal random assumption. We also emphasize that the "rand.+syst." case refers to the worst case scenario, since all sources should in reality have a random part. The distinction between both components is currently not addressed [18,33] and should be investigated in the next phase. The reality probably lies between the ideal random assumption and the worst-case scenario. Quantifying the detailed structure of the uncertainties is a key scientific challenge that should be addressed by coordinated effort among SVC teams and metrology institutes part of the future OC-SVC network.

In summary, the present analysis based on two years of OLCI data demonstrates that the Lampedusa location (either east or north-west, the latter not shown here) is theoretically suitable for an SVC site, in comparison with the existing site in Hawaii. Clearly, the screening criteria currently defined for an ideal location such as Hawaii removes a lot of data in the Mediterranean Sea (about 48% removal at Lampedusa versus less than 9% at Hawaii, starting from the number of valid data after flag and geometrical screening). However, the amplitude of the marine signal at Lampedusa is below that of the oligotrophic waters of Hawaii (while not being as low as for mesotrophic waters), what advantageously drives the uncertainty propagation from sea level to the SVC gains (term $t\rho_w^t/\rho_{gc}$).

7. Discussion and Conclusions

In this paper, we presented the overall design of the EURYBIA OC-SVC infrastructure, taking into account the Copernicus long-term and multi-mission perspective to deliver OC products and services.

Our strategy, to reach the goal to deliver state-of-the-art Fiducial Reference Measurements (FRM) in a very tight development schedule (expected operational target after 36 months from the beginning of the development), is to operate independently but in compliance with MarONet specification [72]. This solution brings to a network concept (e.g., AERONET-OC, [73]) and takes full advantage of more than 30 years of MOBY experience in operation and technological developments. EURYBIA will be compatible with the MarONet network in terms of radiometric data collection, traceability, and protocols, but autonomous regarding detailed architecture, development, funding, and operations.

Our proposed solution for the European OC-SVC Infrastructure (Figure 1) offers the possibility to develop and deploy a technologically proven system within a realistic timeframe for the Copernicus needs. Moreover, it allows to avoid the compatibility issues experienced in the past two decades due to the adoption of different instrumentation and processing approaches.

The MarONet optical system, constituting the core instrumentation of the infrastructure, has been verified to be fully compliant with the radiometric and spectral requirements ([3], see Table 1). The EURYBIA Field Segment accounts for the instrumentation, infrastructure and operations required to acquire radiometric FRM in the field. The field segment includes also an on-shore laboratory for pre- and post-deployment instrument calibration, characterization, and verification of the calibration sources' stability. Complete SI traceability will be insured by a Metrology Institute. The EURYBIA Ground Segment communicates with the buoys to transmit/receive data and to monitor the infrastructure status ensuring operations on 24/7/365. The processing unit computes data products with their associated uncertainties and performs the quality assurance of results. Data products are distributed within the official Copernicus mechanism through the dissemination unit, which also gives access to the open-source processing code and documentation.

The EURYBIA infrastructure is proposed to be settled in the Strait of Sicily offshore of Lampedusa Island, one of the potential European OC-SVC sites [11–13]. Lampedusa is a small flat island in the central Mediterranean Sea far away from the mainland and main traffic routes, but very well connected to Europe. The island is surrounded by oligotrophic waters and hosts a Marine Protected Area.

Our analysis, based on satellite and in-situ observations, revealed the occurrence of consistent and homogeneous cloud free conditions (> 60% during summer). The median value of the AOD at 870 nm is <0.1, hence below the currently adopted threshold value (0.15) for acceptance of measurements for vicarious calibration. Sporadic larger values of AOT are observed only during desert dust transport events characterized by relatively short time duration (1–2 days) over the whole Mediterranean region [74].

Being Lampedusa a very small island characterized by low and smooth topography, its meteorological conditions are dominated by synoptic scale phenomena, and local effects are negligible, as demonstrated by the absence of a diurnal cycle in cloudiness and aerosol load. Wind patterns are driven by the main synoptic structures, showing prevailing directions from NW, with a limited interannual variability and most frequent wind speed between 4 m/s and 8 m/s. Wave and current conditions are suitable to ensure the required stability of the optical buoy.

The analysis of the optical characteristics of the waters surrounding the Island were compared with those observed at the MOBY site. This reveals that in Lampedusa the annual average chlorophyll is <0.08 mg m^{-3}, with a day-to-day and local spatial variability below 10%. An end-to-end uncertainty budget based on OLCI data adopting EUMETSAT screening protocols for SVC and site characterization demonstrates the relevance of this site. The overall uncertainty of EURYBIA SVC gains is estimated to be about 0.05% per decade in the blue-green wavelengths. This value is comparable with those estimated for the reference site in Hawaii and in compliance with stability requirements for climate studies in this spectral domain.

We propose to settle the location site at more than 15 km off the island of Lampedusa at suitable sea depth conditions (> 300m) to avoid adjacency and bathymetry effects. Two candidate locations (Figure 5), satisfying these geophysical criteria and relatively distant from the main marine traffic routes, have been identified—LMP1 (35.5248°N–12.7667°E) and LMP2 (35.7430°N–12.3579°E). Both candidate sites have been evaluated to be fully compliant with the environmental (e.g., cloudiness, aerosols, surface wind, currents, and waves) and logistical requirements for vicarious calibration identified by Mazeran et al. [3]. Recently, Bulgarelli and Zibordi [13] carried out a complete study on the adjacency effect of the Lampedusa Island to address implications on a hypothetical nearby system vicarious calibration infrastructure for satellite ocean color sensors. Their results, as expected, confirmed that the site should be located at distances larger than approximately 14 km from the coast (requirement satisfied by both LMP1 and LMP2). They also suggested that the region in the direction of the reflected sunbeam should be avoided. Based on this latter recommendation LMP1 location should be preferred. The final selection of the exact location where to deploy the optical buoy will require further studies on logistics and environmental characterization, including analyses of ad-hoc in-situ observation to be carried out around the Island. This will be part of future studies to be carried out in phase 3 of the EUMETSAT Programme.

It is important to underline that the EURYBIA design is modular—any update of individual subsystems will be possible without the need to reconsider the whole system architecture. The proposed infrastructure will be able to host, on the moored buoy, additional instrumentation that can be used also for CAL/VAL of other satellite missions. The scheduled cruises for infrastructure maintenance will give an additional opportunity to collect in-situ measurements needed to improve the current OC product algorithms as well as to develop new ones.

Our EURYBIA design is independent from the proposed installation site, so other potential locations for its deployment could be considered. A preliminary analysis of other potential sites was carried out during the EUMETSAT project (not shown) revealing that Lampedusa is a very good candidate for the installation of the European OC-SVC Infrastructure, also in comparison to other locations. A detailed evaluation of other potential European sites will require a deep inter-comparison of environmental characteristics, including a complete evaluation of end-to-end uncertainty budget as we carried out for Lampedusa and MOBY. This will be part of our future studies.

In conclusion, the logistics, the capability and required expertise to operate the infrastructure will be ensured by the fact that Lampedusa has been hosting a long-term well-established Climate Observatory for more than twenty years. In addition, the presence of the Climate Observatory has the advantage to make available relevant observations on atmospherics variables, including aerosols profiles that could be used to improve the accuracy of system vicarious calibration. Lampedusa, as proposed site, represents a trade-off between ideal conditions reached in Hawaii, for the reference MOBY infrastructure, and the priority to host the EURYBIA system in European waters.

Author Contributions: Conceptualization, R.S., G.L.L., D.D., C.M., and A.d.S.; formal analysis, S.C., C.Y., A.P., C.M., G.V., L.C., D.D., A.d.S., and D.M.; data curation, S.C., G.V., T.K., and F.Z.; writing—original draft preparation, G.L.L., M.Y., K.V., R.B., D.M., D.D., A.d.S. and C.M.; writing—review and editing, G.L.L., C.C., D.D, C.M., K.V., and R.S.; supervision, R.S.; project administration, C.C.; funding acquisition, R.S. All authors have read and agreed to the published version of the manuscript.

Funding: The development of the EURYBIA concept and its preliminary design have been supported by EUMETSAT and the European Union through project Preliminary Design of the Copernicus Ocean Colour Vicarious Calibration under contract EUM/CO/18/4600002161/EJK. MarONet will be supported by NASA, while MOBY-Refresh, on which the optical system is based, is supported by NOAA.

Acknowledgments: The SVC design greatly benefited from comments of the review panel composed by M. Wang, B. C. Johnson, A. Reppucci, G. Zibordi, R. Gilmore, K. N. Babu, and EUMETSAT Team Members composed by E. Kwiatkowska, F. Montagner, I. Cazzaniga, E. Obligis, J. Chimot, and D. Dessailly. The MOBY team (B. C. Johnson, M. Feinholtz, A. Gleason, S. Flora, T. Houlihan, M. Yarbrough, and K. Voss) contributed to the design and construction of MOBY-Refresh. Contributions by C. Bommarito and D. Sferlazzo are gratefully acknowledged. V. Artale and S. Marullo are thanked for making available temperature and salinity data from the Lampedusa Oceanographic Observatory. Analyses of OMI NO_2 data used in this paper were produced with the Giovanni online data system (https://giovanni.gsfc.nasa.gov/giovanni/). developed and maintained by the NASA GES DISC. This study has been conducted using EU Copernicus Marine Service Information.

Conflicts of Interest: The authors declare no conflict of interest.

Abbreviations

ACTRIS	Aerosol, Clouds and Trace Gases
ADCP	Acoustic Current Doppler profiler
ADU	Analog Digital Counts
AE	Adjacency Effect
AERONET	Aerosol Robotic Network
AOD	Aerosol Optical Depth
AOT	Aerosol Optical Thickness
BOUSSOLE	BOUée pour l'acquiSition d'une Série Optique à Long termE
BRDF	Bidirectional Reflectance Distribution Function
BSG	Blue Spectro Graph
C3S	Copernicus Climate Change Service
CCD	Charged Coupled Device
CDOM	Coloured Dissolved Organic Matter
CF	Cloud Fraction
CMEMS	Copernicus Marine Environment Monitoring Service
CTD	Conductivity Temperature Depth
CTP	Cloud Top Pressure

CV	Coefficient of Variation
ESA	European Space Agency
EUB	Estimated Uncertainty Budget
EUMETSAT	European Organisation for the Exploitation of Meteorological Satellites
EURYBIA	EUropean RadiometrY Buoy and InfrAstructure
FOV	Field of View
FRM	Fiducial Reference Measurements
FWHM	Full Width at Half Maximum
GS	Ground Segment
GUM	Guide to the expression of Uncertainty in Measurement
ICOS	Integrated Carbon Observation System
IOCCG	International Ocean Colour Coordinating Group
IOPs	Inherent Optical Properties
LUT	Look-up Tables
MarONet	Marine Optical Network
MC	Monte Carlo
MFRSR	Multi Filter Rotating Shadowband Radiometer
MOBY	Marine Optical BuoY
MODIS	Moderate Resolution Imaging Spectroradiometer
NASA	National Aeronautics and Space Administration
NIR	Near-Infrared
NMI	National Metrology Institute
NOAA	National Oceanic and Atmospheric Administration
NRT	Near Real-Time
OC-SVC	Ocean Colour System Vicarious Calibration
OLCI	Ocean and Land Colour Instrument
OMI	Ozone Monitoring Instrument
PACE	Plankton, Aerosol, Cloud, ocean Ecosystem
PDC	Post-Deployment Calibration
PM10	Particulate Matter smaller than 10μm
QC/QA	Quality Control / Quality Assurance
ROI	Region of Interest
RSEM	Relative Standard Error of the Mean
RSG	Red Spectro Graph
S3	Sentinel-3
SDY	Sequential Day of the Year
SI	International System of Units
SeaWiFS	Sea-viewing Wide Field-of-view Sensor
SRF	Spectral Response Function
SST	Sea Surface Temperature
TCS	Traceable to Calibration Standards
TOA	Top Of Atmosphere
TOC	Total Ozone Column
UTC	Coordinated Universal Time
VIIRS	Visible Infrared Imaging Radiometer Suite
VPH	Volume Phase Holographic

References

1. Ansper, A.; Alikas, K. Retrieval of Chlorophyll a from Sentinel-2 MSI data for the European Union water framework directive reporting purposes. *Remote Sens.* **2018**, *11*, 64. [CrossRef]
2. Groom, S.; Sathyendranath, S.; Ban, Y.; Bernard, S.; Brewin, R.; Brotas, V.; Brockmann, C.; Chauhan, P.; Choi, J.; Chuprin, A.; et al. Satellite ocean colour: Current status and future perspective. *Front. Mar. Sci.* **2019**, *6*, 485. [CrossRef]

3. Mazeran, C.; Brockmann, C.; Ruddick, K.; Voss, K.J.; Zagolsky, F.; Antoine, D.; Bialek, A.; Brando, V.; Donlon, C.; Franz, B.A.; et al. Requirements for Copernicus Ocean Colour Vicarious Calibration Infrastructure; EUMETSAT Study. 2017. Available online: https://www.eumetsat.int/website/wcm/idc/idcplg?IdcService=GET_FILE&dDocName=PDF_COP_OCEAN_COL_CAL&RevisionSelectionMethod=LatestReleased&Rendition=Web (accessed on 28 February 2020).

4. Zibordi, G.; Mélin, F.; Voss, K.J.; Johnson, B.C.; Franz, B.A.; Kwiatkowska, E.; Huot, J.-P.; Wang, M.; Antoine, D. System vicarious calibration for ocean color climate change applications: Requirements for in-situ data. *Remote Sens. Environ.* **2015**, 361–369. [CrossRef]

5. Lerebourg, C.; Vendt, R.; Donlon, C. Fiducial reference measurements for satellite ocean colour (FRM4SOC). In *Proceedings of the D-240 Proceedings of WKP-1 (PROC-1) Report of the International Workshop*; European Space Agency: ESRIN: Frascati, Italy, 2016.

6. Gordon, H.R. In-orbit calibration strategy for ocean color sensors. *Remote Sens. Environ.* **1998**, *63*, 265–278. [CrossRef]

7. IOCCG. *International Network for Sensor Inter-Comparison and Uncertainty Assessment for Ocean Color Radiometry (IN-SITU-OCR)*; International Ocean Color Coordinating Group: Dartmouth, NS, Canada, 2012.

8. Donlon, C. *Sentinel-3 Mission Requirements Traceability Document (MRTD)*; European Space Agency: Noordwijk, The Netherlands, 2011.

9. Clark, D.K.; Yarbrough, M.; Feinholz, M.; Flora, S.; Broenkow, W.; Johnson, C.B.; Brown, S.W.; Yuen, M.; Mueller, J. MOBY, a radiometric buoy for performance monitoring and vicarious calibration of satellite ocean color sensors: Measurement and data analysis protocols. In *Ocean Optics Protocols for Satellite Ocean Color Sensor Validation, Revision 3, Volume 2*; Mueller, J.L., Fargion, G.S., Eds.; NASA Goddard Space Flight Center: Greenbelt, MD, USA, 2002; Volume NASA/TM–2002-21004, pp. 138–170.

10. Antoine, D.; Guevel, P.; Desté, J.-F.; Bécu, G.; Louis, F.; Scott, A.J.; Bardey, P. The "boussole" buoy—A new transparent-to-swell taut mooring dedicated to marine optics: Design, tests, and performance at sea. *J. Atmos. Ocean. Technol.* **2008**, *25*, 968–989. [CrossRef]

11. Zibordi, G.; Mélin, F. An evaluation of marine regions relevant for ocean color system vicarious calibration. *Remote Sens. Environ.* **2017**, *190*, 122–136. [CrossRef]

12. Zibordi, G.; Mélin, F.; Talone, M.; European Commission; Joint Research Centre. *System Vicarious Calibration for Copernicus Ocean Colour Missions: Updated Requirements and Recommendations for a European Site*; Publications Office of the European Union: Brussels, Belgium, 2017; ISBN 978-92-79-75340-4.

13. Bulgarelli, B.; Zibordi, G. Adjacency radiance around a small island: Implications for system vicarious calibrations. *Appl. Opt.* **2020**, *59*, C63–C69. [CrossRef]

14. EUMETSAT. *Sentinel-3 OLCI Marine User Handbook, Version 1H, Ref. EUM/OPS-SEN3/MAN/17/907205*; Technical Report; EUMETSAT: Darmstadt, Germany, 2018.

15. Voss, K.J.; Yarbrough, M.A.; Johnson, B.C.; Feinholz, M.E.; Gleason, A.; Flora, S.J. *Present Status of the Marine Optical Buoy (MOBY) Refresh and MOBY-Net*; MOBY: Dubrovnik, Croatia, 2018.

16. Feinholz, M.; Johnson, C.B.; Voss, K.; Yarbrough, M.; Flora, S. Immersion coefficient for the Marine Optical Buoy (MOBY) radiance collectors. *J. Res. Natl. Inst. Stand. Technol.* **2017**, *122*, 1–9. [CrossRef]

17. Voss, K.J.; Johnson, C.B.; Yarbrough, M.A.; Gleason, A.; Flora, S.J.; Feinholz, M.E.; Peters, D.; Houlihan, T.; Mundell, S.; Yarbrough, S. An overview of the Marine Optical Buoy (MOBY): Past, present and future. In Proceedings of the D-240 FRM4SOC-PROC1 Proceedings of WKP-1 (PROC-1) Fiducial Reference Measurements for Satellite Ocean Colour (FRM4SOC), Tartu, Estonia, 8–13 May 2017.

18. Johnson, B.C.; Voss, K.J.; Yarbrough, M.A.; Flora, S.J.; Feinholz, M.E.; Peters, D.; Houlihan, T.; Mundell, S. MOBY radiometric calibration and associated uncertainties. In Proceedings of the D-240 FRM4SOC-PROC1 Proceedings of WKP-1 (PROC-1) Fiducial Reference Measurements for Satellite Ocean Colour (FRM4SOC), Tartu, Estonia, 8–13 May 2017.

19. Barden, S.C.; Arns, J.A.; Colburn, W.S. *Volume-Phase Holographic Gratings and Their Potential for Astronomical Applications*; D'Odorico, S., Ed.; SPIE: Kona, HI, USA, 1998; pp. 866–876. [CrossRef]

20. Flora, S.J.; Broenkow, W.W. Data Reduction Algorithms for the Marine Optical Buoy and Marine Optical System; Moss Landing Marine Laboratories Technical Publication 08-X Moss Landing, CA 95039. 2008, p. 16. Available online: http://data.moby.mlml.calstate.edu/timeseries/other_reports/data%20reduction%20algorithms%20for%20the%20marine%20optical%20buoy.pdf (accessed on 28 February 2020).

21. Vabson, V.; Kuusk, J.; Ansko, I.; Vendt, R.; Alikas, K.; Ruddick, K.; Ansper, A.; Bresciani, M.; Burmester, H.; Costa, M.; et al. Laboratory intercomparison of radiometers used for satellite validation in the 400–900 nm range. *Remote Sens.* **2019**, *11*, 1101. [CrossRef]

22. Zibordi, G.; Talone, M.; Jankowski, L. Response to temperature of a class of in-situ hyperspectral radiometers. *J. Atmos. Ocean. Technol.* **2017**, *34*, 1795–1805. [CrossRef]

23. Torrecilla, E.; Pons, S.; Vilaseca, M.; Piera, J.; Pujol, J. Stray-light correction of in-water array spectroradiometers. Effects on underwater optical measurements. In Proceedings of the OCEANS 2008, Quebec City, QC, Canada, 15–18 September 2008; pp. 1–5. [CrossRef]

24. Feinholz, M.E.; Flora, S.J.; Yarbrough, M.A.; Lykke, K.R.; Brown, S.W.; Johnson, B.C.; Clark, D.K. Stray light correction of the marine optical system. *J. Atmos. Ocean. Technol.* **2009**, *26*, 57–73. [CrossRef]

25. Talone, M.; Zibordi, G.; Ansko, I.; Banks, A.C.; Kuusk, J. Stray light effects in above-water remote-sensing reflectance from hyperspectral radiometers. *Appl. Opt.* **2016**, *55*, 3966. [CrossRef] [PubMed]

26. Seckmeyer, G.; Bernhard, G. *Cosine Error Correction of Spectral UV-Irradiances*; Stamnes, K.H., Ed.; SPIE: Tromso, Norway, 1993; pp. 140–151. [CrossRef]

27. Zibordi, G.; Bulgarelli, B. Effects of cosine error in irradiance measurements from field ocean color radiometers. *Appl Opt.* **2007**, *46*, 5529–5538. [CrossRef] [PubMed]

28. Zibordi, G.; Voss, K.J.; Johnson, C.B.; Mueller, J.L. Protocols for satellite ocean colour data validation. *In-Situ Opt. Radiom.* **2019**, *3*. [CrossRef]

29. D'Alimonte, D.; Zibordi, G.; Kajiyama, T. Effects of integration time on in-water radiometric profiles. *Opt. Express* **2018**, *26*, 5908. [CrossRef] [PubMed]

30. Kajiyama, T.; D'Alimonte, D.; Cunha, J.C. A high-performance computing framework for Monte Carlo ocean color simulations. *Concurr. Comput. Pract. Exp.* **2017**, *29*, e3860. [CrossRef]

31. D'Alimonte, D.; Kajiyama, T. Effects of light polarization and waves slope statistics on the reflectance factor of the sea surface. *Opt. Express* **2016**, *24*, 7922. [CrossRef]

32. Mueller, J.L. *Reduced Uncertainties in Measurements of Water-Leaving Radiance, and Other Optical Properties, Using Radiative Transfer Models and Empirical Data Analysis*; CHORS, San Diego State University Research Foundation: San Diego, CA, USA, 2007.

33. Voss, K.J.; Gordon, H.R.; Flora, S.; Johnson, B.C.; Yarbrough, M.; Feinholz, M.; Houlihan, T. A method to extrapolate the diffuse upwelling radiance attenuation coefficient to the surface as applied to the Marine Optical Buoy (MOBY). *J. Atmos. Ocean. Technol.* **2017**, *34*, 1423–1432. [CrossRef]

34. Austin, R.W. The remote sensing of spectral radiance from below the ocean surface. In *Optical Aspects of Oceanography*; Jerlov, N.G., Steemann-Nielsen, E., Eds.; Academic Press Inc.: London, UK, 1974; pp. 317–344.

35. Gordon, H.R.; Morel, A.Y. *Remote Assessment of Ocean Color for Interpretation of Satellite Visible Imagery: A Review*; Springer: New York, NY, USA, 2010; ISBN 978-1-118-66370-7.

36. Voss, K.J.; Flora, S. Spectral dependence of the seawater–air radiance transmission coefficient. *J. Atmos. Ocean. Technol.* **2017**. [CrossRef]

37. Gordon, H.R.; Wang, M. Influence of oceanic whitecaps on atmospheric correction of ocean-color sensors. *Appl. Opt.* **1994**, *33*, 7754. [CrossRef]

38. Morel, A.; Antoine, D.; Gentili, B. Bidirectional reflectance of oceanic waters: Accounting for raman emission and varying particle scattering phase function. *Appl. Opt.* **2002**, *41*, 6289. [CrossRef] [PubMed]

39. Werdell, P.J.; Behrenfeld, M.J.; Bontempi, P.S.; Boss, E.; Cairns, B.; Davis, G.T.; Franz, B.A.; Gliese, U.B.; Gorman, E.T.; Hasekamp, O.; et al. The plankton, aerosol, cloud, ocean ecosystem mission: Status, science, advances. *Bull. Am. Meteorol. Soc.* **2019**, *100*, 1775–1794. [CrossRef]

40. Grasso, M.; Pedley, H.M. The pelagian islands: A new geological interpretation from sedimentological and tectonic studies and its bearing on the evolution of the central Mediterranean Sea (Pelagian Block). *Geol. Romana* **1985**, *24*, 13–34.

41. Ciardini, V.; Contessa, G.M.; Falsaperla, R.; Gómez-Amo, J.L.; Meloni, D.; Monteleone, F.; Pace, G.; Piacentino, S.; Sferlazzo, D.; Sarra, A. di Global and Mediterranean climate change: A short summary. *Annali Dell'Istituto Superiore Di Sanità* **2016**. [CrossRef]

42. di Sarra, A.; Bommarito, C.; Anello, F.; Di Iorio, T.; Meloni, D.; Monteleone, F.; Pace, G.; Piacentino, S.; Sferlazzo, D. Assessing the quality of shortwave and longwave irradiance observations over the ocean: One year of high-time-resolution measurements at the lampedusa oceanographic observatory. *J. Atmos. Ocean. Technol.* **2019**, *36*, 2383–2400. [CrossRef]

43. Casasanta, G.; di Sarra, A.; Meloni, D.; Monteleone, F.; Pace, G.; Piacentino, S.; Sferlazzo, D. Large aerosol effects on ozone photolysis in the Mediterranean. *Atmos. Environ.* **2011**, *45*, 3937–3943. [CrossRef]

44. di Sarra, A.; Fua, D.; Cacciani, M.; Di Iorio, T.; Disterhoft, P.; Meloni, D.; Monteleone, F.; Piacentino, S.; Sferlazzo, D. Determination of ultraviolet cosine-corrected irradiances and aerosol optical thickness by combined measurements with a Brewer spectrophotometer and a multifilter rotating shadowband radiometer. *Appl. Opt.* **2008**, *47*, 6142. [CrossRef]

45. Meloni, D.; Di Biagio, C.; di Sarra, A.; Monteleone, F.; Pace, G.; Sferlazzo, D.M. Accounting for the solar radiation influence on downward longwave irradiance measurements by pyrgeometers. *J. Atmos. Ocean. Technol.* **2012**, *29*, 1629–1643. [CrossRef]

46. Harrison, A.W.; Coombes, C.A. Performance validation of the Gueymard sky radiance model. *Atmos Ocean* **1989**, *27*, 565–576. [CrossRef]

47. Coombes, C.A.; Harrison, A.W. Calibration of a three-component angular distribution model of sky radiance. *Atmos Ocean* **1988**, *26*, 183–192. [CrossRef]

48. Gregg, W.W.; Carder, K.L. A simple spectral solar irradiance model for cloudless maritime atmospheres. *Linmol. Ocean.* **1990**, *35*, 1657–1675. [CrossRef]

49. Volpe, G.; Dionisi, D.; Brando, V.; Colella, S.; Bracaglia, M.; Pitarch, J.; Falcini, F.; Sammartino, M.; Benincasa, M.; Santoleri, R. *IOPs Continuous Measurements for Ocean Monitoring and Calibration and Validation of Satellite Data*; EGU: Vienna, Austria, 2018; Volume 20, p. 16329.

50. Petzold, T.J. *Volume Scattering Functions for Selected Ocean Waters*; Scripps Institution of Oceanography: San Diego, CA, USA, 1972.

51. Maritorena, S.; Morel, A.; Gentili, B. Diffuse reflectance of oceanic shallow waters: Influence of water depth and bottom albedo. *Limnol. Ocean.* **1994**, *39*, 689–703. [CrossRef]

52. Simoncelli, S.; Fratianni, C.; Pinardi, N.; Grandi, A.; Drudi, M.; Oddo, P.; Dobricic, S.; Mediterranean Sea Physical Reanalysis (CMEMS MED-Physics) [Data Set]. Copernicus Monitoring Environment Marine Service (CMEMS). 2019. Available online: https://doi.org/10.25423/MEDSEA_REANALYSIS_PHYS_006_004 (accessed on 28 February 2020).

53. Jouini, M.; Béranger, K.; Arsouze, T.; Beuvier, J.; Thiria, S.; Crépon, M.; Taupier-Letage, I. The sicily channel surface circulation revisited using a neural clustering analysis of a high-resolution simulation: The sicily channel surface circulation. *J. Geophys. Res. Ocean.* **2016**, *121*, 4545–4567. [CrossRef]

54. Copernicus Climate Change Service (C3S). ERA5: Fifth Generation of ECMWF Atmospheric Reanalyses of the Global Climate. Copernicus Climate Change Service Climate Data Store (CDS), June 2019. 2017. Available online: https://cds.climate.copernicus.eu/cdsapp#!/home (accessed on 28 February 2020).

55. Pace, G.; Cremona, G.; di Sarra, A.; Meloni, D.; Monteleone, F.; Sferlazzo, D.; Zaninin, G. Continuous vertical profiles of temperature and humidity at Lampedusa island. In Proceedings of the 9th International Symposium on Tropospheric Profiling, L'Aquila, Italy, 3–7 September 2012.

56. Kotsias, G.; Lolis, C.J. A study on the total cloud cover variability over the Mediterranean region during the period 1979–2014 with the use of the ERA-Interim database. *Theor. Appl. Climatol.* **2018**, *134*, 325–336. [CrossRef]

57. Meloni, D.; di Sarra, A.; Biavati, G.; DeLuisi, J.J.; Monteleone, F.; Pace, G.; Piacentino, S.; Sferlazzo, D.M. Seasonal behavior of Saharan dust events at the Mediterranean island of Lampedusa in the period 1999–2005. *Atmos. Environ.* **2007**, *41*, 3041–3056. [CrossRef]

58. Long, C.N.; Ackerman, T.P. Identification of clear skies from broadband pyranometer measurements and calculation of downwelling shortwave cloud effects. *J. Geophys. Res. Atmos.* **2000**, *105*, 15609–15626. [CrossRef]

59. Trisolino, P.; di Sarra, A.; Anello, F.; Bommarito, C.; Di Iorio, T.; Meloni, D.; Monteleone, F.; Pace, G.; Piacentino, S.; Sferlazzo, D. A long-term time series of global and diffuse photosynthetically active radiation in the Mediterranean: Interannual variability and cloud effects. *Atmos. Chem. Phys.* **2018**, *18*, 7985–8000. [CrossRef]

60. Min, Q.; Wang, T.; Long, C.N.; Duan, M. Estimating fractional sky cover from spectral measurements. *J. Geophys. Res.* **2008**, *113*, D20208. [CrossRef]

61. Pace, G.; di Sarra, A.; Meloni, D.; Piacentino, S.; Chamard, P. Aerosol optical properties at Lampedusa (Central Mediterranean). 1. Influence of transport and identification of different aerosol types. *Atmos. Chem. Phys.* **2006**, *6*, 697–713. [CrossRef]

62. Meloni, D.; di Sarra, A.; Pace, G.; Monteleone, F. Aerosol optical properties at Lampedusa (Central Mediterranean). 2. Determination of single scattering albedo at two wavelengths for different aerosol types. *Atmos. Chem. Phys.* **2006**, *6*, 715–727. [CrossRef]

63. di Sarra, A.; Sferlazzo, D.; Meloni, D.; Anello, F.; Bommarito, C.; Corradini, S.; De Silvestri, L.; Di Iorio, T.; Monteleone, F.; Pace, G.; et al. Empirical correction of multifilter rotating shadowband radiometer (MFRSR) aerosol optical depths for the aerosol forward scattering and development of a long-term integrated MFRSR-Cimel dataset at Lampedusa. *Appl. Opt.* **2015**, *54*, 2725. [CrossRef] [PubMed]

64. Calzolai, G.; Nava, S.; Lucarelli, F.; Chiari, M.; Giannoni, M.; Becagli, S.; Traversi, R.; Marconi, M.; Frosini, D.; Severi, M.; et al. Characterization of PM $_{10}$ sources in the central Mediterranean. *Atmos. Chem. Phys.* **2015**, *15*, 13939–13955. [CrossRef]

65. Bulgarelli, B.; Zibordi, G. Analysis of Adjacency Effects for Copernicus Ocean Colour Missions. 2018. Available online: https://www.semanticscholar.org/paper/Analysis-of-adjacency-effects-for-Copernicus-Ocean-Barbara-Giuseppe/018cffaf395d3cf29f2254073bbc7b5f7b8c4c8e (accessed on 2 April 2020).

66. BIPM; IEC; IFCC; ILAC; ISO; IUPAC; IUPAP; OMIL. *Evaluation of Measurement Data—Guide to the Expression of Uncertainty in Measurement*; GUM 1995 with minor corrections; JCGM: Saint-Cloud, France, 2008.

67. BIPM; IEC; IFCC; ILAC; ISO; IUPAC; IUPAP; OMIL. *Evaluation of Measurement Data—Supplement 2 to the "Guide to the Expression of Uncertainty in Measurement"—Extension to Any Number of Output Quantities*; JCGM: Saint-Cloud, France, 2011.

68. Brown, S.W.; Flora, S.J.; Feinholz, M.E.; Yarbrough, M.A.; Houlihan, T.; Peters, D.; Kim, Y.S.; Mueller, J.L.; Johnson, C.B.; Clark, D.K. The marine optical buoy (MOBY) radiometric calibration and uncertainty budget for ocean colour satellite sensor vicarious calibration. In Proceedings of the SPIE on Optics & Photonics, Sensors, Systems, and Next-Generation Satellites XI, Florence, Italy, 17 October 2007.

69. Zibordi, G.; Melin, F.; Berthon, J.-F. A regional assessment of OLCI data products. *IEEE Geosci. Remote Sens. Lett.* **2018**, *15*, 1490–1494. [CrossRef]

70. Franz, B.A.; Bailey, S.W.; Werdell, P.J.; McClain, C.R. Sensor-independent approach to the vicarious calibration of satellite ocean color radiometry. *Appl. Opt.* **2007**, *46*, 5068. [CrossRef]

71. Antoine, D.; Morel, A. A multiple scattering algorithm for atmospheric correction of remotely sensed ocean colour (MERIS instrument): Principle and implementation for atmospheres carrying various aerosols including absorbing ones. *Int. J. Remote Sens.* **1999**, *20*, 1875–1916. [CrossRef]

72. Voss, K.J.; Johnson, B.C.; Yarbrough, M.A.; Gleason, A. *Moby-Net: An Ocean Colour Vicarious Calibration System*; Earth Science Technology Forum: Pasadena, CA, USA, 2015.

73. Zibordi, G.; Holden, B.; Melín, F.; D'Alimonte, D.; Berthon, J.F.; Slutsker, I.; Giles, D. AERONET-OC: An overview. *Can. J. Remote Sens.* **2010**, 488–497. [CrossRef]

74. Israelevich, P.; Ganor, E.; Alpert, P.; Kishcha, P.; Stupp, A. Predominant transport paths of Saharan dust over the Mediterranean Sea to Europe: Saharan dust transport to europe. *J. Geophys. Res. Atmos.* **2012**, *117*. [CrossRef]

 *remote sensing*

Article

MODIS Aqua Reflective Solar Band Calibration for NASA's R2018 Ocean Color Products

Shihyan Lee [1,2,*], **Gerhard Meister** [2] and **Bryan Franz** [2]

1   Science Applications International Corp., Reston, VA 20190, USA
2   NASA Goddard Space Flight Center, Ocean Ecology Laboratory, Greenbelt, MD 20771, USA;
    gerhard.meister-1@nasa.gov (G.M.); bryan.a.franz@nasa.gov (B.F.)
*   Correspondence: shihyan.lee@nasa.gov; Tel.: 1-301-869-9681

Received: 9 August 2019; Accepted: 12 September 2019; Published: 20 September 2019

Abstract: Remote-sensing ocean color products have stringent requirements on radiometric calibration stability. To address a calibration deficiency in Moderate Resolution Imaging Spectroradiometer (MODIS) Aqua in recent years, the NASA Ocean Biology Processing Group (OBPG) developed a new calibration for reflective solar bands. Prior to the reprocessing of NASA's ocean color products for 2018 (R2018), the OBPG MODIS products had been based on calibration provided by the MODIS Calibration Support Team (MCST). Several modifications were made to the MCST calibration approach to improve the calibration accuracy for ocean color products. These include (1) applying 936-nm detector normalization to solar diffuser stability monitor (SDSM) data to reduce coherent noise; (2) modeling solar diffuser (SD) degradation wavelength dependency to determine SD degradation in near-infrared and shortwave infrared wavelengths; (3) computing detector gains using SD screen-closed data to better match ocean radiance levels in all bands; (4) performing a simple atmospheric correction to reduce bidirectional reflectance distribution function (BRDF) effects in desert trends; (5) estimating and using modulated relative spectral response (RSR) impact on ocean data to adjust the calibration coefficients; (6) using smoothing to characterize the temporal change in calibration; and characterizing response versus scan angle (RVS) changes using 2nd-order polynomials to improve spatial/temporal calibration stability. Relative to the previous R2014 ocean color products, the R2018 calibration removed the suspect late-mission global trends in blue-band water-leaving reflectance and some anomalously large short-term variability (spikes) in the temporal trend of chlorophyll concentration. This paper will describe the OBPG calibration with a focus on the differences between the MCST and OBPG approaches.

Keywords: satellite; calibration; solar diffusor; SDSM; desert trend; lunar calibration; RVS; MODIS; Aqua; ocean color

---

1. Introduction

The Moderate Resolution Imaging Spectroradiometer (MODIS) on the Earth Observing System (EOS) platform has 36 spectral bands to provide near-global observations every 2 days [1]. Of the 36 spectral bands, 20 are reflective solar bands (RSB) with a spectral range of 0.41 to 2.1 μm. Within the RSB, bands 8–16 are optimized to observe ocean biological processes (Table 1) and the rest of the RSB are designed for land and atmosphere applications but can also be used in ocean science application with reduced accuracy.

The MODIS is a key instrument aboard the Terra and Aqua Satellites. The two EOS spacecrafts, Terra and Aqua, were launched on 18 December 1999 and 4 May 2002, respectively, into sun-synchronous polar orbits at an altitude of 705 km. Terra is on a descending node with an equator crossing time of 10:30 AM (local time) and Aqua is on an ascending node with a 1:30 PM equator crossing time.

Since they began operations, both instruments have continuously provided global earth observation data to study land, ocean, and atmospheric processes. However, due to the changing polarization sensitivities in MODIS Terra (MODIST) [2], the Terra ocean products have been relying on Ocean Biology Processing Group (OBPG) crosscalibration to correct the temporal calibration biases [3]. The crosscalibration uses MODIS Aqua (MODISA) global 7-day composites of water-leaving radiance as a truth field, combined with modeled atmospheric path radiance and surface contributions, to estimate the temporal change in gain and polarization as a function of time and scan angle.

Table 1. Moderate Resolution Imaging Spectroradiometer (MODIS) solar diffuser stability monitor (SDSM) detectors and reflective solar bands (RSB) center wavelengths.

SDSM	Wavelength	Land/ Atmosphere	Center Wavalength (nm)	Band Width (nm)	Ocean	Center Wanalength	Band Width (nm)
Detector 1	412	Band 1	645	50	Band 8	412	15
Detector 2	466	Band 2	859	35	Band 9	443	10
Detector 3	530	Band 3	469	20	Band 10	488	10
Detector 4	554	Band 4	555	20	Band 11	531	10
Detector 5	646	Band 5	1240	20	Band 12	551	10
Detector 6	747	Band 6	1640	24	Band 13	667	10
Detector 7	857	Band 7	2130	50	Band 14	678	10
Detector 8	904	Band 17	905	30	Band 15	748	10
Detector 9	936	Band 18	936	10	Band 16	869	10
		Band 19	940	50			
		Band 26	1375	30			

Because of the MODIST-MODISA crosscalibration, the Terra ocean products are not considered an independent science data product and the accuracy of Terra products is dependent on the quality of Aqua products. Therefore, the calibration accuracy of MODISA is crucial, as the entire suite of MODIS ocean color products rely on the quality of Aqua products. Historically, the Aqua ocean color products are produced using the MODIS Calibration Support Team (MCST) radiometric calibration and are fine-tuned by OBPG crosscalibration [4] and vicarious calibration [5]. The Aqua ocean color products have been maintained at science quality until recent years [6]. As it is getting increasingly challenging to calibrate the aging MODISA sensor, a decision was made by OBPG to develop an independent radiometric calibration specifically for the ocean color products.

The main challenge of calibrating remote-sensing ocean color products is its sensitivity to calibration uncertainties. In general, calibration errors are magnified by anywhere from 5 to 100 times in ocean color products [7]. To produce meaningful climate data records, a long-term radiometric calibration stability of about 0.1% is often needed [8]. To achieve this stringent requirement, there are several advantages for an independent, end-to-end calibration solution. First, because it is all "in house", the radiometric calibration process can be efficiently integrated into the data production life cycle (Figure 1). For each version of calibration look-up-table (LUT), corresponding vicarious coefficients and crosscalibration LUT [4,5] are regenerated and a global, life-of-mission test processing is performed to assess the impacts. The uncertainties in radiometric calibration are significantly amplified during ocean color product retrieval [7]. This is further complicated by the subsequent changes in vicarious and crosscalibration in response to given changes made in radiometric calibration. Therefore, it is essential to integrate the testing of a radiometric calibration LUT into the data production life cycle in order to accurately evaluate a radiometric calibration LUT on ocean color products.

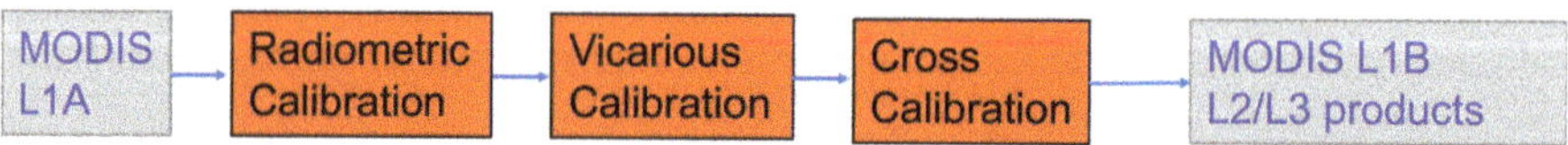

Figure 1. Ocean Biological Processing Group (OBPG) ocean color production process: L1A is the formatted raw instrument data. L1B is the calibrated radiance. L2/L3 are the level 2 and 3 ocean color products [8].

For each version of a calibration LUT, a global mission test reprocessing is performed that includes the first 4-day period of each month over the life of the mission. Each 4-day period is binned into a 9.2-km sinusoidal projection to form a 4-day mean global distribution of ocean color products for each month, which is then separated regionally (e.g., geographically and by water type) and trended with time. Temporal anomaly trends are also computed by subtracting the mean seasonal cycle from the global or regional time series. The ocean color products are reviewed by OBPG staff based on the known ocean biological processes to identify any product deficiency that could have originated from calibration issues. This process is vitally important to achieve the stringent radiometric calibration stability required to maintain ocean color products at science quality [9]. Maintaining radiometric calibration stability is especially difficult for Aqua MODIS as its radiometric response has changed significantly during more than 17 years of operation [10].

An ocean color specific calibration LUT can be optimized to the spectral properties of ocean scenes. For example, the solar calibration for RSB are performed at 2 different solar diffuser (SD) radiance levels (with and without SD screen attenuation). Normally, the land bands are calibrated at high SD radiance and ocean bands are calibrated using low SD radiances to maximize the signal-to-noise ratio (SNR). Since the spectral radiance of the ocean is closer to the low SD radiances, the OBPG solar calibration computes all detectors gains at the low SD radiance levels. Because the MODIS calibration algorithm uses a linear gain, calibrating detector gains at radiance closer to ocean radiance will reduce the impact of nonlinearity in detector responses on ocean color products. This approach, however, will increase nonlinearity impacts on high radiance scenes.

In addition, some radiometric characteristics can be better refined in subsequent OBPG crosscalibration and vicarious calibration and the radiometric calibration strategy can be simplified to improve calibration stability. For example, a previous study showed that the response versus scan angle (RVS) of band 8 (412nm) can be better characterized by a 4th-order fit [11]. However, our preliminary analysis showed that the fit residual reduction by the higher-order fit is insignificant. Compared to using the 2nd-order fit (same as in instrument prelaunch calibration), the 4th-order RVS fit is less stable temporally. Since the OBPG crosscalibration already used 4th-order polynomials to fine-tune the residual RVS uncertainties on ocean color products, using 2nd-order fit for RVS can improve temporal calibration stability while achieving similar RVS calibration accuracies on ocean color products.

The flow chart in Figure 2 summarizes the steps used to generate the OBPG RSB LUT. First, we used the yaw maneuver data to characterize the solar diffuser stability monitor (SDSM)/SD sunscreen transmission functions (tau) and Bidirectional Reflectance Function (BRF) functions and to store them in an internal LUT. In each subsequent SD calibration event, the SDSM calibration computes the relative SD reflectivity (H factor) using SD door open data. After the H factor is determined, the SD calibration computes the detector gain (m1) from the detector response and the SD radiance estimate from H factor and solar incident radiance. The computed m1 is then adjusted by modulated relative spectral response (mRSR) impact on ocean color products based on the typical ocean spectral radiance [12]. The mRSR impact is computed by optical degradation estimated by each detector's radiometric gain (m1) and electronic gain determined by electronic self-calibration (Ecal).

To estimate temporal change in RVS, desert and lunar observations are used to track instrument response changes at various scan angles. At the desert site, the time series of calibrated instrument reflectance, after correcting the atmospheric contributions, are computed as desert trends, which are used to track the RVS change at the 16 repeated observation angles. The calibrated lunar irradiance is computed from the monthly lunar observation. The calibrated lunar irradiances are compared to RObotic Lunar Observatory (ROLO) model predictions to produce lunar trend, which is used to track the RVS at the lunar observation angle.

The raw m1 and RVS values are estimated at the temporal interval of the data collection frequency, i.e., solar, lunar, and desert calibration. A temporal smoothing is applied to generate the time series of m1 and RVS. The smoothed m1 and RVS time series are used to populate the RSB LUT at the predefined time intervals.

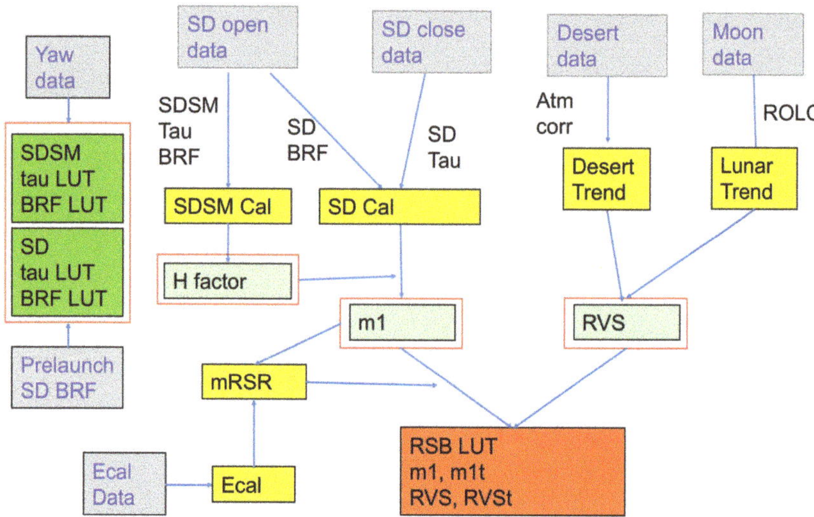

Figure 2. Aqua RSB calibration procedures: The calibration process produces an RSB look-up-table (LUT), which contains temporal gain (m1) and response versus scan angle (RVS) parameters at time steps of m1t and RVSt respectively. The grey boxes are the source data; the yellow boxes are the calibration processes; the intermediate calibration parameters are outlined by red lines; and the red box is the final product of RSB calibration LUT. Ecal: electronic self-calibration; Yaw: spacecraft yaw maneuver.

In this paper, we will describe the calibration methodology and procedures applied by the OBPG in producing the RSB calibration LUT for the R2018 ocean color reprocessing and forward stream processing. The paper will focus on describing the differences between OBPG MODISA calibration and the approach used in the MCST calibration. The principle mechanisms of MODIS RSB calibrations will not be described in detail, as they have been discussed extensively in past literatures during the 19+ years of MODIS operation.

2. MODIS L1B Calibration Algorithm

The MODIS L1B calibration algorithm uses a nominal gain (m1) and response versus scan (RVS) angle function to convert the detector response of each earth view pixel to top-of-atmosphere (TOA) reflectance (Equation (1)) [13].

$$Ref = dn^*(b,ms,d,p) * m1(b,ms,d,t) / rvs(b,ms,p,t) \tag{1}$$

where Ref is reflectance, b is band, ms is mirror side, d is detector, t is time, and p is pixel number. Notice the scan angle in RVS is determined by pixel number and is independent of the detector. Both m1 and RVS are time dependent as both parameters are changing over time. Operationally, the instantaneous m1 and RVS values are interpolated from the calibration LUT that stores m1 and RVS parameters at predefined time intervals.

3. Detector Gain

The MODIS RSB radiometric gain is primarily determined by SD and SDSM calibration assembly (Figure 3). Table 1 lists the center wavelengths of the SDSM detectors and Aqua RSB. During solar calibration events, the SDSM calibration estimates the relative SD reflectivity by the ratio of the measured SDSM detector's sun and SD view responses. The SD radiance during solar calibration can be computed by the solar incident angle on the SD and SDSM-estimated SD reflectivity. Finally, the SD calibration determines the detector gain as the ratio of instrument response and estimated SD radiance [11,13].

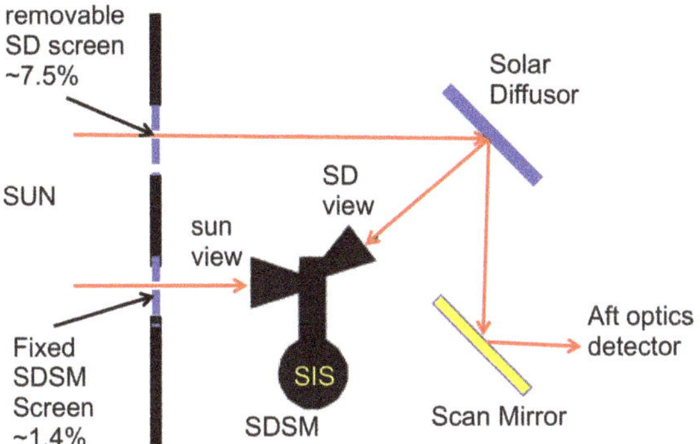

Figure 3. Schematic of solar diffuser (SD)/SDSM calibration assembly. The SD sunscreen has ~7.5% transmission and SDSM sunscreen has ~1.4% transmission.

4. Method

4.1. SDSM Calibration

The MODIS SDSM is designed to track the change of SD reflectivity during on-orbit operation. The knowledge of SD reflectivity is the key to accurately computing detector gain, as the reflectivity of the SD is known to degrade on-orbit due to UV exposure. The MODIS SD degradation is described by the H factor, which is the ratio of instantaneous SD reflectivity to the reflectivity of the pristine diffuser. The OBPG method of estimating SD degradation was described previously in Reference [14]. The notable differences between our method and the one used in the MCST calibration are (1) use of detector 9 normalization to remove the wavelength-dependent coherent noise; (2) use of a wavelength model to estimate SDSM detector 9 degradation; and (3) use of the wavelength model to interpolate and extrapolate SD degradation measured at SDSM detector wavelengths to RSB wavelengths. Figure 4 shows the SDSM H factors and model-estimated SD degradation for bands 5–7.

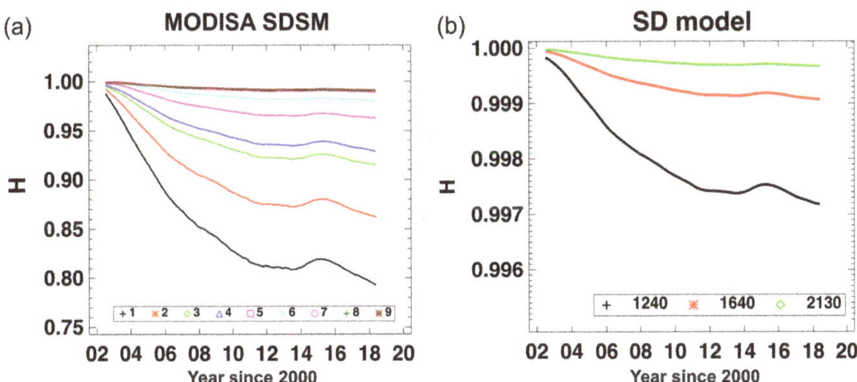

Figure 4. (**a**) SD reflectivity (H) for SDSM detectors 1 to 9 and (**b**) SD wavelength model-estimated SD degradation at MODIS Aqua (MODISA) bands 5 (1240 nm), 6 (1640 nm), and 7 (2130 nm).

4.1.1. Detector Gain: m1

The primary MODIS L1B product for RSB is the calibrated TOA reflectance factor [13]. Operationally, a LUT storing a time series of calibration coefficients is used to convert the instrument responses to TOA reflectance factors (see Equation (1)). The calibration coefficients m_1 for band (b), detector (d), and mirror side (ms) are computed as follows:

$$m_1(b, d, ms) = \frac{BRF_{SD}(b,,) * H(b) * cos\theta}{dn^*_{SD}(b, d, ms) * d^2_{ES}}$$

(2)

where b is band; ms is mirror side; d is detector; dn^*_{SD} is the background-subtracted, temperature-corrected digital count from the SD measurement; dES is the earth–sun distance; θ, γ are the solar zenith and azimuth angles on SD; H is the SD degradation; and BRF_{SD} is the BRF of the SD characterized by yaw maneuver data. For data collected with an SD screen, BRF_{SD} is the combined effects of SD screen transmission function (tau) and SD BRF [14].

The m1 coefficient is computed at each solar calibration event. During each solar calibration event, the earth–sun distance, instrument temperatures, and solar zenith and azimuth angles are computed from spacecraft telemetry. The SD BRF is computed using the solar zenith and azimuth angles, and the SD BRF LUT is derived from yaw maneuver data [14]. The RSB H factor is interpolated from the H factors at SDSM detector wavelengths based on the SD wavelength model [14]. dn*SD is the detector response at SD view subtracted by the space view detector response and then adjusted by temperature effects based on the measured instrument temperatures.

The SD calibration is performed both with and without SD attenuation screen to provide 2 calibration radiance levels. The SD attenuation screen has a transmissivity of ~7.5%, resulting in screen open and closed SD radiance ratios of ~13. The high and low SD radiances are designed to calibrate land and ocean bands, respectively, to maximize the SNR within the band's dynamic range. Figure 5 shows that the typical ocean radiance is between the high and low SD radiances for bands below 600 nm. Above 600 nm, the typical ocean radiance is lower than the screen-closed SD radiance. Since the MODIS L1B calibration assumes a linear function, it is desirable to have the calibrated radiance be similar to the scene radiance to reduce biases due to detector response nonlinearity.

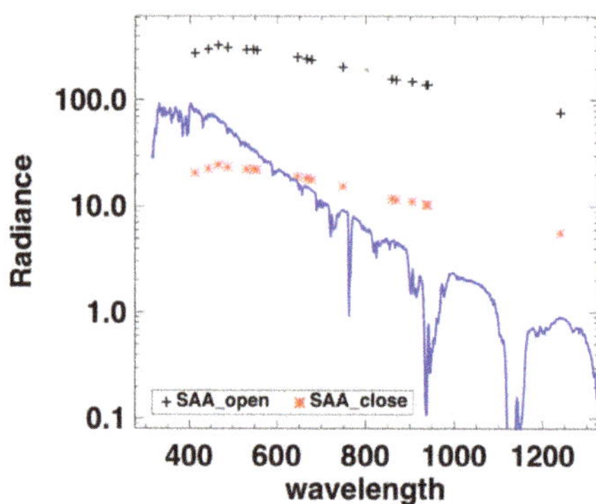

Figure 5. Comparison of typical ocean radiance (blue curve) and average Aqua SD radiance with (SAA_close) and without (SAA_open) SD attenuation screen assembly.

The ocean band detectors saturate at SD radiances without SD screen attenuation. Figure 6 shows MODISA band 9 (443 nm) dn*SD during solar calibration events. Band 9 detectors are saturated when the SD screen is open (Figure 6a). The screen-closed dn*SD shows a decreasing trend and an annual cycle due to seasonal variation in earth–sun distances. Using the screen-closed dn*SD, we compute m1, as shown in Figure 6b.

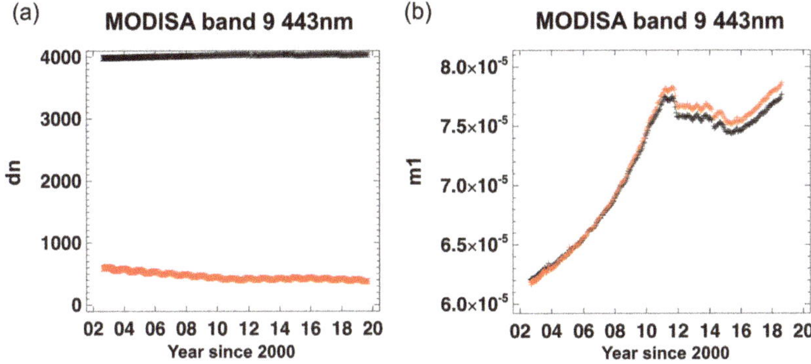

Figure 6. MODISA band 9 SD calibration: (**a**) detector 1 background-subtracted, temperature-corrected dn at SD view for screen open (black) and screen closed (red) and (**b**) computed m1 for detector 1 (black) and detector 10 (red).

For MODIS land bands, the m1 can be computed by either SD screen open or closed data, as both high and low SD radiances are within the detectors' dynamic range. Figure 7 shows that the band 2 detector gains computed using high (screen open) and low (screen closed) SD radiances have up to 1.5% differences. The ratios of m1 computed from high/low SD radiances also changed over time and are detector dependent. Since m1 is computed as the linear fit, the differences in high/low radiance m1 indicate the detector gain is not linear over the dynamic range. The temporal trend in high/low radiance m1 ratios indicates the detector response nonlinearity is changing over time. Figure 7b shows that the temporal change in detector response nonlinearity is detector dependent. Similar detector response nonlinearity behaviors are observed for most land band detectors, and the temporal change of detector response nonlinearities are most prominent for bands 1 and 2. The detector nonlinearity behavior is likely caused by the readout electronics as the detector gain of MODIS high-resolution bands (bands 1–4) have been shown to be subframe and radiance dependent [15]. To optimize calibration for ocean color products, we used low SD radiance data to compute m1 to better match the typical ocean surface radiance.

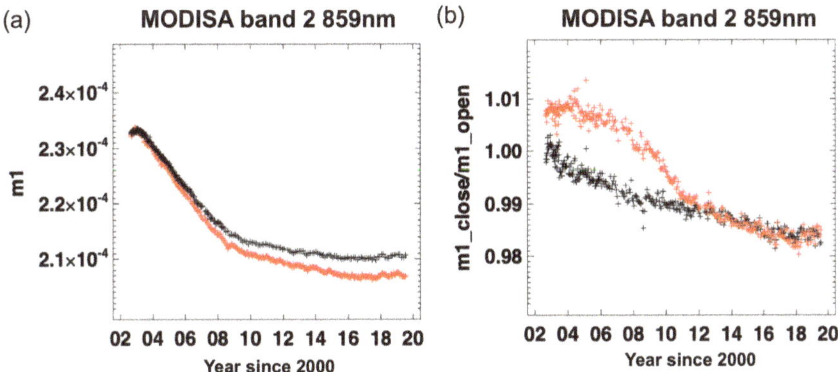

Figure 7. MODISA band 2 SD calibration: (**a**) detector 1 m1 computed from screen open (black) and closed (red) data and (**b**) ratios of m1 computed using SD screen open and closed data for detectors 1 (black) and 10 (red).

4.1.1.1. m1 Time Series

MODISA performs solar calibrations every 3 weeks after the first few years of operation [16]. The computed m1 (Figures 6 and 7) has noise of up to 0.5 %. As the change in detector gain is expected to be gradual, a boxcar smoothing function is used to compute the running average of m1. The smoothing interval is set at 3 months for green and blue bands and 1 year for red and Near-InfraRed (NIR) bands. The smoothing intervals are selected to maximize the temporal m1 smoothness while retaining its temporal trend. In Figure 8, MODISA band 8 shows sharp changes in detector gains between 2011 and 2012. The same features are shown in the other blue and green bands with reduced magnitudes. The shorter smoothing time interval for blue/green bands is used to retain the temporal features. For red/NIR bands, the m1 changes are gradual and do not have distinct features (Figure 8b), and the longer interval is used to generate smoother m1 time series.

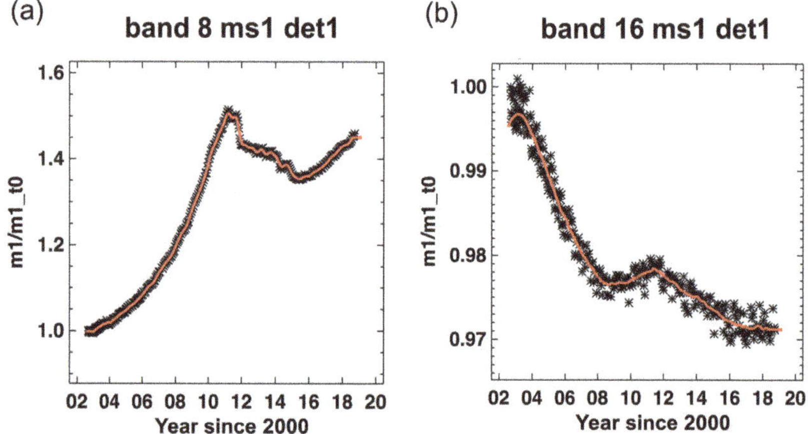

Figure 8. MODISA-normalized m1 computed from SD calibrations (symbols) and the smoothed m1 used for calibration (red curve) for mirror side 1 and detector 1 of (**a**) band 8 and (**b**) band 16.

4.2. RVS

The RVS is the ratio of the detector response at a particular scan angle to the reference scan angle. In MODIS, the RVS is normalized to the SD angle, where the nominal detector gain (m1) is determined. Because the MODIS scan mirror is not protected, the MODISA RVS has changed significantly over time due to nonuniform mirror degradation [17]. We track the temporal RVS change by trending the instrument responses over pseudo-invariant calibration targets, i.e., SD, moon, and desert, observed at different scan angles [17,18]. For MODISA, the SD and moon are always observed at fixed scan angles. The desert can be observed at 16 angles from the repeatable orbital tracks. The SD is observed about every 3 weeks during solar calibration events. The moon is observed monthly except for the summer months when the moon is too close to the horizon. The desert site can be observed daily at one of the 16-day repeat scan angles, if the scene is free of clouds.

By comparing the relative trend difference at different scan angles, we can estimate the temporal change in RVS. The temporal trends at each angle and calibration target are normalized to the first measurement, and the RVS at each scan angle is compared relatively. Therefore, this method determines the change in RVS relative to the first measurement, where the Aqua RVS can be approximated as prelaunch measured values.

One of the limiting factors of this method is that the target brightness needs to be within the dynamic range of the detectors. Among the calibration targets, desert sites are too bright for bands 10 to 16 and certain parts of the moon are also too bright for bands 13 to 16, but the saturated moon pixel radiance can be estimated by the band ratioing method [19]. Therefore, for bands 1–9, 17–19, and 26, desert, moon, and SD are used to track RVS and only moon and SD are used to track RVS for bands 10–16 (Table 2).

Table 2. Calibration targets used to track MODISA temporal RVS change and fitting function used for RVS characterization. The desert site is not used for bands 10–16 due to saturation.

Bands	Calibration Targets	RVS Characterization
Bands 1–9 Bands 17–19 Bans 26	Moon, SD, desert	2nd order polynomial
Bands 10–16	Moon, SD	2nd order polynomial

Ideally, all calibration targets should have the same brightness to reduce the detector response nonlinearity impact. However, the brightness of SD, moon, and desert are significantly different, which adds additional uncertainty to RVS estimates using a combination of calibration targets.

The time-dependent RVS is computed by fitting 2nd-order polynomials over the temporal relative response vs. scan angles from all available calibration targets (Table 2). As described earlier, 2nd-order polynomials are more stable as they are less affected by noise in lunar and desert trends. The residual higher-order RVS behaviors are removed by the OBPG crosscalibration. This process optimizes the overall RVS correction for ocean color products when used in conjunction with OBPG crosscalibration. However, the temporal RVS by itself might not be the best possible RVS characterization for other disciplines without additional product specific corrections.

Since the RVS is normalized at the SD angle, the lunar and desert trends are computed using the time-dependent m1 (see Section 4.1), so the moon and desert trends are essentially the RVS trends at their respective observation angles. Using the time-dependent detector m1 for lunar and desert trends remove the potential uncertainties from temporal change in mirror side and detector-to-detector gain ratios. Using time-dependent detector m1 is especially important for lunar trends, as the moon is not a uniform target and cannot be evenly sampled among detectors. Because the moon image is an ellipsoid, more lunar pixels are sampled by middle detectors than the edge detectors. If the temporal gain ratios between middle and edge detectors changes, this will cause biases in the lunar trend.

4.2.1. Desert Trend

We used the Libya 4 site to track temporal RVS change. Libya 4 is one of the widely used pseudo-invariant sites due to its long-term surface reflectance stability [20]. Despite being a stable target, the satellite observations showed significant bidirectional reflectance distribution function (BRDF) effects with respect to solar incident and instrument view angles. A previous study characterized such effects by simultaneously fitting the solar incident, instrument view angles, and RVS changes [17]. Although this method can reduce the BRDF effects in the observed data, the method is empirical and requires a good assumption of the basic shape of RVS trends.

Since satellite observations are made with different geometrics between the sun and the instrument, the changing atmospheric path radiance from different solar incident and instrument view angles can cause the BRDF effect to be seen in the desert observation. To remove atmospheric contributions, an atmospheric model, 6SV [21], is used to estimate Rayleigh scattering and gaseous absorption. Figure 9 shows the band 8 mirror side 1 RVS characterization process at a −42-degree scan angle. The instrument TOA reflectance (Figure 9a) shows a large annual cycle. Figure 9b shows that the 6SV model estimated atmospheric contribution based on the solar and instrument view geometry. Figure 9b shows that the annual cycle of atmospheric contribution matches well with the annual variations of the TOA reflectance (Figure 9a). Removing the atmospheric contribution from TOA reflectance (Figure 9c) significantly reduces the annual cycles in the desert trending. In Figure 9c, the solar incident power and atmospherically corrected desert trend has a residual noise of a few percent, a significant reduction from the observed data. The residual noise in the desert trend could come from variations in aerosol, polarization effects, and possibly small short-term BRDF changes in the Libya 4 site itself.

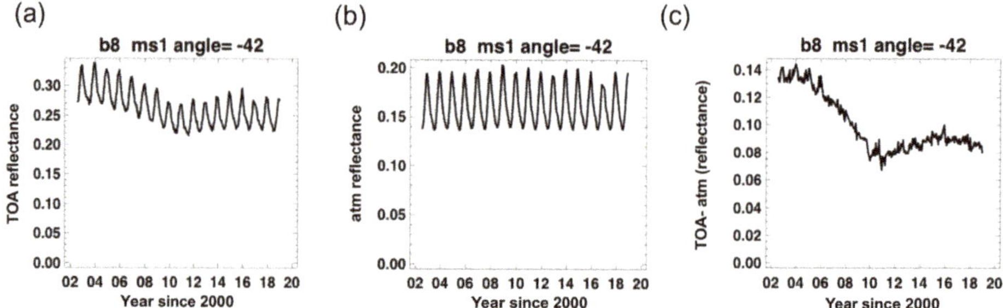

Figure 9. Desert trend for Aqua band 8, mirror side 1 at a scan angle of −42 degrees: (**a**) top-of-atmosphere (TOA) reflectance; (**b**) atmospheric contribution in reflectance; and (**c**) TOA reflectance—atmospheric reflectance. The TOA reflectance is computed using Equation (1) with RVS set to 1.

To further demonstrate the importance of atmospheric correction, we compared the desert site RVS before and after atmospheric correction in January 2003 when the RVS should have minimal deviation from the prelaunch measured values. In Figure 10a, the sensor-observed TOA reflectance showed a 60% variation in response at different scan angles. After correcting for the atmospheric contribution, the RVS variation (Figure 10b) is significantly reduced. Since, the prelaunch measured RVS for band 8 is only about 2%, a 60% variation in RVS is unlikely after only 6 months of on-orbit operation. The results indicate the atmospheric correction removed the bulk of the instrument observation BRDF artifacts to improve RVS characterization accuracy.

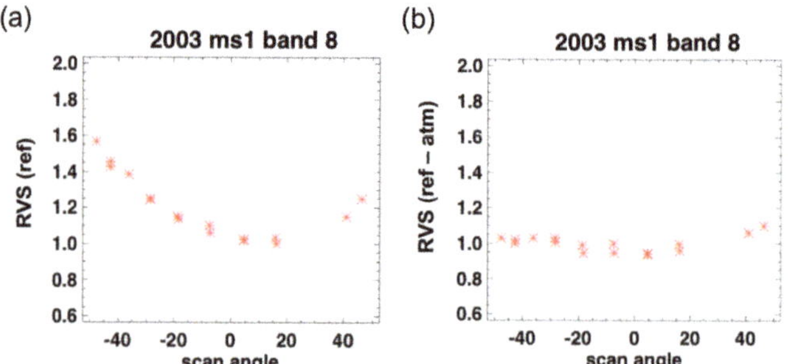

Figure 10. Libya 4 desert observations for MODISA band 8, mirror side 1 during 2003 January: The plots show the normalized reflectance (ref) versus scan angle (RVS) before (**a**) and after (**b**) atmospheric correction. Notice in some angles, 2 clear sky observations were made from the 16-day repeat orbits in January 2003.

To reduce the residual noise in the atmospherically corrected desert trends, a 1-year running average was computed and used to produce smoothed temporal trends at each observation angle. Figure 11 shows examples of the computed desert trends and the smoothed RVS trends at −42 degrees for bands 8 and 9. The smoothed RVS desert trends will later be combined with lunar trends to characterize temporal RVS change.

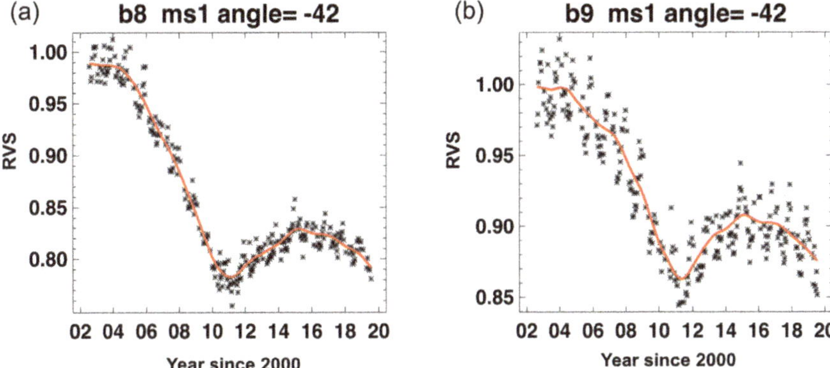

Figure 11. RVS computed from desert trends of MODISA mirror side 1 for (**a**) band 8 and (**b**) band 9 at a −42-degree scan angle: The black symbols are the computed RVS at each desert observation, and the red curve is the temporal smoothed RVS.

4.2.2. Lunar Trend

The moon is an ideal calibration target, as its surface is photometrically stable [22] and it can be observed by MODIS without atmospheric contamination. The geometrically corrected lunar irradiance can be obtained from the US Geological Survey's RObotic Lunar Observatory (ROLO) photometric model [23], using the observation time and the satellite position to determine the lunar phase and libration angles and sun–moon/satellite–moon distances. The lunar trend is computed as the ratio of the sensor-observed and model-predicted lunar irradiances. The temporal changes in the lunar trends provide the temporal RVS change at lunar observation angle [24].

The lunar trends (Figure 12) are normalized to the first observation to remove the residual biases between SD-calibrated and ROLO-predicted lunar irradiance. The normalized lunar trends are the temporal RVS changes at an Space View (SV) angle, which can be up to 20% in band 8 (412 nm). The temporal RVS changes are more prominent in the blue bands (8, 9, 3, and 10) and decrease at longer wavelengths. The temporal RVS trends resemble the m1 trends (Figure 8), where the trends in the blue bands reversed around 2011. For green to NIR bands, the temporal RVS changes are small with no clear reversing trends like the blue bands.

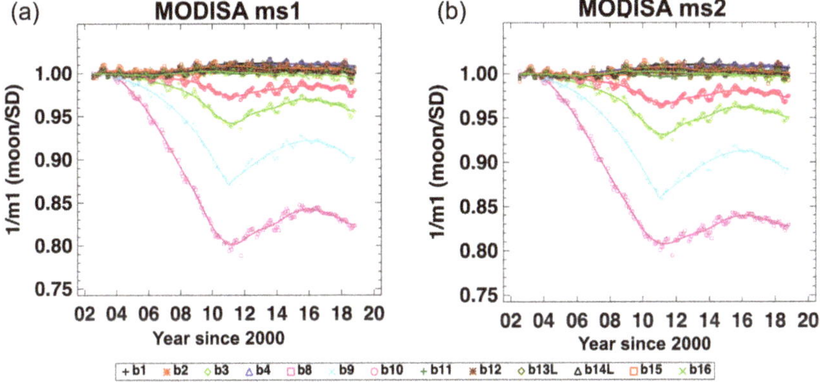

Figure 12. MODISA lunar trends for selected RSBs for (**a**) mirror side 1 and (**b**) mirror side 2: The symbols indicate each lunar calibration data point, and the curves are the fitted lunar trends.

The lunar trends show a quasi-annual cycle, which is likely due to the residual uncertainty in the ROLO model's libration correction [25]. To compute the RVS change at an SV angle, a 1-yr running average was computed to produce smoothed lunar trends, except for bands 8 and 9. A 2-piece fit was used for bands 8 and 9 before and after 2011 to better characterize the reversal in trends.

4.2.3. RVS Characterization

To characterize the temporal RVS, the available desert trends are combined with lunar trends. Since the lunar and desert trends are computed with time-dependent SD m1, the lunar/desert trends are the RVS normalized to SD angles. For bands 1–9, 17–19, and 26, the temporal RVS is estimated as a 2nd-order polynomial fit between the scan angle and the smoothed RVS from lunar and desert trends (Table 2). For bands 10–16, the RVS change is small (<4%) and only lunar trends were used to characterize the temporal RVS change. Since the moon is always observed at the same scan angle, we used the prelaunch estimated RVS curve as the basic shape to characterize the 2nd-order RVS fit from the lunar observation angle alone. This method assumes that small changes in RVS do not significantly change the shape of the RVS.

Figure 13 shows the characterized RVS for selected ocean bands and years. As mentioned earlier, band 8 has the largest temporal RVS change. In 2011, the RVS variation was the largest with the mirror response at the beginning of the scan (−55 degrees) ~30% higher than at the end of the scan (55 degrees). For the green to NIR bands, the RVS and their changes are considerably smaller.

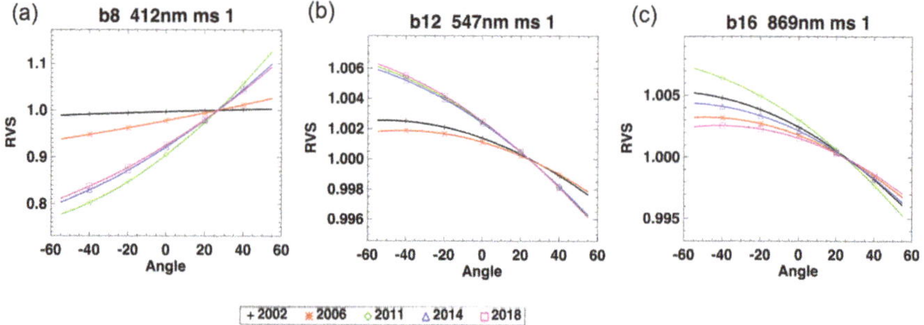

Figure 13. MODISA RVS fitting at selected years for (**a**) band 8 (412 nm), (**b**) band 12 (547 nm), and (**c**) band 16 (869 nm), mirror side 1.

4.3. Relative Spectral Response (RSR) Modulation

The temporal radiometric calibration for Aqua MODIS can be computed from the time-dependent m1 and RVS estimated in Sections 4.1 and 4.2. The calibration coefficients convert the instrument response to TOA reflectance or radiance. However, the Aqua detector RSR has been shown to change over time due to the spectrally dependent optical degradation [12]. The temporal RSR change will cause bias in observed ocean radiances. Applying time-dependent RSR in ocean color product retrieval is not feasible due to complexity and computational time limitations. The current solution is to apply an adjustment in calibration coefficients to correct the estimate of the calibration bias in ocean color products due to RSR modulation.

The derivation and correction of the time-dependent modulated RSR impact on Aqua ocean color products is detailed in our previous study [12]. The calibration bias due to RSR modulation is caused by the spectral differences between calibration targets and ocean scenes (Figure 14a). The RSR modulation impact on ocean color products is the combined effects of biases in observed SD, desert, and ocean scene radiances due to RSR modulation. Figure 14b shows the biases in ocean color product radiances due to modulated RSR computed using the typical spectral radiances in Figure 14a. Figure 14b shows that band 8 is the only band significantly impacted by RSR modulation with up to

1% biases in 2011. The large band 8 RSR modulation is the result of its large out-of-band response convolved with the large spectral radiance difference between calibration targets and the typical ocean scene. For the remaining ocean bands, the impacts from modulated RSR are less than 0.1%.

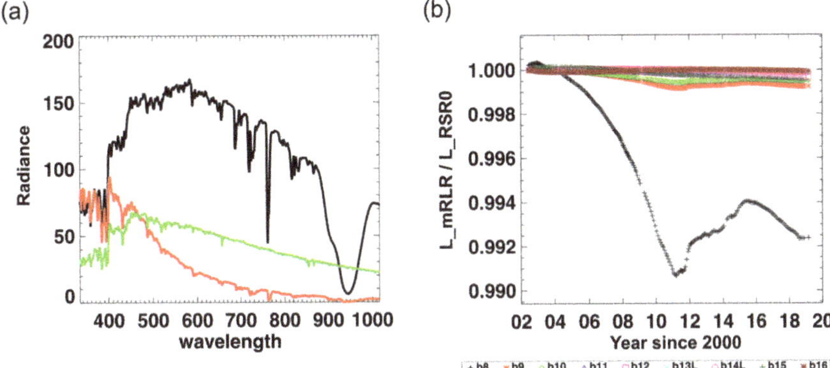

Figure 14. (**a**) Typical radiance for ocean (red), SD (green), and the desert (black) and (**b**) mean calibration bias (L_mRSR/L_RSR0) due to modulated RSR for bands 8 to 16: L_mRSR is the typical ocean radiance computed using modulated RSR (mRSR), and L_RSR0 is the typical ocean radiance computed using prelaunch measured RSR (RSR0).

The biases of modulated RSR are estimated per band for all ocean bands (bands 8 to 16), where RSRs are measured for both in-band and out-of-band regions. Although the modulated RSR impact is dependent on scan angle due to the angular-dependent optical degradation [12], we apply the scan-angle mean impact for simplicity. The correction is meant to correct the bias in global radiometric trends. The angular dependency of modulated RSR impact is small and will be added to residual RVS characterization uncertainty and corrected by OBPG crosscalibration.

5. Results

5.1. OBPG RSB Calibration LUT

The main differences between OBPG and MCST calibration are summarized in Table 3. Additional differences could come from computation methodology and data selection. Figure 15 compares the temporal trends of band 8 m1 and RVS between OBPG and MCST LUT at selected scan angles. Figure 15 shows that the OBPG and MCST gain (m1/RVS) at an SD angle is within 1.5% different but can be up to 4% different at large scan angles. The RVS differences at an SD angle are the modulated RSR effects as described in Section 4.3. The scan-angle-dependent gain differences between OBPG and MCST LUT will change the global trends in ocean color products.

The OBPG RSB calibration LUTs are based on the MCST LUT with the time-dependent m1 and RVS coefficients replaced by the values derived in Sections 4.1 and 4.2. The correction of modulated RSR impact is applied to RVS to keep the m1 coefficients as a pure product of solar calibration for better calibration traceability. Applying the modulated RSR correction to either m1 or RVS produces the same results.

Since the OBPG LUT retains the MCST LUT format, it can be used in MODIS L1B processing outside of OBPG. The OBPG produces a monthly updated LUT, based on the monthly updated MCST LUT to extend m1, RVS, and modulated RSR correction. The crosscalibration LUT (see following section) is also updated monthly after each new RSB LUT is generated to keep the correction up-to-date.

Table 3. Calibration methodology differences between R2014 and R2018.

	R2014 *	R2018
SDSM calibration	Detector 9 normalization Sd/SDSM screen for detector 9 degradation	Detector 9 normalization refined by SD/SDSM screen SDSM wavelength model for interpolation and detector 9 degradation
SD calibration	Use highest SD unsaturated signal to maximize SNR	Use low SD signal for all bands to better match ocean radiance
Temporal RVS characterization	2nd to 4th order fit Remove BRDF effect by fitting observed data vs. view and illumination geometric	2nd order fit for all bands Use atmospheric correction to remove BRDF effect Apply modulated RSR impact

* Based on MCST publications [11,17].

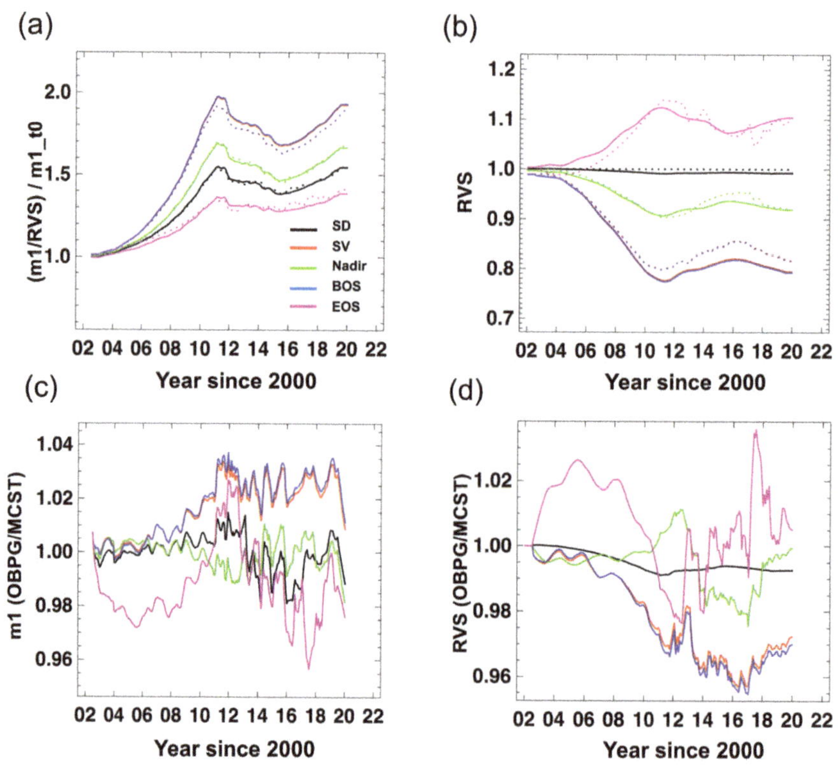

Figure 15. Time series of OBPG and MCST mirror side 1, detector-averaged normalized band 8 for (**a**) m1 and (**b**) RVS and (**c**,**d**) their differences at SD, space view (SV), Nadir, beginning of scan (BOS; −55 degrees), and end of scan (EOS; 55 degrees). Solid line: OBPG, dashed line: MCST. Colors indicate scan angles. The scan angle-specific detector gain is computed as m1/RVS, and m1_t0 is the SD m1 at the beginning of the mission. Note: the legend in Figure 15a is applied to Figure 15b–d.

5.2. Artifact Mitigation

Based on the OBPG radiometric calibration LUT, vicarious and crosscalibrations were performed to correct absolute instrument calibration biases and to flat-field the residual RVS, mirror side, and detector-to-detector uncertainties [4,5]. The vicarious and crosscalibration were performed using Rrs products; therefore, any residual atmospheric correction errors will also be included. Without vicarious and crosscalibration, a perfectly calibrated sensor could still produce biased and/or stripped Rrs images due to residual error from atmospheric correction.

The vicarious calibration produces a constant factor for each band, which has no impact on temporal trends [5]. However, the residual RVS correction can potentially impact global temporal trends if the correction is large [4]. Figure 16 compares the crosscalibration-computed residual RVS correction (M11) between R2014 and R2018 products. The magnitude of R2018 M11 correction is smaller than the R2014 correction after 2011 and similar to before 2011. The R2014 M11 correction varies significantly after 2011, indicating RVS calibration issues. The large M11 correction will change the global temporal trends, as the mean earth scene correction is not 1. This is the case in R2014 products as the scene-averaged M11 correction changes significantly after 2011 (Figure 17).

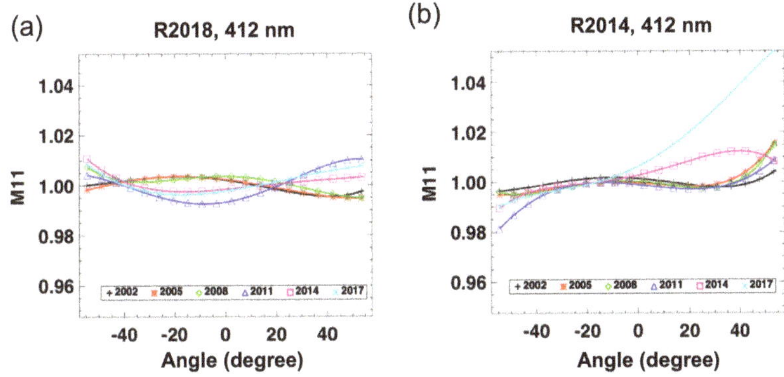

Figure 16. Band 8 (412 nm), mirror side 1, detector 1, crosscalibration correction factors (M11) for select years: (**a**) R2018 calibration and (**b**) R2014 calibration. M11 is the RVS correction (unitless).

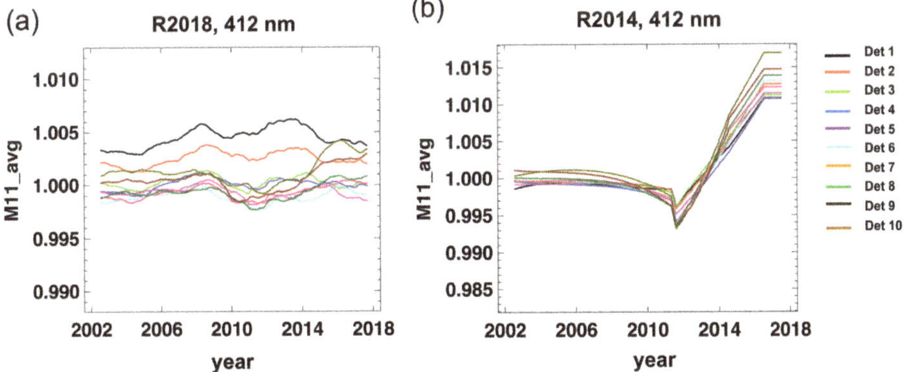

Figure 17. Time series of band 8, mirror side 1, scan-averaged M11 correction for (**a**) R2018 and (**b**) R2014 products: The scan-angle-averaged M11_avg indicates the correction on the global mean. Colors indicate detectors 1–10.

In R2018 products, the M11 is normalized to ensure no temporal trends are introduced by the crosscalibration. Compared to R2014, the R2018 M11 has larger detector spread. This is because the MCST calibration incorporates detector-to-detector gain correction using ocean sites [11] for the calculation of their m1, whereas in R2018, the striping correction is applied only during the crosscalibration. There is no impact of the M11 detector spread on image quality in R2014 vs. R2018 after applying the OBPG crosscalibration.

6. Discussion

To demonstrate the impact of the calibration approach described in this paper, the ocean color products from the R2018 reprocessing are compared with the previous R2014 reprocessing. The only changes between these two reprocessing events were the updated instrument calibration approach and the updated vicarious calibration [26]. Figure 18 shows a comparison of the two reprocessing versions for the time series of spectral water-leaving reflectances (Rrs) and derived chlorophyll concentration, as derived from monthly averages for the globally distributed deepest ocean waters (>1000-m depth). We restrict our analysis to this deep-water subset as it allows for an assessment over the vast majority of the ocean surface while avoiding complex coastal regions with highly variable terrestrial influences. In Figure 18a, the R2014 water-leaving reflectances (solid lines) for the shortest wavelengths (412 nm and 443 nm) show a strong downward trend after 2012, which is not associated with any known geophysical events. This late-mission drift in the blue was significantly reduced in the R2018 products. Similarly, the large seasonal variability in the R2014 chlorophyll product, especially after 2012, is much reduced in the R2018 products (Figure 18b). Such variability will arise if the calibration in the blue is erroneously low, as the observed radiance has a strong seasonal cycle due to variations in solar geometry and associated atmospheric path radiance, and the correction for that variable atmospheric contribution to the TOA radiance will result in an underestimation of water-leaving reflectance in the blue. The chlorophyll concentration is derived from the ratio of blue to green water-leaving reflectances, such that a lower blue–green ratio implies a higher chlorophyll concentration, thus giving rise to seasonal peaks in the R2014 chlorophyll products. There is also an overall bias (increase) in the blue wavelengths between the R2014 and R2018 results, which is caused by an update to the vicarious calibration [26] and which also contributes to reduced seasonal variability in the R2018 chlorophyll product over the full mission lifetime.

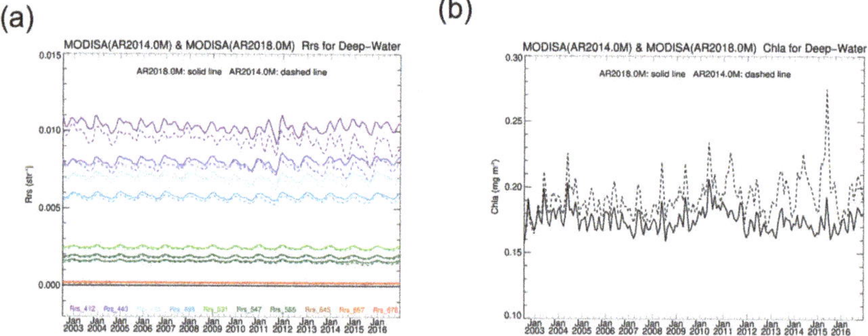

Figure 18. MODIS global deep-water temporal trends for (**a**) water-leaving reflectance (Rrs) and (**b**) chlorophyll. Solid line: R2018, dashed line: R2014 products.

Another very useful subset of the oceans for assessment of product temporal stability is the mean over the clearest ocean gyres, far from terrestrial influences, where chlorophyll concentrations are very low (typically <0.1 mg m-3) and homogenous, and weak variations in phytoplankton abundance or physiology, which modulate the absorption of blue light much more than green light, are the only significant influences on the ocean water optical properties. Temporal anomalies of water-leaving reflectance for the green ocean channel (547 nm) and green land channel (555 nm) were computed by subtracting the mean seasonal cycle from the water-leaving reflectance time series as derived over oligotrophic waters. In the absence of any major geophysical events, we can expect this time series to be relatively stable over time. The temporal anomalies for R2014 (Figure 19a,c) show a long-term trend and shorter-term variabilities that are larger than in the R2018 results (Figure 19b,d). Also notable is that the R2014 results for the two green bands with very similar center wavelengths (Figure 19a,c) are less consistent than the R2018 products (Figure 19 b,d). These results indicate that the R2018 products are more stable and consistent relative to the R2014 products.

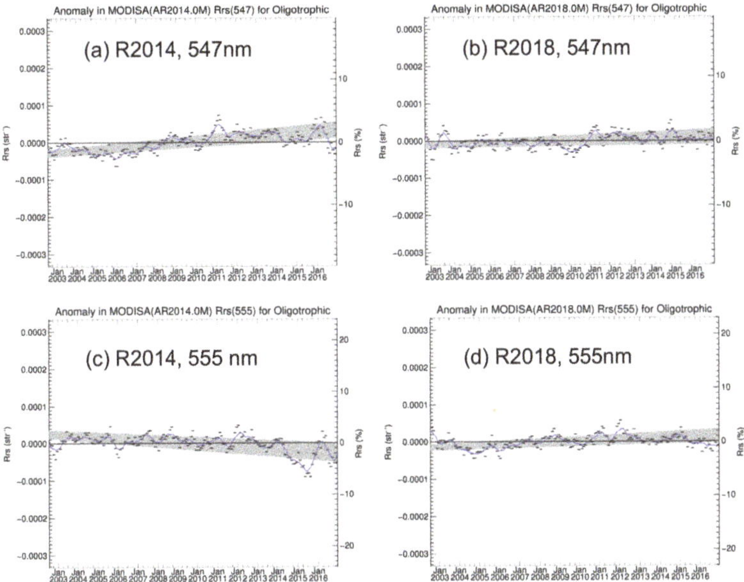

Figure 19. MODIS water-leaving reflectance (Rrs) temporal anomaly in global oligotrophic water: (**a,c**) R2014 products for 547 nm and 555 nm and (**b,d**) R2018 products for 547 nm and 555 nm.

7. Conclusions

In this paper, we described the independently developed MODIS Aqua radiometric calibration used in OBPG R2018 ocean color products. The OBPG calibration is developed based on the general MODIS calibration principles with a focus on ocean color products. The in-house developed calibration process is integrated in the ocean-color-product production cycle which enables fast evaluation. Timely product evaluation facilitates accurate assessment of the combined changes in radiometric calibration and the subsequent OBPG vicarious and crosscalibration. The integrated calibration development process is key to achieving the high radiometric calibration accuracy required for ocean color science.

Because the calibration is developed specifically for OBPG ocean color products, it is optimized for OBPG ocean color processing. For example, the nominal detector gains are calibrated with low SD radiance to better approximate ocean surface radiance. Applying the OBPG calibration for high-radiance targets could result in biases due to detector response nonlinearity. Furthermore, the RVS is characterized by simple 2nd-order fits to improve temporal stability. Applying the OBPG crosscalibration is needed to obtain the optimal RVS characterization.

Using the OBPG-developed radiometric calibration, the R2018 ocean color products show improved quality over the previous products (R2014). The global trends of the R2018 products are more stable and do not have late-mission anomalous behaviors shown in the R2014 products. The R2018 products are more consistent across wavelengths than R2014 products. Compared to the R2014 coefficients, the OBPG crosscalibration in R2018 is more consistent over time, indicating a more stable RVS characterization in R2018. The improvement in R2018 ocean-color-product quality and crosscalibration stability are the results of improvements in R2018 radiometric calibration.

Author Contributions: S.L. development the methodology, performed the analysis and wrote the manuscript. G.M. provided the advice, oversaw the progress and contributed to the writing and revising of the manuscript. B.F. contributed to the ocean color product validation processes and the writing and editing of the manuscript.

Remote Sens. **2019**, *11*, 2187

Funding: This research was funded by "Advancing the Quality and Continuity of Marine Remote Sensing Reflectance and Derived Ocean Color Products from MODIS to VIIRS" (17-TASNPP17-0065), submitted to the Science Mission Directorate's Earth Science Division, in response to NASA Research Announcement (NRA) NNH17ZDA001N-TASNPP: Research Opportunities in Space and Earth Science (ROSES-2017).

Acknowledgments: The authors would like to thank MCST for providing in-depth knowledge of MODIS instrument characteristics and past and current calibration issues and developments.

Conflicts of Interest: The authors declare no conflict of interest.

Glossary

Acronyms	Definition
6SV	Atmospheric correction model. https://6s.ltdri.org/
BOS	Beginning of the Scan
BRDF	Bidirectional Reflectance Distribution Function
BRF	Bidirectional Reflectance Function
Ecal	Electronic self Calibration
EOS	End of the Scan
H factor	Relative SD reflectivity
L1A	Formatted raw instrument data
L1B	Calibrated radiance computed from L1A
L2	Derived per-pixel ocean color products L1B
L3	Global binned ocean color products from L2
LUT	Look-Up-Table
M1	Detector radiometric gain coefficient
M1t	Time stamp of m1
M11	OBPG cross-calibration RVS correction coefficient
MCST	MODIS Calibration Support Team
MODIS	Moderate Resolution Imaging Spectroradiometer
MODISA	MODIS Aqua
MODIST	MODIS Terra
mRSR	Moduated RSR
NIR	Near-InfraRed
OBPG	Ocean Biological Processing Group
R2014	https://oceancolor.gsfc.nasa.gov/reprocessing/r2014/aqua/
R2018	https://oceancolor.gsfc.nasa.gov/reprocessing/r2018/aqua/
ROLO	RObotic Lunar Observatory
RSB	Reflective Solar Bands
RSR	Reflective Spectral Response
Rrs	Remote sensing reflectance
RVS	Response Versus Scan angle
RVSt	Time stamp of RVS
SAA	Solar Attanuation screen Assembly
SD	Solar Diffuser
SDSM	Solar Diffuser Stability Monitor
SIS	Spectral Integrating Sphere
SNR	Signal to Noise Ratio
SV	Space View
Tau	Screen transmission functions
TOA	Top of Atmosphere

References

1. Xiong, X.; Barnes, W.L.; Chiang, K.; Erives, H.; Che, N.; Sun, J.; Isaacman, A.T.; Salomonson, V.V. Status of Aqua MODIS on-orbit calibration and characterization. In Proceedings of the Sensors, Systems, and Next-Generation Satellites VIII, Maspalomas, Spain, 13–15 September 2004.

2. Franz, B.A.; Kwiatowska, E.J.; Meister, G.; McClain, C.R. Moderate Resolution Imaging Spectroradiometer on Terra: Limitations for ocean color applications. *J. Appl. Remote. Sens.* **2008**, *2*, 23525. [CrossRef]

3. Meister, G.; Eplee, R.E.; Franz, B.A. Corrections to MODIS Terra calibration and polarization trending derived from ocean color products. In Proceedings of the Earth Observing Systems XIX, San Diego, CA, USA, 18–20 August 2014.

4. Meister, G.; Franz, B.A. Corrections to the MODIS Aqua Calibration Derived from MODIS Aqua Ocean Color Products. *IEEE Trans. Geosci. Remote. Sens.* **2014**, *52*, 6534–6541. [CrossRef]

5. Franz, B.A.; Bailey, S.W.; Werdell, P.J.; McClain, C.R. Sensor-independent approach to the vicarious calibration of satellite ocean color radiometry. *Appl. Opt.* **2007**, *46*, 5068–5082. [CrossRef] [PubMed]

6. Franz, B.A. Status of MODIS and VIIRS OC & SST Production & Distribution. In Proceedings of the MODIS/VIIRS Science Team Meeting, Silver Spring, MD, USA, 6–10 June 2016.

7. Turpie, K.R.; Eplee, R.E.; Franz, B.A.; Del Castillo, C. Calibration uncertainty in ocean color satellite sensors and trends in long-term environmental records. *SPIE Sens. Technol. Appl.* **2014**, *9111*, 911103.

8. Franz, B.A.; Bailey, S.W.; Meister, G.; Werdell, P.J. Consistency of the NASA Ocean Color Data Record. In Proceedings of the Ocean Optics 2012, Glasgow, UK, 8–12 October 2012.

9. International Ocean-Colour Coordinating Group. *In-Flight Calibration of Satellite Ocean-Colour Sensors*; Report Number 14; International Ocean Colour Coordinating Group: Dartmouth, NS, Canada, 2013.

10. MODIS Characterization and Calibration Team. MODIS Instrument Operations Status. In Proceedings of the MODIS/VIIRS Science Team Meeting, Silver Spring, MD, USA, 15–19 October 2018.

11. Sun, J.; Angal, A.; Chen, H.; Geng, X.; Wu, A.; Choi, T.; Chu, M. MODIS reflective solar bands calibration improvements in Collection 6. In Proceedings of the Earth Observing Missions and Sensors: Development, Implementation, and Characterization II, Bertinoro, Italy, 25–28 June 2012.

12. Lee, S.; Meister, G. MODIS Aqua Optical Throughput Degradation Impact on Relative Spectral Response and Calibration of Ocean Color Products. *IEEE Trans. Geosci. Remote Sens.* **2017**, *55*, 5214–5219. [CrossRef]

13. Xiong, X.; Sun, J.-Q.; Esposito, J.A.; Guenther, B.; Barnes, W.L. MODIS reflective solar bands calibration algorithm and on-orbit performance. In Proceedings of the Optical Remote Sensing of the Atmosphere and Clouds III, Québec City, QC, Canada, 3–6 February 2003.

14. Lee, S.; Meister, G. MODIS solar diffuser degradation determination and its spectral dependency. In Proceedings of the Earth Observing Systems XXIII, San Diego, CA, USA, 21–23 August 2018.

15. Meister, G.; Pan, C.; Patt, F.S.; Franz, B.; Xiong, J.; McClain, C.R. Correction of subframe striping in high-resolution MODIS ocean color products. In Proceedings of the Earth Observing Systems XII, San Diego, CA, USA, 10–14 August 2007.

16. MODIS Characterization Support Team. Available online: https://mcst.gsfc.nasa.gov/ (accessed on 12 September 2019).

17. Sun, J.; Xiong, X.; Angal, A.; Chen, H.; Wu, A.; Geng, X. Time-Dependent Response Versus Scan Anglefor MODIS Reflective Solar Bands. *IEEE Trans. Geosci. Remote Sens.* **2013**, *52*, 3159–3174. [CrossRef]

18. Xiong, X.; Salomonson, V.V.; Chiang, K.-F.; Wu, A.; Guenther, B.W.; Barnes, W. On-orbit characterization of RVS for MODIS thermal emissive bands. In Proceedings of the Passive Optical Remote Sensing of the Atmosphere and Clouds IV, Honolulu, Hi, USA, 9–10 November 2004.

19. Xiong, X.; Geng, X.; Angal, A.; Sun, J.; Barnes, W. Using the Moon to track MODIS reflective solar bands calibration stability. In Proceedings of the Sensors, Systems, and Next-Generation Satellites XV, Prague, Czech Republic, 19–22 September 2011.

20. Helder, D.L.; Basnet, B.; Morstad, D.L. Optimized identification of worldwide radiometric pseudo-invariant calibration sites. *Can. J. Remote. Sens.* **2010**, *36*, 527–539. [CrossRef]

21. Vermote, E.F.; Tanré, D.; Deuzé, J.L.; Herman, M.; Morcrette, J.-J. Second Simulation of the Satellite Signal in the Solar Spectrum, 6S: An Overview. *IEEE Trans. Geosci. Remote Sens.* **1997**, *35*, 675–686. [CrossRef]

22. Kieffer, H.H. Photometric Stability of the Lunar Surface. *Icarus* **1997**, *130*, 323–327. [CrossRef]

23. Kieffer, H.H.; Stone, T.C. The Spectral Irradiance of the Moon. *Astron. J.* **2005**, *129*, 2887–2901. [CrossRef]

24. Sun, J.-Q.; Xiong, X.; Barnes, W.L.; Guenther, B. MODIS Reflective Solar Bands On-Orbit Lunar Calibration. *IEEE Trans. Geosci. Remote. Sens.* **2007**, *45*, 2383–2393. [CrossRef]

25. Bailey, S.W. On-orbit calibration of the Suomi National Polar-Orbiting Partnership Visible Infrared Imaging Radiometer Suite for ocean color applications. *Appl. Opt.* **2015**, *54*, 1984. [CrossRef]

26. Franz, B.A.; Bailey, S.W.; Eplee, R.E., Jr.; Lee, S.; Patt, F.S.; Proctor, C.; Meister, G. NASA Multi-Mission Ocean Color Reprocessing 2018. In Proceedings of the Ocean Optics XXIV, Dubrovnik, Croatia, 8–12 October 2018; Available online: https://www.researchgate.net/publication/331652881_NASA_Multi-Mission_Ocean_Color_Reprocessing_20180 (accessed on 12 September 2019).

Letter

Filling the Gaps of Missing Data in the Merged VIIRS SNPP/NOAA-20 Ocean Color Product Using the DINEOF Method

Xiaoming Liu [1,2,*] and Menghua Wang [1]

[1] National Oceanic and Atmospheric Administration, National Environmental Satellite, Data, and Information Service, Center for Satellite Applications and Research, E/RA3, 5830 University Research Ct., College Park, MD 20740, USA; Menghua.Wang@noaa.gov

[2] Cooperative Institute for Research in the Atmosphere at Colorado State University, Fort Collins, CO 80523, USA

* Correspondence: Xiaoming.Liu@noaa.gov; Tel.: +1-301-683-3341; Fax: +1-301-683-3301

Received: 18 December 2018; Accepted: 16 January 2019; Published: 18 January 2019

Abstract: The Visible Infrared Imaging Radiometer Suite (VIIRS) on the Suomi National Polar-orbiting Partnership (SNPP) and National Oceanic and Atmospheric Administration (NOAA)-20 has been providing a large amount of global ocean color data, which are critical for monitoring and understanding of ocean optical, biological, and ecological processes and phenomena. However, VIIRS-derived daily ocean color images on either SNPP or NOAA-20 have some limitations in ocean coverage due to its swath width, high sensor-zenith angle, high sun glint, and cloud, etc. Merging VIIRS ocean color products derived from the SNPP and NOAA-20 significantly increases the spatial coverage of daily images. The two VIIRS sensors on the SNPP and NOAA-20 have similar sensor characteristics, and global ocean color products are generated using the same Multi-Sensor Level-1 to Level-2 (MSL12) ocean color data processing system. Therefore, the merged VIIRS ocean color data from the two sensors have high data quality with consistent statistical property and accuracy globally. Merging VIIRS SNPP and NOAA-20 ocean color data almost removes the gaps of missing pixels due to high sensor-zenith angles and high sun glint contamination, and also significantly reduces the gaps due to cloud cover. However, there are still gaps of missing pixels in the merged ocean color data. In this study, the Data Interpolating Empirical Orthogonal Functions (DINEOF) are applied on the merged VIIRS SNPP/NOAA-20 global Level-3 ocean color data to completely reconstruct the missing pixels. Specifically, DINEOF is applied to 30 days of daily merged global Level-3 chlorophyll-a (Chl-a) data of 9-km spatial resolution from 19 June to 18 July 2018. To quantitatively evaluate the accuracy of the DINEOF reconstructed data, a set of valid pixels are intentionally treated as "missing pixels", so that reconstructed data can be compared with the original data. Results show that mean ratios of the reconstructed/original are 1.012, 1.012, 1.015, and 0.997 for global ocean, oligotrophic waters, deep waters, and coastal and inland waters, respectively. The corresponding standard deviation (SD) of the ratios are 0.200, 0.164, 0.182, and 0.287, respectively. Gap-filled daily Chl-a images reveal many large-scale and meso-scale ocean features that are invisible in the original SNPP or NOAA-20 Chl-a images. It is also demonstrated that the gap-filled data based on the merged products show more details in the dynamic ocean features than those based on SNPP or NOAA-20 alone.

Keywords: VIIRS; SNPP; NOAA-20; DINEOF; ocean color data; data merging; gap-filling

---

1. Introduction

Ocean color data are critical for monitoring and understanding of water optical, biological, and ecological processes and phenomena, and it is also an important source of input data for physical and biogeochemical ocean models [1]. Since the launch of the Visible Infrared Imaging Radiometer Suite (VIIRS) on the Suomi National Polar-Orbiting Partnership (SNPP) [2,3] in October 2011, ocean color products derived from VIIRS-SNPP have been routinely produced globally [3,4]. On 18 November 2017, the follow-up VIIRS sensor housed in the National Oceanic and Atmospheric Administration (NOAA)-20 satellite was launched as the first of four sensors in the Joint Polar Satellite System (JPSS) satellite series, and global ocean color data from NOAA-20 are also being routinely produced. For both SNPP and NOAA-20, chlorophyll-a (Chl-a) concentration [5–7], normalized water-leaving radiance spectra $nL_w(\lambda)$ [8,9], including new $nL_w(\lambda)$ data using VIIRS imaging bands [10], and water diffuse attenuation coefficient at the wavelength of 490 nm $K_d(490)$ and at the domain of photosynthetically available radiation (PAR) $K_d(PAR)$ [11,12], are all generated as standard VIIRS ocean color products. Some experimental products such as inherent optical properties (IOPs) [13,14], and a newly added quality assurance (QA) score product for measuring data quality [15] are also included for evaluation. In addition, to improve ocean color data quality over turbid coastal and inland waters [16–18], the shortwave infrared (SWIR)-based and near-infrared (NIR)-SWIR combined atmospheric correction algorithms have been used to routinely produce global VIIRS ocean color products for both SNPP and NOAA-20 [19,20]. Furthermore, VIIRS ocean color data processing algorithms have been significantly improved, including improved sensor on-orbit calibration using both solar and lunar approaches [21,22]. However, for either VIIRS-SNPP or VIIRS-NOAA-20, there are always missing pixels in the VIIRS-measured ocean color data imageries due to cloud cover and various other reasons, e.g., strong sun glint contamination, dust storms, very large solar- and sensor-zenith angles, etc. It is certainly useful to fill the gap of missing pixels before being used as input for ocean models and for various other applications.

As a follow-up VIIRS instrument, VIIRS-NOAA-20 is essentially built the same as VIIRS-SNPP. Therefore, the sensor characteristics of the two instruments are very similar. Both SNPP and NOAA-20 operate at the 824-km sun-synchronous polar orbit, which crosses the equator at about 13:30 local time. There is about a 50-minute delay between the paths of NOAA-20 and SNPP, which makes the NOAA-20's path running through the middle of two adjacent SNPP paths, and vice versa. The overlap of the spatial coverages of the two sensors automatically fills each other's gaps caused by high sensor-zenith angles and high sun glint contamination, and it significantly reduces the missing pixels in the merged images. In addition, ocean color products from SNPP and NOAA-20 have the same spatial and temporal resolution, and are processed with the same software package, i.e., the Multi-Sensor Level-1 to Level-2 (MSL12) ocean color data processing system [3]. Therefore, the statistics of their ocean color products are very close to each other, and in fact the data can be directly merged without adjustment for matching up each other's statistical properties. However, there are still lots of missing data in the merged VIIRS SNPP/NOAA-20 products.

To completely fill the gaps of missing pixels in the merged VIIRS SNPP/NOAA-20 ocean color data, the Data Interpolating Empirical Orthogonal Functions (DINEOF) [23,24] method is used in this study to reconstruct the missing data in the ocean color images. The DINEOF exploits the spatio-temporal coherency of the data to infer a value at the missing location and has been successfully adopted in various applications using ocean remote sensing data [25–30]. With more and more availability and usage of ocean color data in recent years, the DINEOF method has also been applied to ocean color data from various sensors including the Moderate Resolution Imaging Spectroradiometer (MODIS) [31–33], the Spinning Enhanced Visible and Infrared Imager (SEVIRI) onboard Meteosat Second Generation 2 [34], and the Korean Geostationary Ocean Color Imager (GOCI) [35]. Most recently, Liu and Wang [36] used the DINEOF to fill the gaps of missing pixels in VIIRS-SNPP global ocean color data. In that study, the DINEOF is applied to VIIRS-SNPP global Level-3 binned ocean color data of 9-km spatial resolution and the DINEOF reconstructed ocean color data are used to fill the gap of missing data. In particular, daily, 8-day, and monthly VIIRS global Level-3 binned ocean color data, including Chl-a concentration, K_d(490), as well as $nL_w(\lambda)$ at the five VIIRS visible bands are tested and evaluated. Results show that the DINEOF method can successfully reconstruct and gap-fill meso-scale and large-scale spatial ocean features in the global VIIRS Level-3 images, as well as capture the temporal variations of these features.

In this study, the DINEOF method is used to reconstruct and gap-fill a merged VIIRS SNPP/NOAA-20 global daily ocean color product, specifically the Chl-a data. With the reconstructed (gap-filled) VIIRS global daily Chl-a images, ocean features can now be well identified and observed both spatially and temporally. Some examples that demonstrate the advantages and usefulness of the gap-filled VIIRS SNPP/NOAA-20 merged global ocean color products are provided. The DINEOF is also applied to ocean color data from a single sensor, VIIRS-SNPP or VIIRS-NOAA-20, and the gap-filled data based on SNPP or NOAA-20 are compared with those based on the two-sensor merged Chl-a product.

2. Data and Methods

2.1. VIIRS SNPP and NOAA-20 Ocean Color Level-2 and Global Level-3 Data

The MSL12 is the official NOAA VIIRS ocean color data processing system for both SNPP and NOAA-20 (and all other follow-on VIIRS sensors), and it has been used for processing VIIRS data from Sensor Data Records (SDR or Level-1B data) to the Environmental Data Records (EDR or Level-2 data) [3,4]. MSL12 was developed for producing consistent ocean color products using the same ocean color data processing system for multiple satellite ocean color sensors [37–39]. In addition to the standard NIR-based atmospheric correction algorithm [8], the current version of MSL12 has been improved to include the SWIR- and NIR-SWIR-based atmospheric correction algorithms [16,19,20], including a recent algorithm improvement using the information from the short blue band, i.e., M1 in Table 1 [40], for improved satellite ocean color data over coastal and inland waters. Indeed, advantages of the SWIR-based ocean color data processing over global highly turbid coastal and inland waters for various applications have been demonstrated in several previous studies [41–43].

Table 1. The VIIRS reflective solar band (RSB) spectral band nominal center wavelength (in nm) for SNPP and NOAA-20.

Sensor	M1	M2	M3	M4	M5	M6	M7	M8	M9	M10	M11	I1	I2	I3
VIIRS-SNPP	410	443	486	551	671	745	862	1238	1378	1610	2250	638	862	1610
VIIRS-NOAA-20	411	445	489	555	667	746	868	1238	1376	1604	2258	642	867	1603

The VIIRS SNPP and NOAA-20 global Level-3 ocean color product data are generated using the spatial and temporal binning from the corresponding Level-2 data [44]. In this study, VIIRS SNPP and NOAA-20 global daily Level-3 binned data of 9-km spatial resolution from 19 June to 18 July 2018 are produced. In the Level-3 data processing, the grid elements (9×9 km^2 bins) are arranged in rows beginning at the South Pole. Each row begins at $180°$ longitude and circumscribes the Earth at a given latitude. For 9-km spatial resolution Level-3 data, there are only three bins in the first row near the South Pole, and the number of bins in each row gradually increases with latitude in the southern hemisphere. The maximum number of bins in a row is reached at the equator, and then the number of bins decreases with latitude in the northern hemisphere. Pixels containing valid Level-2 data are mapped to bins of 9×9 km^2. Within each bin, statistics of mean or median are accumulated for periods of one day for daily ocean color products. Before the binning process, several Level-2 masks and data quality flags from MSL12 (e.g., land, cloud [45], stray-light [46], high sun glint [47], high sensor-zenith angle, high solar-zenith angle, etc.) are applied to VIIRS ocean color Level-2 data. As examples, Figure 1a,b shows the global Level-3 daily Chl-a images of 21 June 2018 for SNPP and NOAA-20, respectively. Obviously, there are many missing pixels in the original Level-3 VIIRS Chl-a images. In fact, there are ~70% missing pixels in these global daily images due to cloud cover, high sun glint contamination, high solar- or sensor-zenith angles, and various other reasons.

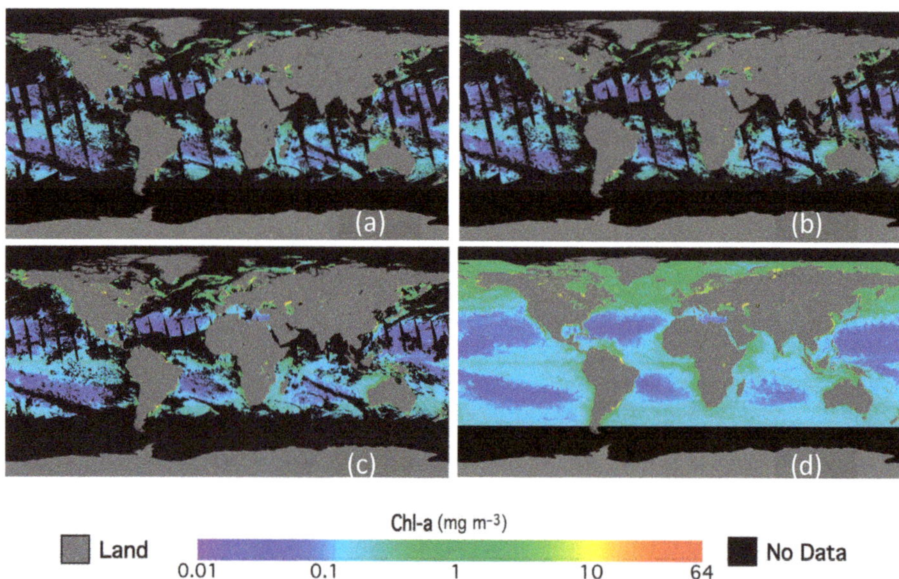

Figure 1. The Visible Infrared Imaging Radiometer Suite (VIIRS)-measured global Level-3 Chl-a image on 21 June 2018 from (**a**) Suomi National Polar-Orbiting Partnership (SNPP), (**b**) National Oceanic and Atmospheric Administration (NOAA)-20, (**c**) merged VIIRS SNPP/NOAA-20 Chl-a image on the same day, and (**d**) gap-filled VIIRS SNPP/NOAA-20 merged Chl-a image on the same day.

2.2. Merging VIIRS SNPP and NOAA-20 Global Level-3 Ocean Color Data

As noted previously, SNPP and NOAA-20 operate at the same sun-synchronous polar orbit of 824-km. However, the positions of the two satellites in the orbit are intentionally arranged so that NOAA-20 leads SNPP by a half orbit, or ~50 minutes. The 50-minute difference between their paths makes the NOAA-20's nadir points always running through SNPP's gaps due to high sensor-zenith angle, and vice versa. This special arrangement of satellite positions in the orbit significantly benefits the observation of the ocean color from the space: the overlap of the spatial coverages of the two sensors fills most of each other's gaps in images due to high sensor-zenith angles and high sun glint contamination, and significantly reduces the missing pixels in the merged images. In addition, ocean biological features are assumed to have little change in ~50 minutes, so that the ocean color data from the two sensors can be directly merged.

The VIIRS SNPP and NOAA-20 instruments both have 14 reflective solar bands (RSBs) with three image bands (I1–I3) and 11 moderate bands (M1–M11) operating in the spectral region of 0.41–2.25 μm. There are slight differences between VIIRS-SNPP and VIIRS-NOAA-20 in the nominal center wavelengths of each band. Table 1 lists the VIIRS RSB spectral band nominal center wavelength for VIIRS-SNPP and VIIRS-NOAA-20. In the ocean color data processing, the VIIRS-NOAA-20 $nL_w(\lambda)$ of five bands (M1–M5) are adjusted to match up with those of VIIRS-SNPP, so that Chl-a, $K_d(490)$, and $K_d(\text{PAR})$ products are equivalent from the two sensors, and can be merged. In this study, only Chl-a products are merged as a preliminary test.

The VIIRS SNPP and NOAA-20 have the same swath width of 3040 km, and horizontal resolution of 750 m for M bands, and 375 m for I bands at nadir. With the same spatial and temporal resolutions, and their ocean color products processed using the same MSL12 software package, Chl-a data from the two instruments are statistically equivalent. Figure 2 shows the scatter plots of VIIRS NOAA-20 versus SNPP Chl-a data on 21 June 2018. The mean ratios of VIIRS NOAA-20/SNPP are 1.069, 1.024, 1.057, and 1.127 in global ocean, global oligotrophic waters (Chl-a < 0.1 mg/m^3), global deep waters (water depth > 1 km), and global coastal and inland waters (water depth $\leq$ 1 km), respectively; and the corresponding standard deviation (SD) values are 0.212, 0.162, 0.175, and 0.336, respectively. Therefore, the statistics of the ocean color products from the two sensors are very close to each other, and they can be directly merged. In this study, the merged VIIRS SNPP/NOAA-20 ocean color products are calculated as weighted averages, where the weight is the number of valid pixels in each bin. The input datasets of the merging process are 30 days of daily SNPP and NOAA-20 global 9-km resolution Level-3 binned Chl-a data from June 19 to July 18, 2018, and the output data are merged SNPP/NOAA-20 global 9-km resolution data in Level-3 binned format.

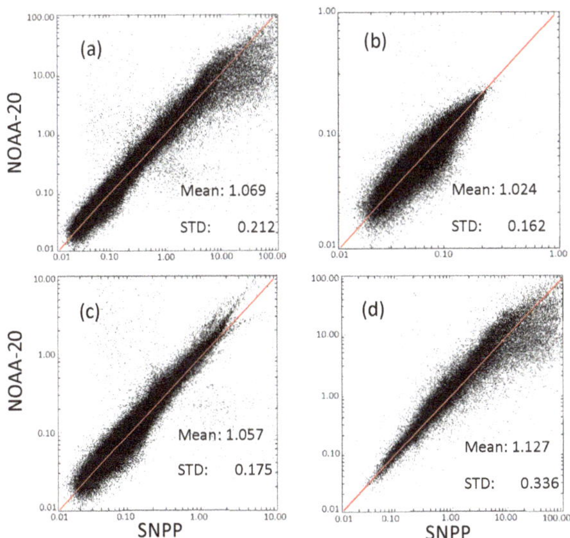

Figure 2. Scatter plots of VIIRS NOAA-20 versus SNPP Chl-a in (**a**) global all regions, (**b**) global oligotrophic waters, (**c**) global deep waters, and (**d**) global coastal and inland waters on 21 June 2018.

2.3. Gap-Filling of the Merged VIIRS SNPP/NOAA-20 Data

The DINEOF method [23,24] is an Empirical Orthogonal Function (EOF)-based technique, which allows the extraction of the dominant spatial patterns observed in a data time series through an iterative approach, while simultaneously filling in the missing data. During the DINEOF process, the original dataset is first stored in a spatio-temporal matrix with $m \times n$ dimensions, where m is the number of grids in the spatial domain and n is the number of time steps in the time series. Initially, the temporal and spatial mean is removed from the data, and all missing values are set to zeroes. The first EOF mode is then calculated by using the singular value decomposition (SVD) technique, and the missing values are replaced with the initial guess by the data reconstruction using the spatial and temporal functions of only the first EOF mode. The first EOF mode is then recalculated iteratively using the previous best guess as the initial value of the missing data for the subsequent iteration until the process converges. Subsequently, the number of EOFs increases one by one and for each EOF mode, the whole reconstruction is operated again until convergence. Then, by using a cross-validation technique, the optimal number of EOF modes can be determined. Finally, the reconstruction procedure is performed again, based on the optimal EOF modes, until convergence is reached. The process to determine the optimum number of EOF modes in the final reconstruction is fully automatic. For example, if the error (reconstructed−original) of the validation pixels decreases gradually from mode 1 to mode 10, but the error starts to increase gradually from mode 11 to mode 13, then the first 10 modes are considered as optimum.

In this study, the DINEOF is utilized to reconstruct (gap-filled) the missing pixels in the merged VIIRS SNPP/NOAA-20 Chl-a data. Specifically, DINEOF is applied to 30 days of daily merged global Level-3 Chl-a data of 9-km spatial resolution from 19 June to 18 July 2018. It is noted that DINEOF is applied directly to the VIIRS Level-3 bin data files, rather than the mapped data files. However, it is a very large dataset for 30 days of global Chl-a data with 9-km spatial resolution, which makes the performance of DINEOF analysis very inefficient. To speed up the DINEOF process, the same procedure used for gap-filling of SNPP ocean color data by Liu and Wang [36] is used. Basically, the global dataset is evenly divided into 16 10°-zonal sections between 80°S and 80°N, and DINEOF is applied to each of the zonal sections separately. In fact, DINEOF analyses on the 16 zonal sections are performed in parallel to further improve data processing efficiency.

3. Results

3.1. Merged VIIRS SNPP/NOAA-20 Products

The VIIRS SNPP and NOAA-20 global daily Level-3 Chl-a data of 9-km spatial resolution, from 19 June to 18 July 2018, were merged. As an example, Figure 1c shows the merged SNPP/NOAA-20 Chl-a concentration on 21 June 2018. The spatial coverage of the merged ocean color data was significantly improved compared with either SNPP (Figure 1a) or NOAA-20 (Figure 1b). Particularly, in Figure 1a,b, the gaps of high sensor-zenith angle occurred in both the northern and southern hemisphere. The gaps of sun glint, however, occurred only to the north of the equator because it was summer (June–July) in the northern hemisphere. In Figure 1c, gaps of missing pixels due to the high sensor-zenith angle and high sun glint contamination from both VIIRS sensors were almost filled, thanks to the overlap of the swath of the two sensors. The remaining missing pixels were due to a very small part of high sensor-zenith angles from the one sensor and a small part of high sun glint contamination from the other. They were significantly reduced. It can be clearly seen in Figure 1c that the gaps of the missing pixels due to the high sensor-zenith angles and high sun glint contamination were much narrower than those in Figure 1a,b. In the southern hemisphere, however, it was completely free of gaps from sun glint contamination and high sensor-zenith angle. In addition, the missing pixels due to cloud cover were also reduced in the merged Chl-a image, because of the shifts of the cloud pixels during the 50-minute period. Overall, the merged VIIRS SNPP/NOAA-20 global Level-3 daily data had ~38% more valid pixels than those from the SNPP or NOAA-20 data alone.

The major gaps of missing pixels in the merged VIIRS SNPP/NOAA-20 data were due to cloud covers. Since there was only a 50-minute delay between the paths of the two satellites, cloud usually had little change in its shape, but could travel a short distance. The merging only reduced some limited pixels on the edges of the cloud cover, but the general pattern of gaps due to cloud cover was still the same in the merged data as in the SNPP or NOAA-20 data. Indeed, the merging could not improve spatial coverage in regions with large areas of thick clouds, such as in the Arabian Sea and Bay of Bengal, equatorial Atlantic Ocean, and high-latitude Pacific and Atlantic Oceans. The summer monsoon usually brings large amounts of moisture and rainfall to the Arabian Sea from June to July.

3.2. Gap-Filled VIIRS SNPP/NOAA-20 Products

The DINEOF method was applied to the merged VIIRS SNPP/NOAA-20 global daily Level-3 Chl-a data from 19 June to 18 July 2018. The DINEOF has two options to reconstruct missing data: fully reconstructed images and filled images. A fully reconstructed image was calculated from the retained EOF modes using the DINEOF method. In the fully reconstructed image, all ocean color data were reconstructed on every ocean pixel (including non-missing pixels). The reconstructed image had no-gaps spatially. However, there were some small differences between reconstructed and original data even for non-missing pixels due to truncated EOF modes in computing all data values. The filled image was a combination of the original image and the reconstructed image, i.e., missing pixels were filled with reconstructed data and original data were kept for non-missing pixels. As an example, the fully reconstructed global Chl-a image of 21 June 2018 is shown in Figure 1d. All gaps of missing pixels due to the high sun glint and large sensor-zenith angles were filled with valid pixels. The gaps of cloud-covered pixels were also reconstructed, especially in the northern Atlantic and Pacific, Arabian Sea and Bay of Bengal, and the equatorial Atlantic Ocean as well. The gaps of missing pixels were all filled very smoothly in Figure 1d. However, the large area of missing pixels in the Southern Ocean close to Antarctica were not reconstructed, since there were no valid pixels in the entire area from 19 June to 18 July 2018 due to high solar-zenith angles during the northern hemisphere summer time.

To evaluate the accuracy of the value from reconstructed pixels, 5% of valid (non-missing) pixels in the original Level-3 data were randomly selected, and intentionally treated as "missing pixels." These pixels were reconstructed and compared with the original data for validation. Figure 3 shows the density scatter plots of the reconstructed data versus original data in the global ocean, oligotrophic waters, deep waters, and coastal and inland waters. In general, most points are close to the 1:1 line in the global ocean, but the oligotrophic waters show the best results (Figure 3b), whilst the highest level of scatter is found in the coastal and inland waters (Figure 3d). Quantitatively, the mean and SD of the reconstructed/original ratios in oligotrophic water are 1.012 and 0.164, respectively. The mean and SD in deep waters are 1.015 and 0.182, respectively, which are slightly higher than those in oligotrophic waters. In the coastal and inland waters, the mean and SD are 0.997 and 0.287, respectively. Overall, for all pixels in the global ocean, the mean and SD of the reconstructed/original ratios are 1.012 and 0.200, respectively. It is noted that the locations of the validation pixels are selected randomly using the random number generator in the Interactive Data Language (IDL). In addition, the DINEOF package includes a built-in function to specify the "size of cloud surface" for cross-validation as a parameter of input. We tested the cross-validation with different cloud patch sizes and found that the cross-validation result does not change significantly with the size of the cloud surface. The main reason is that DINEOF considers not only spatial coherence, but also temporal coherence. Since the location of the cloud patches are selected randomly, even with the large size of cloud patch, the temporal variations in the time-series can still be used to infer the missing values. The advantage of the DINEOF over traditional interpolation is that it utilizes major EOF modes to reconstructs the missing pixels, which capture the major signals of variations in both spatial and temporal domains simultaneously.

3.3. Ocean Features Revealed in the Gap-Filled VIIRS SNPP/NOAA-20 Chl-a Data

Gap-filled daily Chl-a images reveal many large-scale and meso-scale ocean features that are invisible in the merged Chl-a images, or the original VIIRS SNPP, NOAA-20 Chl-a images. As shown in Figure 1d, the most obvious large-scale ocean features are the oligotrophic waters (Chl-a < 0.1 mg/m^3) in the center of the five subtropical ocean gyres, i.e., North Atlantic, South Atlantic, North Pacific, South Pacific, and South Indian Ocean, which cover major parts of the global oceans. Because of the easterly winds near the equator and the equatorial upwelling, there is more enhanced Chl-a concentration (~0.1–0.5 mg/m^3) in the equatorial Pacific Ocean and equatorial Atlantic Ocean regions than in the subtropical gyres. However, no significant Chl-a enhancement is found in the equatorial Indian Ocean. In addition, the Gulf Stream in the North Atlantic Ocean and Kuroshio in the North Pacific Ocean mark a clear boundary in the high latitude ocean regions: high Chl-a concentrations are found to the north of the boundary, while low Chl-a to the south.

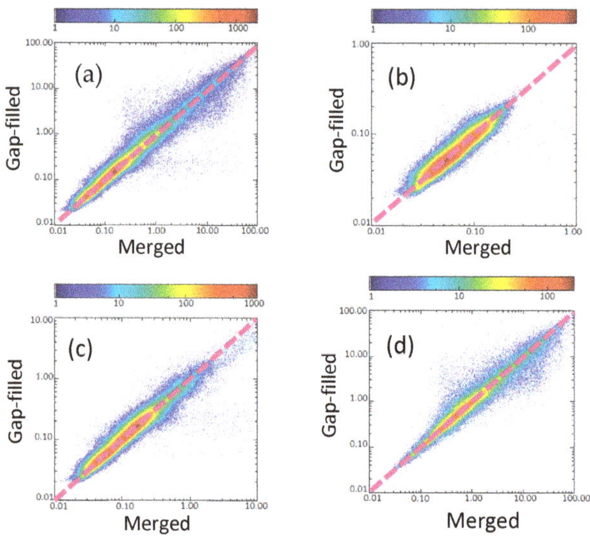

Figure 3. Density scatter plots of gap-filled versus merged Chl-a values for (**a**) all selected validation pixels, (**b**) oligotrophic waters, (**c**) deep waters, and (**d**) coastal and inland waters on 21 June 21 2018.

In addition to these large-scale features, the gap-filled images also reveal many meso-scale ocean features, such as eddies and filaments associated with western boundary currents or coastal currents. Figure 4 shows the transformation of meso-scale eddies in the Gulf of Mexico (GOM) from 19 June to 18 July 2018. Ocean circulation in the GOM is dominated by the loop current (LC), an extension of the western boundary current system of the North Atlantic Ocean that loops into the GOM. The LC enters the Gulf through the Yucatan Channel and exits through the Florida Straits. The LC occasionally extends further northward into the GOM, approaching the northern shelf break. This long extension of the loop will eventually separate to form an anticyclonic loop current eddy (LCE). Formation of an LCE by separation from the LC happens irregularly every several months, and there can be a number of LCEs in the GOM at one time. Figure 4 shows three LCEs in the GOM: two of them (A and B) were aged LCEs that already moved to the west side of GOM, and one (C) was a new LCE that just separated from the LC. Loop current eddy A and B are about 100 km in diameter, and C is about 200 km in diameter. During the one-month period from 19 June to 18 July 2018, A and B were still moving continuously to the west, and A was showing some changes in its shape while interacting with the LC. The LC was very flat till 7 July, when the LC started to bulge northward, getting ready to shed a new LCE. The details of the transformation of these meso-scale ocean features were not available in either SNPP, NOAA-20, or in the merged images, but only revealed in the gap-filled images. The transitions of these meso-scale ocean features were very smooth both temporally and spatially. To compare with the original VIIRS images without gap-fillings, Chl-a global daily images from SNPP, NOAA-20, or merged SNPP/NOAA-20 can be found in the NOAA Ocean Color Team website (https://www.star.nesdis.noaa.gov/sod/mecb/color/), in particular, using the powerful image display tool, i.e., the Ocean Color Viewer (OCView) [48]. With OCView, the user can zoom in to the GOM region (or any other region) with high spatial resolution (2 km) images, and select SNPP, NOAA-20 or the merged SNPP/NOAA-20 images.

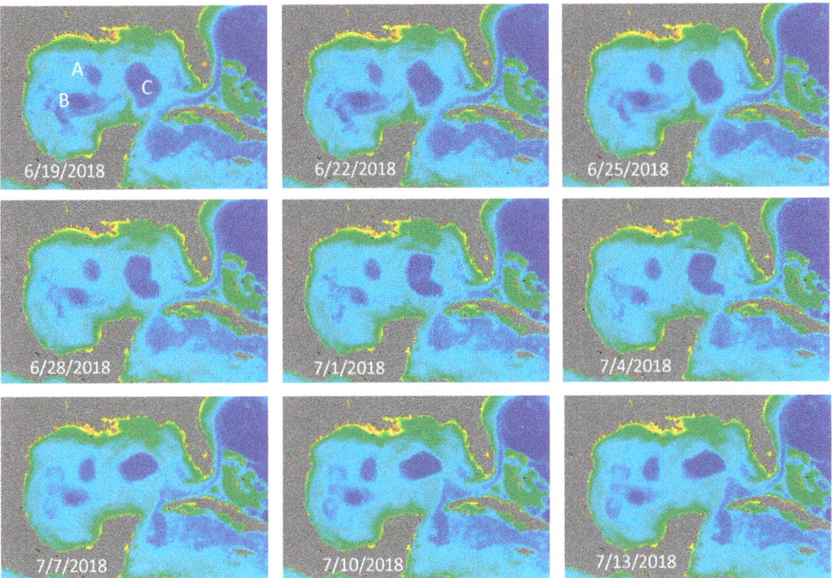

Figure 4. Daily Chl-a images of the Gulf of Mexico (GOM) from 19 June to 13 July 2018 with three loop current eddies (**A**, **B**, and **C**) marked on the 19 June 2018 Chl-a image.

3.4. Comparison with Gap-Filled Data Based on VIIRS SNPP or NOAA-20

The DINEOF was also applied to Chl-a data from a single sensor, VIIRS-SNPP or VIIRS-NOAA-20, and the gap-filled data based on a single sensor were compared with gap-filled data based on the merged products. Figure 5 shows the gap-filled data based on VIIRS-SNPP, VIIRS-NOAA-20, and merged VIIRS SNPP/NOAA-20 products in the GOM on 21 June, 30 June, and 18 July 2018, as examples. It can be seen that, in general, the gap-filled data based on the merged products show the same pattern as those based on SNPP or NOAA-20 alone. However, there are more details in the dynamic features in the gap-filled data based on the merged product than those based on SNPP or NOAA-20 alone, and the interactions of the loop current with surrounding eddies are different in the merged products. On 21 June, the gap-filled data based on merged VIIRS SNPP/NOAA-20 products (Figure 5c) show more details in the coastal eddy feature "A" than VIIRS-SNPP (Figure 5a) or VIIRS-NOAA-20 (Figure 5b) alone. On 30 June 2018, the feature of the big eddy "B" are different in the gap-filled data based on merged products (Figure 5f). From 30 June to 18 July, the eddy "B" was pushed northward by the loop current, and transformed from a north–south oriented shape to a west–east oriented shape. In the animation of the daily images (not shown here), the interactions of the eddy "B" with the loop current were more clearly shown in the merged products. On 18 July 2018, the eddy "C" was seen more clearly in Figure 5i than that in Figure 5g or Figure 5h. Generally, the merged data have more spatial coverage than that from either SNPP or NOAA-20 alone. Indeed, some ocean dynamic features could be partially blocked by pixels with high sensor-zenith angle, sun glint, or cloud in VIIRS SNPP or NOAA-20 images. But, merging data from the two sensors can significantly increase valid pixels in the coverage. Consequently, the gap-filled ocean color products based on the merged data set can capture more details of the ocean dynamic features.

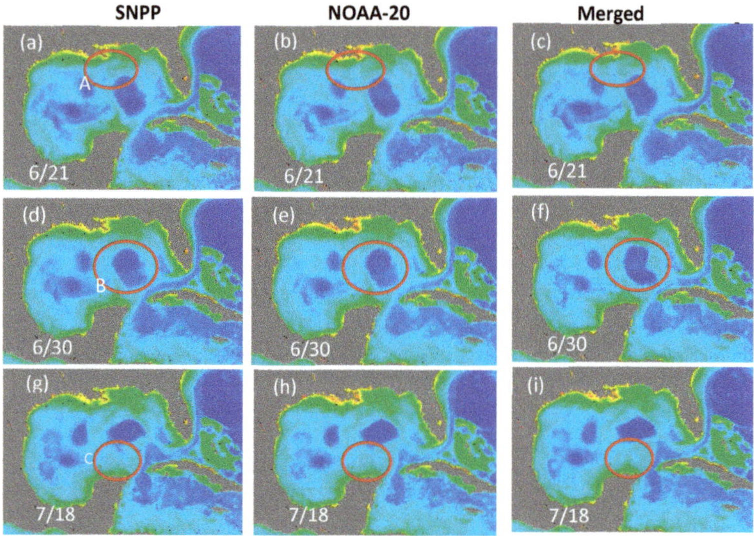

Figure 5. Comparison of the VIIRS-derived ocean features in the GOM in the gap-filled images based on SNPP (left column), NOAA-20 (middle column), and SNPP/NOAA-20 merged (right column) data in 2018 on 21 June (top row) (**a–c**), 30 June (middle row) (**d–f**), and 18 July (bottom row) (**g–i**).

4. Discussion and Conclusions

In a recent study [36], we used the DINEOF to fill the gaps of missing pixels in daily VIIRS-SNPP ocean color data, and it was demonstrated that the DINEOF method can successfully reconstruct and reveal meso-scale and large-scale spatial ocean features in the global VIIRS Level-3 images, as well as capture the temporal variations of these features. Based on the same methodology, in this study, the DINEOF was applied on the daily merged VIIRS SNPP/NOAA-20 global Level-3 ocean Chl-a data. The VIIRS-SNPP and VIIRS-NOAA-20 have very similar sensor characteristics, and the statistics of the ocean color data derived from the two sensors are very close to each other. In addition, global VIIRS SNPP and NOAA-20 ocean color data are derived using the same MSL12 ocean color data processing system. As shown in examples, merging VIIRS-SNPP and VIIRS-NOAA-20 data almost removes the gaps of missing pixels due to high sensor-zenith angles and high sun glint contamination, and also significantly reduces the gaps due to cloud cover. The remaining regular gaps in the merged images are due to small gaps from high sensor-zenith angles in one sensor and sun glint contamination in another. Overall, there are ~38% more valid ocean color pixels in the merged VIIRS SNPP/NOAA 20 global Level-3 daily data than those in the SNPP or NOAA-20 data alone.

In the gap-filled Chl-a data, all missing pixels due to the high sensor-zenith angle and high sun glint were mostly reconstructed. The gaps of cloud-covered pixels were also reconstructed, especially in regions with large areas of thick clouds, such as in the northern Atlantic and Pacific, Arabian Sea and Bay of Bengal, and the equatorial Atlantic Ocean as well. Quantitatively, the mean ratios of the reconstructed/original data are 1.012, 1.012, 1.015, and 0.997 for the global ocean, global oligotrophic waters, global deep waters, and global coastal and inland waters, respectively; and the corresponding SD values are 0.200, 0.164, 0.182, and 0.287, respectively. Gap-filled daily Chl-a images reveal many large-scale and meso-scale ocean features and their variations, which are invisible in the original VIIRS-SNPP or VIIRS-NOAA-20 Chl-a images.

The gap-filled Chl-a data based on two-sensor merged products were compared with gap-filled data based on single sensor VIIRS-SNPP or VIIRS-NOAA-20 alone. It is demonstrated that the gap-filled data based on the merged products show more details in the ocean features than those based on a single sensor alone, and there are differences in the dynamics of ocean features. This is an important message that adding more sensors into the merged products will significantly improve the quality of gap-filled ocean color data. Previous missions such as the Sea-viewing Wide Field-of-view Sensor (SeaWiFS) (1997–2010), MODIS on the Terra (1999–present) and Aqua (2002–present) satellites, and the Medium-Resolution Imaging Spectrometer (MERIS) on the ENVISAT (2002–2012) had provided long-term high-quality global ocean color data to the community. Currently, in addition to the VIIRS on SNPP and NOAA-20, the Ocean and Land Color Instrument (OLCI) on the Sentinel-3A (2016–present) and Sentinel-3B (2018–present) satellites, and the Second-Generation Global Imager (SGLI) on the Global Change Observation Mission-Climate (GCOM-C) (2017–present) are all providing an unprecedented view of ocean optical, biological, and biogeochemical properties on a global scale simultaneously. Adding these sensors in the data merging process will significantly improve the spatial coverage in the merged global ocean color data, and the quality of gap-filled ocean color data as well.

Author Contributions: X.L. carried out the main research work for developing algorithm, obtaining the results, and analyzing the data. M.W. suggested the topic and contributed to the algorithm development and evaluation.

Funding: This work was supported by the Joint Polar Satellite System (JPSS) funding and NOAA Product Development, Readiness, and Application (PDRA)/Ocean Remote Sensing (ORS) Program funding.

Acknowledgments: We thank two anonymous reviewers for their useful comments. The views, opinions, and findings contained in this paper are those of the authors and should not be construed as an official NOAA or U.S. Government position, policy, or decision.

Conflicts of Interest: The authors declare no conflict of interest.

References

1. Yoder, J.A.; Doney, S.C.; Siegel, D.A.; Wilson, C. Study of marine ecosystem and biogeochemistry now and in the future: Example of the unique contributions from the space. *Oceanography* **2010**, *23*, 104–117. [CrossRef]

2. Goldberg, M.D.; Kilcoyne, H.; Cikanek, H.; Mehta, A. Joint Polar Satellite System: The United States next generation civilian polar-orbiting environmental satellite system. *J. Geophys. Res. Atmos.* **2013**, *118*, 13463–13475. [CrossRef]

3. Wang, M.; Liu, X.; Tan, L.; Jiang, L.; Son, S.; Shi, W.; Rausch, K.; Voss, K. Impact of VIIRS SDR performance on ocean color products. *J. Geophys. Res. Atmos.* **2013**, *118*, 10347–10360. [CrossRef]

4. Wang, M.; Shi, W.; Jiang, L.; Voss, K. NIR- and SWIR-based on-orbit vicarious calibrations for satellite ocean color sensors. *Opt. Express* **2016**, *24*, 20437–20453. [CrossRef] [PubMed]

5. O'Reilly, J.E.; Maritorena, S.; Mitchell, B.G.; Siegel, D.A.; Carder, K.L.; Garver, S.A.; Kahru, M.; McClain, C.R. Ocean color chlorophyll algorithms for SeaWiFS. *J. Geophys. Res.* **1998**, *103*, 24937–24953. [CrossRef]

6. Hu, C.; Lee, Z.; Franz, B.A. Chlorophyll a algorithms for oligotrophic oceans: A novel approach based on three-band reflectance difference. *J. Geophys. Res.* **2012**, *117*, C01011. [CrossRef]

7. Wang, M.; Son, S. VIIRS-derived chlorophyll-a using the ocean color index method. *Remote Sens. Environ.* **2016**, *182*, 141–149. [CrossRef]

8. Gordon, H.R.; Wang, M. Retrieval of water-leaving radiance and aerosol optical thickness over the oceans with SeaWiFS: A preliminary algorithm. *Appl. Opt.* **1994**, *33*, 443–452. [CrossRef]

9. Wang, M. (Ed.) *Atmospheric Correction for Remotely-Sensed Ocean-Colour Products*; Reports of International Ocean-Colour Coordinating Group IOCCG: Dartmouth, NS, Canada, 2010. [CrossRef]

10. Wang, M.; Jiang, L. VIIRS-derived ocean color product using the imaging bands. *Remote Sens. Environ.* **2018**, *206*, 275–286. [CrossRef]

11. Wang, M.; Son, S.; Harding, J.L.W. Retrieval of diffuse attenuation coefficient in the Chesapeake Bay and turbid ocean regions for satellite ocean color applications. *J. Geophys. Res.* **2009**, *114*, C10011. [CrossRef]

12. Son, S.; Wang, M. Diffuse attenuation coefficient of the photosynthetically available radiation Kd(PAR) for global open ocean and coastal waters. *Remote Sens. Environ.* **2015**, *159*, 250–258. [CrossRef]

13. Lee, Z.P.; Carder, K.L.; Arnone, R.A. Deriving inherent optical properties from water color: A multiple quasi-analytical algorithm for optically deep waters. *Appl. Opt.* **2002**, *41*, 5755–5772. [CrossRef] [PubMed]

14. Shi, W.; Wang, M. Characterization of particle backscattering of global highly turbid waters from VIIRS ocean color observations. *J. Geophys. Res. Oceans* **2017**, *122*, 9255–9275. [CrossRef]

15. Wei, J.; Lee, Z.; Shang, S. A system to measure the data quality of spectral remote-sensing reflectance of aquatic environments. *J. Geophys. Res. Oceans* **2016**, *121*, 8189–8207. [CrossRef]

16. Wang, M.; Shi, W. Estimation of ocean contribution at the MODIS near-infrared wavelengths along the east coast of the U.S.: Two case studies. *Geophys. Res. Lett.* **2005**, *32*, L13606. [CrossRef]

17. Wang, M.; Shi, W.; Tang, J. Water property monitoring and assessment for China's inland Lake Taihu from MODIS-Aqua measurements. *Remote Sens. Environ.* **2011**, *115*, 841–854. [CrossRef]

18. Wang, M.; Son, S.; Shi, W. Evaluation of MODIS SWIR and NIR-SWIR atmospheric correction algorithm using SeaBASS data. *Remote Sens. Environ.* **2009**, *113*, 635–644. [CrossRef]

19. Wang, M. Remote sensing of the ocean contributions from ultraviolet to near-infrared using the shortwave infrared bands: Simulations. *Appl. Opt.* **2007**, *46*, 1535–1547. [CrossRef]

20. Wang, M.; Shi, W. The NIR-SWIR combined atmospheric correction approach for MODIS ocean color data processing. *Opt. Express* **2007**, *15*, 15722–15733. [CrossRef]

21. Sun, J.; Wang, M. Radiometric calibration of the VIIRS reflective solar bands with robust characterizations and hybrid calibration coefficients. *Appl. Opt.* **2015**, *54*, 9331–9342. [CrossRef]

22. Sun, J.; Chu, M.; Wang, M. On-orbit characterization of the VIIRS solar diffuser and attenuation screens for NOAA-20 using yaw measurements. *Appl. Opt.* **2018**, *57*, 6605–6619. [CrossRef] [PubMed]

23. Alvera-Azcarate, A.; Barth, A.; Rixen, M.; Beckers, J. Reconstruction of incomplete oceanographic data sets using Empirical Orthogonal Functions. Application to the Adriatic Sea. *Ocean Model.* **2005**, *9*, 325–346. [CrossRef]

24. Beckers, J.; Rixen, M. EOF calculations and data filling from incomplete oceanographic data sets. *J. Atmos. Ocean Technol.* **2003**, *20*, 1839–1856. [CrossRef]

25. Ganzedo, U.; Alvera-Azcarate, A.; Esnaola, G.; Ezcurra, A.; Saenz, J. Reconstruction of sea surface temperature by means of DINEOF. A case study during the fishing season in the Bay of Biscay. *Int. J. Remote Sens.* **2011**, *32*, 933–950. [CrossRef]

26. Mauri, E.; Poulain, P.M.; Juznic-Zontac, Z. MODIS chlorophyll variability in the northern Adriatic Sea and relationship with forcing parameters. *J. Geophys. Res.* **2007**, *112*, C03S11. [CrossRef]

27. Mauri, E.; Poulain, P.M.; Notarstefano, G. Spatial and temporal variability of the sea surface temperature in the Gulf of Trieste between January 2000 and December 2006. *J. Geophys. Res.* **2008**, *113*, C10012. [CrossRef]

28. Nechad, B.; Alvera-Azcarate, A.; Ruddick, K.; Greenwood, N. Reconstruction of MODIS total suspended matter time series maps by DINEOF and validation with autonomous platform data. *Ocean Dyn.* **2011**, *61*, 1205–1214. [CrossRef]

29. Sirjacobs, D.; Alvera-Azcarate, A.; Barth, A.; Lacroix, G.; Park, Y.; Nechad, B.; Ruddick, K.; Beckers, J. Cloud filling of ocean color and sea surface temperature remote sensing products over the Southern North Sea by the data interpolating empirical orthogonal functions methodology. *J. Sea Res.* **2011**, *65*, 114–130. [CrossRef]

30. Volpe, G.; Nardelli, B.B.; Cipollini, P.; Santoleri, R.; Robinson, I.S. Seasonal to interannual phytoplankton response to physical processes in the Mediterranean Sea from satellite observations. *Remote Sens. Environ.* **2012**, *117*, 223–235. [CrossRef]

31. Li, Y.; He, R. Spatial and temporal variability of SST and ocean color in the Gulf of Maine based on cloud-free SST and chlorophyll reconstructions in 2003–2012. *Remote Sens. Environ.* **2014**, *144*, 98–108. [CrossRef]

32. Moradi, M.; Kabiri, K. Spatio-temporal variability of SST and chlorophyll-a from MODIS data in the Persian Gulf. *Mar. Pollut. Bull.* **2015**, *98*, 14–25. [CrossRef] [PubMed]

33. Hilborn, A.; Costa, M. Applications of DINEOF to Satellite-Derived Chlorophyll-a from a Productive Coastal Region. *Remote Sens.* **2018**, *10*, 1449. [CrossRef]

34. Alvera-Azcarate, A.; Vanhellemont, Q.; Ruddick, K.G.; Barth, A.; Beckers, J.-M. Analysis of high frequency geostationary ocean colour data using DINEOF. *Estuarine Coast. Shelf Sci.* **2015**, *159*, 28–36. [CrossRef]

35. Liu, X.; Wang, M. Analysis of diurnal variations from the Korean Geostationary Ocean Color Imager measurements using the DINEOF method. *Estuarine Coast. Shelf Sci.* **2016**, *180*, 230–241. [CrossRef]

36. Liu, X.; Wang, M. Gap filling of missing data for VIIRS global ocean color product using the DINEOF method. *IEEE Trans. Geosci. Remote Sens.* **2018**, *56*, 4464–4476. [CrossRef]

37. Wang, M. A sensitivity study of SeaWiFS atmospheric correction algorithm: Effects of spectral band variations. *Remote Sens. Environ.* **1999**, *67*, 348–359. [CrossRef]

38. Wang, M.; Franz, B.A. Comparing the ocean color measurements between MOS and SeaWiFS: A vicarious intercalibration approach for MOS. *IEEE Trans. Geosci. Remote Sens.* **2000**, *38*, 184–197. [CrossRef]

39. Wang, M.; Isaacman, A.; Franz, B.A.; McClain, C.R. Ocean color optical property data derived from the Japanese Ocean Color and Temperature Scanner and the French Polarization and Directionality of the Earth's Reflectances: A comparison study. *Appl. Opt.* **2002**, *41*, 974–990. [CrossRef]

40. Wang, M.; Jiang, L. Atmospheric correction using the information from the short blue band. *IEEE Trans. Geosci. Remote Sens.* **2018**, *56*, 6224–6237. [CrossRef]

41. Shi, W.; Wang, M. Satellite views of the Bohai Sea, Yellow Sea, and East China Sea. *Prog. Oceanogr.* **2012**, *104*, 35–45. [CrossRef]

42. Shi, W.; Wang, M.; Jiang, L. Spring-neap tidal effects on satellite ocean color observations in the Bohai Sea, Yellow Sea, and East China Sea. *J. Geophys. Res.* **2011**, *116*, C12032. [CrossRef]

43. Shi, W.; Wang, M. Ocean reflectance spectra at the red, near-infrared, and shortwave infrared from highly turbid waters: A study in the Bohai Sea, Yellow Sea, and East China Sea. *Limnol. Oceanogr.* **2014**, *59*, 427–444. [CrossRef]

44. Campbell, J.W.; Blaisdell, J.M.; Darzi, M. Level-3 SeaWiFS Data Products: Spatial and Temporal Binning Algorithms. *Oceanogr. Lit. Rev.* **1996**, *9*, 952.

45. Wang, M.; Shi, W. Cloud masking for ocean color data processing in the coastal regions. *IEEE Trans. Geosci. Rem. Sens.* **2006**, *44*, 3196–3205. [CrossRef]

46. Jiang, L.; Wang, M. Identification of pixels with stray light and cloud shadow contaminations in the satellite ocean color data processing. *Appl. Opt.* **2013**, *52*, 6757–6770. [CrossRef] [PubMed]

47. Wang, M.; Bailey, S.W. Correction of the sun glint contamination on the SeaWiFS ocean and atmosphere products. *Appl. Opt.* **2001**, *40*, 4790–4798. [CrossRef] [PubMed]

48. Mikelsons, M.; Wang, M. Interactive online maps make satellite ocean data accessible. *EOS* **2018**, *99*. [CrossRef]

 *remote sensing*

Article

Consistency of Radiometric Satellite Data over Lakes and Coastal Waters with Local Field Measurements

Krista Alikas *, Ilmar Ansko, Viktor Vabson, Ave Ansper, Kersti Kangro, Kristi Uudeberg and Martin Ligi

Tartu Observatory, University of Tartu, Observatooriumi 1, Tõravere, 61602 Tartumaa, Estonia;
ilmar.ansko@ut.ee (I.A.); viktor.vabson@ut.ee (V.V.); ave.ansper@ut.ee (A.A.); kersti.kangro@ut.ee (K.K.);
kristi.uudeberg@ut.ee (K.U.); martin.ligi@ut.ee (M.L.)
* Correspondence: krista.alikas@ut.ee

Received: 26 December 2019; Accepted: 7 February 2020; Published: 12 February 2020

Abstract: The Sentinel-3 mission launched its first satellite Sentinel-3A in 2016 to be followed by Sentinel-3B and Sentinel-3C to provide long-term operational measurements over Earth. Sentinel-3A and 3B are in full operational status, allowing global coverage in less than two days, usable to monitor optical water quality and provide data for environmental studies. However, due to limited ground truth data, the product quality has not yet been analyzed in detail with the fiducial reference measurement (FRM) dataset. Here, we use the fully characterized ground truth FRM dataset for validating Sentinel-3A Ocean and Land Colour Instrument (OLCI) radiometric products over optically complex Estonian inland waters and Baltic Sea coastal areas. As consistency between satellite and local data depends on uncertainty in field measurements, filtering of the in situ data has been made based on the uncertainty for the final comparison. We have compared various atmospheric correction methods and found POLYMER (POLYnomial-based algorithm applied to MERIS) to be most suitable for optically complex waters under study in terms of product accuracy, amount of usable data and also being least influenced by the adjacency effect.

Keywords: atmospheric correction; Sentinel-3 OLCI; Copernicus; measurement uncertainty; validation

---

1. Introduction

There is a growing constellation of satellite sensors providing Earth observation data for monitoring aquatic ecosystems. These ecosystems provide complex conditions for optical remote sensing in terms of optical water types. The adjacency to land also causes bias due to multiple scattering of the light detected by the sensor [1] which needs to be removed when the aim is to detect optical properties of the water.

Sentinel-3 (S3) is an ocean and land mission that consists of three satellites S3A, S3B and S3C, providing environmental monitoring data under the Copernicus program until 2030 [2,3]. There are four different instruments onboard S3 satellites, from which Ocean and Land Colour Instrument (OLCI) is a medium-resolution imaging spectrometer aimed for monitoring optical water quality. OLCI fulfills many of the mission objectives, e.g., measuring ocean and land surface color, monitoring seawater quality and pollution and monitoring land use change. OLCI is also the main contributor to monitoring inland waters [4]. One of the S3 mission requirements is that the measurements and products shall include uncertainty estimates [4]. The uncertainties shall be within 5% for the radiometric data. Currently in OLCI-A open water products, the water-leaving reflectance $\lfloor \rho_w \rfloor_N$ partly meets the S3 mission requirements at averaged global and temporal scales [5] where bands at 490, 510 and 560 nm are within the 5% mission requirement uncertainty for all water types; bands 400, 412 and 442 are within 5–10% depending on water type and bands at longer wavelengths having uncertainties higher or depending on a water type. OLCI-A Level-2 Ocean Colour (OC) operational products provide additional per-pixel uncertainty estimates but they have not been verified yet as they do not include uncertainty estimate from the Level-1 products [2]. In parallel, active research is ongoing to better quantify in situ measurement uncertainties in controlled (laboratory) [6] and in natural measurement (outdoors) [7] conditions. This enables collecting fiducial reference measurements (FRM) that are traceable to SI standards and can be further used with confidence for the validation of satellite products. This FRM data improves the comparability of radiometric in situ measurements with ocean color data, e.g., S3 OLCI data to analyze the performance of the algorithms, estimate the errors and characterize the main sources of uncertainties.

From the S3 OLCI constellation, both S3 A and B are on-orbit and provide daily measurements. The products of S3 A and B OLCI sensor are slightly different due to (1) instrument radiometric response; (2) system vicarious calibration; (3) modeling of the temporal evolution at Level-1 products; (4) spectral response functions; (5) radiometric and geometric behaviors of both instruments [2]. In this study, we focused solely on S3A OLCI products to exclude conclusions based on the differences between the sensors and their data processing.

The objectives of this work are: (1) to analyze the in situ measured radiometric data in order to find the sources and variability of uncertainties; (2) to provide uncertainty budgets for in situ datasets which can be used to reveal and explain differences with the S3A OLCI data over optically complex waters; and (3) to identify the most suitable atmospheric correction (AC) algorithm.

2. Materials and Methods

2.1. Study Sites

Field measurements were carried out in large Estonian lakes Peipsi and Võrtsjärv and in Baltic Sea coastal waters (Table 1). Lake Peipsi and Võrtsjärv are large, shallow and well-mixed Estonian lakes. Lake Peipsi is a transboundary waterbody between Russia and Estonia. It is divided into three parts: in the north mesotrophic Lake Peipsi *sensu stricto* (s.s.), in the south hypertrophic Lake Pihkva and their connection and river-like eutrophic Lämmijärv. Different water types can be observed in this lake: clearest and deepest water in Peipsi s.s., turbid and brownish water near the river Suur Emajõgi delta and phytoplankton rich Lämmijärv and L. Pihkva. In the summertime, cyanobacterial blooms may be present in the entire lake. Lake Võrtsjärv is a very turbid, well-mixed, eutrophic and non-stratified lake with a high absorption coefficient of colored dissolved organic matter at wavelength 442 nm (a_{CDOM}), phytoplankton and concentration of total suspended matter (TSM). Water level variations control nutrient and phytoplankton availability. The Baltic Sea is a semi-enclosed sea with relatively high freshwater input by rivers. The ranges of the optical parameters over the Baltic Sea areas are wide [8], and a_{CDOM} is the dominant optical component [9] with a strong relationship with salinity [10].

Table 1. Morpho-dynamical data and bio-optical parameters of the study sites (minimum, maximum and median value).

Parameter	Lake Peipsi	Lake Võrtsjärv	Baltic Sea Coastal
Mean depth, m	7	2.8	7.5
Max depth, m	15.3	6	12
Chla, mg·m^{-3}	2.7–122 (25.1)	3.4–72.2 (40.4)	0.7–10.7 (4.1)
C_{TSM}, g·m^{-3}	1.3–61.3 (8.8)	1.6–52.7 (16)	5.0–24.3 (10)
a_{CDOM} (442), m^{-1}	1.2–7 (2.9)	1.9–8.9 (2.9)	0.6–3.7 (0.9)
Secchi depth, m	0.4–3.6 (1.5)	0.2–2.6 (0.75)	0.5–4.3 (1.3)

Note: TSM: total suspended matter; a_{CDOM}: absorption coefficient of colored dissolved organic matter at 442 nm; Chla: chlorophyll-a.

2.2. In Situ Measurements

The above-water radiometric measurements were performed from the research vessel (approximately 6 m long) from about 2 m height from the water surface. In each station, the vessel was anchored and radiometric measurements were recorded at least for 15 min. The water-leaving reflectance spectra were calculated from the well-synchronized time series measured with the three above water TriOS-RAMSES hyperspectral radiometers following the protocol of REVAMP [11]. Calculations included the following steps: firstly, all measured radiance and irradiance spectra were corrected for the stray light [6,12]; secondly, spectral response functions of OLCI bands were used to convolve spectra into OLCI band values; thirdly, the time series of water-leaving reflectance $\lfloor\rho_w\rfloor_N$ was calculated as

$$\lfloor\rho_w\rfloor_N = \pi R_{rs}(\lambda) = \pi\frac{L_u(\lambda) - \rho(W)L_d(\lambda)}{E_d(\lambda)} \tag{1}$$

where $R_{rs}(\lambda)$ is the remote sensing reflectance, $L_u(\lambda)$ is the upwelling radiance from the sea, $L_d(\lambda)$ is the downwelling radiance from the sky, $E_d(\lambda)$ is the downwelling irradiance and $\rho(W)$ is the sea surface reflectance as function of wind speed (W, m·s^{-1}), calculated as

$$\rho(W) = 0.0256 + 0.00039W + 0.000034W^2 \tag{2}$$

[11]. Next, the Near-Infrared (NIR) similarity correction with $\lambda_1 = 720$ nm, $\lambda_2 = 780$ nm and $\alpha = 2.35$ was applied to the water-leaving reflectance according to Ruddick et al. [13]. The constant parameter α of the NIR similarity correction [14] is determined in [13] and depends on the choice of wavelengths λ_1 and λ_2; $\alpha = 2.35$ for the $\lambda_1 = 720$ nm and $\lambda_2 = 780$ nm.

After that, for each measurement station, the median of $R_{rs}(560)$ (OLCI band value with center at 560 nm) was calculated. Any spectrum deviating from the mode more than ±10% was excluded from further analysis in order to eliminate outliers due to changing measurement and illumination conditions. Finally, the mean water-leaving reflectance with uncertainty [15] was calculated to each measurement station.

Additionally, environmental parameters such as wind speed, cloudiness, sun condition, solar elevation angle, wave height, Secchi depth and concentrations of optically significant constituents (OSC, e.g., concentration of chlorophyll-a, concentration of total suspended matter and absorption coefficient of colored dissolved organic matter at wavelength 442 nm) were measured. The wind speed was measured with a handheld mechanical anemometer. The overall sky cloudiness, the presence of clouds in front of the sun and the wave height were estimated by visual inspection. For cloudiness, a 100-point scale was used with 0 for clear sky to 100 for fully covered. Sun condition was classified into four groups: clear, partially covered, through optically thin clouds and fully covered. For Secchi depth, the white disk with 30 cm diameter was used and measurements were performed on the shaded side of the vessel. The solar elevation angle was calculated based on the measurements time and geographic coordinates. For concentrations of OSC, the water samples were collected from the water surface (up to 0.5 m depth) and analyzed using the methods of Lindell et al. [16]. Chlorophyll-a (Chl a) was measured spectrophotometrically with a Hitachi U-3010 spectrophotometer and calculated according to the method of Jeffrey and Humphrey [17]. TSM was measured gravimetrically. Lastly, a$_{CDOM}$(442) was derived by water sample filtering through a filter with a pore size of 0.2 μm and measured in a 5 cm optical cuvette against distilled water with a Hitachi U-3010 spectrophotometer.

2.3. Uncertainty Budget

The methodology used for the uncertainty evaluation is consistent with the ISO Guide to the Expression of Uncertainty in Measurement (GUM) [18]. The evaluation is based on the measurement model, which describes the output quantity Y as a function f of input quantities X_i: $Y = f(X_1, X_2, X_3, \ldots)$. For example, for remote sensing reflectance $R_{rs}(\lambda)$, Equations (1) and (2) are used. For every input quantity X_i, respectively, estimate x_i and standard uncertainty $u(x_i)$ are evaluated which are considered as parameters of probability distribution describing the X_i. The combined standard uncertainty $u_c(y)$ for output estimate is calculated from the standard uncertainties associated with each input estimate x_i, using a first-order Taylor series of $y = f(x_1, x_2, x_3, \dots)$. There are two types of standard uncertainties: Type A is of statistical origin; Type B is determined by other means. Both types of uncertainties are indicated as standard deviation, denoted correspondingly by s and u. In calibration of array spectrometers, the uncertainty contributions arising from averaging of a large number of repeatedly measured spectra is considered as of Type A. Contributions from calibration certificates (standard lamp, diffuse reflectance panel, multimeter, current shunt, etc.), but also from instability and spatial non-uniformity of the lamp are considered for Type B.

Radiometric calibration of the irradiance and radiance sensors and their uncertainty budgets are described in [19]. The uncertainty of radiometric calibration stated in [19] has been successfully verified in international comparison between four participants (Tartu observatory of University of Tartu, National Physical Laboratory, The Joint Research Centre, TriOS) in 2016, and since 2018, respective calibration services are accredited by Estonian Accreditation Centre (EAK). Information about the use of calibrated sensors for laboratory and field measurements, about the evaluation of corrections due to different effects and/or respective uncertainty contributions without corrections applied are given in [6,7]. Additional information about the long-time instability of sensors can also be found in [6].

In the three-radiometer system used for the determination of $R_{rs}(\lambda)$ in this paper, the same standard lamp was used for the calibration of all three sensors measuring, respectively, E_d, L_u and L_d. Therefore, the system calibration accounts for mechanical alignment of the lamp, plaque and sensors, for inadequate baffling, only for the short time instability of the irradiance standard, and for the uncertainty of the diffuse reflectance plaque. The contribution of the lamp calibration uncertainty cancels almost fully out.

The following components were included to the uncertainty budget of the in situ $R_{rs}(\lambda)$: radiometric calibration of the three-radiometer system, responsivity drift of sensors after calibration, temperature effects, interpolation to the common wavelength scale, angular response of the irradiance sensor, wind speed uncertainty, uncertainty in the stray light correction and contribution due to polarization effects (all Type B estimates), repeatability of recorded time series, corrected for lag-1 autocorrelation and uncertainty in the NIR similarity correction (both Type A estimates).

Due to the unstable nature of natural illumination in individual time series recorded during field measurements, often a rather strong autocorrelation was visible. Consequently, besides white noise, a relatively high contribution of $1/f$ type noise can be expected, and the effective number of repetitions will be substantially reduced. Thus, in this case, the effective number of independent measurements has to be considered instead of actual number of data points in the recorded series [20]:

$$n_e \approx n_t \frac{1 - r_1}{1 + r_1} \tag{3}$$

where r_1 is the lag-1 autocorrelation coefficient of the analyzed time series.

If the averaged values of different radiometers are used in calculations, then due to these random drifts in time series, zero correlation between the signals of different radiometers cannot be expected, and respective correlations shall always be estimated and accounted for. The calculation scheme of $R_{rs}(\lambda)$ used in this article is based on example H2.4 of (ISO GUM) [18], where the output time series is determined from three sets of simultaneously obtained observations. By using this approach, the combined uncertainty of $R_{rs}(\lambda)$ is estimated from the time series of $R_{rs}(\lambda)$ output spectra, and the evaluation of correlations between input quantities is not needed. As the values of the time series are statistically dependent, for uncertainty of the averaged value, the effective sample size is calculated by using Equation (3). NIR similarity correction is calculated for every spectrum of water-leaving reflectance $\lfloor \rho_w \rfloor_N$, and then as the average of these values. For uncertainty of the averaged NIR similarity correction value, the effective sample size estimated from Equation (3) has also been used.

Finally, the spectra of the relative uncertainty components were convolved to the OLCI band values used for comparison.

2.4. Atmospheric Correction Processors for OLCI Data

S3A OLCI L1 and L2 Full Resolution Non Time Critical data were downloaded from databases CODAREP (period 2016–2017, baseline 2.23) and CODA (2018, baseline 2.42). Same-day match-ups were used and the distance from the shoreline and the time difference between the satellite overpass and in situ measurements were derived. A 1×1 pixel area was used as a satellite match-up point.

L1 data were processed with AC processors POLYMER (v4.10), Case 2 Regional CoastColor (C2RCC) Processor (v1) and Alternative Neural Net (v1) using SNAP (v6) software developed by Brockmann Consult, Array Systems Computing and Communication and Systémes (C-S).

S3 OLCI L1 are geo-located top-of-atmosphere radiance products, which have passed quality checks and radiometric calibration with pixel classification, correction for atmospheric gasses and smile effect correction. S3 OLCI L2 are atmospherically corrected products produced by using two different AC methods (Baseline Atmospheric Correction (BAC) and Alternative Atmospheric Correction) in parallel for ensuring similarity and consistency to MERIS products. BAC is based on previously developed AC for MERIS [21], which includes also a Bright Pixel Correction [22]. BAC is based on a coupled atmosphere–hydrological model using spectral optimization inversion for outputting water-leaving reflectances. It includes sun glint and white gaps correction, which is determined by certain thresholds for glint detection on a pixel. For detecting the correct band for further AC procedure, Case 2 NIR reflectance estimation process is performed based on radiometry, which includes aerosol and Rayleigh correction. Uncertainties of L1 and L2 products do not contain the full uncertainty budget at the moment, therefore these are suggested by the developers only for qualitative analyses [23]. Quality control was done by excluding pixels flagged as: WQSF_lsb_CLOUD, WQSF_lsb_CLOUD_AMBIGUOUS, WQSF_lsb_CLOUD_MARGIN, WQSF_lsb_COSMETIC, WQSF_lsb_SUSPECT, WQSF_lsb_HISOLZEN, WQSF_lsb_SATURATED, WQSF_lsb_HIGHGLINT, WQSF_lsb_OCNN_FAIL, WQSF_lsb_AC_FAIL and including flagged as WQSF_lsb_BPAC_ON.

POLYMER (POLYnomial-based algorithm applied previously to MERIS) is an AC processor originally developed for MERIS products to remove the sun glint effects of ocean waters, however further development made it applicable to optically complex waters. The AC procedure uses a spectral matching method based on polynomial the atmospheric model and bio-optical water reflectance model, which use all the spectral bands in the visible spectrum. The models are adjusted to obtain the best spectral fit for optimizing the parameters into both models. Unlike other AC processors, POLYMER is based only on NIR bands, which makes the processor able to derive the water-leaving reflectance in the presence of sun glint. For processing the data for the analysis, default parameters were used [24,25]. Quality control was done by including pixels where layer "Bitmask" had values 0 and 1024.

Case 2 Regional CoastColor (C2RCC), originally developed by Doerffer and Schiller [26], is an atmospheric correction processor for optically complex Case 2 waters, which is trained and able to work in extreme conditions of scattering and absorption. It is based on inversion by the neural network technology, which uses a large database of radiative transfer simulations of water-leaving reflectance and top-of-the-atmosphere TOA radiances, taking into account certain water parameters (temperature, salinity) and atmospheric conditions (ozone, air pressure) [27]. For processing the data for the analysis, default parameters (except salinity for inland waters 0.0001) were used. Alternative Neural Net (ALTNN) is a combined AC processor of C2RCC and Case-2 Extreme. Alternative Neural Net is based on the same neural network system as C2RCC, but it has been improved and revised for more accurate results. It is a test version for OLCI and MERIS data with an extended training range and a larger number of training samples, which reduces noise in results and the gives opportunity to derive more reliable results [27]. For both C2RCC and ALTNN, the pixels were excluded from the analyses, which were flagged as: Rhow_OOS, Cloud_risk, Rhow_OOR, Rtosa_OOR, Rtosa_OOS, quality_flags_sun_glint_risk.

2.5. Data Analysis

First, the in situ dataset was analyzed with Principal Component Analysis (PCA) to visualize the variation present in the in situ radiometric dataset in relation to the uncertainty budget. PCA analyses were performed with R software using the ggbiplot package and prcomp function. Inputs for the PCA were the concentrations of (1) Chl a (mg·m^{-3}), (2) TSM (g·m^{-3}), (3) $a_{CDOM}(442)$ (m^{-1}), (4) wind speed (m·s^{-1}), (5) cloudiness (from 0 to 100), (6) Secchi depth (m), (7) solar elevation angle (degree), (8) R_{rs} at 560 (sr^{-1}), (9) wave height (m), (10) presence of clouds in front of the sun in the scale of 1–4 (1, clear; 2, fully covered; 3, through a thin cloud; 4, partially covered).

Second, the accuracy of the satellite-derived Remote Sensing Reflectance $R_{rs}(\lambda)_{olci,i}$ was then compared against the in situ measured $R_{rs}(\lambda)_{insitu,i}$ values. Mean Absolute Percentage Difference (MAPD) was applied to investigate dispersion and Mean Percentage Difference (MPD) to investigate bias:

$$MAPD_\lambda = \frac{\sum_{i=1}^{n} 100 \left| \frac{R_{rs}(\lambda)_{olci,i} - R_{rs}(\lambda)_{insitu,i}}{R_{rs}(\lambda)_{insitu,i}} \right|}{n} \tag{4}$$

$$MPD_\lambda = \frac{\sum_{i=1}^{n} 100 \left(\frac{R_{rs}(\lambda)_{olci,i} - R_{rs}(\lambda)_{insitu,i}}{R_{rs}(\lambda)_{insitu,i}} \right)}{n} \tag{5}$$

Here, $R_{rs}(\lambda)_{insitu,i}$ and $R_{rs}(\lambda)_{olci,i}$ are, respectively, in situ and OLCI-derived values for the band λ and match-up i.

3. Results

3.1. In Situ Dataset in Terms of Associated Uncertainties

Figure 1 shows the in situ measured water-leaving reflectance processed to OLCIs wavelengths (a), corresponding uncertainty estimates (b) and the relationship between the R_{rs} and uncertainty (c) on selected wavelengths.

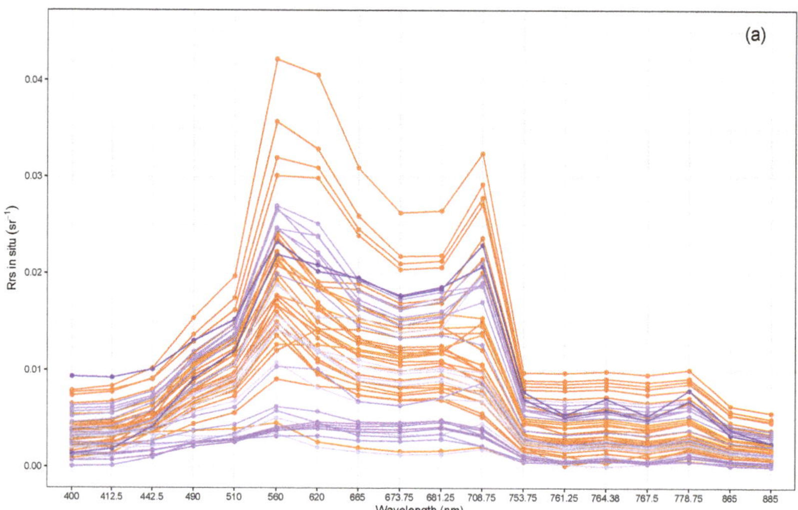

Figure 1. *Cont.*

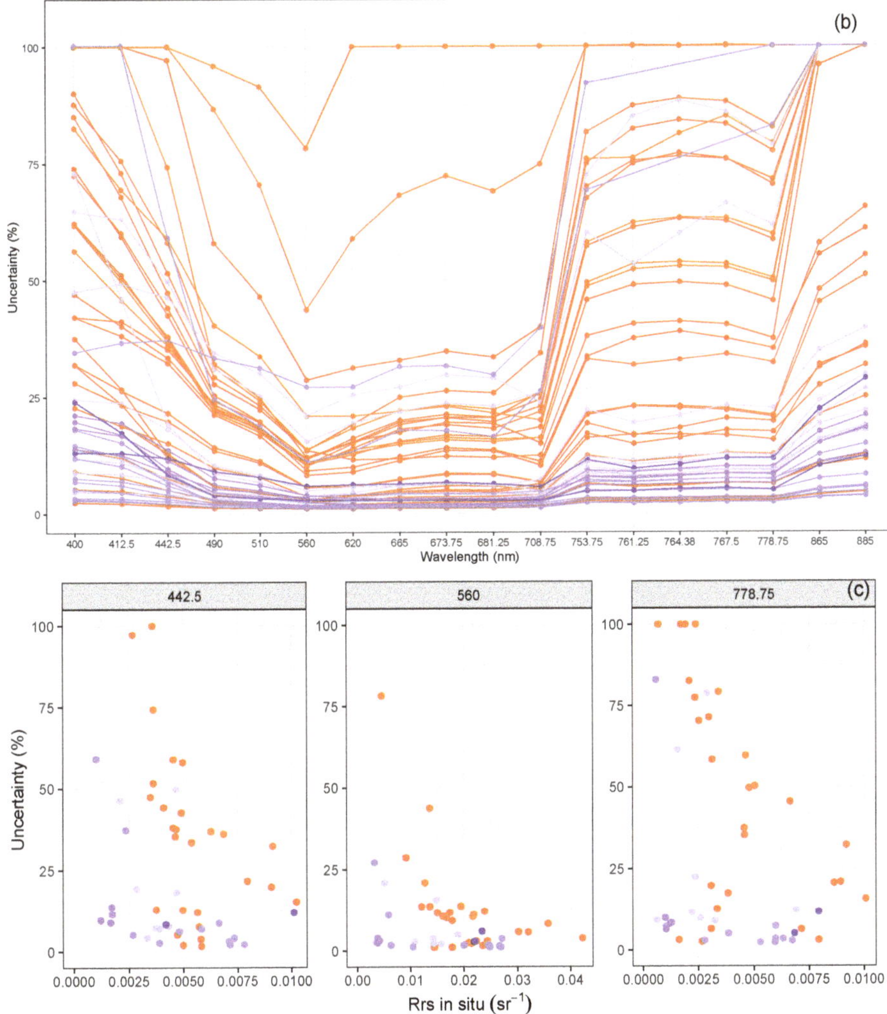

Figure 1. In situ measured water-leaving reflectance after processed with (**a**) S3A Ocean and Land Colour Instrument (OLCI) SRF, (**b**) calculated full uncertainty budget for each band and relationship between (**c**) R_{rs} and uncertainty in selected bands.

The uncertainties are highest in short visible bands from 400 nm and decreasing toward green wavelengths (560 nm). For bands starting from 753.75 nm, the uncertainties are increasing toward longer wavelengths. Table A1 shows that the median uncertainty is less than 10% for bands 490–708.75 with the lowest median uncertainty of 3.9% for the 560 nm band. For bands 510–708.75, 50% of the measurements were obtained with <5% uncertainty, whereas for bands 400–442.5 and 753.75–885, less than 15% of the measurements were obtained with <5% uncertainty (Table A1).

Figure 1c shows elevated uncertainty for some measurements in case of a weaker signal from the water; however, there is still a cluster of measurements with lower uncertainties independent from the signal strength.

To show the contribution of various components to the full uncertainty budget, data from two stations with similar optical water quality but having different environmental conditions (wind speed, cloudiness) were analyzed.

The left panel in Figure 2 shows spectra measured at one station in challenging conditions (upper panel): wind speed 2 m·s^{-1}, overall cloudiness 90%, sun partially covered during the measurements (ID 839 in Table A2) and in good conditions (lower panel): wind speed 1 m·s^{-1}, overall cloudiness 5%, no clouds in front of the sun (ID 786 in Table A2). For both stations, the median hyperspectral spectra and the ones calculated on OLCIs wavelengths (blue) have similar shape and magnitude, although they greatly differ based on the uncertainty estimates (gray line). The uncertainty budget (on the right panel in Figure 2) shows the main contribution comes from environmental conditions for the upper panel due to the high standard deviation in the station spectra caused by changing and challenging environmental conditions during the measurements. The good measurement conditions resulted in a lower standard deviation and there is no domination of one certain component to the full uncertainty budget (Figure 2, lower panel).

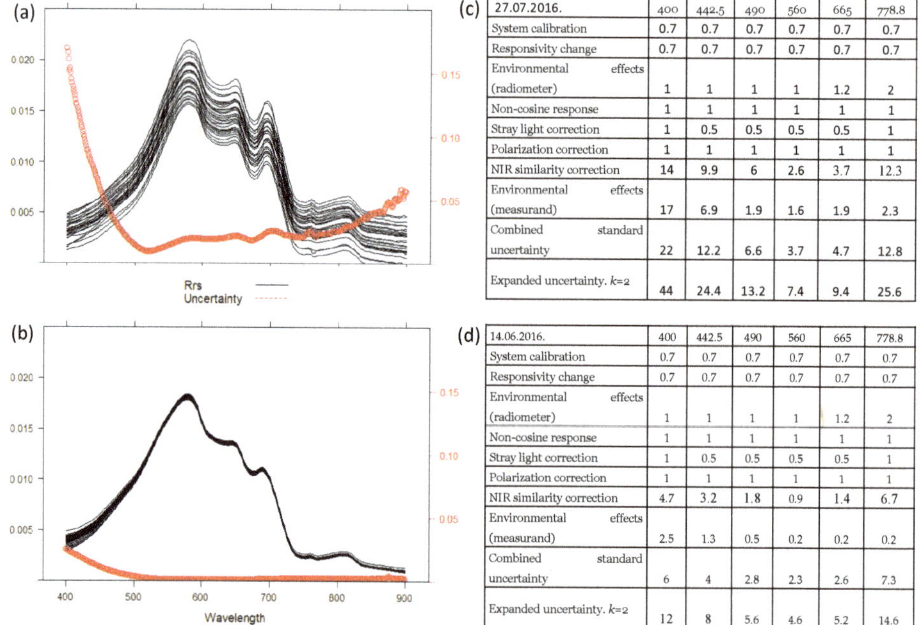

(c)

27.07.2016.	400	442.5	490	560	665	778.8
System calibration	0.7	0.7	0.7	0.7	0.7	0.7
Responsivity change	0.7	0.7	0.7	0.7	0.7	0.7
Environmental effects (radiometer)	1	1	1	1	1.2	2
Non-cosine response	1	1	1	1	1	1
Stray light correction	1	0.5	0.5	0.5	0.5	1
Polarization correction	1	1	1	1	1	1
NIR similarity correction	14	9.9	6	2.6	3.7	12.3
Environmental effects (measurand)	17	6.9	1.9	1.6	1.9	2.3
Combined standard uncertainty	22	12.2	6.6	3.7	4.7	12.8
Expanded uncertainty. $k=2$	44	24.4	13.2	7.4	9.4	25.6

(d)

14.06.2016.	400	442.5	490	560	665	778.8
System calibration	0.7	0.7	0.7	0.7	0.7	0.7
Responsivity change	0.7	0.7	0.7	0.7	0.7	0.7
Environmental effects (radiometer)	1	1	1	1	1.2	2
Non-cosine response	1	1	1	1	1	1
Stray light correction	1	0.5	0.5	0.5	0.5	1
Polarization correction	1	1	1	1	1	1
NIR similarity correction	4.7	3.2	1.8	0.9	1.4	6.7
Environmental effects (measurand)	2.5	1.3	0.5	0.2	0.2	0.2
Combined standard uncertainty	6	4	2.8	2.3	2.6	7.3
Expanded uncertainty. $k=2$	12	8	5.6	4.6	5.2	14.6

Figure 2. Example of radiometric data measured in two stations with different environmental conditions. Spectra measured during one station and derived uncertainty (**a,b**) and calculated uncertainty budget on selected bands (**c,d**). The upper row represents challenging conditions (ID 839 in Table A2) and panel (**b**) good conditions (ID 786 in Table A2).

3.2. PCA

To study the factors associated with different levels of uncertainty in the R_{rs} data (Figure 2a,b), a PCA was applied on the full in situ dataset. Ten input parameters were used for the PCA: TSM, Chl a, a$_{CDOM}$(442), Secchi depth, R_{rs}(560), solar elevation angle, overall sky cloudiness, wind speed, wave height, sun conditions. This resulted in 10 principal components and the contribution from each parameter is shown in Table 2.

Based on the calculated uncertainty budget on the 442 nm band, four categories based on the level of uncertainty were derived: (1) $u(442) < 5\%$, (2) $5\% \leq u(442) < 40\%$, (3) $40\% \leq u(442) < 70\%$, (4) $u(442) > 70\%$. This was used as an additional layer of information for each point in interpreting the PCA results.

Table 2. Principal Component Analysis (PCA) components (PC1–PC10) and contribution by measured parameters. Parameters with highest contribution to first four principal components are marked in bold. The lowest row shows the cumulative proportion (%) of each component to the total variation.

Parameters	PC1	PC2	PC3	PC4	PC5	PC6	PC7	PC8	PC9	PC10
TSM $(g \cdot m^{-3})$	**0.53**	−0.09	0.15	0.01	0.07	0.01	0.19	0.12	−0.55	0.57
Chl a $(mg \cdot m^{-3})$	**0.41**	−0.29	0.11	0.31	0.27	−0.22	0.33	0.32	0.13	−0.54
$a_{CDOM}(442)$ (m^{-1})	0.09	−0.31	−0.45	0.38	−0.59	−0.03	−0.25	−0.02	−0.33	−0.18
Secchi (m)	**−0.49**	0.12	−0.20	0.06	0.40	0.11	0.14	0.23	−0.64	−0.22
$R_{rs}(560)$ (sr^{-1})	0.31	0.31	0.34	−0.36	−0.22	0.32	−0.32	0.16	−0.25	−0.47
Solar elevation angle (°)	0.04	**0.51**	−0.17	-0.19	−0.39	−0.46	0.55	−0.04	−0.08	−0.10
Overall sky cloudiness (%)	0.09	**0.46**	0.11	**0.41**	0.22	−0.52	−0.52	−0.01	−0.06	0.06
Wind speed $(m \cdot s^{-1})$	−0.38	−0.04	**0.46**	0.21	−0.39	−0.05	0.05	0.63	0.07	0.19
Wave height (m)	−0.22	−0.31	**0.57**	−0.01	−0.08	−0.29	0.07	−0.56	−0.30	−0.17
Sun conditions *	0.04	0.37	0.16	**0.61**	−0.08	0.52	0.29	−0.31	0.04	−0.02
Cumulative Proportion	31	57	73	85	91	94	97	98	99	100

* Clear—1; fully covered—2; partially covered—3; through thin cloud—4.

The first principal component (PC1) described 30.8% of variance in the dataset and had the highest contribution from TSM, Chl a and Secchi depth (Table 2, Figure 3). Therefore, PC1, the group explaining the highest variation in the in situ data, can be associated with the optical properties of the water. Figure 3 shows how each in situ measurement is positioned in terms of principal components and associated level of uncertainty estimated at the 442 nm band. There is no association between the level of the uncertainty and the variables contributing most to PC1 (Figure 3). PC2, describing 26.4% of the variance, can be associated mainly with changes in the solar elevation angle and overall sky cloudiness (Table 2). Based on the PCA results, neither PC2 can be used to differentiate points with different levels of measurement uncertainty.

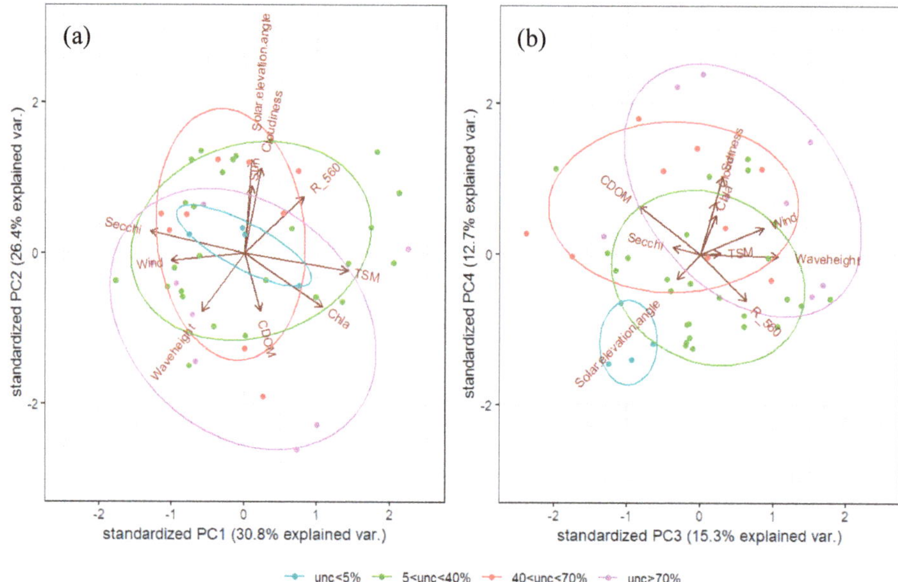

Figure 3. Each data point from the in situ dataset is positioned in terms of the first four principal components (PC) and associated level of uncertainty estimated at the 442 nm band. The clusters based on the uncertainty are shown on (**a**) for PC1 and PC2 and on (**b**) for PC3 and PC4.

Based on PC3 and PC4, different clusters were formed in terms of low (<5% uncertainty at the 442 nm band) and high (>70% uncertainty at the 442 nm band) measurement uncertainty (Figure 3). Both PC3 and PC4 are determined by environmental conditions. The main contribution to PC3 comes from wave height and wind speed, to PC4 from sun conditions (if there are clouds in front of the sun or not) and overall sky cloudiness (Table 2). Based on the PC3 and PC4, measurements with a lower level of uncertainty are associated with lower wave height and wind speed (PC3) and also good illumination conditions (clear sky and no clouds in front of the sun) (PC4).

3.3. Spatial and Temporal Effects on Combining Satellite and In Situ Data

The S3A OLCI image (Figure 4) coupled with in situ sampling dataset from 14 June 2016 was analyzed for the performance of various AC processors in comparison with in situ data in terms of changing temporal and spatial conditions.

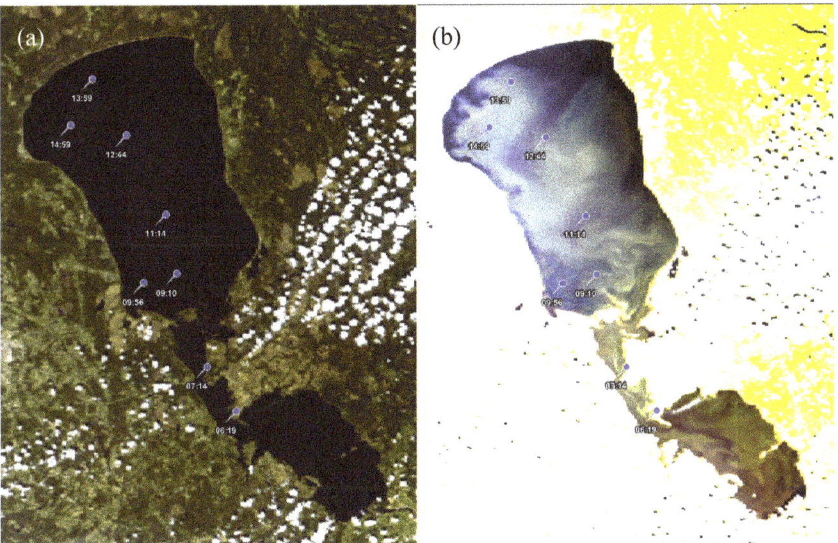

Figure 4. S3A/OLCI L1 image as RGB (**a**) and with modified RGB histogram (**b**) from 14 June 2016 with overpass 8:25 UTC. The in situ sampling stations with measurement time (UTC) on the same day are marked with pins.

The reference measurements, derived R_{rs} from AC processors and environmental conditions for each station, are shown in Figure 5. Each measurement was performed in conditions where no clouds were in front of the sun (Figure A1). The overall sky cloudiness was higher in the first two stations (from 30% to 10%, respectively) but stayed constant at 5% for the following stations (Figure A1). The clouds were Cirrus and Cirrostratus for the three first stations, and later Cumulus and Cirrostratus. Wind speed changed from 0 to 5.5 m/s and wave height from 0 to 0.3 m, both increasing toward the evening (except at station #5, at 11:14 UTC, Figure A1). The in situ measured $R_{rs}(\lambda)$ with associated uncertainties compared to quality-controlled AC retrievals (Figure 5) show that, although the image was cloud-free (Figure 4), the number of retrievals varies station-by-station. The lower two panels (Figure 5) shows the ratio of the AC processor-derived $R_{rs}(\lambda)$ to the in situ measured $R_{rs}(\lambda)$ with respective uncertainty. The uncertainties are higher in the first and last three stations, while stations #2 to #5 have very low uncertainties as the measurements have been performed in conditions with low wave height and wind speed in combination with a solar elevation angle above 40 degrees. The optical properties of water are similar in the first two stations (Figure A1), where the main changes are in the overall sky cloudiness (decrease from 30% to 10%), solar elevation angle (from 36 to 43 degrees) and a slight increase in wind speed (from 0 to 2.5 m/s) and wave height (from 0 to 0.05 m).

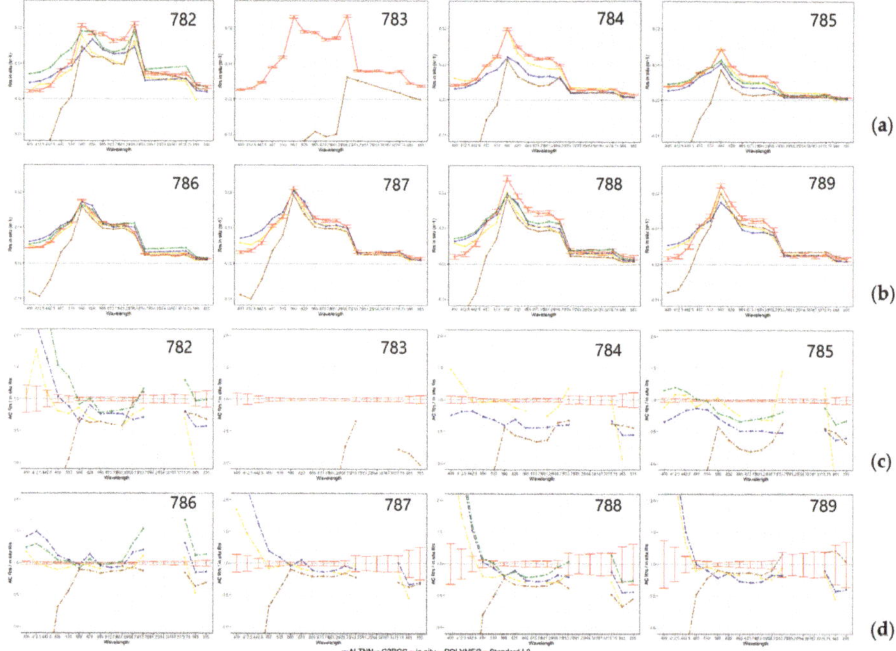

Figure 5. The comparison of in situ and atmospheric correction (AC) processor-derived R_{rs} (**a**,**b**) and the ratio of the AC-derived R_{rs} to the in situ measured R_{rs} (**c**,**d**) in 8 match-up stations on 14 June 2016. Red error bars show the uncertainty derived from in situ measurements. For each point (ID 782 – ID 789 in Table A2), the changes in the sky conditions, the temporal and spatial changes in the measurement conditions and the optical properties of water can be found in Figure A1.

In the first two stations, the Chl a, TSM and a_{CDOM} absorption are the highest and decrease in the following stations, where they stay fairly the same (Figure A1).

The best performance from each AC processor is in the case of station #5 measured at 11:14 UTC where all processor-derived R_{rs} pass the quality control by flags. This station is about 13 km away from the coast. The most complex conditions were for station #2 of the day (at 7:14 UTC), where measurements were performed in a very narrow part of the lake (Figure 4), just 0.8 km from the nearest coast.

Figure 5a shows the changes in the R_{rs} uncertainty during one measurement campaign. Uncertainty was higher during early morning and evening measurements which can be linked with the sun elevation angle (Figure A1) and also with the wind speed in the evening measurements (3.5–5 m·s^{-1}). During the midday station, in good measurement conditions (Figure A1), the uncertainties stayed low in the whole spectrum, e.g., station #4, where uncertainty was 1.1–1.5% for bands 490–708.75, and between 2% and 6.5% for shorter and longer wavelengths. In contrast, in the second to last station, #7, the uncertainty was 3–5.9% for bands 490–708.75, and between 12.1% and 33% for shorter and longer wavelengths, which can be explained by changes in the environmental conditions, e.g., lower sun elevation angle (36.7 degrees) and higher wind speed (5.5 m·s^{-1}).

Although the uncertainties for the first three OCLI bands (up to 442.5 nm) are higher, the AC processor-derived R_{rs} is only derived in the limits of uncertainty by POLYMER AC in few cases; all other AC results tend to under- or overestimate. While the standard AC strongly underestimates bands up to 560 nm, its accuracy is comparable with other processors for longer wavelengths. For bands 665, 673.75 and 681.25, AC-derived R_{rs} is slightly (about 20–30%) underestimated. While POLYMER tends to show the most accurate spectra, it has the highest inaccuracies at the 865 nm band, where all other AC processors perform better.

3.4. Validation of AC Processors on All Match-Ups

Visual observation of spectra from all match-up stations showed that in productive waters (Chl a > 20 mg·m^{-3}), the products of C2RCC and ALTNN (Figure 6a) tend to give peak reflectance at 620 nm instead of 560 nm as measured in situ. C2RCC and ALTNN products are often flagged out over absorbing waters (a$_{CDOM}$ > 1.5 m^{-1}) with Chl a level < 20 mg·m^{-3} (Figure 6d, ID 2228-2288 in Table A2) regardless of the distance from the shore. It was also noted that the agreement between the processors increases further away from the shore (Figure 6b), except the bands up to 510 nm in case of standard product. In the vicinity of land (Figure 6c), the discrepancies between the various AC processors become higher, although the POLYMER tends to be least affected by the adjacency effect.

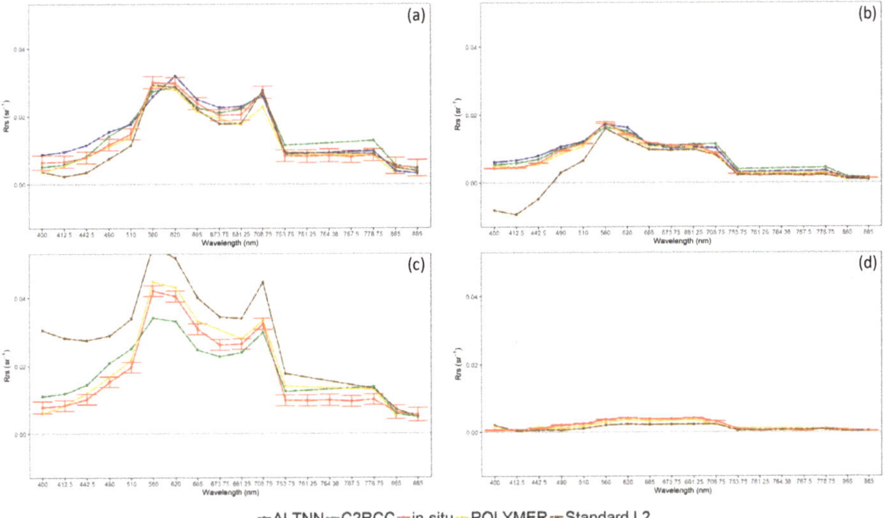

Figure 6. Comparison of all AC processors at four selected match-up station: (**a**) ID833, (**b**) ID786, (**c**) ID835 and (**d**) ID 2288 (in Table A2). Red denotes in situ measurements and corresponding uncertainty at every wavelength.

Table A2 shows the overview of bio-optical properties, environmental variables and adjacency to land for each match-up point with a reference to R_{rs} at 560 nm, measured in situ or derived by AC processors. Combining the results from regression analyses (Figures A2–A5) and Table A2, good conditions for satellite retrievals can be associated with high distance from the land (>7 km,, e.g., ID 786, 787, 842, 1985 and 1986 from Table A2) in combination with good illumination conditions (clear in front of sun, wind speed 0.5 to 5.5 m·s^{-1}, wave height < 0.15 m). The uncertainties at 560 nm stayed under 14%. In these cases, the difference between the processors was usually < 10% and deviation from the in situ < 15% (Table A2). The highest errors between in situ and satellite-derived R_{rs} (ID 903, 901 in Table A2) can be associated with low solar elevation angle (<30 degrees) in combination with high Chl a (>34 mg·m^{-3}) which results in high uncertainty for the in situ measurements (78.3% and 20.9% at 560 nm, respectively). Additionally, measurements performed with high wind speed (5 m·s^{-1}) in combination with high wave height (0.4 m) are associated with higher uncertainties and the errors are high between the AC processor retrievals and situ data (20% to 50%) but low between the processors (2–13%, ID 894, 898 in Table A2).

3.5. Filtering the In Situ Data Based on the Uncertainty

The difference between the in situ measured R_{rs} and AC derived is the highest for the 400 and 412.5 nm bands in the case of each processor (Figures 7 and A2–A4 and Figure A5a). Based on the statistics, the POLYMER-derived R_{rs} is the most accurate for all bands, except at 865 nm. POLYMER-AC-derived R_{rs} values are well aligned around at in 1:1 line (Figure A2) with a relatively smaller bias compared to other processors.

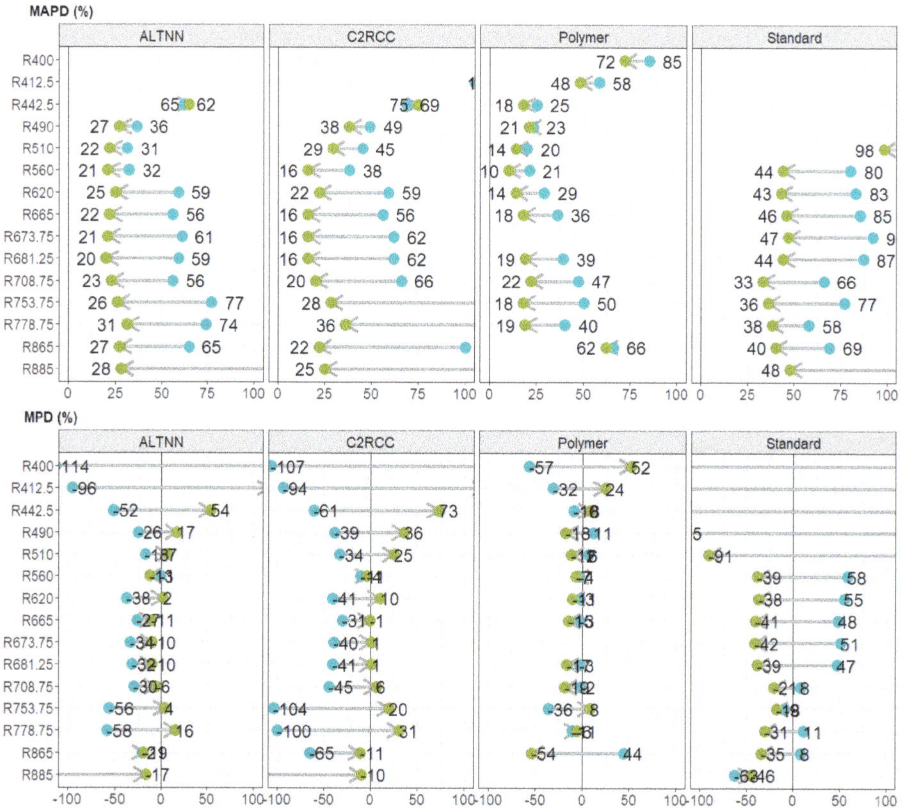

Figure 7. Change in the accuracy of AC processors retrievals on all match-up data (blue dots) and after filtering the in situ data based on the 5% uncertainty threshold at least in one band. The number of match-up points for all data and after filtering: POLYMER (41 → 24), Case 2 Regional CoastColor (C2RCC) (27 → 15), Alternative Neural Net (ALTNN) (29 → 17), Standard L2 (49 → 26). The limits for the *x*-axis have been set to ±100%.

Bands up to 510 nm are overestimated by ALTNN and C2RCC (MPD 114% and 107%, respectively, at 400 nm), slightly overestimated by POLYMER (MPD 57%) and strongly underestimated by standard products (MPD −463%). Improved accuracy is obtained for longer wavelengths by all processors. For POLYMER, ALTNN and C2RCC, the band at 560 nm is derived with the highest accuracy, e.g., MAPD 21% (POLYMER), 32% (ALTNN), 28% (C2RCC). In general, for all AC processors, the derived R_{rs} for bands from 753 nm onwards show higher scatter compared to green bands, although the majority of the estimates are in the limits of the associated in situ measurement uncertainty (Figure 7, and more in detail for each processor in Figures A2–A5).

As the median uncertainty for the in situ R_{rs} varies greatly on different wavelengths (Appendix A, Table A1), the filtering of data was made based on the up to 5% uncertainty criteria at least in one band. This eliminated the in situ data measured in not optimal environmental or measurement conditions and improved the accuracy of the AC processor retrievals (Figures 7 and 8 and for each processor separately; Figures A2–A5). This decrease in match-ups was about 55% for each processor: POLYMER (41 → 24), C2RCC (27 → 15), ALTNN (29 → 17), Standard L2 (49 → 26).

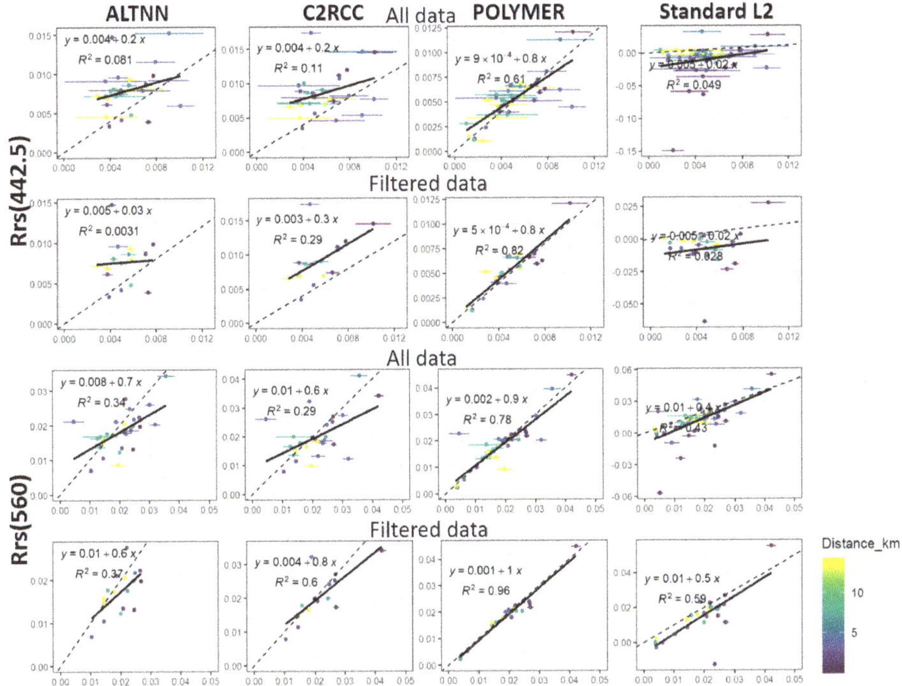

Figure 8. Correlation between in situ measured (x-axis) vs. satellite-derived rho (y-axis) on OLCI bands at 442.5 and 560 nm from C2RCC, ALTNN, POLYMER and standard AC. For both bands, upper row shows validation results on all data and lower after filtering the in situ data based on the 5% criteria at least in one band. For each match-up point, the color shows the distance from the shore.

To estimate the reason for outliers in the correlation plots (Figures A2–A4 and Figure A5b), the distance from the 1:1 line was measured for every point and analyzed against different variables. The outliers were most evident in the adjacency to land in the first five kilometers, and the errors between the AC-derived and in situ measured R_{rs} decrease with increasing distance from the shore (Figure 9). It is especially pronounced in the standard OLCIs R_{rs} products, then C2RCC and ALTNN. In the case of POLYMER, the adjacency has the lowest impact on the quality of the retrievals compared to other processors. As the match-up dataset contains data pairs with variable time difference between the in situ measurements and satellite overpass, Figure 9 shows that over these inland and coastal waters, the time difference between −3 h up to +6 h (in situ minus satellite overpass) can be associated with outliers only in few cases.

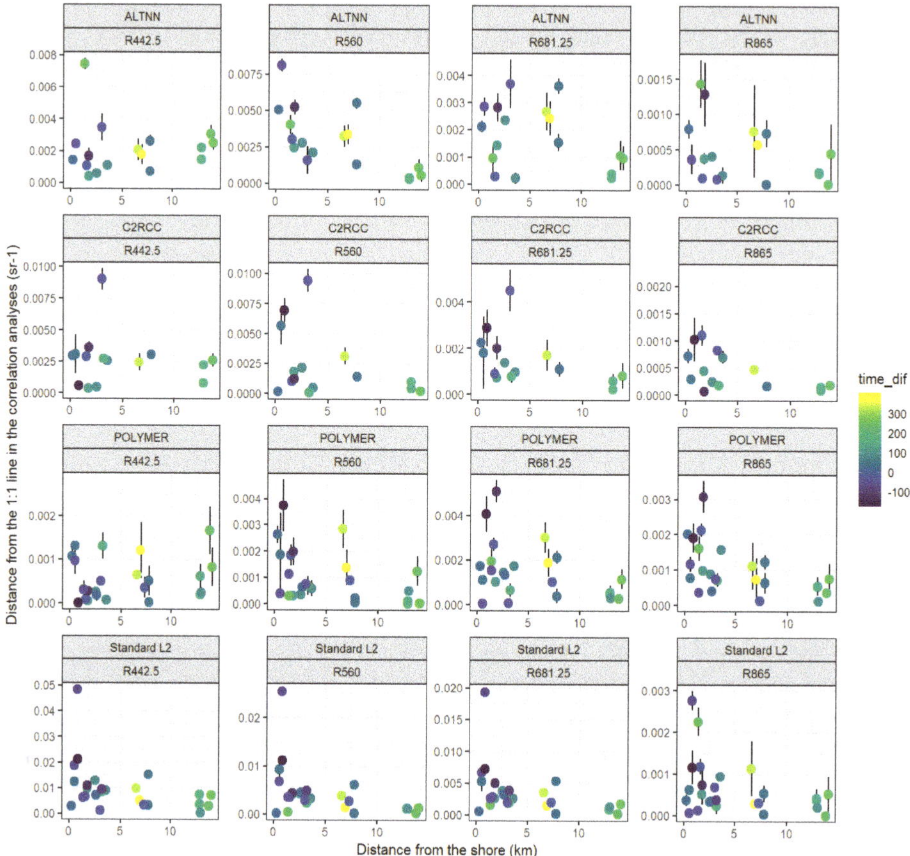

Figure 9. The distance from the 1:1 line in the correlation plots (*y*-axis) and the distance from the shore (*x*-axis) for each match-up point. For each match-up point, the color shows the time difference in minutes between the in situ measurement and the satellite overpass, where the negative values indicate in situ measurement was done prior to satellite overpass.

3.6. Comparability between AC Processors

To exclude the conclusions on AC performances due to the different number of match-up points and processor-based flagging, the statistics were calculated only for these stations when all four processors derived quality control retrievals (Table 3). These eliminated all the match-up stations representing a_{CDOM}-rich waters because none of the C2RCC and ALTNN processor results passed the quality control by flags for these conditions.

Table 3 shows POLYMER retrievals have the smallest dispersion over all OLCIs' bands except for 865 nm, and compared to other processors it has substantially smaller errors in the blue bands (400–490). For bands from 560 nm onwards, the retrievals of all processors are relatively close, while the C2RCC derived R_{rs} have the highest errors. However, the products of C2RCC and ALTNN tend to give a peak in reflectance at 620 nm instead of 560 nm in case of highly productive waters (strong absorption in 681 nm and scatter at 709 nm) and also overestimate R_{rs} at the 778.75 nm band. Although standard L2 products give systematically negative reflectance at shorter wavelengths, the products show comparable accuracy with other processors for bands from 560 nm onwards (Table 3, Figure 10).

Table 3. Statistics on match-up data for tested AC processors only for 22 match-up points when each AC processor (Alternative Neural Net—ALTNN, Case 2 Regional CoastColor Processor—C2RCC) resulted in quality-controlled retrieval.

	MAPD (%)				MPD (%)			
Name	Polymer	Standard	ALTNN	C2RCC	Polymer	Standard	ALTNN	C2RCC
400	**50**	353	109	136	**25**	−276	97	120
412.5	**43**	336	93	116	**22**	−312	83	103
442.5	**22**	194	56	74	**6**	−192	50	64
490	**22**	82	35	52	**−5**	−82	28	40
510	**24**	56	31	48	**−2**	−51	21	36
560	**29**	38	35	42	14	**−2**	7	16
620	**40**	49	71	69	16	**4**	52	51
665	**51**	62	67	64	20	**12**	41	39
673.75	NA	**68**	73	72	NA	**14**	48	49
681.25	**56**	66	70	72	21	**16**	46	50
708.75	**64**	73	**68**	78	**24**	40	42	55
753.75	**75**	97	93	138	**55**	58	72	122
778.75	**56**	71	88	125	**24**	25	71	114
865	**70**	90	76	119	−50	**32**	**32**	81
885	NA	268	241	421	NA	214	198	386

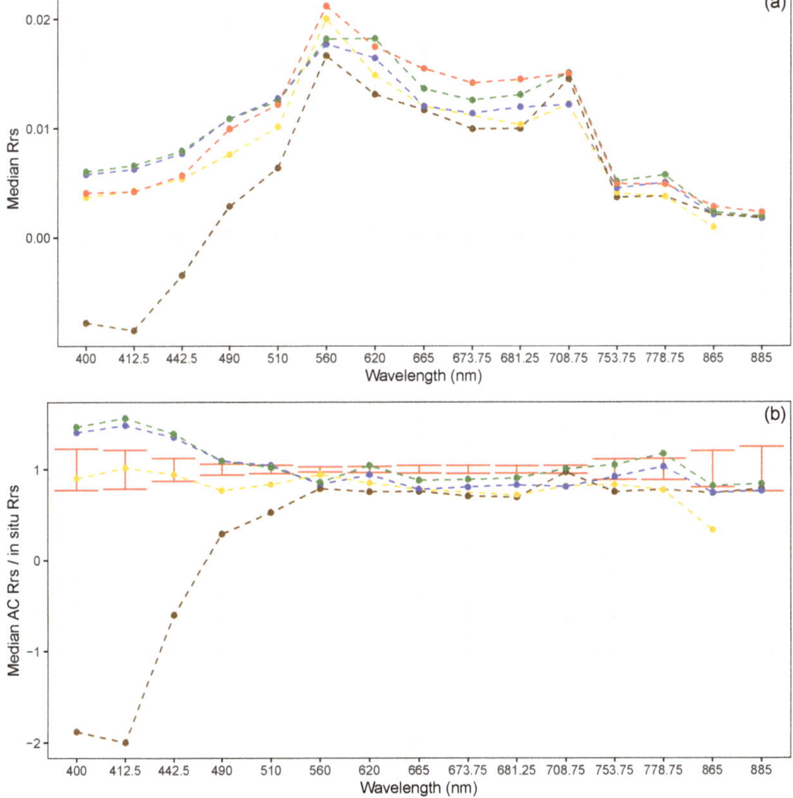

Figure 10. Median R_{rs} for each dataset (**a**) after including match-up stations only when all four AC processors resulted in quality-controlled retrieval. The lower panel (**b**) is the ratio between median AC R_{rs} to median in situ R_{rs} and the median in situ uncertainty at every band.

As the uncertainty of OLCI products was not available at the time of writing, we decided to include the standard deviation between outputs from different AC processors to indicate the accuracy of satellite-driven reflectance (Figure 10).

The highest discrepancies for the AC processors are in the blue wavelengths (up to 442.5 nm). Between these bands, POLYMER median R_{rs} has the closest match to in situ R_{rs} and within the limits of the in situ measurement uncertainty, while C2RCC and ALTNN overestimate about 50% and standard L2 products strongly underestimate R_{rs} values (Figure 10). For the 490 and 510 nm bands, the consistency increases, and for bands 560 to 778, all AC processors derive comparable results and deviate up to 30% from the in situ R_{rs} spectra. The R_{rs} at 665 to 681 nm are systematically underestimated by all processors.

4. Discussion

In order to improve the comparability between field-measured and satellite-derived radiometric data, the uncertainty budget associated with each dataset has to be considered. This enables to assess the uncertainty originating from each component and therefore allows to detect components with the highest uncertainty. This knowledge can direct future research to focus on decreasing the uncertainty on components that impede the meaningful comparison between satellite and in situ data.

4.1. Sources and Variation of Uncertainties on the In Situ Measured Radiometric Data

SI-traceable radiometric calibration of sensors before the field measurements is indispensable in order to guarantee the S3 mission uncertainty requirement of 5% for the in situ results. Without such calibration, only relative variability between the same type sensors measuring the same object was from 6% to 10%, as revealed during the laboratory inter-comparison of radiometers used for satellite validation [6]. The largest differences between sensors (about 10%) were evident for blue bands. After uniform SI-traceable calibration, the variability between sensors measuring the same stable source in laboratory conditions was well under 1% in the whole range from 400 nm to 900 nm.

Unfortunately, the radiometric calibration of sensors still may be insufficient for producing firm SI-traceable in situ results, as conditions during radiometric calibration and conditions during actual field measurements can be significantly different and therefore not entirely covered in the traceability chain [7,28]. Although traceability may be well in place under "normal" operating conditions, this may not be the case under the full range of realistic operating conditions, which may prevail during field measurements [7,28,29]. Major differences during calibration and in-field use arise from various temperature conditions, from differences in the spectral and spatial distributions of measured radiation, to much larger instability/variability of recorded field signals. All these aspects do contribute to the substantial increase in the uncertainty of the field results [7].

In this study, a large part of the in situ measurements did not meet the S3 mission uncertainty requirement. As seen from analyzing the uncertainty budget for the all in situ dataset, R_{rs} at OLCI bands shorter than 442 nm and longer than 753.75 nm are obtained with less than 5% uncertainty only in about 10% of the cases. The smallest relative uncertainty of in situ results is usually achieved for green bands where the signal level during calibration and during field use is the largest. For the red and NIR bands, the contribution of temperature effects due to the temperature sensitivity of Si-based sensors may be rather large. At the same time, the expected signal level in this range is low. Therefore, relative effects due to different corrections (NIR similarity correction, stray light correction, etc.) can be substantial. For blue bands, the signal level of a reference radiation source used for calibration is low, and as a consequence, calibration uncertainties usually the largest. Relative effects due to different corrections for this range are also rather large. For blue bands, the high uncertainty can be due to the contamination by sun glint in case of above-water measurements, which should be corrected [30,31]. Therefore, it is a challenge to collect in situ data with low uncertainty especially for blue and red and NIR bands, which would be suitable for the validation of satellite data. The large uncertainties in blue bands are also present in the data collected by AERONET-OC stations, especially over the Baltic Sea area [32].

Laboratory and field intercomparison of ocean color radiometers [6,7] has clearly demonstrated that the uncertainty of in situ results cannot be reduced only with enhanced accuracy of radiometric calibration. For achieving better SI-traceability of field measurements, besides crucial radiometric calibration, additional characterization of individual sensors including temperature dependence, nonlinearity, angular response of all radiance and irradiance sensors, spectral stray light and wavelength scale effects is very important. The results of these tests enable considering all possible effects arising from specific field conditions during the in situ measurements.

It was shown in the uncertainty budget of in situ measurements performed during changing environmental conditions that a substantial increase of contributions from the repeatability of recorded time series and from uncertainty in the NIR similarity correction was present, and the increase of some systematic effects due to environmental effects can be expected. Therefore, it is crucial to perform measurements used for validation in optimal weather conditions.

4.2. Benefits of Including Uncertainty Budget for the In Situ Dataset

The PCA analyses showed that the level of uncertainty in the in situ data is associated with various environmental conditions. Based on the dataset used here, it was possible to classify data either with low or high uncertainty based on the wave height and wind speed and also the illumination conditions (visibility of sun and overall sky cloudiness). It was shown that the solar elevation angle is an important factor, especially while performing measurements over phytoplankton rich waters. This kind of analysis should be expanded to cover different optical water types to be able to set thresholds for data acquisition to have confidence in the in situ data which will be used to validate ocean color products.

In this study, we set a 5% uncertainty threshold at least in one band to filter the in situ data. This filtering decreased the dispersion and bias for most of the AC processor-derived results. This eliminates the conclusions made possibly based on the in situ data collected in poor environmental conditions. It was shown that over multiple cases, the performance of the AC processors was poor although the level of uncertainty in the in situ data was low. For these cases, neither the environmental conditions nor time difference between satellite overpass and in situ measurement had an effect. Instead, for each processor and for every band, the distance from the shore was then the parameter explaining most often the deviations between the in situ measured and satellite-derived radiometric data. The adjacency effect correction is an issue in the case of each tested AC processor but was less pronounced in the POLYMER products.

Bulgarelli and Zibordi [1] showed the adjacency perturbations can reach up to 36 km from land in case of OLCI data. The extent and degree are sensitive to land cover and slightly also to water type (at blue wavelengths), depending also on the signal-to-noise ratio level of the sensor. It was shown that perturbations induced by adjacency effect at NIR and visible wavelengths might compensate each other and biases are not directly linked with the intensity of the reflectance of the nearby land. For our test sites, both poorly (green vegetation) and highly reflective (sand beaches) land cover was present. Bulgarelli and Zibordi [1] found that adjacency effect increases with water absorption in case of highly reflective land covers (e.g., sand), and over bare soil and green vegetation (which was the majority of land cover in this study), the impact might be limited to few first kilometers for the visible wavelengths. We found that the outliers could be explained by the vicinity to land up to 5 km. Thus, the adjacency effect is an important factor contributing to the accuracy of the satellite-derived products over coastal and enclosed water bodies. Various optical water types, adjacent to a composite of multiple land covers, can result in various perturbations and should be accounted for while deriving accuracy estimates on satellite-derived radiometric data.

4.3. Comparison of Atmospheric Correction Algorithms over Optically Complex Waters

To get accurate atmospheric correction over lakes and coastal waters is much more difficult than in a marine environment due to larger non-uniformity and instability of atmospheric parameters (atmospheric water vapor, temperature, etc.), due to different water types, shallow water and/or vicinity to land effects.

For the satellite-derived radiometric product, the highest deviations from the in situ data are for the blue bands. This is a challenge partly due to the higher uncertainties in the field data [33] and also due to the optimal parametrization of the AC processor [34,35]. The green bands are estimated most accurately and the bands in the red and NIR wavelengths are often estimated within the limit of in situ uncertainty. C2RCC and ALTNN products often showed R_{rs} peak at 620 nm over phytoplankton rich waters and also a majority of data was flagged out over absorbing waters (high a_{CDOM} in combination with low Chl a). Standard L2 product underestimates strongly R_{rs} up to 560 nm but for longer wavelengths shows comparable results with other processors. This can be partly explained by the system vicarious calibration (SVC) gains, which mitigate the radiometric biases in L1b, but they are currently not optimal for optically complex waters. For the standard L2 products, there are many ongoing activities (bright pixel correction) and planned activities on SVC (recomputation of gains for OLCI-A, computation for OLCI-B) with potential perspective to derive SVC gains specific to complex water products and also additional improvements on cloud flags (cloud risk, cloud shadow, snow/ice cloud). This should all increase the accuracy of the OLCI-A operational products for the Case 2 waters.

In general, the flagging criteria for satellite data used in this study do not eliminate invalid pixels for the standard L2 products. For the products of POLYMER, ALTNN and C2RCC, the flagging criteria seem reasonable as the few clear outliers were associated with higher uncertainty in the in situ data or the vicinity of the land.

Currently, the POLYMER AC tends to give the most accurate results for the radiometric data over tested inland and coastal waters. It tends to account for the adjacency effect better compared to other processors. It shows the lowest bias compared to in situ values in all bands except 865 nm and performs equally well over different water types.

5. Conclusions

Copernicus program has ensured a growing constellation of Sentinel satellites usable to monitor optical water quality from regional to global applications. Consistency between radiometric satellite data and local field measurements has to be obtained in order to have the assurance on the remote sensing derived products. The traceability chain for both measurements provides knowledge about the components associated with the highest uncertainty which should be accounted for and decreased, if possible, to improve the accuracy and precision of the measurements.

The calibration and characterization of the radiometers are essential to know the measurements' uncertainty in controlled conditions. However, the environmental conditions tend to differ from the laboratory conditions causing effects that cannot be properly accounted for. In situ radiometric measurements should be performed in optimal conditions to reduce noise in the data due to poor environmental conditions (high wind speed, wave height, low solar elevation angle). This allows obtaining in situ data with low uncertainty which can be further used to validate the satellite-derived radiometric data.

From the tested AC processors, POLYMER radiometric products show the best agreement with in situ data, being least influenced by the adjacency effect. The C2RCC and ALTNN processors tend to depend on the level of OAS in water, being not suitable for highly absorbing waters with low Chl a content. OLCI-A operational L2 products strongly underestimate the blue wavelengths, although, from the 560 nm band onwards, the results from all AC processors are relatively similar.

Author Contributions: Conceptualization, K.A., I.A. and V.V.; methodology, K.A., I.A. and V.V.; software, I.A.; investigation, K.A., I.A., V.V.; data curation, K.A., I.A., V.V., A.A., K.K., K.U. and M.L.; writing—original draft preparation, K.A.; writing—review and editing, K.A., I.A., V.V., A.A., K.K., K.U. and M.L.; visualization, K.A. and I.A.; funding acquisition, K.K., M.L. All authors have read and agreed to the published version of the manuscript.

Funding: This research was funded by (1) European Space Agency project Fiducial Reference Measurements for Satellite Ocean Color (FRM4SOC), contract no. 4000117454/16/I-Sbo; (2) EU's Horizon 2020 research and innovation programme (grant agreement no. 730066, EOMORES) and; (3) Estonian Research Council grants PSG10 and PUTJD719.

Acknowledgments: We are grateful to the Centre for Limnology, Estonian University of Life Sciences for possibility to have joint field work cruises that enabled us to collect the data used in this manuscript. We would also like to thank three reviewers for the constructive feedback to our manuscript.

Conflicts of Interest: The authors declare no conflicts of interest.

Appendix A

Table A1. Cumulative frequency of measurements with categorized uncertainty estimate, $u(R_{rs})$, for every OLCI band. The lowest row shows the median uncertainty for every band. Total 59 in situ measurements.

$u(R_{rs})$	400	412.5	442.5	490	510	560	620	665	673.75	681.25	708.75	753.75	778.75	865	885
<5	6	7	9	21	26	31	31	29	28	28	28	11	10	4	4
<10	12	12	21	32	33	37	36	35	35	35	35	24	23	11	11
<20	22	26	33	35	45	51	50	46	44	45	43	32	32	25	22
<50	37	40	48	53	56	58	57	57	57	57	56	42	41	35	33
50–100	59	59	59	59	59	59	59	59	59	59	59	59	59	59	59
Median	31.9	26.4	17.5	8.7	6.8	3.9	4.2	5.2	6	5.9	5.7	16.2	15.7	27.4	31.7

Table A2. Overview of match-up dataset. The bio-optical properties, recorded environmental conditions and radiometric data at the 560 nm band, measured in situ and obtained from tested AC processors.

ID	Secchi (m)	TSM (gm^3)	Chla (mg m^3)	a_{CDOM} (m^{-1})	Sun *	Solar .elev	Cloud (%)	Wind (m s)	Wave Height	Dist (km)	Unc (%)	In Situ R_{rs}_560	Sat_med R_{rs}_560	Difference (%) from the AC Median R_{rs}					Difference (%) from the In Situ R_{rs}			
														Poly	L2	ALT	C2R	insitu	Poly	L2	ALT	C2R
735	0.7	10.8	34.7	2.5	1	51	90	3.4	0.05	4.8	9.3	0.018	0.026	21	−21	NA	NA	32	−16	−78	NA	NA
782	0.75	15.3	36.3	3.8	1	36	30	0	0	1.8	2.5	0.021	0.016	−10	10	18	−17	−27	13	29	35	8
783	0.8	15.3	37.2	3.2	1	43	10	2.5	0.05	0.8	1.6	0.023	−0.013	NA	0	NA	NA	286	NA	154	NA	NA
784	1.4	9.7	15.7	1.7	1	53	5	1.5	0.05	7.8	1.8	0.020	0.012	−65	7	0	NA	−64	0	43	39	NA
785	1.6	6.6	10.3	1.5	1	55	5	3.5	0.1	2.6	1.1	0.014	0.011	−37	23	4	−4	−32	−4	41	27	21
786	1.4	8.6	13.8	1.5	1	53	5	1	0	12.9	1.1	0.018	0.017	−6	5	−3	3	−5	0	10	3	8
787	1.4	6.3	9.2	1.8	1	45	5	3.5	0.15	14.1	2.3	0.021	0.021	−4	5	0	NA	−4	0	9	4	NA
788	1.4	7.1	14.0	1.5	1	37	5	5.5	0.3	6.6	3	0.024	0.020	−3	5	0	0	−23	17	23	19	18
789	1.4	6.3	10.6	1.6	1	29	7	5	0.2	7.0	3	0.022	0.020	−1	0	13	NA	−10	9	9	21	NA
790	1.8	8.8	20.2	1.4	3	37	60	5	0.2	7.3	11.8	0.017	0.016	−11	11	NA	NA	−11	0	20	NA	NA
791	1.6	5.8	19.2	1.5	3	45	60	5	0.2	6.6	11.7	0.015	0.014	−11	11	NA	NA	−6	−4	16	NA	NA
792	1.7	7.7	19.9	1.3	3	51	90	5.5	0.15	14.1	10.3	0.017	0.013	−7	7	NA	NA	−30	18	28	NA	NA
794	1.4	5.8	17.8	4.2	3	47	100	4.5	0.15	2.7	28.6	0.009	−0.010	NA	0	NA	NA	194	NA	206	NA	NA
832	0.45	16.3	31.9	2.5	1	46	30	0	0.15	3.0	5.7	0.032	0.019	−4	4	−5	36	−65	37	42	37	62
833	0.5	24.7	36.1	2.4	1	48	35	0.5	0.05	3.0	5.8	0.030	0.028	−3	−3	9	3	−6	3	3	14	9
834	0.5	16.7	35.5	2.0	1	51	50	0	0.05	4.8	8.3	0.036	0.040	2	−3	15	−2	12	−11	−17	4	−15
835	0.4	20.0	30.1	2.3	1	53	80	0	0.005	0.5	3.8	0.042	0.045	0	−24	NA	24	6	−6	−31	NA	19
838	0.9	12.5	28.2	2.8	3	47	100	3	0	3.0	12.1	0.024	0.019	NA	55	0	0	−28	NA	65	22	22
839	1.35	9.1	15.0	2.3	3	50	90	2	0.05	0.9	10.7	0.017	0.002	NA	0	NA	NA	−905	NA	90	NA	NA
841	0.65	9.7	22.0	4.7	3	50	80	2.5	0.05	0.7	13.6	0.012	−0.024	NA	0	NA	NA	150	NA	300	NA	NA
842	1.7	4.3	9.5	1.6	1	47	80	5.5	0.15	7.4	13.6	0.014	0.014	−3	45	0	0	0	−3	45	0	0
894	1.8	7.7	28.5	1.8	2	36	60	5	0.4	7.8	43.7	0.014	0.018	−5	7	5	−13	23	−36	−21	−23	−47
898	1.1	6.8	20.5	1.2	1	31	25	5	0.4	13.6	13.7	0.019	0.009	2	−2	9	−4	−107	52	51	56	50
899	0.9	10.5	25.5	1.6	1	35	10	6	0.4	12.9	10.5	0.022	0.019	−7	−2	2	4	−14	6	10	14	16
900	0.8	12.5	24.8	1.9	1	33	10	4	0.4	2.4	11.1	0.022	0.018	−16	2	−2	25	−24	6	20	18	40
901	0.7	12.0	34.3	4.8	1	29	5	3	0.4	7.3	20.9	0.013	0.015	10	32	−10	−12	14	−5	21	−28	−31

Table A2. *Cont.*

ID	Secchi (m)	TSM (gm³)	Chla (mg m³)	a_{CDOM} (m⁻¹)	Sun *	Solar .elev	Cloud (%)	Wind (m s)	Wave Height	Dist (km)	Unc (%)	In Situ R_{rs}_560	Sat_med R_{rs}_560	Difference (%) from the AC Median R_{rs}					Difference (%) from the In Situ R_{rs}			
														Poly	L2	ALT	C2R	insitu	Poly	L2	ALT	C2R
903	0.6	17.5	55.3	3.4	1	20	10	2	0.3	3.1	78.3	0.005	0.022	−3	6	3	−19	79	−402	−355	−369	−478
1981	0.7	16.0	29.1	3.5	3	35	60	1	0.01	0.6	21	0.005	−0.057	NA	0	NA	NA	109	NA	1224	NA	NA
1982	0.7	17.5	48.5	3.4	3	41	70	2	0.1	3.1	5	0.019	0.022	9	−4	4	−46	14	−6	−21	−12	−71
1983	1.2	7.7	17.5	1.8	1	44	65	3	0.1	0.9	15.6	0.015	−0.003	NA	0	NA	NA	646	NA	118	NA	NA
1985	1.1	8.4	19.6	1.9	1	46	60	2	0.1	7.8	2.2	0.016	0.017	5	8	−5	−5	6	−2	2	−12	−13
1986	1.3	6.0	25.4	1.7	1	46	30	2	0.1	12.9	2.3	0.014	0.014	3	13	−3	−3	1	1	11	−4	−4
1987	1.4	7.3	24.6	1.7	3	40	25	4	0.2	13.8	3.9	0.014	0.015	−5	9	−4	4	6	−12	3	−11	−2
2205	0.9	12.7	16.7	NA	4	29	60	2	0.01	0.9	3.7	0.027	0.017	−26	35	NA	0	−57	20	58	NA	36
2207	0.9	14.6	27.4	NA	4	42	50	2	0.15	0.5	1.8	0.025	0.015	−60	0	13	NA	−63	2	39	47	NA
2208	0.9	14.5	26.1	NA	4	46	40	1	0.1	1.6	1.5	0.027	0.023	−4	7	4	−9	−15	10	19	16	5
2209	0.9	10.8	20.3	NA	4	50	80	0.5	0.05	0.3	1.2	0.027	0.025	7	−7	21	−9	−8	14	1	27	−1
2210	0.9	8.7	13.9	NA	4	52	100	1	0.1	3.6	1.2	0.025	0.023	−5	11	5	−6	−8	3	18	12	3
2211	0.9	9.4	27.0	NA	4	51	80	3	0.2	1.8	1.3	0.010	0.007	−35	37	6	−6	−41	4	55	34	24
2212	1	7.4	9.6	NA	4	50	80	4	0.2	3.2	1.8	0.020	0.020	−6	16	NA	0	−1	−5	17	NA	1
2228	2	1.5	6.8	3.1	1	50	10	5	0.3	2.7	1.8	0.006	0.003	−105	105	NA	NA	−139	14	102	NA	NA
2230	1.6	4.0	9.0	3.2	1	43	40	3	0.1	13.9	27.2	0.003	0.003	24	−24	NA	NA	3	22	−27	NA	NA
2232	2.3	3.5	8.5	2.1	1	36	70	1	0	8.5	11.1	0.006	0.006	2	−2	NA	NA	1	1	−3	NA	NA
2284	1.2	5.7	15.1	5.7	1	43	0	4	0.15	3.2	3.2	0.004	−0.003	NA	0	NA	NA	248	NA	168	NA	NA
2285	1.2	5.1	13.1	5.8	1	46	0	4	0.2	1.4	3.3	0.004	0.001	−243	243	NA	NA	−479	41	125	NA	NA
2286	1.8	6.0	11.1	4.1	1	48	3	4	0.3	7.3	4	0.004	0.001	−102	102	NA	NA	−204	34	101	NA	NA
2288	1.4	3.8	8.5	4.0	1	49	20	4	0.2	13.0	2.5	0.004	0.003	−18	18	NA	NA	−45	19	44	NA	NA
2578	0.7	22.3	34.7	3.2	2	37	80	4.5	0.4	1.4	2.9	0.022	0.023	1	0	−23	NA	3	−2	−3	−26	NA
2579	0.5	23.0	49.8	3.2	2	43	80	4	0.2	3.2	6	0.023	0.020	−11	33	−25	11	−17	5	43	−6	24

* Clear—1; fully covered—2; partially covered—3; through thin cloud—4.

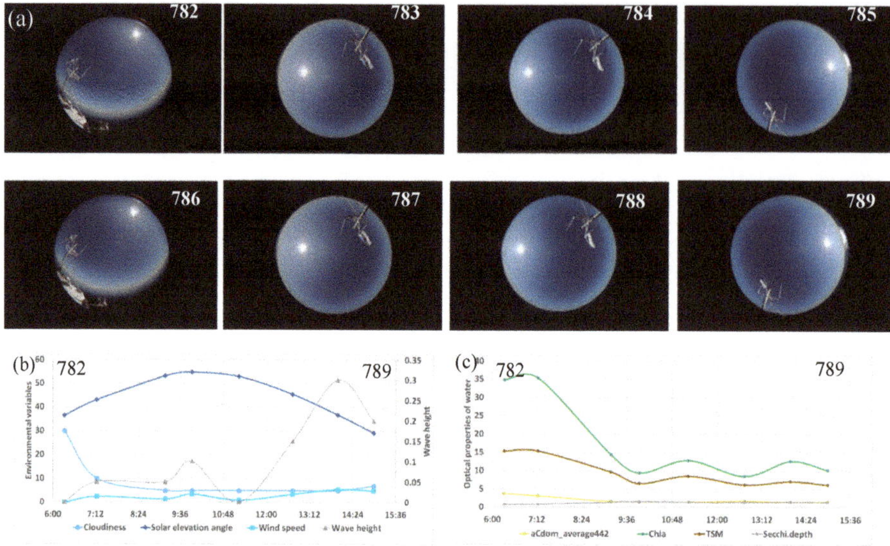

Figure A1. The changes in the sky conditions (**a**), the temporal and spatial changes in the measurement conditions (**b**) and optical properties of water (**c**) on 14th June 2016.

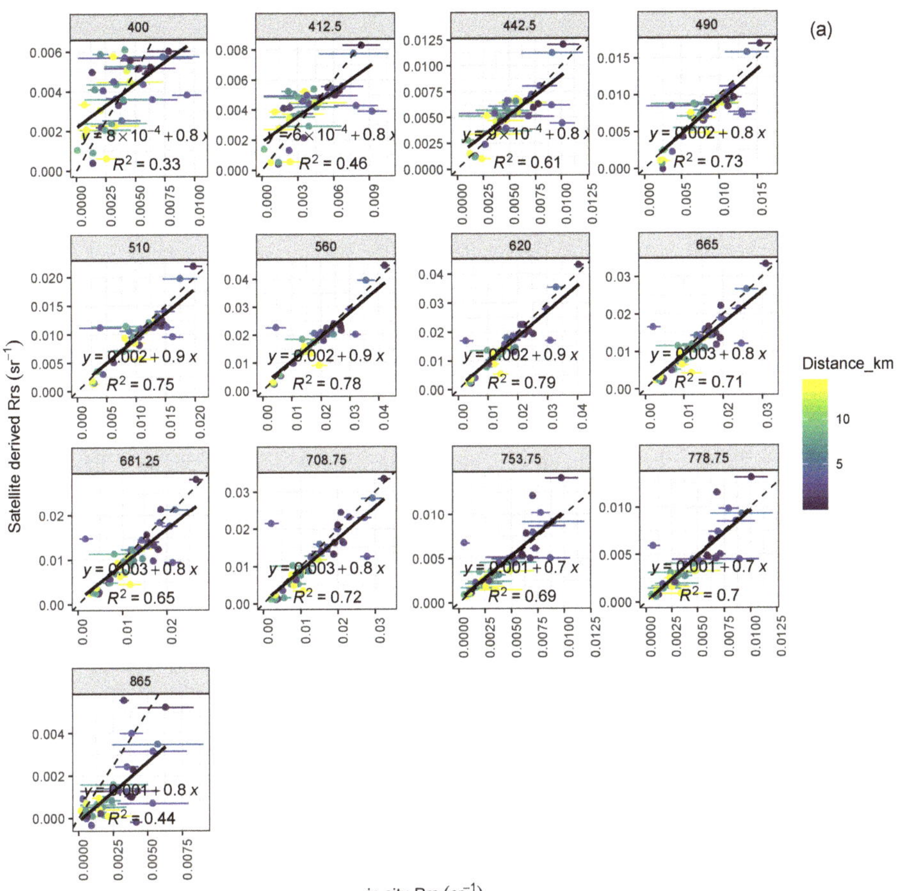

Figure A2. *Cont.*

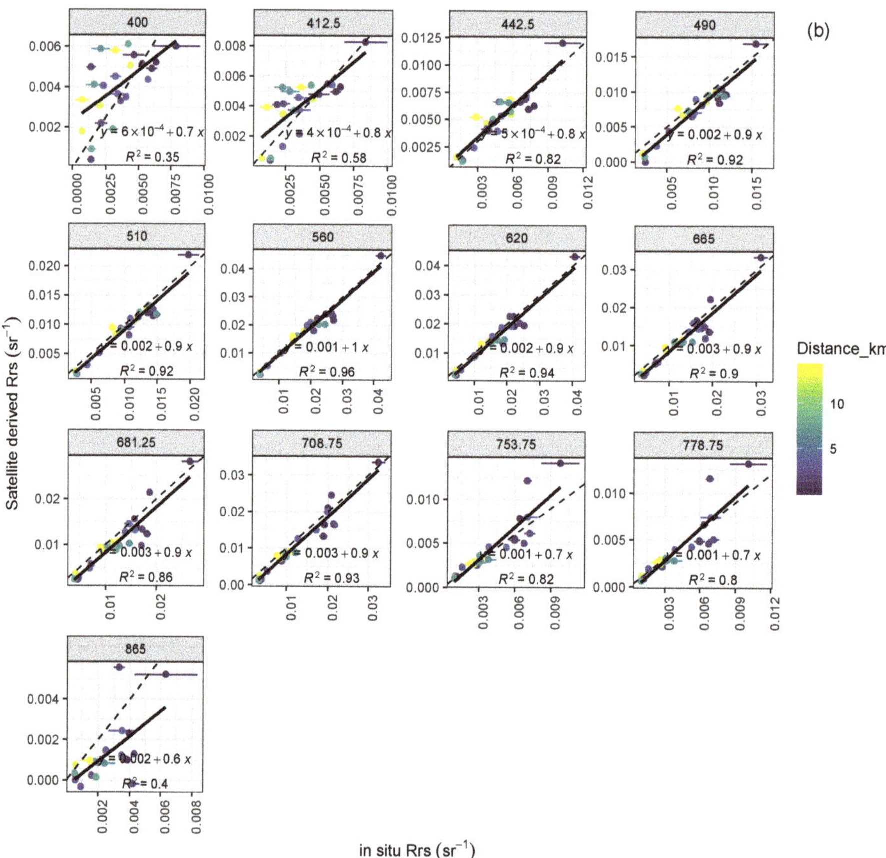

Figure A2. Correlation between in situ measured (*x*-axis) vs. POLYMER-derived rho (*y*-axis) on OLCI bands. Upper figure (**a**) shows validation results on all data and lower (**b**) after filtering the in situ data based on the 5% criteria at least in one band. For each match-up point, the color shows the distance from the shore.

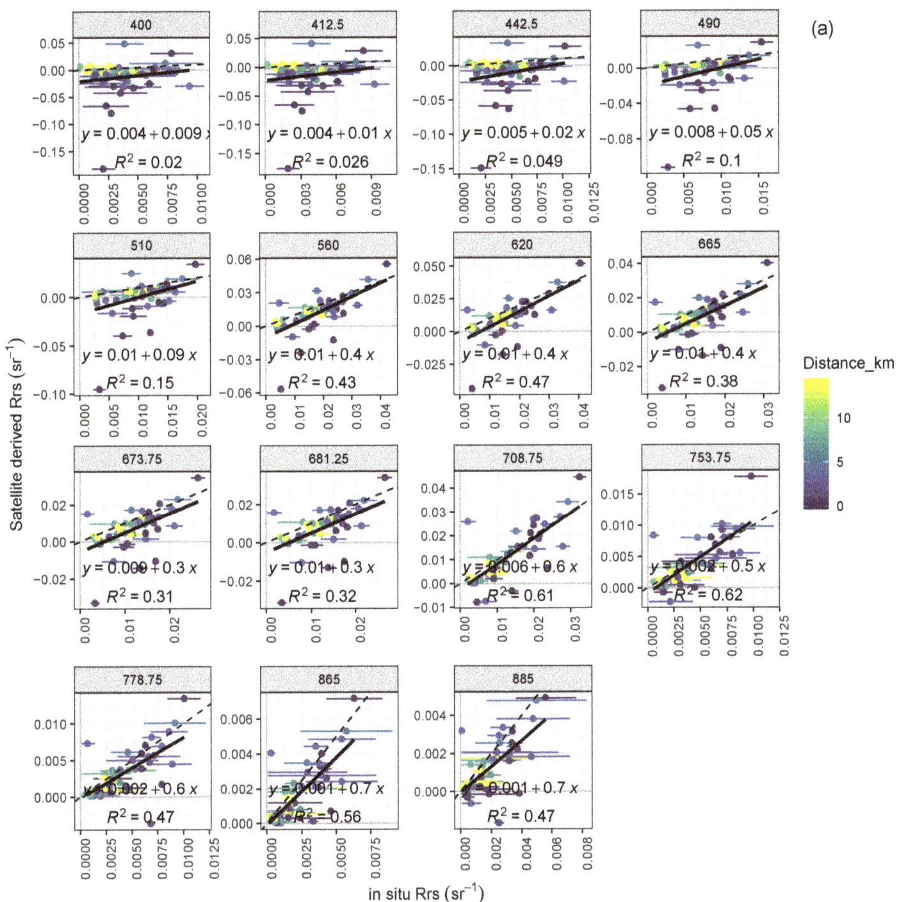

Figure A3. *Cont.*

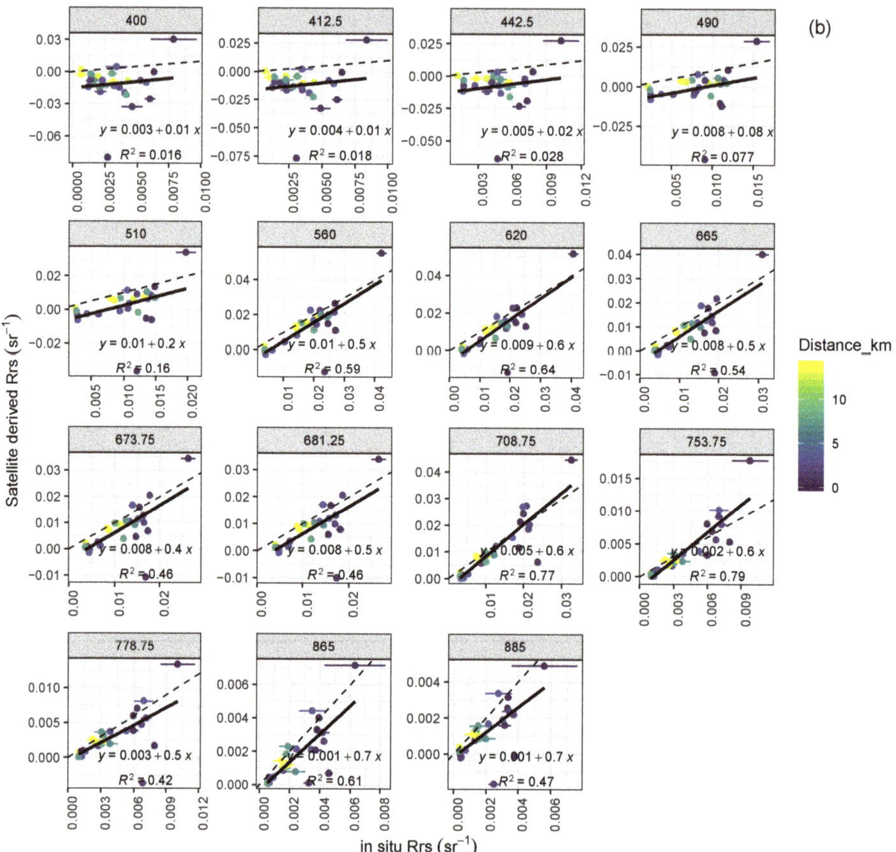

Figure A3. Correlation between in situ measured (*x*-axis) vs. standard L2-derived rho (*y*-axis) on OLCI bands. Upper figure (**a**) shows validation results on all data and lower (**b**) after filtering the in situ data based on the 5% criteria at least in one band. For each match-up point, the color shows the distance from the shore.

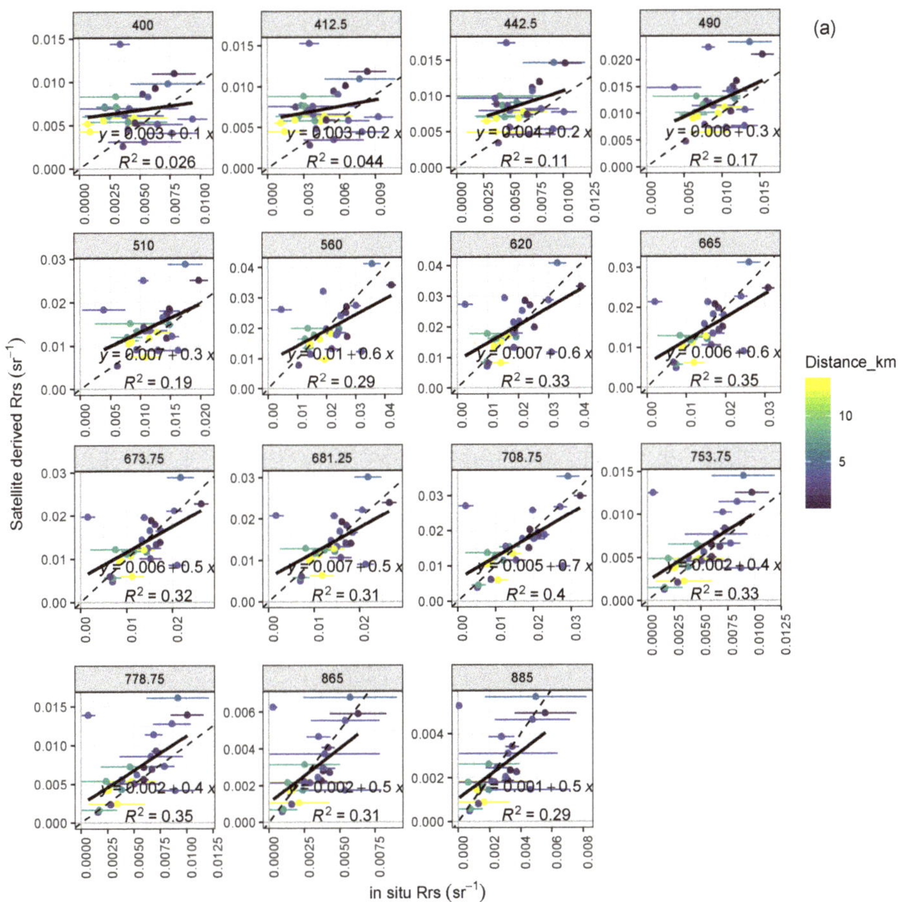

Figure A4. *Cont.*

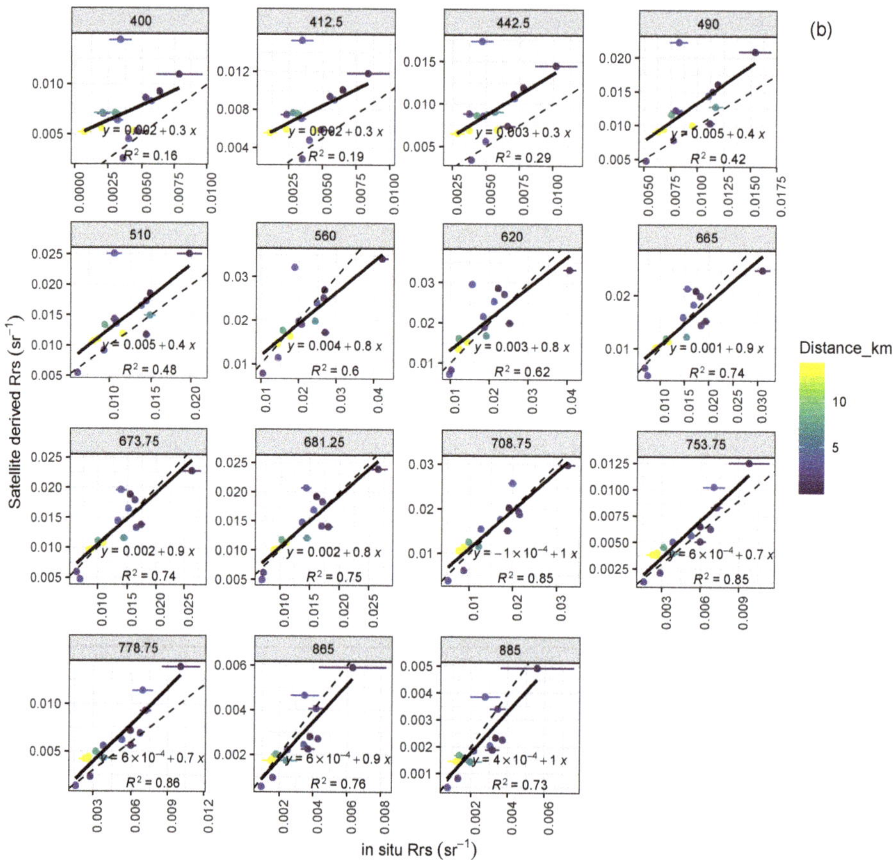

Figure A4. Correlation between in situ measured (*x*-axis) vs. C2RCC-derived rho (*y*-axis) on OLCI bands. Upper figure (**a**) shows validation results on all data and lower (**b**) after filtering the in situ data based on the 5% criteria at least in one band. For each match-up point, the color shows the distance from the shore.

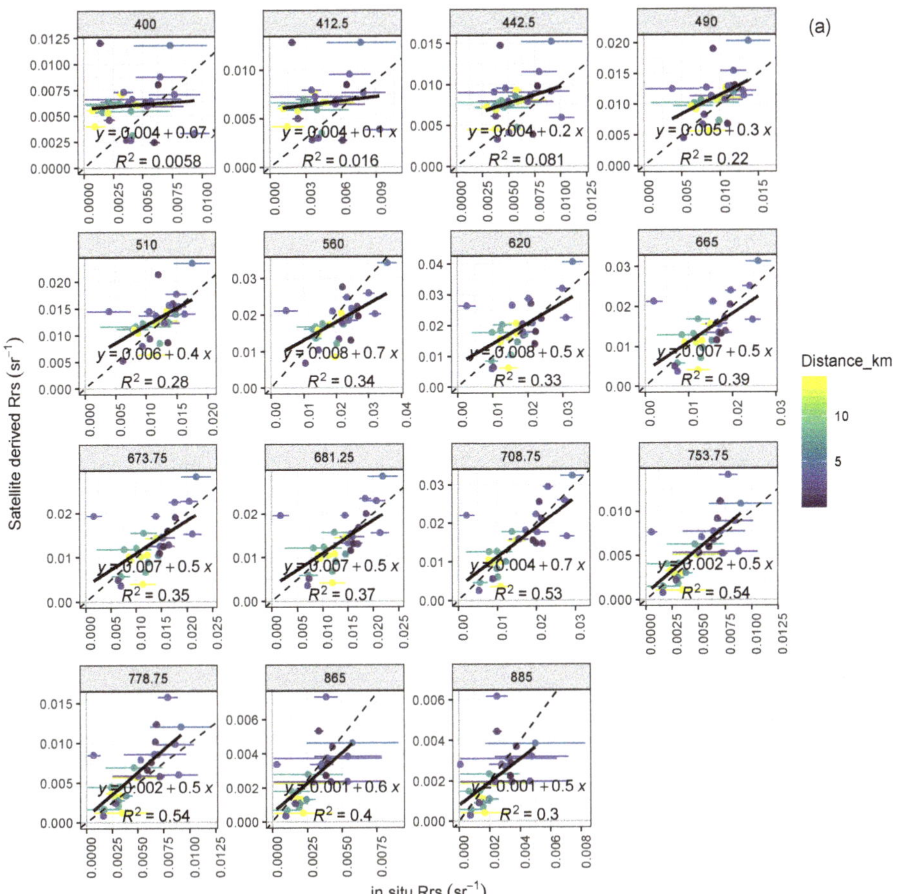

Figure A5. *Cont.*

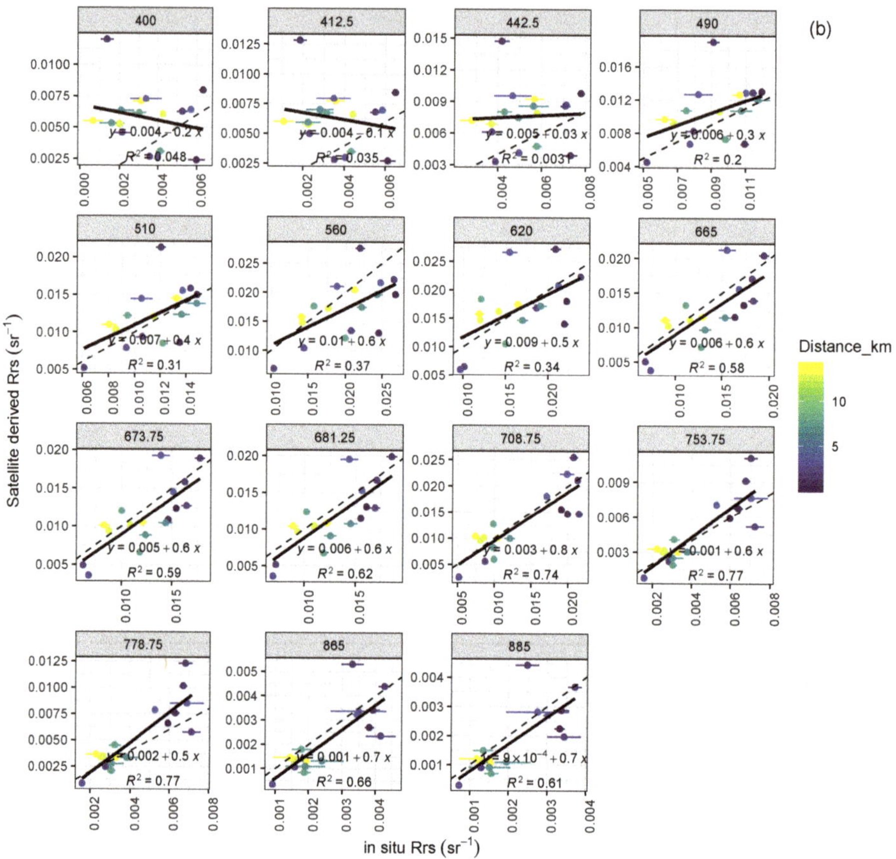

Figure A5. Correlation between in situ measured (*x*-axis) vs ALTNN-derived rho (*y*-axis) on OLCI bands. Upper figure (**a**) shows validation results on all data and lower (**b**) after filtering the in situ data based on the 5% criteria at least in one band. For each match-up point, the color shows the distance from the shore.

References

1. Bulgarelli, B.; Zibordi, G. On the detectability of adjacency effects in ocean color remote sensing of midlatitude coastal environments by SeaWiFS, MODIS-A, MERIS, OLCI, OLI and MSI. *Remote Sens. Environ.* **2018**, *209*, 423–438. [CrossRef] [PubMed]

2. EUMETSAT Mission Management. *Sentinel-3 Product Notice—OLCI Level-2 Ocean Colour, v1*; EUMETSAT: Darmstadt, Germany, 2019.

3. ESA, Sentinel-3 Family Grows. Available online: https://www.esa.int/Applications/Observing_the_Earth/Copernicus/Sentinel-3/Sentinel-3_family_grows (accessed on 21 December 2019).

4. ESA. *Sentinel-3 Mission Requirements Traceability Document (MRTD), EOP-SM/2184/CD-cd*; ESA: Paris, France, 2011.

5. ESA. *Sentinel-3 Mission Requirements Document (MRD), EOPSMO/1151/MD-md*; ESA: Paris, Franch, 2007; Available online: https://earth.esa.int/c/document_library/get_file?folderId=13019&name=DLFE-799.pdf (accessed on 20 December 2019).

6. Vabson, V.; Kuusk, J.; Ansko, I.; Vendt, R.; Alikas, K.; Ruddick, K.; Ansper, A.; Bresciani, M.; Burmester, H.; Costa, M.; et al. Laboratory Intercomparison of Radiometers Used for Satellite Validation in the 400–900 nm Range. *Remote Sens.* **2019**, *11*, 1101. [CrossRef]

7.	Vabson, V.; Kuusk, J.; Ansko, I.; Vendt, R.; Alikas, K.; Ruddick, K.; Ansper, A.; Bresciani, M.; Burmester, H.; Costa, M.; et al. Field Intercomparison of Radiometers Used for Satellite Validation in the 400–900 nm Range. *Remote Sens.* **2019**, *11*, 1129. [CrossRef]

8.	Kratzer, S.; Moore, G. Inherent Optical Properties of the Baltic Sea in Comparison to Other Seas and Oceans. *Remote Sens.* **2018**, *10*, 418. [CrossRef]

9.	Kowalczuk, P.; Stedmon, C.A.; Markager, S. Modelling absorption by CDOM in the Baltic Sea from season, salinity and chlorophyll. *Mar. Chem.* **2006**, *101*, 1–11. [CrossRef]

10.	Harvey, E.T.; Kratzer, S.; Andersson, A. Relationships between colored dissolved organic matter and dissolved organic carbon in different coastal gradients of the Baltic Sea. *AMBIO* **2015**, *44*, 392–401. [CrossRef] [PubMed]

11.	Tilstone, G.H.; Moore, G.F.; Doerffer, R.; Røttgers, R.; Ruddick, K.G.; Pasterkamp, R.; Jørgensen, P.V. Regional Validation of MERIS Chlorophyll products in North Sea REVAMP Protocols Regional Validation of MERIS Chlorophyll products. In Proceedings of the Working Meeting on MERIS and AATSR Calibration and Geophysical Validation (ENVISAT MAVT-2003), Frascati, Italy, 20–24 October 2003.

12.	Vabson, V.; Ansko, I.; Alikas, K.; Kuusk, J.; Vendt, R.; Reinat, A. Improving Comparability of Radiometric In Situ Measurements with Sentinel-3A/OLCI Data. Presented at the Fourth S3VT Meeting, Darmstadt, Germany, 13–15 March 2018. [CrossRef]

13.	Ruddick, K.; De Cauwer, V.; Park, Y.; Moore, G. Seaborne measurements of near infrared water-leaving reflectance: The similarity spectrum for turbid waters. *Limnol. Oceanogr.* **2006**, *51*, 1167–1179. [CrossRef]

14.	Ruddick, V.; De Cauwer, V.; Van Mol, B. Use of the near infrared similarity reflectance spectrum for the quality control of remote sensing data. *Remote Sens. Coastal Ocean. Environ.* **2005**, *5885*, 588501. [CrossRef]

15.	Alikas, K.; Ansko, I.; Vabson, V.; Ansper, A.; Kangro, K.; Randla, M.; Uudeberg, K.; Ligi, M.; Kuusk, J.; Randoja, R.; et al. Validation of Sentinel-3A/OLCI data over Estonian inland waters. Presented at the AMT4 Sentinel FRM Workshop; Plymouth, UK: 20–21 June 2017. [CrossRef]

16.	European Commission; Joint Research Centre (European Commission). *Manual for Monitoring European Lakes Using Remote Sensing Techniques*; Lindell, T., Pierson, D., Premazzi, G., Zilioli, E., Eds.; Office for Official Publications of the European Communities: Luxembourg, 1999; pp. 29–69.

17.	Jeffrey, S.W.; Humphrey, G.F. New spectrophotometric equations for determining chlorophylls a, b, c1 and c2 in higher plants, algae and natural phytoplankton. *Biochem. Physiol. Pflanz.* **1975**, *167*, 191–194. [CrossRef]

18.	*Guide to the Expression of Uncertainty in Measurement*; International Organization for Standardization (ISO): Geneva, Switzerland, 1995.

19.	Kuusk, J.; Ansko, I.; Vabson, V.; Ligi, M.; Vendt, R. *Protocols and Procedures to Verify the Performance of Fiducial Reference Measurement (FRM) Field Ocean Colour Radiometers (OCR) Used for Satellite Validation; Technical Report TR-5*; Tartu Observatory: Tõravere, Estonia, 2017.

20.	Santer, B.D.; Wigley, T.M.L.; Boyle, J.S.; Gaffen, D.J.; Hnilo, J.J.; Nychka, D.; Parker, D.E.; Taylor, K.E. Statistical significance of trends and trend differences in layer-average atmospheric temperature time series. *J. Geophys. Res. Atmos.* **2000**, *105*, 7337–7356. [CrossRef]

21.	Antoine, D.; Morel, A. A multiple scattering algorithm for atmospheric correction of remotely sensed ocean colour (MERISinstrument): Principle and implementation for atmospheres carrying various aerosols including absorbing ones. *Int. J. Remote Sens.* **1999**, *20*, 1875–1916. [CrossRef]

22.	Moore, G.F.; Aiken, J.; Lavender, S.J. The atmospheric correction of water colour and the quantitative retrieval of suspended particulate matter in Case II waters: Application to MERIS. *Int. J. Remote Sens.* **1999**, *20*, 1713–1733. [CrossRef]

23.	EUMETSAT. *Sentinel-3 Marine User Handbook, v1H e-Signed*; EUMETSAT: Darmstadt, Germany, 2018.

24.	Steinmetz, F.; Deschamps, P.Y.; Ramon, D. Atmospheric correction in presence of sun glint: Application to MERIS. *Opt. Express* **2011**, *19*, 9783–9800. [CrossRef] [PubMed]

25.	Steinmetz, F.; Ramon, D. Sentinel-2 MSI and Sentinel-3 OLCI consistent ocean colour products using POLYMER. In Proceedings of the SPIE 10778, Remote Sensing of the Open and Coastal Ocean and Inland Waters, Honolulu, HI, USA, 24–26 September 2018; p. 107780E. [CrossRef]

26.	Doerffer, R.; Schiller, H. The MERIS case 2 water algorithm. *Int. J. Remote Sens.* **2007**, *28*, 517–535. [CrossRef]

27.	Brockmann, C.; Doerffer, R.; Marco, P.; Stelzer, K.; Embacher, S.; Ruescas, A. Evolution Of The C2RCC Neural Network For Sentinel 2 and 3 For The Retrieval of Ocean Colour Products in Normal and Extreme Optically Complex Waters. In Proceedings of the Living Planet Symposium, Prague, Czech Republic, 9–13 May 2016.

28. International Vocabulary of Metrology—Basic and General Concepts and Associated Terms (VIM), 3rd ed.; JCGM 200: 2012. Available online: http://www.bipm.org/vim (accessed on 17 December 2019).

29. Possolo, A.; Iyer, H.K. Invited Article: Concepts and tools for the evaluation of measurement uncertainty. *Rev. Sci. Instrum.* **2017**, *88*, 011301. [CrossRef]

30. Kutser, T.; Vahtmäe, E.; Paavel, B.; Kauer, T. Removing glint effects from field radiometry data measured in optically complex coastal and inland waters. *Remote Sens. Environ* **2013**, *133*, 85–89. [CrossRef]

31. Lee, Z.; Ahn, Y.-H.; Mobley, C.; Arnone, R. Removal of surface-reflected light for the measurement of remote-sensing reflectance from an above-surface platform. *Opt. Express* **2010**, *18*, 26313–26324. [CrossRef]

32. Zibordi, G.; Berthon, J.-F.; Mélin, F.; D'Alimonte, D.; Kaitala, S. Validation of satellite ocean color primary products at optically complex coastal sites: Northern Adriatic Sea, Northern Baltic Proper and Gulf of Finland. *Remote Sens. Environ.* **2009**, *113*, 2574–2591. [CrossRef]

33. Antoine, D.; d'Ortenzio, F.; Hooker, S.B.; Bécu, G.; Gentili, B.; Tailliez, D.; Scott, A.J. Assessment of uncertainty in the ocean reflectance determined by three satellite ocean color sensors (MERIS, SeaWiFS and MODIS-A) at an offshore site in the Mediterranean Sea (BOUSSOLE project). *J. Geophys. Res. Ocean.* **2008**, *113*, C07013. [CrossRef]

34. Li, J.; Jamet, C.; Zhu, J.; Han, B.; Li, T.; Yang, A.; Guo, K.; Jia, D. Error Budget in the Validation of Radiometric Products Derived from OLCI around the China Sea from Open Ocean to Coastal Waters Compared with MODIS and VIIRS. *Remote Sens.* **2019**, *11*, 2400. [CrossRef]

35. Gossn, J.I.; Ruddick, K.G.; Dogliotti, A.I. Atmospheric Correction of OLCI Imagery over Extremely Turbid Waters Based on the Red, NIR and 1016 nm Bands and a New Baseline Residual Technique. *Remote Sens.* **2019**, *11*, 220. [CrossRef]

 *remote sensing*

Article

A Virtual Geostationary Ocean Color Sensor to Analyze the Coastal Optical Variability

Marco Bracaglia [1,2,*], **Rosalia Santoleri** [1], **Gianluca Volpe** [1], **Simone Colella** [1], **Mario Benincasa** [1] **and Vittorio Ernesto Brando** [1]

[1] Istituto di Scienze Marine (CNR-ISMAR), Via Fosso del Cavaliere 100, 00133 Rome, Italy; rosalia.santoleri@cnr.it (R.S.); gianluca.volpe@cnr.it (G.V.); simone.colella@cnr.it (S.C.); mario.benincasa@artov.ismar.cnr.it (M.B.); vittorio.brando@cnr.it (V.E.B.)

[2] Dipartimento di Scienze e Tecnologie, Università degli Studi di Napoli Parthenope, Via Amm. F. Acton 38, 80133, Naples, Italy

* Correspondence: marco.bracaglia@artov.ismar.cnr.it

Received: 14 April 2020; Accepted: 6 May 2020; Published: 12 May 2020

Abstract: In the coastal environment the optical properties can vary on temporal scales that are shorter than the near-polar orbiting satellite temporal resolution (~1 image per day), which does not allow capturing most of the coastal optical variability. The objective of this work is to fill the gap between the near-polar orbiting and geostationary sensor temporal resolutions, as the latter sensors provide multiple images of the same basin during the same day. To do that, a Level 3 hyper-temporal analysis-ready Ocean Color (OC) dataset, named Virtual Geostationary Ocean Color Sensor (VGOCS), has been created. This dataset contains the observations acquired over the North Adriatic Sea by the currently functioning near-polar orbiting sensors, allowing approaching the geostationary sensor temporal resolution. The problem in using data from different sensors is that they are characterized by different uncertainty sources that can introduce artifacts between different satellite images. Hence, the sensors have different spatial and spectral resolutions, their calibration procedures can have different accuracies, and their Level 2 data can be retrieved using different processing chains. Such differences were reduced here by adjusting the satellite data with a multi-linear regression algorithm that exploits the Fiducial Reference Measurements data stream of the AERONET-OC water-leaving radiance acquired at the Acqua Alta Oceanographic Tower, located in the Gulf of Venice. This work aims to prove the suitability of VGOCS in analyzing the coastal optical variability, presenting the improvement brought by the adjustment on the quality of the satellite data, the VGOCS spatial and temporal coverage, and the inter-sensor differences. Hence, the adjustment will strongly increase the agreement between the satellite and in situ data and between data from different near-polar orbiting OC imagers; moreover, the adjustment will make available data traditionally masked in the standard processing chains, increasing the VGOCS spatial and temporal coverage, fundamental to analyze the coastal optical variability. Finally, the fulfillment by VGOCS of the three conditions for a hyper-temporal dataset will be demonstrated in this work.

Keywords: fiducial reference measurements; hyper-temporal dataset; optical radiometry; coastal environment; observation geometry

---

1. Introduction

Ocean Color Radiometry (OCR) satellite data allow observing the variability of the inherent optical properties (IOPs) and the concentrations of the water components in the coastal and open ocean with a large spatial coverage [1].

Near-polar orbiting OC satellites usually have a spatial resolution that goes from 300 m to 1000 m and provides ~1 image per day of basins located at middle latitudes. These spatial and temporal resolutions are of the same order as the temporal and spatial scales of most of the open ocean bio-optical processes and, consequently, are sufficient to properly characterize such areas [2–4]. On the contrary, in coastal waters, meteorological, marine, and hydrological drivers set up shorter temporal scales for such processes, compared to the open ocean [4]; indeed, in those areas, rapid changes can be observed in the time scale of few hours or less [4–8]. Consequently, an improved temporal resolution than what is currently provided by existing near-polar orbiting sensors is needed to accurately analyze the coastal environment [2–4].

An improved temporal resolution can be provided by OCR geostationary sensors [9–15], which acquire multiple observations of the same area during the same day. Nevertheless, the only currently working OCR geostationary sensor is the Geostationary Ocean Color Imager (GOCI) [16] over the Korean Peninsula, while such sensors are absent over the European Seas.

The first step in improving the near-polar orbiting sensor temporal resolution has been accomplished in [5], where the overlapping scenes of the Visible Infrared Imaging Radiometer Suite (VIIRS), mounted on the SUOMI-NPP satellite, have been exploited to observe the short term variability of some OC quantities in the Gulf of Mexico. This approach has then been used in [6] to observe the particulate backscattering (b_{bp}) variability in the North Adriatic Sea (NAS, Figure 1), adjusting the remote sensing reflectance (R_{rs}) spectra following [17]. This adjustment, based on a multi-linear regression (MLR) procedure and the AERONET-OC Acqua Alta Oceanographic Tower (AAOT) [18,19] in situ radiometric data (Figure 1), was used to reduce the uncertainties introduced by the different viewing geometry of different VIIRS images and to use data generally masked in the standard processing chains. The AERONET-OC in situ data are part of the Fiducial Reference Measurements (FRM) data stream as they guarantee traceability, accuracy, long-term stability, and cross-site consistency [20]. Those aspects are central for system vicarious calibration, product validation, uncertainty estimation, and development and assessment of bio-optical models for the generation of derived high-level products [21–26].

The present work aims to reduce the gap between near-polar orbiting and geostationary mission temporal resolutions, creating a Level 3 hyper-temporal analysis-ready OC dataset [27,28], containing observations provided by the currently functioning near-polar orbiting OC imagers.

This dataset is defined as an analysis-ready dataset as it is conceived to provide to users, without the expertise in pre-processing procedures, "ready to use" OC data to analyze the coastal optical variability at the high temporal resolution, similarly to the CEOS Analysis Ready Data for Land (CARD4L) [27].

The definition of a hyper-temporal dataset is based on the three conditions provided in [29], for which a hyper temporal dataset must:

(a) Be univariate (i.e., with multiple images of the same parameter only).
(b) Contain a set of time slices, all of which must be precisely coregistered (with image-to-image pixels perfectly aligned spatially).
(c) Exhibit radiometric consistency between images (i.e., they are measured using the same sensors or inter-validated sensor systems, and exhibit a degree of normalization between time slices).

This dataset is named Virtual Geostationary Ocean Color Sensor (VGOCS). It contains multiple images acquired during the same day by the currently functioning near-polar orbiting OC imagers over the NAS study area, allowing approaching the temporal resolution of a geostationary satellite. For each sensor, the time series starts for the first day of acquisition until 31 October 2019.

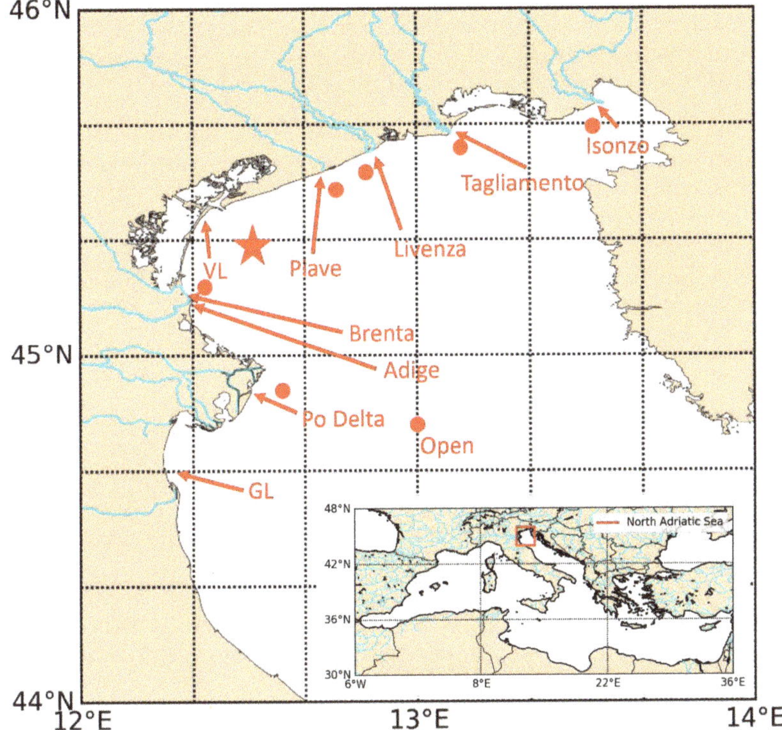

Figure 1. Map of the NAS with the rivers of the basin. The red star identifies the Acqua Alta Oceanographic Tower (AAOT), VL the Venice Lagoon, GL the Goro Lagoon, the red dots the virtual buoys (Section 2.7). In the box on the bottom right the Mediterranean Sea, with the NAS identified by the red square.

The creation of this dataset started from the Level 2 (L2) R_{rs} spectra provided by National Aeronautics and Space Administration (NASA) and the European Organisation for the Exploitation of Meteorological Satellites (EUMETSAT); the radiometric data of all sensors have been preserved at their native spectral resolution and re-projected on a common 1 km × 1 km equirectangular grid. The grid spatial resolution of 1 km is probably too coarse for the short spatial scale processes that can take place in the coastal environment [4,30–33]; nevertheless, this is the best that is possible to achieve for the creation of a hyper-temporal dataset created with several OCR sensors. After the re-projection, the R_{rs} spectra, adjusted following [17], have been used to calculate various IOPs by the Quasi Analytical Algorithm version 6 (QAA) [34,35]. This algorithm has been chosen as it has been demonstrated suitable for the b_{bp} retrieval in the NAS [6,36,37].

The adjustment presented in [17] is here exploited to use data from different sensors, as its application to the R_{rs} spectra can reduce the artifacts that can be observed between different satellite images. Indeed, the use of data from different sensors introduces an additional source of uncertainty, besides the one due to the different viewing geometry of each satellite image [6,38]. Hence, the sensors have different spatial and spectral resolutions, their calibration procedures can have different accuracies, and their L2 data can be retrieved using different processing chains. Consequently, looking at two scenes acquired during the same day by different sensors, it is not easy to discern how much of the observed difference is due to a real process or the artifacts, introduced by the different data uncertainties [8,39].

The objective of this work is to prove:

- The suitability of VGOCS in analyzing the coastal optical variability, by the study of the effect of the adjustment on the quality of the satellite data, the VGOCS spatial and temporal coverage, and the intersensor differences.
- The fulfillment of the three conditions for a hyper-temporal dataset [29].

The manuscript is structured as follows: in the next section the study area, the material, and the methods used in this work will be presented, while in Section 3 the obtained results will be analyzed. In the latter section, the results of the match-up analyses between different satellite and in situ data before and after the application of the adjustment will be presented. Then the spatial and temporal coverage and the inter-sensor differences of the adjusted VGOCS data will be analyzed. In the last two sections, the results will be discussed and the conclusions and the future perspective for this work will be presented.

2. Materials and Methods

2.1. Study Area

The NAS is characterized by a large number of rivers (listed in Figure 1) that can discharge waters with different biogeochemical and optical properties [40–42]. This basin is located in the North-Eastern Mediterranean (Figure 1) and its main rivers are shown in Figure 1. All those rivers are relevant for the optical dynamics of the basin [40,43,44], but Po and Adige are those that provide most of the NAS freshwater [45,46].

The NAS optical variability is strongly influenced by meteo-marine and hydrological forcings, and by the basin shallow bathymetry, which has an average depth smaller than 35 m [41,43–50]. Due to this complexity, several works studied the bio-optical properties and the dynamics of the NAS using several kinds of data and methods such as in situ, ship-board, and satellite observations and oceanographic models [51–56].

2.2. VGOCS Sensors

VGOCS is composed of the observations acquired over the NAS by the current functioning near-polar orbiting OC sensors. Those involved in the dataset are

- The NASA Moderate-Resolution Imaging Spectroradiometer (MODIS) sensors mounted on the AQUA and TERRA satellites, here simply referred to as AQUA and TERRA respectively.
- The National Oceanic and Atmospheric Administration (NOAA) VIIRS sensors mounted on the SUOMI-NPP and NOOA-20 (previously JPSS-1) satellites, here referred to as VIIRSN and VIIRSJ respectively.
- The European Space Agency (ESA)/EUMETSAT Ocean and Land Color Instrument (OLCI) mounted on Sentinel 3A, here referred simply as OLCI.

The two MODIS sensors, mounted on the AQUA and TERRA satellites, are the only sensors of the dataset that have the same technical characteristic and resolutions; indeed, the two VIIRS sensors are very similar to each other but differ for the spectral resolution [57]. The latter is depicted for each sensor in Figure 2 [57–60].

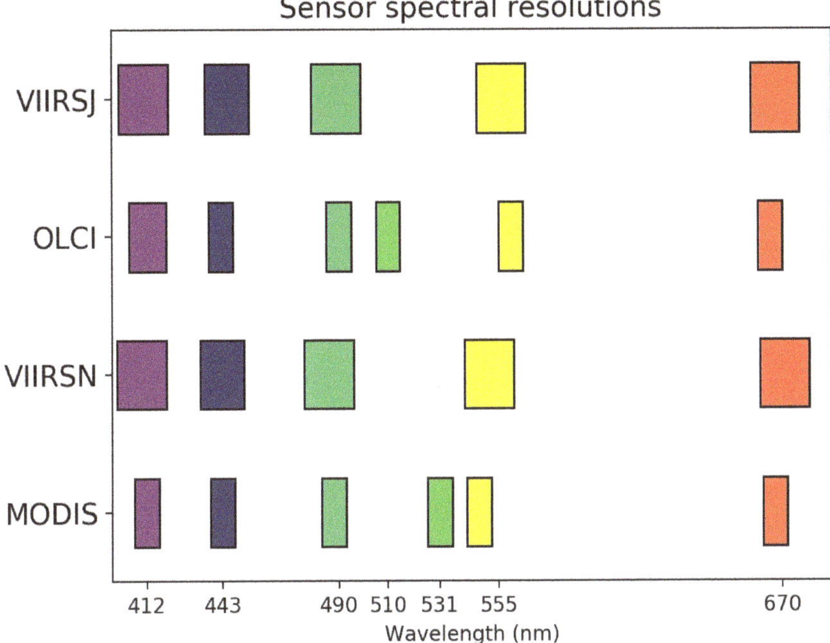

Figure 2. Spectral resolutions of the sensor involved in the Virtual Geostationary Ocean Color Sensor (VGOCS).

The MODIS OC bands have a 1000 m spatial resolution [60], the two VIIRS sensors a 750 m one [58], while the OLCI data are provided at two different resolutions: Full resolution (FR, 300 m) and Reduce Resolution (RR, 1200 m) [59]. In this study, the RR products are exploited, as the FR data suffer from a salt and pepper effect that is due to atmospheric correction problems (i.e., a lot of valid observations are surrounded by masked pixels, not shown).

OLCI, which is the only ESA/EUMETSAT sensor used in VGCOS, is the one with the lower revisiting frequency. This sensor has an orbital cycle of 27 days [59], while for the NASA/NOAA (N/N) sensors this is of 16 days [58,60].

OLCI has also a different viewing geometry and engineering than the N/N sensors. Indeed, the swath of the MODIS and VIIRS sensors is centered on the nadir and a rotating mirror reflects the radiation onto a set of Charge-Coupled Device (CCD) detectors. In OLCI, 5 CCD detectors are mounted in a fan arrangement on a common optical bench and the radiance arrives at them after passing a calibration assembly [59]. The sensor has a total field of view of 68.5°, while each detector has a field of view 14.1° with a slight overlap of 0.6° between each other [59]. Moreover, to reduce the sun-glint effect, which impacted a large number of the MEdium Resolution Imaging Spectrometer (MERIS) observations at sub-tropical latitudes [59,61], the OLCI swath is not centered at nadir but is tilted 12.6° westwards [59].

2.3. VGOCS Dataset and Processing Chain

The VGOCS dataset is created starting from the standard L2 R_{rs} spectra provided by EUTMESAT (OLCI, processing baseline v2.23) [62–64] and NASA (N/N sensors, processing version R2018.0) [65], that are stored in the archive of CNR ISMAR in Rome [39].

The VGOCS processing chain (Figure 3) is mostly the same as the one used in [6], which was based on the standard Copernicus Marine Environment Monitoring System (CMEMS) operational processing chain for the Mediterranean Sea [39], with some changes. The main difference is that CMEMS provides a single daily image, derived from the merging of data from different sensors [39], while in the VGOCS processing chain the merging procedure is not performed. Hence, all the images are retained to have multiple satellite observations of the basin during the same day.

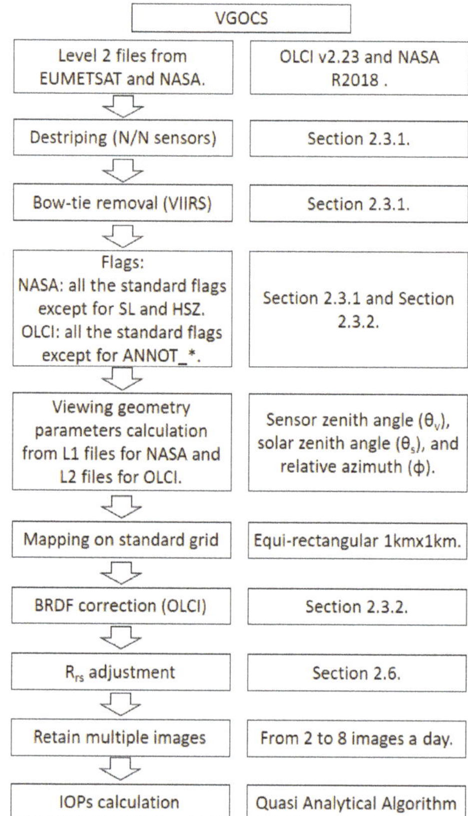

Figure 3. On the left the flow chart of the VGOCS processing chain, on the right some additional information for each step.

VGOCS contains observations of the NAS from 2002 (AQUA and TERRA operating) to 31 October 2019 (all sensors operating), providing 2 to 8 images a day depending on the number of functioning sensors and on the day of the satellite orbital cycles. The latter parameter is useful as for each sensor the number of images and their viewing geometry is the same for the same orbital cycle day, except for some minor changes in the ground-track and overpass times. The orbital cycle day is calculated in the same way for all the N/N sensors and it is a number that goes from 1 to 16. Hence, all the N/N sensors involved in VGOCS have an orbital cycle of 16 days [58,60]. For each file, defining as NDT the number of days between the satellite observation and the first TERRA acquisition (8 February 2000), the orbital cycle day can be defined as the remainder of NDT/16 plus one. The orbital cycle of OLCI is of 27 days, thus the orbital cycle day is defined as the remainder of NDO/27 plus one (where NDO is the number of days between the satellite observation and the first OLCI acquisition, 26 April 2016).

To use the same grid for each image of the dataset, all of them are re-projected on a standard 1 km × 1 km equirectangular grid by the nearest neighbor approach.

The name format of each file is VGOCS_YYYYJJJHHMMSS_X.nc, with Y, J, H, M, S (year, Julian day of the year, hour, minute, second) referring to the start time of the sensor acquisition, while X is the label which identifies the sensor (J: VIIRSJ, V: VIIRSN, A: AQUA, T: TERRA, O: OLCI).

The parameters that can be found in each VGOCS file are stored in different groups:

- The R_{rs}_data group, where the R_{rs} spectra at the native sensor resolution are stored.
- The IOP_data group, where all the parameters (listed below) calculated with the QAA can be found [34,35]:

  - reference wavelength (λ_0)
  - total absorption at the reference wavelength ($a(\lambda_0)$)
  - particulate backscattering at the reference wavelength ($b_{bp}(\lambda_0)$)
  - b_{bp} spectral slope (η)
  - b_{bp} at 443 nm ($b_{bp}(443)$)
  - absorption by nonalgal particles and dissolved matter at 443 nm ($a_{dg}(443)$)
  - a_{dg} spectral slope (s)
  - absorption by phytoplankton at 443 nm ($a_{phy}(443)$)

- The Atmospheric_data group contains some atmospheric parameters, such as aerosol optical thickness and angstrom coefficients, extracted from the L2 files.
- The Geo_data group contains information about the applied flags, extracted from the L2 files, and about some viewing geometry parameters, such as the sensor zenith angle (θ_v), the solar zenith angle (θ_s), and the relative angle between the solar and sensor azimuth angle (φ). Those parameters for the OLCI sensor are extracted from the L2 file, for the VIIRS sensors from the L1 GEO files, while for the MODIS sensors they are retrieved using the L1B files as the input of l2gen.

Moreover, latitude, longitude, and the satellite orbital cycle day are provided as stand-alone variables.

2.3.1. NASA/NOAA Sensors

The processing chain used for the N/N sensors is the same reported in [6] for the VIIRSN data. All the N/N sensor images have been destriped following [66,67], while, for the two VIIRS sensors, the bow-tie missing data have been filled by linear interpolation [58]. Not all the standard L2 flags [68] have been applied; hence, the data have been unmasked for the High Sensor Zenith (HISATZEN, here HSZ) and the stray-light flag (STRAYLIGHT, here SL). The first one masks data acquired at a $\theta_v > 60°$ [69,70], while the second one masks data that could be affected by the adjacency effect of clouds and lands [71–74].

The HSZ flag has not been applied because a large number of images are acquired at $\theta_v > 60°$, and unmasking for this flag makes available a larger number of valid pixels. Using the HSZ data is fundamental for the aim for which VGOCS is conceived, as, to analyze the daily optical variability, it is important to have at disposal as many full images as possible. In this study, an image will be considered as "full" if the sensors detected at least 90% of the water pixels of the NAS, while it will be considered as "partial" if the sensor detected between 30% and 90% of the water pixels.

In Figure 4 the number of images for each day of the orbital cycle is presented for the N/N sensors considering as masked the HSZ observations. As can be seen, only for days 5, 8, 13, and 16 there are five full images (in green), while for the other days there are maximum one partial (in black) and four full images.

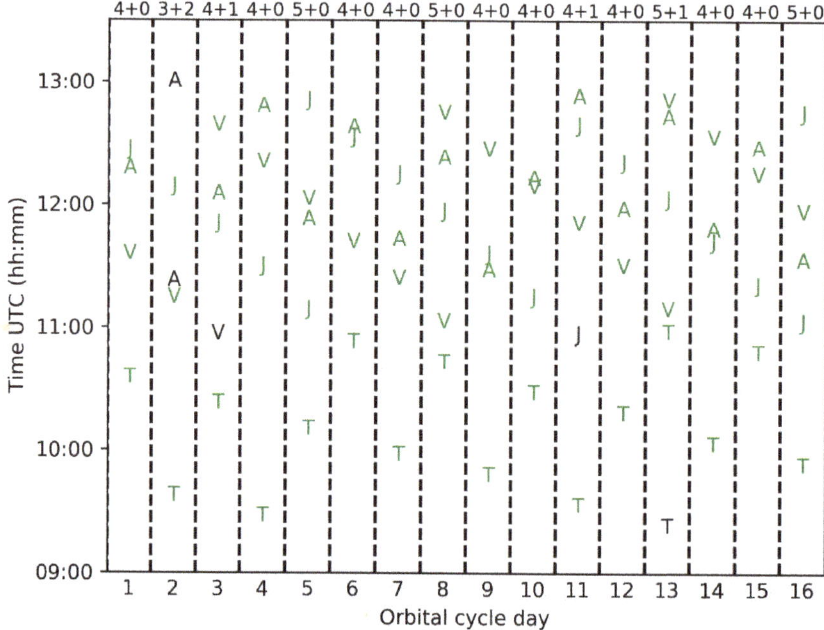

Figure 4. Available images for the N/N sensors for each day of the orbital cycle, with the application of the High Sensor Zenith (HSZ) flag. On the x-axes the orbital cycle day, on the y-axes the scan start time, with the different letters identifying the different sensors. In green the full images, in black the partial images. On the top for each orbital cycle day the number of full images plus the number of the partial ones.

Figure 5 is the same of Figure 4, but with the pixels unmasked for the HSZ flag. The HSZ pixels are 22.9% of the pixels scanned by the N/N sensors in the NAS, and unmasking them provides a larger number of full and partial images, in comparison with Figure 4. Now the minimum number of available images is reached on day 3 with 5 full images and the maximum on day 4 with 7 full and 1 partial image. Particularly:

- For VIIRSN there are 9 full orbit overlaps in comparison with the two available maskings for HSZ.
- For VIIRSJ there are 8 full and 2 partial overlaps in comparison with the two available maskings for HSZ.
- For AQUA and TERRA, there are 3 full orbit overlaps (previously zero), and some additional partial images.

Thus, the unmasking for this flag strongly improves the spatial and temporal coverage of VGOCS, making available up to 3 additional full images a day.

It is important to remind here that OLCI has not been taken into account in Figures 4 and 5 as it has a different orbital cycle and no HSZ flag. This sensor can provide one additional complete image of the NAS per day except for days 7, 11, 15, 19, 23, and 27 of its orbital cycle when the NAS is not overpassed by Sentinel-3A.

The SL flag masks all the pixels that could be affected by the light reflected by the land surface that can be forward scattered into the sensor field of view [71–74]. Nevertheless, the application of this flag results in the masking of a large part of the coastal area (here identified in red in Figure 6) [6,75], as can be noted in Figure 7. Here two $b_{bp}(443)$ maps retrieved from the original VIIRSN R_{rs} during 21 March 2013 12:48 UTC are presented: on the left the SL flag is active, and a large portion of the coastal area results masked; on the right, the SL flag is not applied and the majority of the coastal pixels are now available.

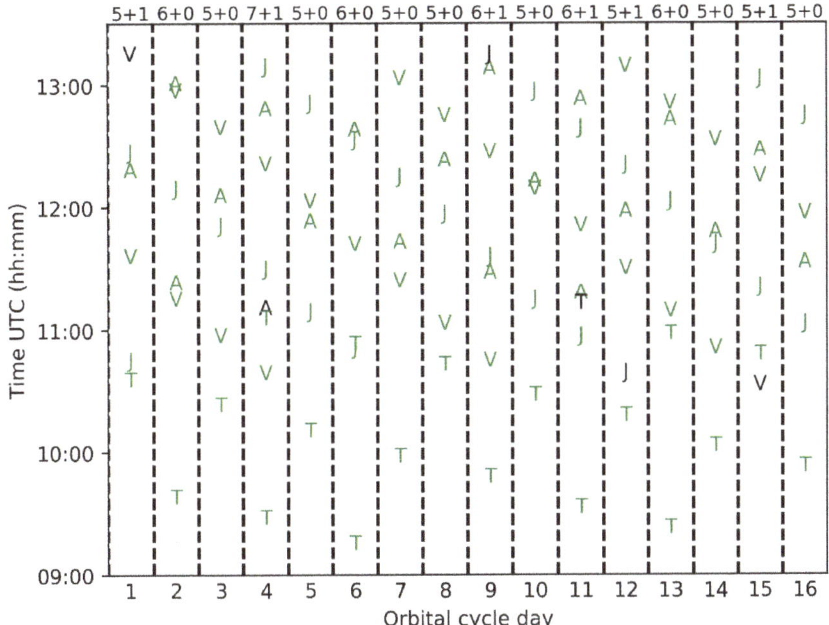

Figure 5. Available images for the N/N sensors for each day of the orbital cycle, without the application of the HSZ flag. On the x-axes the orbital cycle day, on the y-axes the scan start time, with the different letters identifying the different sensors. In green the full images, in black the partial images. On the top for each orbital cycle day the number of full images plus the number of the partial ones.

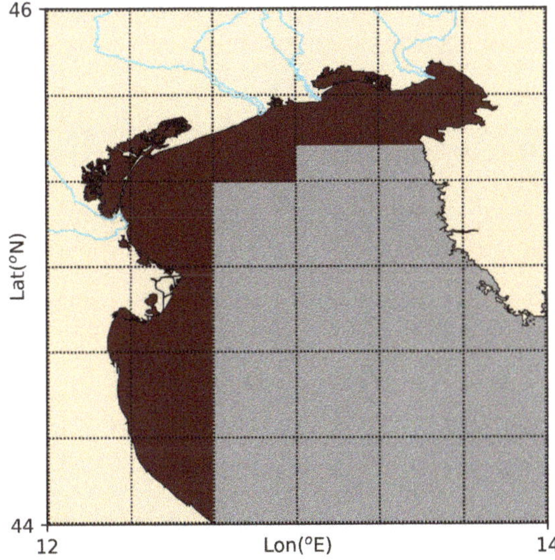

Figure 6. Map of the North Adriatic Sea (NAS), in red the coastal area used to quantify the stray-light flag (STRAYLIGHT, here SL) and ANNOT_* effect on the VGOCS spatial coverage.

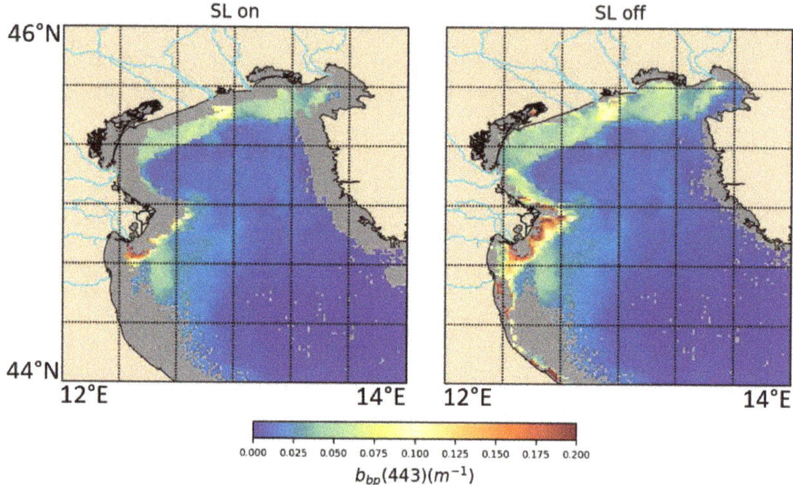

Figure 7. Maps of $b_{bp}(443)$ retrieved from the original VIIRSN R_{rs} for 21 March 2013 12:48 UTC. On the left, the SL flag is active (SL on), while the pixels are unmasked (SL off) on the right.

The behavior observed in Figure 7 is common to a large number of images of the dataset, as the mean percentage of the valid pixels flagged by SL in the coastal area is >40% for every single N/N sensor and the entire VGCOS N/N time series (Table 1). Hence, to analyze the coastal optical variability it is necessary to unmask such pixels.

Table 1. The mean percentage of SL flagged pixels in comparison with the total number of coastal valid pixels, for each N/N sensor and for the entire N/N VGOCS time series.

Sensor	Percentage of SL Masked Pixels
TERRA	42.0%
AQUA	41.2%
VIIRSN	45.1%
VIIRSJ	44.7%
VGOCS (no OLCI)	42.4%

The larger uncertainties generally observed for the HSZ and SL data can be reduced by the adjustment procedure [6]; consequently, the agreement of such data with the in situ observations will be analyzed in Section 3.2.

2.3.2. OLCI Sensor

The EUMETSAT OLCI data have different L2 processing, leading to files with structures and variables differing than those for the N/N sensors. Particularly, a different atmospheric correction algorithm is applied to the TOA radiances; hence, for OLCI, instead of the algorithm from [76], used in the L2 NASA processing, the Baseline Atmospheric Correction (BAC) algorithm [77,78] is exploited. Moreover, the R_{rs} provided in the OLCI L2 files are not corrected for the bi-directional effect. For this reason, for the OLCI data, an additional step is needed in the processing chain (Figure 3) to be coherent with the N/N data. Hence, the OLCI R_{rs} spectra are corrected with the bi-directional reflectance distribution function (BRDF), following the procedure outlined in the works of Morel and his co-workers from 1991 [69,78–81].

The EUMETSAT OLCI L2 flags are different than those used in the NASA L2 processing. In this work, in agreement with [82] and the latest EUMETSAT indication (I. Cazzaniga, pers. comm.) [83], the ANNOT_DROUT flag is not applied, as in the latest version of CMEMS [84]. This flag masks pixel where the value of the residual for the surface reflectance at 510 nm is above a certain threshold, defined in the climatology. Moreover, in this study, the ANNOTMIXR1, ANNOT_TAU6, and ANNOT_ABSO_D named annotation flags (here ANNOT_*), have not been applied in the match-up analysis to analyze their effect on the VGOCS spatial coverage and their agreement with in situ data. Hence, all those flags concern the quality of the atmospheric correction in open waters, and the application to optically complex waters could be questionable [82,83].

In Figure 8, the maps of b_{bp}(443) retrieved from the original OLCI R_{rs} during 21 September 2017, with and without the application of the ANNOT_DROUT and ANNOT_* flags, are presented. The mean percentage of valid pixels masked by the ANNOT_DROUT flag in the coastal area (Figure 6) is around 40.9% for the OLCI time series, the ANNOT_MIXR1 flag masks the 20.1% of them, while the other two ANNOT_* flags mask the 0.8% of the coastal pixels.

Also for the ANNOT_DROUT and ANNOT_* data, the effect of the adjustment on their agreement with in situ observations will be analyzed in Section 3.2.3 to justify the decision for unmasking data traditionally masked in the standard processing chains.

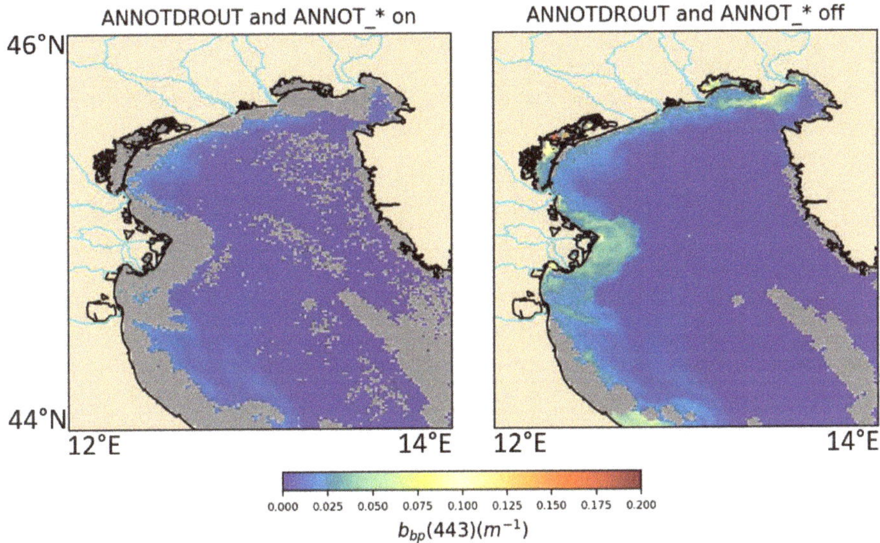

Figure 8. Maps of b_{bp}(443) retrieved from the original OLCI R_{rs} for 21 September 2017 09:40 UTC. On the left, the ANNOT_ * flags are active (ANNOT_* on), while the pixels are unmasked on the right (ANNOT_* off).

2.4. In Situ Data

AAOT, part of the AERONET-OC network [18,19], is located in the NAS, 8 nautical miles south-east of the Venice Lagoon (12.509° E, 45.314° N, the red star in Figure 1). The in situ data here acquired has been used to validate and adjust the satellite R_{rs} spectra of each sensor.

AERONET-OC can be considered as the archetype FRM for the validation of satellite remote-sensing radiometric data over water [20,85]. Hence, to compare the AAOT data with the satellite R_{rs} spectra, the normalized water-leaving radiances here acquired have been divided by the extra-solar irradiance (F0) [86]. Particularly, in this study the Version 3 L2 AAOT data (cloud-screened and quality assured data) have been exploited and integrated with the L1.5 data (only cloud-screened) for periods where the L2 data were not still processed.

The exploitation of these data to adjust the satellite spectra is fundamental for this study; indeed, AAOT is located in an area with a large optical variability that can be considered as representative of the basin variability [19,87]. Consequently, the adjustment coefficients here retrieved can be used to adjust the R_{rs} spectra in the entire basin [6].

2.5. Match-up Analyses

To assess the quality of the satellite R_{rs} spectra, pairwise match-up analyses were performed using the AAOT in situ data as a reference. Those analyses were conducted for every sensor of VGOCS separately, because different sensors can have different sources of uncertainties and, consequently a different agreement with the in situ spectra.

The match-up procedure is consistent with the one of [6], where for each couple of satellite/in situ observations acquired at $\Delta t < 30$ min, a 3×3 box around the AAOT area has been chosen, keeping only boxes with at least 3 valid pixels. The value of the in situ R_{rs} is the mean of all the AAOT data acquired between t (time of the satellite observation)-Δt (30 min) and t+Δt, while the value of the satellite R_{rs} is the mean of the 3x3 box data For each valid couple of observations, the in situ R_{rs} spectrum has been band-shifted to the satellite wavelengths, following [88].

To test the agreement between the satellite and in situ data, different statistical parameters have been calculated and their formulas have been presented in Table 2.

Table 2. Formulations of the statistical parameters used in this study to assess the agreement between two datasets, where n is the number of concurrent observations of the match-up, x the in situ measurement, and y the satellite observation.

Statistical Parameter	Formulation
Determination coefficient (r^2)	$r^2 = \dfrac{\sum_{i=1}^{n}(x_i-\bar{x})(y_i-\bar{y})}{\sqrt{\sum_{i=1}^{n}(x_i-\bar{x})^2}\sqrt{\sum_{i=1}^{n}(y_i-\bar{y})^2}}$
Mean Absolute Difference (MAD)	$MAD = \dfrac{\sum_{i=1}^{n}|y_i-x_i|}{n}$
Root Mean Squared Difference (RMSD)	$RMSD = \sqrt{\dfrac{\sum_{i=1}^{n}(y_i-x_i)^2}{n}}$
Mean absolute percentage difference (MAPD)	$MAPD(\%) = \dfrac{100}{n}\sum_{i=1}^{n}\left|\dfrac{y_i-x_i}{x_i}\right|$

In the R_{rs} match-up analyses, the entire dataset (ED) of each sensor was divided into two subsets: training (TD) and validation dataset (VD). The first was used to retrieve the coefficients of the adjustment (presented in Section 2.6), while the second one was used to validate the satellite data before and after the adjustment procedure (Section 3). For each N/N sensor (except for VIIRSJ), VD is composed of all the data acquired between 1 September 2015 and 31 October 2019, while the remaining data are part of TD. The split of the dataset is not applied to the OLCI and VIIRSJ data, as they are the youngest sensors between those used in VGOCS and they have a lower number of match up points in comparison with the other sensors. Hence, the splitting may not allow capturing all the optical variability of the basin; consequently, for those two sensors, the data used to retrieve the adjustment coefficients are the same used to validate the dataset (TD=VD=ED).

Match-up analyses have been performed also for data traditionally masked in the standard processing chains. In Section 3.2 match-up analyses for the HSZ, SL, ANNOT_DROUT, and ANNOT_* data have been performed. For the HSZ flag, observation is considered as flagged if the pixel collocated with AAOT is scanned at $\theta v > 60°$; for SL, ANNOT_DROUT, and ANNOT_*, an observation is identified as flagged if the considered flag is active for at least 1/3 of the box valid pixels.

2.6. Satellite R_{rs} Adjustment

To reduce discrepancies and biases between in situ and satellite measurements, to exploit data usually masked in the standard processing chains, and to reduce the inter-sensor differences, a multi-linear regression algorithm (MLR), like the one presented in [17], is exploited to adjust the satellite R_{rs} spectra. As different sensors have different uncertainties, temporal coverages, and agreements with the in situ observations, a different set of adjustment coefficient is calculated for each of them.

Particularly, for each couple of observations of the TD the difference between the in situ and satellite measurement is evaluated for each band:

$$\Delta R_{rs}(\lambda) = R_{rs}^{is}(\lambda) - R_{rs}^{or}(\lambda), \tag{1}$$

where $R_{rs}^{is}(\lambda)$ is the in situ R_{rs} and $R_{rs}^{or}(\lambda)$ is the original satellite R_{rs}. The multilinear regression scheme is

$$\langle \Delta R_{rs}(\lambda) \rangle = a_0^{sat} + \sum_{i=1}^{n} a_i^{sat} R_{rs}^{or}(\lambda_i) + a_6^{sat}\theta_v + a_7^{sat}\theta_s + a_8^{sat}\varphi, \tag{2}$$

The input vectors of the adjustment are the original R_{rs} spectra, and the values of the three viewing geometry parameters (θ_v, θ_s, and φ) corresponding to the pixel collocated with AAOT.

The coefficients a_i^{sat} ($i = 0, \ldots, 8$) were calculated performing the MLR scheme between $\Delta R_{rs}(\lambda)$ and the input vectors. Then the adjusted R_{rs} can be retrieved with:

$$R_{rs}^{adj}(\lambda) = R_{rs}^{or}(\lambda) + \langle \Delta R_{rs}(\lambda) \rangle, \tag{3}$$

This set of fixed coefficients have been used to adjust the R_{rs} spectra of each image of the single sensor dataset, as the AAOT in situ data are representative of the optical variability of the basin. Finally, the adjusted R_{rs} have been exploited as the input of the QAA to calculate different IOPs [34,35].

2.7. Inter-Sensor Differences

As stated in the previous sections, the adjustment is applied to reduce the R_{rs} inter-sensor differences that are due to different observation geometries, calibration accuracies, data processing, and resolutions of the sensors. To quantify such effect r^2 and MAPD have been calculated in different locations of the NAS for images acquired in temporal proximity by different sensors ($\Delta t < 10$ min). This has been done starting from the hypothesis that in a time range of 10 min no considerable oceanographic processes can occur and that consequently most of the differences observed between two images are likely due to artifacts. The chosen locations are here named virtual buoys (red dots in Figure 1) and they are situated:

- At AAOT, to analyze the effect of the adjustment in an area with large optical variability.
- Close to the Livenza, Brenta-Adige, Piave, Tagliamento, and Isonzo river mouths to analyze the effect of the adjustment in optically complex waters.
- In an off-shore location (here named simply OPEN), to analyze the effect of the adjustment in open waters.
- Close to the Po river mouth to analyze the effect of the adjustment in the area where the largest amount of fresh-water is usually available [45,46], due to the high river discharges.

Hence, the last two virtual buoys represent the two optical extreme situations for open and optically complex waters respectively.

As stated earlier, to calculate r^2 and MAPD only couples of images acquired at $\Delta t < 10$ min have been selected and for each virtual buoy a 3×3 box around its location has been chosen. Then only the couples of images with both boxes with at least 3 valid R_{rs} pixels have been maintained, and r^2 and MAPD have been calculated for each band of the R_{rs} spectrum, similar to a match-up analysis. As the sensors have a different spectral resolution, in this analysis all the R_{rs} spectra have been band-shifted to the MODIS bands, using the approach from [88].

3. Results

3.1. Match-Up Analyses

In this section, the agreement between the satellite and in situ R_{rs} spectra before and after the application of the adjustment will be evaluated. In Table 3 the numbers of match-up points for each VGOCS sensor are shown.

Table 3. The number of match-up points for each VGOCS sensor. Between parenthesis, for the N/N sensors the number of match-up points for the ED SL affected data, for OLCI those flagged by the ANNOT_DROUT/ANNOT_* flags.

Sensor	Number of Match-Up Points
TERRA	218 (178)
AQUA	357 (230)
VIIRSN	449 (200)
VIIRSJ	216 (48)
OLCI	64 (37/20)

Previous studies demonstrated how the MLR adjustment strongly increases the agreement between the in situ and satellite spectra [6,17]. In this work the effect of the adjustment on all the VGOCS sensors will be evaluated but, for the safe of brevity, the scatterplots will be presented only for OLCI. For the N/N sensors, only the spectra of r^2 and MAPD before and after the application of adjustment are shown (Figures 9 and 10) and the other statistical parameters (RMSD and MAD) are presented from Tables 4–7).

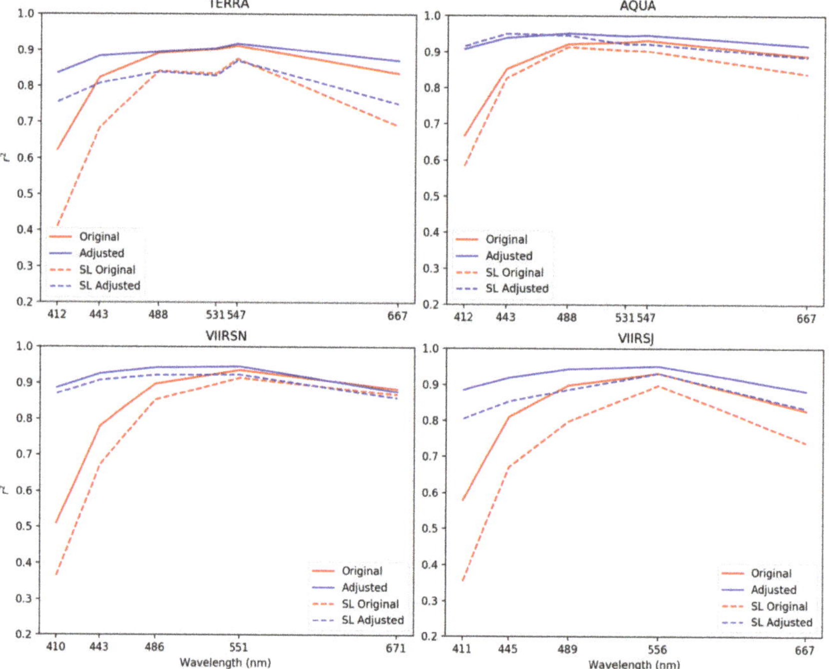

Figure 9. Spectra of r^2 before (red) and after (blue) the adjustment procedure for each N/N sensor. The continuous lines identified the results for VD, the dotted ones for the ED SL affected data.

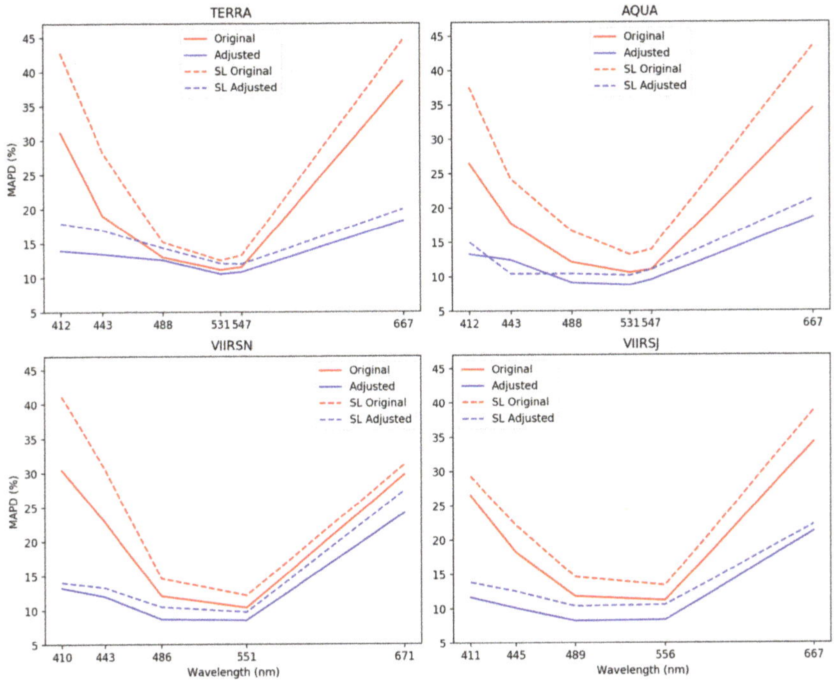

Figure 10. Spectra of MAPD before (red) and after (blue) the adjustment procedure for each N/N sensor. The continuous lines identified the results for VD, the dotted ones for the ED SL affected data.

Table 4. Statistical parameters for the match-up analysis between the AAOT in situ R_{rs} and the original (adjusted) VD TERRA R_{rs}.

TERRA Bands	r^2	MAD (sr^{-1})	RMSD (sr^{-1})	MAPD (%)
$R_{rs}(412)$	0.62 (0.84)	1.3×10^{-3} (6.6×10^{-4})	1.6×10^{-3} (1.0×10^{-3})	31.2 (14.0)
$R_{rs}(443)$	0.82 (0.89)	9.1×10^{-4} (6.3×10^{-4})	1.2×10^{-3} (9.0×10^{-4})	19.1 (13.5)
$R_{rs}(488)$	0.90 (0.90)	9.2×10^{-4} (8.1×10^{-4})	1.3×10^{-3} (1.2×10^{-3})	13.1 (12.6)
$R_{rs}(531)$	0.90 (0.90)	8.5×10^{-4} (7.6×10^{-4})	1.3×10^{-3} (1.2×10^{-3})	11.2 (10.6)
$R_{rs}(547)$	0.91 (0.92)	8.4×10^{-4} (7.6×10^{-4})	1.2×10^{-3} (1.2×10^{-3})	11.6 (10.9)
$R_{rs}(667)$	0.83 (0.87)	4.1×10^{-4} (2.3×10^{-4})	5.2×10^{-4} (3.5×10^{-4})	38.6 (18.3)

Table 5. Statistical parameters for the match-up analysis between the AAOT in situ R_{rs} and the original (adjusted) VD AQUA R_{rs}.

AQUA Bands	r^2	MAD (sr^{-1})	RMSD (sr^{-1})	MAPD (%)
$R_{rs}(412)$	0.67 (0.91)	1.1×10^{-3} (6.1×10^{-4})	1.6×10^{-3} (7.9×10^{-4})	26.5 (13.3)
$R_{rs}(443)$	0.85 (0.94)	8.7×10^{-4} (5.7×10^{-4})	1.2×10^{-3} (7.4×10^{-4})	17.8 (12.5)
$R_{rs}(488)$	0.92 (0.95)	8.8×10^{-4} (6.4×10^{-4})	1.2×10^{-3} (9.1×10^{-4})	12.1 (9.1)
$R_{rs}(531)$	0.93 (0.95)	8.1×10^{-4} (6.6×10^{-4})	1.2×10^{-3} (9.9×10^{-4})	10.6 (8.8)
$R_{rs}(547)$	0.93 (0.95)	8.0×10^{-4} (6.9×10^{-4})	1.1×10^{-3} (1.0×10^{-3})	11.0 (9.6)
$R_{rs}(667)$	0.89 (0.92)	3.5×10^{-4} (2.1×10^{-4})	4.4×10^{-4} (2.9×10^{-4})	34.5 (18.7)

Table 6. Statistical parameters for the match-up analysis between the AAOT in situ R_{rs} and the original (adjusted) VD VIIRSN R_{rs}.

VIIRSN Bands	r^2	MAD (sr^{-1})	RMSD (sr^{-1})	MAPD (%)
$R_{rs}(410)$	0.51 (0.89)	1.5×10^{-3} (6.6×10^{-4})	2.1×10^{-3} (8.8×10^{-4})	30.5 (13.3)
$R_{rs}(443)$	0.78 (0.93)	1.2×10^{-3} (6.0×10^{-4})	1.7×10^{-3} (8.3×10^{-4})	22.9 (12.1)
$R_{rs}(486)$	0.90 (0.94)	9.2×10^{-4} (6.6×10^{-4})	1.3×10^{-3} (1.0×10^{-3})	12.2 (8.8)
$R_{rs}(551)$	0.94 (0.95)	7.7×10^{-4} (6.1×10^{-4})	1.1×10^{-3} (9.7×10^{-4})	10.5 (8.6)
$R_{rs}(671)$	0.88 (0.88)	3.1×10^{-4} (2.6×10^{-4})	4.1×10^{-4} (3.7×10^{-4})	29.8 (24.2)

Table 7. Statistical parameters for the match-up analysis between the AAOT in situ R_{rs} and the original (adjusted) VD VIIRSJ R_{rs}.

VIIRSJ Bands	r^2	MAD (sr^{-1})	RMSD (sr^{-1})	MAPD (%)
$R_{rs}(411)$	0.58 (0.89)	1.4×10^{-3} (6.0×10^{-4})	1.9×10^{-3} (7.9×10^{-4})	26.6 (11.7)
$R_{rs}(445)$	0.81 (0.92)	1.0×10^{-3} (6.0×10^{-4})	1.3×10^{-3} (8.2×10^{-4})	18.3 (10.2)
$R_{rs}(489)$	0.90 (0.94)	9.5×10^{-4} (6.4×10^{-4})	1.3×10^{-3} (9.0×10^{-4})	11.9 (8.3)
$R_{rs}(556)$	0.93 (0.95)	7.9×10^{-4} (5.6×10^{-4})	1.1×10^{-3} (8.0×10^{-4})	11.3 (8.4)
$R_{rs}(667)$	0.83 (0.88)	3.3×10^{-4} (1.9×10^{-4})	4.2×10^{-4} (2.7×10^{-4})	34.3 (21.3)

After the adjustment MAPD is spectrally reduced for each N/N sensor and r^2 is spectrally increased except for the 671 nm band for VIIRSN (nearly constant, Figures 9 and 10). Particularly, r^2 is now close to or larger than 0.90 for each band of each sensor, with just the exception of the first TERRA band for which r^2 is 0.84. For MAPD, a lower decrease is observed in the green part of the spectrum, since the green bands were already in good agreement with the in situ observations before the adjustment. The blue and red bands show the larger improvement: for the first original blue band, MAPDs were larger than 25% for all sensors, while after the adjustment they are lower than 15%; for the red band MAPDs were between 25.0% and 30.0%, and after the adjustment, these decreased to values that go from 18.3% (TERRA) to 24.2% (VIIRSN). Hence, after the adjustment, the agreement between the satellite and in situ spectra is strongly improved, and this is confirmed also by the other statistical parameters (from Tables 4–7) that are spectrally reduced.

The agreement with the in situ spectra differs from one sensor to another. Particularly, TERRA is the one that shows the larger discrepancies with the in situ data both before and after the adjustment procedure. This can be due to the serious degradation of the TERRA mirror; indeed, for this sensor, not even the vicarious calibration has led to the retrieval of reliable OC products, and a calibration based on other functioning OC sensor observations is applied to the TERRA data [3,89]. This degradation can also explain the lower number of valid observations and match-up points for TERRA (218), in comparison with AQUA (357) and VIIRSN (449). Despite that, the adjustment has strongly improved the quality of the TERRA R_{rs} spectra, enabling a more reliable usage of the data from this sensor.

Some differences in the statistics for different sensors can also be partly due to the different spectral resolutions; indeed, in the band shifting procedure the spectral distances between the native in situ (AAOT) and target bands (satellite) are different for different sensors, and larger spectral distances can increase the uncertainty of the band-shifting [88]. This validates the choice to keep the spectral resolution of each VGOCS sensor and to do not band-shift all of them to a set of predefined wavelengths. Indeed, the AAOT central wavelengths varied during the years, and choosing a set of predefined wavelengths for the satellite sensors could lead to the need to band-shift both in situ and satellite data to this set, increasing the uncertainty.

In Figures 11 and 12, the scatterplots for the original and adjusted OLCI R_{rs} in comparison with the in situ data are shown, with the statistical parameters spectra shown in Figure 13 and listed in Table 8. The number of match-up points is only 67 due to the short OLCI time series, the larger revisiting time of Sentinel 3A in comparison with the other VGOCS satellites, and the lack of AAOT data from the end of June to the beginning of October of 2017. The uncertainty observed for $R_{rs}(412)$ is larger than those observed for the other OC sensors of the constellation. Despite that, the adjustment largely reduces this uncertainty, showing similar results to those of the previous analyses (Figure 13). Indeed, after the adjustment r^2 is 0.89 for the red band and larger than 0.9 for the other ones, while MAPD is $\approx$23% for the red band, 10.3% for the 412 nm band, and lower than 10.4% for the other ones.

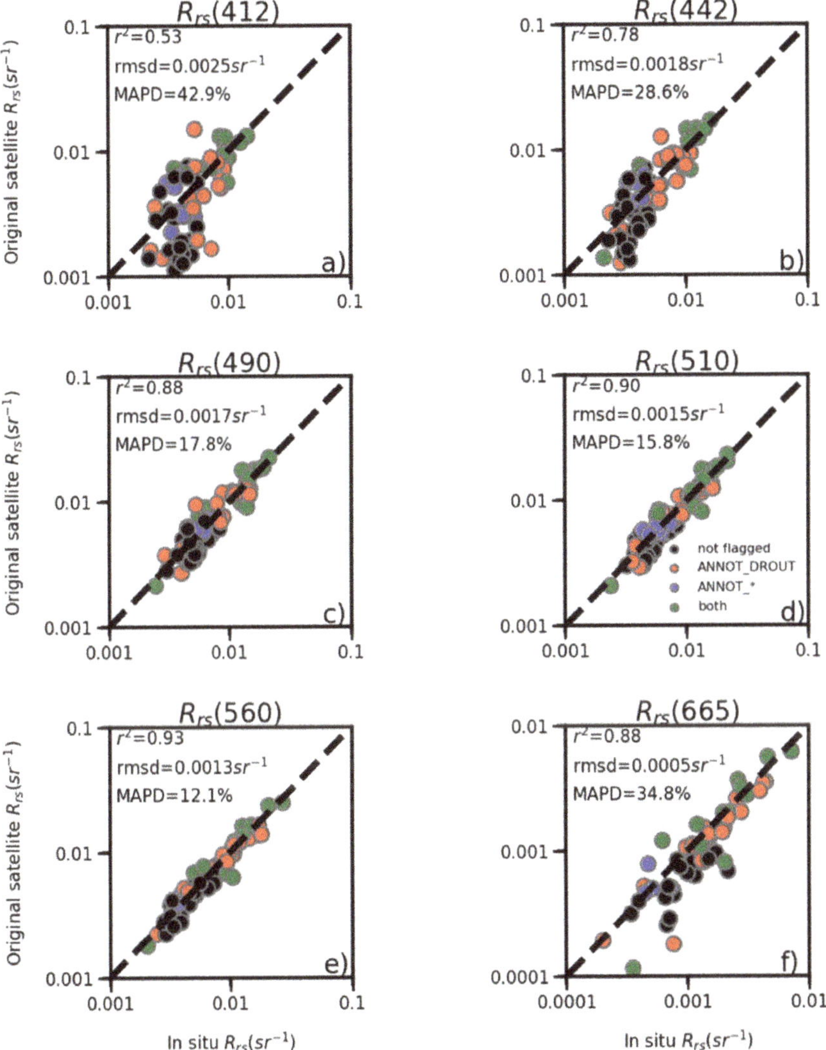

Figure 11. Match-up analysis between the in situ R_{rs} and the original VD OLCI R_{rs}. In the logarithmic scale on the x-axes the in situ R_{rs} and on the y-axes the original OLCI R_{rs} for each band (**a–f**). The red points are the observations where the ANNOT_DROUT flag is active, the blue points those where the ANNOT_* flags are active, the green ones those where both flags are active, and the black points where none of those flags is active.

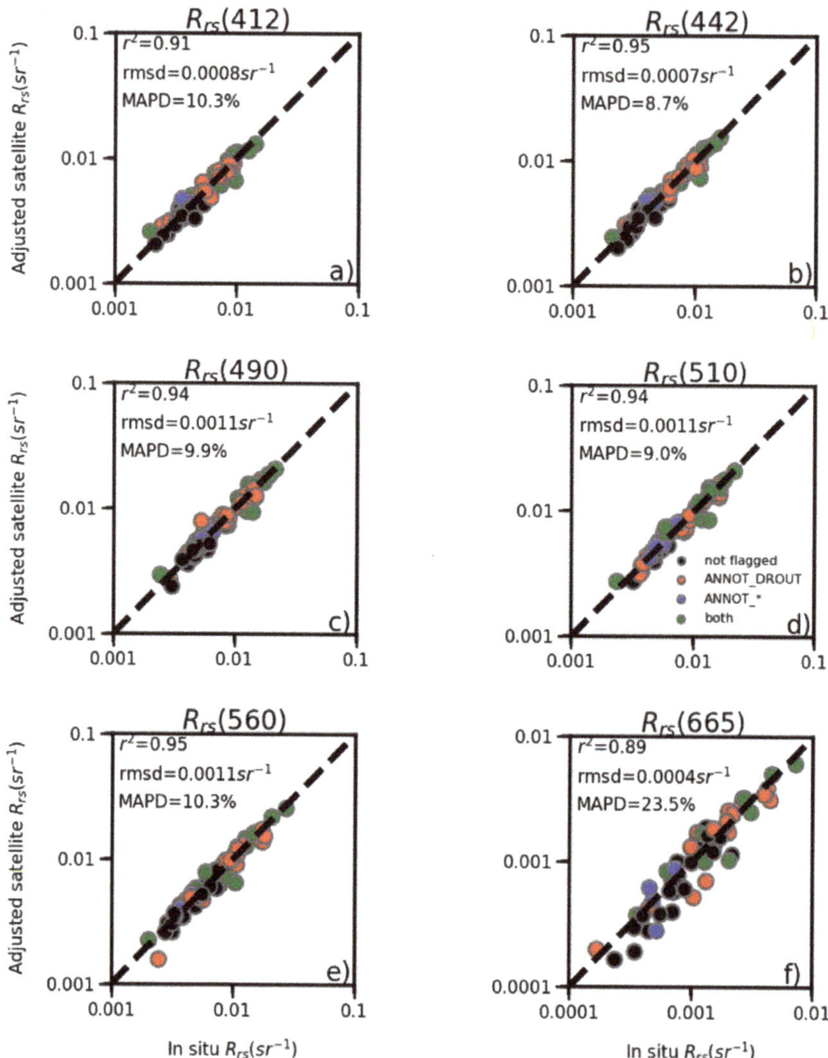

Figure 12. Match-up analysis between the in situ R_{rs} and the adjusted VD OLCI R_{rs}. In the logarithmic scale on the x-axes the in situ R_{rs} and on the y-axes the adjusted OLCI R_{rs} for each band (**a–f**). The red points are the observations where the ANNOT_DROUT flag is active, the blue points those where the ANNOT_* flags are active, the green ones those where both flags are active, and the black points where none of those flags is active.

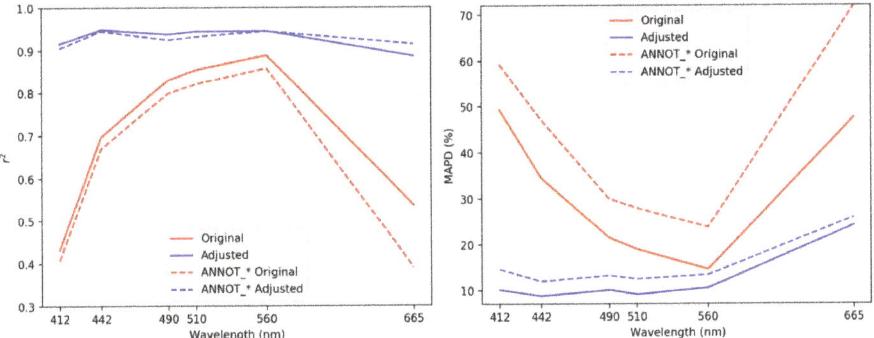

Figure 13. Spectra of r^2 (left) and MAPD (right) before (red) and after (blue) the adjustment procedure for OLCI. The continuous lines identified the results for VD, the dotted ones for the ANNOT_* data.

Table 8. Statistical parameters for the match-up analysis between the AAOT in situ R_{rs} and the original (adjusted) VD OLCI R_{rs}.

OLCI Bands	r^2	MAD (sr^{-1})	RMSD (sr^{-1})	MAPD (%)
$R_{rs}(412)$	0.43 (0.92)	2.1×10^{-3} (5.6×10^{-4})	2.8×10^{-3} (7.9×10^{-4})	49.4 (13.3)
$R_{rs}(442)$	0.70 (0.95)	1.6×10^{-3} (5.0×10^{-4})	2.2×10^{-3} (7.4×10^{-4})	34.6 (12.5)
$R_{rs}(490)$	0.83 (0.94)	1.4×10^{-3} (7.7×10^{-4})	1.9×10^{-3} (9.1×10^{-4})	21.4 (9.1)
$R_{rs}(510)$	0.85 (0.95)	1.3×10^{-3} (7.2×10^{-4})	1.8×10^{-3} (9.9×10^{-4})	19.0 (8.8)
$R_{rs}(560)$	0.89 (0.95)	1.0×10^{-3} (7.7×10^{-4})	1.6×10^{-3} (1.0×10^{-3})	14.6 (9.6)
$R_{rs}(665)$	0.54 (0.89)	4.9×10^{-4} (3.1×10^{-4})	1.0×10^{-3} (2.9×10^{-4})	47.6 (18.7)

The better agreement between the satellite and in situ spectra strongly increases also the quality of the IOPs retrieved by the QAA [6]. Indeed, the quality of the QAA outputs is dependent on the input R_{rs} spectrum uncertainty; consequently, the reduction of the discrepancies between the satellite and in situ R_{rs} bring to a better agreement between the satellite and in situ QAA parameters [6].

3.2. VGCOS Spatial and Temporal Coverage

As it is shown in Section 2.3, the unmasking of the HSZ, SL, ANNOT_DROUT, ANNOT_* data leads to an increase of the VGOCS spatial and temporal coverage. In this section, the agreement of those data with the in situ spectra will be analyzed, to evaluate if they can be unmasked in the VGOCS dataset.

3.2.1. HSZ Flag

The quality of the OC data can be dependent on the viewing geometry and, particularly, on θ_v. The HSZ flag masks data acquired at θ_v larger than 60° since they can be affected by larger uncertainties in the atmospheric correction and the normalization of the water leaving radiance [69,70]. Nevertheless, also data acquired at large θ_v, but lower than 60° (for example 50° < θv < 60°) can be affected by larger discrepancies in comparison with in situ or other satellite data [6,38].

To evaluate this θ_v dependence for the N/N sensors, MAPD has been calculated for different ranges of θ_v with a step of 10° (Figure 14) for the match-up analyses of the previous section. Before the adjustment, the VIIRSN and VIIRSJ R_{rs} with $\theta_v > 50°$ show larger MAPDs in comparison of those acquired at $\theta_v < 50°$, especially for the 411/2 nm band and the 667/671 ones. After the adjustment MAPD is lower in each bin for each sensor, reducing a general bias with the in situ data, confirming the aggregated results presented in Section 3.1. MAPDs for data acquired at $\theta_v > 50°$ are still slightly larger than those acquired at lower θ_v, but the dependence is now consistently reduced. For example, for the original VIIRSJ 411 nm band MAPD is between 18.1% and 28.5% for the first five bins, while it

is larger than 34% for the last two; after the adjustment, MAPD is between 10.4 and 13% for θv < 50°
and it is <14.0% for θv > 50°, confirming the reduced θv dependence.

AQUA and TERRA show similar behavior of the one observed for VIIRSJ/N, but for them also
the original data acquired at 40° < θ_v < 50° show larger discrepancies with the in situ data, but they
are strongly reduced with the adjustment. This larger θ_v dependence for AQUA and TERRA can be
due to the pixel growth effect [90,91], which can be more evident in the MODIS sensors, as this is
reduced in VIIRS by its aggregation algorithm [91]. The main result of this analysis is that now most
of the observations acquired at θ_v > 50° have MAPD lower of those acquired at θ_v < 50° before
the adjustment.

Consequently, data previously masked by the HSZ flag (θ_v > 60°) can now be exploited, making
available a large number of data, usually discarded in standard applications.

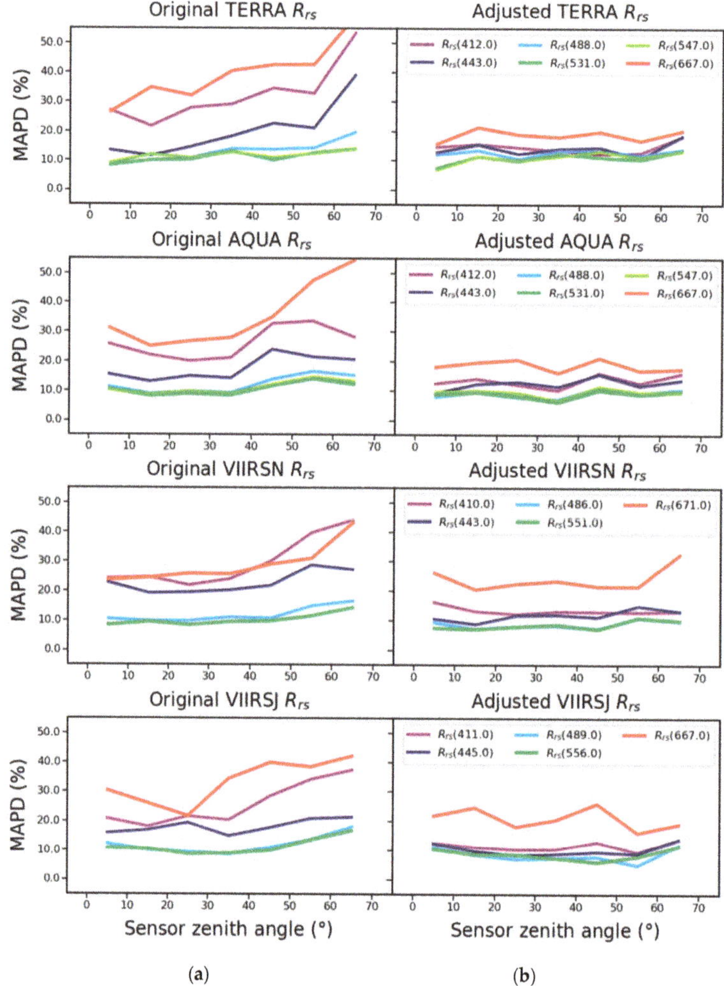

Figure 14. Analysis of the MAPD dependence from θ_v before (left, column **a**) and after the adjustment
(right, column **b**). On the x-axes the wavelength, on the y-axes MAPD, with the different colours of the
lines identifying the different R_{rs} bands.

As already shown in Figures 4 and 5, unmasking for the HSZ flag provides up to 3 additional full images a day of the NAS, strongly increasing the VGOCS spatial and temporal coverage, which is fundamental for the aim for which VGOCS is conceived.

3.2.2. SL Flag

The SL affected pixels can have large sources of uncertainties [6,71–74], and, before using them, it is important to test their agreement with the in situ observations. For this reason, match-up analyses for the original and adjusted R_{rs} of the N/N sensors have been performed, considering only the SL affected data (spectra of r^2 and MAPD in Figures 9 and 10). For those analyses, for each sensor ED (TD+VD) has been used, to have a larger number of match-up points.

For the original SL affected data (red dotted line), the values of r^2 and MAPD are considerably lower and larger respectively than the values observed for VD (red continuous line), confirming that the stray-light effect can significantly lower the quality of the OC data. Nevertheless, after the adjustment r^2 and MAPD are considerably closer to 1 and 0 respectively. The behavior of each sensor is similar to the one observed in Section 3.1, with TERRA that still shows larger discrepancies in comparison with the in situ data, due to the degradation of its mirror [3,89]. There are still some residuals due to the stray-light effect, as the MLR algorithm can only minimize and not eliminate the larger uncertainties introduced by this effect. Nevertheless, the values of the statistical parameters (blue dotted lines) are now closer to those observed for VD after the adjustment (blue continuous line). Moreover, the values of MAPD and r^2 of the adjusted SL data are respectively lower and larger than those observed for the original R_{rs} in VD.

Hence, as for the HSZ flag (Section 3.2.1), the adjustment makes available such data, allowing to analyze the coastal optical variability and further improving the VGOCS spatial coverage close to shore (Section 2.3.1).

3.2.3. ANNOT_DROUT and ANNOT_* Flags

The ANNOT_DROUT and the ANNOT_* flags have not been applied in the OLCI processing chain, as they mask a large number of pixels close to the coast (Section 2.3.2).

As stated earlier, recently EUMETSAT has removed the ANNOT_DROUT flag from the standard L2 flags (I.Cazzaniga, pers.comm.) [83]. The validity of this choice is confirmed by the match-up analyses of Figures 11 and 12, where the comparison between the in situ and original/adjusted OLCI R_{rs} is presented. In these figures, the green and red data are those for which the ANNOT_DROUT flag is active. Not considering them, the number of match-up points is drastically reduced, as only 30 observations are now available in comparison with the 67 of Section 3.1. Masking for ANNOT_DROUT data, MAPD is larger and r^2 and RMSD lower (not shown) for all the bands in comparison with the match-up analyses accomplished considering the entire OLCI VD (Section 3.1). The RMSD reduction is not due to a better agreement of this dataset with the in situ data, but simply to the order of magnitude of the masked data. Indeed, only one observation with $R_{rs}(412) > 0.01$ sr^{-1} is not masked by ANNOT_DROUT (blue and black points, Figure 11), leading to a lower value of RMSD, which is an absolute difference, and it is dependent from the order of magnitude of the observations, while MAPD is a relative difference. Hence, the application of the ANNOT_DROUT flag does not allow capturing all the optical variability of the basin. For those reasons, this should not be applied in optically complex waters, as previously stated in [82].

In the OLCI match-up analysis, all the data flagged as ANNOT_* (blue and green points in Figures 11 and 12, with the r^2 and MAPD spectra in Figure 13) are due to the ANNOT_MIXR1 flag, as this is the only one which usually results to be applied in coastal waters (Section 2.3.2). The ANNOT_MIXR1 flag data show larger MAPDs (red dotted line) in comparison with those of the entire dataset (red continuous line), but those are strongly reduced after the adjustment (blue dotted line). Still slightly larger discrepancies are shown in comparison with those of the adjusted VD R_{rs}

(blue continuous line), but their agreement with in situ data is better than the one of the original R_{rs} of VD (red continuous line), similarly to the SL data.

Hence, the adjusted ANNOT_MIXR1 data should not be masked in the NAS coastal waters, for the OLCI processing baseline v2.23. On the contrary, the ANNOT_TAU6, and ANNOT_ABSO_D data have been finally masked in the VGOCS dataset. Hence, they are not usually applied in coastal waters and the use of such data could result in large uncertainties in open waters, as those data have not been characterized by a match-up analysis.

3.3. OLCI Camera Dependence

As stated earlier, the quality of the OLCI R_{rs} spectra can be dependent on the different cameras mounted on the optical bench. Indeed, the presence of different camera implies that different ranges of θ_v correspond to different detectors; consequently, a plot similar to Figure 14 is presented in Figure 15 but with MAPD calculated for the different camera observations (C1: Camera1, C2: Camera2 ...).

To distinguish between the different cameras, the detector_index field provided in the OLCI L2 files is exploited. Indeed, this field goes from 0 to 3699 (that are the total number of columns of a single L1 file) and each bin of 740 columns corresponds to a different camera (e.g., from 0 to 739 Camera1, from 740 to 1479 Camera 2 and so on ...).

The observations from each camera (from C1 to C5) are 17,19,17,8, and 6 respectively. Such numbers of observations are too low to characterize OLCI camera dependence. Nevertheless, using the available match-up points it is possible to observe that the value of MAPDs for the original R_{rs} are consistently different for different cameras. Those differences are strongly reduced after the adjustment, with MAPD showing similar values for different cameras. Hence, also if it is not possible to analyze the OLCI camera dependence, the observed reduction of this effect brought by our adjustment is sufficient for the aim of our study.

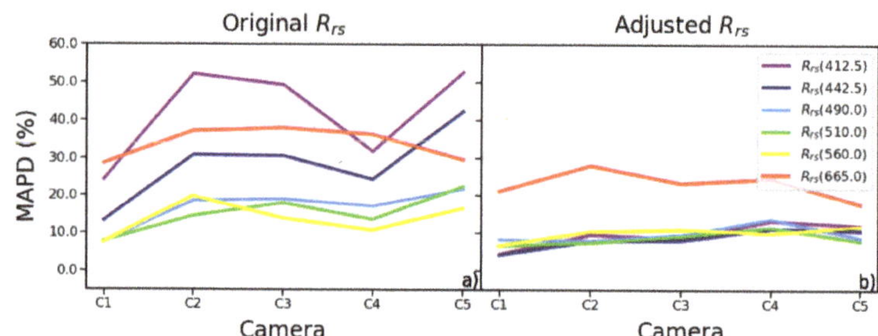

Figure 15. Analysis of the dependence of MAPD from the different OLCI cameras for the (**a**) original and (**b**) adjusted R_{rs}. On the x-axes the camera (from C1 to C5), on the y-axes MAPD, with the different colours of the lines identifying the different R_{rs} bands.

3.4. Inter-Sensor Differences

As described in Section 2.7, to qualify the residual inter-sensor differences, MAPD and r^2 have been calculated at the virtual buoys for couples of images acquired in temporal proximity.

In Figures 16 and 17, the spectra of r^2 and MAPD at the virtual buoys are presented for the original and adjusted R_{rs}, band-shifted to the MODIS bands. r^2 is everywhere increased for the two blue bands, while for the other bands it remains almost constant (or slightly reduced), except for the red band at OPEN, where this parameter is reduced. For what concerns MAPD, this is reduced for each band in each location, with a larger decrease observed for the two blue bands and the red band, and a slight increase in the green part of the spectrum at OPEN.

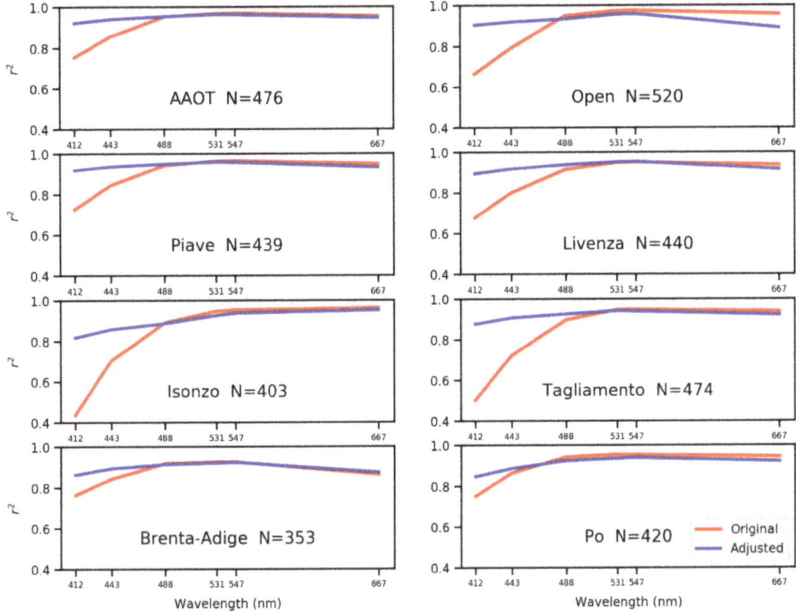

Figure 16. The r^2 calculated at the virtual buoys for the original (red) and adjusted R_{rs} (blue) for images acquired in temporal proximity by the different VGOCS sensors. Inside the box the virtual buoy name with the number of the couple of images used in the calculation of r^2.

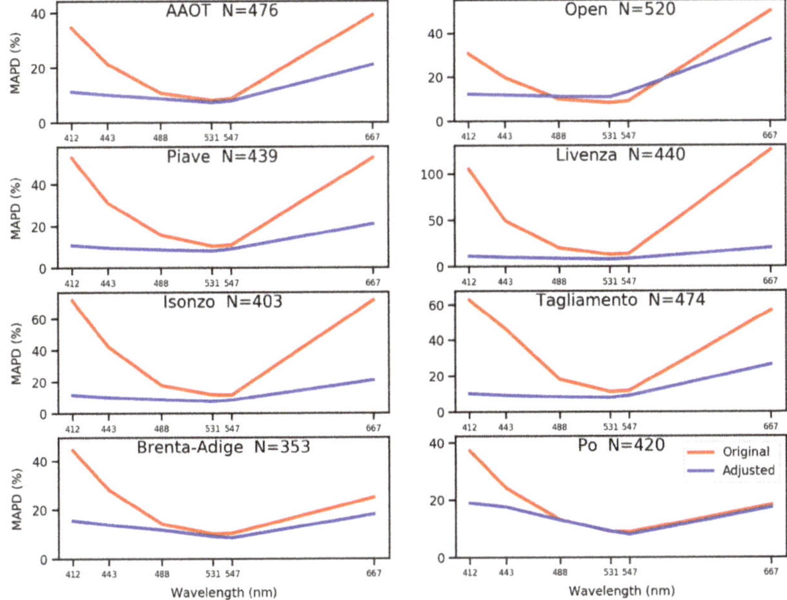

Figure 17. Mean absolute percentage difference (MAPD) calculated at the virtual buoys for the original (red) and adjusted R_{rs} (blue) for images acquired in temporal proximity by the different VGOCS sensors. Inside the box the virtual buoy name with the number of the couple of images used in the calculation of MAPD.

The lower reduction/increase of MAPD/ r^2 in the 488–547 nm range is due to the good agreement of these bands also for the original R_{rs}; thus, the adjustment does not influence too much the R_{rs} retrieval for them, as the original bands are already in good agreement with in situ observations.

The MAPD decrease observed at AAOT in Figure 17 after the adjustment is consistent with the behavior observed for the match-up analyses of Section 3.1, performed used the AAOT in situ data as a reference. For the original sensor match-up analyses, MAPD for the first blue band was between 26.5% and 42.9%, while for the second blue band this was between 17.8% to 28.6%; after the adjustment, those ranges decreased to 10.3–14.0% and 8.7–13.5% respectively. Similarly, for AAOT in the inter-sensor analysis, the values of MAPD decreased from 34.8% and 21.4% to 11.2% and 10.1% for the 412 and 443 nm bands respectively. Moreover, after the adjustment, the MAPD values for the green bands are slightly lower than 10.0%, while the one for the red band is around 20.0%, similarly to the results of Section 3.1.

At the other virtual buoys, the MAPD spectra after the adjustment are similar to the one observed for AAOT, while for the original R_{rs} very different spectra were observed for each of them. Indeed, for the original R_{rs}, the ranges of MAPD for the blue bands were between 30.7% (OPEN) and 105.8% (Livenza) and between 19.9% (OPEN) and 49.6% (Livenza). After the adjustment, for the first band, MAPD is between 10.4% (Tagliamento) and 19.0% (Po), while for the second one it is between 9.4% (Tagliamento) and 17.4% (Po). Not considering Po and Brenta-Adige, the superior limits of the two ranges reduce to 12.4% and 11.9% (OPEN), which are values similar to those observed for AAOT in both match-up and inter-sensor analysis. Similarly, for the green bands, all values of MAPD are around 10.0%, while for the red band MAPD is always around 20.0%, except for OPEN, where an MAPD of 37.3% is observed.

Different values of MAPD are observed for the blue bands at Po/Brenta-Adige and the red band at OPEN in comparison with the other virtual buoys since such locations represent the extreme situation for optically complex and open waters respectively, and consequently, the adjustment can be less effective for such type of waters at certain wavelengths.

Similar values of the statistical parameters for the match-up and inter-sensor analysis imply that the residual differences now observed between different satellite images are mostly due to the uncertainty in the atmospheric correction and the absolute calibration of the space sensors; consequently, those that are due to the different technical characteristics, resolutions, viewing geometry, and processing chains are notably reduced. Moreover, these findings confirm the underlying assumption of this study, i.e., that the AAOT in situ data can be exploited to adjust the R_{rs} spectra of the entire basin [6,87].

4. Discussion

This study tested the effect of an adjustment on the quality of the satellite data, the VGOCS spatial and temporal coverage, and the inter-sensor differences to demonstrate the suitability of VGOCS in analyzing the coastal optical variability.

Several match-up analyses between the original R_{rs} of different satellite sensors and the AAOT R_{rs} data were performed. The adjustment brought to a significant improvement of the satellite R_{rs} retrieval with a large spectral increase of r^2 and a spectral reduction of the other statistical parameters for all the VGOCS sensors. The better agreement between the satellite and in situ spectra strongly increases also the quality of the IOPs retrieved by the QAA, which are needed to analyze the coastal optical variability [6].

The adjustment was exploited also to make available data generally masked in the standard processing chains. Indeed, data acquired at $\theta v > 60°$ are usually masked by the application of the HSZ flag, due to larger uncertainties [38]. This is confirmed by the analysis of the dependence of the quality of the original R_{rs} spectra on θ_v; indeed, for the original R_{rs} larger discrepancies were observed for high θ_v, confirming the results of previous works [6–38]. Nevertheless, the adjustment reduced the biases between the satellite and in situ data for all ranges of θ_v, especially for those acquired at $\theta_v > 50°$, leading to a strong reduction of the θ_v dependence. This enabled to use also data usually masked by the HSZ flag, improving the VGOCS temporal and spatial coverage.

The VGOCS spatial coverage is further increased by the use of data usually masked by the SL flag for the N/N sensors and by the ANNOT_DROUT and ANNOT_MIXR1 flags for OLCI.

The adjustment strongly reduces the uncertainty for the SL affected pixels and it allows to have more valid pixels close to the shore, as the application of this flag leads to the masking of the 40% of the coastal pixels. In the future, to additionally reduce the SL data uncertainty, the calculation of a different set of adjustment coefficient could be considered, using in the TD only SL data.

The analysis of the ANNOT_DROUT data confirms the findings of [82] and the validity of the choice by EUMETSAT to remove this flag from the list of the standard L2 flags (I. Cazzaniga, pers.comm.) [83], as they result to be in good agreement with in situ data in the coastal area and they allow to analyze the coastal optical variability.

The behavior of the ANNOT_MIXR1 data, which are 20.1% of the OLCI coastal pixels, is similar to the SL ones, thus they have been unmasked. On the contrary, as it is not possible to characterize the ANNOT_TAU6 and ANNOT_ABSO_D data, they have been masked in our dataset, as they are not encountered in the coastal area and they can show large uncertainties in open waters.

The VGOCS spatial and temporal coverage will be further increased in the future. Hence, in the dataset, only the sensors that are currently functioning were included, as the present days are those with a larger number of sensors on orbit. Nevertheless, in the future also other sensors, that are not anymore functioning, such as the Sea-Viewing Wide Field-of-View Sensor (SeaWiFS) mounted on the satellite SeaStar and MERIS mounted on the Envisat satellite, will be included. Moreover, when the vicarious calibration process will be accomplished, also the OLCI Sentinel 3B sensor will be added to the constellation.

The analysis of the θ_v dependence was not performed for OLCI as different ranges of θ_v are scanned by different cameras. The low number of match-up points does not allow to deeply analyze the camera dependence, but different values of MAPD were observed for different cameras. After the adjustment, the difference between different cameras is strongly reduced and MAPD is similar for each of them. In the future, to deeply characterize this effect, the analysis can be performed simply enlarging the Δt of the match-up. Indeed, in this study Δt has been chosen as very narrow (30 min) to take into account the large temporal variability of the coastal area.

Starting from the hypothesis that no considerable oceanographic processes can occur in 10 min, the differences between R_{rs} spectra measured by different sensors have been estimated using images acquired in temporal proximity. The values of MAPD observed in this analysis after the adjustment, similar to those observed in the comparison with the in situ data, proved that the MLR scheme notably reduced the inter-sensor differences, which can lead to a misleading interpretation of the basin optical variability when two images from different sensors are compared.

In the introduction, the VGOCS has been defined as a hyper-temporal dataset, but to be coherent with this definition, the dataset must fulfill the three conditions defined in [29] and reported in Section 1:

(a) The variables stored in the dataset must be univariate. As the R_{rs} data are provided at their native spectral resolution to reduce the uncertainty in the band-shifting procedure, the variables stored in the R_{rs}_data group are different for each sensor and this condition seems to be not fulfilled. Nevertheless, the R_{rs} data group is mainly provided for post-processing procedures and it allows the users to calculate different IOPs and water component concentrations using different algorithms [92–95]. The parameters that are needed to analyze the coastal optical variability, the aim for which VGOCS is conceived, are the IOPs, that are the same for each sensor, fulfilling the first requirement.

(b) All the images must be time referenced and the same grid should be used for each image of the dataset. This is fulfilled because the name of each file gives information about the acquisition time, and each image of the dataset is reprojected on a standard 1 km × 1 km equirectangular grid.

(c)   All the sensors must be inter-validated between each other. This condition is also reached, as they are all adjusted on the AERONET-OC AAOT in situ data and the adjustment significantly reduced the inter-sensor differences.

5. Conclusions

This work aims to fill the gap between the near-polar orbiting and geostationary sensor temporal resolutions over the European coastal basins by the creation of a hyper-temporal Ocean Color analysis-ready dataset, named Virtual Geostationary Ocean Color Sensor (VGOCS). The latter fulfills the conditions needed to be defined as a hyper-temporal dataset and contains observations acquired from several near polar-orbiting Ocean Color imagers over the North Adriatic Sea, allowing approaching the geostationary sensor temporal resolution.

The problem in using data from different sensors is that they are characterized by different uncertainty sources that can introduce artifacts between different satellite images. For this reason, a multi-linear regression adjustment, based on a Fiducial Reference Measurements data-stream, has been applied to the remote sensing reflectance spectra of the VGOCS sensors. Hence, this adjustment, besides improving the agreement between the satellite and in situ data, reduced the inter-sensor differences, allowing the use of data from different sensors in analyzing the coastal optical variability. Moreover, the adjustment made available data traditionally masked in the standard processing chains, strongly increasing the VGOCS spatial and temporal coverage in the coastal area, an improvement that is fundamental to properly capture most of the coastal optical variability.

The approach here presented can be extended to all the coastal areas where one or more automatic radiometric stations, that represent the entire basin optical properties, are present. In European waters, this approach could be extended to:

- The Western Black Sea where the Gloria, Section-7, and Galata Platform AERONET-OC stations are present [96].
- In the Northern Sea where the Zeebrugge-MOW1 and Thornton_C-Power, part of the AERONET-OC network, are located [97].
- In the Baltic Sea where the Helsinki Lighthouse, Irbe Lighthouse, and Gustav Dalen Tower AERONET-OC stations are present [19]. As this basin is located at higher latitudes than the NAS, the satellite overpasses above such area are more frequent, making the approach also more powerful. Nevertheless, due to a large amount of CDOM present in these waters [98], a different atmospheric correction is needed to retrieve reliable Rrs spectra [99–101].
- The North-Western Mediterranean where the BOUee pour l'aquiSition d'une Serie Optique a Long termE mooring (BOUSSOLE) [102] and the Casablanca AERONET-OC [103] station are located. Particularly, BOUSSOLE has been used for the system vicarious calibration of the OLCI sensor [104]; hence, it could be interesting to test the effect of the adjustment using such data as input of the MLR algorithm.
- The areas where the different WATERHYPERNET network sites are located [105]. This network, which will be developed in the next years, is based on the concept of AERONET-OC [26], but with the essential advantage of the exploitation of a hyperspectral radiometer [106]. The use of such an instrument will allow validating each band of each satellite mission [106], reducing the uncertainties that could be introduced by band-shifting procedures [88,107].

The use of the Fiducial Reference Measurements data stream for the creation of this dataset for different coastal basins can be a reliable and unique way to compensate for the lack of an OCR geostationary sensor. Hence, the VGOCS dataset, available upon request to the authors, allows non-expert users to have at the disposal "ready to use" OC data to analyze the coastal optical variability at high temporal resolution. Moreover, expert users can perform analysis of exploiting products usually not provided in standard L3 files, such as flags and viewing geometry parameters.

Author Contributions: The general conception of this work was developed by M.B. and V.E.B.; M.B., G.V., M.B.E, and S.C. designed and implemented the processing chain. M.B. and V.E.B. designed the data analysis. M.B. processed the satellite imagery and performed the data analysis. M.B. and V.E.B. wrote the manuscript. S.C., G.V., and R.S. contributed to the interpretation of the results. All co-authors provided critical comments to the manuscript. All authors have read and agreed to the published version of the manuscript.

Funding: This research was supported by the European Union's Copernicus Marine Environment and Monitoring Service (grant number: 77-CMEMS-TAC-OC) and the European Union's Horizon 2020 research and innovation programme (HYPERNETS project, Grant agreement n° 775983).

Acknowledgments: The AERONET Team is acknowledged for the continuous effort in supporting the AERONET-OC sub-network. Giuseppe Zibordi from the Joint Research Center of the European Commission is acknowledged for establishing and maintaining the AAOT AERONET-OC site. Vega Forneris and Flavio Lapadula from ISMAR-CNR are acknowledged for advice on the implementation of the processing chain. Ilaria Cazzaniga provided relevant information enabling the OLCI flag and camera dependence analysis. Francesco Bignami, Davide D'Alimonte, Davide Dionisi, Federico Falcini, Michela Sammartino, Enrico Zambianchi, and Giuseppe Zibordi, provided useful comments on earlier versions of this manuscript.

Conflicts of Interest: The authors declare no conflict of interest.

References

1. IOCCG. *Why Ocean Colour? The Societal Benefits of Ocean-Colour Technology*; Reports of the International Ocean-Colour Coordinating Group, No. 7; IOCCG: Bedford, NS, Canada, 2008.

2. IOCCG. *Mission Requirements for Future Ocean-Colour Sensors*; Reports of the International Ocean-Colour Coordinating Group, No. 13; IOCCG: Bedford, NS, Canada, 2012.

3. McClain, C.R.; Meister, G.; Monosmith, B. Satellite Ocean Color Sensor Design Concepts and Performance Requirements. In *Experimental Methods in the Physical Sciences*; Academic Press: Cambridge, MA, USA, 2014; Volume 47, pp. 73–119.

4. Mouw, C.B.; Greb, S.; Aurin, D.; DiGiacomo, P.M.; Lee, Z.; Twardowski, M.; Binding, C.; Hu, C.; Ma, R.; Moore, T.; et al. Aquatic color radiometry remote sensing of coastal and inland waters: Challenges and recommendations for future satellite missions. *Remote Sens. Environ.* **2015**, *160*, 15–30. [CrossRef]

5. Arnone, R.A.; Vandermeulen, R.A.; Soto, I.M.; Ladner, S.D.; Ondrusek, M.E.; Yang, H. Diurnal changes in ocean color sensed in satellite imagery. *J. Appl. Remote Sens.* **2017**, *11*, 032406. [CrossRef]

6. Bracaglia, M.; Volpe, G.; Colella, S.; Santoleri, R.; Braga, F.; Brando, V.E. Using overlapping VIIRS scenes to observe short term variations in particulate matter in the coastal environment. *Remote Sens. Environ.* **2019**, *233*, 111367. [CrossRef]

7. Chen, Z.; Hu, C.; Muller-Karger, F.E.; Luther, M.E. Short-term variability of suspended sediment and phytoplankton in Tampa Bay, Florida: Observations from a coastal oceanographic tower and ocean color satellites. *Estuar. Coast. Shelf Sci.* **2010**, *89*, 62–72. [CrossRef]

8. Qi, L.; Hu, C.; Barnes, B.B.; Lee, Z. VIIRS captures phytoplankton vertical migration in the NE Gulf of Mexico. *Harmful Algae* **2017**, *66*, 40–46. [CrossRef]

9. Choi, J.K.; Park, Y.J.; Ahn, J.H.; Lim, H.S.; Eom, J.; Ryu, J.H. GOCI, the world's first geostationary ocean color observation satellite, for the monitoring of temporal variability in coastal water turbidity. *J. Geophys. Res. Ocean.* **2012**, *117*. [CrossRef]

10. Lavigne, H.; Ruddick, K. The potential use of geostationary MTG/FCI to retrieve chlorophyll-a concentration at high temporal resolution for the open oceans. *Int. J. Remote Sens.* **2018**, *39*, 2399–2420. [CrossRef]

11. Neukermans, G.; Ruddick, K.; Bernard, E.; Ramon, D.; Nechad, B.; Deschamps, P.Y. Mapping total suspended matter from geostationary satellites: A feasibility study with SEVIRI in the Southern North Sea. *Opt. Express* **2009**, *17*, 14029–14052. [CrossRef]

12. Choi, J.K.; Park, Y.J.; Lee, B.R.; Eom, J.; Moon, J.E.; Ryu, J.H. Application of the Geostationary Ocean Color Imager (GOCI) to mapping the temporal dynamics of coastal water turbidity. *Remote Sens. Environ.* **2014**, *146*, 24–35. [CrossRef]

13. Concha, J.; Mannino, A.; Franz, B.; Kim, W. Uncertainties in the Geostationary Ocean Color Imager (GOCI) Remote Sensing Reflectance for Assessing Diurnal Variability of Biogeochemical Processes. *Remote Sens.* **2019**, *11*, 295. [CrossRef]

14. Ruddick, K.; Neukermans, G.; Vanhellemont, Q.; Jolivet, D. Challenges and opportunities for geostationary ocean colour remote sensing of regional seas: A review of recent results. *Remote Sens. Environ.* **2014**, *146*, 63–76. [CrossRef]

15. Wang, M.; Son, S.; Jiang, L.; Shi, W. Observations of ocean diurnal variations from the Korean geostationary ocean color imager (GOCI). In *Ocean Sensing and Monitoring VI*; International Society for Optics and Photonics: Bellingham, WA, USA, 2014; Volume 9111, p. 911102.

16. Ryu, J.H.; Han, H.J.; Cho, S.; Park, Y.J.; Ahn, Y.H. Overview of geostationary ocean color imager (GOCI) and GOCI data processing system (GDPS). *Ocean Sci. J.* **2012**, *47*, 223–233. [CrossRef]

17. D'Alimonte, D.; Zibordi, G.; Melin, F. A statistical method for generating cross-mission consistent normalized water-leaving radiances. *IEEE Trans. Geosci. Remote Sens.* **2008**, *46*, 4075–4093. [CrossRef]

18. Zibordi, G.; Holben, B.; Hooker, S.B.; Mélin, F.; Berthon, J.F.; Slutsker, I.; Giles, D.; Vandemark, D.; Feng, H.; Rutledge, K.; et al. A network for standardized ocean color validation measurements. *EOS Trans. Am. Geophys. Union* **2006**, *87*, 293–297. [CrossRef]

19. Zibordi, G.; Mélin, F.; Berthon, J.F.; Holben, B.; Slutsker, I.; Giles, D.; D'Alimonte, D.; Vandemark, D.; Feng, H.; Fabbri, B.E.; et al. AERONET-OC: A network for the validation of ocean color primary products. *J. Atmos. Ocean. Technol.* **2009**, *26*, 1634–1651. [CrossRef]

20. Zibordi, G.; Voss, K.J. In situ optical radiometry in the visible and near infrared. In *Experimental Methods in the Physical Sciences*; Academic Press: Cambridge, MA, USA, 2014; Volume 47, pp. 247–304.

21. Zibordi, G.; Voss, K.J. Requirements and strategies for in situ radiometry in support of satellite ocean color. In *Experimental Methods in the Physical Sciences*; Academic Press: Cambridge, MA, USA, 2014; Volume 47, pp. 531–556.

22. Banks, A.C.; Vendt, R.; Alikas, K.; Bialek, A.; Kuusk, J.; Lerebourg, C.; Ruddick, K.; Tilstone, G.; Vabson, V.; Donlon, C.; et al. Fiducial Reference Measurements for Satellite Ocean Colour (FRM4SOC). *Remote Sens.* **2020**, *12*, 1322. [CrossRef]

23. Alikas, K.; Ansko, I.; Vabson, V.; Ansper, A.; Kangro, K.; Uudeberg, K.; Ligi, M. Consistency of Radiometric Satellite Data over Lakes and Coastal Waters with Local Field Measurements. *Remote Sens.* **2020**, *12*, 616. [CrossRef]

24. Liberti, G.L.; D'Alimonte, D.; di Sarra, A.; Mazeran, C.; Voss, K.; Yarbrough, M.; Bozzano, R.; Cavaleri, L.; Colella, S.; Cesarini, C.; et al. European Radiometry Buoy and Infrastructure (EURYBIA): A Contribution to the Design of the European Copernicus Infrastructure for Ocean Colour System Vicarious Calibration. *Remote Sens.* **2020**, *12*, 1178. [CrossRef]

25. Vabson, V.; Kuusk, J.; Ansko, I.; Vendt, R.; Alikas, K.; Ruddick, K.; Ansper, A.; Bresciani, M.; Burmester, H.; Costa, M.; et al. Field intercomparison of radiometers used for satellite validation in the 400–900 nm range. *Remote Sens.* **2019**, *11*, 1129. [CrossRef]

26. Zibordi, G.; Mélin, F.; Voss, K.J.; Johnson, B.C.; Franz, B.A.; Kwiatkowska, E.; Huot, J.; Wang, M.; Antoine, D. System vicarious calibration for ocean color climate change applications: Requirements for in situ data. *Remote Sens. Environ.* **2015**, *159*, 361–369. [CrossRef]

27. Lewis, A.; Lacey, J.; Mecklenburg, S.; Ross, J.; Siqueira, A.; Killough, B.; Szantoi, Z.; Tadono, T.; Rosenavist, A.; Goryl, P.; et al. CEOS Analysis Ready Data for Land (CARD4L) Overview. In Proceedings of the IGARSS 2018 IEEE International Geoscience and Remote Sensing Symposium, Valencia, Spain, 22–27 July 2018; pp. 7407–7410.

28. Scarrott, R.G.; Cawkwell, F.; Jessopp, M.; O'Rourke, E.; Cusack, C.; de Bie, K. From land to sea, a review of hypertemporal remote sensing advances to support ocean surface science. *Water* **2019**, *11*, 2286. [CrossRef]

29. Piwowar, J.M.; Peddle, D.R.; LeDrew, E.F. Temporal mixture analysis of arctic sea ice imagery: A new approach for monitoring environmental change. *Remote Sens. Environ.* **1998**, *63*, 195–207. [CrossRef]

30. Bissett, W.P.; Arnone, R.A.; Davis, C.O.; Dickey, T.D.; Dye, D.; Kohler, D.D.; Gould, R.W., Jr. From meters to kilometers: A look at ocean-color scales of variability, spatial coherence, and the need for fine-scale remote sensing in coastal ocean optics. *Oceanography* **2004**, *17*, 32–43. [CrossRef]

31. Aurin, D.; Mannino, A.; Franz, B. Spatially resolving ocean color and sediment dispersion in river plumes, coastal systems, and continental shelf waters. *Remote Sens. Environ.* **2013**, *137*, 212–225. [CrossRef]

32. Lee, Z.; Hu, C.; Arnone, R.; Liu, Z. Impact of sub-pixel variations on ocean color remote sensing products. *Opt. Express* **2012**, *20*, 20844–20854. [CrossRef] [PubMed]

33. Kutser, T. Quantitative detection of chlorophyll in cyanobacterial blooms by satellite remote sensing. *Limnol. Oceanogr.* **2004**, *49*, 2179–2189. [CrossRef]

34. Lee, Z.; Carder, K.L.; Arnone, R.A. Deriving inherent optical properties from water color: A multiband quasi-analytical algorithm for optically deep waters. *Appl. Opt.* **2002**, *41*, 5755–5772. [CrossRef]

35. Lee, Z.; Carder, K.L.; Arnone, R.A. "Quasi-Analytical Algorithm v6 Update". 2014. Available online: http://www.ioccg.org/groups/Software_OCA/QAA_v6_2014209.pdf (accessed on 5 May 2020).

36. Pitarch, J.; Bellacicco, M.; Volpe, G.; Colella, S.; Santoleri, R. Use of the quasi-analytical algorithm to retrieve backscattering from in-situ data in the Mediterranean Sea. *Remote Sens. Lett.* **2016**, *7*, 591–600. [CrossRef]

37. Pitarch, J.; Bellacicco, M.; Organelli, E.; Volpe, G.; Colella, S.; Vellucci, V.; Marullo, S. Retrieval of Particulate Backscattering Using Field and Satellite Radiometry: Assessment of the QAA Algorithm. *Remote Sens.* **2020**, *12*, 77. [CrossRef]

38. Barnes, B.B.; Hu, C. Dependence of satellite ocean color data products on viewing angles: A comparison between SeaWiFS, MODIS, and VIIRS. *Remote Sens. Environ.* **2016**, *175*, 120–129. [CrossRef]

39. Volpe, G.; Colella, S.; Brando, V.E.; Forneris, V.; La Padula, F.; Di Cicco, A.; Sammartino, M.; Bracaglia, M.; Artuso, F.; Santoleri, R. Mediterranean ocean colour Level 3 operational multi-sensor processing. *Ocean Sci.* **2019**, *15*, 127–146. [CrossRef]

40. Solidoro, C.; Bastianini, M.; Bandelj, V.; Codermatz, R.; Cossarini, G.; Canu, D.M.; Ravagnan, E.; Salon, S.; Trevisani, S. Current state, scales of variability, and trends of biogeochemical properties in the northern Adriatic Sea. *J. Geophys. Res. Oceans* **2009**, *114*. [CrossRef]

41. Bignami, F.; Sciarra, R.; Carniel, S.; Santoleri, R. Variability of Adriatic Sea coastal turbid waters from SeaWiFS imagery. *J. Geophys. Res. Oceans* **2007**, *112*. [CrossRef]

42. Zavatarelli, M.; Raicich, F.; Bregant, D.; Russo, A.; Artegiani, A. Climatological biogeochemical characteristics of the Adriatic Sea. *J. Mar. Syst.* **1998**, *18*, 227–263. [CrossRef]

43. Brando, V.E.; Braga, F.; Zaggia, L.; Giardino, C.; Bresciani, M.; Matta, E.; Bellafiore, D.; Ferrarin, C.; Maicu, F.; Benetazzo, A.; et al. High-resolution satellite turbidity and sea surface temperature observations of river plume interactions during a significant flood event. *Ocean Sci.* **2015**, *11*, 909–920. [CrossRef]

44. Marini, M.; Jones, B.H.; Campanelli, A.; Grilli, F.; Lee, C.M. Seasonal variability and Po River plume influence on biochemical properties along western Adriatic coast. *J. Geophys. Res. Oceans* **2008**, *113*. [CrossRef]

45. Cozzi, S.; Giani, M. River water and nutrient discharges in the Northern Adriatic Sea: Current importance and long term changes. *Cont. Shelf Res.* **2011**, *31*, 1881–1893. [CrossRef]

46. Falcieri, F.M.; Benetazzo, A.; Sclavo, M.; Russo, A.; Carniel, S. Po River plume pattern variability investigated from model data. *Cont. Shelf Res.* **2014**, *87*, 84–95. [CrossRef]

47. Degobbis, D.; Precali, R.; Ivancic, I.; Smodlaka, N.; Fuks, D.; Kveder, S. Long-term changes in the northern Adriatic ecosystem related to anthropogenic eutrophication. *Int. J. Environ. Pollut.* **2000**, *13*, 495–533. [CrossRef]

48. Tesi, T.; Miserocchi, S.; Acri, F.; Langone, L.; Boldrin, A.; Hatten, J.A.; Albertazzi, S. Flood-driven transport of sediment, particulate organic matter, and nutrients from the Po River watershed to the Mediterranean Sea. *J. Hydrol.* **2013**, *498*, 144–152. [CrossRef]

49. Malačič, V.; Viezzoli, D.; Cushman-Roisin, B. Tidal dynamics in the northern Adriatic Sea. *J. Geophys. Res. Oceans* **2000**, *105*, 26265–26280. [CrossRef]

50. Wang, X.H.; Pinardi, N.; Malacic, V. Sediment transport and resuspension due to combined motion of wave and current in the northern Adriatic Sea during a Bora event in January 2001: A numerical modelling study. *Cont. Shelf Res.* **2007**, *27*, 613–633. [CrossRef]

51. Barale, V.; Jaquet, J.M.; Ndiaye, M. Algal blooming patterns and anomalies in the Mediterranean Sea as derived from the SeaWiFS data set (1998–2003). *Remote Sens. Environ.* **2008**, *112*, 3300–3313. [CrossRef]

52. Bellafiore, D.; Ferrarin, C.; Braga, F.; Zaggia, L.; Maicu, F.; Lorenzetti, G.; Manfè, G.; Brando, V.E.; De Pascalis, F. Coastal mixing in multiple-mouth deltas: A case study in the Po delta, Italy. *Estuar. Coast. Shelf Sci.* **2019**, *226*, 106254. [CrossRef]

53. Benincasa, M.; Falcini, F.; Adduce, C.; Sannino, G.; Santoleri, R. Synergy of Satellite Remote Sensing and Numerical Ocean Modelling for Coastal Geomorphology Diagnosis. *Remote Sens.* **2019**, *11*, 2636. [CrossRef]

54. Ferrarin, C.; Davolio, S.; Bellafiore, D.; Ghezzo, M.; Maicu, F.; Mc Kiver, W.; Drofa, O.; Umgiesser, G.; Bajo, M.; De Pascalis, F.; et al. Cross-scale operational oceanography in the Adriatic Sea. *J. Oper. Oceanogr.* **2019**, *12*, 86–103. [CrossRef]

55. Braga, F.; Zaggia, L.; Bellafiore, D.; Bresciani, M.; Giardino, C.; Lorenzetti, G.; Maicu, F.; Manzo, C.; Riminucci, F.; Ravaioli, M.; et al. Mapping turbidity patterns in the Po river prodelta using multi-temporal Landsat 8 imagery. *Estuar. Coast. Shelf Sci.* **2017**, *198*, 555–567. [CrossRef]

56. Boldrin, A.; Carniel, S.; Giani, M.; Marini, M.; Bernardi Aubry, F.; Campanelli, A.; Grilli, F.; Russo, A. Effects of bora wind on physical and biogeochemical properties of stratified waters in the northern Adriatic. *J. Geophys. Res. Oceans* **2009**, *114*. [CrossRef]

57. Liu, X.; Wang, M. Filling the gaps of missing data in the merged VIIRS SNPP/NOAA-20 ocean color product using the DINEOF method. *Remote Sens.* **2019**, *11*, 178. [CrossRef]

58. Cao, C.; Xiong, X.; Wolfe, R.; De Luccia, F.; Liu, Q.; Blonski, S.; Lin, G.G.; Nishihama, M.; Pogorzala, D.; Oudrari, H.; et al. *Visible Infrared Imaging Radiometer Suite (VIIRS) Sensor Data Record (SDR) User's Guide*; NOAA Technical Report NESDIS: College Park, MD, USA, 2013.

59. Donlon, C.; Berruti, B.; Buongiorno, A.; Ferreira, M.H.; Féménias, P.; Frerick, J.; Goryl, P.; Klein, U.; Laur, H.; Mavrocordatos, C.; et al. The global monitoring for environment and security (GMES) sentinel-3 mission. *Remote Sens. Environ.* **2012**, *120*, 37–57. [CrossRef]

60. Masuoka, E.; Fleig, A.; Wolfe, R.E.; Patt, F. Key characteristics of MODIS data products. *IEEE Trans. Geosci. Remote Sens.* **1998**, *36*, 1313–1323. [CrossRef]

61. Kay, S.; Hedley, J.; Lavender, S. Sun glint correction of high and low spatial resolution images of aquatic scenes: A review of methods for visible and near-infrared wavelengths. *Remote Sens.* **2009**, *1*, 697–730. [CrossRef]

62. ESA. 2020. Available online: https://scihub.copernicus.eu/userguide/BatchScripting (accessed on 5 May 2020).

63. EUMETSAT. 2020. Available online: https://coda.eumetsat.int/ (accessed on 5 May 2020).

64. EUMETSAT. 2020. Available online: https://codarep.eumetsat.int/ (accessed on 5 May 2020).

65. NASA. 2020. Available online: https://oceancolor.gsfc.nasa.gov/reprocessing/r2018/ (accessed on 5 May 2020).

66. Bouali, M.; Ignatov, A. Adaptive reduction of striping for improved sea surface temperature imagery from Suomi National Polar-Orbiting Partnership (S-NPP) visible infrared imaging radiometer suite (VIIRS). *J. Atmos. Ocean. Technol.* **2014**, *31*, 150–163. [CrossRef]

67. Mikelsons, K.; Wang, M.; Jiang, L.; Bouali, M. Destriping algorithm for improved satellite-derived ocean color product imagery. *Opt. Express* **2014**, *22*, 28058–28070. [CrossRef]

68. NASA. 2020. Available online: https://oceancolor.gsfc.nasa.gov/atbd/ocl2flags (accessed on 5 May 2020).

69. Morel, A.; Antoine, D.; Gentili, B. Bidirectional reflectance of oceanic waters: Accounting for Raman emission and varying particle scattering phase function. *Appl. Opt.* **2002**, *41*, 6289–6306. [CrossRef] [PubMed]

70. Morel, A.; Mueller, J.L. Normalized water-leaving radiance and remote sensing reflectance: Bidirectional reflectance and other factors. *Ocean Opt. Protoc. Satell. Ocean Color Sens. Valid.* **2002**, *2*, 183–210.

71. Kaufman, Y.J. Atmospheric effect on spatial resolution of surface imagery. *Appl. Opt.* **1984**, *23*, 3400–3408. [CrossRef]

72. Sterckx, S.; Knaeps, E.; Ruddick, K. Detection and correction of adjacency effects in hyperspectral airborne data of coastal and inland waters: The use of the near infrared similarity spectrum. *Int. J. Remote Sens.* **2011**, *32*, 6479–6505. [CrossRef]

73. Bulgarelli, B.; Zibordi, G.; Mélin, F. On the minimization of adjacency effects in SeaWiFS primary data products from coastal areas. *Opt. Express* **2018**, *26*, A709–A728. [CrossRef]

74. Van Mol, B.; Ruddick, K. Total Suspended Matter maps from CHRIS imagery of a small inland water body in Oostende (Belgium). In Proceedings of the 3rd ESA CHRIS/Proba Workshop, Frascati, Italy, 21–23 March 2005; pp. 21–23.

75. Barnes, B.B.; Cannizzaro, J.P.; English, D.C.; Hu, C. Validation of VIIRS and MODIS reflectance data in coastal and oceanic waters: An assessment of methods. *Remote Sens. Environ.* **2019**, *220*, 110–123. [CrossRef]

76. Gordon, H.R.; Wang, M. Retrieval of water-leaving radiance and aerosol optical thickness over the oceans with SeaWiFS: A preliminary algorithm. *Appl. Opt.* **1994**, *33*, 443–452. [CrossRef]

77. Lavender, S.J.; Pinkerton, M.H.; Moore, G.F.; Aiken, J.; Blondeau-Patissier, D. Modification to the atmospheric correction of SeaWiFS ocean colour images over turbid waters. *Cont. Shelf Res.* **2005**, *25*, 539–555. [CrossRef]

78. Moore, G.F.; Aiken, J.; Lavender, S.J. The atmospheric correction of water colour and the quantitative retrieval of suspended particulate matter in Case II waters: Application to MERIS. *Int. J. Remote Sens.* **1999**, *20*, 1713–1733. [CrossRef]

79. Morel, A.; Gentili, B. Diffuse reflectance of oceanic waters: Its dependence on Sun angle as influenced by the molecular scattering contribution. *Appl. Opt.* **1991**, *30*, 4427–4438. [CrossRef]

80. Morel, A.; Gentili, B. Diffuse reflectance of oceanic waters. II. Bidirectional aspects. *Appl. Opt.* **1993**, *32*, 6864–6879. [CrossRef]

81. Morel, A.; Gentili, B. Diffuse reflectance of oceanic waters. III. Implication of bidirectionality for the remote-sensing problem. *Appl. Opt.* **1996**, *35*, 4850–4862. [CrossRef]

82. Zibordi, G.; Mélin, F.; Berthon, J.F. A regional assessment of OLCI data products. *IEEE Geosci. Remote Sens. Lett.* **2018**, *15*, 1490–1494. [CrossRef]

83. *Sentinel-3 Product Notice—OLCI Level-2 Ocean Colour*; EUMETSAT: Darmstadt, Germany, 2019.

84. Volpe, G.; Colella, S.; Brando, V.E.; Benincasa, M.; Forneris, V.; Bracaglia, M.; Di Cicco, A.; D'Alimonte, D. *Ocean Colour Production Centre. Ocean Colour Mediterranean and Black Sea Observation Product*; Copernicus Publications: Göttingen, Germany, 2019.

85. Ruddick, K.G.; Voss, K.; Boss, E.; Castagna, A.; Frouin, R.; Gilerson, A.; Hieronymi, M.; Johnson, B.C.; Kuusk, J.; Lee, Z.; et al. A Review of Protocols for Fiducial Reference Measurements of Water-Leaving Radiance for Validation of Satellite Remote-Sensing Data over Water. *Remote Sens.* **2019**, *11*, 2198. [CrossRef]

86. Thuillier, G.; Floyd, L.; Woods, T.N.; Cebula, R.; Hilsenrath, E.; Hersé, M.; Labs, D. Solar irradiance reference spectra for two solar active levels. *Adv. Space Res.* **2004**, *34*, 256–261. [CrossRef]

87. Mélin, F.; Vantrepotte, V.; Clerici, M.; D'Alimonte, D.; Zibordi, G.; Berthon, J.F.; Canuti, E. Multi-sensor satellite time series of optical properties and chlorophyll-a concentration in the Adriatic Sea. *Prog. Oceanogr.* **2011**, *91*, 229–244. [CrossRef]

88. Mélin, F.; Sclep, G. Band shifting for ocean color multi-spectral reflectance data. *Opt. Express* **2015**, *23*, 2262–2279. [CrossRef]

89. Lee, S.; Meister, G.; Franz, B. MODIS Aqua Reflective Solar Band Calibration for NASA's R2018 Ocean Color Products. *Remote Sens.* **2019**, *11*, 2187. [CrossRef]

90. Cao, C.; De Luccia, F.J.; Xiong, X.; Wolfe, R.; Weng, F. Early on-orbit performance of the visible infrared imaging radiometer suite onboard the Suomi National Polar-Orbiting Partnership (S-NPP) satellite. *IEEE Trans. Geosci. Remote Sens.* **2014**, *52*, 1142–1156. [CrossRef]

91. Pahlevan, N.; Sarkar, S.; Franz, B.A. Uncertainties in coastal ocean color products: Impacts of spatial sampling. *Remote Sens. Environ.* **2016**, *181*, 14–26. [CrossRef]

92. Brando, V.E.; Dekker, A.G.; Park, Y.J.; Schroeder, T. Adaptive semianalytical inversion of ocean color radiometry in optically complex waters. *Appl. Opt.* **2012**, *51*, 2808–2833. [CrossRef]

93. Loisel, H.; Stramski, D.; Dessailly, D.; Jamet, C.; Li, L.; Reynolds, R.A. An Inverse Model for Estimating the Optical Absorption and Backscattering Coefficients of Seawater From Remote-Sensing Reflectance Over a Broad Range of Oceanic and Coastal Marine Environments. *J. Geophys. Res. Ocean.* **2018**, *123*, 2141–2171. [CrossRef]

94. Werdell, P.J.; Franz, B.A.; Bailey, S.W.; Feldman, G.C.; Boss, E.; Brando, V.E.; Dowell, M.; Hirata, T.; Lavender, S.J.; Lee, Z.; et al. Generalized ocean color inversion model for retrieving marine inherent optical properties. *Appl. Opt.* **2013**, *52*, 2019–2037. [CrossRef]

95. Werdell, P.J.; McKinna, L.I.; Boss, E.; Ackleson, S.G.; Craig, S.E.; Gregg, W.W.; Lee, Z.; Maritorena, S.; Roesler, C.S.; Rousseaux, C.S.; et al. An overview of approaches and challenges for retrieving marine inherent optical properties from ocean color remote sensing. *Prog. Oceanogr.* **2018**, *160*, 186–212. [CrossRef]

96. Zibordi, G.; Mélin, F.; Berthon, J.F.; Canuti, E. Assessment of MERIS ocean color data products for European seas. *Ocean Sci.* **2013**, *9*, 521–533. [CrossRef]

97. Van der Zande, D.; Vanhellemont, Q.; De Keukelaere, L.; Knaeps, E.; Ruddick, K. Validation of Landsat-8/OLI for ocean colour applications with AERONET-OC sites in Belgian coastal waters. In Proceedings of the Ocean Optics Conference, Victoria, BC, Canada, 23–28 October 2016.

98. Berthon, J.F.; Zibordi, G. Optically black waters in the northern Baltic Sea. *Geophys. Res. Lett.* **2010**, *37*. [CrossRef]

99. Darecki, M.; Stramski, D. An evaluation of MODIS and SeaWiFS bio-optical algorithms in the Baltic Sea. *Remote Sens. Environ.* **2004**, *89*, 326–350. [CrossRef]

100. IOCCG. *Atmospheric Correction for Remotely-Sensed Ocean-Colour Products*; Reports of International Ocean-Color Coordinating Group; IOCCG: Bedford, NS, Canada, 2010; p. 10.

101. Zibordi, G.; Berthon, J.F.; Mélin, F.; D'Alimonte, D.; Kaitala, S. Validation of satellite ocean color primary products at optically complex coastal sites: Northern Adriatic Sea, Northern Baltic Proper and Gulf of Finland. *Remote Sens. Environ.* **2009**, *113*, 2574–2591. [CrossRef]

102. Antoine, D.; Chami, M.; Claustre, H.; d'Ortenzio, F.; Morel, A.; Bécu, G.; Gentili, B.; Louis, F.; Ras, J.; Roussier, E.; et al. *BOUSSOLE: A Joint CNRS-INSU, ESA, CNES, and NASA Ocean Color Calibration and Validation Activity*; National Aeronautics and Space Administration: Washington, DC, USA, 2006.

103. Talone, M.; García-Ladona, E. *AERONET-OC-Bioptical Survey of the Ebro's Shelf*; Instituto de Ciencias del Mar (ICM): Barcelona, Brazil, 2020.

104. Lamquin, N.; Bourg, L.; Lerebourg, C.; Martin-Lauzer, F.R.; Kwiatkowska, E.; Dransfeld, S. System Vicarious Calibration of Sentinel-3 OLCI. In Proceedings of the CALCON Technical Meeting 2017, Logan, UT, USA, 23 August 2017.

105. HYPERNETS. Available online: https://www.hypernets.eu/from_cms/test_sites (accessed on 5 May 2020).

106. Vansteenwegen, D.; Ruddick, K.; Cattrijsse, A.; Vanhellemont, Q.; Beck, M. The pan-and-tilt hyperspectral radiometer system (PANTHYR) for autonomous satellite validation measurements—Prototype design and testing. *Remote Sens.* **2019**, *11*, 1360. [CrossRef]

107. Pahlevan, N.; Smith, B.; Binding, C.; O'Donnell, D.M. Spectral band adjustments for remote sensing reflectance spectra in coastal/inland waters. *Opt. Express* **2017**, *25*, 28650–28667. [CrossRef]

MDPI
St. Alban-Anlage 66
4052 Basel
Switzerland
Tel. +41 61 683 77 34
Fax +41 61 302 89 18
www.mdpi.com

Remote Sensing Editorial Office
E-mail: remotesensing@mdpi.com
www.mdpi.com/journal/remotesensing